BIOCHEMISTRY
A Short Course

BIOCHEMISTRY
A Short Course

Harry R. Matthews
Department of Biological Chemistry
University of California School of Medicine
Davis, California

Richard A. Freedland
Department of Molecular Biosciences
University of California School of Veterinary Medicine
Davis, California

Roger L. Miesfeld
Department of Biochemistry
University of Arizona
Tucson, Arizona

WILEY-LISS
A JOHN WILEY & SONS, INC., PUBLICATION
New York • Chichester • Weinheim • Brisbane • Singapore • Toronto

Address all Inquiries to the Publisher
Wiley-Liss, Inc., 605 Third Avenue, New York, NY 10158-0012

Printed in the United States of America.

The text of this book is printed on acid-free paper.

Library of Congress Cataloging-in-Publication Data

Matthews, Harry Roy, 1942–
Biochemistry : a short course / Harry R. Matthews, Richard A. Freedland, Roger L. Miesfeld.
p. cm.
Includes Index.
ISBN 0-471-02205-5 (pbk. : alk. paper)
1. Biochemistry. I. Freedland, Richard A. (Richard Allan)
II. Miesfeld, Roger L. III. Title.
QP514.2.M387 1996
574.19′2—dc20 96-16469

Printed in the United States of America

10 9 8 7 6 5 4 3 2 1

This book is dedicated to our wives, Iris, Beverly, and Elizabeth,
for their patience and support during all the evenings and weekends
we spent working on this project.

CONTENTS

CHAPTER 4: MEMBRANES

CHAPTER 5: NUCLEOPROTEIN COMPLEXES

CHAPTER 6: HEMOGLOBIN ILLUSTRATES PROTEIN BEHAVIOR

CHAPTER 7: INTRODUCTION TO ENZYMES

CHAPTER 8: REGULATION OF ENZYME ACTIVITY BY NONCOVALENT CHANGES

CHAPTER 13: GLUCONEOGENESIS

CHAPTER 14: FATTY-ACID METABOLISM: CATABOLISM

CHAPTER 15: LIPID METABOLISM: SYNTHESIS AND TRANSPORT

CHAPTER 16: HORMONES

CHAPTER 17: AMINO-ACID METABOLISM: UREA, GLUTAMATE, GLUTAMINE, AND ASPARTATE

CHAPTER 18: METABOLISM OF OTHER ALIPHATIC AMINO ACIDS

CHAPTER 19: METABOLISM OF AROMATIC AMINO ACIDS AND PROTEIN ECONOMY

CHAPTER 20: PURINES, PYRIMIDINES, AND HEME METABOLISM

CHAPTER 21: GENOME ORGANIZATION

CHAPTER 22: DNA REPLICATION

CHAPTER 30: MOLECULAR BIOLOGY OF ANIMAL VIRUSES

CHAPTER 31: ONCOGENES AND HUMAN CANCER

APPENDIX: MENDELIAN GENETICS 477

GLOSSARY 479

INDEX 489

PREFACE

The science of biochemistry is rich in textbooks. Admirable scholarship and artistry have gone into preparing and publishing several of the comprehensive biochemistry texts which are invaluable reference works. There is no need for another of these books—our colleagues have already done a great job. This text was developed to fill a need that arose directly from our experiences teaching biochemistry to undergraduate, graduate, medical, and veterinary students. We have recognized the demands on students' time and listened to their concerns about focusing on the primary concepts of a subject—the ideas and approaches that will remain when the details are forgotten. A primary task of a teacher is to identify key concepts and explain them clearly to the students. As Nicholas Allison put it (*Aldus Magazine,* **1,** 16):

> The word, "education," comes from the Latin, "educere," meaning, literally, "to lead out." Despite the common notion that a teacher's job is to stuff things in—knowledge, manners, behaviors—most great educators agree that "leading out" is the true task, helping students develop an appetite for knowledge and an appreciation for its fruits, and guiding them to where and how it can be found. Education is empowerment.

Identification of key concepts and rejection of extraneous material were critical tasks in preparing this text. In our teaching, we have made decisions, sometime heart-wrenching, to omit details and topics that may have been critical to our own understanding or are just too elegant or fascinating for most authors to omit. *Biochemistry—A Short Course* builds on our insights and emphasizes clear explanation, illustrated with many figures, and made palatable by good design and white space. This is not a condensed text or just a summary. The small size of the book arises from its logical development of exclusively those concepts and approaches that are critical to modern biochemistry.

We have also listened to students who appreciate the relevance of biochemistry to human health and disease and we have often—although not always—chosen the human example over a sometimes more complete story from a lower organism. Biochemistry, including, as our short course does, molecular biology, is both vital for the practice of modern medicine and the basis on which the medicine of the future is being built. We have a duty to convey to our students and readers the critical relevance of biochemistry to modern life and death.

Because we believe biochemistry is so important that its study should appeal to many students, we have tried to minimize the amount of prior knowledge the reader needs to bring to the book. There is a short appendix on Mendelian Genetics and Chapter One includes brief introductions of critical elements of structural biology and cell biology that are necessary for later discussions. However, a basic prior knowledge of organic chemistry will be helpful for understanding particularly the chapters on metabolism. Focus on concepts has not resulted in a diminution of scope, and the book covers both traditional biochemistry and areas that have emerged more recently, often called molecular biology. The unity of these areas is illustrated on the front cover of our book by a model of part of a protein—called the glucocorticoid receptor. In appropriate cells, this protein links a hormone signal, originating outside the cell, with changes in metabolism inside the cell, using existing genetic information.*

Happily, some students will have the curiosity and determination to delve more deeply into their subject. To encourage these students, we have included a reference section at the end of each chapter. These readings develop the key ideas presented in the chapter and provide a path to the primary literature. As additional aids to study, there are questions at the end of each chapter and short discussions of the answers.

The sheer size or density of most biochemistry textbooks intimidates students and much goes unread. *Biochemistry—A Short Course* is more student-friendly and will help students learn by raising their self-confidence and defining their objectives more clearly.

Computers provide unique methods of visualizing three-dimensional structure and unique ways of describing biochemical processes with interactive cartoons. One of us, Harry R. Matthews, maintains a site for this purpose at http://moby.ucdavis.edu/HRM.

ACKNOWLEDGMENTS

We are particularly grateful to Dr. Dmitry Bochkariov, who crafted the illustrations with outstanding skill and patience. We are also extremely grateful to those who contributed specific illustrations—Dr. Fern Tablin for her electron micrograph of a megakaryocyte, Dr. Grace Rosenquist for an example of a hydropathy plot, Dr. Mark Chapman for a sample DNA sequencing gel, Dr. Stephen Rundlett for DNA footprinting and electrophoretic mobility shift data, and Sharon Pascoe for the RT–PCR data of the human androgen receptor.

We are grateful to our colleagues who read and commented on parts of the manuscript including Drs. Tom Jue, Joachim Schnier and D. A. Walsh.

* The protein structure model, based on X-ray crystallography, represents the 66 amino-acid DNA binding domain of the glucocorticoid receptor (Luisi, B. F., Xu, W. X., Otwinowski, Z., Freedman, L. P., Yamamoto, K. R., and Sigler, P. B. "Crystallographic analysis of the interaction of the glucocorticoid receptor with DNA." *Nature* **352:**497–505, 1991). In Chapter 5, we discuss the biochemical characteristics of several DNA binding proteins, one of which is the so-called "zinc-finger" domain of the glucocorticoid receptor shown on the cover. Chapters 11–20 describe cellular control of key metabolic processes, and one example is the role the glucocorticoid receptor plays in modulating glucose production by the liver. Finally in chapter 29, we show how the glucocorticoid receptor is activated by hormone binding to function as a eukaryotic transcription factor capable of altering the expression of specific genes. Therefore, rather than separating the underlying principles into biochemical and molecular biology concepts, *Biochemistry—A Short Course* uses integrated examples, like the glucocorticoid receptor, to emphasize key molecular principles at the cellular level.

BIOCHEMISTRY
A Short Course

CHAPTER 1

MOLECULES AND CELLS

INTRODUCTION

Molecules, and ions, are the basic building blocks of all living structures and food that is taken in is broken down to simple molecules, but not further, before being used. Molecules may be energy stores, transducers of energy into work and vice versa, internal signals, or *receptors* for external signals such as light or smell. Modern discoveries are leading to important new approaches to human disease, including *rational drug design* and *gene therapy.* Drugs range from simple calcium carbonate pills to synthetic proteins such as *tissue plasminogen activator*—used to help the body dissolve *blood clots.* The vast majority of drugs are molecules, although we are learning how to make macromolecular complexes—synthetic viruses—to increase the specificity and efficacy of pharmacological intervention. Thus, in addition to its practical value, the study of biological molecules provides wonderful insights into how we function. Biochemistry (and molecular biology) is the study of those processes, occurring in living organisms, that can be understood at the molecular level.

This chapter introduces the important prerequisite topics of *cell biology* and *intermolecular forces* (the forces between molecules) to provide the background necessary to appreciate what follows. Genetics is more important for later chapters and its basics are provided in the Appendix.

MOLECULES AND CELLS

A molecule is a specific group of atoms connected together by *covalent bonds*—shared electron clouds. Molecules in the body may be relatively simple, such as the oxygen molecule, O═O, or highly complex, such as the massive ***glycoproteins*** that insulate cells from mechanical shock. Biological molecules interact with one another primarily through *noncovalent interactions.* In general, such interactions are individually weak and nonspecific, but the body harnesses these seemingly innocuous forces to generate extraordinary specificity and strength. Many individual interactions are brought together in highly specific ways to stabilize the ***transition states*** of biochemical reactions, pass sophisticated messages, and generate complex macromolecular and cellular structures.

Scientists like to group and classify objects of study as a preliminary to understanding them. The study of living organisms is no exception, and a complex organism such as the human body is described in terms of hierarchical structures, starting with molecules and continuing with macromolecular structures, cells, organs or tissues, the whole organism, and then interactions within and between groups of organisms and between the groups and their environment. Cells are assemblies of *macromolecular structures*—assemblies of molecules—including a *plasma membrane* that encloses the cell and structures that maintain the cell, allow it to reproduce, and carry out specific cellular functions.

Subcellular Structure

Subcellular structure is in the realm of cell biology, but we need an overview to place our biochemical and molecular biological studies in their context in the living organism. Cells are subclassified into ***eukaryotic*** and ***prokaryotic*** cells. Eukaryotic cells occur in higher organisms from *yeast* to *human* and have characteristic subcellular structures including ***mitotic chromosomes*** and a *nuclear membrane.* Prokaryotic cells include *bacteria* and other simple organisms; they lack much of the eukaryotic cell structure; in particular, the *nucleus* is absent. In all cells, the solution within the plasma membrane is the ***cytoplasm.*** The *cytoplasm* includes a very concentrated aqueous solution of small and large molecules crowded together with, in *mammalian cells,* a dynamic structure—the ***cytoskeleton***—providing three-dimensional shape and order within the cell.

Within the cytoplasm of a eukaryotic cell, the **nucleus** is usually the most prominent substructure or *organelle* (Figs. 1.1 and 1.2). The nucleus has its own membrane—the nuclear membrane—surrounding the **nucleoplasm.** Within the *nucleoplasm* are found:

1. The *nuclear lamina* beneath the nuclear membrane
2. The *nuclear matrix* organizing the *chromosomes* within the nucleus
3. The ***nucleolus*** where *ribosomal RNA* molecules are made (Chapter 24)

The nucleus is both the site of almost all *nucleic acid synthesis* in the cell (Chapter 24) and the location of most of the *hereditary material* of the cell. In the molecular biology section of this book (chapters 21–31, we will see how the nucleus plays a critical role in separating ***transcription*** (*RNA synthesis*) from ***translation*** (*protein synthesis*).

The nuclear membrane is continuous with the ***endoplasmic reticulum*** **(ER),** a membrane structure that divides the cytoplasm into two topologically separate domains. One side of the ER is the true "inside" of the cell; the other side, called the "lumen," is *topologically equivalent* to the outside of the cell. In this context, topologically equivalent domains are those between which molecules can move, or be transported, without having to pass through a *membrane* (Chapter 4). Thus, a protein that is made in the cytoplasm can freely diffuse throughout the cytoplasm, but cannot pass to the outside of the cell or into the nucleus without going through a membrane. There are mechanisms for passing a protein from the cytoplasm through the *ER* or through the nuclear membrane. The process of placing each molecule of the cell into its appropriate *subcellular compartment* is critical to the ordered functioning of the cell.

Mitochondria are *topologically independent* and are found within the cytoplasm. They cooperate with the cytoplasm and use molecular oxygen to "burn" molecules derived from food, producing ***adenosine triphosphate (ATP),*** which provides energy for cellular processes. Other subcellular organelles occur, in specific cells, such as the *chloroplasts* found in *plant cells.*

Electron microscopy reveals smaller features such as ***ribosomes*** (the site of protein synthesis) or *centrioles* (organizers of *cell division*), which are examples of macromolecular structures. Current research suggests that there may be many other *macromolecular structures,* often highly dynamic, such as assemblies of ***receptor protein kinases*** with substrates and *adaptor molecules,* or components of the *cytoskeleton* or *nuclear matrix.*

The cellular structure is dynamic. The most dramatic changes occur when a cell divides into two daughter cells as part of the process of *cell proliferation.* Cell division in eukaryotes involves the process of *mitosis* in which the nuclear membrane dissolves, the *chromosomes condense* and separate into two groups, and two nuclear membranes reassemble around the chromosomes. Finally, in the process of *cytokinesis,* the cell membrane contracts in the middle to form a shape like the character "8" and the two halves separate to form independent cells. Although it is the most dramatic of the microscopically visible events, the division itself is just part of a complete cycle of molecular events in which the cell prepares for division and then carries it out (Fig. 1.3).

Cells fresh from a division enter a phase of the *cell cycle* called *G1 phase,* where most of the control of cell proliferation occurs. When the cell receives the signal to leave G1 phase, it passes to *S phase,* during which the chromosomes are replicated. When *replication* is complete, the cell enters *G2 phase* and carries out molecular preparations for mitosis, which then follows. Although the key control points are usually in G1 phase, the transition from G2 phase to mitosis is a key *control point* in some cells and in others the G1 phase is

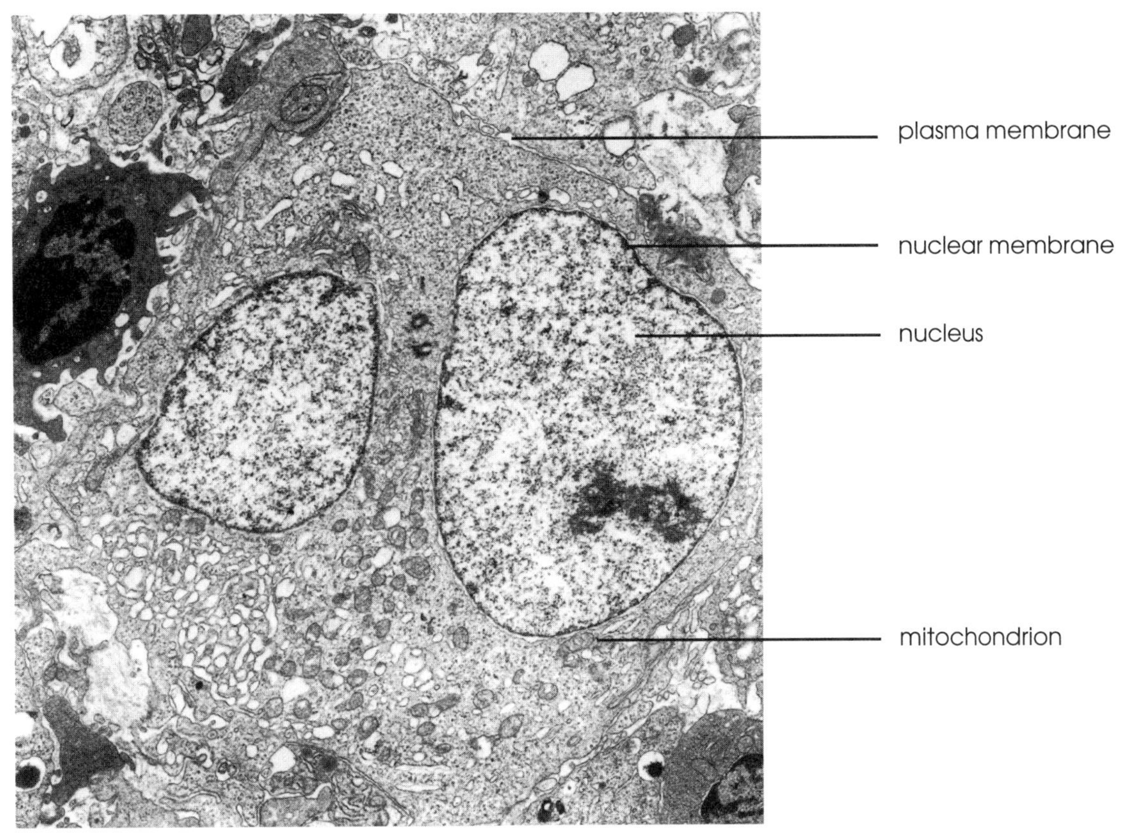

Figure 1.1.
Electron micrograph of a eukaryotic cell. This cell is a mammalian megakaryocyte, a cell found in the blood. It contains the usual organelles found in eukaryotic cells, shown in diagrammatic form in Figure 1.2, except that this particular cell has two nuclei—common for megakaryocytes—whereas most cell types have only one nucleus, illustrating the diversity as well as the similarities of cellular organization. The original was graciously provided by Dr. Fern Tablin.

missing. In general, the sequence of cell cycle phases is remarkably conserved.

Supracellular Structures

In sufficiently complex organisms, cells are assembled into *organs* or organ systems including the blood, brain, liver, kidney, skin, and many others. Although it is hard to deal with organs at the molecular level, we will touch on the ways in which molecules provide the communications within and between organs.

CELLULAR PROCESSES NEED ENERGY SUPPLIED BY ATP

Cells are anathema to the *Second Law of Thermodynamics* and must continually expend energy to maintain their ordered structures and to carry out their specific functions. With the exception of geothermal sources, our energy comes from the sun. Heat can be absorbed directly, reducing the load on the body's internal heating system, but most of the body's energy comes via the transduction of the sun's radiation into chemical energy by plants through the process of *photosynthesis.* When animals eat plants and breath oxygen, the *chemical energy* thus acquired is converted to a form accessible to the cell.

There are usually several stages to energy production in the body. The food is first digested to simple molecules as already mentioned, and the energy-rich molecules—mostly simple ***sugars***—are transported via the bloodstream to the cells of the body. Excess energy is stored by individual cells or by a supracellular energy storage system from which energy can be returned to the blood stream, when needed, in the form of **ketone bodies.** Cells take up the sugars or *ketone bodies* and use molecular oxygen to synthesize *ATP,* which can be

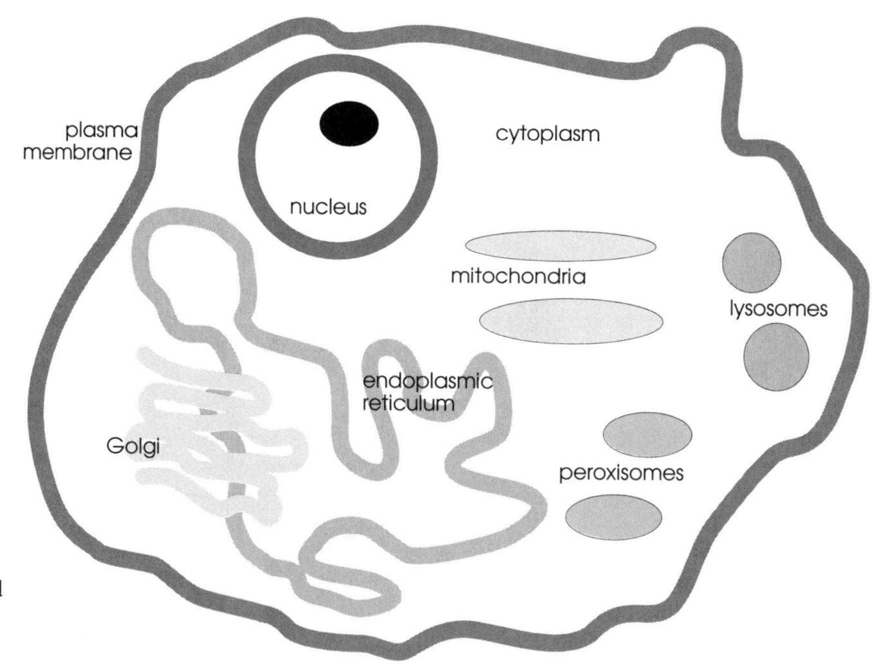

Figure 1.2.
Diagrammatic representation of a eukaryotic cell and its major organelles. Although cells vary widely in their detailed appearance, most cells in the body contain the organelles shown here.

regarded as the energy currency within the cell. ATP is synthesized in the cytoplasm and *mitochondria* and used throughout the cell as energy is needed. These processes are discussed in more detail in Chapter 11.

Many of the reactions that the cell carries out are energetically unfavorable—that is, the product has more internal energy than the initial compounds. The cell complies with the Laws of Thermodynamics by coupling such reactions to hydrolysis of ATP so that the overall final state (including hydrolyzed ATP) has less internal energy than the overall initial state (including ATP).

FORCES BETWEEN MOLECULES

Forces acting between molecules are, by definition, noncovalent. A full discussion of intermolecular forces is well beyond the scope of this book, but some of the more significant features are important for understand-

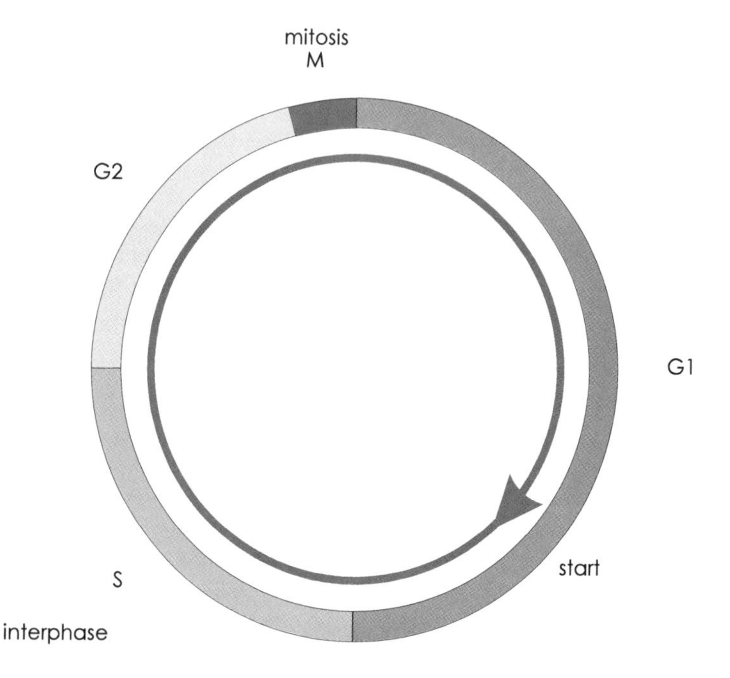

Figure 1.3.
Cell cycle outline. The cell cycle begins at a point in G1 phase and moves clockwise through S phase, G2 phase, and mitosis. The cells then divide and reenter G1 phase.

ing macromolecular structures and interactions which are central to a molecular view of life. In addition, noncovalent forces act between adjacent parts of an individual macromolecule, determining its preferred shape. Forces between separate molecules are *inter*molecular; the same forces acting within one macromolecule are *intra*molecular.

The major noncovalent molecular forces are:

1. ***Electrostatic interaction*** (also called ***ionic bonds*** or ***salt links***)
2. ***Hydrogen bonds***
3. *Steric hindrance*
4. Van der Waals attraction
5. ***Hydrophobic*** *interaction*

1. *Electrostatic interactions* take place between chemical groups that carry an *electrical charge,* as in interactions between ions where like charges repel and unlike charges attract. The interactions are strong, but fall off with the square of the distance between the charges. In some cases, although the molecule does not have fully charged groups, there may be a partial charge separation—as in ***α-helices*** (Chapter 3)—creating an *electric dipole* that can interact electrostatically with other charges or dipoles. Electrostatic interactions are very sensitive to the environment. They are weakened by a water environment and further weakened by the presence of small ions such as Na^+ or Cl^-. Thus, changes in salt concentration, including those containing the divalent ions Ca^{2+} or Mg^{2+}, can have dramatic effects on macromolecular structure.

2. *Hydrogen bonds* are noncovalent interactions in which a hydrogen atom is shared between two other atoms, as in >N—H . . . O═C< where the hydrogen is shared by the nitrogen and oxygen atoms. In this example, the shared hydrogen is more closely associated with the nitrogen and thus the nitrogen is the hydrogen donor and the oxygen is the hydrogen acceptor. In other examples, oxygen atoms, as well as nitrogen atoms, may act as hydrogen donors and oxygen or nitrogen may be hydrogen acceptors. Hydrogen bonds have optimum lengths, and are directional, the shared hydrogen preferring to lie on the straight line joining the donor and acceptor atoms. Thus, the atoms involved have to be placed precisely with respect to one another, making the hydrogen bond an important feature in determining *specificity* in molecular interactions.

The energy associated with one hydrogen bond is comparable with thermal energies at body temperature so that any one hydrogen bond is weak and easily broken. However, macromolecules typically form many hydrogen bonds and the cumulative effect can be decisive. Macromolecules may exist in many different shapes, or conformations, with similar energies. The formation of hydrogen bonds can be a major factor, determining which conformation is preferred.

3. By far the strongest noncovalent interaction is the repulsion that occurs when two atoms get close enough for their electron clouds to interact directly. This repulsion is very strong, although it drops off extremely rapidly as the atoms move further apart. The force—often called *steric hindrance*—prevents spatial overlap of any atoms that are not covalently bonded.

4. *Van der Waals forces* occur between atoms that are close together but not "overlapping" and provide a weak attraction. Although Van der Waals forces are individually weak, biological macromolecules are generally very dense structures with many close atomic contacts. The sum of the Van der Waals forces is a significant factor in the stability of biological structures.

5. *Hydrophobic interactions* are explained in more detail in Chapter 3 in the context of protein structure. Simply stated, atoms or groups of atoms that cannot form hydrogen bonds tend to aggregate in aqueous solutions to minimize their exposure to water, whose structure they disrupt.

SUMMARY

1. The body is organized in a hierarchy of structures starting with molecules as the smallest unit—except for some ions—and progressing to macromolecules, macromolecular complexes, organelles, cells, tissues and organs, the complete organism, and groups of organisms.

2. Cells are classified as either prokaryotic, simple cells, like bacterial cells with no separate cell nucleus, or eukaryotic, more complex cells such as those found in the human body.

3. Eukaryotic cells have many separate organelles. Mitochondria are the site of ATP synthesis; nuclei contain most of the nucleic acid of eukaryotic cells, including the chromosomes; the endoplasmic reticulum divides the cell into different topological domains.

4. Cells require energy to maintain their or-

dered state. Most of this energy is used in the form of ATP, which is readily broken down to release energy for cellular needs.

5. Molecules are held together at the most basic level by covalent bonds.

6. Intermolecular interactions are key to a cell's function and they occur through an array of many small noncovalent forces including electrostatic interactions, hydrogen bonds, steric hindrance, Van der Waals interactions, and hydrophobic interactions.

REFERENCES

Learning Biochemistry, Richard F. Ludueña, 1995, Wiley, New York, NY.

Introduction to Cell and Molecular Biology, Stephen L. Wolfe, 1995, Wadsworth Publishing Co., Belmont, CA.

Medical Cell Biology, Steven R. Goodman, 1994, J. B. Lippincott, Philadelphia, PA.

Principles of Protein Structure, G. E. Schultz and R. H. Schirmer, 1979, Springer, New York, NY.

Principles that determine the structure of proteins, C. Chothia, 1984, *Annu. Rev. Biochem.,* **53:**537–572.

REVIEW QUESTION

1. Which of the following statements about the forces involved in macromolecular structure is false?
   a) Covalent bonds are directional and of a fixed average length.
   b) Hydrogen bonds are directional and of a fixed average length.
   c) Electrostatic interactions may be either attractive or repulsive.
   d) Hydrophobic interactions are directional and of a fixed average length.
   e) Steric interactions are very strongly distance dependent.

ANSWER TO REVIEW QUESTION

1. **d** The correct response is d because the statement is false. Hydrophobic interactions tend to exclude water, rather than provide a specific "bond" like covalent, hydrogen, or electrostatic bonds. Steric interactions are very strongly distance-dependent because they provide a very strong repulsion when two atoms try to overlap in space, but there is essentially no steric interaction if the atoms are further apart.

CHAPTER 2

NUCLEIC ACIDS

INTRODUCTION

Deoxyribonucleic acid (**DNA**) is the hereditary material of all cells—genes are made of *DNA*. The last 30 years have seen enormous strides in our understanding of DNA, and in our ability to manipulate it, which have led to the biotechnology industry and to the beginning of a new level of understanding of cell biology. Technological progress has already led to important applications (e.g., *human growth factor,* tissue plasminogen activator, new clinical *diagnostic tools,* and genetic engineering of plants) and gene therapy is in the *clinical trial* stage. But we are still only at the initial stages of applying this powerful new knowledge. The next 30 years will see equally large strides in the application of our knowledge of DNA to clinical and other problems. We have learned much by using these new tools and have developed many new concepts. However, it is probably fair to say that we have barely begun to explore this new level of understanding and that the promise of molecular genetics and its application to diseases like cancer still requires many years of painstaking but exciting work, including but not limited to the study of genes and DNA sequences in the *Human Genome Project. Ribonucleic acid* (**RNA**) is involved in many different processes in the cell related to the conversion of the information in the genes into active gene products.

STRUCTURE

Chemical (Primary) Structure of DNA and RNA

The gross chemical structure of DNA led to a simple conceptual understanding of genetic processes in the 1950s. We are now beginning to understand the structure and function of DNA in more detail and the Human Genome Project (Chapter 21) is attempting to formulate a complete chemical description of human DNA by early in the next century.

Both DNA and RNA are nucleic acids, that is, chains, or polymers, of *nucleotides.* Each **nucleotide** found in DNA or RNA comprises three chemical units: a base, a sugar, and a phosphate group (Fig. 2.1). One base with its sugar is called a **nucleoside.** In RNA the sugar is a 5-carbon molecule, *ribose,* present as a ring structure; in DNA the 2′ carbon of ribose is reduced by the loss of an oxygen to give *deoxyribose.*

The bases found in DNA are either *purines* (*adenine* or *guanine*) or *pyrimidines* (*cytosine* or *thymine*), although some modified bases are found, such as *5-methyl cytosine.* Ribonucleic acid contains adenine, cytosine, and guanine as in DNA, but thymine is usually replaced by *uracil,* which lacks the 5-methyl group present in thymine (Fig. 2.2). In some types of RNA, unlike DNA, modified bases are common. In both DNA and RNA the base is attached through a nitrogen to the

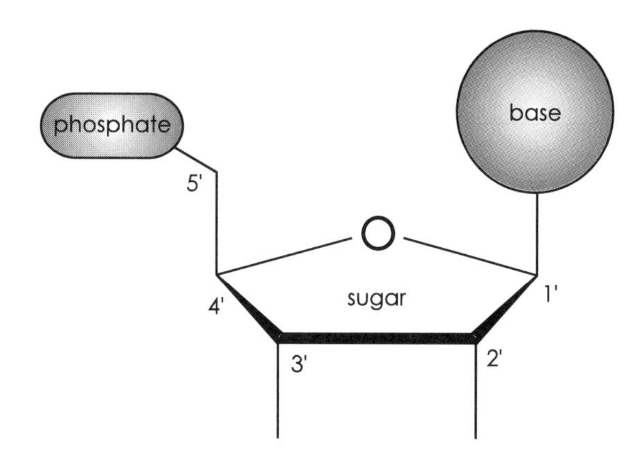

Figure 2.1.
Chemical structures of DNA and RNA nucleotides. The three components of a nucleotide—base, sugar, and phosphate—are shown diagrammatically.

1′ carbon of the sugar to give the nucleoside. Nucleosides are linked by a phosphate group joining the 5′ carbon of one nucleoside to the 3′ carbon of the next. Thus, nucleic acids have a direction: when the chain is written with a free 5′-OH on the left and a free 3′-OH on the right, the left to right direction is 5′ to 3′. The backbone of the polymer is formed from the sugar-phosphates and the bases form side-chains, as shown in Figure 2.2.

Some of the base sequences found in eukaryotic DNA are simple, that is, multiple *tandem repeats* of a short sequence such as TTAGGG found repeated at the ends (*telomeres*) of human chromosomes. Some of the sequence is accounted for by highly repetitive elements, such as the million copies of a nucleotide sequence 123 base pairs in length, known as the *Alu sequence,* which are found dispersed throughout the *human genome.* Other sequences can be recognized, such as specific protein binding sites, and may occur often, as in the case of the *TATAA sequence* found near the start of many genes. Much of the sequence, however, is complex, reflecting the complexity of protein sequences or the random nature of evolutionary changes.

The human genome comprises the information contained in one set of human chromosomes which themselves contain about 3 billion *base pairs* (***bp***) of DNA in 24 chromosomes (22 *autosomes* + 2 *sex chromosomes*; chromosomes are discussed in Chapter 5). Each DNA molecule is a linear, unbranched, polymer of great length (on the order of centimeters) in comparison to its width of 2 nanometers (nm). The great length of a DNA molecule is illustrated by calculating the length of DNA present in one adult human. The length is (length of 1 bp)(number of bp per cell)(number of cells in the body) which is (0.34 nm)(6×10^9)(10^{13}) or 2×10^{10} km, about as far as the distance from the earth to the sun and back. Molecules of RNA are shorter than chromosomal DNA, from less than 100 nucleotides up to several thousand nucleotides long.

Nucleic Acid Secondary Structure

In a DNA molecule, two sugar–phosphate backbones are twisted around one another with the bases inside. The bases stack to shield their hydrophobic faces from the solvent and form hydrogen bonds in two specific patterns, as shown in Figure 2.3, adenine pairing with thymine (or uracil) and guanine pairing with cytosine. This specific base-pairing scheme is known as *Watson–Crick base pairing* in recognition of its discovery by *James Watson* and *Francis Crick* in Cambridge, England in 1953, based on experimental data from *Wilkins, Chargaff,* and others. Both Watson–Crick and other forms of base pairing are found in parts of RNA molecules. One alternative form of base pairing is known as *Hoogsteen pairing* and other forms are generally known as *non-Watson–Crick base pairs.*

In DNA, essentially every base is part of a base pair; in RNA only a portion of the bases are paired. Most of the DNA in cells is in the structure known as **B-form DNA** (Fig. 2.4). *B-form DNA* is actually a family of similar structures that vary in their detailed conformation depending on the precise sequence of individual bases along the chain. B-form DNA comprises two sugar–phosphate chains with *complementary sequences* such that when the two chains are wound round one another, in opposite (*antiparallel*) directions, to form a right-handed *double helix,* each base is opposite its Watson–Crick partner and a base pair is formed (Fig. 2.5). The helix is said to be *right-handed* because the residues spiral clockwise as you look forward along the DNA chain. The phosphate groups are on the out-

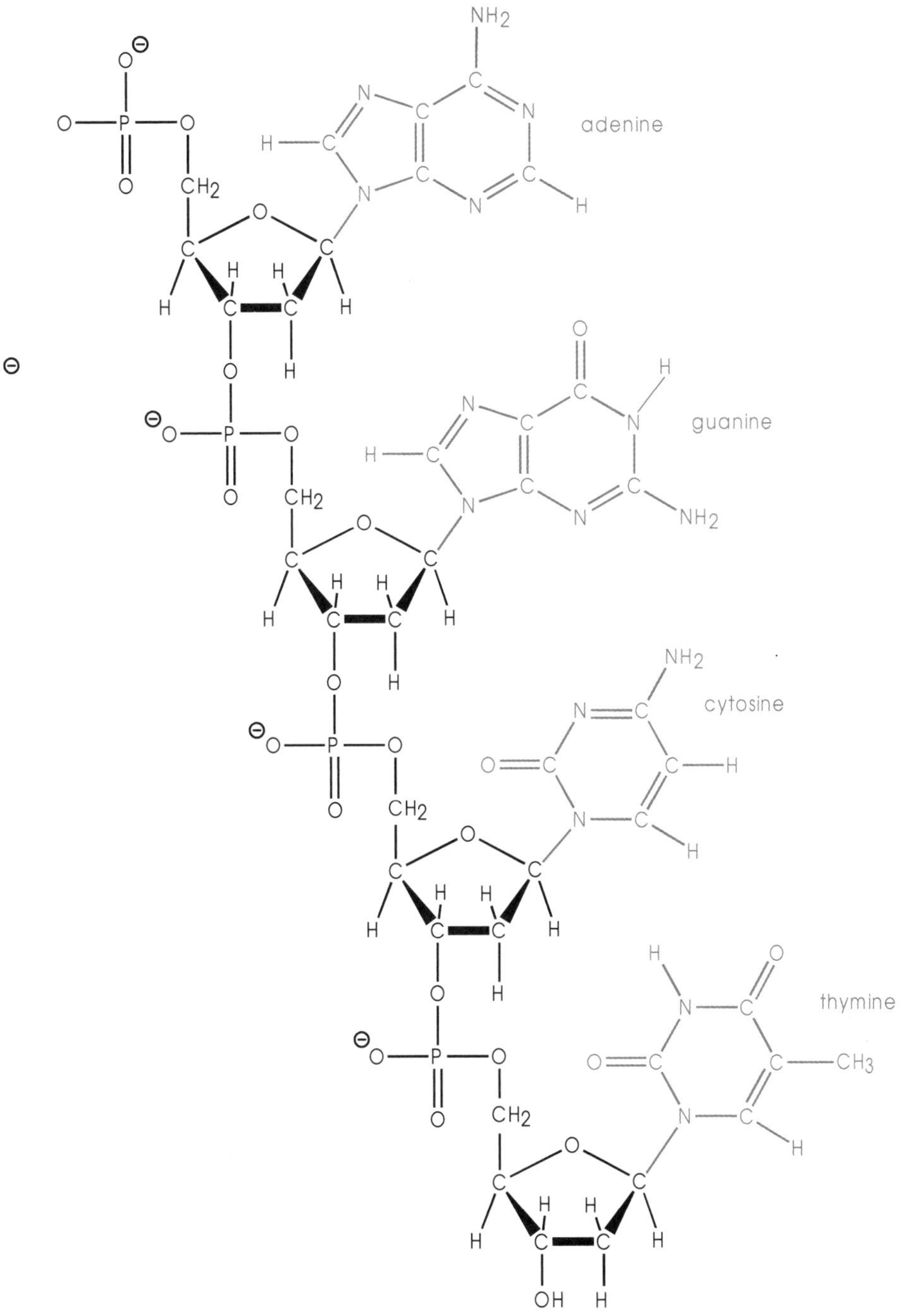

Figure 2.2.
Chemical structure of a residue in a polynucleotide. The chemical formula of a single strand of DNA, four bases long, is shown. The bases are picked out in purple and their names are given. The sugar–phosphate "backbone" is black.

side of the structure, where solvent ions can interact with the charges, and there are "grooves" between the phosphate chains—a ***minor groove*** and a ***major groove.*** The edges of the base pairs are accessible to the solvent in the grooves and provide regions where specific protein binding can occur in a sequence-dependent fashion. The B-DNA double helix has 10–10.5 bp per turn of the helix, a rise per residue of about 0.34 nm, and a diameter of about 2 nm. The DNA molecules can be "seen" in conventional electron microscopy if they are stained or shadowed to increase contrast, or "felt" in an *atomic force microscope.*

Under appropriate circumstances, DNA can adopt radically different secondary structures such as *A-DNA, H-DNA, Z-DNA,* or *cruciform DNA.* A-DNA is a Watson–Crick base-paired double helix that is shorter and fatter than B-DNA. The structure is found in RNA and in situations where hybrid double helices form with one RNA strand and one DNA strand; the 2′-OH found in RNA is inconsistent with a B-DNA

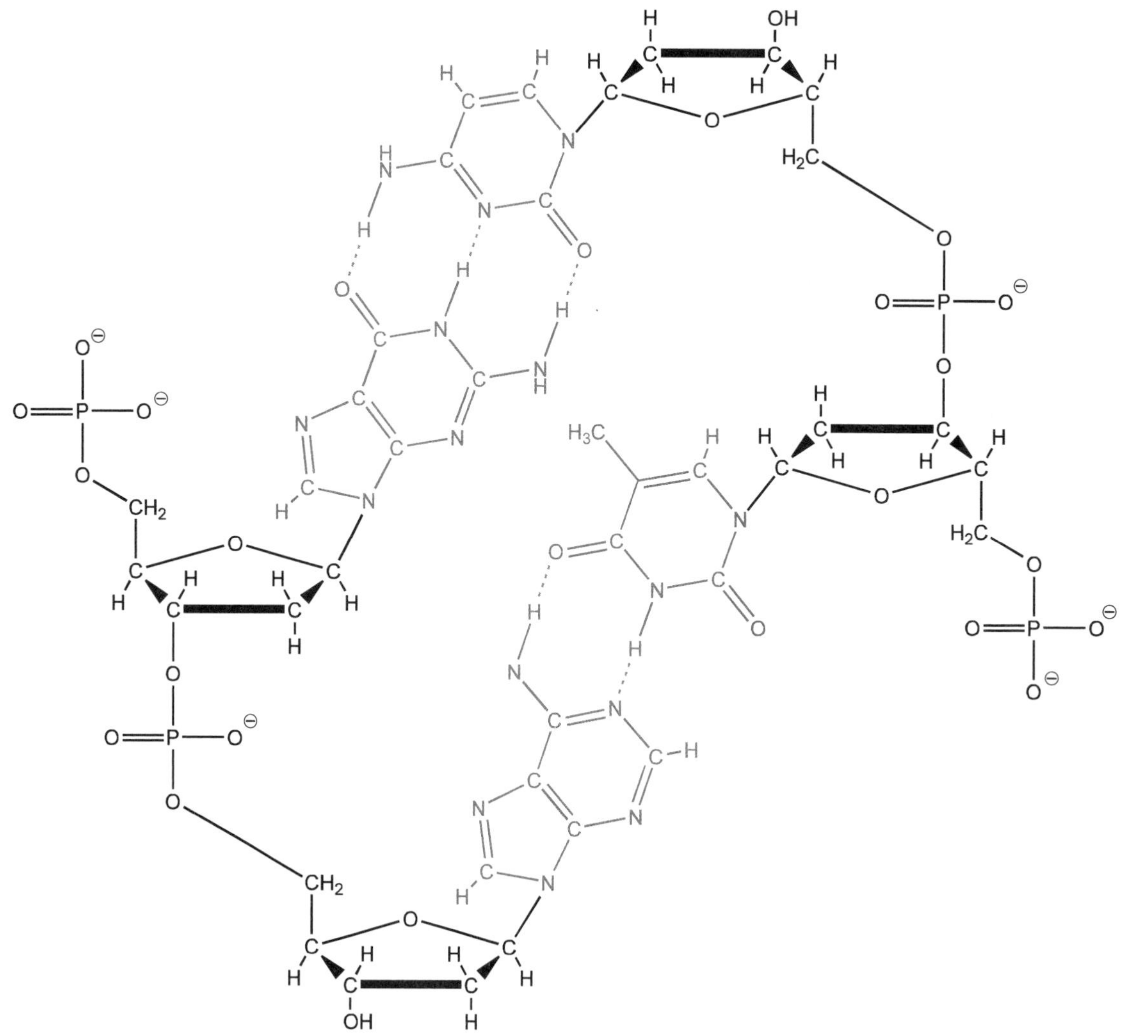

Figure 2.3.

Chemical structures of base pairs. The chemical formula of a double strand of DNA, 2 base pairs long, is shown. The base pairs are picked out in purple. The upper base pair is a G–C pair and the lower one is an A–T pair. The hydrogen bonds are shown as gray, broken lines. The sugar–phosphate backbone is black.

structure because of steric hindrance. H-DNA includes a *triple-stranded* region, Z-DNA is a *left-handed double helix* of Watson–Crick base pairs, and a cruciform structure is mainly B-DNA in which part of the structure is formed by intrastrand base pairing where part of each strand loops back on itself. Quadruplex DNA contains 4 strands wound round each other.

In all the double-stranded forms, the two strands of DNA are wound together in an antiparallel manner. Hence, the two single strands, 5′-ATTCGAAT-3′ and 5′-ATTCGAAT-3′, will come together to form the double strand:

```
ATTCGAAT
TAAGCTTA
```

In the double strand, the top strand is normally written in the *5′ to 3′ direction.* The direction of the chain can be specified explicitly as follows: 5′-ATTCGAAT-3′. The nature of the end groups can also explicitly be added; for example, 5′ppp-ATTCGAAT-3′OH has a 5′ triphosphate on one end and a hydroxyl on the other.

The RNA molecules may also occur as complete double helices containing two antiparallel RNA strands, similar to a DNA double helix in the A-DNA conformation. In many RNA molecules, however, double-stranded structures are formed, but are interspersed with single-stranded regions. The RNA double helices may be intramolecular, that is, the RNA strand folds back on itself. To get Watson–Crick base pairing in an intramole-

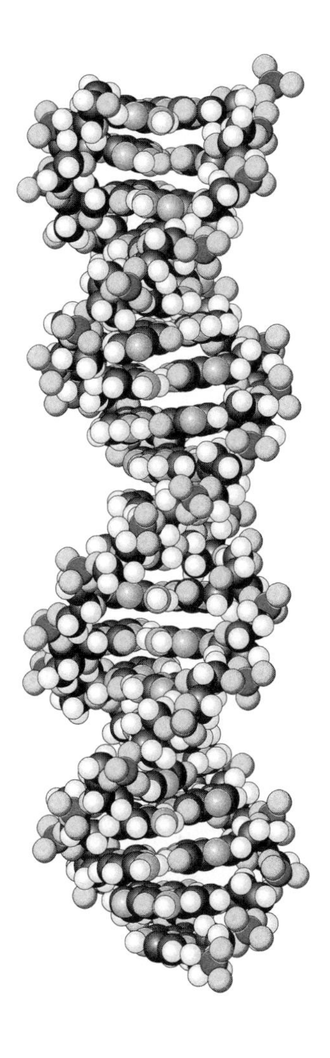

Figure 2.4.

B-form DNA molecular model. A space-filling model of B-DNA is shown with the helix axis vertical. The phosphate atoms are picked out in dark purple, nitrogens in pale purple, oxygens in purple-grey, carbons in black and hydrogens in light gray. The B-DNA structure may be visualized interactively on the World Wide Web at http://moby.ucdavis.edu/HRM/Biochemistry/molecules.htm.

cular double helix, the nucleotide sequence on one side of the helix must be an **inverted repeat** of the sequence on the other side. For example, in the sequence **ACUGGAAU**CCUC**AUUCCAGU,** the second bold sequence is an *inverted repeat* of the first bold sequence. This is hard to see unless the complimentary strand is drawn in:

ACUGGAAUCCUC**AUUCCAGU**
UGACCUUAGGAG**UAAGGUCA**

Now you can see that the sequence on the top strand reading from left to right starts with **ACUGGAAU** and the sequence on the bottom strand reading from right to left is **ACUGGAAU.** We use the top strand for reading left to right and the bottom strand for reading right to left so that we are always going in the **5′ to 3′ direction** along the nucleic acid strand. This inverted repeat is interrupted by the arbitrary sequence, CCUC, which can form a loop allowing the second copy to fold back and base pair with the first copy of the repeat (Fig. 2.6). Cruciform structures (Fig. 2.7) are formed in DNA in a similar way.

Contiguous inverted repeats (i.e., inverted repeats with no spacer between the repeats) are examples of **palindromes.** A *palindrome* looks the same read from left to right or from right to left, as in the word "deed." A palindromic DNA or RNA sequence can fold back to form a "*hairpin*" double-stranded structure (Fig. 2.6).

Repeats that are not inverted are *direct repeats* or *tandem repeats.* **ATTCGAAT**ATTCGAAT contains a contiguous direct repeat (the first copy is in bold characters to emphasize the repeat).

Virtually all the nucleotides in DNA are in double-stranded secondary structures, but this is not true for RNA. In addition, RNA double helices may be less regular than B-DNA with some *pyrimidine–pyrimidine base pairs* and sometimes a *looped-out base* (Fig. 2.8). The determination of which nucleotides in an RNA molecule are in double-stranded structures involves both ***phylogenetic*** and biochemical data. In the phylo-

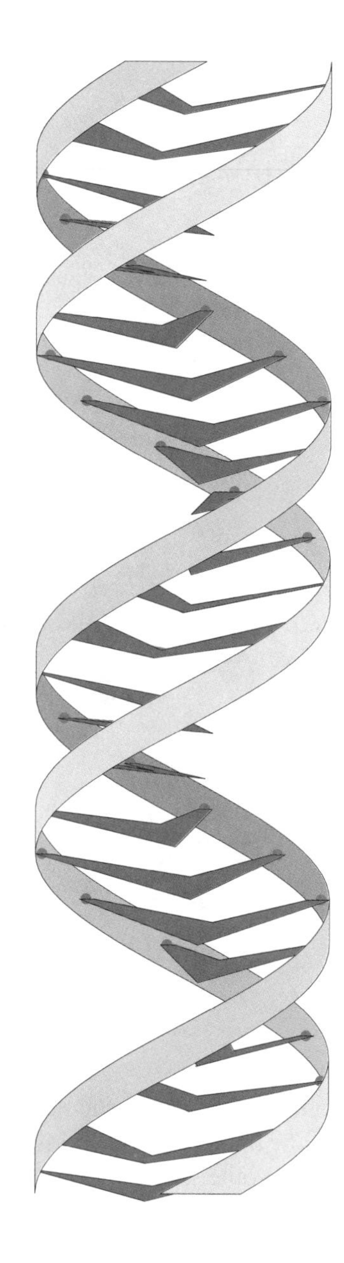

Figure 2.5.

B-form DNA cartoon. B-form DNA is shown in cartoon form, emphasizing the two sugar–phosphate backbones as ribbons connected by base pairs. An interactive ribbon view of B-DNA is on the World Wide Web at http://moby.ucdavis.edu/HRM/Biochemistry/molecules.htm.

genetic approach, models of base pairing are tested by looking for paired *mutations* in homologous RNA molecules from different species. If a mutation occurs in a required double-stranded region, then a ***complementary mutation*** should be found on the opposite side of the helix so that satisfactory base pairing is maintained in the mutant. Biochemical data involves reagents that distinguish between bases that are in double-stranded secondary structures and bases that are in single-stranded regions. The combination of these approaches has led to extensive secondary structure maps of *ribosomal RNA* molecules and other RNA molecules. The deter-

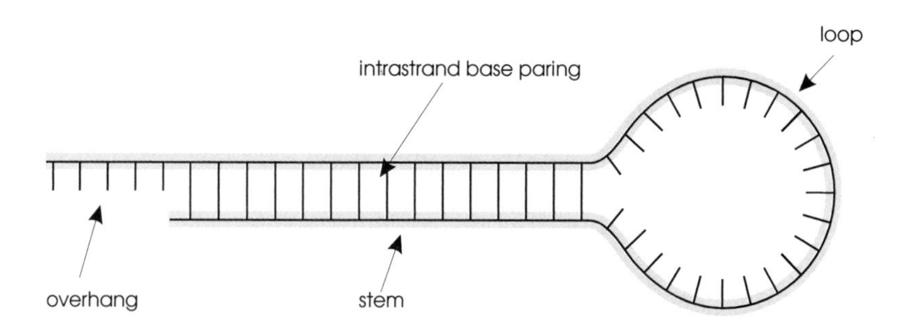

Figure 2.6.

A hairpin loop. A molecule of RNA is shown, in diagrammatic form, in a hairpin-loop structure. The sugar–phosphate backbone is highlighted in purple and the individual bases are indicated by short lines perpendicular to the backbone.

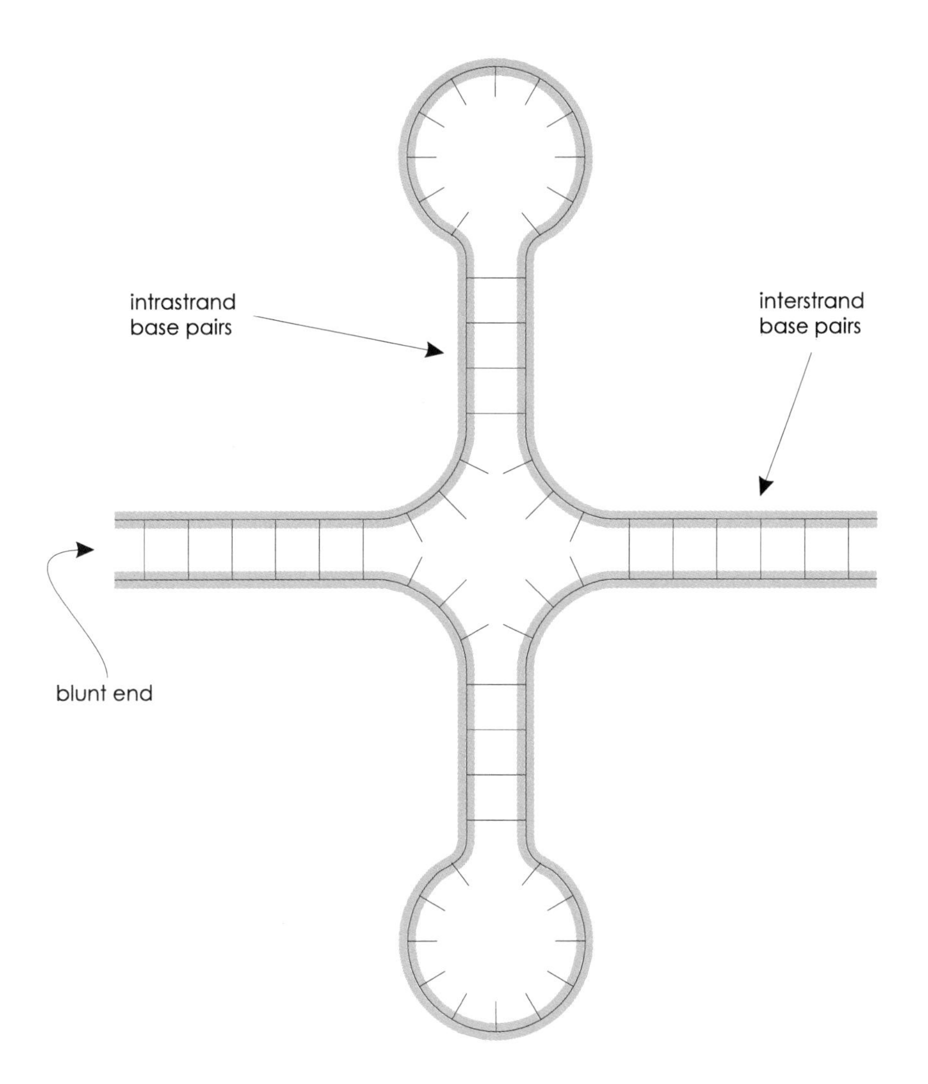

Figure 2.7.
A cruciform. A nucleic acid molecule is shown in a cruciform structure. Two strands with complementary base sequence are shown with the backbones highlighted in purple, the individual bases are the short lines perpendicular to the backbones. Base pairing occurs wherever a base is shown joined to its partner. Note that both interstrand and intrastrand base pairing occur.

mination of the *tertiary structure,* however, remains a challenge in most cases.

Nucleic Acid Tertiary Structure

Pure DNA folds randomly in solution depending on its natural stiffness and the tendency of some sequences to produce bends. B-DNA is quite flexible and a sequence of 200 bp can readily flex through 90°. Stable DNA bends are a common feature of DNA bound to proteins and, in general, when complexed with proteins, DNA takes up specific folded structures which are essential to its function. Examples include the folding of DNA in the *nucleosome* (Chapter 5) and its interaction with *recombination enzymes* (Chapter 23). Also, DNA tertiary structure may be important in the *control of transcription* (Chapter 29). The folding of DNA in chromosomes is a remarkable feat and is discussed further in Chapter 5.

In some cases, more is known about tertiary structure in RNA than in DNA. The full three-dimensional structure is known for some ***tRNA*** molecules from ***X-ray crystallography*** data. The structure involves several *stem-loop* regions of secondary structure which interact with one another to give a specific tertiary structure. The tRNA structure shown in Figure 2.8 includes stem-loop structures, unusual bases, pyrimidine–pyrimidine base pairs and looped-out bases. The inherent structure of RNA is clearly critical for its function, as shown by the conservation of secondary structure in ribosomal RNA (below) and by studies of *catalytic RNA* and ***RNA splicing*** (Chapter 25).

DOMAINS AND SUPERCOILING

The discussion above assumes that the nucleic acid has free ends. In practice, for DNA, this is not a good approximation to the situation in the cell. ***Plasmids*** and

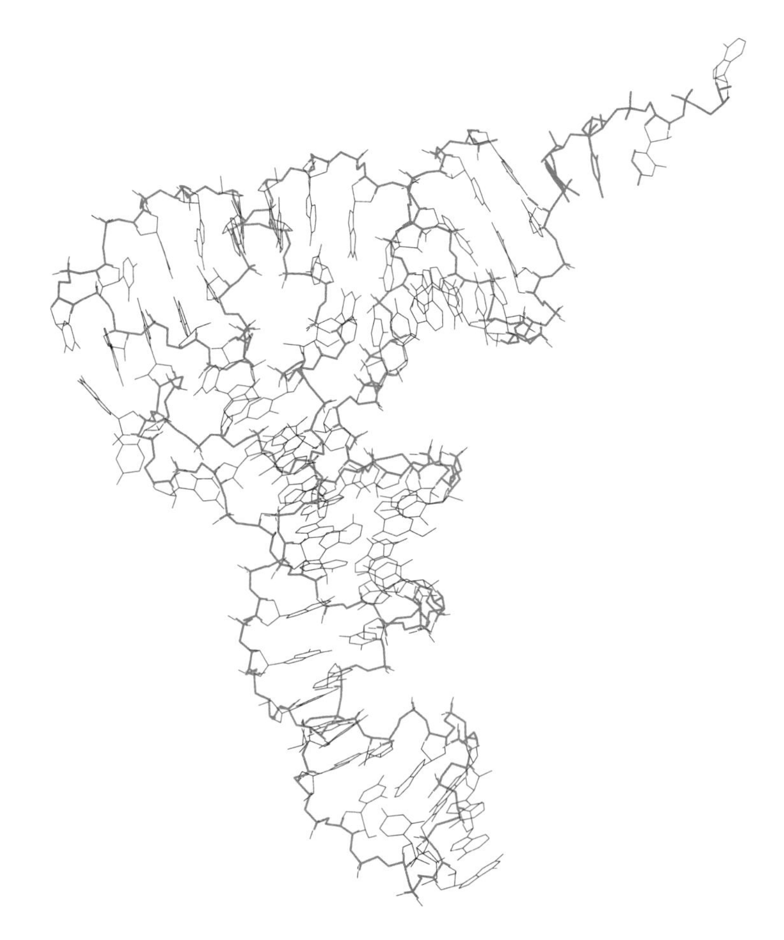

Figure 2.8.

tRNA tertiary structure. A tRNA molecule is shown in the structure determined by X-ray crystallography. The backbone is picked out in purple and the bases and parts of the sugars are in black. Note the single-strand 5′ end at the upper right, and the base-paired stems along the top and leading to the anticodon loop at the bottom of the figure. The other two stems are folded on top of one another in the center and are difficult to distinguish in this view of the whole molecule. An interactive copy of this structure is available on the World Wide Web at http://moby.ucdavis.edu/HRM/Biochemistry/molecules.htm.

bacterial chromosomes are circular, and *eukaryotic chromosomes,* although linear, have their DNA anchored at points 50–100 kb (1 kb = 1000 bp) apart, thus separating the chromosomal DNA into ***topological domains*** (Fig. 2.9). The ends of a domain are not free to move, but are associated with a matrix structure in the nucleus which includes an enzyme, *DNA topoisomerase II,* which is discussed further below.

In a section of DNA with free ends, if the center is twisted, the twist will propagate in both directions along the DNA and "fall off" at the ends. The *twist* that moves in one direction will be right-handed, and in the other direction, left-handed. These additional twists to the DNA are known as positive and negative **supercoiling.** Positive *supercoiling* tends to increase the twist between one nucleotide and the next, negative supercoiling reduces it. In a circular DNA, the supercoils will "cancel out" when they meet opposite the original twist. The subject of supercoiling in DNA may be referred to as *DNA topology.*

If a piece of DNA with fixed ends is twisted in the center, the supercoiling can propagate only as far as the fixed ends, and if the original twist is maintained, the *supercoils* will be maintained. In bacteria, supercoils can be generated by an enzyme, *DNA gyrase,* which has not been found in eukaryotes. In all cells, supercoils are generated by *DNA polymerase* and *RNA polymerase* as they move along the DNA strand, "reading" the base se-

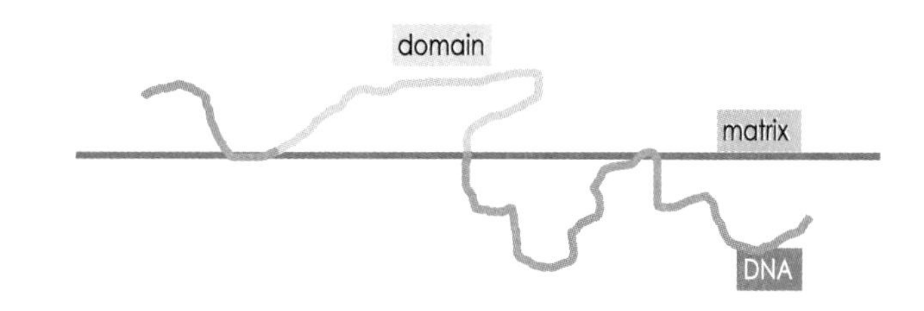

Figure 2.9.

Chromosomal domains. The straight gray line represents part of the nuclear matrix and the purple line represents part of a chromosome. The chromosome is attached to the matrix where the lines meet or cross. The light purple part of the chromosome represents one domain.

quence (Fig. 2.10). Other proteins may also induce supercoiling, as *histones* do when DNA is wound into **nucleosomes** (Chapter 5). When DNA is isolated and separated from the nuclear matrix and binding proteins, it will relax to a nonsupercoiled form unless it is circular, in which case the supercoiling will be retained.

In the cell, supercoiling can be relaxed only by cutting the DNA. Cutting one strand is sufficient, since rotation may then occur around the remaining sugar–phosphate bond to remove the supercoil. This untwisting reaction is catalyzed by *DNA topoisomerase I,* which also seals the single-stranded gap after the supercoil has relaxed. DNA topoisomerase II can also relax supercoiled DNA, but it does so by making a temporary double-strand cut instead of a temporary single-strand cut. DNA topoisomerases are used in the cell to allow DNA to be unwound for ***replication*** (Chapter 22) and *transcription* (Chapter 24), and DNA topoisomerase II is used in mitosis to "untangle" DNA and allow it to fold into the special structure adopted by chromosomes during the process of mitosis.

DNA DENATURATION

The DNA double strand can be denatured by raising the pH or the temperature, or drastically lowering the ionic strength. Denatured DNA consists of the two single strands free in solution. Since these experiments may be carried out by raising the temperature of a solution of DNA, the *denaturation* process is often referred to colloquially as "melting" the DNA.

The strength of the double-strand structure depends strongly on the *ionic strength* of the solution and its pH and on base composition: GC base pairs, with their 3 hydrogen bonds, are more stable than AT base pairs, with their two hydrogen bonds. Hence, a DNA double strand denatures in stages with the AT-rich regions denaturing first and the GC-rich regions denaturing last. This can be seen either by capturing the intermediate state with "bubbles" in the AT-rich regions by electron microscopy or, in solution, by measuring the increase in absorbance of *ultraviolet light* (reflecting loss of *base stacking*) that accompanies denaturation (Fig. 2.11).

Reassociation

When denatured DNA is returned to ***physiological conditions*** (the conditions found in the body), the single strands come back together to regenerate B-form DNA in a process called *renaturation* or *reassociation.* Renaturation requires the base pairs to match up. Hence, a given single strand will only reassociate with a single strand that has a complementary sequence. The rate of renaturation depends on the concentration of the complementary sequences. In part, this is just a simple matter of how much DNA is present per unit volume of solution, a critical parameter in reassociation (and *hybridization*—see Section "Hybridization"). If the concentration of complementary strands is high, they will renature rapidly; if the concentration is low, they will renature slowly. Thus, in laboratory experiments it is important to have

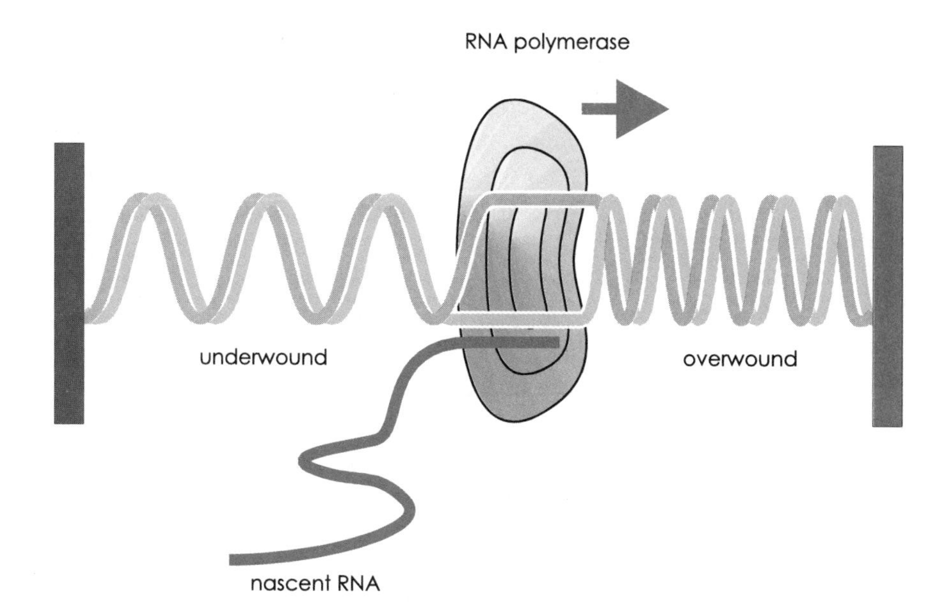

Figure 2.10.

Supercoiling by RNA polymerase. The purple and gray lines represent the two strands of the DNA template, anchored to the nuclear matrix indicated by the hatched boxes. The ellipse represents RNA polymerase synthesizing RNA (the solid purple line).

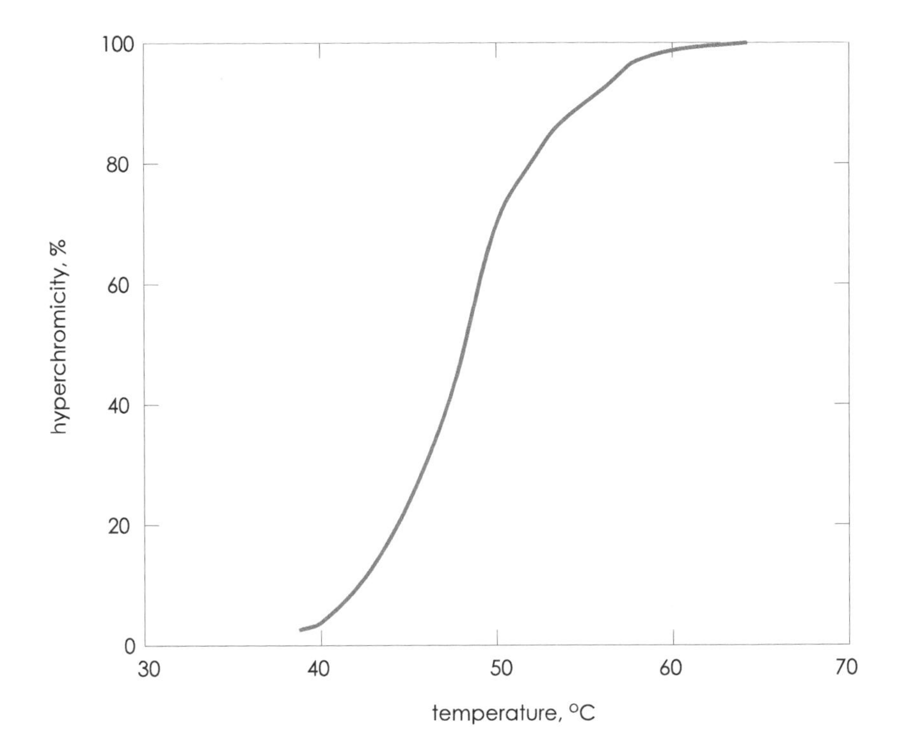

Figure 2.11.
Thermal denaturation of DNA (UV melting curve). A solution of double-stranded DNA was placed in a spectrophotometer and the change in absorbance at 260 nm was measured as the temperature of the sample was increased.

the concentrations of the species that are to renature as high as possible. In the body, chromosome structures and specific proteins help to keep separated DNA strands close together and hence in high local concentration.

The effect of concentration is complicated by the *complexity of DNA*. At one extreme, if many fragments of a simple *homopolymer* made from poly dA and poly dT are denatured and then renatured, the renaturation will be very fast since any poly dA strand can pair with any poly dT strand. Consider a bacterial DNA broken into many fragments of 200–500 bp and denatured. When this mixture is renatured, the process will be much slower because each bacterial DNA fragment will have to "search" for its complement rather than renaturing with the first fragment with which it collides.

Human cells contain much more DNA than bacteria and so fragmented human DNA should reassociate more slowly than fragmented bacterial DNA. For a large part of the human DNA that is true, and a very slow rate of reassociation is seen. For a significant fraction of human DNA, however, much faster reassociation rates are seen. Reassociation rates for eukaryotic DNA are divided into four classes (Fig. 2.12):

1. Zero-time reassociation comes from palindromic sequences that can form "*snap-back*" or "hairpin" structures. Such molecules reassociate very rapidly, at a rate independent of their concentration, because they do not need to interact with another molecule to form a double-stranded structure.
2. Very rapidly reassociating sequences come from a simple sequence that is repeated many times (up to a million times per cell) in clusters of **tandem repeats** of the same sequence. Such DNA is also called *satellite DNA* since it forms separate, satellite peaks in density gradients due to its uniform base composition. One satellite DNA is found at the telomeres (ends) of chromosomes; another is at the *centromeres.*
3. *Intermediate reassociating sequences* come from heterogeneous groups of sequences that are repeated 10–10,000 times per cell. Many of these sequences are interspersed—scattered—throughout the genome. Others represent tandem repeats of highly repeated genes, such as the genes for ribosomal RNA or histones (Chapter 21).
4. *Slowly reassociating sequences* come from unique sequences that are not repeated, which includes most genes.

The classes overlap to some extent, especially when the conditions of reassociation are varied. Under

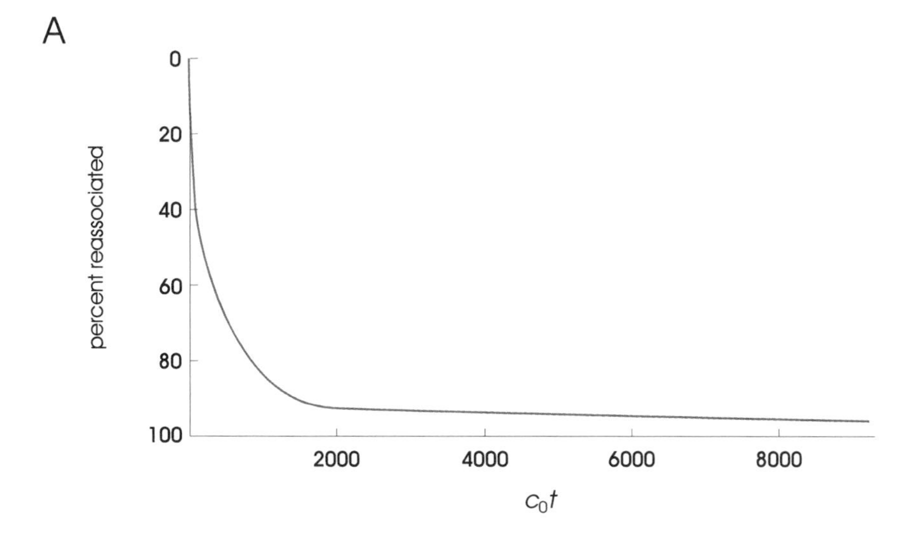

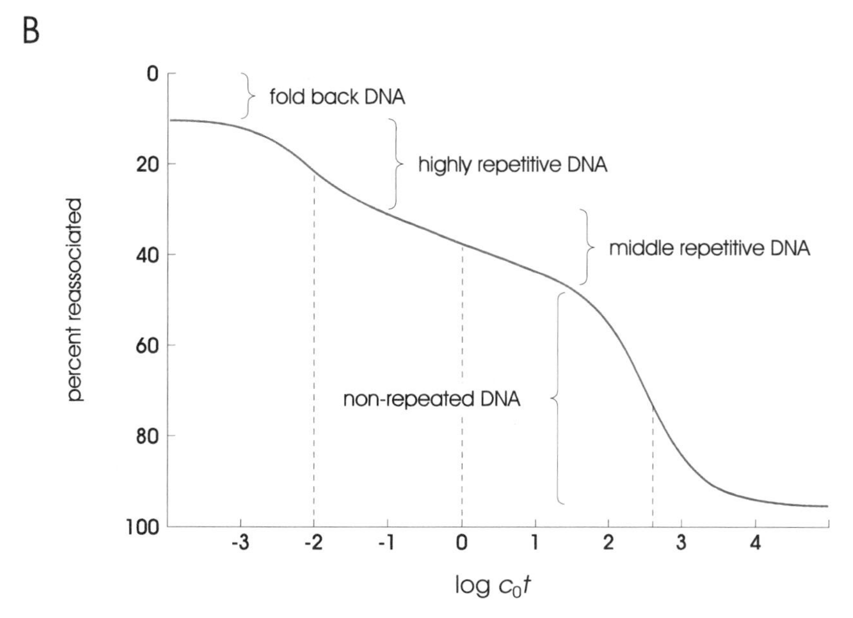

Figure 2.12.

Reassociation kinetics. (A) The reassociation of denatured DNA, initial concentration c_0, is followed as a function of time t on a linear scale. (B) The same data is shown on a logarithmic scale.

"stringent" conditions, only exact matches of base pairs will renature. Under less-stringent conditions, a certain amount of mismatch can be tolerated and closely related sequences will renature.

Hybridization

Additional DNA strands, or RNA strands, can be added to a denatured DNA sample and allowed to participate in the renaturation reaction. When the added DNA or RNA associates with one of the original DNA single strands this is called hybridization. Hybridization is a very powerful technique for detecting particular DNA or RNA sequences, especially in conjunction with *electrophoresis* (***Southern blotting*** and ***northern blotting***) or in primer annealing in the ***polymerase chain reaction*** (***PCR,*** Chapter 22).

In the case of Southern or northern blotting, a mixture of DNA (for Southern blotting) or RNA (for northern blotting) is subjected to **gel electrophoresis** (Fig. 2.13). In this technique, a solution containing the mixture of nucleic acids is placed in a well in a slab of agarose or other gel material. An electric field is used to drive the molecules through the gel and when the molecules are dispersed throughout the gel, the electric field is switched off. The rate at which the molecules move through the gel during the electrophoresis stage depends on the charge on the molecule and on its size. DNA has a uniform charge density and flexible rod shape and thus moves through the gel according to its size, smaller molecules moving faster. RNA folds to give a variety of shapes and charge densities, but in a strongly denaturing solution it unfolds and also runs according to size. Thus, in both cases the nucleic acid molecules are displayed in the gel separated according to size, usually expressed as bp (DNA) or bases (RNA) (Fig. 2.14).

DNA is denatured in the gel (RNA is denatured before electrophoresis) and the face of the gel slab is placed in contact with a membrane made of *nitrocellulose, nylon,* or another material. The single-stranded nucleic acid molecules are then transferred from the gel to the membrane by *capillary transfer* (blotting) or other methods. The membrane containing the blotted nucleic acids is called the *blot.*

To detect a nucleic acid of a specific sequence, a nucleic-acid **probe** having a sequence complementary to the sequence of the sought-after nucleic acid is prepared. The probe may be either radioactive or chemically tagged with a fluorescent or chemiluminescent label. The blot is incubated with the *probe* under conditions that allow hybridization so that the probe reassociates with its complementary sequence on the blot and becomes bound to the blot. After sufficient time for reassociation has elapsed, the unbound probe is washed away and the presence and location of the hybridized probe can be determined from its radioactivity or other tag, indicating the presence of the complementary sequence on the original blot (Fig. 2.15).

In practice, it is important to use high enough concentrations of DNA on the filter and/or probe in so-

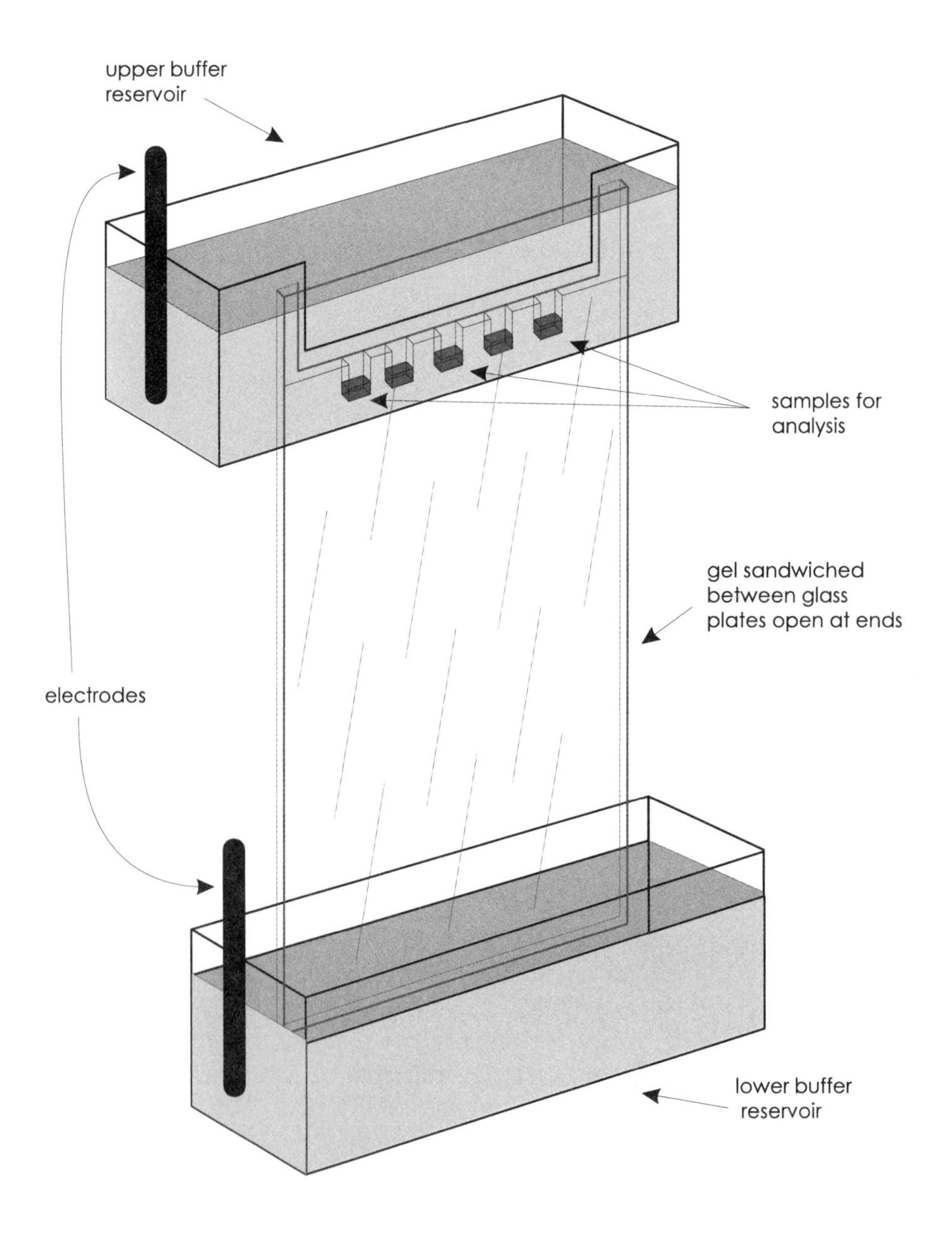

Figure 2.13.
DNA electrophoresis apparatus.

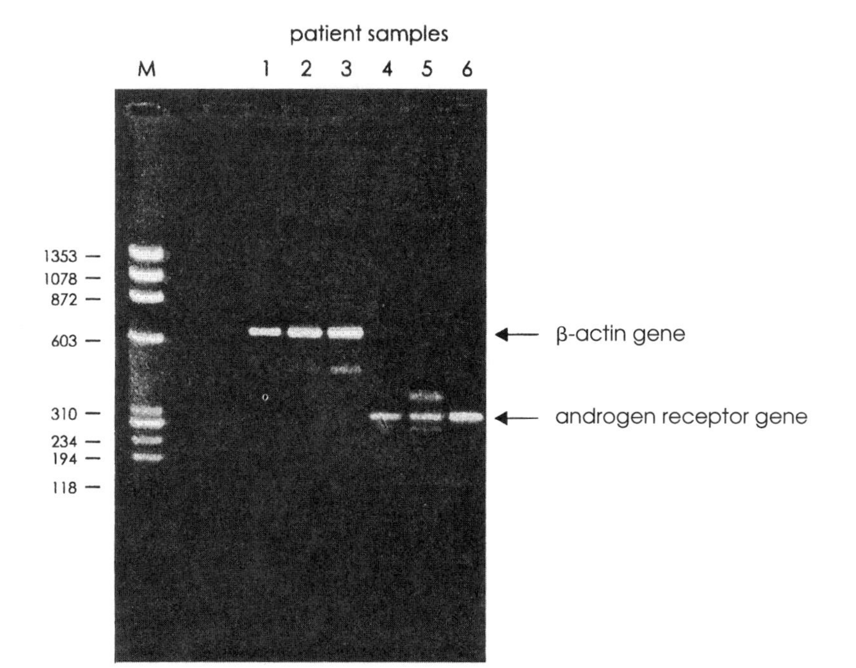

Figure 2.14.

DNA electrophoresis results. A gel after DNA electrophoresis. Lane M is molecular weight markers. The numbers along the left-hand side are the sizes, in bp. at the marker molecule.

lution. When using *genomic DNA,* an average *single-copy gene* fragment on the membrane is approximately 1 pg (10^{-12} g) of DNA, whereas the specific probe for that gene is present at approximately 1 ng/mL (10^{-9} g/mL) solution. Therefore, based on both the *complexity* of sequences in the genomic DNA, and on the small amount of specific sequence, hybridization needs to be carried out for a sufficient amount of time (about 24 hours). In contrast, when cloned DNA is fixed to the membrane, the amount of gene-specific DNA on the filter is approximately 1 μg (10^{-6} g); moreover, the complexity of other sequences on the membrane, such as plasmid DNA, is lower. In this case, the filter only needs to hybridize for 1 hour or less, using the same probe to get a comparable result. The important thing to remember is that nucleic acid hybridization, as exemplified by Southern blotting, is dependent on sequence complexity, nucleic acid concentration, and time. The lower the sequence complexity and the higher the nucleic acid concentration, the less time it takes for hybrid formation.

The term *Southern blotting* is used because the method was developed by *E. M. Southern* and the term *northern blotting* was introduced for RNA blots to distinguish them from DNA blots. More recently, protein chemists have gotten into the act by blotting electrophoresed proteins to a membrane and detecting specific protein molecules with antibodies, which is known as western blotting. Finally, southwestern blots (a renatured protein blot and a DNA probe) are used to study *protein–DNA interactions.*

NUCLEIC ACID SEQUENCING

The most important structural feature of DNA is the sequence of base pairs along the molecule. For RNA, both base sequence and higher-level structures are critical. Direct *sequencing of RNA* is difficult and will not be dealt with here. RNA can be *reverse transcribed* into DNA and the DNA sequence determined (Chapter 22). *DNA sequencing* is an evolving technology and currently based on the principles developed by *Sanger* and his colleagues and by *Maxam* and *Gilbert.* DNA sequence is determined in blocks of 200–600 bases and these blocks of sequence are then assembled into ever-larger blocks or "*contigs*" to give very long sequences, hundreds of thousands of bases, and eventually millions of bases. The principle of the Sanger method is outlined below.

DNA SEQUENCING BY SANGER'S METHOD

The *Sanger sequencing* method is outlined here. The principle depends on the ability of an enzyme, DNA polymerase, to make a very accurate complementary

1. break DNA into fragments by cutting at specific places

2. separate fragments by size using gel electrophoresis

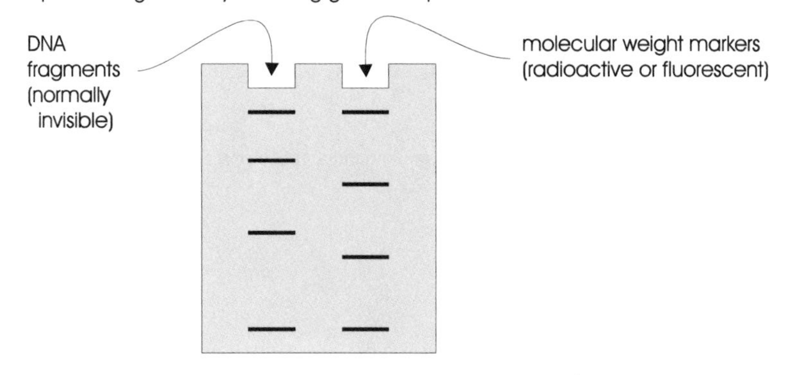

3. denature with alkali; blot DNA fragments from gel to nitrocellulose

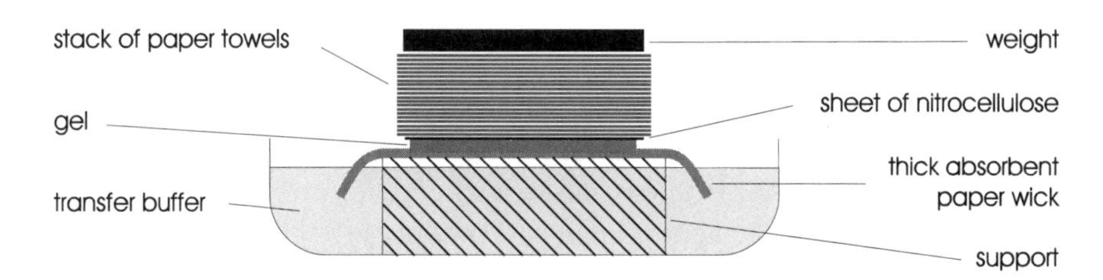

4. hybridize with radioactive (or fluorescent) probe; wash off unhybridized probe; determine position of radioactivity

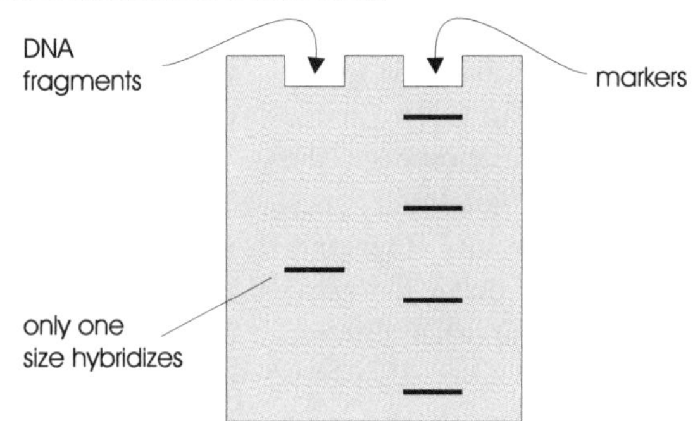

Figure 2.15.

Southern blotting. A schematic of the process of Southern blotting.

copy of one strand of DNA (the *template strand*). The polymerase needs a starting point on the template strand and a supply of the *nucleoside triphosphates* that are the substrates for DNA polymerase. For the starting point, chemical synthesis is used to make a *primer* molecule that comprises one strand of a short sequence of DNA complementary in sequence to part of the template strand and ending at the required starting point. The primer is mixed with the template strand and allowed to hybridize so that the primer binds to the complementary sequence on the template. One end of the primer then presents a free 3′-OH group for DNA polymerase to add a new nucleotide (Fig. 2.16).

The polymerase begins at the provided starting point and continues for an extremely long distance along the DNA strand, synthesizing the complementary copy as it goes by joining the 5′-phosphate of a new nucleotide to the free 3′-OH of the strand being synthesized. All four nucleotides must be provided or the synthesis will stop when the missing nucleotide is required. Synthesis will also stop if the 3′-OH is missing. In the Sanger method, four synthesis reactions are run in parallel, all with the same starting point. Each reaction contains a 2′-,3′-dideoxynucleotide (ddA in reaction 1, ddC in reaction 2, and so on for ddG and ddT) in addition to the normal four 2′-deoxynucleotides. Synthesis continues normally in each reaction until, by chance, a dideoxynucleotide is incorporated. When this happens, the chain can no longer elongate because there is no 3′-OH. Thus, each reaction will finish with a mixture of synthesized DNA strands all starting at the same point, but finishing at any of the points where a dideoxynucleotide could be incorporated. Thus, the reaction containing ddA contains only DNA sequences ending in ddA (corresponding to a T on the template strand), the reaction containing ddC contains only DNA sequences ending in ddC, and similarly for the reactions containing ddG or ddT.

The DNA molecules in each reaction are then separated according to their length. Since they all have the same starting point, the length of each fragment is the distance from the starting nucleotide to the finishing nucleotide. In the reaction containing ddA, all the finishing bases are A, and so the lengths of the fragments from this reaction give the positions of all the As. Similarly, the positions of all the Cs are given by the reaction containing the ddC, and similarly for the reactions containing ddG and ddT. Usually, the reaction

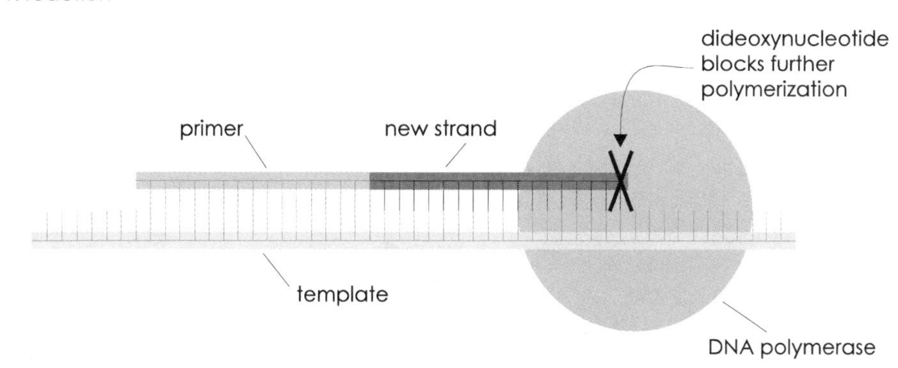

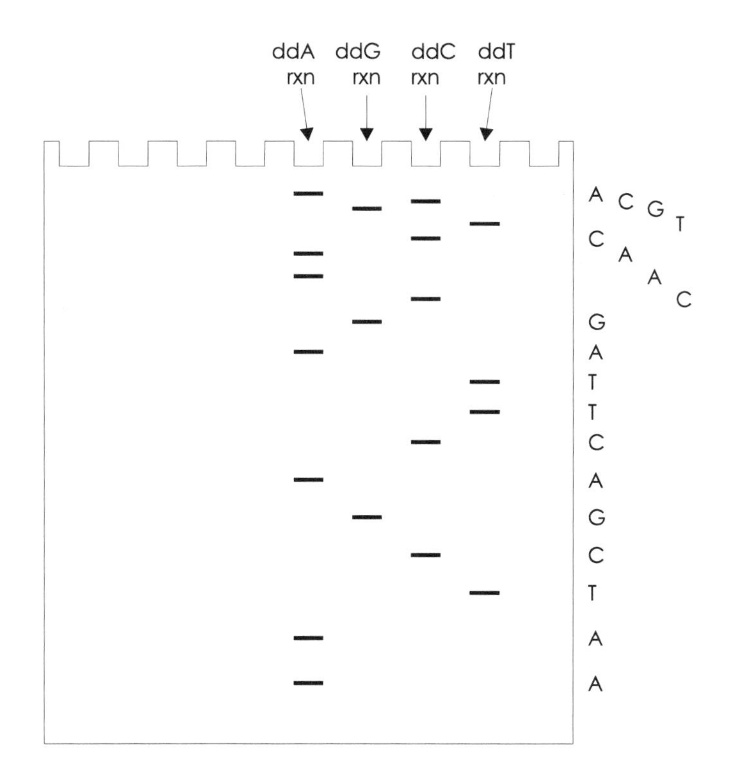

Figure 2.16.

Primer and template in Sanger sequencing. The cartoon, panel 1, shows a long single strand of DNA (labeled "template") annealed to a primer that is complementary in sequence to part of the template. DNA polymerase has already synthesized a "new strand" of DNA, starting from the 3′ end of the primer and continuing along the template strand until a dideoxynucleotide was incorporated. The polymerase is now stopped as there is no 3′-OH to allow synthesis to continue.

The lower panel shows an idealized sketch of the results of gel electrophoresis of many DNA molecules derived as shown in the top panel. The lane labeled "ddA rxn" contains DNA from a reaction mixture containing dideoxyadenosine triphosphate. Similarly, the ddG, ddC and ddT lanes had the indicated dideoxynucleotide.

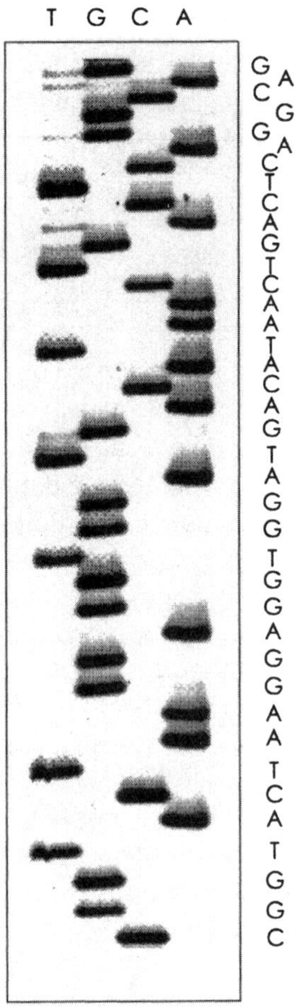

Figure 2.17.

Sanger sequencing gel result. An autoradiograph of a gel from a Sanger sequencing experiment. The nucleotide sequence can be read directly from the gel, as shown on the right of the gel.

mixtures are separated on adjacent lanes of a *polyacrylamide gel* and the sequence can be "read off" by looking for the shortest fragment. If it is in the lane containing the ddA reaction, it indicates the position of an A. The lane containing the next shortest fragment indicates the base following A and so on for as long as the gel can be read (Fig. 2.17).

SUMMARY

1. Nucleic acids store and process the genetic information with the help of proteins.

2. Nucleic acids are long polymers with a repeating structure comprising sugars joined through their 3′ and 5′ carbons by phosphate groups. Each sugar also has a base attached to the 1′ carbon.

3. DNA contains only deoxyribose sugars while RNA contains only ribose. The major bases found in DNA are adenine, guanine, cytidine, and thymine. RNA contains mostly adenine, guanine, cytidine, and uracil. The bases pair with one another to form A-T, G-C, or A-U base pairs.

4. Both nucleic acids can form antiparallel double-stranded structures with the bases stacked and paired in the center and the phosphates wound around the outside. In DNA the predominant structure in the body is B-form DNA, or B-DNA, a structure with bases perpendicular to the helix and about 10 base pairs per turn of the helix. In RNA, much of the structure is in single-stranded conformations with substantial regions of intramolecular double strands.

5. Many DNA molecules are torsionally constrained at their ends, either by being circular, and so lacking ends, or by being attached to a proteinaceous structure, the nuclear matrix. In eukaryotic cells, a region bounded by adjacent matrix attachment sites is a topological domain.

6. The processes of nucleic acid synthesis introduce supercoiling into DNA. This is relaxed in the

cell by the action of topoisomerases which break the DNA temporarily to allow relaxation and then rejoin the strand(s).

7. The two strands of a DNA double helix are easily separated by denaturing DNA by raising the temperature or the pH. The strands may reassociate when conditions return to physiological. The rate of reassociation depends on the concentrations of the complementary strands.

8. An additional nucleic acid sequence, a probe, may be included in a reassociation reaction and form double strands with components of the initial DNA. This hybridization reaction is the basis of the Southern and northern blotting techniques used to determine if a known nucleic-acid sequence is present in a mixture of nucleic-acid sequences.

9. The sequence of nucleotides contains the information carried by the molecule. Sequences are determined by the "Sanger" sequencing technique involving DNA synthesis from a known starting point to termination at a known base and measurement of the length synthesized.

REFERENCES

Understanding DNA, C. R. Calladine and Horace R. Drew, 1992, Academic Press, San Diego, CA.

What Mad Pursuit, F. H. C. Crick, 1988, Weidenfeld and Nicholson, London.

"DNA Topoisomerase Poisons as Anti-Tumor Drugs," I. F. Liu, 1989, *Ann. Rev. Biochem.,* **58:**351–375.

World Wide Web at http://moby.ucdavis.edu/HRM/Biochemistry/molecules.htm.

REVIEW QUESTIONS

1. Each of the following sequences represents one strand of a double-stranded oligonucleotide. Which oligonucleotide contains an inverted repeat?
   a) ATTGGCATGCG
   b) ATTGGTATTGG
   c) ATTGGTGGTTA
   d) ATTGGTCCAAT
   e) AAGTAAGTAAG

2. Two single-stranded DNA probes of the same length and concentration are added to an unfractionated mixture of DNA fragments from a human cell line. One probe is complementary to a region of ribosomal RNA; the other is complementary to a region of globin mRNA. The mixture is heated to denature the DNA fragments and then allowed to cool. Why does the ribosomal RNA probe form double-stranded structures before the globin mRNA probe does?
   a) The A+T content of the ribosomal RNA probe is higher.
   b) The G+C content of the ribosomal RNA probe is higher.
   c) There is more ribosomal RNA in the cell.
   d) The ribosomal RNA genes are repeated.
   e) The sequence of the ribosomal RNA is simpler.

ANSWERS TO REVIEW QUESTIONS

1. ***d*** You need to write in the lower strand to see which contains the inverted repeat. The lower strand for oligonucleotide d is 3′-TAACCAGGTTA-5′. In this case, the 5′-ATTGG sequence on the upper strand is an inverted repeat of the 5′-ATTGG sequence on the lower strand.

2. ***d*** The ribosomal RNA (rRNA) genes are greater in number than the globin genes, in any human cell, so the chance of a rRNA probe meeting a complementary human DNA sequence is greater than the chance of a globin probe meeting its complementary sequence. Hence, during renaturation/hybridization, the rRNA probe will form hybrid double helices before the globin probe does.

CHAPTER 3

PROTEINS

INTRODUCTION

Proteins are extraordinarily versatile molecules. They are found within cells and in extracellular locations. They may serve structural roles, as in ***collagen,*** or they may be functional as in enzymes, *transporters,* muscles, *hormones,* and *receptors.* Proteins are synthesized as linear polymers of 19 *amino acids* and one *imino acid* (commonly referred to slightly inaccurately as 20 amino acids). After synthesis of the initial polymer, or *polypeptide,* the protein may be ready for its function, or additional chemical modification of the structure may occur. In virtually all cases, specific folding of the protein into a fixed three-dimensional structure is also required. Proteins, together with RNA and *polysaccharides,* play a major role in their own synthesis, modification, and folding.

AMINO ACIDS

The amino acids that make up proteins each have a central carbon atom, the *α-carbon,* to which is attached a hydrogen atom, the *α-amino group,* the *α-carboxylate group,* and a variable side chain. The 19 different amino acid side chains are shown in Figure 3.1. Except in the case of *glycine,* where the side-chain is a hydrogen atom, the α-carbon is asymmetric, making the amino acids *optically active* (i.e., their solutions rotate the plane of *polarized light*). The amino acids found in proteins are of the L- configuration, although D- amino acids can be found elsewhere in nature. The twentieth residue found in proteins comes from *proline,* an imino acid, in which the side-chain folds back to bond to the α-amino group forming a ring (Fig. 3.1).

The properties of an amino acid are determined partly by its α-amino ($-NH_3^+$) and α-carboxylate ($-COO^-$) groups and partly by its side-chain. In proteins it is mainly the side-chain properties that are important, together with the *peptide bond* (see below). However, many amino acids occur free in body fluids and the analysis of proteins usually involves breaking them down into amino acids, so the properties of free amino acids are important. Determination of the concentrations of free amino acids in *blood plasma* or *urine* is used as a diagnostic indicator (e.g., measurement of *phenylalanine* concentration is used in the *diagnosis* of *phenylketonuria*).

The most important properties of amino acids, for the purpose of understanding protein structure and function, are electrostatic charge, hydrophilic/hydrophobic properties, and chemical reactivity, which are discussed below.

Electrostatic Properties

The folding of a protein and its interactions with other molecules are affected by the *electrostatic interactions*

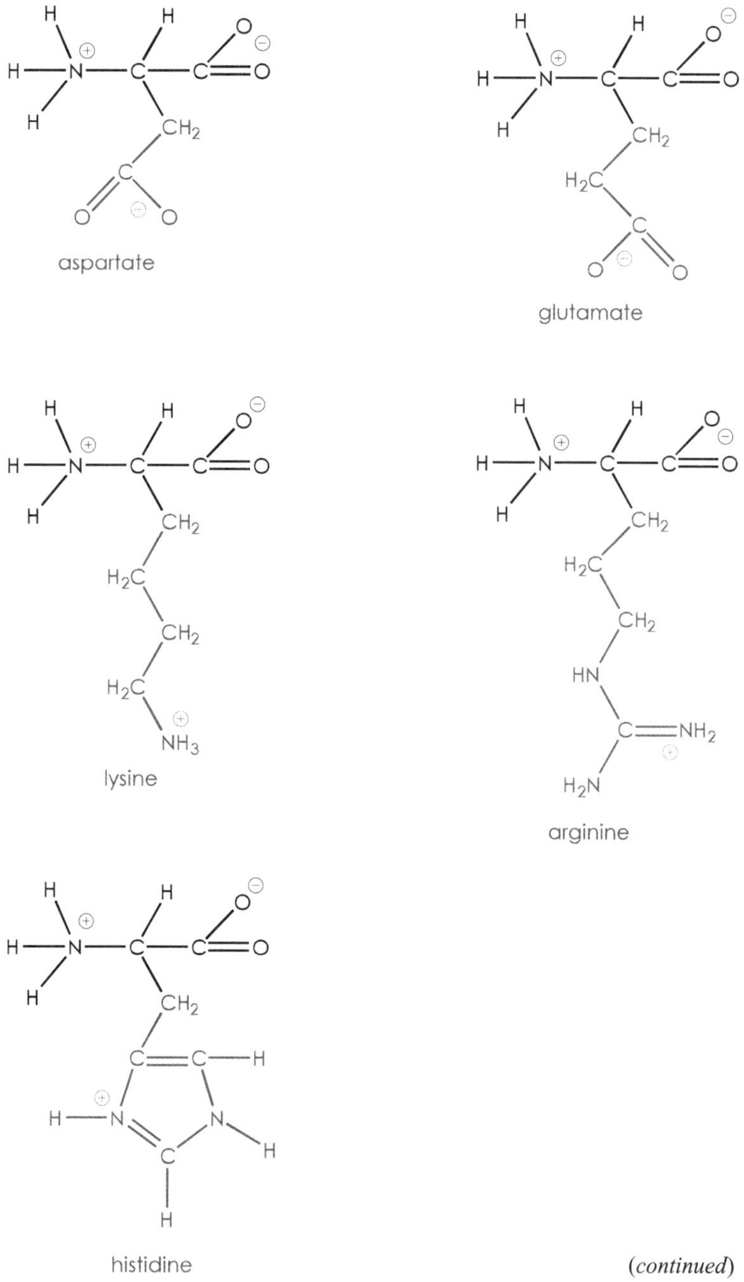

aspartate

glutamate

lysine

arginine

histidine

(continued)

Figure 3.1.

Amino acids found in proteins; the side-chains are shown in purple.

(a) Amino acids with side-chains that are usually charged under physiological conditions.

that occur between charged groups. Thus, it is important to understand the charges that can be present on amino acids and proteins. The text box at the end of the chapter, p. 64, defines *pK* (equivalent to pK_a) and *pI.*

All amino acids have charged α-amino and α-carboxylate groups. In a protein or polypeptide, the formation of the peptide bond (described below) removes these charges, except at the ends of the *polypeptide chain.* Some amino acids have charged side-chains in addition to the α-amino and α-carboxylate groups. In biochemical parlance, an acidic side-chain is negatively charged except at pH values far below the p*K*; a basic side-chain is positively charged except at pH values far above the p*K*. The magnitude of the charge found on an amino-acid side-chain depends on the p*K* of the group and the pH of the solution surrounding the group.

When such an amino acid is incorporated into a protein, the local chemical environment in the protein may affect the p*K* of a group and thus alter its charge. Such changes can be very important in the function of proteins (e.g., the *Bohr effect* in hemoglobin depends on structural changes that affect the p*K* of *histidine*

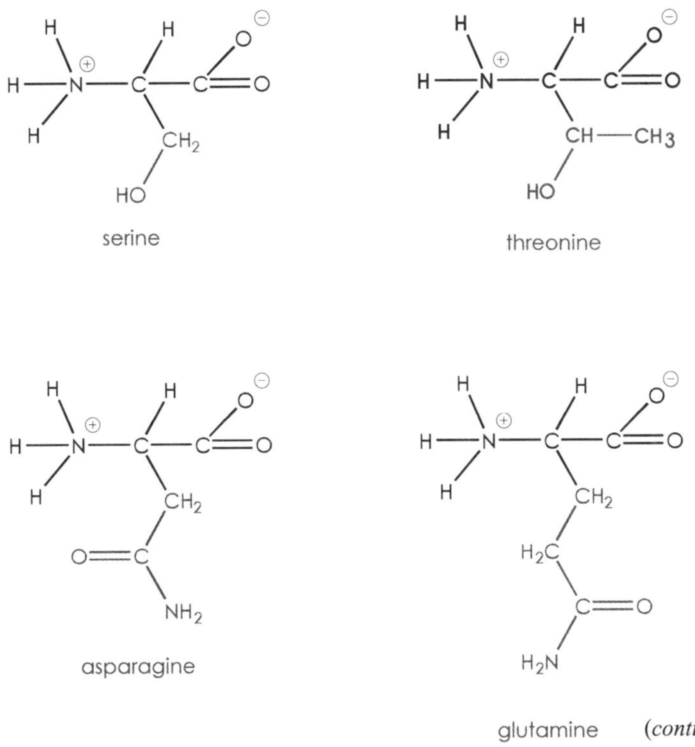

serine

threonine

asparagine

glutamine (*continued*)

Figure 3.1.(*continued*)

(b) Amino acids with side-chains that are polar but not charged under physiological conditions.

residues). In most cases, the side-chains that are significantly charged at pH 7.5 (with the charge in parentheses) are *aspartate* (–1), *glutamate* (–1), *arginine* (+1), *lysine* (+1), and *histidine* (between 0 and +1), while at this pH the α-carboxylate group has a charge –1 and the α-amino group in a protein has a charge that can vary between 0 and +1, depending on the protein and its environment (Table 3.1).

Hydrophobicity

As a solvent, water is quite highly structured, owing to hydrogen bond interactions. Pure water is relatively disordered (high *entropy*) because the water molecules within the solution are indistinguishable. When solute is added, some water molecules will be adjacent to a solute molecule. If the solute can hydrogen bond to these water molecules, many of them will remain indistinguishable from those in the bulk of the solvent. Thus, the water will be only slightly more ordered when the solute is added. This is a *hydrophilic* solute. On the other hand, if the solute cannot form hydrogen bonds, many of the water molecules adjacent to the solute molecule will be distinguishable (by unsatisfied hydrogen bonding possibilities) from water molecules in the bulk of the solvent. This is a **hydrophobic** solute. A large decrease in entropy (increase in order) of the water occurs when a *hydrophobic* solute is added to pure water. Since entropy tends to increase, hydrophobic molecules in water will tend to aggregate so as to minimize the surface area exposed to the water molecules.

The tendency of hydrophobic groups to aggregate is a powerful force driving the structure and interactions of proteins, *lipids,* nucleic acids, and other molecules. Many amino-acid side-chains have some hydrophilic character and some hydrophobic character. Hydrocarbon groups are invariably hydrophobic, whereas *polar groups,* that can form hydrogen bonds, and charged groups are hydrophilic. Thus, *glycine, cysteine,* and *alanine* have hydrophobic side-chains, but they are not strongly hydrophobic because of their small size. Proline is hydrophobic and the larger hydrocarbon side-chains, found in *leucine, isoleucine,* and *valine,* are strongly hydrophobic, as is *methionine.*

Phenylalanine is hydrophobic and additional interactions involving the π electrons above and below the ring encourage stacking of ring structures in proteins and nucleic acids. The other amino-acid ring structures, *tyrosine* and *tryptophan,* can also stack, but have both hydrophobic parts (the ring structures) and groups that can form hydrogen bonds, –OH or >NH respectively.

The charged amino-acid side-chains, listed above, are primarily hydrophilic, but the "linker" between the

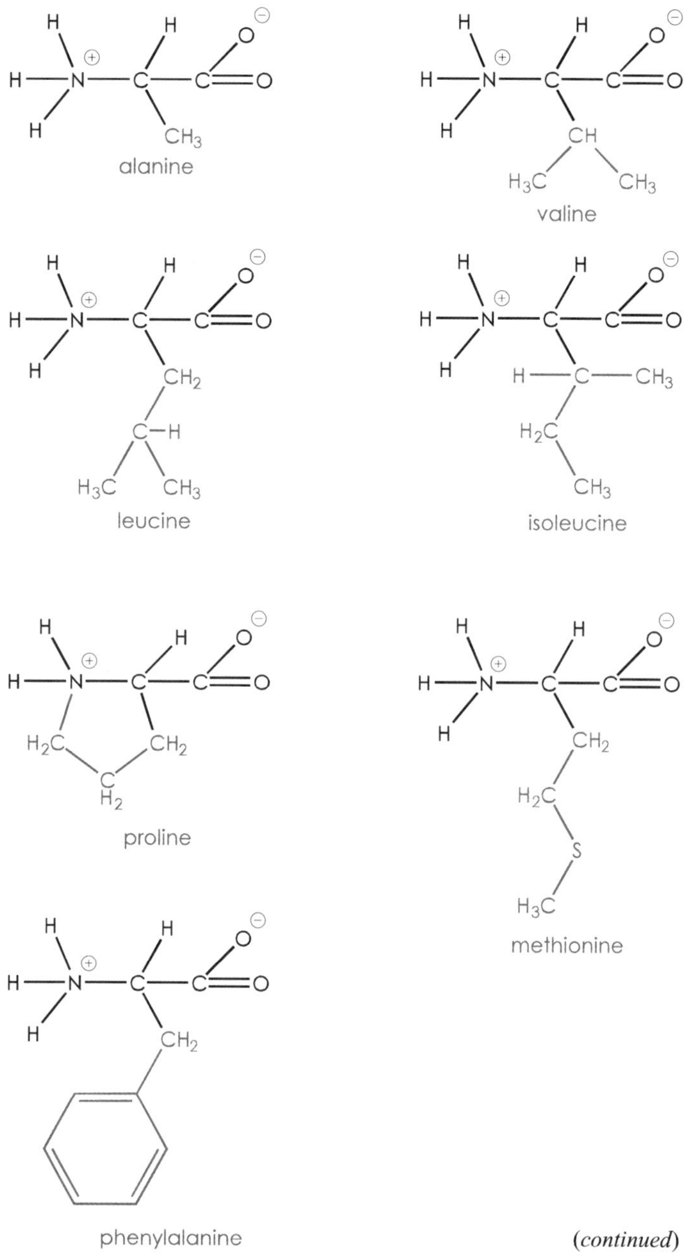

alanine

valine

leucine

isoleucine

proline

methionine

phenylalanine

Figure 3.1.(*continued*)
(c) Amino acids with hydrophobic side-chains.

(*continued*)

charge group and the α-carbon, especially in lysine and arginine, has hydrophobic character. *Asparagine, glutamine, serine,* and *threonine* are also primarily hydrophilic, although not charged (Table 3.2).

Chemical Reactivity

All the amino acids incorporated into proteins in the body are stable under physiological conditions. However, some are easily modified by chemical or enzymatic reactions while others are difficult to change. The more reactive amino acids are those with acidic or basic groups on the side-chain (aspartate, glutamate, arginine, histidine, lysine) or sulfhydryl (Fig. 3.2) or hydroxyl groups (cysteine, serine, threonine). The more reactive amino acids are often part of *enzyme active sites* (Chapter 7) where they participate in the reaction catalyzed by the enzyme.

POLYPEPTIDES

Amino acids can be linked together indefinitely in a continuous unbranched sequence by joining amino and carboxylate groups together with the elimination of wa-

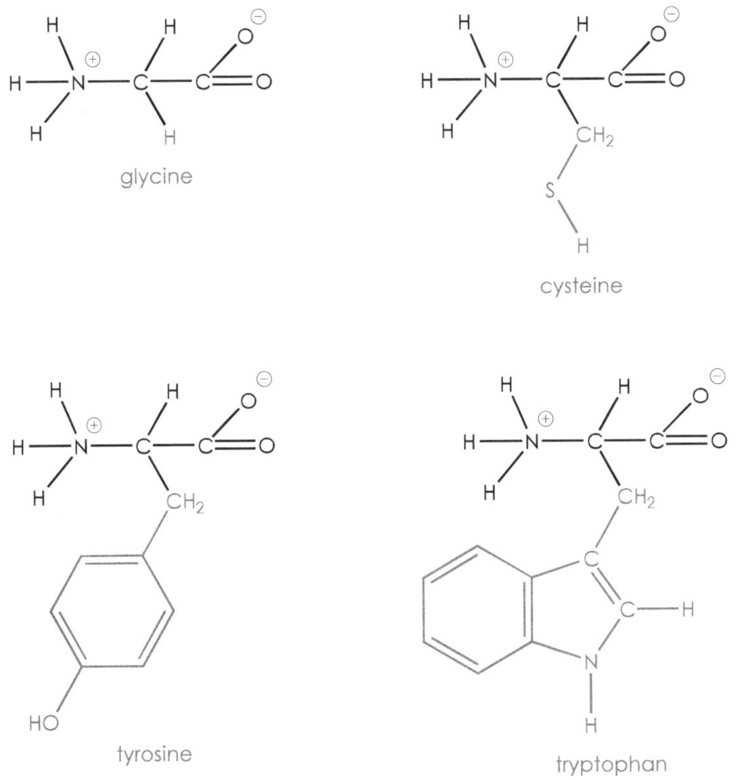

Figure 3.1.(*continued*)
(d) Additional amino acids.

ter (Fig. 3.3) to produce an *amide* and a *carbonyl* group joined by a *peptide bond.* In the cell, this is accomplished by a complex mechanism involving ribosomes and many other complex molecules. The complexity is required because of the need for *fidelity,* that is, the correct amino acid must be added at each point in the synthesis. Peptide synthesis can also be carried out chemically, to a valuable but limited extent, by the *Merrifield* synthesis technique.

Either process gives a chain, the polypeptide "backbone" that has lost its charged α-amino and α-carboxylate groups, except at the very ends of the chain. Each peptide group forms good hydrogen bonds, the >N—H (amide group) being cationic and the >C=O (carbonyl group) anionic through resonance. The chain has a unique direction and is conventionally written with the amino terminus (N-terminus) on the left and the carboxylate terminus (C-terminus) on the right:

$$NH_3^+\text{–}\underset{}{\overset{R}{\overset{|}{CH}}}\text{–CO–NH–}\underset{\underset{R}{|}}{CH}\text{–CO–NH–}\overset{R}{\overset{|}{CH}}\text{–}COO^-$$

Proteins have a sequence of amino-acid side-chains that is unique to each type of protein. The sequence is written from the amino terminus to the carboxylate terminus, as in the sequence of the A chain of insulin: NH_3^+-Gly-Ile-Val-Glu-Gln-Cys-Cys-Ala-Ser-

Table 1.1. Important Acidic and Basic Groups in Proteins

p*K*	Chemical group	Usual status in the body
3.0–3.2	*C*-terminal carboxyl	Negative charge
3.0–4.7	Aspartate side-chain carboxyl	Negative charge
~4.5	Glutamate side-chain carboxyl	Negative charge
5.6–7.0	Histidine *imidazole*	Partly charged (positive)
7.8–8.4	*N*-terminal amino	Partly charged (positive)
9.8–10.4	Tyrosine hydroxyl	Uncharged
9.1–10.8	Cysteine sulfhydryl	Uncharged
9.4–10.6	Lysine side-chain amino	Positive charge
11.6–12.6	Arginine guanidyl	Positive charge

Table 3.2. Classification of Amino Acids as Mainly Polar or Nonpolar[a]

Mainly polar (hydrophilic)	Mainly nonpolar (hydrophobic)
Aspartate	Alanine
Glutamate	Valine
Lysine	Isoleucine
Arginine	Leucine
Histidine	Phenylalanine
Serine	Methionine
Threonine	Proline
Asparagine	Tryptophan
Glutamine	

[a]Note that many amino acids have both polar and nonpolar parts.

Val-Cys-Ser-Leu-Tyr-Gln-Leu-Glu-Asn-Tyr-Cys-Asn-COO^-. When only the composition is known (i.e., the sequence is unknown), the amino acids are separated by commas and enclosed in parentheses, for example, (Ala,Cys_2,Gly) means a peptide containing one Ala, two Cys, and one Gly in an unknown order. A "polypeptide chain" is, by convention, a continuous chain linked only by peptide bonds. A protein may have only one polypeptide chain and then the polypeptide and the protein are synonymous. In other cases, a protein may have more than one polypeptide chain, as does insulin. In such cases, the different peptide chains within a protein may be held together by noncovalent forces, as in the case of hemoglobin, sometimes supplemented by covalent cystine cross links, as in the case of insulin. In many cases, the noncovalent interactions allow more than one conformation and a protein may switch from one conformation to another as part of its function.

The overwhelming majority of peptide bonds in proteins occur between α-amino and α-carboxylate groups, giving a single linear unbranched chain. In rare, but important, cases, side-chain amino and carboxylate groups can be joined by an isopeptide bond to create a branch (as in joining the C-terminus of the small protein ubiquitin to another protein) or to provide cross links (as in fibrin polymers in a blood clot).

Within an organism, a particular protein has a fixed sequence, derived from the gene. Proteins with similar functions in different species usually have related but different sequences depending on the evolutionary separation of the species and protein concerned. Changes in an amino acid that do not dramatically change the character of the residue, such as Val→Ile, are most common. These are ***conservative replacements***. Many proteins show regions of sequence that are highly conserved during evolution and such sequences often represent structural domains with a conserved shape and/or function. We are beginning to learn to recognize these regions, or domains, and to deduce function from amino-acid sequence, in some cases. In some proteins, the structure is conserved, although the sequence is not.

Protein Structure

The chemical groups in proteins and the covalent bonds between them (the primary structure) are only part of the story since each protein generally takes up a specif-

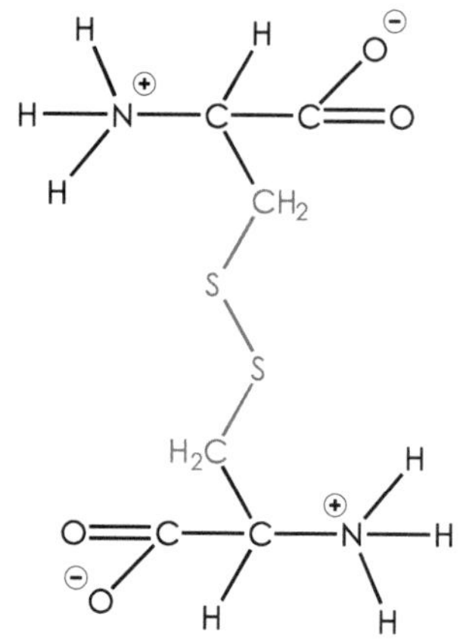

Figure 3.2.
Cystine. Two cysteine side-chains can cross-link to form cystine.

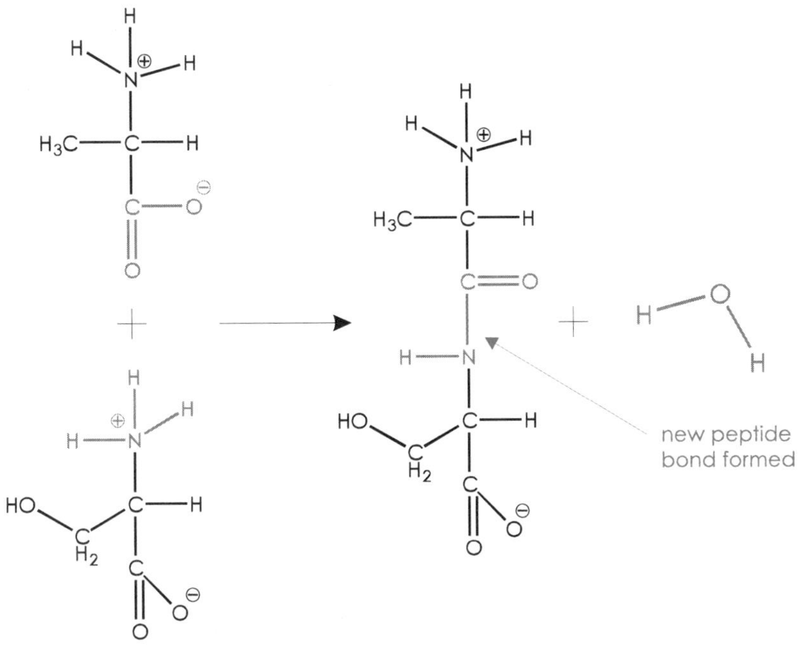

new peptide bond formed

Figure 3.3.

Chemical structure of peptide bond formation. The peptide bond is shown in purple.

ic shape or folding pattern during or after synthesis. It is this specific shape, or structure, that brings together the parts of the protein that are needed for its function. Proteins fold into complex and unique shapes, and the techniques of *X-ray crystallography* and *magnetic resonance* have revealed the structures of many proteins in great detail. Several common recurring themes help us to understand how proteins fold and how folding affects their function. Conformation is the term generally used for spatial arrangements that are determined by individually weak, noncovalent, bonds.

Proteins that cause *disease* may do so by having an incorrect structure, as we will see later with an *inborn error* in hemoglobin where a small amino acid, glycine, is replaced by a large amino acid, phenylalanine, that cannot fit into the small space available within the normal hemoglobin structure. In this case, the structure of hemoglobin is destabilized and anemia results.

Peptides and proteins can form very complex shapes by rotation of the backbone about the single bonds on either side of the α-carbon atom of each residue. The angles of rotation are called ψ (psi) and ϕ (phi) and may have different values for each α-carbon atom.

A polypeptide chain will form many different structures, but the stability of these structures is strongly dependent on the side-chains. The side-chains interact with one another, with the backbone, and with the solvent, producing many noncovalent interactions that determine the shape of the protein. In proteins exported from the cell or with extracellular domains, covalent interactions between pairs of cysteine residues often produce disulfide bridges that strongly constrain the shape of the protein. The actual folding pattern is thus fixed by the environment of the protein and its amino-acid sequence. The folding of the main chain can be described in terms of the two angles, ψ and ϕ, for each peptide bond because the angle of the . . . N—C . . . bond of the peptide group is fixed by resonance between the . . . C═O and . . . C—N . . . bonds, giving both of them a partial double-bond character and causing the . . . C—N . . . bond to be rigid (no rotation) and in the plane of the . . . C═O bond (Fig. 3.4). Full atomic coordinates are needed to describe the positions of the side chains, which are flexible.

The structure of any one protein is described in terms of a hierarchy of structures: primary, secondary, tertiary, and *quaternary.* The primary structure is the sequence of amino acids along the polypeptide chain. This is generally synonymous with the covalent structure of the protein, except that any disulfide bridges may alternately be classified as parts of the tertiary or quaternary structure. The secondary structure refers to the local folding of the macromolecular backbone and can be specified by the angles between adjacent peptide groups. Much of the secondary structure found in proteins falls into well-defined structures that are found in many proteins. Tertiary structure refers to the next order of folding of the polypeptide chain in which ele-

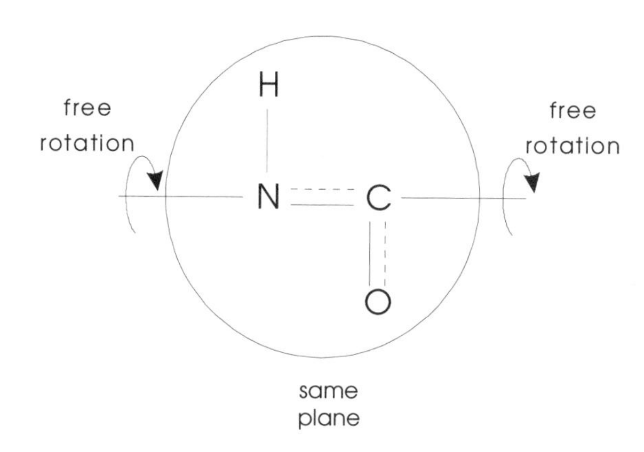

Figure 3.4.

Planar peptide bond. The peptide bond is circled. Free rotation is indicated for the bonds on either side of the peptide bond, but the peptide bond, itself, is planar due to resonance as indicated by the broken lines. In one resonance form, the N═C bond is double while, in the other, the C═O bond is double. An interactive view of a peptide bond is on the World Wide Web at http://moby.ucdavis.edu/HRM/Biochemistry/molecules.htm.

ments of secondary structure are arranged with respect to one another. Only proteins with more than one polypeptide chain have quaternary structure, which refers to the number, type, and arrangement of the different polypeptide chains in the protein.

Secondary Structure

Protein secondary structures may be uniform or irregular. In this context, a uniform structure is one in which each amino acid is displaced from the previous one in the same way. The uniform structures are particularly striking in structural proteins, especially when the amino-acid sequence is repetitive (e.g., *keratin,* silk *fibroin, collagen*) and they also occur in short segments of globular proteins (e.g., *hemoglobin,* enzymes). Most amino-acid sequences are theoretically capable of forming several types of secondary structure, but specific structures are found in practice because interactions of side chains can either stabilize or destabilize particular structures. The structure that is actually observed is determined by the amino-acid sequence and the solution conditions. There are two common uniform structures found in proteins, *α-helices* and *β-sheets,* and a third group of structures that appears frequently, the *reverse turns.*

The most common helix structure is the *alpha-helix* (α-helix). The polypeptide chain is coiled like a right-handed screw to form a compact cylinder. The amino-acid side-chains extend outward from the axis to coat the cylinder. The rise per amino-acid residue is 0.15 nm, giving a pitch (the pitch of a helix is the distance along the axis that corresponds to one turn of the helix) of 0.54 nm so the coil is fairly tight. The diameter of the coiled polypeptide backbone is approximately 0.7 nm, but this is increased by the side-chains. Each residue is rotated 100° relative to the previous one, giving 3.6 residues per turn of 360°. This brings the >C═O of one residue in line (in the direction of the helix axis) with the >N—H of the fourth residue ahead, and so on, and hydrogen bonds are formed between every backbone >C═O . . . H—N< pair in the helix except at the ends. This means that every peptide >N-H and >C═O hydrogen bonds to another peptide in the same chain (intrachain hydrogen bonds). The α-helix is particularly stable because the line of the hydrogen bond (>N—H . . . O═LC<) is nearly straight. This line runs parallel to the helix axis for all the backbone hydrogen bonds in an α-helix (Fig. 3.5).

A polypeptide made from L-amino acids can theoretically form either right- or left-handed α-helices, but the right-handed α-helix is more stable and is the form found in proteins. Other types of helix are possible, but are only found as short segments at the ends of α-helical regions in proteins (although some synthetic polypeptides will form unusual helices).

The polar nature of the amide and carbonyl groups means that there tends to be an excess positive charge at one end of an α-helix and an excess negative charge at the other end, since the hydrogen bonding pattern leaves unbonded amides at one end and unbonded carbonyl groups at the other end. This means that an α-helix forms an electric dipole that can participate in electrostatic interactions.

Some amino-acid residues are more likely to form α-helices than others. In particular, proline is known as a "helix-breaker" because it prevents the polypeptide chain folding into an α-helix. From examination of α-helices in known protein structures, a table of the probability of a residue being found in α-helix, or other secondary structure, can be constructed. When a protein of unknown structure is studied, the amino-acid sequence can be compared with the table of probabilities and re-

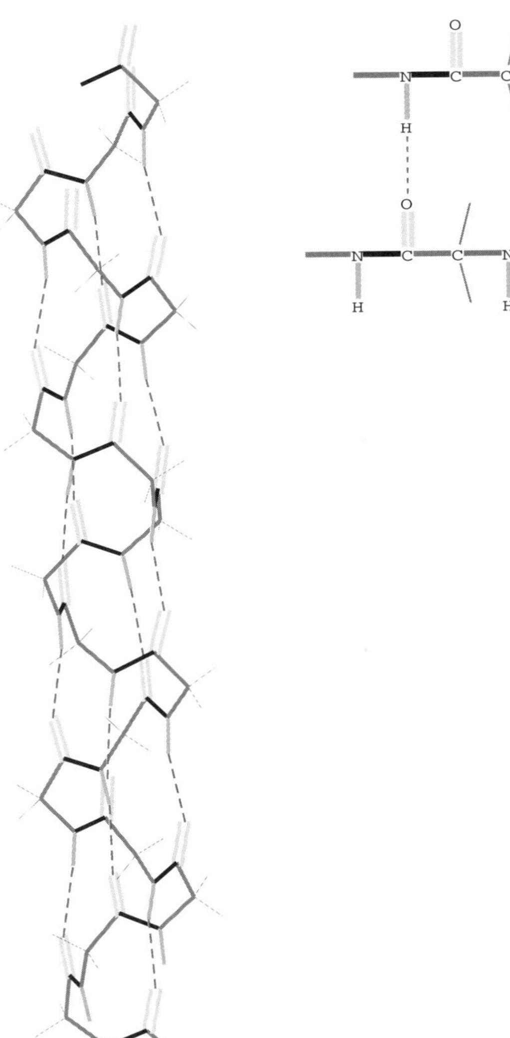

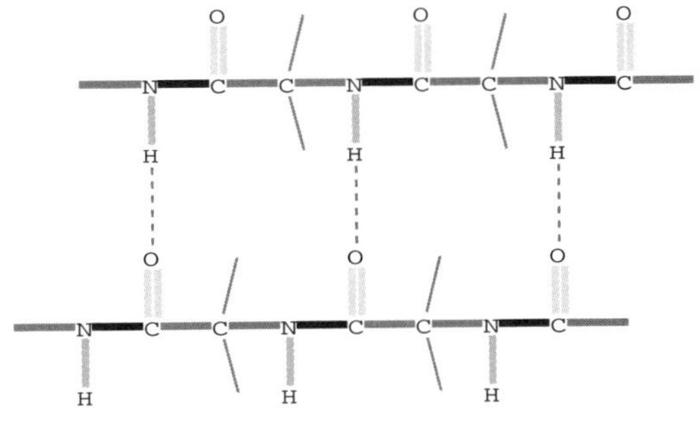

Figure 3.5.

α Helix. The peptide backbone of an α-helix is shown on the left, in black and gray with the C═O and N—H bonds of the peptide groups picked out in purple. The broken purple lines indicate the hydrogen bonds between C═O and N—H groups in adjacent turns of the helix. Side-chains are not shown. This structure is based on X-ray diffraction data. The inset at the top right shows a highly diagrammatic representation of the hydrogen bonding reduced to two dimensions. The black and gray peptide backbones shown are parts of the same chain, on adjacent turns of the helix. In this view, you can see how one end of the helix has all exposed C═O groups while the other end has all exposed N—H groups. Within the helix, neither C═O nor N—H groups are exposed because they are hydrogen bonded to each other. An interactive α helix is on the World Wide Web at http://moby.ucdavis.edu/HRM/Biochemistry/molecules.htm.

gions of structure with many residues frequently found in α-helices can be predicted to have an α-helical structure in the new protein. Taken together with other factors, such as the clustering of hydrophobic residues on one face of the helix, these predictive methods can be useful, although not definitive.

The second common group of secondary structures is the beta-structure (β-structure), often called a *β-sheet* or *β-barrel.* In these structures the polypeptide chain is much more extended, with 0.35 nm between adjacent peptide groups compared with 0.15 nm for the α-helix. The rotation per residue is 180° and there are 2 residues per "turn." The structure also features hydrogen bonds (>N—H . . . O═C<) between all the backbone >N—H and >C═O groups except at the edges. In the β-structure, however, the hydrogen-bonded >N—H and >C═O groups are on separate chains (or different parts of the same chain, looped back). The hydrogen bonds are perpendicular to the direction of the polypeptide chains. An indefinite number of chains can be hydrogen bonded together to form either a sheet, the β-sheet, or a barrel-shaped structure, the β-barrel. The sheet is not quite flat since the α-carbons are alternately slightly above or below the plane of the sheet. Thus, the β-sheet may be called a pleated sheet or a *β-pleated sheet.* The peptide groups are in the plane of the "pleats," but the side-chains project above and below the plane of the sheet (Fig. 3.6).

Beta-structures may have adjacent chains parallel or antiparallel and both types are found in proteins. In the best example of β-sheets, the silk-protein fibroin, the chains are antiparallel. Fibroin has an amino-acid sequence that resembles . . .—(Gly—Ala)$_n$—. . . . In the β-sheet this gives just —H as the —R group on one

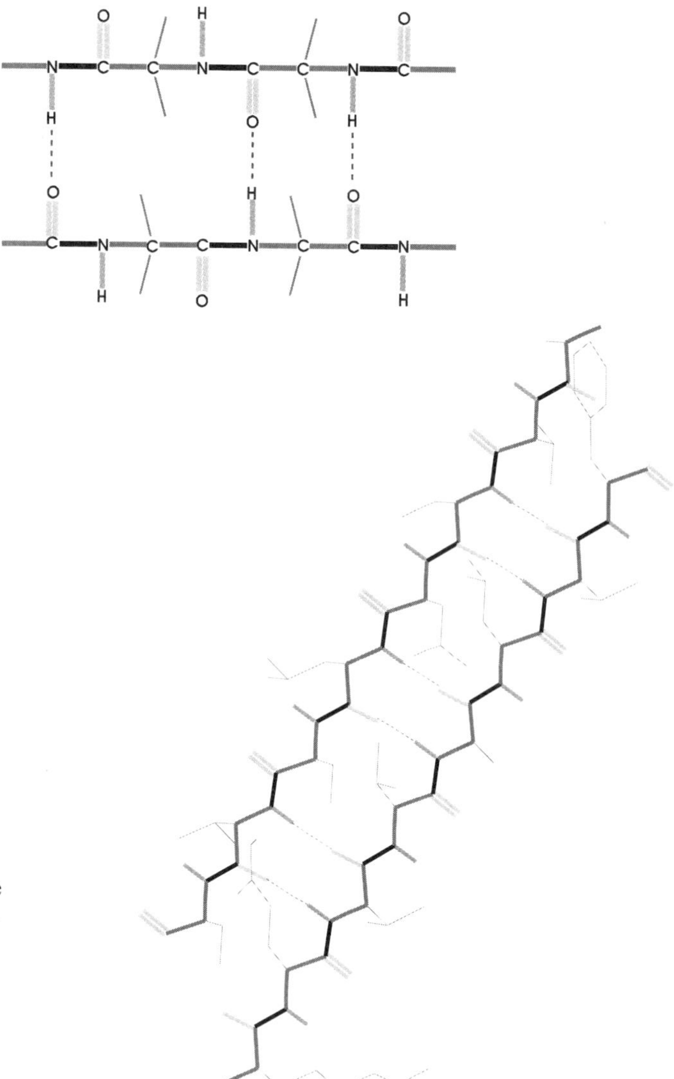

Figure 3.6.

β Sheet. The main part of the figure (lower right) shows two antiparallel polypeptide strands in a β-sheet structure. The polypeptide backbone is in black and gray, the side-chains are in light gray, and the C═O and N—H groups of the peptide bonds are picked out in purple. Broken purple lines show the hydrogen bonds between the chains.

The inset, top left, shows a highly diagrammatic representation of the hydrogen bonding pattern, in two dimensions. It shows that, in this antiparallel arrangement, each edge of the sheet has both N—H and C═O groups exposed. An interactive model of a β-sheet is on the World Wide Web at http://moby.ucdavis.edu/HRM/Biochemistry/molecules.htm.

side of the sheet and —CH_3 as the —R group on the other side of the sheet. These very small —R groups are particularly suitable for β-structures because larger —R groups tend to interact unfavorably and prevent the backbone folding correctly. The occurrence of β-structures in more complex protein sequences can be predicted from data on known protein structures, as described for α-helices.

The third group of conformations is **reverse turns.** These are important and common conformational elements in globular proteins that enable the chain to reverse direction and fold back on itself. Proline is often used to cause the structure to reverse direction. Reverse turns can also be stabilized by hydrogen bonding between the first and the third residue. The chain can fold in six different ways to make this hydrogen bond, two of which are like part of a modified α-helix called a 3_{10} helix (3_{10} bends). Some other forms have been called β-bends because the H-bonding pattern resembles that in a β-sheet (Fig. 3.7).

The remaining residues in proteins are in one of two general types of structure. The first type is found in *globular proteins* and is a fixed but unique or nonrepetitive structure. The second type is found in nonglobular proteins or nonglobular parts of globular proteins. In this structure, the conformation of the polypeptide chain is not rigidly fixed and in solution the angles are continually changing as the chain moves around in the solution. This lack of structure is called a **random coil.** In some cases, at least, the *random coil* form found in the isolated protein may be bound to another structure (e.g., nucleic acid) in the cell and so have a fixed con-

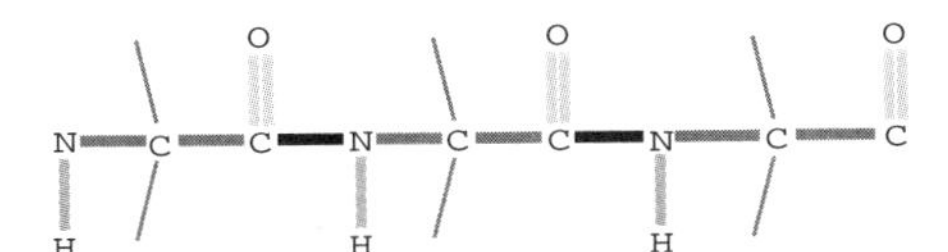

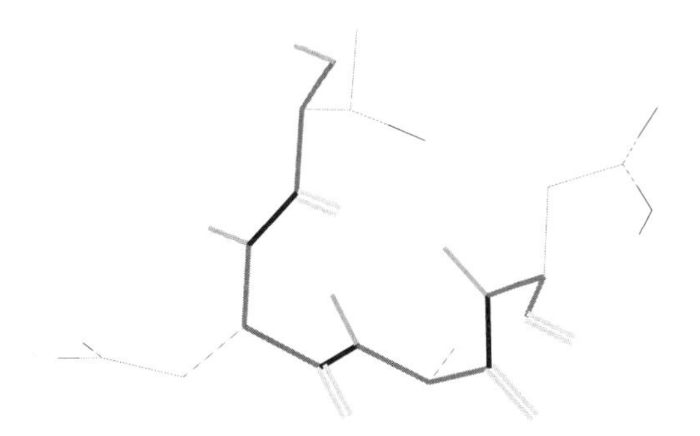

Figure 3.7.

Reverse turn. The structure, lower right, shows an example of a sharp reverse turn in a polypeptide chain. The backbone is shown in black and gray with the sidechains in light gray and the C═O and N—H groups of the peptide bonds picked out in purple. Depending on the specific configuration of the reverse turn, hydrogen bonding may occur between backbone groups or between backbone groups and side-chains. No hydrogen bonding is shown in this example (see the inset, top left).

formation. Random coils are not found to a very large extent in enzymes, but they are an important feature of some other proteins, for example, the histone proteins found in chromosomes.

Functional Motifs and Domains

The terms primary, secondary, tertiary, and quaternary apply to the structural hierarchy in proteins. Increasingly, however, we are also thinking about proteins in terms of functional *motifs* and *domains.* A functional motif or domain is a part of a protein that serves a particular function. A considerable number of functional motifs and domains is now recognized and the group is increasing in size. In general, a motif is a short section of amino-acid sequence with a recognized function, while a functional domain is a larger section of amino-acid sequence. The size distinction is not clearly drawn, but sequences of less than 25 amino acids with a recognized function would normally be referred to as motifs, whereas sequences larger than 50 amino acids with a recognized function would normally be referred to as domains.

A motif may be as small as two amino acids. For example, the *proline-directed* ***protein kinases*** recognize the sequence —Ser/Thr—Pro— and phosphorylate the serine or threonine. Not all the residues in the motif need to be defined. For example, the recognition signal for addition of *N*-linked *oligosaccharide* is —Asn—Xxx—Ser/Thr— where the symbol "Xxx," like "Any," stands for any amino acid. More often, the motifs are somewhat longer, as in the recognition signal for the *cyclic AMP-dependent protein kinase* where three or four amino acids in the motif are recognized (. . . Phe . . . 7 amino acids . . . Arg—Arg—Any—Ser—hydrophobic— . . .), or in protein-targeting signals. Motifs may be defined in terms of structure as well as sequence. For example, the *zinc finger* motif is a family of structures that binds zinc and participates in the sequence-specific binding of proteins to DNA.

Functional domains are substantially larger parts of proteins. For example, an *SH2 domain* is a sequence of around 100 amino acids (Fig. 3.8) that forms a functional unit that binds specifically to *phosphotyrosine*; similarly, an *SH3 domain* forms a functional unit that binds to *polyproline* sequence motifs. Functional domains have been identified in proteins that bind DNA and affect the synthesis of RNA. The *DNA-binding domain* of the protein, in some cases, can be separated from the *transcription-modulating domain.* In the case of *steroid hormone receptor,* there are three functional domains, a *gene-activating domain,* a DNA-binding domain, and a *hormone-binding domain.* Domains often provide catalytic activity in proteins; for example, just two classes of *protein kinase* catalytic domains are found in over 1000 protein kinases.

Structural Domains

Domains are often conserved in structure as well as function and may be thought of as either functional or structural domains according to the context.

One of the simplest examples of structural do-

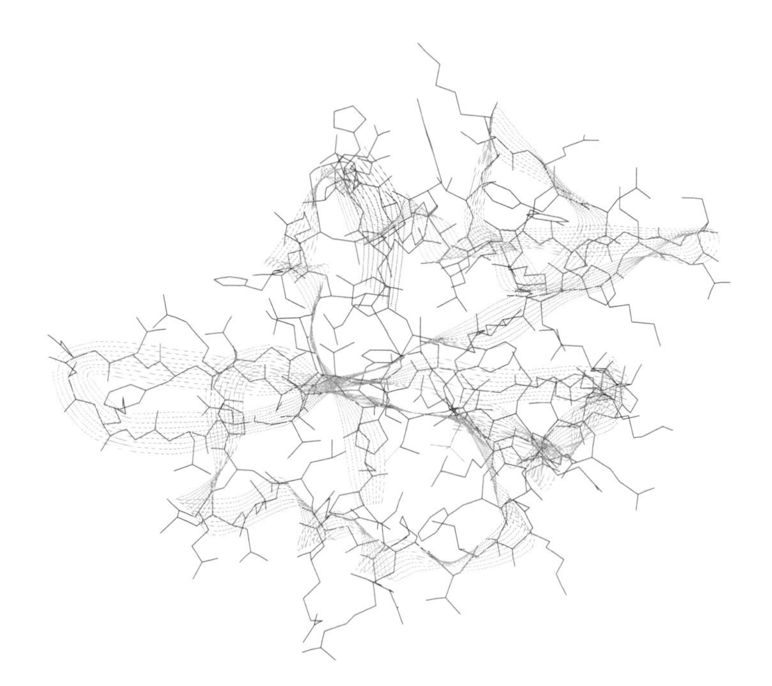

Figure 3.8.

A protein domain. This is an SH2 domain whose function is to bind phosphotyrosine. The bonds are shown in black and the purple ribbons are used to indicate the direction of the backbone. An interactive model of an SH2 domain is on the World Wide Web at http://moby.ucdavis.edu/HRM/Biochemistry/molecules.htm.

mains occurs in membrane-spanning proteins. Usually, such proteins have a sequence of domains in which a *membrane-spanning domain* follows an *extracellular domain* and precedes an *intracellular domain.* There may be several repetitions of this sequence in one protein. In these cases, the domains can be distinguished as follows: the extracellular domain usually contains O-linked oligosaccharide chains and cysteines oxidized to cystine, forming *disulfide bridges*; the membrane-spanning domain is usually in the form of an α-helix with predominantly hydrophobic amino-acid side-chains, to interact with the hydrophobic core of the membrane. The *immunoglobulin fold* is a widespread structural domain found within the extracellular domains of membrane-spanning proteins.

Another important example is the so-called "*nucleotide fold*." This is a domain that binds nucleotides and is found in many proteins that bind ATP or GTP. The amino-acid sequence of the nucleotide fold domain is sufficiently conserved that it can be recognized even in a protein of unknown function.

Many DNA-binding proteins fall into identified classes with recognizable domains. For example, many bacterial DNA-binding proteins use the "*helix-turn-helix*" structural motif and many DNA-binding proteins from higher organisms have another structural motif called "zinc fingers."

As the number of recognizable domains with known functions increases, it is becoming more feasible to deduce at least part of a protein's *function,* as well as its structure, from its amino-acid sequence. Protein kinases provide an excellent example.

Tertiary Structures

Tertiary structure is the level above secondary structure in the hierarchy of structures. For example, in the zinc-finger motif mentioned above, there are regions of secondary structure, α-helix, and β-sheet. These interact to form a specific higher-order structure, the tertiary structure.

A substantial number of tertiary structures has been determined by analysis of protein crystals using X-ray diffraction and, more recently, by analysis of proteins in solution using magnetic resonance techniques. In some cases it has been possible to show that the crystal structure is closely related to the structure in solution. For example, it is possible to demonstrate that some enzymes (*ribonuclease, chymotrypsin, carboxypeptidase*) are catalytically active even when fixed in the crystal conformation by chemical cross-linking. In the case of *lysozyme* it has been possible to confirm, by very detailed magnetic resonance spectroscopy measurements, that the solution conformation is similar to the crystal structure. Developments in magnetic resonance spectroscopy are providing the ability to determine the structures of small proteins or separate domains in solution directly.

Structural domains can recur both in proteins with

similar functions, such as *myoglobin* and hemoglobin, and in otherwise dissimilar proteins. However, there is a tremendous variety of structures in proteins and detailed understanding or predicting of tertiary structure is well beyond current knowledge. Structurally related proteins often have related amino-acid sequences, but similar structures may also be formed by unrelated sequences.

Some generalizations, while helpful for giving a general understanding of protein tertiary structures, are not universally true:

1. The conformations are very compact, leaving a minimum of empty space inside and relatively few sites for internal bound water.
2. All the charged side-chains are found on the surface of the molecule where they interact with the solvent.
3. Large hydrophobic side-chains tend to be buried in the interior of the molecule.
4. Most of the amino-acid residues are in one of the three major forms of secondary structure.
5. The tertiary structure is stabilized mainly by a combination of cystine cross-links (where present) and hydrophobic forces. Hydrophobic forces are generated by interactions between different regions of the polypeptide chain. For example, two α-helices may interact to form an *α-helical coiled coil* or a β-barrel may provide a core surrounded by α-helices, and so on.
6. The tertiary structure is often clearly related to the function of the protein, as in the *heme-binding* pocket of myoglobin and *substrate clefts* in enzymes.
7. The protein may include several separate domains.

Quaternary Structures

A large number of proteins extend their structures further by packing together two or more polypeptide subunits to make dimeric, trimeric, tetrameric, or larger proteins. A *hetero*dimer has different subunits; a *homo*dimer has identical subunits. For example, the DNA-binding proteins known as *restriction endonucleases* may be symmetrical homodimers. This allows them to approach their recognition sequence from either direction along the DNA chain. The subunits of a protein are held together by noncovalent forces and disulfide bridges similar to the interactions involved in tertiary structures. The net effect of the interactions is a very specific association of the subunits, usually critical to the function of the protein.

Quaternary structure is often intimately related to function. For example, more than one enzyme activity may be present, allowing the product of one reaction to be immediately taken up as the substrate for the next, eliminating the delay that would otherwise be caused by diffusion from one enzyme to another. In many cases, the same ligand binding or enzyme activity may reside on more than one subunit of a protein allowing the quaternary structure to mediate interactions between these subunits. For example, in oxygen binding by hemoglobin (discussed in more detail in Chapter 6), the four subunits each bind oxygen but the affinity of any one subunit depends on the presence or absence of oxygen on the other subunits. Such variable affinity is known as *cooperativity* and is key to oxygen delivery by hemoglobin. Similarly, cooperativity mediated by the quaternary structure of certain enzymes is the basis of *feedback regulation* of *metabolic pathways.*

POSTTRANSLATIONAL MODIFICATIONS

Polypeptide chains are synthesized in the cytoplasm of a cell by the process known as translation. The polypeptide may be ready for use immediately after translation or it may require further maturation steps, such as facilitated folding, complex formation, transport to another cell compartment, or covalent modification of chemical groups on the protein. The term *posttranslational modifications* is used for the last of these processes. Some of the modifications take place in the cytoplasm or nucleus, particularly *phosphorylation* and *acetylation,* and others take place in the endoplasmic reticulum or *Golgi apparatus,* particularly the addition of sugar or *polysaccharide* residues.

Protein phosphorylation is a pervasive posttranslational modification in cells. It is reversible and can dramatically affect the activity of a modified protein. Protein phosphorylation is one of the most important mechanisms used for *signal transduction* by cells. In prokaryotic cells, the best-known reversible protein phosphorylations occur on histidine and aspartate; in eukaryotes the best-known occur on the hydroxyl groups of serine, threonine, and tyrosine, although histidine can also be phosphorylated (Fig. 3.9). Other reversible modifications also occur, such as the acetylation of lysine residues in histone proteins.

Many modifications appear to be irreversible, in-

phosphoarginine

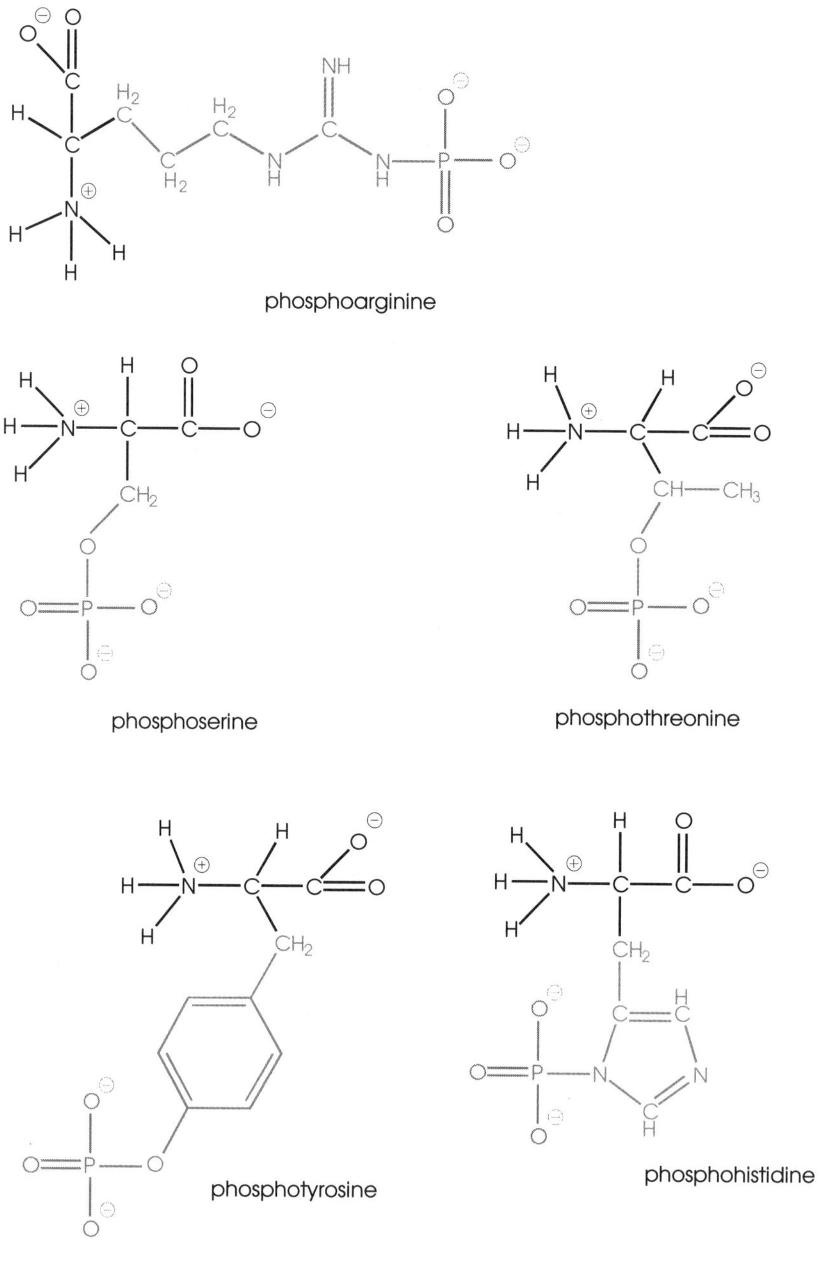

phosphoserine

phosphothreonine

phosphotyrosine

phosphohistidine

Figure 3.9.

Phosphoamino acids. Examples of phosphoamino acids found in proteins. The side-chain is purple and the phosphate group is gray.

cluding the hydroxylation of proline and other modified amino acids found in structural proteins such as collagen, the formation of *γ-carboxyglutamate* in *blood* and *bone* proteins, acetylation of the α-amino group of many proteins, and *isoprenylation* of proteins to anchor them to the cytoplasmic face of the plasma membrane.

PROTEIN PURIFICATION

Proteins are generally present in the body in extremely complex mixtures. Although modern approaches allow many experiments to be carried out in these complex mixtures, critical information can only be obtained from pure proteins. The process of *protein purification* is quite laborious and unpredictable in general. It is necessary to apply several purification procedures serially to the sample. Two of the most powerful and versatile groups of procedures are *column* ***chromatography*** and *gel electrophoresis.*

Column Chromatography

A chromatography column is simply a glass or steel tube packed with a chromatography matrix. The matrix is small beads—typically 10 or 15 μm in diameter—packed tightly into the column, and the nature of these beads determines the nature of the chromatography. The protein mixture to be separated is dissolved in a suitable solvent and pumped through the column where the proteins interact with the chromatography matrix. The matrix will tend to bind to some proteins, thus retarding them so that they elute from the column after proteins that do not bind to the matrix. In this way, proteins can be separated on the basis of their affinity for the chromatography matrix.

The most commonly used chromatography matrices are *affinity, ion–exchange, reverse–phase,* and *molecular sieving* matrices. An affinity matrix contains a specific *ligand* that binds only the protein of interest—a specific enzyme inhibitor or a particular DNA sequence can be used this way. The protein of interest binds to the column and the other proteins in the mixture pass through. The protein is then eluted with a solvent that breaks the interaction between the ligand and the protein so that the protein can be eluted in pure form.

An ion–exchange matrix contains charged groups—either positively charged, a *cation exchanger,* or negatively charged, an *anion exchanger.* Proteins with a charge opposite to that of the chromatography matrix bind, while those of like charge pass straight through. The column is then eluted with an increasing concentration of salt in the eluting solvent. The ions in the salt solution compete with the bound proteins for the charged groups on the resin and as the salt concentration increases, the proteins are displaced from the matrix, or resin. The more weakly bound proteins elute first. In this way, a gradient of salt concentration can be used to separate proteins into many different groups, some of which may contain a single, pure, protein.

A reverse–phase chromatography matrix is a hydrophobic surface, such as *silica.* The protein mixture is loaded in a relatively hydrophilic solvent, so that proteins with hydrophobic patches on their surfaces will bind to the column. The column is then eluted with a solution of increasing hydrophobicity and the proteins are eluted in order of their hydrophobicity as they are displaced by the solvent gradient. This method also separates protein mixtures into many different groups.

A molecular sieve matrix—also known as *gel filtration*—does not bind the proteins being separated, but allows small proteins to enter the beads while large proteins are excluded, owing to the size of the pores in the beads. Thus, small proteins have further to travel through the column and they elute later, after the large proteins. Thus, a protein mixture can be separated into groups according to size.

The equipment for column chromatography can be simple, using open columns and manual elution and collection of the eluent, or quite complex, using FPLC (fast protein liquid chromatography) or HPLC (high-performance liquid chromatography) technology. Although the principles are the same, the FPLC and HPLC equipment is much faster and more reproducible than the manual methods.

Gel Electrophoresis of Proteins

The principles of gel electrophoresis are explained in Chapter 2. Proteins are more complex than nucleic acids, in this respect, since the charge on a protein may be positive, negative, or zero, whereas nucleic acids are always negatively charged. Polyacrylamide is the gel of choice for protein gel electrophoresis. It is possible to electrophorese proteins in the absence of a gel, by doing the experiment in a very narrow tube, to reduce lateral diffusion. This technique is known as *capillary gel electrophoresis* and is gaining popularity for its resolution, speed, and reproducibility, although it requires complex equipment.

Gel electrophoresis of proteins falls into three groups, *native gels, SDS gels,* and *isoelectric focusing.* In native gels, proteins are electrophoresed in a solvent that maintains the native structure of the protein. Thus, the direction of movement will depend on the charge on the protein and the speed will depend on both the charge and the size of the protein. This method is very useful for separating proteins that differ more in charge than in size.

The SDS gels contain the detergent *sodium dodecyl sulfate* (also known as *sodium lauryl sulfate*). The *SDS* forms a "coat" around each protein molecule, denaturing it and shielding its charge. Thus, on an SDS gel all proteins move toward the anode because of the negative charge on the SDS "coat." The uniform charge density of the "coat" means that the proteins move mainly according to their denatured size, which is determined by their *molecular weight.* The SDS gels are useful because they will separate almost any mixture of proteins, including membrane proteins that are insoluble in the absence of detergent, with standard conditions. The SDS gels may also be used to determine the molecular weight of a protein by comparing its migration velocity with that of standard proteins of known molecular weight.

Isoelectric focusing uses a gel with a pH gradient.

The protein mixture is loaded on one end of the gradient and almost all proteins will move into the gel because they will have the same charge at extremes of pH. As the proteins move along the pH gradient, the changing pH will cause the protein charge on each protein to vary and eventually reach zero at the point in the gradient where the pH equals the pI of the protein. Since the charge is zero, the protein will stop moving and eventually all the proteins will be stationary in the gel, having reached their isoelectric points.

Isoelectric focusing and SDS gel electrophoresis can be combined in *two-dimensional gel electrophoresis*. The proteins are first separated by isoelectric focusing and then the tube or lane of the gel is placed along the top of a slab of SDS gel and electrophoresed again. This is a very powerful separation method.

In general, gel electrophoresis is most useful for analyzing small amounts of proteins or for the final stage of purification. Although column chromatography can be used on small amounts of material, it is most useful when handling large amounts of proteins.

SUMMARY

1. Proteins carry out most of the body's molecular activities, sometimes in collaboration with nucleic acids or carbohydrates.

2. Proteins are polymers of amino acids.

3. Amino acids are charged at their primary amino and carboxylate ends and arginine, aspartate, glutamate, histidine, and lysine are charged on their side chains at physiological pH. Charged molecules interact with each other through electrostatic forces.

4. Many amino acids have nonpolar side chains. Such groups cannot fit into the structure of water and so they tend to aggregate in aqueous solution to minimize the area of interaction with water. This tendency is called a hydrophobic interaction.

5. The primary amino and carboxylate groups of amino acids are chemically reactive and many amino acids also have chemically reactive side-chains containing hydroxyl, amine, or carboxylate groups; other side-chains, such as phenylalanine, are not reactive.

6. Proteins are formed by polymerizing amino acids. Each amino acid is joined to the next through the reaction of an amino group with a carboxylate group, resulting in the formation of a peptide bond and the elimination of water.

7. Proteins take up several different secondary structures, including the α-helix, β-sheet, and reverse turns. The shape of the polypeptide chain is constrained by the planar peptide bond and noncovalent interactions.

8. Protein sequences and structures contain many recognizable motifs and domains.

9. Interactions of secondary structures within a single polypeptide chain generate tertiary structures—the complex three-dimensional shapes of proteins. Interactions between separate polypeptide chains in the same protein generate a quaternary structure, which mediates interactions between the subunits. Tertiary and quaternary structures may be stabilized by disulfide bridges between cysteine residues, forming cystines.

10. Protein synthesis in the body is constrained to 20 amino acids (including the imino acid proline), but modifications made after translation greatly extend the range of side chains found in mature proteins. Reversible modifications provide opportunities for regulation of protein function.

11. Proteins in the cell exist in a crowded complex mixture. Purification of individual proteins is achieved by a sequence of purification steps, including column chromatography. Protein samples may be analyzed by polyacrylamide gel electrophoresis.

TEXT BOX: DEFINITION OF P*K* AND PI

In an aqueous solution, pH = $-\log_{10}([H^+])$. For example, at "neutral" pH, $[H^+] = [OH^-] = 10^{-7}$ M and the pH is $-\log_{10}([10^{-7}])$ which is 7. Note that a lower pH corresponds to a higher $[H^+]$. The pH inside a cell or in body fluids, called physiological pH, can vary, but is typically slightly above neutral pH, often around pH 7.6.

At physiological pH, a solution of an amino acid with an uncharged side-chain (e.g., alanine),

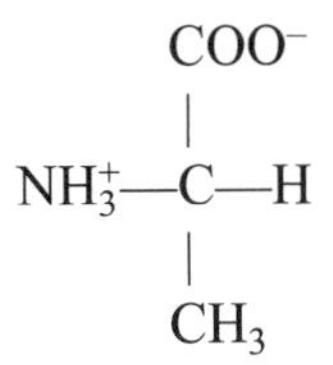

will have both the carboxyl and the amino groups ionized, as shown. If hydrogen ions (H^+) are added to lower the pH, the carboxyl group will lose its ionization:

$$—COO^- + H^+ \Leftrightarrow —COOH$$

if hydroxyl ions (OH^-) are added to raise the pH, the amino group will lose its ionization:

$$OH^- + —NH_3^+ \Leftrightarrow H_2O + —NH_2$$

Amino acids can thus behave either as acids or as bases, which makes them *ampholytes.* At neutral pH, where both amino and carboxyl groups are ionized, the molecule is a dipolar ion or a *zwitterion.*

The **p*K*** of a particular chemical group is the pH at which the group is half ionized. The pH at which half the carboxyl groups are ionized is the pK_1' (e.g., the pK_1' for alanine is 2.3) and the pH at which half the amino groups are ionized is the pK_2' (pK_2' for alanine is 9.6).

The pH of a solution of an amino acid can be calculated from the *Henderson–Hasselbalch* equation: pH = pK' + log [(proton acceptor)/(proton donor)]. For pK_1', the *proton acceptor* form is —COO^- and the *proton donor* form is —COOH. For pK_2' they are —NH_2 and —NH_3^+, respectively. An example at the end of the chapter shows how to use the Henderson–Hasselbalch equation to calculate the charges on a molecule. We use the term pK' for the apparent, or measured, value and the term pK for the value after correction for thermodynamic nonideality. In many circumstances, this distinction is ignored.

There is a pH at which the net charge on a molecule is zero, that is, if we add all the charges on the molecule together, the result is zero. This pH is known as the *isoelectric point,* **pI.** The pI of alanine is about 6.0. The pI of a protein can be used to separate closely related proteins by isoelectric focusing. If the pI is above 7, the protein is said to be basic; if the pI is below 7, the protein is said to be acidic, although all proteins have both acidic and basic groups.

TEXT BOX: FORCES INVOLVED IN DETERMINING PROTEIN STRUCTURE

The main forces involved in determining α-carbon angles are:

1. **Cross-linking by cystine** provides very stable reference points in the structure. Cystine cross-links can be removed by reducing agents or replaced by oxidizing agents.
2. **Hydrophobic interactions.** Hydrophobic interactions are stabilized by salts and destabilized by *guanidinium, urea, thiocyanate,* and other *chaotropic* compounds.
3. **Hydrogen bonds** (H-bonds) are individually weak, but the cumulative effect of the many H-bonds in a typical protein is an important feature of the structure. H-bonds are directional and often contribute to the specificity of interactions within and between proteins and other molecules, including other proteins and nucleic acids. Both backbone and side-chains can H-bond and the cumulative effect is important. H-bonds are destabilized by chaotropic compounds.
4. **Electrostatic forces,** also known as **ionic bonds** or **salt bridges,** occur only between charged chemical groups. Thus, these interactions occur exclusively between the side-chains of residues that are charged or when dipoles are formed. The charges must be close together in space. Electrostatic interactions are stronger in the absence of water (e.g., in the interior of a structure) and are destabilized by salts.
5. Atoms occupy a fixed volume of space that is very difficult to compress, except by covalent bond formation. Thus, atoms cannot overlap in their position. The effect of this on protein structure is called **steric hindrance.** As one would expect, bulky side-chains such as tryptophan, phenylalanine, methionine, and others have quite restricted positions due to steric hindrance, whereas small side-chains such as

glycine, alanine, and serine can take up a wide variety of positions, permitting sharp changes in direction of the polypeptide chain. Proline is a special case in which the folding of the polypeptide chain is restricted by the covalent bond between the side-chain and the α-amino group.

6. **Other interactions.** A variety of additional interactions is important, such as the π–π bond interactions when rings stack, the formation of dipoles in α-helices that generate additional electrostatic interactions, and Van der Waals interactions, which favor close contact between atoms and other forces.

A WORKED EXAMPLE USING THE HENDERSON–HASSELBALCH EQUATION

Consider one of the histidine residues in hemoglobin that is partly responsible for transport of protons in the blood. At a particular position in the capillaries, a hemoglobin molecule releases oxygen, causing the hemoglobin to shift into the taut structure. As a result, our histidine undergoes a change in pK from 7.0 to 7.4. If the blood pH at this point remains constant at pH 7.4, how many protons does the histidine residue take up as a result of the conformational change?

This question comes down to finding the change in the charge on the histidine. When the p*K* rises, the charge increases and the amount of the increase corresponds to the number of protons taken up. (Actually, the answer will be a fraction of a proton. This really means that the histidine, which is dynamically binding and releasing its proton, will be spending more time with the proton bound.)

What is the charge when the p*K* is 7.0? Use the Henderson–Hasselbalch equation to calculate the ratio of the charged and uncharged forms and use the result to calculate the average charge on the histidine as follows:

At pH 7.4 and p*K* 7.0, the Henderson–Hasselbalch equation becomes:

$$7.4 = 7.0 + \log([\text{proton acceptor}]/[\text{proton donor}])$$

In the case of histidine, the proton acceptor form is uncharged and the proton donor form is charged, giving:

$$7.0 = 7.4 + \log([\text{uncharged}]/[\text{charged}])$$

Solving this equation gives:

$$[\text{uncharged}]/[\text{charged}] = 10^{(7.4-7.0)} = 2.5$$

This means that for every charged histidine residue in this situation there must be an additional 2.5 histidine residues uncharged, or for every 3.5 histidine residues, 1.0 is charged. Thus, the average charge is 1.0/3.5, which is 0.29.

Now we need the average charge when the p*K* is 7.4. We could work through the Henderson–Hasselbalch equation again with the new p*K* value, or we can take a short cut by remembering that when the pH is equal to the p*K*, half the histidines are charged, and so the average charge is 0.5.

Thus, when the p*K* changes from 7.0 to 7.4, the charge on the histidine increases from 0.29 to 0.50, a change of 0.21. The answer to the original question is: "0.21 protons are taken up when the p*K* changes from 7.0 to 7.4 at a constant pH of 7.4."

REFERENCES

Introduction to Protein Structure, Carl Branden and John Tooze, 1991, Garland Publishing, New York.

Proteins: Structures and Molecular Properties, 2nd ed., Thomas E. Creighton, 1993, Freeman and Co., New York.

"The Anatomy and Taxonomy of Protein Structure," J. S. Richardson, 1981, *Adv. Prot. Chem.,* **34:**167–339.

"Protein Folding in the Cell," M. J. Gething and J. Sambrook, 1992, *Nature* **355:**33–45.

The "bio" in biochemistry: protein folding inside and outside the cell, R. J. Ellis, 1996, *Science,* **272:**1448–1449.

World Wide Web at http://moby.ucdavis.edu/HRM/Biochemistry/molecules.htm.

REVIEW QUESTIONS

1. Which of the following substitutions in an α-helical part of a protein is most likely to affect the function of the protein?
   a) Glu → Asp
   b) Lys → Arg
   c) Val → Phe
   d) Ser → Cys
   e) Gln → Pro

2. An amino acid has three ionizable groups, the α-amino and α-carboxyl groups and a side chain that can be positively charged. The p*K* values are 7, 3, and 11, respectively. Which of the following pH values is nearest to the pI of this amino acid?
   a) 3
   b) 5
   c) 7
   d) 9
   e) 11

3. Which of the following partial amino-acid sequences is likely to be found in the membrane-spanning domain of a protein with one such domain?
   a) -Leu-Ile-Gln-Asn-Cys-Trp-
   b) -Leu-Asn-Ala-Ser-Val-Phe-
   c) -Leu-Val-Pro-Asn-Ile-Met-
   d) -Leu-Arg-Ile-Asp-Val-Lys-
   e) -Leu-Val-Met-Phe-Ala-Ile-

ANSWERS TO REVIEW QUESTIONS

1. **e** All of the substitutions are conservative, except Gln → Pro. In addition, proline cannot fit into an α-helix, so the structure would certainly be disrupted by a proline.

2. **d** The isoelectric point, pI, of a molecule is the pH at which the overall net charge on the molecule is zero. Calculate the net charge at each of the pH values given as possible answers and see which net charge is closest to zero. At pH 9, the net charge is -1 from the carboxyl group ($—COO^-$), 0 from the α-amino group ($—NH_2$), and +1 from the side-chain ($—NH_3^+$). The net charge is thus 0. Hence the pI is pH 9.

3. **e** All of these sequences, except e, contain at least one polar residue. The sequence that lacks polar residues is the most likely to be found in an isolated membrane-spanning domain.

CHAPTER

4

MEMBRANES

INTRODUCTION

Many molecules are small enough and hydrophilic enough to move freely through the aqueous environment of the cell, even under the *molecular crowding* conditions found in the cytoplasm. Hence, to keep the required molecules in and the unwanted molecules out, all cells are surrounded by a membrane called the plasma membrane. Eukaryotic cells also have extensive intracellular membrane structures to regulate the movement of molecules between intracellular compartments. Since the molecules moving in the aqueous environment are hydrophilic, the barrier used by the cell is a continuous hydrophobic sheet that is impermeable to hydrophilic molecules. To prevent the two-dimensional hydrophobic sheet from folding up into a solid three-dimensional ball, the sheet is surrounded on both sides by a hydrophilic layer that allows the membrane to take up an extended shape. The hydrophobic barrier is made from lipids and the hydrophilic cover is made from the head-groups of the lipid molecules. This structure is called a ***lipid bilayer.*** Since the barrier has to regulate the passage of molecules through it, the membrane contains protein structures that carry ions, molecules, or signals across the membrane.

LIPIDS, PHOSPHOLIPIDS, AND MEMBRANE STRUCTURE

The hydrophobic part of the membrane is made from a class of lipids that comprises *fatty acids* attached covalently to hydrophilic, that is, polar, heads, often joined through a *glycerol* or serine derivative. Thus, *diacyglycerols* and *triacylglycerols* have two or three fatty acids esterified to glycerol, respectively. The main groups of membrane lipids that share this overall structure are *phospholipids* and *glycolipids* (Fig. 4.1). The major constituents of membranes are phospholipids such as *phosphatidyl choline* (Fig. 4.2) *sphingomyelin, phosphatidylserine,* and *phosphatidylethanolamine.* Glycolipids have sugar residues as their polar head groups and include *gangliosides* whose polar groups include *sialic acid.* Glycolipids are found in the extracellular layer of the bilayer in plasma membranes and *inositol phospholipids* are found on the cytoplasmic side and are important in *cell-signaling* mechanisms. Eukaryotic plasma membranes also contain lipids that lack a large hydrophilic part, particularly *cholesterol* (Fig. 4.2b).

A membrane is a large macromolecular complex, about 5 nm thick, in which two sheets of phospholipids aggregate so that their hydrophobic lipid portions are packed together between the outer layers, which are

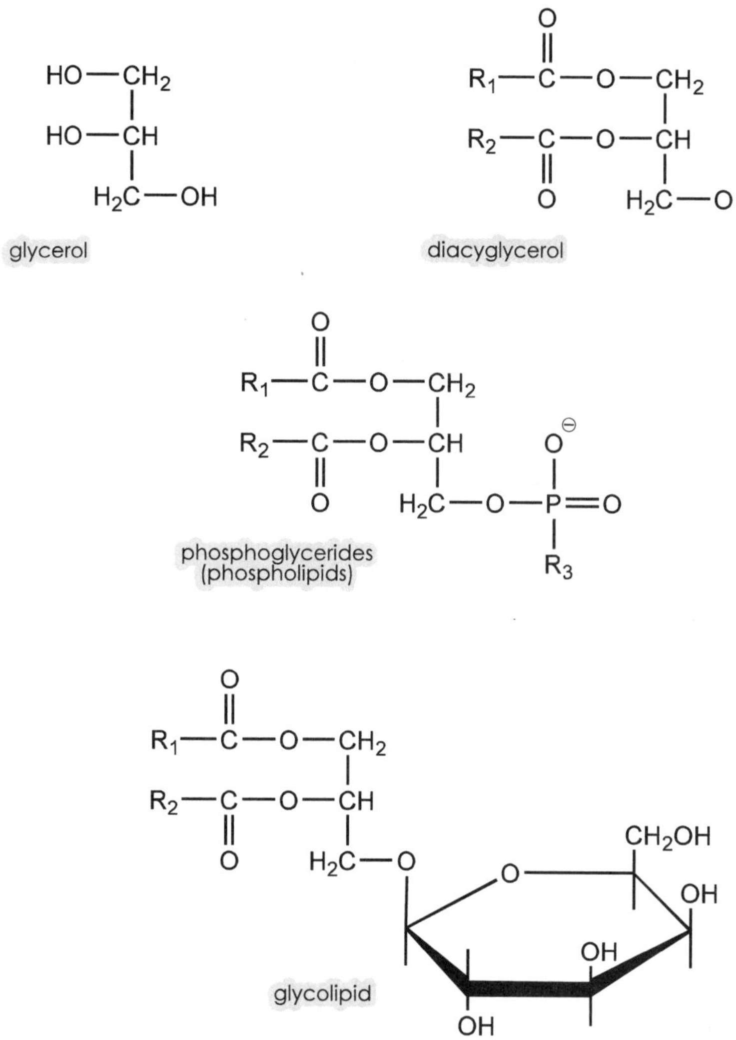

glycerol

diacyglycerol

phosphoglycerides (phospholipids)

glycolipid

Figure 4.1.(a)

Overall structural features of phospholipids, glycolipids and sphingolipids.

Glycerol has a 3-carbon "backbone" with three hydroxyl groups.

Diacylglycerol is derived from glycerol by esterification of the first two hydroxyl groups.

Phosphoglycerides (*phospholipids*) are derived from diacylglycerol by phosphorylation of the remaining hydroxyl group and the addition of a third group, R_3, via the phosphate.

Glycolipids are derived from diacylglycerol by the attachment of a sugar or polysaccharide to the remaining hydroxyl group.

composed of the hydrophilic head groups (Fig. 4.3). This resulting "lipid bilayer" provides an essentially impermeable barrier to the passage of hydrophilic ions and molecules. The membrane is held together by nonspecific hydrophobic interactions in the lipid bilayer and noncovalent polar interactions of the head groups. Unlike the case of globular proteins, where the movement of the atoms in the structure is quite limited, membranes are relatively fluid and individual molecules in the membrane can migrate randomly in two dimensions through the membrane (Fig. 4.3).

Membranes naturally tend to form closed structures, to avoid exposing the hydrophobic ends of lipid bilayers to the solvent. Synthetic closed structures, called *liposomes,* can be made with membrane fragments or synthetic phospholipids. Liposomes can be made to contain compounds buried in the membrane or totally enclosed and are a source of great interest as a vehicle for delivering drugs or genes to diseased cells. Membrane-bound components can be used to target the liposomes to the appropriate cells and fusion with the cell membrane can be used to deliver the contents of the liposome to the cell interior.

Although membranes are stable in the aqueous environment normally found in the body, they can be broken up by organic solvents or by detergents, and it is possible that this accounts for some of the deleterious effects of *ethanol,* for example, on the *pancreas.* Organic solvents, and other substances such as guanidinium salts, destabilize the water structure and so decrease the hydrophobic interactions that hold the lipid bilayer together. Detergents have hydrophobic regions that can penetrate the lipid bilayer and disrupt its structure.

The outer surface of many cellular plasma membranes is exposed to the extracellular environment. In mammalian cells, the major chemical groups on the

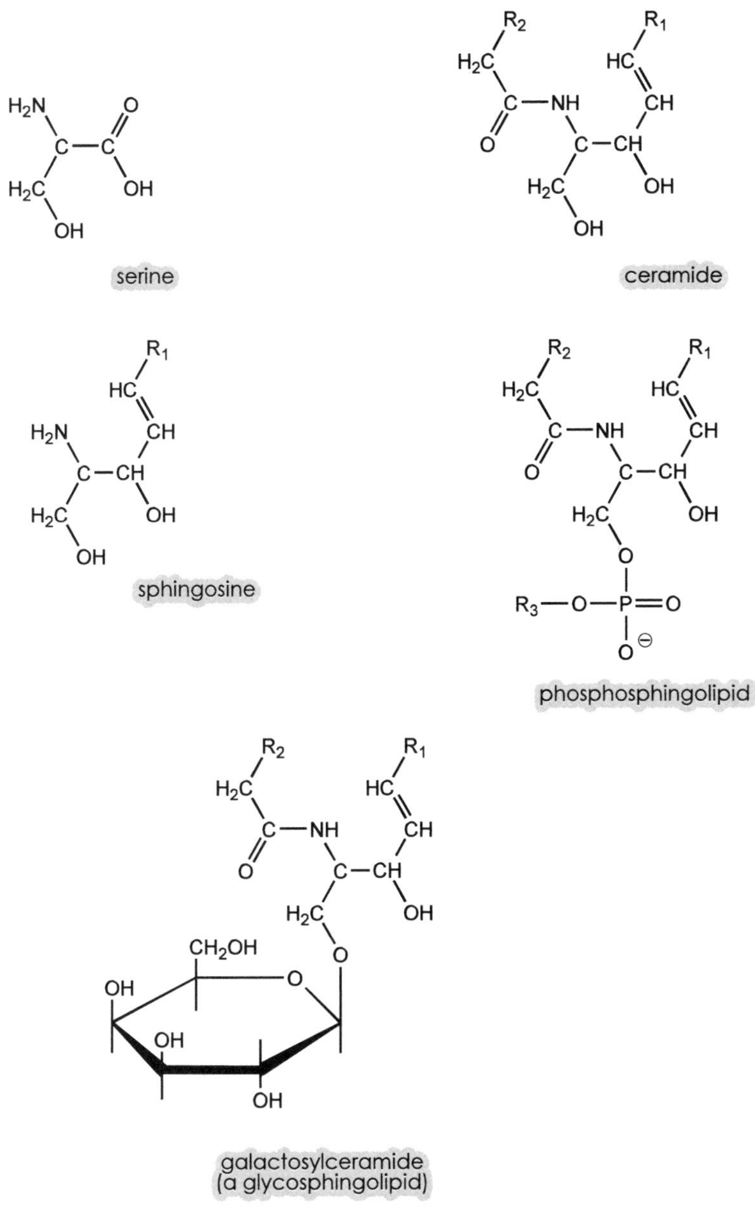

serine

ceramide

sphingosine

phosphosphingolipid

galactosylceramide
(a glycosphingolipid)

Figure 4.1.(b)

Serine is an amino acid, also found in proteins.

Sphingosine is derived from serine by loss of the carbonyl group and the addition of a hydrocarbon side chain in its place.

Ceramide is derived from sphingosine by an additional side chain, attached to the amino group.

A *phosphosphingolipid* is derived from ceramide by addition of a third side chain, to the side chain hydroxyl of the original serine.

A *glycosphingolipid* is derived from ceramide by the addition of a sugar or polysaccharide to the side chain hydroxyl of the original serine.

outer surface are carbohydrates, comprising the polysaccharide parts of integral membrane proteins and glycolipids. Proteins with exceptionally large amounts of covalently bound carbohydrate are called *proteoglycans,* and some of these are bound to the plasma membrane, although they are found mainly in the *extracellular matrix.* This carbohydrate zone on the surface of the cell may be called the *glycocalyx,* and is involved in cell–cell interactions as well as probably providing a protective coat for the cell. In *yeast* and *plant cells,* there is a rigid cell wall surrounding the plasma membrane. In bacteria, such as *E. coli,* there is a double-membrane structure with an aqueous space, the *periplasmic space,* between the outer and inner bilayers.

Drugs that penetrate the plasma membrane and allow the passage of ions can have major effects on cells. For example, the drug *A23187* is an example of the class of drugs known as *ionophores* which permit the passage of specific ions across phospholipid membranes. A23187 allows passage of Ca^{2+} ions and mimics or disrupts normal *cell signaling* events that depend on regulation of $[Ca^{2+}]$. *Gramicidin A* is useful as an *antibiotic*; it allows H^+, Na^+, and K^+ to flow across the phospholipid membranes of gram-positive bacteria, killing them.

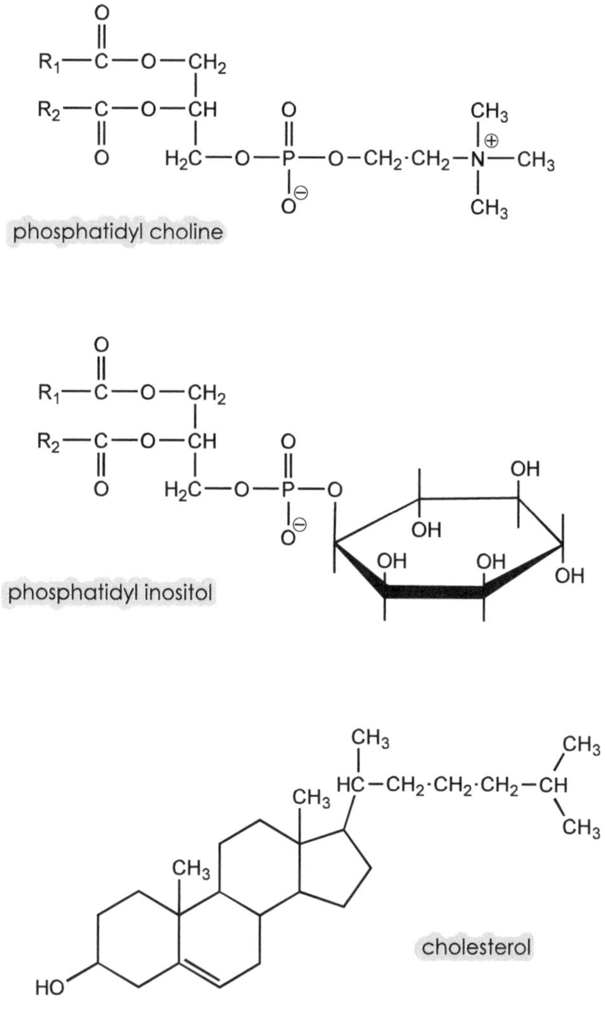

phosphatidyl choline

phosphatidyl inositol

cholesterol

Figure 4.2.

Three important lipids—phosphatidyl choline, phosphatidyl inositol, and cholesterol.

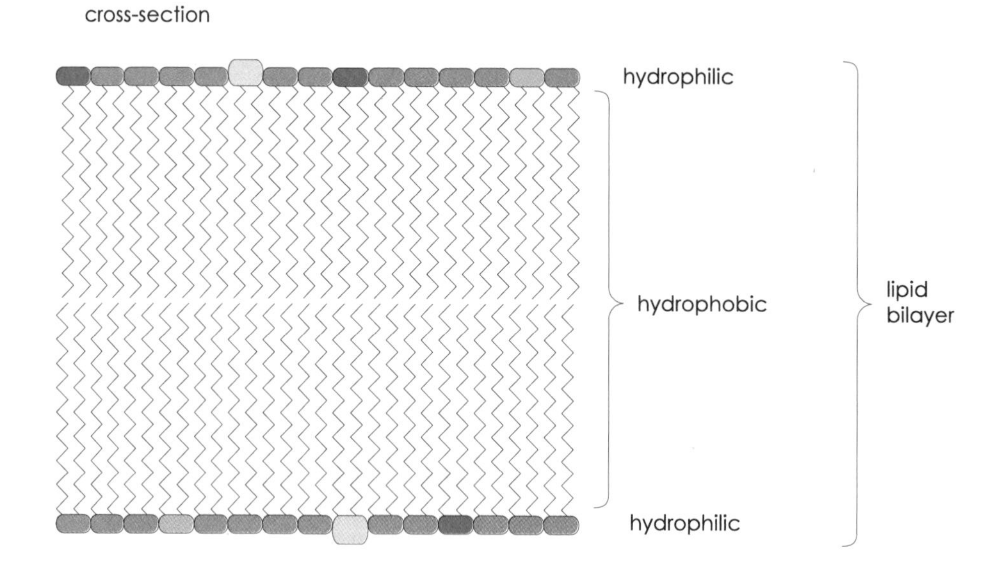

Figure 4.3.(a)

Membrane structure. Lipid bilayer.

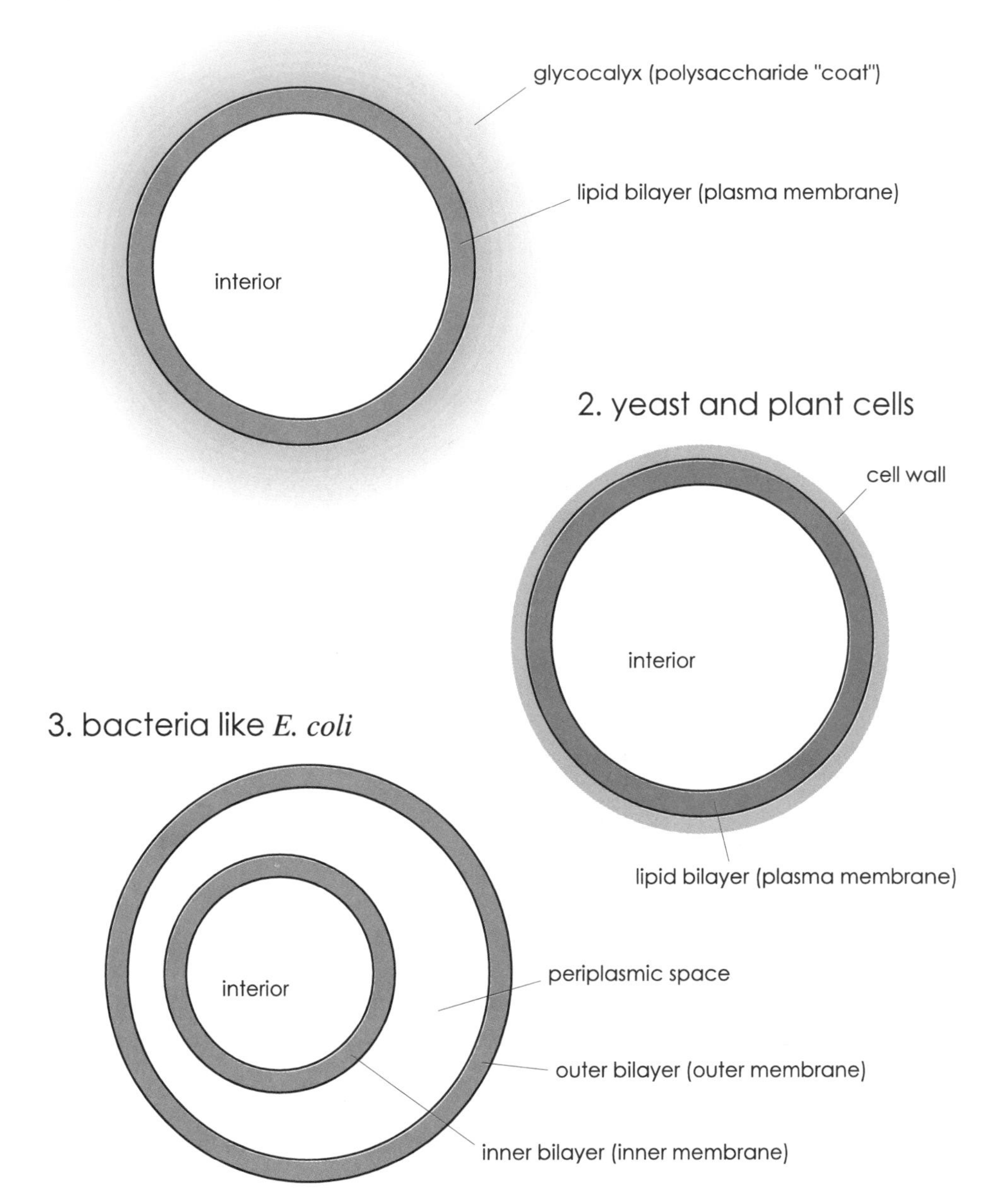

Figure 4.3.(b)
Membrane structure. Organization of plasma membranes in mammalian, yeast, and bacterial cells.

PROTEINS IN MEMBRANES

Passage of polar molecules and ions through the membrane is made possible by the presence of proteins in the membrane and proteins comprise on the order of 50% of the mass of a membrane. A protein can be associated stably with a lipid bilayer membrane in three general ways (Fig. 4.4):

1. The protein can bind to the surface of the membrane (called *peripheral membrane proteins*).
2. The protein can be covalently attached to a lipid prosthetic group, such as a prenyl group, that buries itself in the lipid bilayer (Chapter 3).
3. The protein may cross the membrane (called a *transmembrane protein*) (Fig. 4.4) or be firmly embedded in the bilayer without crossing the membrane (*integral membrane proteins*).

Transmembrane proteins have a special structure where they cross the lipid bilayer. This is usually an α-helix composed of hydrophobic amino-acid side chains,

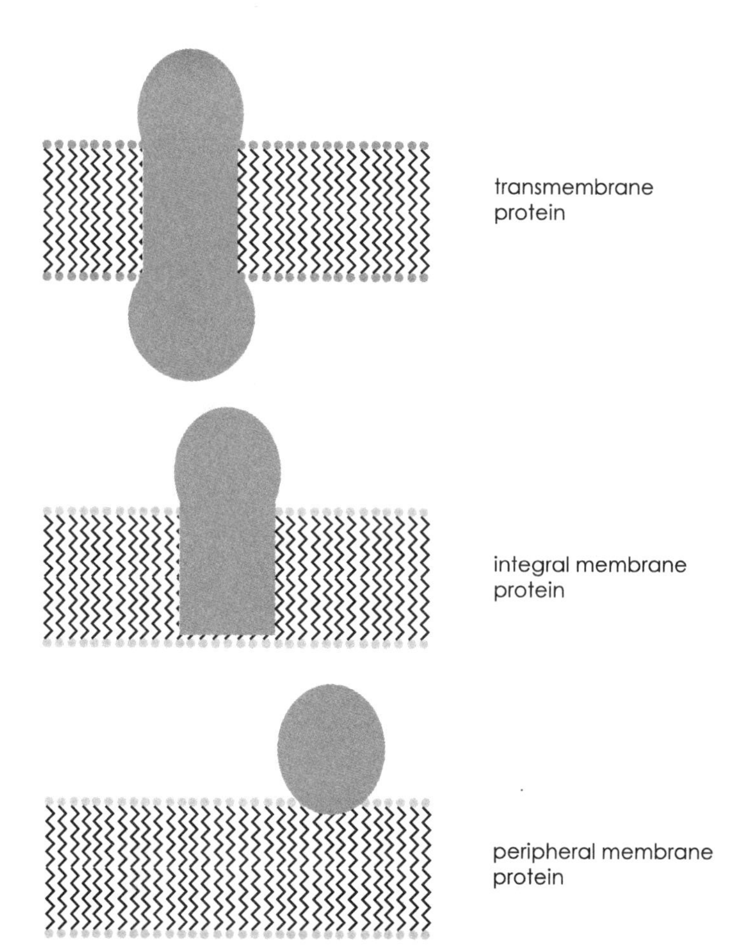

Figure 4.4.
Proteins and a lipid bilayer. Proteins are found in membranes, either spanning the membrane (transmembrane proteins), buried in the membrane (integral membrane proteins), or attached to the membrane surface (peripheral membrane proteins), as shown from top to bottom, respectively.

although a β-barrel structure may also be used. The α-helical structure buries the polar amide groups inside the helix, surrounding them with hydrophobic side-chains, giving a cylinder coated with hydrophobic groups. This hydrophobic α-helix can "dissolve" easily in the hydrophobic lipid bilayer and is known as a *transmembrane helix.* Although one such helix anchors a protein in the membrane, many transmembrane proteins have a bundle of hydrophobic α-helices, with the main chain snaking from side to side of the membrane. Such proteins have a characteristic arrangement of amino acids in their sequence that can be recognized by a *hydropathy plot.*

A **hydropathy plot** shows the average hydrophobicity of successive short sequences of amino acids as a function of position along the protein sequence. For example, the hydropathy index at position 1 in a protein might be the average hydrophobicity of the first 20 amino acids in the sequence. Then, at position 2, the hydropathy index would be the average hydrophobicity of amino acids 2–21; at position 3, it would be the average hydrophobicity of amino acids 3–22; and so on. Regions of high hydrophobicity, at least 20 residues long, indicate transmembrane α-helix structures, while other parts of the protein show slight hydrophobicity or hydrophilicity (Fig. 4.5). While the hydropathy plot is not definitive, it is a very useful method for recognizing an integral membrane protein and predicting its topology.

Integral membrane proteins are oriented in the membrane. In the plasma membrane, the extracellular part of the protein is often characterized by the presence of O-linked polysaccharide groups and disulfide (cystine) bridges which are absent from the cytoplasmic side. Integral membrane proteins may permit and/or regulate the passage of hydrophilic ions and molecules, they may transmit information across the membrane, or they may play structural roles, linking the membrane to the *cytoskeleton* or extracellular matrix or linking two cells together in a ***tight junction.***

The extensive hydrophobicity of integral membrane proteins makes them insoluble in most aqueous environments. However, they may be dissolved in the presence of *detergents* such as SDS (sodium dodecyl

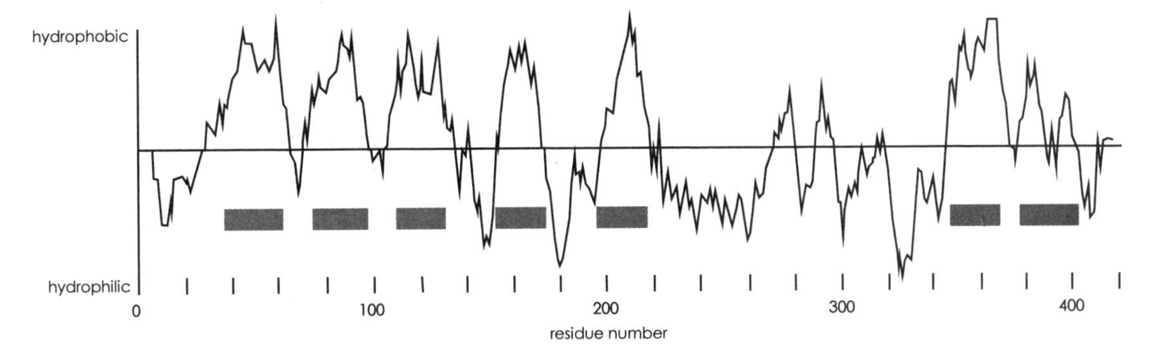

Figure 4.5.

Hydropathy plot. The hydropathic index is plotted on the *y* axis, with hydrophobic domains at the top and hydrophilic domains at the bottom. The *x* axis shows the amino-acid residue number from the amino to the carboxy terminus. The protein shown is a 7-transmembrane receptor (the serotonin receptor), analogous to the epinephrine (adrenalin) receptor. The approximate positions of the seven transmembrane segments are indicated by the purple bars. (Original plot graciously provided by Dr. Grace Rosenquist.)

sulfate) which is a charged (anionic) detergent, or *Triton X-100,* a neutral detergent.

MEMBRANE TRANSPORT PROTEINS

Transport of molecules across plasma membranes is a critical function of all organisms. A single genetic mutation in a transport protein can give rise to *disease.* In *cystinuria,* for example, dietary cysteine can be taken up in the *intestine* in the form of small polypeptides and subsequently released into the bloodstream as the free amino acid which would normally be taken up by the tissues. In cystinuria, however, the import of cystine into cells cannot occur due to a *genetic defect* in the *transport protein* and the cystine in the blood is passed into the *kidney* where *renal reabsorption* is defective and the cystine precipitates as *kidney stones.*

In the case of molecules, such as molecular oxygen, that are hydrophobic enough to diffuse through the lipid bilayer without help, the net transport occurs from high concentration to low concentration—*passive diffusion.* In the case of uncharged polar molecules that diffuse extremely slowly through the lipid bilayer, a protein that opens a *channel* or carries molecules across the membrane in the same direction is just speeding up the diffusion process—*facilitated diffusion.* In the case of charged molecules, the direction of net diffusion depends on both the concentration gradient and the electric potential gradient. The combination of these factors is the ***electrochemical gradient*** and, as with neutral polar molecules, diffusion of charged molecules and ions down the electrochemical gradient can be facilitated by passive transport proteins that open a channel or carry molecules across the membrane. Channels allow much faster diffusion than *carriers,* but both can be regulated to control the rate of diffusion.

In other cases, the cell requires molecules to be moved against their concentration or electrochemical gradient. This is achieved by carrier proteins or "pumps" that use ***active transport.*** The movement of a molecule is coupled to another event, such as the hydrolysis of ATP, that provides the energy to move the molecule against the gradient. When just one type of molecule is transported in this way, the carrier protein is called a *uniporter.*

The versatility of carrier proteins is increased, in some cases, by the coupling of the transport of one type of molecule with the transport of another molecule, either in the same direction, a *symport,* or in opposite directions, an *antiport.*

There are several *superfamilies* of membrane-transport proteins. The largest superfamily, the *ABC transporter superfamily,* is of major *clinical importance* because cells can use members of this superfamily to develop *multidrug resistance.* This is a particular problem in *chemotherapy* where *cytotoxic* drugs are used to kill cancer cells. A cancer-cell population develops that overexpresses the multidrug resistance (*MDR*) protein. The MDR protein transports hydrophobic drugs out of the cell as fast as they diffuse in, thus negating the *cytotoxic* activity of the drug. The MDR protein is capable of transporting a wide range of hydrophobic drugs; once resistance has been developed to one drug, a wide variety of other drugs is also ineffective. *Cystic fibrosis,* a common *genetic disorder,* is due to a defect in the gene for an ABC transporter that functions as a *chloride channel* in ***epithelial cells,*** such as those in the *lung.* This is of particular interest because of the possibility of delivering the correct gene, packaged in a liposome or an inactivated virus, to the lungs by nasal spray.

Nerve impulses are transmitted between cells by the release of a *neurotransmitter* by one cell. The neurotransmitter diffuses across the *synapse* and interacts with an ion channel protein on the target cell. The interaction of the neurotransmitter opens the *ion channel,* passing the signal to the target cell. A *transmitter-gated ion channel* is such an ion channel, and these channels are targets of a variety of drugs, such as *succinylcholine,* a competitive inhibitor of *acetylcholine,* used in *surgery* to relax skeletal muscles, and *barbiturates* and *tranquilizers* such as *Valium* (*diazepam*) and *Librium* (*chlordiazepoxide*) which bind to *γ-aminobutyric acid* **receptors.**

MEMBRANE-BOUND RECEPTORS AND OTHER INTRACELLULAR SIGNALING MOLECULES

The plasma membrane plays a critical role in intercellular communication. Such communication is critical to the survival of complex multicellular organisms. Communication occurs between neighboring cells through direct interactions or between more distant cells through the use of extracellular signaling molecules that are secreted by one cell or group of cells and recognized by other cells, the target cells. *Extracellular signaling molecules* are of two main types: those that can pass through the plasma membrane and interact directly with a receptor inside the *target cell*; and those that cannot pass through the plasma membrane. Examples of the former group are the *steroid*

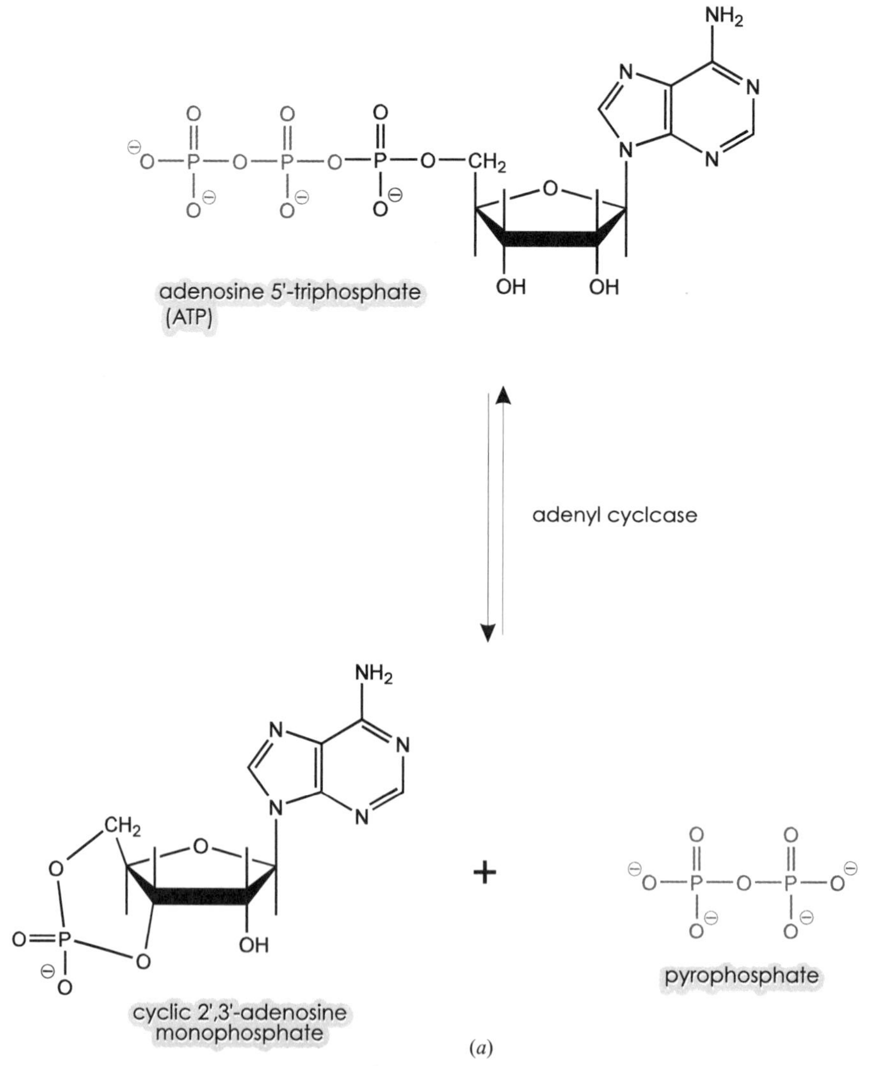

adenosine 5'-triphosphate (ATP)

adenyl cyclcase

cyclic 2',3'-adenosine monophosphate

pyrophosphate

(*a*)

Figure 4.6.(a)

Formation of second messengers. (a) Cyclic AMP is formed from ATP by the action of adenyl cyclase. The reaction is driven by the hydrolysis of the pyrophosphate released in the reaction. Adenyl cyclase is a membrane-bound protein and the reaction occurs in the cytoplasm, close to the membrane.

hormones and of the latter group are *peptide hormones* and molecules bound to the membrane of the signaling cell.

The plasma membrane does not have a direct role in the response to a steroid hormone, other than allowing the steroid to enter the cell. The plasma membrane is the initial point of contact for all the other extracellular signals. The plasma membrane contains transmembrane proteins that have a *receptor domain* on the extracellular surface and a *response domain* on the intracellular surface—the cytoplasmic domain. The extracellular signaling molecule, the ligand, binds to the receptor domain and causes changes—such as *dimerization* of two receptors—that lead to changes in the cytoplasmic domain(s). These changes lead to the generation of a ***second messenger***—an intracellular signaling molecule—or to *protein phosphorylation.* These processes trigger a ***signal transduction*** pathway that carries the signal away from the membrane to the appropriate parts of the cell, such as the nucleus, where *gene expression* can be changed, the mitochondria, where [Ca^{2+}] can be regulated, or the cytoskeleton.

The generation of second messengers may be direct, as in ligand-gated ion channels, or indirect, as in the production of *inositol phosphates* or *cyclic nucleotides* (Fig. 4.6). The indirect production of second messengers is mediated through a membrane protein that has membrane and cytoplasmic domains but no extracellular domain: a *heterotrimeric* ***G protein.*** Binding of ligand to receptor causes the receptor-G protein interaction to change, separating the *G protein* into its α subunit and a heterodimer of the β and γ subunits. The separated subunits can then activate other molecules such as *adenyl cyclase* or a *phospholipase* (Fig. 4.7) to produce second messengers such as *cyclic AMP* or *inositol phosphates* and *diacylglycerol.*

Protein phosphorylation can occur:

1. Directly through protein kinase activity of the receptor itself
2. As a result of second-messenger activation of protein kinases
3. Through activation of cytoplasmic protein kinases by other mechanisms

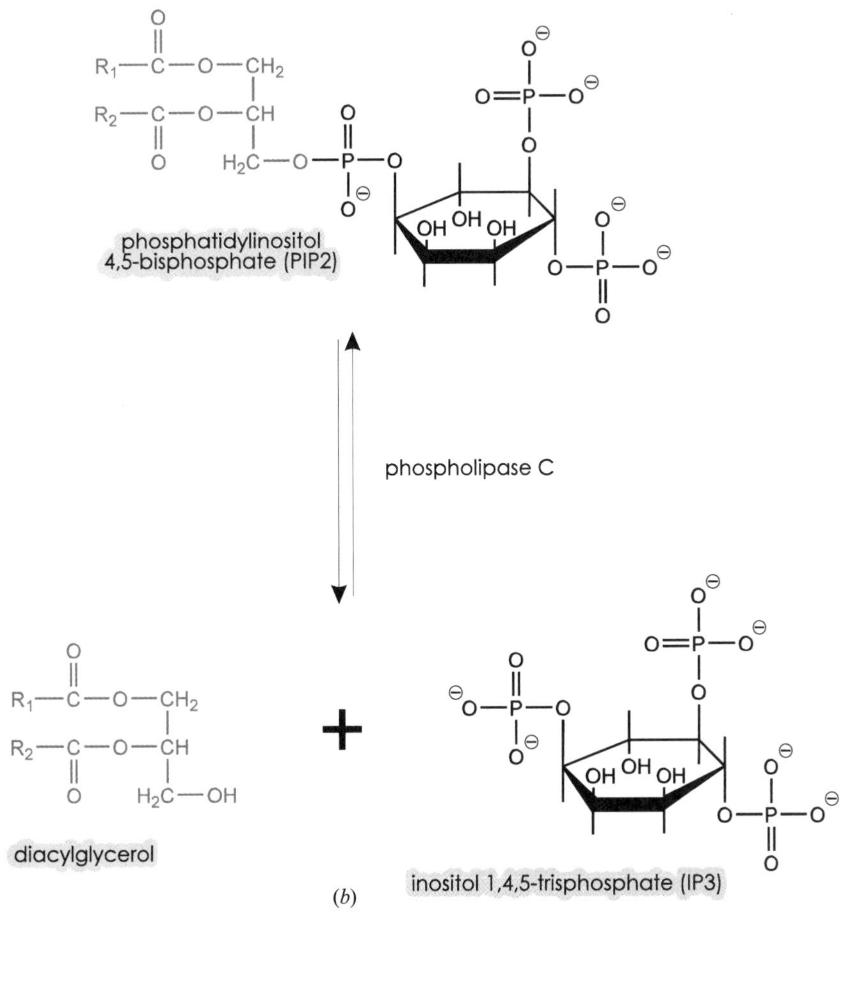

phosphatidylinositol 4,5-bisphosphate (PIP2)

phospholipase C

diacylglycerol

inositol 1,4,5-trisphosphate (IP3)

(*b*)

Figure 4.6.(b)
Formation of second messengers. Phosphatidyl inositol, a membrane phospholipid, is cleaved by phospholipase C, releasing inositol phosphate into the cytoplasm and diacylglycerol, which remains in the membrane.

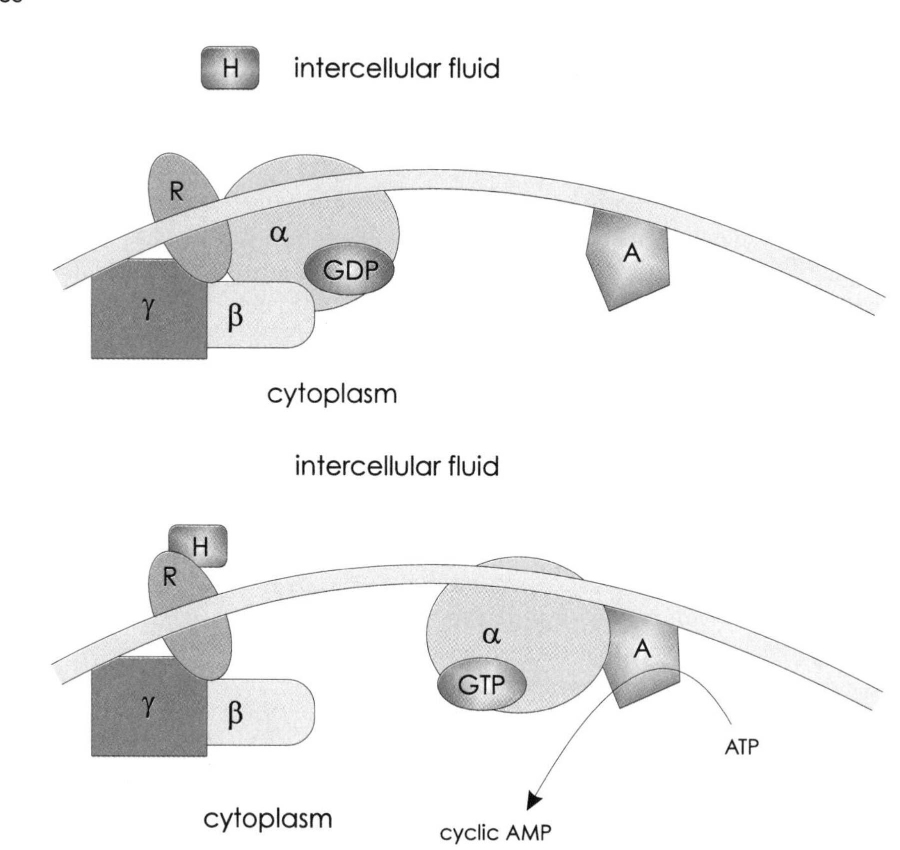

Figure 4.7.

Signal transduction across a membrane mediated by a heterotrimeric G protein. The upper panel shows the resting state. The receptor, R, is not occupied by ligand; the heterotrimeric G protein subunits, αβγ, are associated with one another and with the receptor. The α-subunit has GDP bound. Adenyl cyclase, A, is inactive. The hormone, H, is shown free in the intercellular fluid. The lower panel shows the activated state. The hormone, or ligand, is bound to the receptor. Activation of the receptor has caused the exchange of GDP for GTP on the α-subunit of the heterotrimeric G protein. The α-subunit has dissociated from the receptor complex and is now interacting with adenyl cyclase. Adenyl cyclase is activated, by its interaction with the GTP-bound α-subunit, and catalyzes the conversion of ATP to cyclic AMP.

Some transmembrane receptors have cytoplasmic ***protein phosphatase*** activity. Soluble protein phosphatases, such as *calcineurin, PP1, PP2A, PP2C,* and *protein tyrosine phosphatases,* also mediate intracellular signals.

SUMMARY

1. Lipids are hydrocarbon chains with a polar head group.

2. In aqueous solution, they aggregate with their hydrocarbon chains packed inside, surrounded by the polar head groups. Cellular membranes are constructed this way, with a pair of layers of side-by-side lipids facing each other.

3. Much of the mass of a membrane is protein, either integral membrane, membrane-attached, or membrane-associated proteins.

4. Proteins in membranes may function as transporters, to allow or actively encourage molecules to pass through the membrane. Active transport can maintain a flow against a concentration gradient.

5. Protein receptors in the plasma membrane bind extracellular signaling molecules such as hormones and pass the signal to the interior of the cell.

REFERENCES

Introduction to Protein Structure, Carl Branden and John Tooze, 1991, Garland Publishing, New York.

Biomembranes: Molecular Structure and Function, R. B. Gennis, 1989, Springer-Verlag, New York.

Membranes and Their Cellular Functions, J. B. Finean and R. Coleman, 1984, Blackwell Scientific, Oxford, UK.

A collection of review articles in *Science* **258:**917–969 (1992).

REVIEW QUESTIONS

1. Which of the following statements about phosphatidylinositol is false?
   a) Is a type of phospholipid
   b) Is a derivative of diacylglycerol
   c) Contains three fatty-acid chains
   d) Is found in the plasma membrane
   e) Is a precursor for two important second messengers

2. Kidney stones are a major symptom of cystinuria. What is the basis of this disease?
   a) Dormant forms of yeast accumulate in the kidney
   b) None of the cystine or cysteine in food gets into the bloodstream
   c) Lack of disulfide isomerase
   d) Lack of an amino acid transporter
   e) Dietary deficiency

ANSWERS TO REVIEW QUESTIONS

1. ***c*** Phosphatidyl inositol is a phospholipid containing a sugar (inositol) attached through a phosphate group to diacylglycerol. Thus, a and b are correct, and c is the false statement. Phosphatidyl inositol is indeed found in the plasma membrane, where it can be cleaved by phospholipase to produce the second messengers, diacylglycerol and inositol trisphosphate, so d and e are also correct statements.

2. ***d*** Cystinuria is due to the loss of the amino-acid transporter protein that transports cysteine and cystine into cells. The stones are due to precipitation of cystine in the kidney. Cystine gets into the bloodstream from digestion of food because oligopeptides are taken up into the cells lining the intestine and some of those peptides yield cystine, which is exported to the bloodstream. Cystinuria has nothing to do with yeast, disulfide isomerase, or dietary deficiencies; it is a genetic disease.

CHAPTER
5

NUCLEOPROTEIN COMPLEXES

INTRODUCTION

Although proteins and nucleic acids have well-separated functions in many instances, they also work intimately together in specific complexes containing both nucleic acid and protein. Some *nucleoprotein complexes* are very stable, some are transitory, and others have an intermediate stability. The protein component may provide a structural support for the nucleic acid, but in many cases, the two types of molecule both contribute directly to the function of the complex. Although cases of enzyme action by pure RNA molecules are rare, RNA molecules often act catalytically in nucleoprotein complexes. The chromosome was the first nucleoprotein complex to be discovered and is discussed first. Ribosomes have been studied intensively for many years and contain most of the RNA in the cell. More recently, nucleoprotein structures such as *telomerase, spliceosomes,* and *signal-recognition particles* have illustrated the versatility of nucleoprotein complexes.

CHROMOSOMES

Chromosomes are nucleoprotein complexes that contain the genes—DNA—found in the cell nucleus. The term was introduced when chromosomes were seen in dividing eukaryotic cells and is usually reserved for the nuclear DNA–protein complexes of eukaryotic cells. The term "*bacterial chromosomes*" may, however, be used for the DNA–protein complexes found in bacterial cells. Bacterial "chromosomes" differ from eukaryotic chromosomes in their protein content and structure.

A normal human *somatic cell* has 46 chromosomes, divided into the autosomes and the sex-specific chromosomes known as X and Y; there are 44 autosomes+2X (female) or 44 autosomes+X+Y (male). The autosomes are in pairs, with one of each pair being derived from each parent. Hence, there are 22 different autosomes.

Some *diseases* are due to changes in chromosome structure or gene expression. Large-scale changes in chromosomes can occur, including loss or gain of chromosomes, changes in the number of chromosomes per cell, or *translocations* (exchange of genetic material) between chromosomes. Such changes may cause disease: *trisomy* (3 copies) of *chromosome 21* causes *Down's syndrome* (*Mongolism*); certain translocations between chromosomes 22 and 9 are associated with *chronic myelogenous leukemia*; translocations between chromosomes 8 and 14 are associated with *Burkitt's lymphoma.* More subtle changes in gene expression may also cause disease. For example, the loss of a single gene, the *retinoblastoma gene,* is associated with *retinoblastoma,* an intraocular *cancer* of the developing eye. Increased expression or structural changes in specific genes, called *protooncogenes,* contribute to other cancers.

The chromosomes are located in the nucleus, except during **mitosis** or ***meiosis*** when the nuclear membrane dissociates and the chromosomes condense into "*metaphase chromosomes*" and separate into the two sets destined for the "daughter" cells. Eukaryotic chromosomes are visible as well-defined structures in the *light microscope* when cells are in mitosis or meiosis. In *interphase,* these compact structures disperse throughout the nucleus and can no longer be discerned clearly by light microscopy.

Each chromosome consists of DNA systematically folded by interactions with the proteins of the chromosome. Folding occurs at several levels, or hierarchies. In the lowest level, DNA is wound around protein cores to form nucleosomes, which are described in more detail below. In the next hierarchy, strings of nucleosomes are coiled into a helix known as the *30-nm fiber* because of its apparent width (30 nm) when viewed by *electron microscopy.* The 30-nm fiber is then coiled again and, perhaps, yet again to form the fully condensed metaphase chromosome. During interphase, some of these coils unwind to allow the DNA to function. These structural hierarchies are summarized in Figure 5.1 and described in more detail below.

Chromosomes are associated with the following functions:

1. Storage and *transmission of genetic information.* Storage involves packing the chromosomes up into structures small enough to fit in the cell nucleus. Transmission involves separating duplicated chromosomes into complete sets and separating them so that each new nucleus receives one of each chromosome.
2. Expression of genetic information. Expression involves many steps. At the chromosome level, the gene and its associated regulatory sequences must be made available for binding *transcription factors* (see Chapter 29) and enzymes and they must be allowed to synthesize RNA.
3. Maintenance of genetic information. Maintenance is mainly a matter of repairing the *damage* that occurs to DNA because of chemical attack, especially from *carcinogenic compounds* or *radiation,* such as *sunlight* or *X-rays.*
4. *Recombination* of genetic information. When *reproduction* occurs, genes, or parts of genes, may be exchanged between chromosomes, to generate new combinations and continue *evolution.*

Chromatin

Chromosomes are made of *chromatin* which consists mainly of DNA and proteins. RNA is associated with chromatin that is involved in transcription. As mentioned in Chapter 2, the DNA in chromosomes is primarily in a right-handed double helical structure (B-DNA). Chromosomal RNA is mostly the product of ongoing transcription. Chromatin is a complex, dynamic structure determined mainly by protein binding. Many DNA-binding proteins bind independently of the DNA sequence, to a first approximation; others have strong preferences for particular sequences.

Chromosomal Proteins

The most common structural chromosomal proteins are the histones. Other common chromosomal proteins are called "*nonhistone proteins*" and include some other structural proteins and specific enzymes such as *RNA polymerases* and *DNA topoisomerases.* Histones and many of the enzymes have little or no sequence specificity in their DNA binding properties. That is, they bind equally well to almost all DNA molecules, regardless of their nucleotide sequence. Many regulatory proteins bind to DNA in a sequence-specific manner; that is, they bind much more tightly when they "find" the appropriate nucleotide sequence. Other regulatory proteins do not bind directly to DNA but bind to DNA–protein complexes. Regulatory proteins may be present in very low concentration.

The histone proteins comprise half or more of the mass of chromosomal proteins. In most eukaryotic cells, there are 5 histones: *H1, H2A, H2B, H3,* and *H4.* H1 is called the "*linker" histone*; the others are called "*core" histones,* reflecting their roles in nucleosomes—the first hierarchy of chromosome structure. H1 has a molecular weight of about 21,000 and comprises three structural domains, a central structured domain and two terminal random-coil domains. The core histones are small proteins, with molecular weights of 11,000–15,000, and each has an N-terminal random-coil domain with the rest of the molecule structured, except for a few residues at the C-termini of H3 and H4. The histones are very basic, owing to their high content

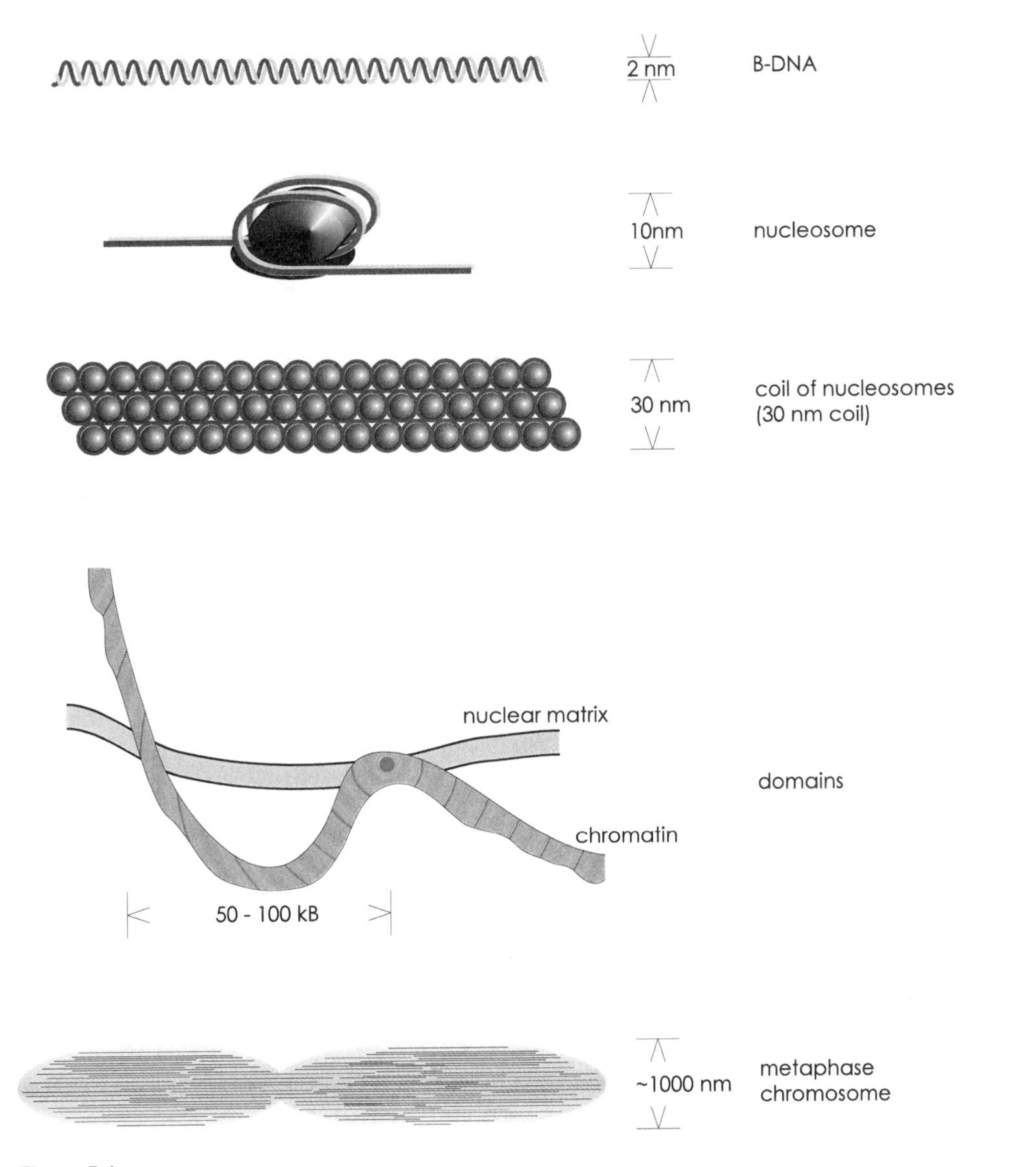

Figure 5.1.
Hierarchy of chromatin structure. The figure illustrates the various known levels of DNA folding in the chromosome. The first level (top) is the coiling of the two strands of DNA around each other in the B-DNA double helix. The double helix is then coiled again, around the histone octamer, to form nucleosome core particles. In inactive regions of the chromatin, the nucleosome core particles are coiled again into the 30-nm coil to produce a fiber. The fiber, and/or strings of nucleosomes, are attached to the nuclear matrix at points about 50–100 kb apart, creating a domain of chromatin between each pair of adjacent attachment points. In metaphase chromosomes, the 30-nm fiber is further coiled to produce the fully condensed chromosome (bottom).

of the basic amino acids lysine and arginine, which are concentrated in the random-coil terminal domains.

The core histones occur in equimolar amounts, approximately two of each per 200 bp of DNA. Histone H1 is present at about half the molar concentration of the core histones.

Except for histone H4, each histone exists in several closely related subfractions or variants that have slightly different amino-acid sequences. Different tissues have different proportions of the various subfractions, but it is not known if this tissue specificity has functional significance. A substantial *variant of histone H1,* termed *H1*^{0}, has been associated with cessation of cell growth, although what role, if any, it plays in the regulation or maintenance of the state of low growth is not known. In some cases, specific variants are under different cell-cycle controls, so that one group is synthesized continuously through the interphase part of the cell cycle, while another group is synthesized together with DNA specifically in S phase.

Histones are also subject to reversible post-synthetic modifications, including phosphorylation, acetylation, and *poly ADP-ribosylation.* The functions of these modifications remain to be fully established, but there are good indications that:

1. Phosphorylation of H1 is associated with *chromosome condensation* in mitosis and meiosis.
2. Acetylation of histones H3 and H4 is associated with enhanced transcription.
3. Acetylation of all core histones is associated with *assembly of chromosomes* as part of their replication.
4. Poly ADP-ribosylation is associated with *DNA repair.*

Nucleosomes

The core histones form a multi-subunit complex, an octamer, containing two of each core histone with their structured domains forming an approximately cylindrical core and the random coil domains in solution. The cylindrical core is composed of a tetramer of H3 and H4 flanked by a dimer of H2A and H2B on each end. Then 146 bp of DNA wraps around the core histone octamer, forming 1.7 turns of a superhelix. This particle is known as the "*nucleosome core particle.*"

Most nucleosome core particles are associated with one molecule of histone H1 per core particle. The H1 molecule organizes an additional 22 base pairs of DNA, forming a total of two superhelical turns. This particle is termed the "*chromatosome,*" although the term is not widely used (Fig. 5.2).

In chromatin, a stretch of "*linker*" *DNA* (0–80 base pairs, typically 30 base pairs) is found between adjacent chromatosomes. The complete repeating structure is called a nucleosome which contains 8 core histones + histone H1 + 168 base pairs of histone-bound DNA + linker DNA. In the electron microscope, nucleosomes can be "seen" as particles arranged along the thin DNA molecule and the structure is named, after its appearance, "*beads on a string.*" In some regions of chromatin, histone H1 may be absent, leaving nucleosome core particles connected by linker DNA.

In a nucleosome, approximately half of the negative electrical charge on the phosphate groups in DNA is neutralized by the positive charges of histidine, lysine, and arginine residues in the histones. The remainder of the phosphate charge is available for interactions with other proteins or with ions or *polyamines.* The vast majority of cellular DNA is in a nucleosome structure, although small regions are free of nucleosomes owing to the binding of nonhistone proteins.

Higher-Order Chromatin Structures

In chromosomes, the nucleosomes are further folded in at least three additional hierarchies of structure. These are called higher-order structures.

In repressed, or inactive, regions of chromatin, ar-

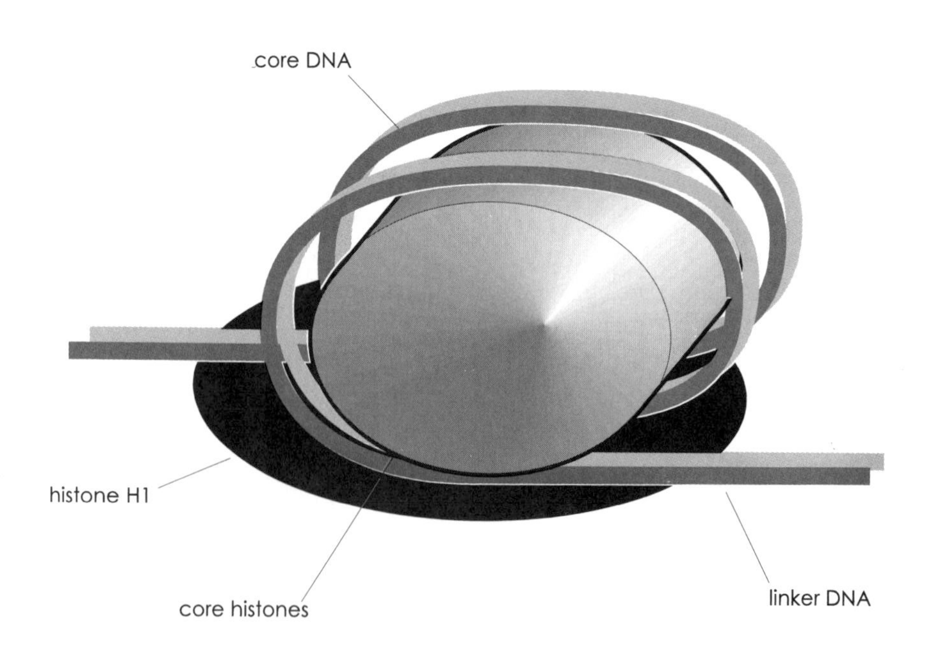

Figure 5.2.

Nucleosome structure. The DNA double helix is shown as a pair of lines in purple and gray. The octamer of core histones is the shaded gray disk and the black ellipse is histone H1. This is a diagrammatic representation showing the components and their approximate relationship to one another.

rays of nucleosomes can form a coil known as the 30-nm fiber or *solenoid.* This appears to have about six nucleosomes per turn, but the path of the DNA between nucleosome units is unknown. Additional levels of packing, beyond the 30-nm fiber, are needed to attain the level of compaction seen in a chromosome.

There is good evidence for independent chromosomal domains of about 50- to 100-thousand base pairs (50–100 kb) of DNA. In such a domain, the ends of the domain are anchored so that the domain may be *torsionally constrained,* even though the chromosome is linear. One domain can contain any or all of the elements of chromatin structure already mentioned. The current hypothesis is that the ends of the domains are anchored to a large, stable protein structure, called the nuclear matrix (in interphase) or the *nuclear scaffold* (in metaphase). In some cases, one domain may represent one independently controlled transcriptional unit. The possibility of DNA becoming supercoiled within a torsionally constrained domain is discussed in Chapter 2.

Structure of Active Chromatin

Active chromatin is a vague term for the parts of a chromosome that are participating in some kind of enzymatic activity (Fig. 5.3). Frequently, transcriptional activity is meant, but sometimes the term is used more generally. Cell biologists use the term *euchromatin* to refer to dispersed chromatin, which may roughly be equivalent to "active" chromatin. Some parts of chromosomes, in specific cells, are folded very tightly and are inaccessible to enzymes and proteins. This inaccessible chromatin represents inactive chromatin and may be equivalent to the material termed *heterochromatin.* Other regions of chromosomes are accessible, during interphase, and may be involved in activities like transcription or replication, or may be available, or "competent," to respond directly to protein signals that would switch on transcription. Hence, there are three overall levels of chromosome activity:

1. Inactive or repressed regions
2. Potentially active or competent regions
3. Active regions

On a large-scale level, these regions can be distinguished by their accessibility to nucleases.

Nucleosomes protect DNA from *nucleases* to differing extents, depending on the nucleases and how closely the nucleosomes are packed together. *Deoxyribonuclease-I* (*DNAase-I* or DNAse-I) is a nuclease that makes single-stranded cuts in DNA with little sequence specificity and eventually reduces a naked DNA molecule to mono- or small oligonucleotides.

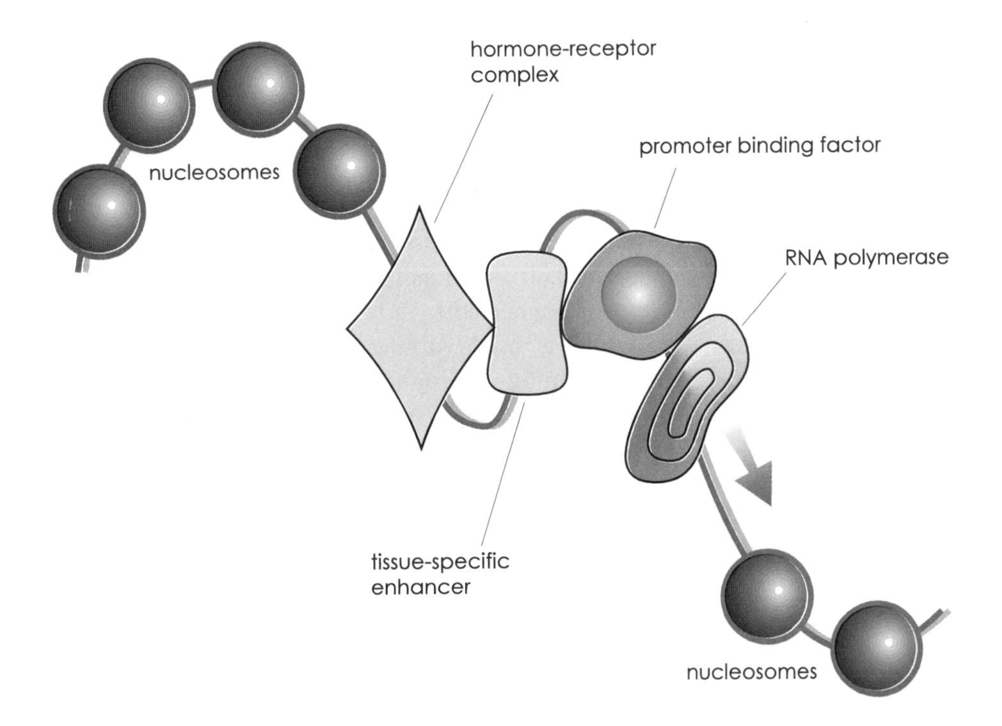

Figure 5.3.
Active chromatin. The DNA is shown as a pair of lines, in purple and gray. The nucleosomes are indicated by small, shaded gray spheres. The other shapes represent RNA polymerase and three transcription factors interacting with one another along the DNA. The diagram is not to scale, and transcription complexes probably contain more components than shown, but this illustrates the principle of multiple factors interacting with one another and with DNA and RNA polymerase.

Partial digestion by DNAase-I is widely used in the study of chromatin structure. *Micrococcal nuclease* (also called *Staphyloccocal nuclease*) makes double-stranded cuts initially in the linker DNA between nucleosomes, useful for mapping the positions of nucleosomes. Micrococcal nuclease digests must be interpreted carefully because this nuclease shows some sequence specificity when partial digestion of purified DNA is carried out as a control. In addition to these enzyme probes, there are chemical reagents that cut DNA. These reagents are much less sequence-dependent than even DNAase-I. Accessible DNA can also be labeled by *methylating enzymes.*

Three levels of sensitivity of the DNA to DNAase-I are found in chromatin:

1. In repressed chromatin, the sensitivity is very low (DNAase-resistant).
2. In transcriptionally competent regions, and the bulk of active regions, the sensitivity is about 10 times higher (DNAase-sensitive).
3. At specific sites in or near *transcriptionally competent* or active regions, the sensitivity approaches that of free DNA (*hypersensitive sites*).

The number and locations of hypersensitive sites in the chromatin fiber are specific to individual genes and also vary depending on whether the gene is active or merely competent.

The precise cause of the difference in structure between the resistant and DNAase-sensitive structures is unknown; it may be related to acetylation of core histones which occurs in the same regions, and/or to a lack of histone H1. Modifications to DNA also occur, mainly *methylation* of cytosine in inactive regions of DNA (Chapter 28).

The DNAase-hypersensitive sites are small and dispersed throughout the chromatin; they are interpreted as being due to the absence of histones or nucleosomes. Often, the binding of other proteins can be detected in the hypersensitive sites, using more sophisticated "*footprinting*" techniques. There is growing evidence for the existence of multi-subunit complexes of proteins bound at hypersensitive sites which may control the initiation of transcription.

In many cases, nucleosomes form almost at random over the DNA sequences they protect. In other cases, a nucleosome may be specifically *positioned*—or *phased*—over a particular DNA sequence. In these cases, nucleosomes can play more specific roles in the regulation of transcription. For example, particular nucleosomes may compete with specific DNA-binding proteins for a particular DNA sequence element. The roles of nucleosomes in transcriptional control are discussed further in Chapter 28.

Sequence-Specific DNA-Binding Proteins

Sequence-specific binding may be seen with special structural proteins or specific enzyme activities, for example, at the ends of chromosomes (termed *telomeres*), where special DNA sequences bind both specific structural proteins and, in some cells, the enzyme, *telomerase,* which maintains the ends of linear chromosomes. Other sequence-specific binding occurs on DNA sequences that regulate the transcription of genes. Some of these sequences are associated with many genes, for example, the sequences that help define the beginning and end of a transcribed sequence; others are more specific and are found associated with groups of genes whose transcription may be regulated coordinately. Although there are many such regulatory transcription factors, there are structural motifs common to groups of factors. The *helix–turn–helix motif* is used for sequence-specific DNA binding in many prokaryotic and some eukaryotic DNA-binding proteins; the *zinc-finger motif* is important in many *eukaryotic transcription factors*; and the *leucine–zipper motif* is important for *protein–protein interactions.*

Protein–protein interactions are a critical part of protein–DNA binding. Many specific transcription factors bind to DNA as homodimers or heterodimers. The *cooperativity* introduced by the dimer structure strengthens the binding and improves the ability to regulate gene transcription by regulating the activity of these transcription factors.

The helix–turn–helix motif is comprised of two α-helices connected by a turn. One α-helix, the *recognition helix,* binds in the *major groove* of B-form DNA, forming hydrophobic contacts with the base pairs and electrostatic interactions between the *α-helix dipole* and the DNA phosphates. The amino-acid side-chains of the recognition helix may form hydrophobic, hydrogen and electrostatic bonds with the DNA. Often the DNA is bent to optimize these interactions. The interactions of the amino-acid side-chains in the recognition helix require a specific DNA sequence, giving the sequence-specificity of binding, but weak binding to other DNA sequences occurs, allowing the protein to "find" its preferred binding site by sliding along the DNA in a two-dimensional search pattern. In the dimer, the two monomers are bound together so that their recognition helices bind to adjacent turns of the DNA major groove. Since the protein–protein interaction domains point toward each other, the protein monomers are pointing in opposite directions. Hence, to bind a homodimer, the specific DNA sequence has to be palindromic—a characteristic of many specific protein binding sites on DNA. While helix–turn–helix proteins have carefully been characterized in prokaryotic systems, they also occur in eukaryotes, for example, *homeobox proteins.*

The zinc-finger motif is a completely different structure, found in some eukaryotic DNA-binding proteins. The number of zinc fingers found in one protein varies substantially. The zinc-finger protein, *TFIIIA,* regulates the transcription of the small ribosomal RNA, *5s* ***rRNA,*** by binding to a control region within the transcribed part of the gene. TFIIIA has nine zinc fingers, each of which binds one zinc ion through coordination with two histidines and two cysteines in each "finger." These are "Cys–His" fingers; a related structure uses four cysteines to bind the Zn^{2+}.

The leucine-zipper motif is an α-helical sequence with a leucine at every seventh position. Two such α-helices, from separate proteins or subunits, coil around each other so that the leucines stack on the inside of the coil. This is an example of the α-helical coiled-coil structure found in fibrous proteins as well as in transcription factors. Since the repeat for an α-helix is 3.4 residues per turn, every seventh residue in a straight helix is almost, but not quite, in the same angular direction around the helix. By coiling the helix slightly, every seventh residue is made to align more precisely. Thus, the leucines in adjacent strands interdigitate and provide hydrophobic stacking interactions to hold the two α-helices together. This motif is used by a variety of eukaryotic transcription factors to produce dimer structures for binding to DNA.

RIBONUCLEOPROTEIN COMPLEXES

The process of gene expression in eukaryotes, which is discussed in more detail in Chapter 29, involves a series of nucleoprotein complexes, starting with chromatin and moving through spliceosomes, ribosomes, and signal-recognition particles. Gene expression begins with the assembly of the correct proteins (transcription factors) on chromatin in the region containing the gene to be expressed. Assembly of the transcription complex results in synthesis of RNA, using the sequence of the DNA as a template, in a process known as transcription. The RNA produced is known as the *transcript*; in its initial form, it is known as the *nascent transcript.* The nascent transcript is quickly packaged into *nucleoprotein particles* (*hnRNPs*). These interact with *small nuclear ribonucleoproteins* (*snRNPs,* pronounced "snurps") that process the transcript to produce *mRNA* (*"messenger" RNA*) which then leaves the nucleus. In the cytoplasm, the mRNA provides a template for protein synthesis on ribosomes. Finally, signal-recognition particles pick out those proteins that need to pass through the endoplasmic reticulum during their synthesis.

Nuclear Ribonucleoprotein Complexes

Heterogeneous nuclear ribonuclear protein particles form on nascent transcripts as soon as they are synthesized. Each hnRNP particle contains about 500 bp of transcript and a variable set of proteins; several hnRNPs form on one nascent transcript. The hnRNP particle packages the transcript until it is further processed to form mRNA.

Small nuclear ribonucleoproteins contain small nuclear RNAs (snRNAs)—known as U1, U2, and so on—which are complexed with proteins to form the snRNPs. The snRNPs recognize specific nucleotide sequences on the nascent transcript and bind to the transcript. Those snRNPs that are involved in *RNA splicing* form a complex called a spliceosome that removes *introns* from the nascent transcripts. The spliceosome is nearly as large as a ribosome. The roles of individual snRNPs in the spliceosome are being elucidated; both RNA and protein have functional roles.

Ribosomes

Ribosomes are much larger nucleoprotein structures than nucleosomes and contain RNA and protein rather than DNA and protein. Although ribosomes are partly assembled in the nucleus, they carry out most of their functions in the cytoplasm. Ribosomes comprise one each of two different nucleoprotein subunits named for their *sedimentation coefficient,* which is a measure of the rate at which they sediment in an *ultracentrifuge.* (The units of sedimentation rate are *svedbergs,* abbreviated "S"; one svedberg is 10^{-13} sec.) Prokaryotic ribosomes have 30-S and 50-S subunits, which together make a *70-S ribosome.* Eukaryotic cells, in their mitochondria, have ribosomes that are synthesized using mitochondrial and nuclear genes and resemble prokaryotic ribosomes. Eukaryotic ribosomes are found in the cytoplasm and are made from 40- and 60-S subunits which combine to make an *80-S ribosome.* Ribosomes bind to mRNA and, during protein synthesis, many ribosomes are distributed along each mRNA molecule in a structure known as a *polysome.* Polysomes are responsible for all cellular protein synthesis. Many additional molecules become associated with the ribosomes, including *tRNA, nascent polypeptides,* accessory protein factors, and cofactors.

The essential ribosome can be isolated free of the additional molecules and is a well-defined particle containing both RNA and proteins. A prokaryotic ribosome contains three RNA molecules and 57 protein molecules. Fifty-three of the protein molecules occur once per ribosome and two occur twice. A eukaryotic ribosome contains four RNA molecules, three of which are homologous to prokaryotic ribosomal RNAs, and one of which is found only in eukaryotic ribosomes. There are about 82 proteins per eukaryotic ribosome.

Ribosomes are too large for the application of the techniques that are used for determining the molecular structures of enzymes and smaller complexes, such as nucleosomes. Thus, atomic-level information is not available. However, a number of ingenious experimental approaches are being used to determine the structure of the RNA in the ribosome and the locations of the proteins in the overall structure defined by electron microscopy. Some of the structural features are described in Chapter 26.

Signal-Recognition Particles

The signal-recognition particle (*SRP*) recognizes the *N*-terminus of specific proteins and helps them pass through the ER for further processing and transport. The *N*-terminal amino-acid sequence recognized by the SRP is known as the *signal sequence.* Each SRP contains one small RNA molecule, called SRP RNA, and six protein subunits. Two of the SRP's functional domains are located at one end of the complex and the third functional domain is at the other end of the complex (Fig. 5.4). The function of the first domain is to bind the signal sequence; this occurs free in the cytoplasm. The domain at the other end of the SRP then interacts with the polysome to cause a pause in protein synthesis. The paused structure then diffuses to the ER, where the third functional domain binds to an *SRP receptor,* an integral membrane protein in the ER. At this point, the SRP is displaced, leaving the polysome attached to the ER and protein synthesis resumes, with the nascent protein being threaded through the ER as it is synthesized.

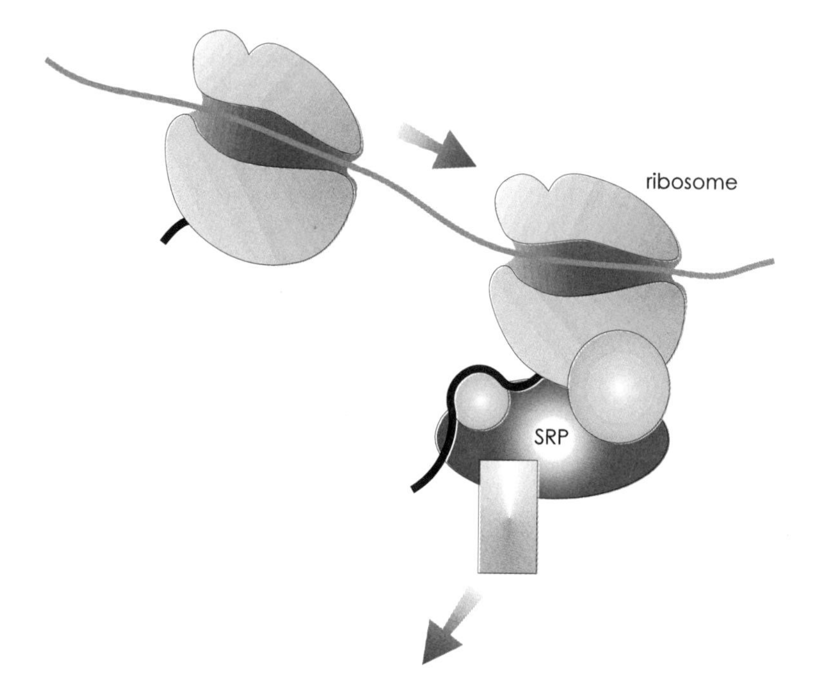

Figure 5.4.

Signal-recognition particle domains. Domains of the signal-recognition particle. The SRP is the purple complex, with the three domains in purple. One domain binds the nascent polypeptide chain (black line), and another binds the polyribosome (gray) and pauses translation. The third domain will bind to the SRP receptor when the complex reaches the ER (not shown). This is very diagrammatic and the SRP is shown larger than it should be in relation to the polyribosome. The function of the SRP can be seen in an animation available on the World Wide Web at http://moby.ucdavis.edu/HRM/Biochemistry/animations.htm.

SUMMARY

1. Many structures in the cell contain both nucleic acid and protein. Such nucleoprotein complexes are generally concerned with the storage and use of genetic information.

2. Nuclear DNA is packaged with proteins to form chromosomes. Chromatin is the nucleoprotein material that makes up chromosomes.

3. The proteins of chromatin are either histones or nonhistones. There are five different types of histones, H1, H2A, H2B, H3, and H4. H1 is called the "linker" histone while the others are "core" histones. In chromatin, two H2A/H2B dimers and one H3/H4 tetramer form an octamer around which DNA is wound in about 1.7 turns of a supercoil, to give a nucleosome core particle. Addition of histone H1 and more DNA to the core particle forms the nucleosome, the major repeating substructure of chromatin. Adjacent nucleosomes are linked by DNA—the linker DNA—typically 30 bp in length.

4. Chromatin structure is classified into three groups—transcriptionally inactive, transcriptionally competent, and transcriptionally active—which can be distinguished by nuclease digestion experiments. The distribution of DNA sequences among these groups varies between tissues reflecting tissue-specific patterns of gene expression.

5. Active chromatin has structural characteristics such as a deficit of histone H1 and acetylation of specific-lysine residues in histones H3 and H4. There are also short regions of DNA that are free from nucleosomes, usually due to the binding of specific nonhistone proteins—transcription factors—to the DNA, forming transcription complexes.

6. Nucleosome arrays can further coil into a higher-order structure with about six nucleosomes per turn. This "30-nm fiber" is further organized and packed to form the higher-order chromosomal structures—heterochromatin, euchromatin, metaphase chromosomes.

7. Regulation of gene expression at the level of transcription is achieved through the action of proteins that recognize and bind to specific DNA sequences. Many such proteins have recurring motifs, such as the helix–turn–helix or zinc finger, that provide the framework for recognition of specific DNA sequences.

8. Several subsequent steps of gene expression, RNA processing, synthesis of the peptide bond, and protein sorting involve ribonucleoprotein complexes, spliceosomes, ribosomes, and signal-recognition particles.

REFERENCES

Introduction to Protein Structure, Carl Branden and John Tooze, 1991, Garland Publishing, New York.

Understanding DNA, C. R. Calladine and Horace R. Drew, 1992, Academic Press, San Diego, CA.

Chromatin, K. E. van Holde, 1988, Springer-Verlag, New York.

Chromatin unfolds, G. Felsenfeld, 1996, *Cell,* **86:**13–19.

Targeting chromatin disruption: Transcription regulators that acetylate histones, A. P. Wolfe and D. Pruss, *Cell,* **84:**817–819.

World Wide Web at http://moby.ucdavis.edu/HRM/Biochemistry/animations.htm.

REVIEW QUESTIONS

1. In the following lists, a length of DNA is paired with a structural component of the chromosomes. Which pairing is the least likely to be correct?
   a) 20 bp 1 turn of the DNA double helix
   b) 146 bp nucleosome core particle
   c) 168 bp chromatosome
   d) 195 bp nucleosome
   e) 100 kbp topologically constrained domain

2. Which of the following particles contains a type of nucleic acid that is different from that found in the others?
   a) Signal-recognition particle
   b) Spliceosome
   c) Nucleosome
   d) Ribosome
   e) Telomerase

ANSWERS TO REVIEW QUESTIONS

1. ***a*** The DNA double helix has about 10 bp per turn in B-DNA, the major form of DNA in chromosomes. Even in Z-DNA, the number of base pairs per turn is only 12. Hence, a is the incorrect pairing. The others are all correct.

2. ***c*** All of the above, except the nucleosome, contain RNA; the nucleosome contains DNA. Telomerase was not discussed in this chapter, but is treated in Chapter 22, DNA replication, since telomerase catalyzes the template-independent synthesis of DNA telomeres.

CHAPTER

6

HEMOGLOBIN ILLUSTRATES PROTEIN BEHAVIOR

INTRODUCTION

Sickle cell anemia, the *thalassemias,* and several other *inherited disorders* are due to genetic changes in hemoglobin. A number of environmental factors, such as *smoking, exercise, carbon monoxide* concentrations in air, *high altitude,* and impaired *lung function* also affect hemoglobin function. Understanding the function of hemoglobin and myoglobin helps with managing and treating these *clinical* and *environmental* factors. The depth of knowledge of hemoglobin also makes it a valuable example of the ways in which proteins work.

Myoglobin is a protein that stores and transports oxygen in mammals. Myoglobin occurs inside muscle cells that need to store oxygen. Hemoglobin occurs in *red blood cells,* also called *erythrocytes.* Myoglobin and hemoglobin have many similarities in structure and mechanism for oxygen binding as well as important individual features that are required for their specific functions.

HEMOGLOBIN

Heme

Both myoglobin and hemoglobin contain a *prosthetic group* (a nonpolypeptide part of a protein), namely *heme.* Prosthetic groups remain bound to the protein permanently, by covalent bonds, noncovalent bonds, or both. The protein without the prosthetic group is called an *apoprotein.* When an enzyme is involved, the apoprotein plus prosthetic group may be called the *holoenzyme.*

apoprotein + prosthetic group → protein (holoenzyme)

The heme prosthetic group contains a reduced iron (Fe^{2+}) atom in a largely hydrophobic, planar, *porphyrin* ring. The four nitrogen atoms of the porphyrin ring bind the iron atom nearly in the plane of the ring and leave the two other coordination positions of the reduced iron available for bonding perpendicular to the plane of the heme (Fig. 6.1).

Globin Structure

The apoprotein, *globin,* of myoglobin is a single polypeptide chain of molecular weight 17,800. The chain is folded into eight α-helical regions, labeled A through H, and these regions interact with one another to give a specific tertiary structure forming a single domain. An interactive model of myoglobin is on the World Wide Web at http://moby.ucdavis.edu/HRM/Biochemistry/molecules.htm.

The structure is approximately spherical and very compact except for a deep hydrophobic cleft or hole.

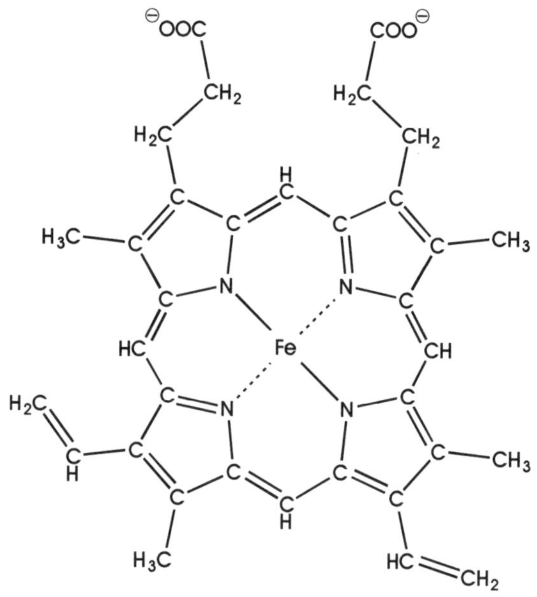

Figure 6.1.

Heme structure. The chemical structure of heme shows the ferrous iron in the center with the hydrophobic surroundings and the two polar propionic acid groups at the top.

This is the heme binding site. It is lined with hydrophobic residues to provide the hydrophobic environment for the heme, which fits in the pocket with its charged carboxylate groups facing the outside.

In addition to the hydrophobic lining, the heme binding site contains two histidine residues that play key roles in oxygen binding. The first, His-F8 (the eighth residue of the F-helix), is known as the ***proximal histidine*** and its imidazole ring is close to the iron atom and bonded covalently to the fifth coordination position of the iron. The second histidine, His-E7, is on the opposite side of the heme group and further from it. It is known as the ***distal histidine*** and is not bonded to the iron atom.

Mechanism of Oxygen Binding

Oxygen will bind to free heme groups, but they are immediately oxidized to give the ferric form (Fe^{3+}) of the heme group, which no longer binds oxygen. In myoglobin the heme group always has its iron in the ferrous form (Fe^{2+}). Oxidation of the heme group is decreased, in myoglobin, by the steric hindrance of the polypeptide framework. Oxidation is increased by a heme–heme intermediate and the polypeptide prevents two heme groups from getting close enough to interact.

Steric considerations are also important in reducing the affinity of myoglobin for carbon monoxide (CO). Free heme has a 25,000-fold greater affinity for CO than for O_2 and CO is present in the atmosphere and in tissues where heme is broken down, so free heme would quickly be used up in binding irreversibly to CO. In myoglobin, the presence of the distal His prevents high-affinity binding of CO to the heme group and reduces the increased affinity for CO to 200-fold (Fig. 6.2).

Quaternary Structure of Hemoglobin

Hemoglobin contains four polypeptide subunits, two α- and two β-subunits, to give a tetrameric structure of $\alpha_2 \cdot \beta_2$. The α- and β-chains are quite similar to one another and to myoglobin in amino-acid sequence and secondary and tertiary structure. However, in hemoglobin, the α- and β-subunits interact specifically to form tetramers. One molecule of hemoglobin thus contains four heme groups and can bind up to four molecules of oxygen. The heme binding pockets in hemoglobin are very similar to those in myoglobin, especially the two histidines (E7 and F8) and two key hydrophobic residues (Phe-CD1, the first residue in the part of the chain joining the C and D helices, and Leu-F4). Notice how these key features of the heme binding site come from residues that are widely spaced along the polypeptide chain but are brought together by folding of the chain. As in myoglobin, the heme binding site allows oxygen to bind reversibly to the heme without oxidizing it and reduces the affinity of heme for CO.

The quaternary structure of hemoglobin generates two other major structural features that are important for the function of hemoglobin. The first is another ligand binding site, in the center of the molecule, where *2,3-bisphosphoglycerate* (*BPG*) binds. The second is the area of interaction between subunits, particularly the α-β interaction.

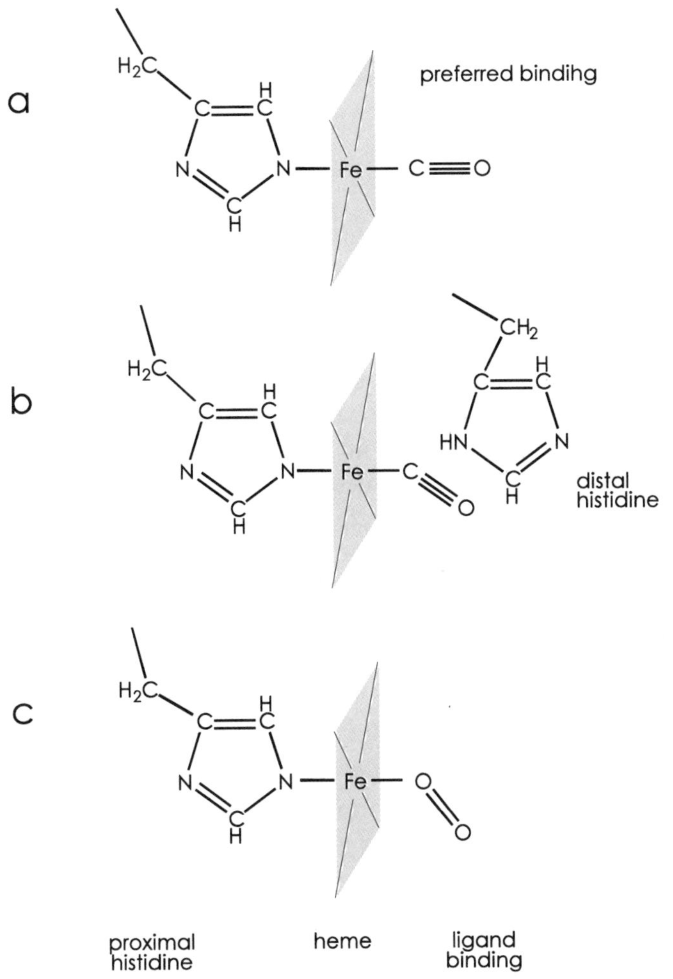

Figure 6.2.

Distal His and CO binding. The figure illustrates the effect of the distal histidine in providing steric hindrance to the binding of carbon monoxide, reducing the affinity of hemoglobin for carbon monoxide. The heme group is shown in purple in each panel, with the proximal histidine on the left and the distal histidine on the right. (a) Imaginary binding of carbon monoxide, as if the distal histidine were absent. (b) Actual binding of carbon monoxide, with the distal histidine present. Steric hindrance reduces the strength of the binding. (c) The preferred angle of oxygen binding avoids steric hindrance from the distal histidine.

Bisphosphoglycerate (BPG)

Bisphosphoglycerate (also called *diphosphoglycerate*) is a highly charged molecule and the BPG binding site in hemoglobin is lined with six charged amino-acid residues and two α-amino groups (Fig. 6.3). The BPG binding site is made from parts of both β-chains and there is only one BPG molecule bound per hemoglobin tetramer. The BPG binding to hemoglobin provides a cross-link between the two β-chains and stabilizes the *taut structure* of hemoglobin.

Bisphosphoglycerate binding controls the *affinity of hemoglobin for oxygen.* It is synthesized in the erythrocyte from an intermediate in a major metabolic pathway and defects in this pathway sometimes produce important consequences for *oxygen transport* by changing the level of BPG in the erythrocyte.

The pathway involved is *glycolysis,* the conversion of *glucose* to *pyruvate.* The initial step is phosphorylation of glucose by the enzyme *hexokinase* and the final step is formation of pyruvate from *phosphoenolpyruvate* by *pyruvate kinase* (Chapter 11). The BPG is made from an intermediate compound, 1,3-bisphosphoglycerate, by the enzyme *bisphosphoglycerate mutase.*

In most cells, the BPG concentration is low, but in erythrocytes it is high, in normal individuals. When an inherited defect in hexokinase occurs, however, the concentrations of the intermediates in the glycolytic pathway drop and so the concentration of BPG drops. Conversely, when there is a pyruvate kinase deficiency, the glycolytic intermediates pile up and the concentration of BPG rises. These diseases thus affect the function of hemoglobin through the effects of BPG.

Bisphosphoglycerate is also responsible for the apparent change in hemoglobin (increased *oxygen affinity*) when blood is stored in an acid–citrate–dextrose medium. Under these conditions, glycolysis stops and the synthesis of BPG stops. However, the degradation of BPG continues, with a resultant drop in BPG concentration. The situation cannot be reversed by

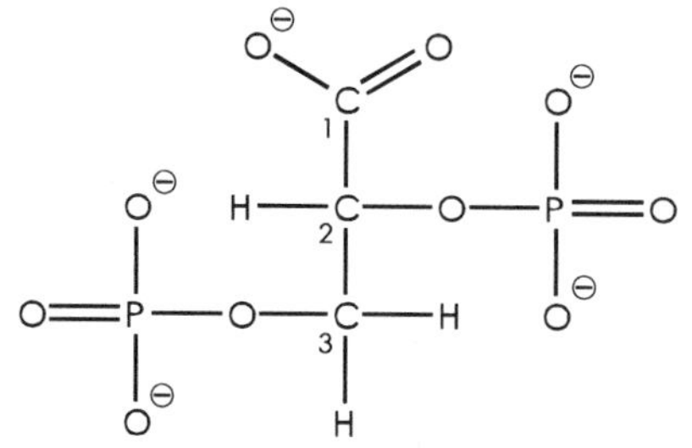

Figure 6.3.

BPG structure. The chemical structure of bisphosphoglycerate shows the three carbons (1–3) of the parent glycerate with the two added phosphates. Note the highly charged nature of this molecule under physiological conditions.

adding BPG directly because this charged molecule will not penetrate the red cell membrane. However, it can be prevented by adding *inosine,* which is converted to BPG by the red blood cell.

The other new region provided by the tetrameric structure of hemoglobin is the interface between unlike subunits, the α–β interface. At this interface, the two subunits are held together by a number of salt links—electrostatic bonds—and by hydrogen bonds.

A key feature of the subunit interactions is the existence of two conformations:

1. The taut or T structure, in which there is a maximum number of salt links.
2. The relaxed or R structure, in which some salt links are broken, leaving parts of the protein free to rotate.

One example of the salt links is Lys at position C5 in the α-chain. This Lys is linked to the carboxy-terminus of the β-chain in the taut (T) structure only. The two structures also have different hydrogen bonds, for example, a Tyr from the α-chain (C7) bonds to an Asp in the β-chain (G1) in the T structure and to an Asn in the α-chain (G4) in the relaxed (R) structure.

The R and T structures are vital to the function of hemoglobin. The binding of oxygen is greatly enhanced in the *R state.* Conversely, the binding of BPG, CO_2, and H^+ are enhanced in the *T state.* Thus, the presence of BPG in the erythrocyte will tend to force hemoglobin into the T, cross-linked, state and force off any bound oxygen which can then be taken up by the tissues.

OXYGEN TRANSPORT

Molecular oxygen is required by the mitochondria of cells for generation of molecules such as ATP. Adenosine triphosphate provides the chemical energy to drive the chemical reactions that cells use to carry out their functions such as macromolecular synthesis, *vision,* or *muscle contraction.* The basic process whereby oxygen reaches mitochondria from the atmosphere is **diffusion.** *Diffusion* is a random process in which molecules move independently of one another. Diffusion results in a net transport of molecules from high- to low-concentration regions and so oxygen diffuses from air to mitochondria on its own. However, the rate at which it diffuses on its own is very slow, owing to the low solubility of molecular oxygen in blood. The rate of diffusion depends on:

1. The concentration gradient
2. The number of molecules on the move
3. The speed at which the molecules move

The concentration of oxygen in the air is fixed and the mitochondria require a certain concentration of oxygen. Hence, the concentration gradient has an upper limit and the only way to increase the rate of diffusion is to get more molecules on the move. Myoglobin and hemoglobin achieve this by being very soluble and binding oxygen. This increases the apparent solubility of oxygen and so increases the number of oxygen molecules on the move.

The path of oxygen from air to mitochondria involves a number of steps (Fig. 6.4):

A. Absorption of oxygen to hemoglobin in erythrocytes in the lungs
B. Blood flow to the tissues
C. Diffusion of oxygen from erythrocyte to cells within a tissue
D. Diffusion to mitochondria within cells (may be facilitated by myoglobin in muscle cells)

Oxygen Storage by Myoglobin

Oxygen diffuses into cells through the plasma membrane and, in muscle cells, binds to myoglobin. The oxy-

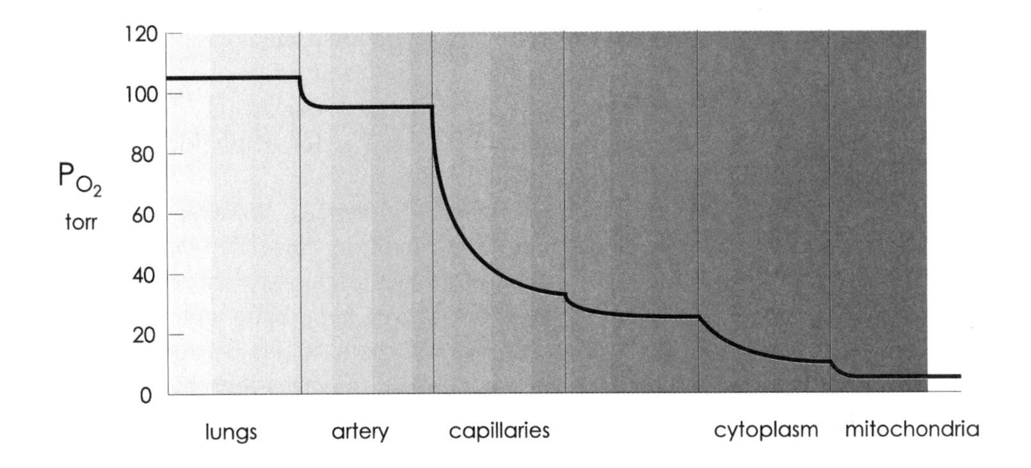

Figure 6.4.
Oxygen concentration gradient in the body. The partial pressure of oxygen is high in the lungs and low in the mitochondria. Oxygen flows down the concentration gradient.

genated myoglobin—oxymyoglobin—then diffuses through the cytoplasm to the mitochondria. Here the oxygen concentration is lower, causing the release of oxygen to give deoxymyoglobin, which diffuses back to the plasma membrane where the high oxygen concentration converts it back to oxymyoglobin, and the process is then repeated. Under normal conditions of adequate oxygen supply, most of the myoglobin remains in the oxygenated form, providing a storage medium that can release oxygen quickly if the demand suddenly increases. Myoglobin is found only in muscle cells.

Myoglobin binds oxygen by a simple chemical equilibrium:

$$\mathrm{Mb} + \mathrm{O_2} \leftrightarrow \mathrm{MbO_2}$$

where Mb represents deoxymyoglobin and MbO_2 represents oxymyoglobin.

The equilibrium constant K is given by

$$K = [\mathrm{Mb}] \cdot [\mathrm{O_2}]/[\mathrm{MbO_2}]$$

where the brackets represent concentrations.

Consequently, the ratio of oxymyoglobin to deoxymyoglobin is just proportional to the oxygen concentration:

$$[\mathrm{MbO_2}]/[\mathrm{Mb}] = [\mathrm{O_2}]/K$$

Another way to look at it is to calculate the fraction of myoglobin that is in the oxygenated form, or the *fractional saturation* of myoglobin:

$$\textit{Saturation} = [\mathrm{MbO_2}]/([\mathrm{Mb}] + [\mathrm{MbO_2}])$$
$$= [\mathrm{O_2}]/([\mathrm{O_2}] + K)$$

This equation for the saturation of myoglobin gives a **hyperbolic curve** for the oxygenation of deoxymyoglobin (Fig. 6.5), very similar to the *hyperbolic curve* found for enzymes that follow the Michaelis–Menten equation (Chapter 7).

The oxygen concentration, or *oxygen tension,* is often quoted as the partial pressure pO_2 of a gas phase in equilibrium with the solution. Pressure can be measured in *torr* (millimeters of mercury), dynes per square

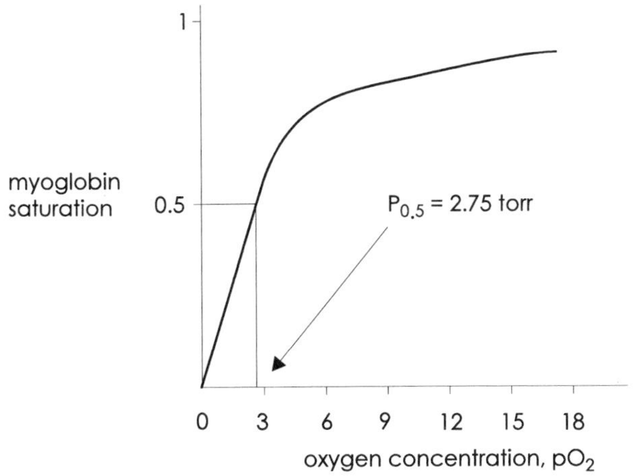

Figure 6.5.
Myoglobin saturation curve. The saturation of myoglobin is shown as a function of oxygen concentration.

centimeter (millibars), or Newtons per square meter (Pascals). 1 torr = 1.333 mbar = 133.3 Pa.

The *partial pressure* that produces 50% saturation of myoglobin is called the P_{50} or $P_{0.5}$. The pressure $P_{0.5}$ is related to the equilibrium constant because, at 50% saturation, the equation above reduces to $0.5 = P_{0.5}/(P_{0.5}+K)$, since saturation = 0.5 and $[O_2] = P_{0.5}$. Therefore, $K = P_{0.5}$.

The oxygen concentration in mitochondria ranges from 0 to 10 torr and in venous blood it is 15 torr or more. The partial pressure for half saturation of myoglobin, the $P_{0.5}$, is 2.75 torr; therefore, under most conditions, myoglobin is quite highly oxygenated and provides a store of oxygen that is immediately available if the mitochondrial level falls because of muscle contraction. The increased oxygenation also facilitates intracellular transport from the inner cell surface to the mitochondria, down the concentration gradient from about 10 torr at the inner cell surface (myoglobin is 80% saturated) to about 1 torr at the mitochondria (myoglobin is about 25% saturated), when oxygen demand is high.

Oxygen Transport by Hemoglobin

Hemoglobin is found in high concentration in red blood cells. It binds oxygen at the heme, to become oxyhemoglobin, when the oxygen concentration is high, in the lungs, and releases oxygen to become deoxyhemoglobin when the oxygen concentration is lower. Oxygen release occurs over a range of $[O_2]$, from 30 to 40 torr in blood capillaries feeding tissues with moderate oxygen consumption, down to 10 torr in tissues with very high oxygen consumption. Of course, hemoglobin with its four subunits can bind up to four molecules of oxygen per hemoglobin, but it has a lower affinity for oxygen than does myoglobin. The $P_{0.5}$ for hemoglobin in erythrocytes is about 27 torr.

Another major difference, however, between myoglobin and hemoglobin concerns the *shape* of the binding curve. A plot of saturation against oxygen tension of hemoglobin is not hyperbolic but *sigmoidal* (Fig. 6.6). The *sigmoidal* shape occurs because binding of oxygen by one subunit enhances the ability of the remaining subunits to bind oxygen. Consequently, at low oxygen tension very little oxygen is bound but, as the oxygen tension gets higher, oxygenation proceeds to increase the binding affinity so that much oxygen is bound. This enhances hemoglobin's transport efficiency because at high oxygen tension it behaves like a molecule with high oxygen affinity and at low oxygen tension it behaves like a molecule with low oxygen affinity. This behavior is called positive cooperativity.

In the blood stream, hemoglobin in red blood cells encounters a high oxygen tension (100 torr) in the lungs which causes it to become 97% saturated with an average $0.97 \times 4 = 3.88$ molecules of oxygen per hemoglobin. Little oxygen is released until the erythrocytes encounter lower oxygen tensions in the capillaries of peripheral tissues. In tissues of moderate oxygen consumption, the oxygen tension will be 30–40 torr and about 1 molecule of oxygen per molecule of hemoglobin will be released. This makes the remaining oxygens very sensitive to further small decreases in oxygen tension, so the next molecules of oxygen are given up relatively easily.

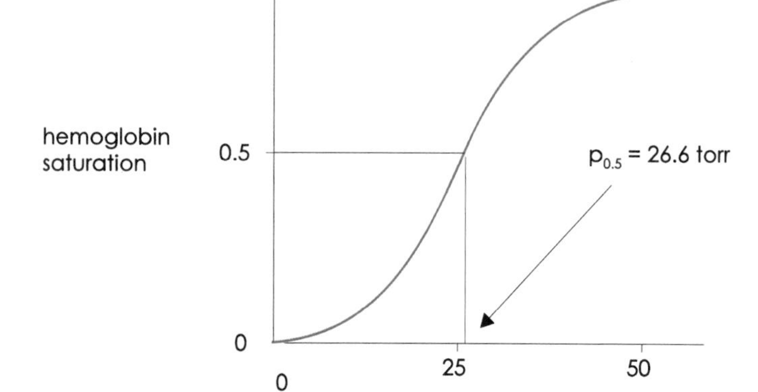

Figure 6.6.

Hemoglobin saturation curve. The saturation of hemoglobin is shown as a function of oxygen concentration.

Oxygen Transport from Lungs to Muscle Mitochondria

In Figure 6.7, notice how the steep part of the myoglobin curve falls in its working range and the steep part of the hemoglobin curve falls in *its* working range. Hemoglobin uses positive cooperativity to achieve this.

In tissues with high rates of oxygen consumption, such as the heart or rapidly working skeletal muscles, the oxygen tension can be as low as 15 torr and this causes hemoglobin to fall to 20% saturation, releasing about 3.1 molecules of oxygen per molecule of hemoglobin (assuming it started at 97% saturation). The small further drop in oxygen tension from 15 to 10 torr causes the release of another 0.5 molecule oxygen per molecule hemoglobin. Thus hemoglobin provides a store of oxygen that can be released with a small drop in oxygen tension.

The Hill Coefficient

The binding of oxygen to myoglobin is described by the equation

$$K = [Mb] \cdot [O_2]/[MbO_2]$$

where K is the equilibrium constant. Taking logarithms of both sides and rearranging them gives

$$\log([MbO_2]/[Mb]) = \log[O_2] - \log K$$

The *Hill plot* is obtained by plotting $\log([MbO_2]/[Mb])$ as a function of $\log[O_2]$. For myoglobin this gives a straight line of slope equal to 1, the *Hill coefficient* (Fig. 6.8).

For hemoglobin, successive binding sites have different equilibrium constants and so the above equation has to be modified. It is found that the Hill plot for hemoglobin has a maximum slope of 2.8, giving a Hill coefficient of 2.8 (Fig. 6.8). A Hill coefficient greater than 1.0 indicates positive cooperativity.

Allosteric Effects

An allosteric effect occurs when binding of a ligand to a particular site on a protein molecule produces an effect at a separate site on the molecule. This is usually achieved through a conformational change in the molecule.

In many cases, including hemoglobin, there are two possible states of an allosteric molecule—R (relaxed) and T (taut). For hemoglobin the conditions for formation of the two forms are:

R: oxygen bound; BPG released
T: oxygen released; BPG bound

In addition, H^+ and CO_2 bind to the T form and stabilize it.

The binding of oxygen to the deoxy form of hemoglobin (deoxyhemoglobin) causes a reduction in diameter of the iron atom which can then relax into the plane of the porphyrin ring. This motion is transferred to the polypeptide through the proximal histidine and eventually causes sliding of the subunits across one an-

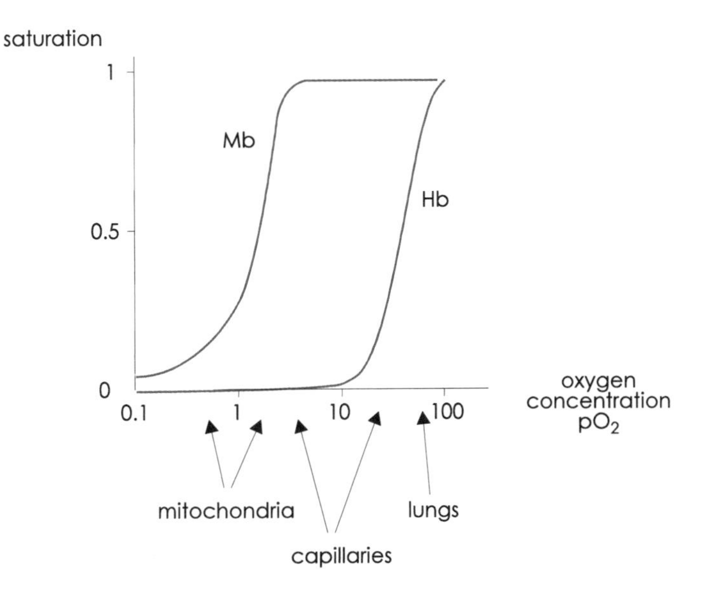

Figure 6.7.

Oxygen transport. The saturation of myoglobin (Mb) and hemoglobin (Hb) are shown as a function of oxygen concentration. To show the respective functions clearly, the oxygen concentration has been plotted on *a logarithmic scale* which stretches out the lower end of the oxygen concentration scale and compresses the higher end. The oxygen concentration found in the lungs, capillaries, and mitochondria are indicated on the oxygen-concentration axis.

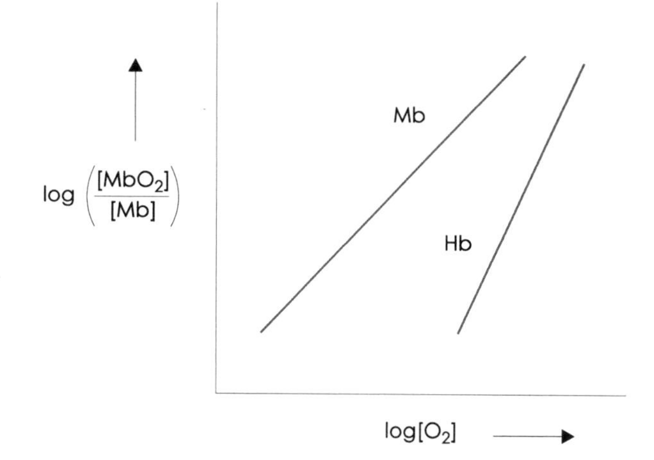

Figure 6.8.

Hill plot. A Hill plot is shown for myoglobin (Mb); it is a straight line of slope equal to 1. A line for hemoglobin is also shown (Hb); in this case, the *y* axis would be the logarithm of the ratio of the number of occupied sites to the number of unoccupied sites, referring to the oxygen binding sites on hemoglobin.

other, breakage of salt links at the α-β interface, and "squeezing out" of BPG. These changes make the remaining heme groups in the molecule more accessible for O_2 binding; therefore, the remaining oxygens bind more easily.

This cooperativity of O_2 binding is known as heme–heme interaction.

2,3-Bisphosphoglycerate

Since binding BPG to the BPG binding site displaces O_2 from the hemes, this is an *allosteric* effect. Bisphosphoglycerate stabilizes the taut form.

We have already mentioned that disturbance of BPG metabolism can affect oxygen binding by hemoglobin. Figure 6.9 shows the effect on oxygen saturation curves.

In hexokinase deficiency, [BPG] is low and so oxygen binding increases and O_2 is not released to the tissues properly. In pyruvate kinase deficiency, [BPG] is high and so oxygen binding is impaired and hemoglobin will not carry O_2 to peripheral tissues.

Bisphosphoglycerate is also involved in high-altitude adaptation. When a person moves from low to high altitude, such as at a ski area, the lowered $[O_2]$ in the atmosphere leads to reduced oxygenation of hemoglobin and reduced supply of oxygen to the tissues. This results in heavy breathing, even for light exercise, and other symptoms such as *headaches* and *nausea.* The body adapts in several ways, including increasing the blood flow, the number of red blood cells, and the [BPG]. The [BPG] can be changed significantly in about 24 hours, much faster than the number of red blood cells can be increased, which takes several weeks. Paradoxically, increasing [BPG] further reduces the oxygenation of hemoglobin in the lungs, although the effect is fairly small. More important, the increased

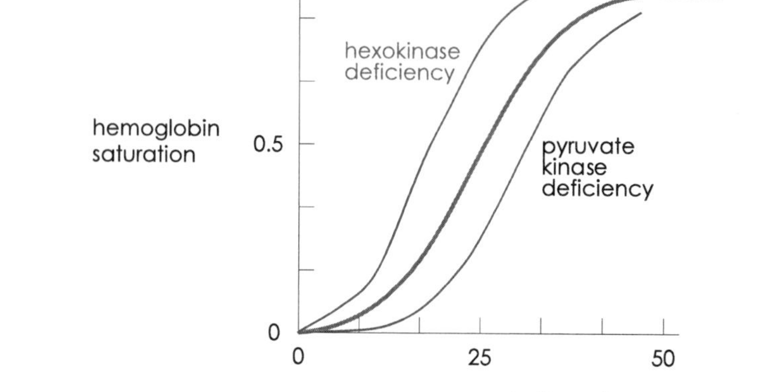

Figure 6.9.

Hexokinase and pyruvate kinase deficiency. The oxygen uptake curves for patients with a deficiency in either hexokinase or pyruvate kinase are shown in purple and black, respectively. The normal uptake curve is shown in gray for comparison.

[BPG] causes more oxygen to be released in the tissues by unloading the hemoglobin more completely than usual (Fig. 6.10). Although this is quite effective at providing a good supply of oxygen under normal high-altitude conditions of low to moderate exercise, the reserve of oxygen bound to hemoglobin in the blood is lost and therefore is not available for more strenuous exercise. The lowering of the oxygen content of the blood can cause serious consequences for people with sickle cell anemia or, in some cases, with *sickle cell trait,* as discussed in more detail below.

Bohr Effect

Both H^+ and CO_2 are produced by metabolism in tissues. More rapidly metabolizing tissues have a greater need for oxygen and also produce more H^+ and CO_2. The higher $[H^+]$—lower pH—and higher $[CO_2]$ are a signal to hemoglobin to release more oxygen so that hemoglobin can "ration" its oxygen supply, giving out more to those tissues that need it the most. The H^+ and/or CO_2 bind to hemoglobin and act allosterically to reduce the affinity of hemoglobin for oxygen. Hence, oxygen is released.

The effect of pH on the oxygen affinity of hemoglobin is known as the *Bohr effect"* and it helps hemoglobin deliver oxygen where it is most required (Fig. 6.11). The H^+ and CO_2, bound to hemoglobin, are transported back to the lungs where the high oxygen tension displaces them, restoring the higher oxygen affinity of hemoglobin, and they are lost into the atmosphere.

The CO_2 is bound covalently to the free α-amino groups of hemoglobin to form a *carbamate.* This re-

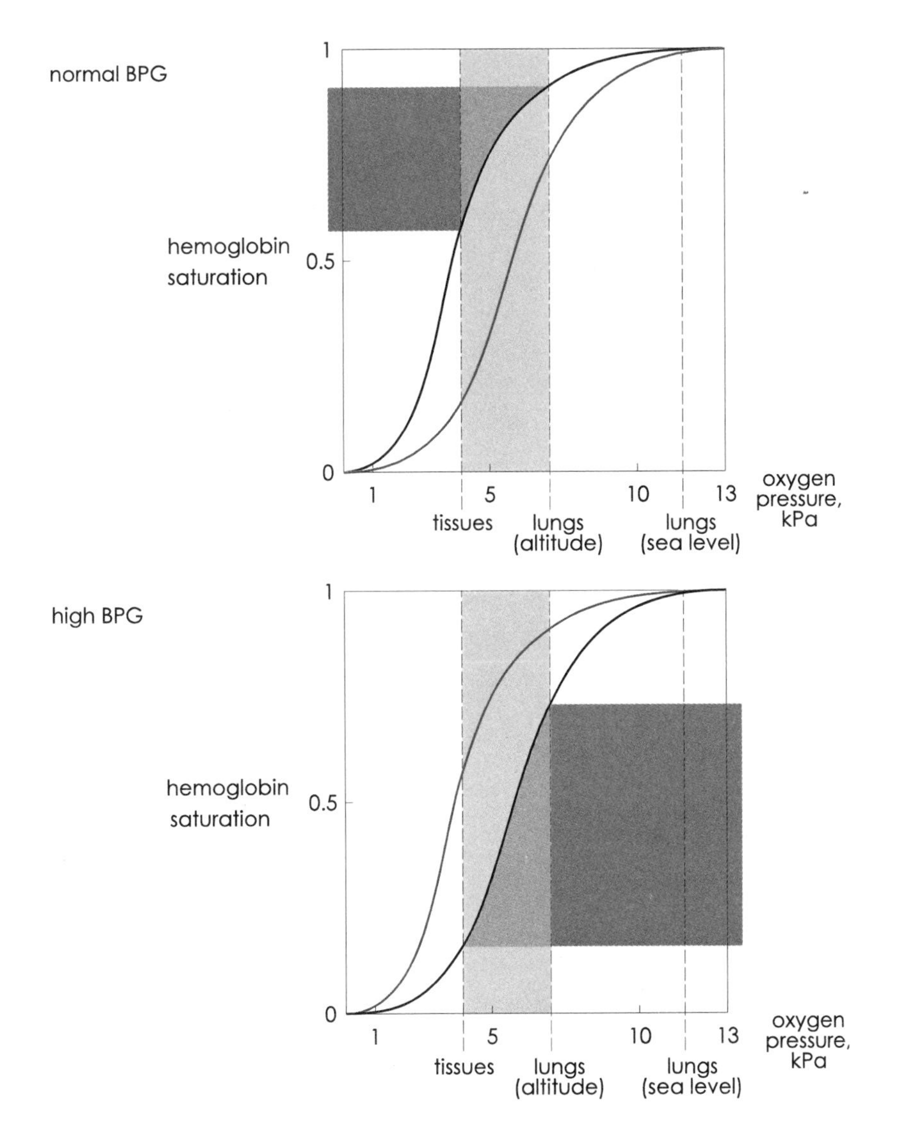

Figure 6.10.

High-altitude adaptation. The figure summarizes oxygen transport at high altitude. The upper panel (normal BPG) shows the situation before adaptation and the lower panel (high BPG) shows it after adaptation. In both panels, the relevant oxygen uptake curve for hemoglobin is shown in black. The oxygen gradient between the lungs and the tissues is shown by the purple shaded area. The gray shaded area indicates the change in oxygen saturation of hemoglobin as it moves from the lungs to the tissues.

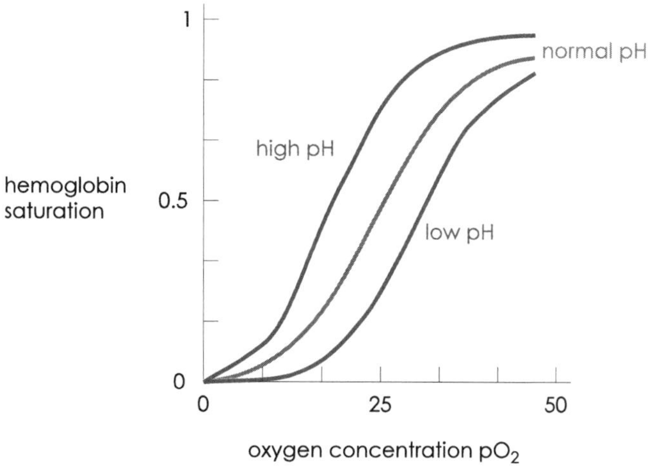

hemoglobin saturation
1
0.5
0
high pH
normal pH
low pH
0
25
50
oxygen concentration pO_2

Figure 6.11.

Bohr effect. The oxygen uptake curves for hemoglobin at pH values above and below normal blood pH are shown in purple and black, respectively. The uptake curve at the normal pH is shown in gray, for comparison.

places the partial positive charge on the free α-amino group with the negative charge on the carbamate group, which participates in additional salt bridges, stabilizing the T form of hemoglobin:

$$R\text{—}NH_2 + CO_2 \Leftrightarrow R\text{—}NHCOO^- + H^+$$

Here H^+ binds to groups (a pair of terminal amino groups and two pairs of histidines) whose pKs are raised by their environment in the T form. The binding of H^+ and CO_2 provides buffering capacity to the blood and helps maintain the blood pH. (CO_2 transport is only partly due to hemoglobin; *carbonic anhydrase* converts CO_2 to bicarbonate, which is very soluble and is carried in solution in the blood after being transported out of the red blood cell.)

NORMAL ADULT HEMOGLOBINS

The major adult hemoglobin is *Hb A* (also called Hb A_0), with subunit composition $\alpha_2{\cdot}\beta_2$. During the life of the erythrocyte, small amounts of modified forms of Hb A, including Hb A_{1a}, Hb A_{1b}, and Hb A_{1c}, are produced by occasional chemical reactions with glucose and other compounds. The *Hb A_{1c}* is a glucosylated form of Hb A, and its formation is related to the blood glucose concentration [Glc]. Consequently, the concentration of Hb A_{1c} can be used to monitor patients with *diabetes mellitus*. Since the normal regulation of [Glc] does not work in patients with diabetes mellitus, patients must take care to control their sugar intake directly. When the patient's [Glc] is high, the rate of formation of Hb A_{1c} is high and the level of Hb A_{1c} increases. Since the hemoglobin has a long lifetime, the proportion of hemoglobin present as Hb A_{1c} is a measure of the average [Glc] over the lifetime of the erythrocyte. Thus, even with tests spaced a month apart, a physician can tell how well blood sugar has been controlled between office visits (Fig. 6.12).

The form *Hb A_2*, subunit composition $\alpha_2{\cdot}\delta_2$, is a minor component of normal adult hemoglobin. The α-chains of Hb A and Hb A_2 are identical. The δ-chains are β-like chains produced at a slow rate from separate genes.

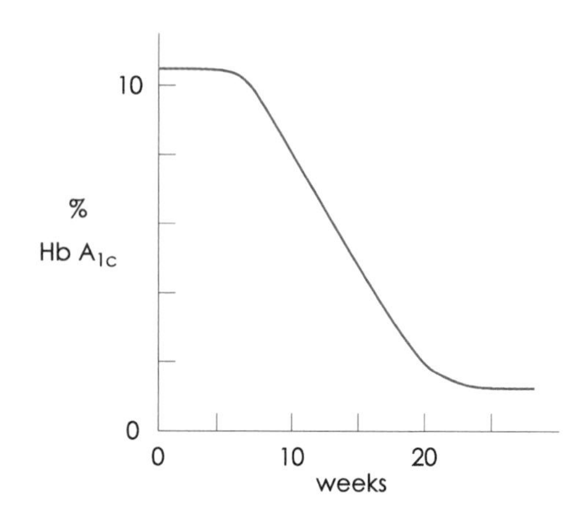

Figure 6.12.

Time course of Hb A1c in treatment of diabetes. The concentration of hemoglobin A1c in the blood of a diabetic patient is shown as a function of time after treatment was started.

Table 6.1. Normal Functional Hemoglobins in Humans

Developmental stage[a]	Name	α or α-like chain	β or β-like chain	Subunit composition
Embryo		ζ	ϵ	$\zeta_2{\cdot}\epsilon_2$
Fetus	Hb F	α	γ	$\alpha_2{\cdot}\gamma_2$
Birth to death	Hb A	α	β	$\alpha_2{\cdot}\beta_2$
Birth to death	Hb A_2	α	δ	$\alpha_2{\cdot}\delta_2$

[a]There is considerable overlap between the stages.

Hemoglobin in Development

Different normal hemoglobins are found during early *development* in humans. They are due to the use of different genes at different stages of development. At least seven different genes are used in humans: α, β, $^A\gamma$, $^G\gamma$, δ, ϵ, and ζ. The *fetal hemoglobin, Hb F,* is $\alpha_2{\cdot}\gamma_2$. ($^A\gamma$ and $^G\gamma$ are both found in fetal hemoglobin; they differ only in the inconsequential change of alanine to glycine at position 136.) *Embryonic hemoglobins* are $\zeta_2{\cdot}\epsilon_2$.

Fetal Hemoglobin, Hb F

The *fetus* requires a different hemoglobin, Hb F, because it has to absorb oxygen from the *placenta,* where the oxygen tension is lower than in the lungs. The increased oxygen affinity required is achieved by having lower affinity for BPG, thus automatically raising the oxygen affinity.

The lower affinity for BPG arises from changes in the BPG binding site (Fig. 6.13). In particular, residue H21 in the γ-chain is Ser, replacing His in the β-chain and in addition, a small proportion (~10%) of the γ-chains have their α-amino group (on the amino-terminus) modified by acetylation, which also reduces the positive charge in the BPG binding site and hence reduces the affinity for BPG.

A reduced affinity for BPG in Hb F gives it an increased oxygen affinity. This allows O_2 to transfer from Hb A to Hb F in the placenta and so provides good oxygenation to fetal tissues.

HEMOGLOBIN IN DISEASE

Abnormalities in hemoglobin are caused by mutation and then spread through the population by *inheritance.* This is the basis of evolution, but many mutations are harmful and these result in *inherited diseases.* Except for genes carried on the sex chromosomes in males, each individual gene occurs at least twice in each cell, one copy coming from each of the parents. Consequently, if a defective mutant gene is inherited from only one parent, the condition is ***heterozygous,*** and if the consequence (the *phenotype*) is *recessive,* the person has no or very mild symptoms. If defective genes are inherited from both parents, the condition is ***homozygous*** and the symptoms are fully expressed, whether the defect is recessive or not.

Diseases involving defective hemoglobin fall into two groups, ***hemoglobinopathies,*** which are due to a *change* in either the α- or β-chains, and ***thalassemias,*** which are due to the *absence* of either the α- or β-chains.

α-Thalassemia. An individual with no functional α-chains (owing to mutation in the α-chain genes or in the control of their expression) makes neither Hb F nor Hb A. The fetus may survive by making a hybrid fetal-embryonic hemoglobin, but it usually dies before term or shortly after delivery (*hydrops fetalis*). The hemoglobin found in such cases is called *Hb-Portland* ($\zeta_2 \cdot \gamma_2$). Such individuals usually have normal production of β-, γ-, and δ-chains, and these can then associate to form homotetramers β_4 (*Hb H*), γ_4 (*Hb Barts*), or δ_4. Homotetramers are very rare in normal individuals, but can be found in heterozygotes or "carriers." The homotetramers will bind O_2, but show none of the cooperativity or allosteric effects of hemoglobin and so cannot function in a useful way. The cooperativity and allosteric effects come from the interaction of unlike subunits.

The heterozygous state is also known as *α-thalassemia trait.* The α-chain genes are present as two copies per chromosome, meaning that a person may have 4, 3, 2, 1, or 0 normal α-chain genes. Four genes is normal, two or three genes is asymptomatic, one gene may show mild symptoms (α-thalassemia trait), and no genes is fatal (α-thalassemia disease).

β-Thalassemia. If no functional β-chains are produced (through mutation of the β-chain genes or their expression), β-thalassemia occurs. Normally, if the β-chain is absent, the γ- and δ-chains (and α-chains) are present, but the δ-chain gene is close to the β-chain gene on the chromosome and in some instances

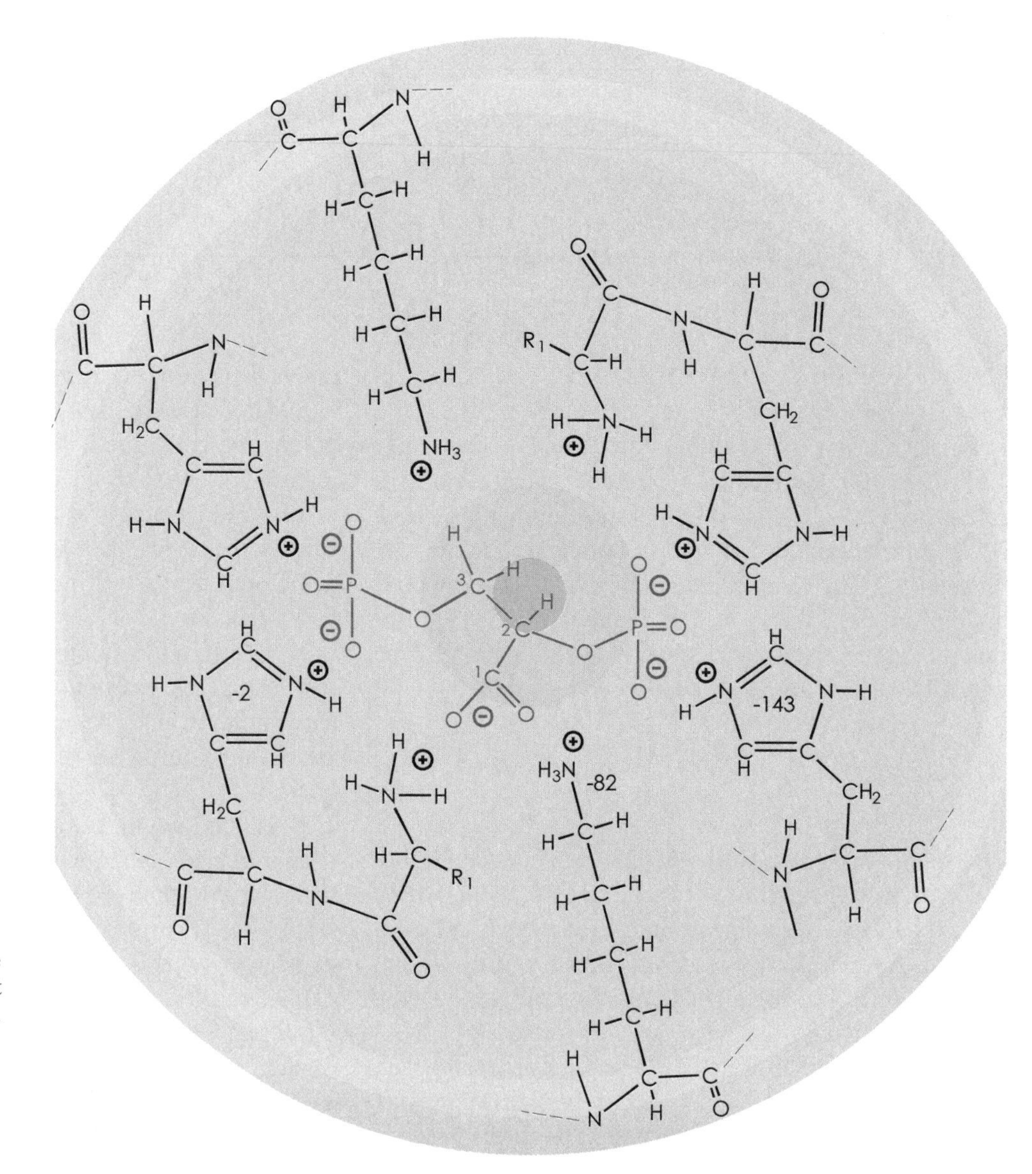

Figure 6.13.

BPG binding site in Hb A. BPG is shown in dark purple and key residues of the β-chains of hemoglobin are shown in black. The key residues of the β-chain shown on the right (top to bottom) are the α-amino group, histidine at position 2, histidine at position 143 and lysine at position 82, in the amino acid sequence. The equivalent residues on the second β-chain are shown on the left.

both are missing. Beta-thalassemia doesn't affect the fetus, but homozygotes are *anemic* after birth. Their blood contains increased amounts of Hb F ($\alpha_2 \cdot \gamma_2$) and Hb A_2 ($\alpha_2 \cdot \delta_2$).

The β-chain genes are present in only one copy per chromosome, so the homozygous state is either two copies of the gene (normal) or no copies of the gene (β-thalassemia disease), and the heterozygous state is one copy of the gene (*β-thalassemia trait*).

Both α- and β-thalassemia trait usually confer some protection from *malaria* on the heterozygote; hence, the mutation survives and can be found in areas of high malaria incidence.

Hemoglobinopathies. Mutations of hemoglobin affect one of four key areas in the hemoglobin molecule as follows:

1. The exterior surface of the complete molecule may be altered. Frequently, such mutations are nearly neutral in effect, except when they reduce the solubility of hemoglobin, as in sickle-cell anemia.
2. The oxygen binding site (sometimes called the active site) is sensitive to mutations that allow oxidation of the heme iron and destroy its ability to bind oxygen. Hemoglobins with this type of mutation are called *ferric-hemoglobins* or *met-hemoglobins* (*Hb M*).
3. The tertiary structure can be prevented from forming correctly by specific mutations, for example, those where steric hindrance is produced by a bulky side-chain. The tertiary

structure is essential for the heme environment and for producing the quaternary structure and heme–heme interaction.

4. Subunit interactions in the quaternary structure are vital for allosteric control of O_2 binding by hemoglobin and this can be affected by mutations of residues in the subunit interfaces. The result is usually an abnormal O_2 affinity.

Examples of each of these effects include:

1. *HbS*, $\alpha_2 \cdot \beta_2^{6Glu \rightarrow Val}$. The nomenclature means that the sixth residue in the β-chain (normally a Glu) has been replaced by Val. *Hemoglobin S* is found in people with sickle-cell anemia. As expected for a polar residue, Glu-6 is on the outside of the molecule. Insertion of a valine here would be expected to reduce the solubility of hemoglobin and encourage aggregation.

As expected, sickle deoxyhemoglobin precipitates as long thin crystals in the erythrocyte. The crystal formation pushes against the plasma membrane, causing it to buckle and become sickle shaped instead of the normal disc shape, hence the name sickle-cell anemia. However, *oxy*hemoglobin is still soluble, which means that if the oxygen tension remains fairly high, sickle hemoglobin can still transport oxygen. In the lungs, the high oxygen tension means that there is almost no deoxyhemoglobin and thus no sickled cells. In the tissues, the lower oxygen concentration (and higher $[H^+]$ and $[CO_2]$) cause release of oxygen and the formation of deoxyhemoglobin which begins to precipitate. If only part of the hemoglobin is deoxygenated, the sickling process is slow and, before there is severe sickling, blood flow returns the erythrocyte to the lungs, where deoxyhemoglobin becomes oxygenated again. In tissues with a high oxygen demand, however, more deoxyhemoglobin will be formed and *sickling* will occur. In this case, the erythrocytes will cycle between a normal shape, when they contain oxyhemoglobin, in the arteries, and a sickle shape, when they contain deoxyhemoglobin, in the veins. This process damages the plasma membrane and shortens the life of the erythrocyte considerably (typically from 120 to 17 days), reducing the number of erythrocytes in the blood and giving rise to anemia.

Sickle cells do not have the smooth shape and flexibility of normal erythrocytes and thus have more difficulty penetrating the capillaries. If the oxygen tension is low enough to cause severe localized sickling, some sickle cells may jam in the blood vessel, restricting the blood flow. Reduction of blood flow causes the oxygen tension in the tissue to fall further, creating more deoxyhemoglobin, and a buildup of sickle cells follows, leading to loss of blood supply to the tissue affected. A crisis situation can build and result in premature death.

Heterozygotes have *sickle-cell trait.* The concentration of Hb S in their blood is rarely sufficient to give clinical symptoms but, at high altitudes and under conditions of prolonged exercise, sickling and clinical symptoms can occur.

Sickling can be prevented by treatment of Hb S with *potassium cyanate.* The cyanate ion reacts irreversibly with amino termini of the hemoglobin chains, which reduces the binding of CO_2 and BPG to hemoglobin (Fig. 6.14). The result is increased O_2 affinity, less deoxyhemoglobin, and hence less sickling. Cyanate has toxic side effects, so a more specific Hb-modifier is being sought.

Other approaches to drug therapy for sickle-cell anemia are to develop competitive inhibitors of sickling, to switch on the fetal Hb genes in the adult, and to develop ways of enlarging the erythrocytes to reduce the concentration of hemoglobin.

2. HbM, for example, *HbM* Osaka, $\alpha_2^{58\,His \rightarrow Tyr} \cdot \beta_2$; *HbM Milwaukee,* $\alpha_2 \cdot \beta_2^{67\,Val \rightarrow Glu}$. His-58 is His F8, the proximal histidine, and in HbM Osaka the change to Tyr results in heme binding in the Fe^{3+} state, which then binds a molecule of water instead of O_2.

Val-67 is one of the hydrophobic contacts between the globin chain and the heme group. Its replacement, in HbM Milwaukee, by a polar residue impairs the binding of heme and may allow water to enter the heme pocket.

These mutants, in the homozygous state, have hemoglobin molecules with two normal chains (the β-chains in HbM Osaka; the α-chains in HbM Milwaukee) and two mutant chains. The mutant chains will not bind O_2 and so will not switch from the T to the R state.

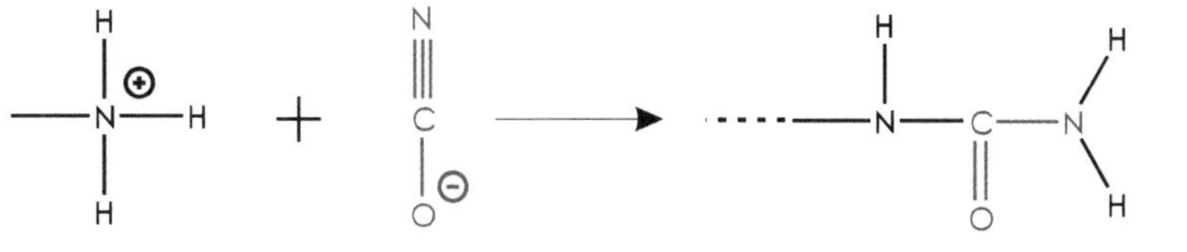

Figure 6.14.
Cyanate reacts with the N-terminus of Hb. Cyanate reacts with the amino terminus of a hemoglobin chain, removing its charge.

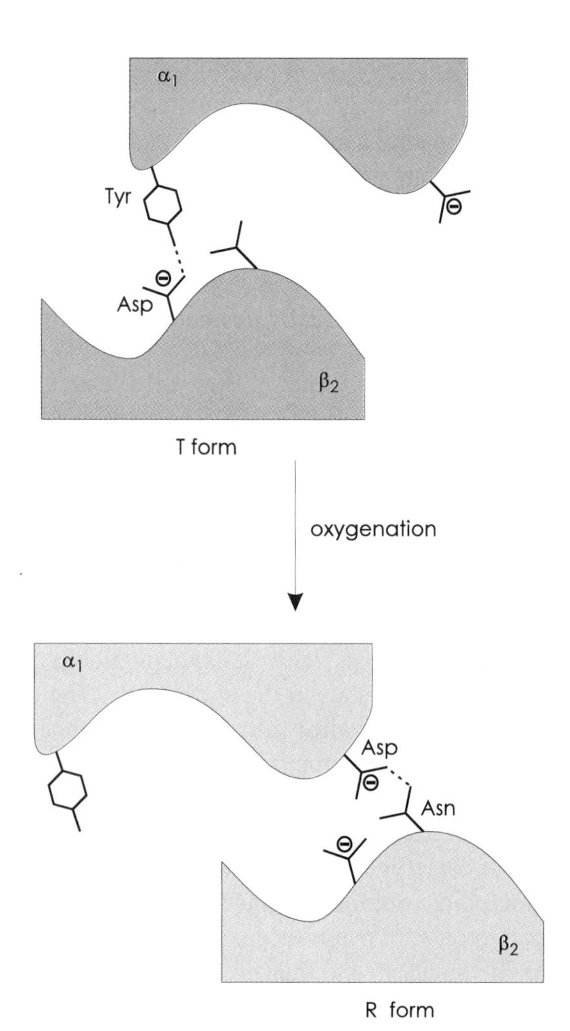

Figure 6.15.
Hydrogen bonding between unlike subunits. The curved shapes represent part of the surface of two subunits of hemoglobin. In the upper panel, the hemoglobin is in the deoxy- state (T) and a hydrogen bond is shown between a tyrosine and an aspartate on two different subunits. In the lower panel, the hemoglobin is in the oxygen-bound state (R) and the subunits have moved with respect to each other, breaking the tyrosine–aspartate hydrogen bond and allowing the formation of a new hydrogen bond, this time between an asparagine and another aspartate on the different subunits.

Consequently, there is no cooperativity of oxygen binding and the O_2 affinity of the two normal chains is too high to be of much use.

3. Hb Riverdale-Bronx, $\alpha_2{\cdot}\beta_2^{25\,Gly\rightarrow Arg}$. The tertiary structure is stabilized by hydrophobic interactions between α-helices. This requires a densely packed structure with little room for water. In this example, a glycine is replaced by arginine. There is not enough space to accommodate this bulky polar side-chain and so the tertiary structure is distorted and unstable.

4. Hb Kempsey, $\alpha_2{\cdot}\beta_2^{99\,Asp\rightarrow Asn}$. The Asp at position 99 in the β-chain is involved in stabilizing the T state, deoxyhemoglobin, through a hydrogen bond with Tyr at C7 on the α-chain. The Asn at position 99 cannot form this bond and so the T state is harder to create. Thus, *Hb Kempsey* has an abnormally high O_2 affinity and doesn't release O_2 at the tissues. The opposite effect occurs in *Hb Kansas,* where Asn 102 is replaced by Thr, destabilizing the R state, and a very low O_2 affinity results.

SUMMARY

1. Hemoglobin is a useful model of protein function. It is found in erythrocytes—red blood cells—where it binds and releases oxygen.

2. Hemoglobin is an example of a protein that contains a nonpolypeptide part, the prosthetic group. Heme is the prosthetic group for hemoglobin and a variety of other proteins.

3. The protein part of hemoglobin, the apoprotein, is folded into several α-helices that are packed together to form a compact molecule with a deep pocket in which heme is bound.

4. Oxygen is bound to the iron atom in heme, which remains in the ferrous state because the globin structure prevents interaction of the oxygen with more than one heme at a time.

5. Hemoglobin is a tetramer of globin chains,

each with one bound heme. Oxygen binding at one or more of the hemes causes a conformational change that increases the affinity for binding subsequent oxygen molecules. This change is inhibited by the binding of hydrogen ions, carbon dioxide, and BPG.

6. In the lungs, the high oxygen concentration forces off hydrogen ions, carbon dioxide, and BPG, and hemoglobin becomes almost saturated with oxygen. As blood flow takes the saturated hemoglobin to the tissues, the blood oxygen concentration drops, releasing some oxygen and allowing binding of hydrogen ions, carbon dioxide, and BPG, thus releasing more oxygen. These cooperative allosteric effects allow hemoglobin to deliver oxygen efficiently to the tissues.

7. Hemoglobin releases more oxygen where the blood pH is lower.

8. Hemoglobin changes during development. Fetal hemoglobin has two α-chains—identical to adult α-chains—and two γ-chains—specific to the fetus. Adults have two α-chains and two β-chains, with a small proportion of β-like δ-chains.

9. Many inherited forms of hemoglobin are known. Some of these cause disease, such as sickle-cell anemia.

REFERENCES

Protein Structure: New Approaches to Disease and Therapy, Max Perutz, 1992, Freeman, New York.

Mechanisms regulating the reactions of human hemoglobin with oxygen and carbon monoxide, 1990, *Ann. Rev. Phys.,* **52**:1–25.

The Molecular Basis of Blood Diseases, G. Stamatoyannopoulos, A. W. Nienhuis, P. Leder, and P. W. Majerus, 1987, W. B. Saunders, Philadelphia, PA.

Thalassemia—a global public health problem, 1996, D. J. Weatherall and J. B. Clegg, *Nature Medicine,* **2**:847–849.

Cuts and scrapes? Plasmin heals! J. D. Vassalli and J. H. Seurat, 1996, *Nature Medicine,* **2**:284–285.

REVIEW QUESTIONS

1. Which statement is incorrect? The prosthetic group of myoglobin
   a) is mainly hydrophobic
   b) contains two negative charges
   c) contains iron in the reduced form
   d) is covalently bound to the proximal histidine
   e) is modified in some common hemoglobinopathies

2. Which statement is false? Oxygen transport in humans
   a) depends on blood flow
   b) depends on variations in pH in the vicinity of various tissues
   c) is linked with pH buffering in the blood
   d) involves active transport of oxygen across cell membranes
   e) requires BPG in the erythrocytes

3. What causes a sickle cell crisis?
   a) The presence of glutamate at position 6 of the β-chains of hemoglobin
   b) Increased concentration of deoxyhemoglobin
   c) Increased blood flow
   d) Shortened life of erythrocytes
   e) High oxygen concentration

ANSWERS TO REVIEW QUESTIONS

1. *e* The prosthetic group of myoglobin is heme and it has all of the properties indicated in a–d. Hemoglobinopathies, however, are due to genetic changes in the globin part of hemoglobin and never to changes in the heme prosthetic group.

2. ***d*** Without blood flow, oxygen would move too slowly through the body. Variations in pH are critical for directing oxygen to the tissues that need it most (Bohr effect). Oxygen release is linked with uptake of hydrogen ions. Without BPG, the oxy-

gen affinity of hemoglobin would be too high. However, d is false; oxygen molecules are hydrophobic and diffuse readily through the plasma membrane—no active transport is necessary, or found to exist.

3. ***b*** At low oxygen tensions, deoxyhemoglobin precipitates in erythrocytes, causing it to assume a sickle shape. The sickled cells can jam in the blood vessel and accumulate, cutting off the oxygen supply to tissues, causing more sickling, and building to cause a sickle-cell crisis. Note that glutamate at position 6 of the β-chains of hemoglobin is found in normal hemoglobin; sickle hemoglobin has valine at this position. The shortened life of the erythrocyte causes anemia, which certainly increases the probability of a sickle-cell crisis, but is not the immediate cause.

CHAPTER 7

INTRODUCTION TO ENZYMES

INTRODUCTION

Enzymes are protein and/or RNA molecules that catalyze chemical reactions. Chemical reactions in biological systems are almost always catalyzed, otherwise they would occur too slowly. The catalysts that are used are proteins (except for a few instances of catalysis by RNA) although other molecules (prosthetic groups, coenzymes) may also be required. The main functional characteristics of enzymes compared with chemical catalysts are their high efficiency, their specificity, and their capacity for regulation. Enzyme-catalyzed reactions can lead to the transformation of energy from one state to another. Enzyme-catalyzed reactions conform to the same laws of thermodynamics as chemical reactions and engines.

Some of the reactions catalyzed are very complex; others are very simple. Many occur at very rapid rates. For example, nearly half the CO_2 produced by tissues is carried to the lungs by dissolving in the bloodstream (the rest is carried by hemoglobin). The rate of solution of CO_2 in water is much too slow for this process and so it is catalyzed by an enzyme called carbonic anhydrase:

$$CO_2 + H_2O \leftrightarrow H_2CO_3$$

Carbonic anhydrase raises the rate of this interconversion 10 million times. Notice that the enzyme itself does not appear as part of the chemical equilibrium, although its name is sometimes written above the arrows. During the reaction, each enzyme molecule combines with its substrates, CO_2 and H_2O, and converts them to product, H_2CO_3. The enzyme and product then separate and the enzyme can bind additional substrate molecules and continue to catalyze further conversions. In this way, one molecule of carbonic anhydrase can hydrate 100,000 molecules of CO_2 per second.

Enzymes do not change the *equilibrium of a chemical reaction*—just the rate. Many reactions in the body can go in either direction, as indicated by the use of a double-headed arrow (↔) in equations. In some cases, the concentrations of the reactants in the body are such that the reaction always proceeds only in one direction. Such reactions are said to be *physiologically irreversible* and are written with a single-headed arrow (→). This is not a property of the enzyme, but depends on the reaction and the reactant concentrations.

The cell operates through chemical reactions. However, much of the interaction of an organism with its environment involves types of energy that are not chemical. Enzymes and enzyme-like proteins can transform chemical energy into other forms of energy and vice versa. The easiest is to transform chemical energy into heat for maintaining *body temperature*. There are also enzymatic mechanisms for cooling, as in secretion of sweat.

Movement is achieved through *muscle contraction* which involves the conversion of the chemical energy stored in ATP into physical work. Sunlight is converted into chemical energy by *photosynthesis* and light is converted into chemical signals by *rhodopsin* in the *eye.* Small ions and molecules are moved against concentration gradients across membranes. Information is stored chemically in chromosomes and in the brain. Enzymes are required for transmission of nerve impulses.

Enzyme specificity is always very high by comparison with chemical catalysts, but varies widely between enzymes. Usually the specificity involves two components: choice of reactant(s) and the reaction catalyzed. The reactant(s) are called *substrate*(s) and the *products* are called products:

$$\text{substrate(s)} \xrightleftharpoons{\text{Enzyme (cofactors)}} \text{product(s)}$$

As an example, consider the specificity of *trypsin,* an enzyme that cleaves protein chains by hydrolyzing the peptide bond. Trypsin requires a basic amino acid, Lys or Arg, and splits the peptide on the carbonyl side of Lys or Arg as shown in Figure 7.1.

The substrate, the peptide, is split into two products. Trypsin can also catalyze hydrolysis of an ester bond. Hence, trypsin has a well-defined specificity that covers a small range of substrates and reactions. Other proteolytic enzymes may be more or less specific. For example, *subtilisin* catalyzes the same reactions as trypsin, but it will hydrolyze the peptide bond between any amino acid. In contrast, *thrombin* will only hydrolyze the peptide bond between arginine and glycine.

Another example of a highly specific enzyme is *glucokinase.* This acts on *D-glucose* and phosphorylates it to give *glucose-6-phosphate.* Glucokinase is important in sugar metabolism in the liver. It is deficient in patients suffering from *diabetes mellitus* in which there is a high blood sugar concentration as a consequence of failure to secrete the *pancreatic hormone* insulin. In muscle the same reaction is carried out by a much less specific enzyme, *hexokinase.*

A very important consequence of enzyme specificity is that it is often possible to detect a particular enzyme and even measure its catalytic activity in the presence of many other enzymes and molecules. (Of course, in the cell the enzyme must be able to carry out its function in the presence of many other enzymes and substrates.) A simple enzyme can often be assayed in a cell or unpurified *extract of a cell* by adding a substrate and measuring the amount of product released. Several conditions must carefully be controlled, to take account of the enzyme kinetics and the presence of regulatory substances and cofactors, but under the right conditions, *enzyme assay* in body fluids or unpurified extracts is a valuable clinical and research tool.

CHARACTERISTICS OF ENZYME-CATALYZED REACTIONS

An enzyme is a *catalyst* and as such it does not alter a *reaction equilibrium. At equilibrium,* the ratio of the substrate concentration(s) to the product concentration(s) is the same, whether or not enzyme is present. The effect of an enzyme is to speed up the *rate* of the reaction so that equilibrium is attained rapidly. If an enzyme-catalyzed reaction requires 1 sec to reach equilibrium in the presence of enzyme, it might require 10^8 sec (about 3 yr) to reach equilibrium in the absence of enzyme.

Enzymes achieve the increase in reaction rate by lowering the *activation energy* of the reaction. The activation energy for a simple chemical reaction can be shown by a graph of *free energy* of the system as a function of the progress of the reaction (the *reaction coordinate*) (Figure 7.2). In this case there is a net fall in free energy for the whole reaction, but in order to get from the initial state to the final state, the system must pass through the *transition state.* The energy for this comes from thermal energy.

In an enzyme-catalyzed reaction, the reaction occurs in several steps, each one of which has a low activation energy; thus, many more molecules have suffi-

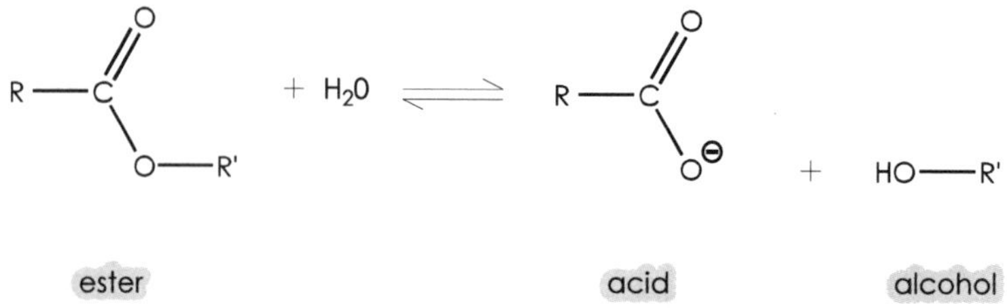

$$R{-}C({=}O){-}O{-}R' + H_2O \rightleftharpoons R{-}C({=}O){-}O^{\ominus} + HO{-}R'$$

ester　　acid　　alcohol

Figure 7.1.

Trypsin specificity. In addition to breaking peptide bonds, trypsin can break ester bonds, producing a carboxylic acid and an alcohol.

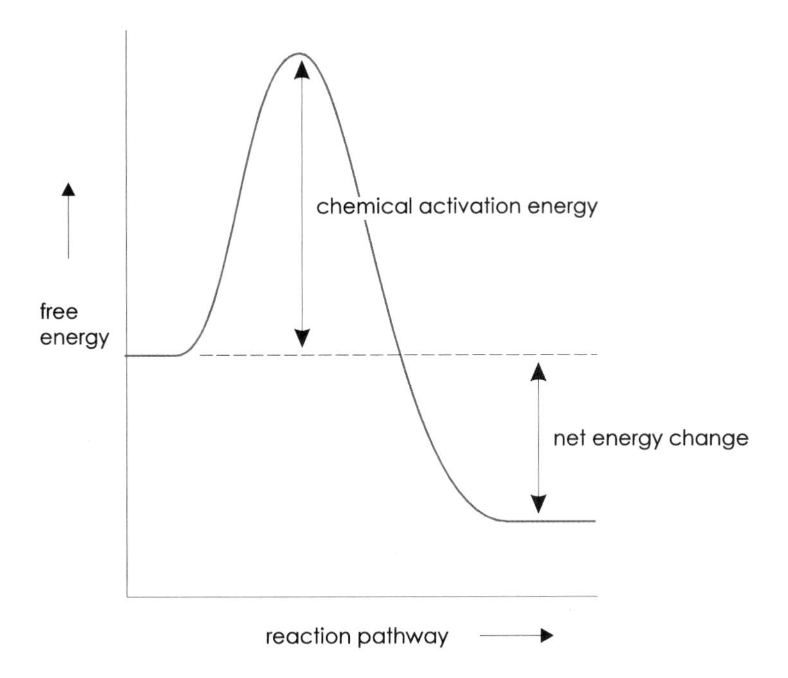

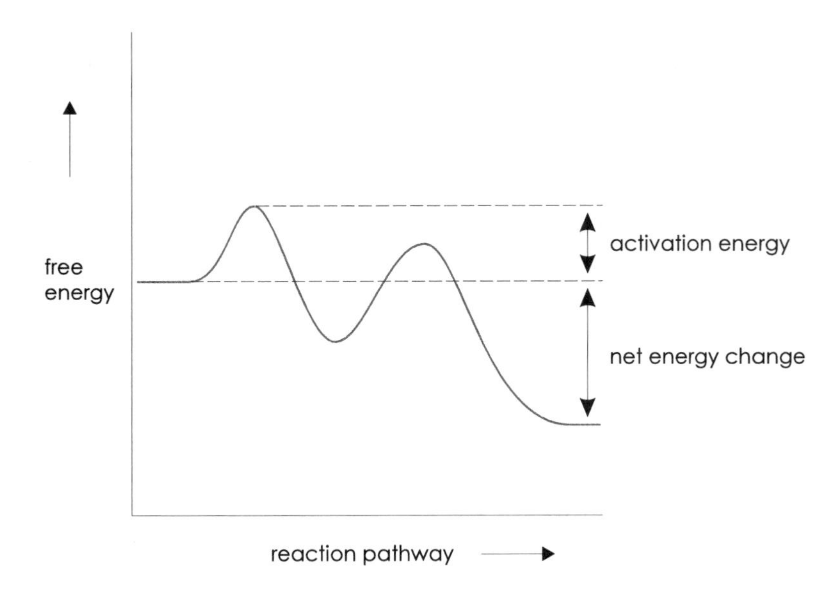

Figure 7.2.
Free energy as a function of the reaction coordinate. In both panels, the purple curve shows how the free energy of a system of reactants (substrate plus enzyme, if present, plus product) changes as substrate is converted to product. In the upper panel, there is no catalysis and the curve shows a high transition-energy barrier inhibiting the conversion of substrate to product. In the lower panel, enzyme catalysis has broken the reaction into steps so that the transition energy for any one step is much less.

cient thermal energy to overcome the activation energy barriers and the reaction goes faster. The simplest case is an enzyme with a single substrate and single product.

Active Sites

Catalysis by an enzyme occurs at the *active site* of the enzyme where the substrate(s) is (are) bound. The main general features of active sites are:

1. The active site takes up a relatively small portion of the total volume of the enzyme.
2. The active site is a three-dimensional entity.
3. The specificity of binding depends on the precisely defined arrangement of atoms in the active site giving a *direct fit* or an *induced fit.*
4. Most substrates are bound to enzymes by relatively weak forces.
5. Active sites are clefts or crevices.

Usually the active-site cleft is formed by folding the polypeptide chain into a specific shape. The parts of the protein not contributing directly to the active site may be required to form or stabilize the structure. Other functions for the rest of the enzyme include regulatory functions (particularly in allosteric enzymes) and signals directing the enzyme to the correct part of the cell (Chapter 27). In some enzymes the active site is a **direct fit** in terms of shape and bonding interactions to the substrate. In other enzymes, the active-site structure has to be modified to bind the substrate, which is called **induced fit.** Hexokinase provides an excellent example of induced fit—the two lobes of the enzyme close around the substrate.

Mechanisms

The mechanism by which an enzyme catalyzes its specific reaction is known in varying levels of detail in a number of cases, for example, *ribonuclease, lysozyme, serine proteases, carboxypeptidase, aspartate amino transferase,* and *lactate dehydrogenase.* The mechanisms differ, but some useful general principles have emerged from studies of individual enzymes. Enzymes use the normal principles of organic chemistry.

The major factors that contribute to enzyme catalysis are:

1. The enzyme binds the substrate(s) (and cofactor) in such a way that the susceptible bonds are (a) in close proximity to the catalytic group on the active side (called the entropy effect) and (b) so oriented in relation to the catalytic group that the transition state is readily formed (orbital steering).
2. Some enzymes may combine with the substrate to form an unstable covalent intermediate that more readily undergoes reaction to form the products.
3. By providing functional groups capable of acting as proton donors or acceptors, the enzyme may bring about general-acid or general-base catalysis.
4. The enzyme may induce *strain* or distortion in the susceptible bond of the substrate, making the bond easier to break.
5. Binding groups in the active site may be positioned so as to stabilize a reactive intermediate (propinquity effect).

Entropy effect and orbital steering. The groups in the active site that bind the substrate increase the effective concentration of substrate because they have an affinity for the substrate, and having "caught" a substrate molecule they hold it in position for the catalytic group to act. If it were not for this active site binding, catalysis would depend on random collisions with the substrate and the time of contact would rarely be long enough for catalysis to occur. The substrate binding site also relies on random collision, but it may actually attract the substrate through electrostatic forces, and once a collision occurs the substrate is bound by an array of noncovalent bonds in the active site. This effect of holding the substrate near the catalytic site is called the *entropy effect.* In addition to binding the substrate, the active site orients it so that it is in the best orientation relative to the catalytic group. This enhances the entropy effect and is called *orbital steering* (Fig. 7.3).

Covalent catalysis. The entropy effect and orbital steering are used by all enzymes. Covalent catalysis is used by some enzymes, those that form covalent enzyme–substrate intermediates. The intermediate compound is highly reactive or unstable and has a high probability of entering the transition state and completing the reaction. The serine proteases—a group of enzymes that have a reactive serine in the active site—are an example of a group of enzymes that use covalent catalysis. Other enzymes also use a covalent bond to serine or cysteine or histidine or lysine.

Acid–base catalysis. Acidic or basic groups within the protein provide a catalytic mechanism that can function instead of *covalent catalysis* or together with covalent catalysis. A general acid is a functional group

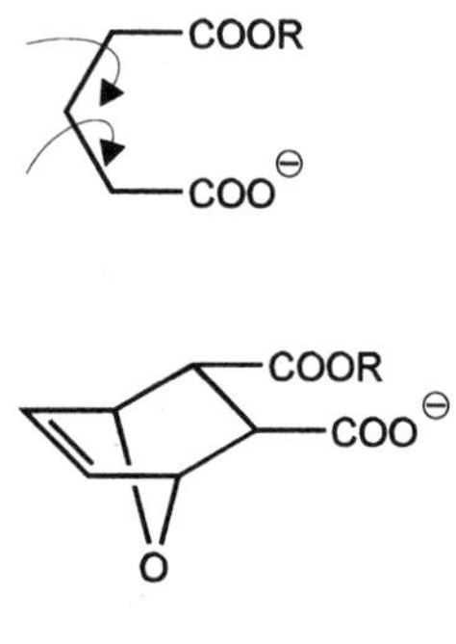

Figure 7.3.

Chemical catalysis. In the molecule shown, the R group can exchange between the two carbons shown. In the upper molecule, free rotation, shown by the arrows, means that the two carbons are rarely close together as the molecule twists around. In the lower molecule, the structure holds the two carbons in a fixed position favorable to the exchange reaction.

capable of donating a proton and a general base accepts a proton. The most important general acids and bases in proteins are shown in Figure 7.4.

One of the most common acid–base catalysts is histidine. Its pK is in the range 5.6–7, depending on its environment in the protein, so it readily accepts or donates a proton.

Distortion of substrate. Many active sites do not fit the substrate directly, but have to be distorted to bind the substrate (*induced fit*). Once bound, the enzyme can then distort the substrate and so weaken existing bonds and promote the formation of new ones. One general theory is that the enzyme-active site is designed to fit the transition state of the substrate. This would help account for the lowering of the activation energy for passing through the transition state because the increase in energy required to change the substrate to the transition state is offset by the decrease in energy of the enzyme in going to the transition state. A good experimental argument in favor of this model comes from the finding that molecules which resemble the transition state of the substrate, but cannot form the product (*transition-state analogs*), often bind very strongly to the enzyme and inhibit it extremely effectively. In general, *transition-state analogs* make the most powerful drugs.

Propinquity effect. Catalysis is enhanced by chemical groups in the enzyme-active site that stabilize a reactive intermediate, helping the enzyme progress from binding of the substrate to catalyzing the correct specific reaction at a high rate.

One other general characteristic is that any particular enzyme will make use of a whole array of different methods and interactions to achieve its purpose.

ENZYME KINETICS

Kinetics—rates of reactions—are very important for the body that maintains its steady state by adjusting reaction rates in response to the environment and to hormonal controls. We begin by considering the rates of chemical reactions in isolated systems and then consider how to use this understanding to work with enzymes in the body.

Chemical reactions are classified according to the number of reactants. In the simplest case, the first-order reaction, there is just one reactant and the rate of the reaction is proportional to the concentration of that reactant. Hence, a graph of reaction rate against reactant concentration is a straight line for a first-order chemical reaction (Fig. 7.5).

In the case of **enzyme-catalyzed reactions,** the kinetics are less straightforward, except under carefully chosen conditions. The time course of an enzyme-catalyzed reaction is generally complex and enzyme reactions are usually characterized by measuring the *initial rate of the reaction,* V_0, which is the reaction rate at

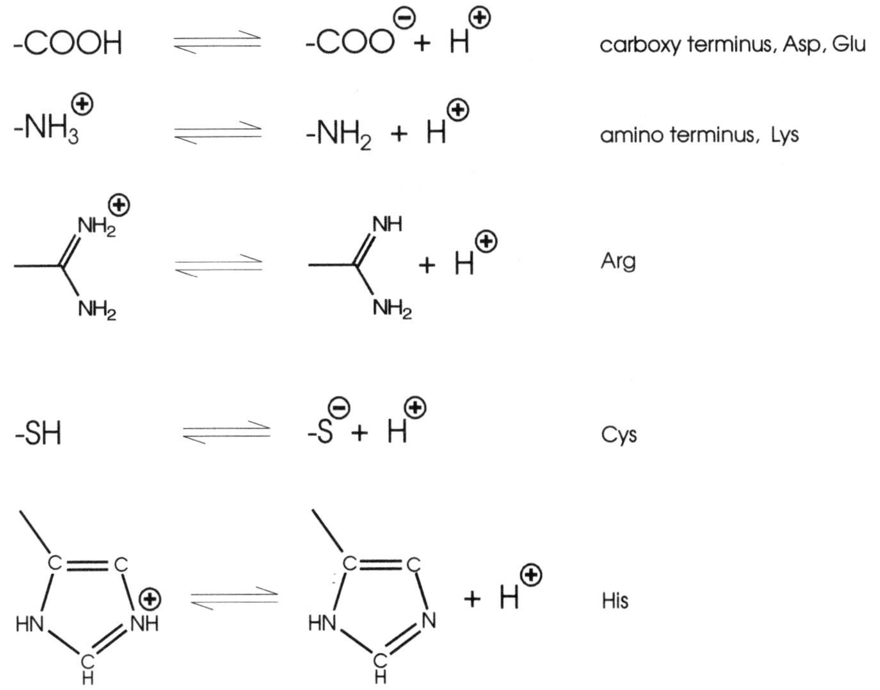

$$-COOH \rightleftharpoons -COO^{-} + H^{+}$$ carboxy terminus, Asp, Glu

$$-NH_3^{+} \rightleftharpoons -NH_2 + H^{+}$$ amino terminus, Lys

$$-C(=NH_2^{+})NH_2 \rightleftharpoons -C(=NH)NH_2 + H^{+}$$ Arg

$$-SH \rightleftharpoons -S^{-} + H^{+}$$ Cys

Imidazolium (HN–CH=NH⁺ ring) ⇌ imidazole (HN–CH=N ring) + H^{+} His

Figure 7.4.

Acidic and basic groups in proteins. Each of the groups shown is found in unmodified proteins and may contribute to the charge on the protein. The dissociated forms are shown on the right.

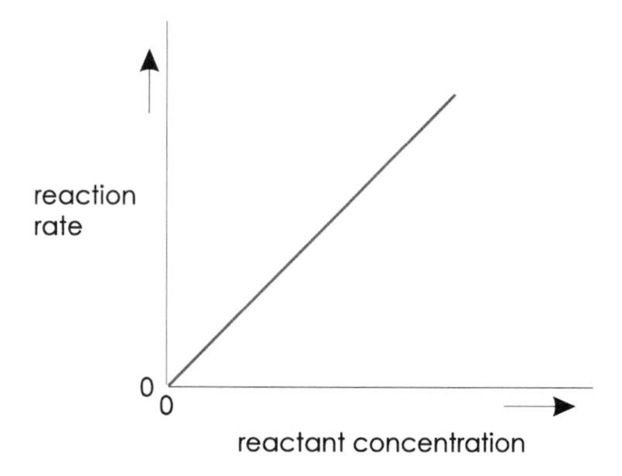

Figure 7.5.
First-order chemical reaction. The purple line shows that the reaction rate (y axis) is simply proportional to the substrate concentration (x axis).

time zero. In a closed system, such as a test tube, an enzyme reaction can be started by adding the enzyme to a solution of the substrate and other molecules that may be required. The reaction will start at a high rate and slow down over time for several reasons:

1. The substrate will be used up and the reaction rate will slow down as each enzyme molecule spends more time diffusing through the solution before it collides with a new substrate molecule.
2. As the reaction proceeds, product will accumulate which may tend to inhibit the enzyme.
3. The enzyme molecules may gradually lose activity owing to random hydrolysis or *denaturation.*

To simplify the kinetics, we will start by considering only the reaction rate right at the beginning of the reaction. The reaction rate at time 0, V_0, is the initial slope of the graph of product concentration against time (Fig. 7.6).

We can measure V_0 for a variety of initial substrate concentrations to understand how the reaction rate will change in response to changes in substrate availability. That is, the initial reaction rate V_0 can be studied as a function of substrate concentration in a simplified laboratory experiment. At very low substrate concentrations, a graph of reaction velocity against substrate concentration (Fig. 7.7) will be a straight line. However, for an enzyme-catalyzed reaction, even with only one substrate and one product, this graph will curve at higher substrate concentrations and will level off at very high substrate concentration. Think of it this way: when the substrate concentration is low, it is the limiting factor, and so an increase in substrate concentration produces a proportional increase in reaction rate. However, as the substrate concentration increases, the enzyme concentration becomes the limiting factor and then a further increase in substrate produces a less-than-proportional increase in reaction rate. Eventually, the enzyme becomes "saturated" and the reaction rate reaches a constant value that doesn't increase significantly as yet more substrate is added. The maximum reaction rate that is achieved is called the V_{max}.

For most enzymes, this graph (Fig. 7.7) has a fair-

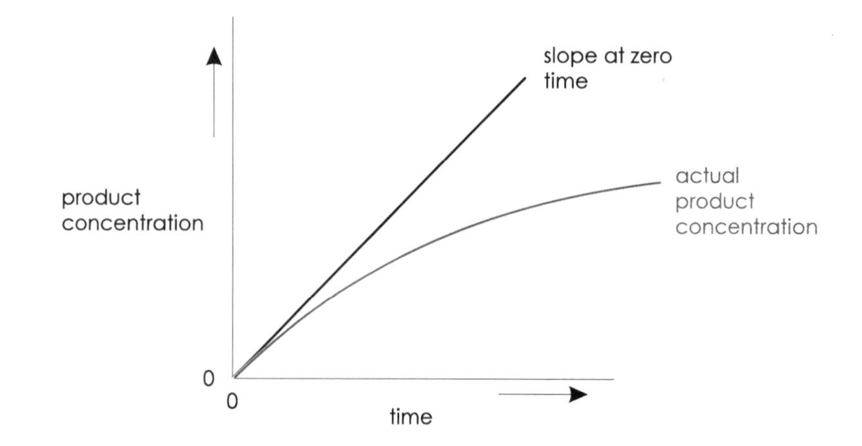

Figure 7.6.
Time course for enzyme reaction. The purple line shows the product concentration (y axis) as a function of time (x axis) for an enzyme reaction. The black line indicates the initial slope of the curve. This initial slope is the initial velocity of the reaction.

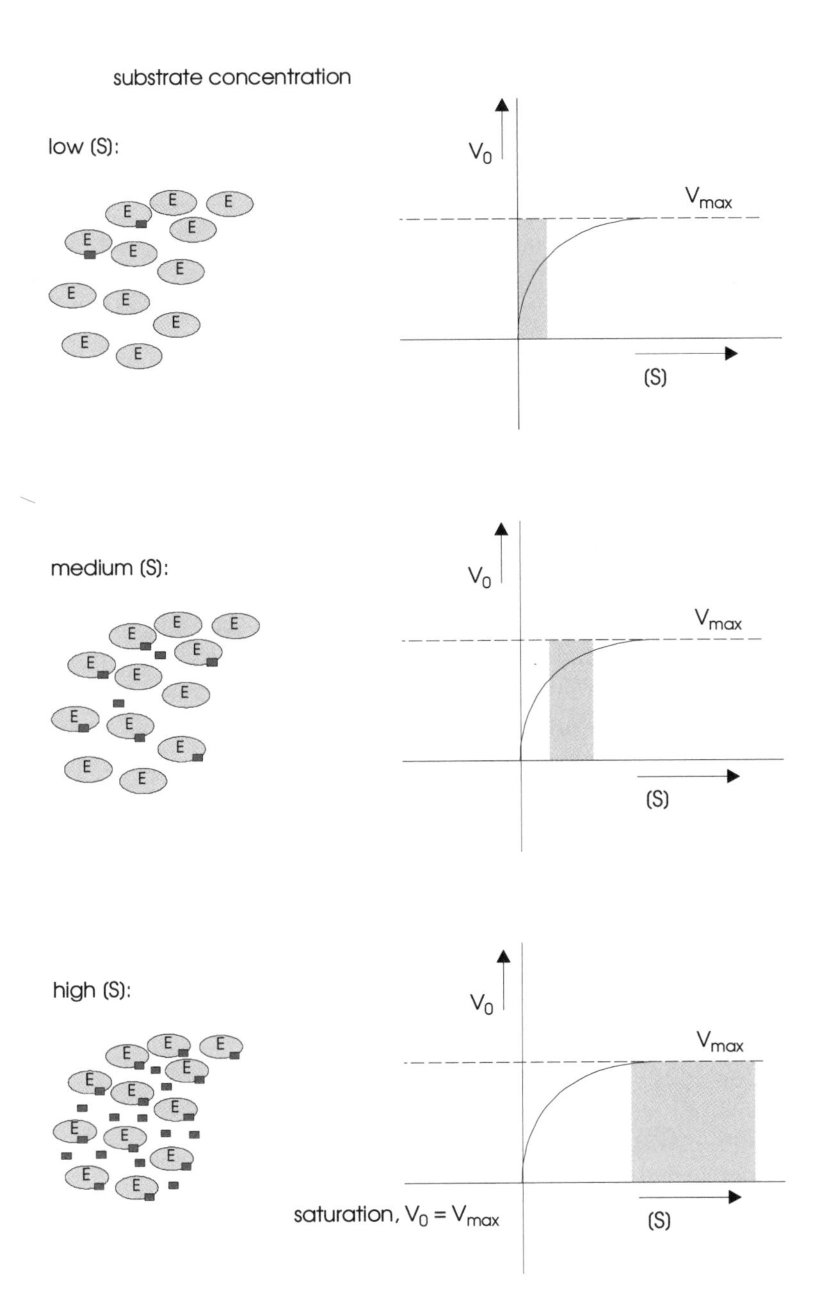

Figure 7.7.
V_0 vs. [S] for an enzyme-catalyzed reaction. In the cartoons on the
left, the gray ellipses, labeled E, represent enzyme molecules. The smaller purple rectangles represent substrate molecules. In the top panel, there are few substrate molecules and they are all bound to the enzyme molecules. In the central panel, there are more substrate molecules and some of them are free in solution, not bound to an enzyme molecule. In the bottom panel, there are many substrate molecules and all the enzyme molecules have a bound substrate. The number of enzyme molecules with bound substrate is an indication of the reaction rate because those enzymes have the opportunity to "act" and convert the substrate to product. The initial reaction rate (*y* axis) is shown as a function of the substrate concentration (*x* axis) in the graphs on the right. The purple highlight indicates the region of the graph that corresponds to the cartoon at its left.

ly simple mathematical shape—a *hyperbola*—which is useful because it allows us to easily calculate enzyme reaction velocities without having to refer to the graph every time. The equation that describes this graph is called the *Michaelis–Menten equation.* It involves two constants (specific for each enzyme) called V_{max} and K_M, the *Michaelis–Menten constant,* which are explained further below. The Michaelis– Menten equation is $V_0 = V_{max} \cdot [S]/(K_M + [S])$ where [S] is the substrate concentration.

Here V_{max} is the maximum velocity seen in the graph of V_0 against [S]. Where does K_M come in? It is the substrate concentration that produces half the maximum velocity (Fig. 7.8). If K_M is large, the reaction rate at low substrate concentration is relatively low. Alternatively, if K_M is small, the reaction rate at low substrate concentration is relatively high. The curve of velocity against substrate will rise quickly if K_M is low and the curve will soon approach V_{max}; if K_M is high, the curve will appear to spread out toward the right, approaching V_{max} only at high substrate concentration. Sometimes, K_M is written as $[S]_{0.5}$.

If we know K_M and V_{max} for a given enzyme, we can use the Michaelis– Menten equation to calculate the reaction rate, V_0. Hence, this is a very useful equation.

reaction rate

V_{max}

$\frac{1}{2}$ V_{max}

substrate concentration

K_M

Figure 7.8.

Definition of K_M. The curve shows how the reaction rate (y axis) for an enzyme-catalyzed reaction changes with the substrate concentration. The lower broken line is drawn at a reaction rate of one half of the maximum rate for this reaction. This line intersects the curve at the substrate concentration equal to K_M.

Alternate Forms of the Michaelis–Menten Equation

The shape of the graph of initial velocity against substrate concentration is hyperbolic. It is hard to draw accurately and to analyze. For example, just where is V_{max}?

The graph can be converted to a linear form by plotting $1/V_0$ against 1/[S] and the result is called a *double reciprocal plot* or a *Lineweaver–Burke plot.* It is widely used for characterizing enzymes. The equation of the line is: $1/V_0 = (K_M/V_{max}) \cdot (1/[S]) + 1/V_{max}$. Hence, the slope of the Lineweaver–Burke plot is K_M/V_{max}, and the intercept on the $1/V_0$ axis is $1/V_{max}$ (Fig. 7.9). The fact that the plot gives a straight line is a useful indicator that the Michaelis–Menten equation does actually describe the kinetics of the enzyme under study. The Michaelis–Menten equation can also be "linearized" as the *Eadie–Hofstee plot* (Fig. 7.10). This plot of V_0 against V_0/[S] gives K_M as $-1 \times$ (the slope of the graph) and V_{max} as the intercept on the V_0 axis.

Which of these plots should be used? To generally understand the behavior of enzymes, use the simple graph of initial velocity against substrate concentration. The linearized forms are useful for calculation of K_M and V_{max}. The Lineweaver–Burke plot is useful for distinguishing between types of inhibition (Chapter 8). The Eadie–Hofstee plot is better than the Lineweaver–Burke plot at picking up deviations from the Michaelis–Menten equation.

Enzyme reactions are more complicated than this simple description, but the Michaelis–Menten equation provides a very useful empirical tool for describing enzymes.

Theoretical Basis for the Michaelis–Menten Equation

The Michaelis–Menten equation can be derived from the chemical kinetics of a very simplified enzyme reaction. The main assumption is that there is only one intermedi-

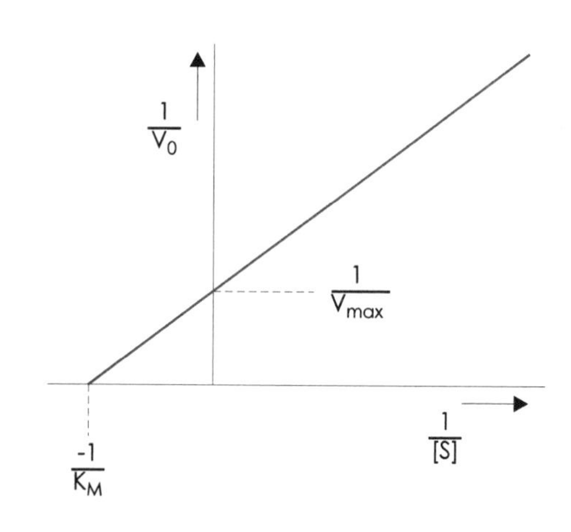

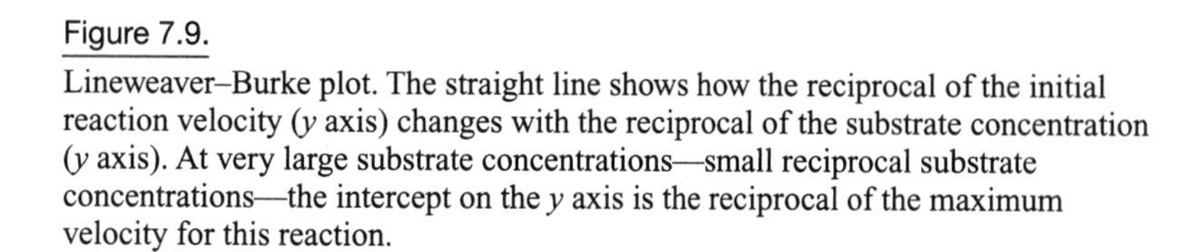

Figure 7.9.

Lineweaver–Burke plot. The straight line shows how the reciprocal of the initial reaction velocity (y axis) changes with the reciprocal of the substrate concentration (y axis). At very large substrate concentrations—small reciprocal substrate concentrations—the intercept on the y axis is the reciprocal of the maximum velocity for this reaction.

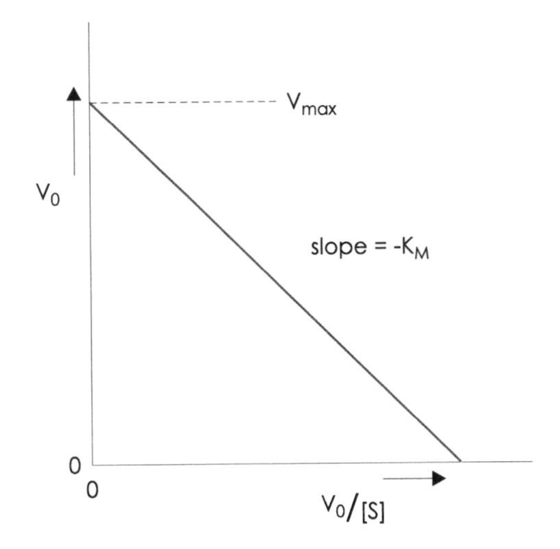

Figure 7.10.
Eadie–Hofstee plot. The straight line shows how the velocity of an enzyme-catalyzed reaction varies with the velocity/substrate ratio to give a line with slope equal to $-K_M$.

ate state, the *enzyme–substrate complex.* Since we are dealing with *initial* velocities, it is not necessary to consider the back reaction (conversion of product to substrate). Thus, the chemical equation for the reaction is:

$$S + E \underset{K_2}{\overset{K_1}{\rightleftharpoons}} ES \overset{K_3}{\longrightarrow} E + P$$

The Michaelis–Menten equation can be derived from this chemical equation, but the details need not concern us here.

The "Meaning" of V_{max} and K_M

The velocity of an enzyme reaction increases as [S] increases. When [S] becomes much larger than K_M, the velocity approaches a constant, that is, as $[S] \rightarrow \infty$, $V_0 \rightarrow V_{max}$. Hence, V_{max} is the maximum velocity that can be achieved with a given total enzyme concentration and it can be measured as the limiting (maximum) velocity obtained as [S] is increased; V_{max} is proportional to enzyme concentration. Thus, V_{max} is a property of the enzyme and its substrate and also depends on enzyme concentration and on the conditions of temperature, pH, and ionic strength in the solution.

The V_{max} of an enzyme is a measure of how fast the reaction it catalyzes can proceed once the enzyme–substrate complex is formed. This is related to the turnover number—the number of substrate molecules converted into product per active site at very high substrate concentration. *Turnover numbers* vary from very high (e.g., 600,000 sec^{-1} for carbonic anhydrase) to relatively slow (e.g., 0.5 sec^{-1} for lysozyme). Another name for turnover number is $\boldsymbol{k_{cat}}$, which is related to V_{max} and the total enzyme concentration, [E] (measured as number of active sites per unit volume of solution), as follows: $V_{max} = k_{cat} \times [E]$.

The Michaelis–Menten constant, K_M, depends on the particular enzyme and substrate being used and on the conditions of temperature, pH, and ionic strength in the solution. Note that K_M is independent of the enzyme concentration, in contrast to V_{max}. Theory shows that K_M is approximately equal to the *dissociation constant* for the enzyme and the substrate, provided that the enzyme–substrate complex reverts to enzyme and substrate much more often than it goes on to generate product. Hence, a low K_M corresponds to tight binding of enzyme to substrate and vice versa. The units of K_M are moles per liter or just M.

Enzymes have higher rates of reaction when the substrate concentration is greater than K_M. This principle is illustrated by the use of *asparaginase* in the treatment of *leukemia.* In leukemia, white blood cells proliferate, out of control, but it was found that their growth depended on the presence of asparagine in the blood. Asparaginase hydrolyzes asparagine to aspartate, and when administered in the bloodstream it removed the asparagine and slowed the growth of malignant white cells. However, it was found that asparaginase from different sources had widely different effects which were eventually traced to differences in K_M. The ineffective asparaginases had a value of K_M that was too high for the low asparagine concentration in the blood.

Enzymes do not Change the Equilibrium

All enzyme-catalyzed reactions are reversible and if the concentration of product is high, the reverse reaction will compete with the forward reaction for enzyme. The V_{max} and K_M values for the reverse reaction are usually different from those of the forward reaction. The characteristics of the forward and back reactions are related by the *Haldane equation*:

$$K_{eq} = V_F \cdot K_B/(V_B \cdot K_F)$$

where K_{eq} is the equilibrium constant of the reaction, V_F and V_B are the V_{max} values for the forward and back reactions, and K_B and K_F are the Michaelis–Menten constants for the back and forward reactions respectively. Enzymes do not change K_{eq}, therefore, the Haldane equation shows that enzymes can have K_M and V_{max} designed to allow the enzyme to work equally well in both directions or can have a large V_{max} in the forward direction and a small V_{max} in the reverse direction, but this has to be balanced by a large K_M in the forward direction and a small K_M in the reverse direction.

For a number of enzymes, the body requires different kinetic parameters at different times and at different places. This can be achieved by regulation of a single enzyme or by having multiple enzymes. Where multiple enzymes occur, they are usually very similar and are called *isozymes* or *isoenzymes.*

Two-Substrate Reactions

Many enzyme reactions in the body involve two substrates. In these cases, the simple reaction scheme used to describe *one-substrate reactions* is inapplicable. However, the concepts of K_M and V_{max} can still be applied to *two-substrate reactions* as follows. If there are two substrates, A and B, and two products, P and Q, the overall reaction is:

$$A + B \Leftrightarrow P + Q$$

The rate of formation of P + Q can be measured as a function of concentration of substrate A with substrate B held constant. In many cases, the Michaelis–Menten equation can be used to describe the results if a superscript A is used to indicate that K_M^A and V_{max}^A apply when [A] is changing with [B] held constant:

$$V_0 = V_{max}^A \cdot [A]/(K_M^A + [A])$$

Similarly, if [B] is varied while [A] is held constant, we get K_M^B and V_{max}^B:

$$V_0 = V_{max}^B \cdot [B]/(K_M^B + [B])$$

Of course, the value of K_M^A depends on the concentration of B that was used. Essentially, the K_M^A and V_{max}^A values can be used to predict the effect of changing just the concentration of A.

Both K_M^A and K_M^B can be found from double reciprocal plots as for single substrate reactions. Multiple plots can be obtained if the experiment is repeated for multiple values of [B] (Fig. 7.11).

Use of the Michaelis–Menten Equation

In a *metabolic pathway,* the initial enzyme is regulated by *allosteric* or other mechanisms as discussed below. The subsequent enzymes in the pathway obey *Michaelis–Menten kinetics.* If the initial enzyme is down-regulated (its reaction rate falls), the concentration of its product falls. Since this product is the substrate for the second enzyme, its reaction rate will fall according to the Michaelis–Menten equation. Hence, its production concentration will fall, and so on down the whole pathway.

This neat control system fails in one important situation. First, consider what happens if one of the enzymes is at V_{max}: any increase in substrate concentration will have no effect, and a decrease will only have an effect if it takes the concentration below that which produces V_{max}. In the body, enzymes are not usually at V_{max} but this situation will occur if an enzyme is severely inhibited, for example, by a drug. The reaction rate will drop, causing the substrate concentration to increase and the inhibited reaction will reach the apparent V_{max}. The pathway will respond by raising the substrate concentration still further but, of course, to no avail. This effect will propagate back through the pathway, raising the concentration of all the intermediates before the block. Of course, the concentration of all intermediates after the block and the final product will be very low, set by the V_{max} of the inhibited enzyme.

It is important to realize that a drug will not only prevent the synthesis of product but will raise the concentrations of previous intermediates to nonphysiological levels. This can cause severe side effects.

Consider another example. If the substrate concentration [S] is below K_M for the enzyme and this substrate, then changes in substrate concentration will directly affect the velocity V_0 of the reaction. For exam-

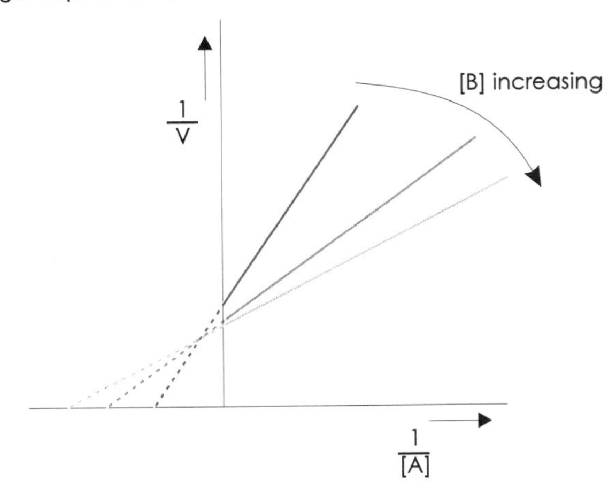

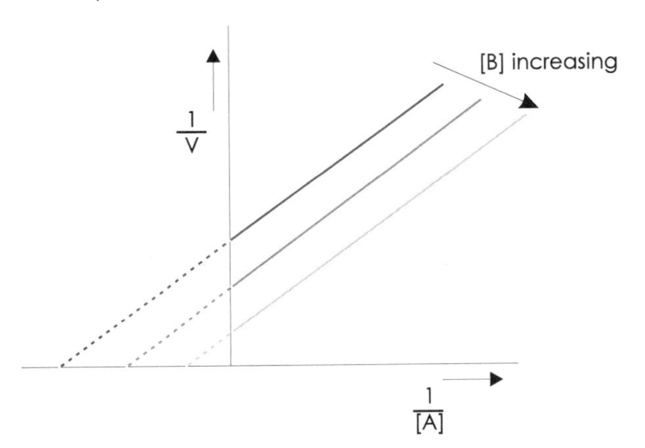

Figure 7.11.

Double-reciprocal plot for a two-substrate enzyme. Double-reciprocal plots are shown for two examples of enzymes with two substrates, each, *A* and *B*. One type of two-substrate enzyme—a single displacement enzyme—typically gives plots like those shown in the top panel. Another type of two-substrate enzyme—a double displacement enzyme—typically gives plots like those shown in the bottom panel. In each panel, there are three lines. For each line, the concentration of one substrate, *B*, is held constant while the other substrate concentration, [*A*], is varied. In each panel, each different line corresponds to a different value of the concentration of the substrate *B*, [*B*].

ple, if [S] starts as K_M and then suddenly doubles, the velocity V_0 will increase from $V_{max}/2$ to $2 \times V_{max}/3$, an increase of about 33%. Similarly, if [S] drops to $K_M/2$, then V_{max} drops to $V_{max}/3$, a drop of about 33%. If [S] $\gg K_M$, then V_0 is insensitive to changes in [S]. These situations can be seen in the metabolism of glucose by hexokinase:

$$\text{ATP} + \text{glucose} \Leftrightarrow \text{ADP} + \text{glucose-6-phospate}$$

A supply of glucose from the blood is essential to brain function and so the brain hexokinase must provide a constant rate of reaction even though blood glucose concentrations vary normally between 3 and 9 mM. Brain hexokinase has a K_M for glucose of 0.05 mM, so the phosphorylation of glucose by brain hexokinase is almost independent of the glucose concentration, in the normal range. In contrast, liver normally synthesizes glucose and exports it to the blood, but the liver does take up glucose when the blood glucose concentration is very high. Liver hexokinase has a K_M for glucose of 2 mM; thus, it is sensitive to changes in blood glucose levels.

ISOZYMES IN DIAGNOSIS

Human blood contains a very complex and carefully balanced array of cells, *platelets,* macromolecules, small molecules, and ions. The cells and platelets can be removed by centrifugation, leaving the *plasma.* If plasma is allowed to clot, the remaining solution is the *serum.* The serum normally contains few enzymes, but damage or disease of a tissue often releases enzymes from the cells of the tissue into the blood. Some of these enzymes are tissue-specific, and identification of *abnormal serum enzymes* has a general utility in diagnosis. The ideal would be to have a specific "marker" enzyme for each tissue, then a serum test for marker enzymes would identify damaged or diseased cells. At this time, markers are not available for all cell types. However, some important examples are available.

Diagnosis of Myocardial Infarction

A patient presents at the emergency room with sudden severe chest pain. Is the pain due to a heart attack or damage to another internal organ or structure? Damage to the heart will release a number of enzymes into the bloodstream, including *creatine kinase* (CK, also called *CPK*) and *lactate dehydrogenase* (*LDH*). Elevated levels of these enzymes are indicators of a *myocardial infarction.* Figure 7.12 shows some changes in blood enzyme levels following myocardial infarction (MI). These enzymes are measured in the serum. It is very important to obtain serum without lysing the cells in the blood, because these cells release enzymes that interfere with the assay of true serum enzymes.

Isoenzymes

These simple measurements of enzyme activity may not be completely specific. For example, damage to the liver or skeletal muscle will also raise LDH in serum. However, the LDH released from heart muscle is different from that released by the liver or skeletal muscle. The different enzymes are called isoenzymes, that is, they catalyze the same reaction. In the case of LDH, the isoenzymes (or isozymes) are regulated differently and match the metabolic requirements of the appropriate tissues.

Lactate Dehydrogenase Isoenzymes

The enzyme LDH is a tetramer of two different subunits, called H and M. The predominant subunit in heart cells is H and the predominant subunit in skeletal muscle or liver is M. The isoenzymes contain different proportions of the two subunits, that is, H_4, H_3M, H_2M_2, HM_3, and M_4. The isoenzymes can be separated electrophoretically.

In normal blood serum, LDH-2 (H_3M) is the main isoenzyme. After myocardial infarction, LDH-1 (H_4) is

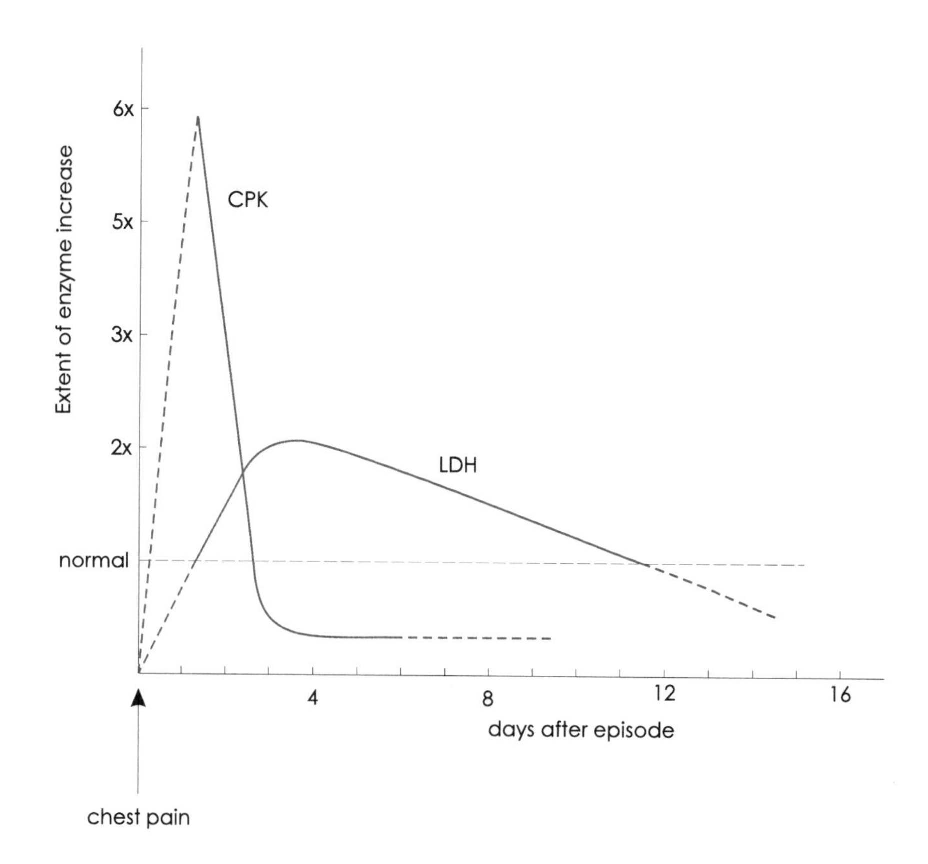

Figure 7.12.

Blood enzymes following myocardial infarction. The activities of two enzymes, creatine kinase (CPK) and lactate dehydrogenase (LDH) were measured in the blood plasma of a patient at various times following a myocardial infarction.

the main enzyme. Hence the relation [LDH-1]<[LDH-2] is "flipped" after myocardial infarction to give [LDH-1]>[LDH-2]. Diseases of liver or skeletal muscle, on the other hand, raise the level of LDH-5 (M_4).

Creatine Kinase Isoenzymes

The enzyme CK is a dimer of two subunits, called M (found in skeletal muscle) and B (found in brain). Heart muscle, the myocardium, contains both subunits and the principal isoenzyme in myocardium is MB in contrast to MM, found in skeletal muscle and BB, found in brain. After a myocardial infarction, where part of the myocardium is damaged by low oxygen due to reduced blood flow (*ischemia*), it is specifically the MB isoenzyme that increases in the blood.

Enzyme Assay in the Clinical Laboratory

The assay of enzyme activity is an important diagnostic and screening procedure.

To obtain results that really measure the enzyme concentration and are independent of substrate concentration and other factors, it is necessary to measure V_{max} under carefully controlled conditions. This can be done by measuring the initial velocity V_0 as a function of [S] and then fitting the results to the Michaelis–Menten equation.

For clinical purposes, a simpler procedure is required, and usually V_{max} is measured directly as the reaction velocity in the presence of a vast excess of substrate ($[S] > 10 \cdot K_M$). If the enzyme has a very high K_M for product (K_M^P), that is, for the back reaction, V_{max} can be obtained from the amount of product formed after a fixed reaction time. If K_M for the product is not high enough, the product will slow the reaction and the apparent V_{max} will be too low.

This situation can be avoided by including additional reactants that remove the product as fast as it is formed, thus maintaining the condition of $[P] \ll K_M^P$. Such a reaction is called a *coupled reaction.* Under these conditions ($[S] \gg K_M^S$; $[P] \ll K_M^P$) the rate of reaction is measured from the formation of P in a fixed time. The simplest way to determine P is to use a spectrophotometer. For example, in the reaction catalyzed by lactate dehydrogenase (lactate + $NAD^+ \leftrightarrow$ pyruvate + $NADH + H^+$), NADH absorbs light at 340 nm (near *ultraviolet*) where none of the other components absorb and so the increase in *absorbance* at 340 nm measures production of NADH. A coupled reaction can be used to generate a spectrophotometric change.

For example, pyruvate kinase can be assayed by the coupled reaction:

phosphoenolpyruvate + H^+ + ADP $\Leftrightarrow$ pyruvate + ATP
(pyruvate kinase reaction)
pyruvate + NADH + H^+ $\Leftrightarrow$ lactate + NAD^+
(LDH reaction)

in which lactate dehydrogenase is added in excess and removes pyruvate as it is formed, oxidizing one molecule of NADH for each molecule of pyruvate used. The reaction can be followed by the fall in absorbance at 340 nm.

The *units of enzyme activity* are micromoles of substrate consumed per minute (U or mmol $\cdot$ min^{-1} at defined temperature, pH, and other conditions). Enzyme concentration = enzyme activity/turnover number. Specific activity = enzyme activity/total protein.

The conditions of the reaction are important, particularly the concentration of cofactors, divalent ions, monovalent ions, temperature, and pH. Enzyme activity generally increases with temperature up to the point where thermal denaturation destroys its activity. Most enzymes have an *optimum pH* which commonly has a value between 6 and 8. The shape of the pH versus activity curve depends strongly on the enzyme.

ENZYME NOMENCLATURE

A large number of different enzymes has been described. Their nomenclature has been rather haphazard, although many enzymes are named by placing -ase after their substrate; others are called product synthetase. Enzymes that transfer phosphate from ATP to a substrate are called substrate kinase. An international enzyme commission (EC) has developed a classification number, a *systematic name,* and a *recommended name.* Recommended names are widely used, but the EC numbers and systematic names are increasingly used to avoid confusion, often in a footnote.

ENZYME COFACTORS OR COENZYMES

Vitamins and Trace Elements

A *cofactor* or *coenzyme* is a non-polypeptide part of an enzyme. It may be tightly bound to the enzyme at all

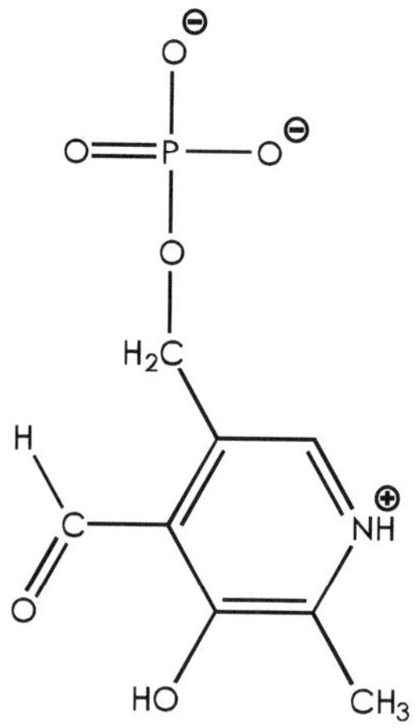

Figure 7.13.
Pyridoxal phosphate. The structure of a cofactor.

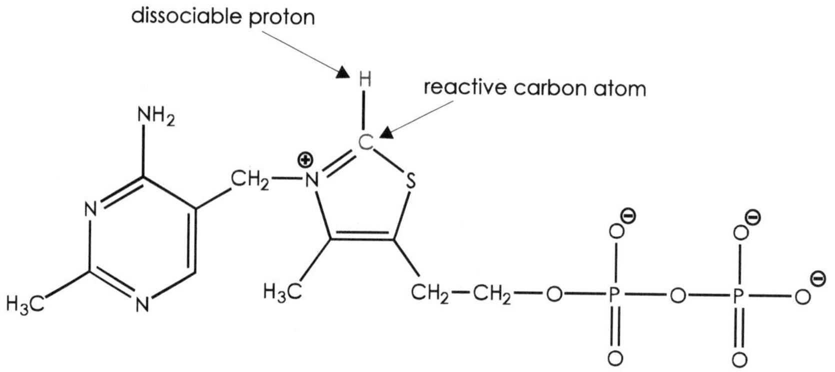

Figure 7.14.
Thiamine pyrophosphate (TPP). The purple color highlights the parts of this cofactor that are directly involved in enzyme reactions.

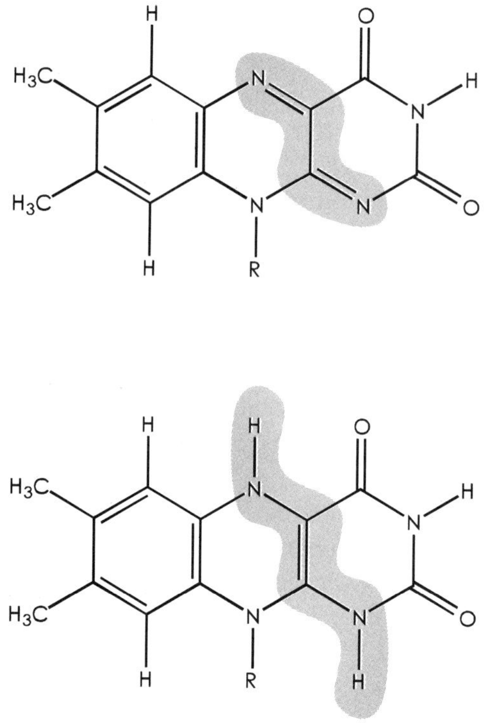

Figure 7.15.
FAD and $FADH_2$. The two structures shown for this cofactor both occur during enzyme reactions in which FAD is a cofactor; the upper form, FAD, becomes reduced to the lower form, $FADH_2$. The changes in the molecule are highlighted in purple.

times, in which case it may also be called a prosthetic group. Alternatively, it may behave more like a second substrate of the enzyme.

Many cofactors are metal ions, for example, Zn^{2+} in carboxypeptidase. These metal ions may be required as *trace elements* in the *diet.* Many other cofactors are specific organic molecules, often derived from *vitamins.* For example, *pyridoxal phosphate* (Fig. 7.13) is derived from B_6 *vitamins*. Some vitamins provide cofactors that the body cannot synthesize. In these cases, the vitamin is an essential component of the diet. Only small amounts are required because, as part of an enzyme, the cofactor is reused many times.

Thiamin (*vitamin* B_1) is necessary in the diet of most vertebrates. Intake of thiamin by humans is close to the minimum requirement and so some foods are supplemented with thiamin. Reactions requiring thiamin are more critical in some tissues than others and the rate of turnover of thiamin also varies, producing variation in the severity of the effects in different tissues. The severity can vary between individuals so that thiamin deficiency (*beri-beri*) can show symptoms of an abnormal nervous system leading to paralysis and wasting of the limbs (dry beri-beri) or, alternatively, the cardiovascular system may suffer the greater deficiency, giving congestive heart failure with a seepage of liquid into the tissues (an edema) so that they become puffy (wet beri-beri). These diseases normally occur when thiamin deficiency

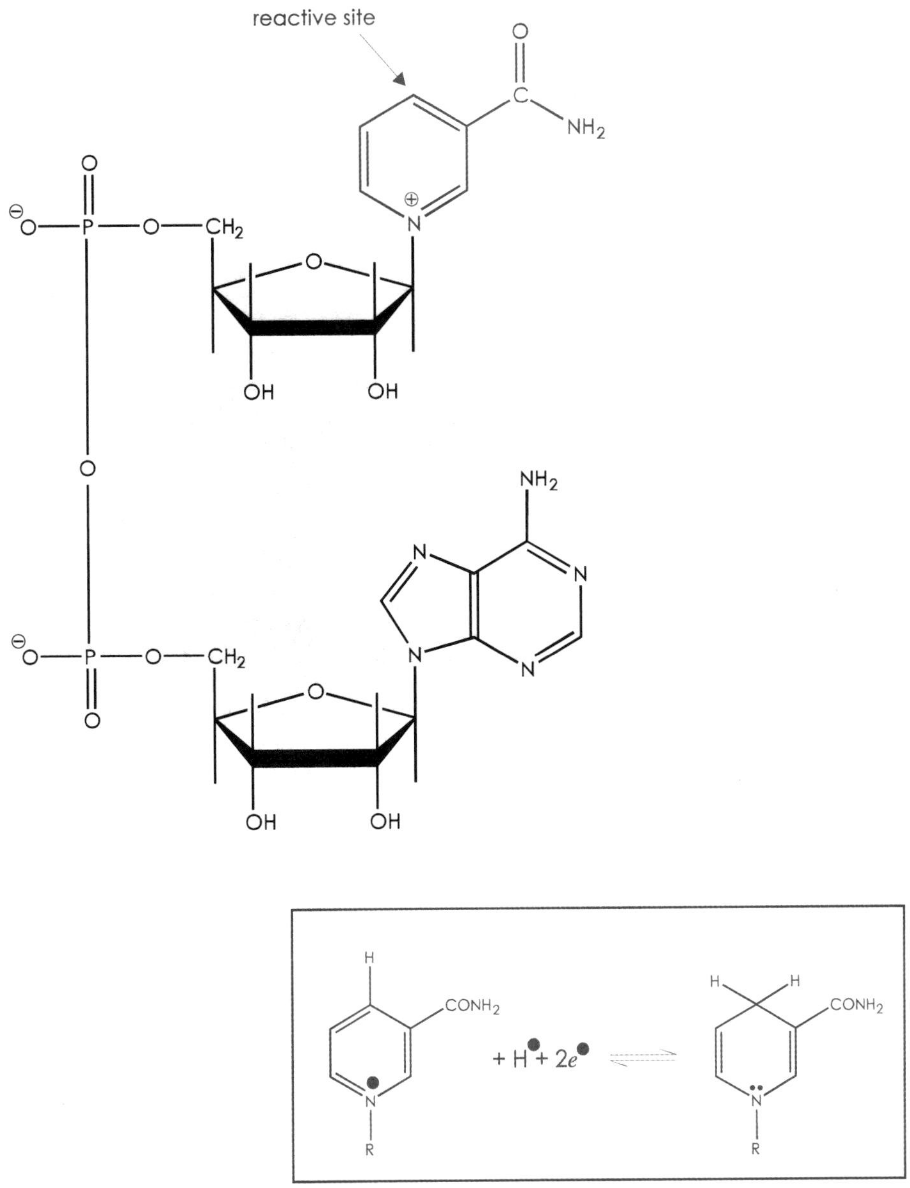

Figure 7.16.
NADH and NAD^+. The full structure of the cofactor, NAD^+, is shown in black and purple. The inset shows how the nicotinamide part, highlighted in purple, changes during an enzyme-catalyzed reaction in which NAD is a cofactor.

is due to lack of material from the outer parts of wheat or rice where the thiamin is concentrated. Thiamin deficiency can also occur among *alcoholics* where it is called *Wernicke's encephalopathy* and usually appears as defects of the central nervous system. It also develops if patients are fed intravenous glucose without vitamin supplements.

Thiamin contains a substituted thiazole group which is involved in the decarboxylation of pyruvate by *pyruvate decarboxylase.* Thiamin occurs in cells largely as its active coenzyme form *thiamin pyrophosphate* (TPP) (Fig. 7.14).

Other key coenzymes are *FMN, FAD, NAD,* and *CoA.*

Both FMN and FAD occur as tightly bound coenzymes (Fig. 7.15). They participate in oxidation/reduction reactions in which the *riboflavin* part is oxidized or reduced complementary to the reduction or oxidation of the substrate. The enzymes are called *flavoproteins* and the oxidized forms have an intense color (yellow, red, or green), although the reduced forms are colorless. *NADH dehydrogenase* is an important example of a flavoprotein; other flavoproteins are involved in oxidative degradation of pyruvate, fatty acids, and amino acids.

Nicotinic acid (niacin) is a precursor of *nicotinamide adenine dinucleotide* (NAD) which plays a central role in metabolism as a coenzyme. It can be synthesized from tryptophan. *Tryptophan and nicotinate deficiency* lead to *pellagra,* which is characterized by *diarrhea, dermatitis,* and *dementia.* Maize-rich diets

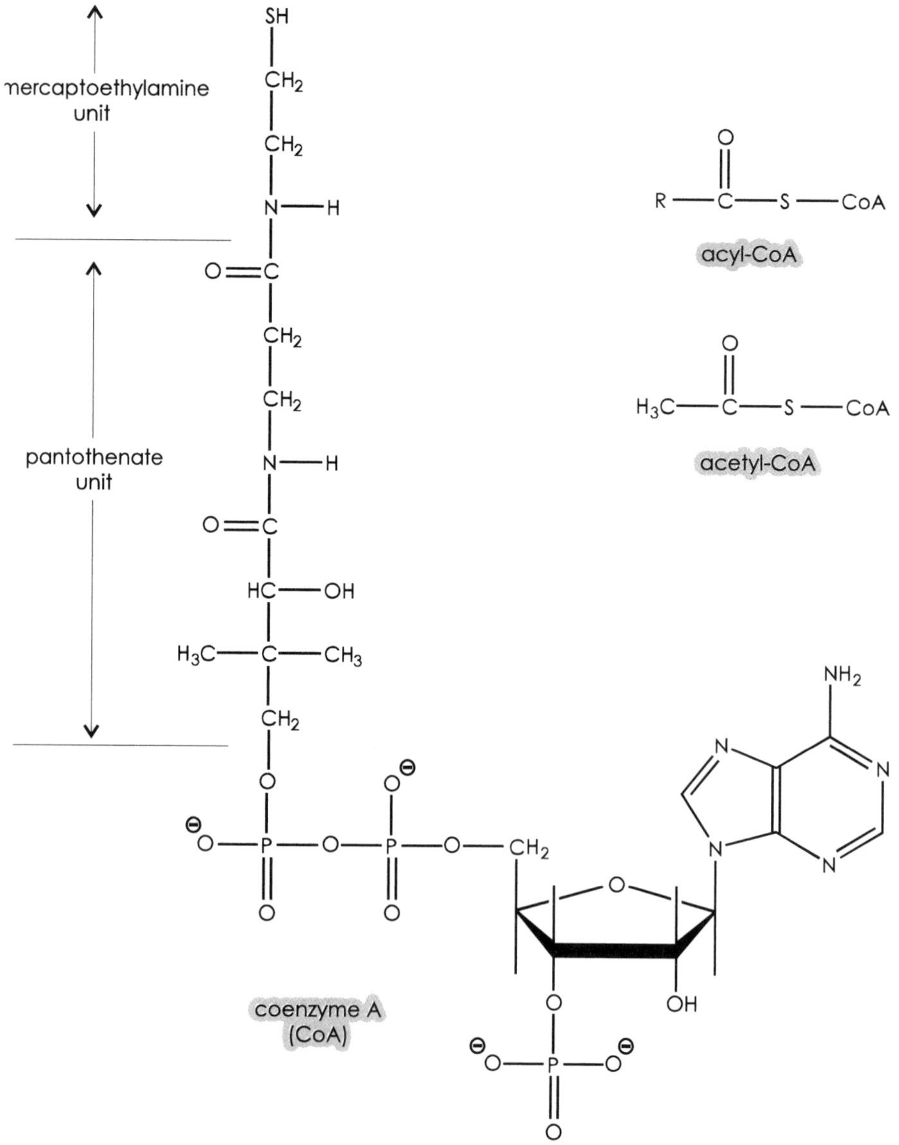

Figure 7.17.

Coenzyme A. The full structure of coenzyme A is shown, with two modified forms, acetyl CoA and acyl CoA, at the top right. In the modified forms the bulk of the coenzyme A molecule is represented by CoA and just the sulfur is shown linking coenzyme A to the acetyl or acyl group, R.

can lead to this problem. Large doses of nicotinate are toxic although nicotinamide is not. Nicotinic acid and nicotinamide are used in two coenzymes, NAD and *NADP* (NAD phosphate) (Fig. 7.16) and catalyze oxidation/reduction reactions. In the reduction reaction, the substrate loses two hydrogen atoms. Both the electrons go to the NAD^+ and the remaining proton goes to the solvent. The resulting reduced form of NAD^+ is NADH.

Pantothenic acid is needed for *coenzyme A* (CoA) (Fig. 7.17). Deficiencies of CoA are not known. It is a widespread coenzyme which serves as a carrier of acyl groups (acetyl-CoA) by linking acetyl to the sulfhydryl group of CoA.

SUMMARY

1. Enzymes are catalysts. They do not affect the equilibrium of a chemical reaction, but they do affect the rate at which the equilibrium is approached. Enzymes are usually quite specific in the reaction(s) they catalyze.

2. Substrate binding and catalysis take place in a restricted portion of an enzyme—the active site. The three-dimensional structure of this site is critical to enzyme activity.

3. Enzyme catalysis is achieved by a variety of mechanisms including both noncovalent and covalent interactions. Enzymes reduce the activation energy for the reaction they catalyze.

4. For a given amount of enzyme, the reaction rate increases as substrate is added up to the point where the enzyme is saturated with substrate, the V_{max}. For most enzymes, the shape of the curve of reaction rate as a function of substrate concentration is hyperbolic, as described by the Michaelis–Menten equation.

5. The K_M of an enzyme for a substrate often reflects the affinity of the enzyme for the substrate.

6. The blood plasma contains different enzymes from those in cells. Damage to cells can cause release of cellular enzymes into the bloodstream, providing an important diagnostic tool. In many cases, the diagnosis can be focused by distinguishing between isoenzymes, which are different enzymes that catalyze the same reaction.

7. Many enzymes use cofactors or coenzymes. These are small molecules that behave like a substrate of the enzyme, but they are recycled by other reactions in the cell and used many times over.

REFERENCES

Enzymes in Metabolic Pathways: A Comparative Study of Mechanism, Structure, Evolution and Control, Milton H. Saier, Jr., 1987, Harper and Row, New York.

Structure and catalysis of enzymes, W. N. Lipscomb, 1983, *Ann. Rev. Biochem.*, **52:**17–34.

Enzyme Structure and Mechanism, 2nd ed., Alan Fersht, 1985, Freeman, New York.

REVIEW QUESTIONS

1. A single subunit enzyme somewhere near the middle of a metabolic pathway in the body is found to be mutated in the patient so that the KM for the enzyme's only substrate is twice the normal value. If nothing else changes, how much must the substrate concentration change in order to maintain the same rate of flow through the pathway?
   a) Exactly double
   b) Exactly half
   c) More than double
   d) Less than half
   e) Between half and double

2. Asparaginase is used to reduce the level of asparagine in the blood in one treatment for leukemia. Which form of asparaginase would be most useful if the blood asparagine level is 0.2 mM?
   a) $K_M = 0.2$ mM; $V_{max} = 0.1$ mM/hr
   b) $K_M = 0.2$ mM; $V_{max} = 0.5$ mM/hr
   c) $K_M = 2.0$ mM; $V_{max} = 0.1$ mM/hr

d) $K_M = 0.1$ mM; $V_{max} = 0.5$ mM/hr
e) $K_M = 0.1$ mM; $V_{max} = 0.1$ mM/hr

3. Which of the following statements about isoenzymes is false?

a) Important in the diagnosis of myocardial infarction
b) Important in the diagnosis of prostate cancer
c) Are tissue-specific
d) Depend on genetic differences between individuals for their clinical value
e) Only exist for certain enzymes

ANSWERS TO REVIEW QUESTIONS

1. **a** The reaction velocity is given by the Michaelis–Menten equation: $V = V_{max} [S]/(K_M + [S])$. If we put in $2 \times [S]$ and $2 \times K_M$, we get: $V = V_{max} \cdot 2 \cdot [S]/(2 \cdot K_M + 2 \cdot [S])$. The twos cancel out and we are left with the original velocity. Thus, a is the correct answer.

2. **d** This enzyme is best because it gives the fastest reaction rate. You can calculate the rate using the Michaelis–Menten equation, or you can simply see that this option has both the highest V_{max} and the lowest K_M of the possible answers and thus it must have the fastest reaction rate.

3. **d** Within each individual, there are many isoenzymes created by a variety of mechanisms. It is the tissue-specific presence of different isoenzymes within each individual that makes them clinically useful. There is no need for genetic variability to make isoenzymes useful, clinically, so d is false.

CHAPTER

8

REGULATION OF ENZYME ACTIVITY BY NONCOVALENT CHANGES

INTRODUCTION

The body produces many enzymes. Some are required almost all the time, just to keep us "ticking over," such as those used for energy production. These enzymes, which are needed most of the time in most cells, are sometimes termed *housekeeping enzymes.* Although housekeeping enzymes are generally present, their activity is regulated up or down according to the needs of the moment in order to keep the body's *metabolism* in balance. For example, enzymes involved with energy *production* will increase when energy is required, as for exercise, and enzymes involved with energy *storage* will increase when energy is in excess, as after a meal. Increases in enzyme activity are called *up-regulation* and decreases are *down-regulation.*

Other enzymes are required only in specific situations. For example, the enzymes for DNA replication are required only when a cell begins its proliferation cycle. Such enzymes may be present at very low levels or be completely absent when they are not required and dramatically increased when they are required. In such cases, the enzyme is said to be "switched on" or "switched off."

ENZYME REGULATION

The body regulates enzyme activity in many ways. It changes the amount of enzyme protein present by changing the rate of synthesis or degradation. In other cases, a covalent modification of the enzyme protein will cause the activity to increase or decrease. Alternatively, other molecules may bind reversibly to the enzyme to change its activity. We will discuss examples of each type of enzyme regulation as they occur in the body.

Some of the most important factors that affect enzyme activity are:

- Change of substrate concentration
- Change of product concentration
- Inhibition by end-product of a pathway (*negative feedback*)
- *Allosteric* activation and inhibition
- *Irreversible* inhibition
- Association/dissociation of a repressor subunit or a specificity-modifier subunit
- Cooperativity

Change of pH
Change of ionic conditions
Reversible *covalent modification* of enzyme
Irreversible covalent modification of enzyme
Change of enzyme synthesis
Change of enzyme degradation

Most of these methods change the rate of the reaction. Sometimes the specificity of the reaction changes.

CHANGE OF SUBSTRATE CONCENTRATION. At very low substrate concentrations, there is plenty of enzyme to interact with substrate; thus, the reaction rate is approximately proportional to substrate concentration. At higher substrate concentrations, there is proportionately less enzyme available; therefore, the reaction rate increases more slowly than the substrate concentration. Eventually, the amount of enzyme becomes the limiting factor, and further increase in substrate concentration has little or no effect on the reaction rate, as described in more detail in Chapter 7 (Fig. 7.7). At the maximum reaction rate, V_{max}, the enzyme is said to be "saturated" with substrate. This simple mechanism of regulating reaction rate is widely used by the enzymes of the body. Additional methods are used, however, to provide more specific, or finer, controls and feedback control.

CHANGE OF PRODUCT CONCENTRATION (PRODUCT INHIBITION). Product will compete with substrate for the active site of the enzyme, depending on the relative affinities for product and substrate. If the affinity for product is in the range of product concentration, the enzyme will remain bound to product for a significant amount of time. During this time, the enzyme is not available for substrate to bind and the reaction velocity falls as product concentration increases. This can result in simple feedback inhibition, but the body usually employs a more sophisticated form of feedback inhibition, involving allosteric enzymes, as well.

NEGATIVE FEEDBACK. Frequently, a particular reaction is catalyzed by a linear series of enzyme reactions called a pathway. The final product of the pathway may act as an inhibitor of the first enzyme in the pathway. Inhibition of the first enzyme reduces the rate of formation of its product, which is the substrate for the next enzyme in the pathway. Thus, the reaction rate for the second enzyme falls and this continues through the pathway until all the reaction rates fall. Hence, the flow of material along the pathway is regulated by the concentration of the final product. The inhibition may be caused by an allosteric effect, that is, the *inhibitor* binds to a site on the enzyme different from the *active site* and causes a conformational change in the enzyme to an inactive or less-active state. We return to this important mechanism later in this chapter.

ALLOSTERIC ACTIVATION OR INHIBITION. Allosteric effects can also be positive in that a particular molecule can cause the formation of an active conformation. For example, the reaction catalyzed by *phosphofructokinase*:

fructose 6-phosphate + ATP
→ *fructose 1,6-bisphosphate* + ADP

is responsive to the concentration of AMP and is increased when the AMP concentration increases, through binding of AMP to the enzyme.

In negative feedback, allosteric regulation results in inhibition.

IRREVERSIBLE INHIBITION. A good example of irreversible inhibition is the inactivation of serine enzymes by reagents that modify the active-site serine. For example, *malathion* is an *insecticide* related to nerve poisons that inactivate the serine enzyme *acetylcholinesterase* (discussed in more detail below). *Phenylmethylsulphonyl fluoride* (*PMSF*) is used to inhibit *serine proteases.*

SUBUNIT ASSOCIATION/DISSOCIATION. As an example, the protein kinase that regulates glycogen metabolism is regulated by a *repressor subunit* that is displaced by cyclic AMP.

$$2R + 2C \Leftrightarrow R_2C_2$$
$$R_2C_2 + 4 \text{ Cyclic AMP} \Leftrightarrow 2C + 2(\text{R.Cyclic AMP}_2)$$

where C is the *catalytic subunit* that is active on its own and R_2C_2 is the inactive complex with the repressor subunit R.

As another example, the *σ subunit* of bacterial RNA polymerase provides specificity for particular substrate (DNA) sequences. Also, *lactose synthetase* in the mammary gland has its reaction altered by binding of a protein subunit.

COOPERATIVITY. Oxygen binding by hemoglobin is an example of cooperativity. In a similar way, many regulated enzymes have multiple substrate binding sites and bind substrate cooperatively. Cooperativity is discussed further in the section on allosteric enzymes.

CHANGE OF PH. Changes of pH in cells are known to occur; for example, pH changes in the blood are important in hemoglobin action. Such changes may have multiple effects because most enzymes are sensitive to pH. *Pepsin* is a highly active protease at low pH (in the *stomach*), but is inactive at neutral pH in the cells where it is synthesized.

CHANGE OF IONIC CONDITIONS. Overall changes in ionic conditions will affect most enzymes, particularly if they have a requirement for the involvement of a particular ion, as in the case of Mg^{2+} and protein kinases. [Ca^{2+}] has specific effects that are mediated by the calcium-binding protein ***calmodulin.***

COVALENT MODIFICATIONS OF ENZYMES. This includes both reversible changes like phosphorylation of specific serines in the enzymes of glycogen metabolism and irreversible changes like *zymogen* activation by proteolysis in *digestion* and *blood clotting.* These mechanisms are considered in more detail below.

CHANGE OF ENZYME SYNTHESIS. Rates of synthesis affect enzyme concentration and hence reaction velocity. Changes in enzyme synthesis may result from changes in mRNA synthesis, stability, or utilization by ribosomes.

CHANGE OF ENZYME DEGRADATION. Degradation rates also affect enzyme concentration and hence reaction rates. One of the most highly regulated enzymes in the body, *ornithine decarboxylase,* is regulated, in large part, by control of the amount of enzyme present. We now examine some of these processes in more detail.

REVERSIBLE ENZYME INHIBITION

The inhibition of certain enzymes by specific metabolites is an important element in the regulation of *intermediary metabolism* and most often occurs with cooperative enzymes that are regulated allosterically. Inhibition of enzymes that obey the Michaelis–Menten equation, noncooperative enzymes, is more commonly used by pharmacists to alter a patient's metabolism. Reversible inhibition of noncooperative enzymes is classified into three groups which can be distinguished kinetically and which have different mechanisms and effects when administered. The classes are called *competitive, uncompetitive,* and *noncompetitive inhibition. Mixed inhibition* also occurs. In all these types of inhibition, the inhibitor (usually a small molecule) binds reversibly and rapidly with the enzyme.

In **competitive inhibition** the inhibitor, I, binds reversibly to the active site of the enzyme. Consequently, the inhibitor competes with the substrate for the active site. As more substrate is added, at constant inhibitor concentration, the inhibitor is displaced and the reaction rate approaches the same maximum value as in the absence of inhibitor. However, more substrate is required to achieve any given reaction velocity in the presence of inhibitor than in its absence. The amount of inhibition depends on the inhibitor constant, K_I, which is also the dissociation constant for the binding of the inhibitor to the enzyme:

$$E + I \Leftrightarrow EI$$
$$K_I = [E] \times [I]/[EI]$$

The graph of reaction velocity, V_0, against substrate concentration, [S], shows extensive inhibition at low substrate concentration and loss of inhibition at high substrate concentration. This graph (Fig. 8.1) is characteristic of competitive inhibition. A competitive inhibitor effectively reduces the apparent affinity of the enzyme for its substrate and so the apparent K_M is affected. At high substrate concentration, the substrate effectively blocks the inhibitor from interacting with the enzyme and thus the eventual maximum velocity, V_{max}, is not affected by a competitive inhibitor.

Another characteristic of competitive inhibition is structural similarity between substrate and competitive inhibitor. The similarity allows the inhibitor to bind in the active site of the enzyme. For example, the enzyme *dihydrofolate reductase* (*DHFR*) is subject to competitive inhibition by *methotrexate* and other compounds. Methotrexate is used in *cancer chemotherapy* because DHFR is required for the synthesis of *thymidine triphosphate,* a specific precursor of DNA (Fig. 8.8). The normal substrate for DHFR is *folate* and methotrexate is closely related (Fig. 8.2).

In **uncompetitive inhibition** the inhibitor binds only to the enzyme–substrate complex. It does not affect the binding of enzyme to substrate, but it does prevent the complex dissociating to give product. Thus, the uncompetitive inhibitor tends to stabilize the

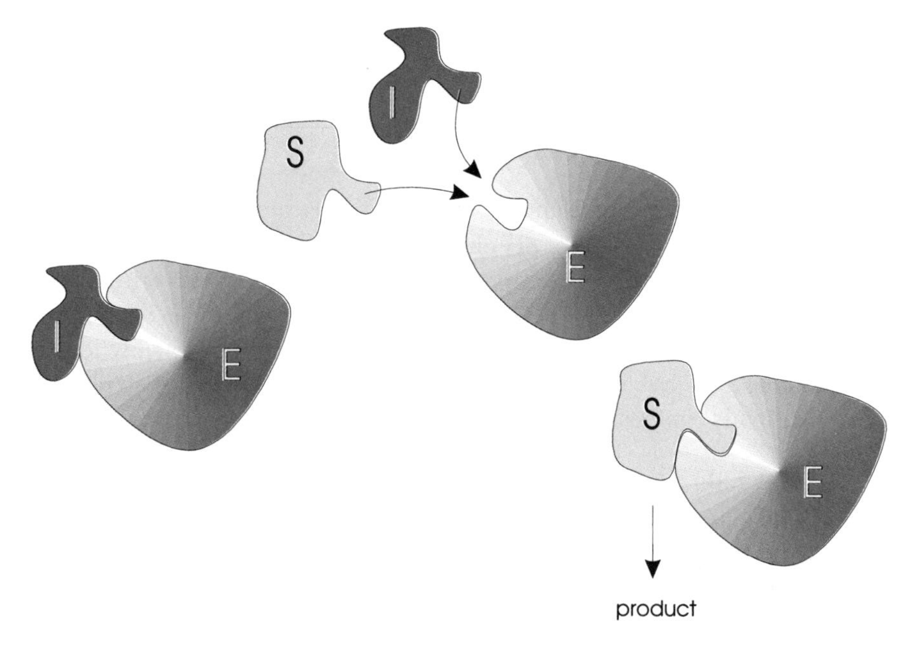

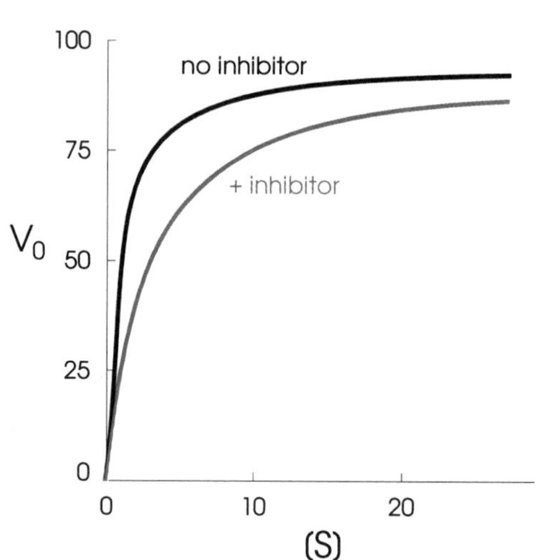

Figure 8.1.

Competitive inhibition. The black and gray shaded objects labeled "E" represent enzyme molecules. The light purple objects labeled "S" represent substrate molecules. The dark purple objects labeled "I" represent inhibitor molecules. The substrate-binding site is represented as an indentation in the enzyme. Either the substrate or the inhibitor can bind to this site in this case.

The graph shows how the velocity of the reaction catalyzed by the enzyme varies with the substrate concentration. The black line shows data obtained in the absence of inhibitor; the purple line shows data obtained in the presence of a competitive inhibitor, as illustrated in the cartoon.

enzyme–substrate complex and this shows up as a decrease in the apparent K_M. Normally, decreased K_M gives rise to increased reaction velocity; in this case, the decrease in K_M is offset by a decrease in V_{max} and the net effect is inhibition. Actually, the ratio of K_M to V_{max} remains constant. As more substrate is added, the reaction velocity will increase, but will always be less than the reaction velocity in the absence of inhibitor, in contrast to competitive inhibition (Fig. 8.3). The inhibitor constant, K_I, is K_I = [ES]×[I]/[ESI] where ES represents the enzyme–substrate complex. Uncompetitive inhibition is common for two-substrate reactions.

In **noncompetitive inhibition** the inhibitor can bind to both the enzyme and the enzyme–substrate complex. It binds at a site separate from the active site and modifies the enzyme conformation to inhibit the formation of product, thus reducing V_{max}. Usually, the inhibitor does not affect substrate binding and so it does not affect K_M. As in uncompetitive inhibition, the reaction rate will increase as more substrate is added, but the rate will always be less than in the absence of inhibitor, at any given substrate concentration. In most cases, the inhibition constant, K_I, can be defined as the dissociation constant for reaction of inhibitor with either free enzyme or the enzyme–substrate complex, which are usually the same. The graph of reaction rate against substrate concentration is a slightly different shape from that for uncompetitive inhibition (Fig. 8.4). In some cases, the binding of inhibitor affects the binding of substrate and thus K_M as well as V_{max}. This is more complicated and is referred to as **mixed inhibition.** Like uncompetitive inhibition, noncompetitive

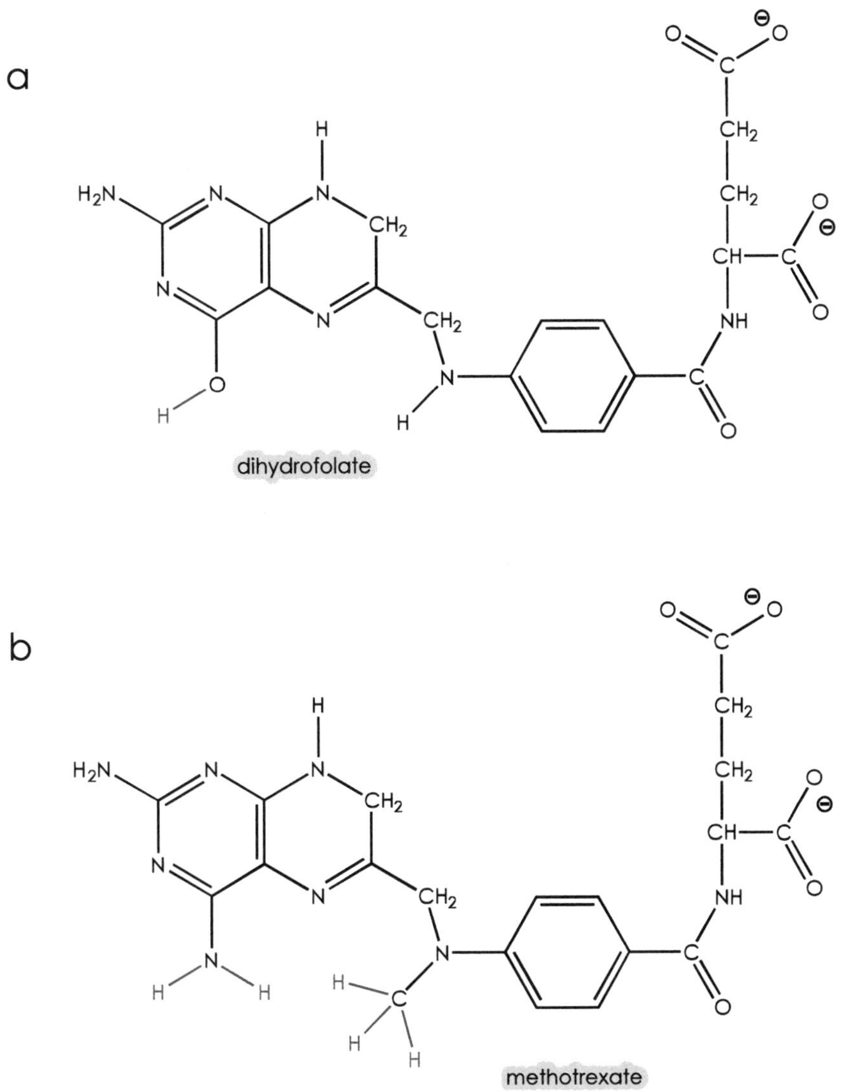

Figure 8.2.
Inhibition of dihydrofolate reductase. (a) The upper molecule, dihydrofolate, is the natural substrate of dihydrofolate reductase; (b) methotrexate, the lower molecule, is a competitive inhibitor. The parts of the molecules that differ are shown in purple.

and mixed inhibition are characteristic of enzymes with more than one substrate.

These types of reversible inhibition are readily distinguished using a double reciprocal—Lineweaver–Burke—plot. Enzyme reaction rate is measured as a function of substrate concentration at several different inhibitor concentrations and the double reciprocal plot is drawn. For each inhibitor concentration, a straight line is drawn through the experimental points. In the case of competitive inhibition, the lines will intersect at the $1/V$ axis, where 1/[S] is zero, since V_{max} is not affected by competitive inhibition. Uncompetitive inhibition gives parallel lines, because both K_M and V_{max} change but their ratio does not. Noncompetitive inhibition gives lines that intersect on the 1/[S] axis, where $1/V$ is zero, because K_M is unaffected but V_{max} changes (Fig. 8.5).

These types of inhibition are shown by simple enzymes (also called Michaelis–Menten enzymes). Some enzymes show much more complex behavior toward inhibitors and may also respond to activators. Such enzymes do not yield such simple graphs of reaction rate against substrate concentration.

EXAMPLES OF THE APPLICATION OF ENZYME INHIBITORS

Treatment for Poison Victims

This is an example of a clinical use of a competitive inhibitor. *Ethylene glycol* is used in *antifreeze* and is not, in itself, poisonous. However, in the body, it is converted to *oxalic acid* by an enzyme pathway that begins with the enzyme *alcohol dehydrogenase.* This reaction

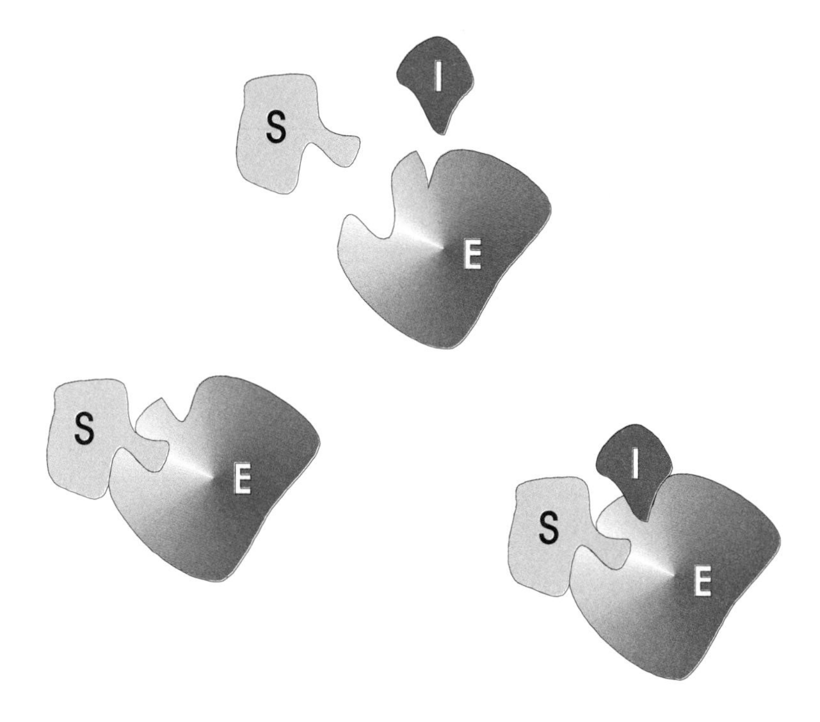

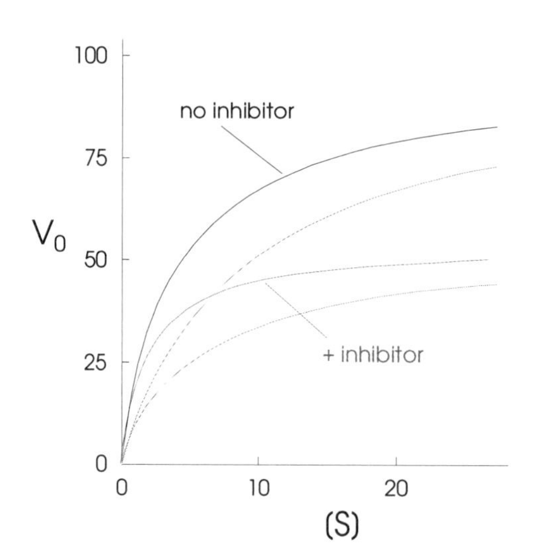

Figure 8.3.

Uncompetitive inhibition. The black and gray shaded objects labeled "E" represent enzyme molecules. The light purple objects, labeled "S" represent substrate molecules. The dark purple objects labeled "I" represent inhibitor molecules. The substrate binding site and the inhibitor binding site are represented as separate indentations in the enzyme.

The graph shows how the velocity of the reaction catalyzed by the enzyme varies with the substrate concentration. The black line shows data obtained in the absence of inhibitor; the purple line shows data obtained in the presence of an uncompetitive inhibitor, as illustrated in the cartoon. The upper gray line is for data obtained in the presence of a competitive inhibitor (cf. Fig. 8.1) and the lower gray line is for data obtained in the presence of a noncompetitive inhibitor (cf. Figure 8.4).

can be inhibited by the competitive inhibitor *ethanol,* CH_3–CH_2OH; thus, a nearly intoxicating, intravenous dose of ethanol competes with the ethylene glycol, which is then excreted harmlessly (Fig. 8.6). A similar procedure is used for *methanol* poisoning.

Inhibitors of Acetylcholinesterase

In the transmission of nerve impulses across a junction between cells—a synapse—the transmitting cell releases *acetylcholine* into the synapse. Acetylcholine diffuses to the receiving cell, where it binds to acetylcholine receptor molecules in the membrane of the receiving cell. Acetylcholine is then degraded by acetylcholinesterase, to free the receptors so that they can respond to the next signal.

Acetylcholinesterase works through a serine residue that forms a covalent intermediate with acetylcholine. The intermediate is then hydrolyzed to give free enzyme, acetate, and *choline.*

Succinylcholine is a competitive inhibitor of acetylcholinesterase and the acetylcholine receptor. In the presence of succinylcholine, the *acetylcholine receptors* remain blocked, giving rise to muscular relaxation that can be useful during *surgery.* Because suc-

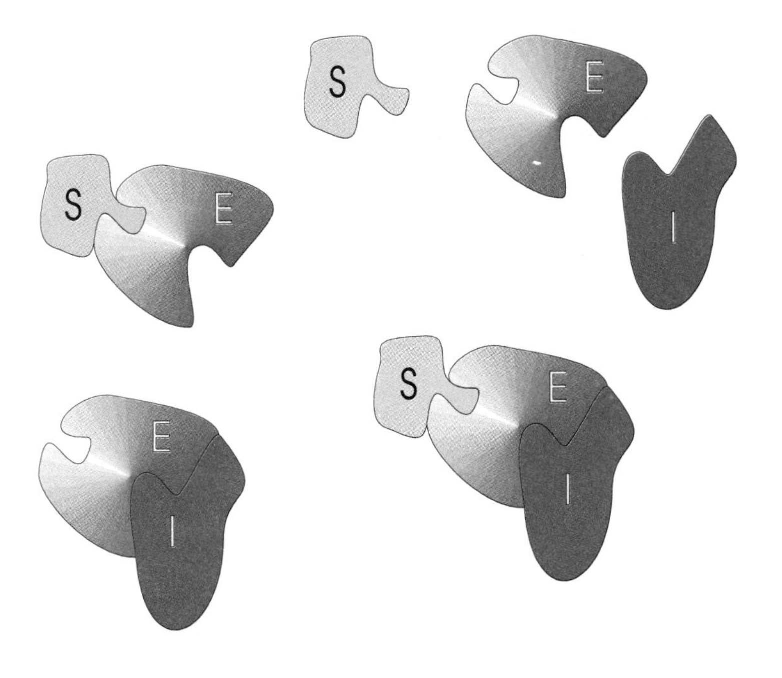

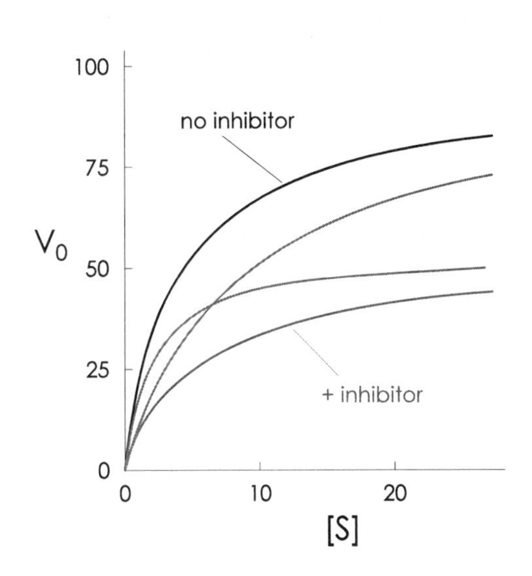

Figure 8.4.

Noncompetitive inhibition. The black and gray shaded objects labeled "E" represent enzyme molecules. The light purple objects labeled "S" represent substrate molecules. The dark purple objects labeled "I" represent inhibitor molecules. The substrate binding site and the inhibitor binding site are represented as separate indentations in the enzyme.

The graph shows how the velocity of the reaction catalyzed by the enzyme varies with the substrate concentration. The black line shows data obtained in the absence of inhibitor; the purple line shows data obtained in the presence of a noncompetitive inhibitor, as illustrated in the cartoon. One of the gray lines is for data obtained in the presence of a competitive inhibitor (cf. Fig. 8.1) and the other gray line is for data obtained in the presence of an uncompetitive inhibitor (cf. Fig. 8.3).

cinylcholine is a competitive inhibitor, it can be overcome by higher than normal levels of acetylcholine. The inhibition is reversible, and the effects wear off when the succinylcholine has been degraded (it is slowly hydrolyzed by acetylcholinesterase) or diffused away from the synapse.

Other types of inhibitors may not be so tolerable. *Organophosphorus* compounds, used in *nerve gases* and *weed killers* (e.g., *parathion*), form a covalent irreversible bond with the active serine and permanently inactivate acetylcholinesterase. This is a type of ***suicide inhibition*** because the inhibitor reacts with the enzyme much like a substrate, but becomes blocked in the intermediate state where the enzyme–phosphoryl bond is stable, in contrast to the hydrolyzable enzyme–acetyl bond. These compounds are life-threatening.

A successful antidote to organophosphorus compounds must hydrolyze the enzyme–phosphoryl bond to release the acetylcholinesterase. The simplest compound that will do this is *hydroxylamine.* Unfortunately, hydroxylamine is toxic at the concentrations needed. However, derivatives can be prepared that work at much lower concentrations and can be used clinically. One such derivative is *pyridine aldoxime methiodide* (Fig. 8.7). In this derivative, the hydroxylamine group is fixed in position relative to the pyridine ring. The pyri-

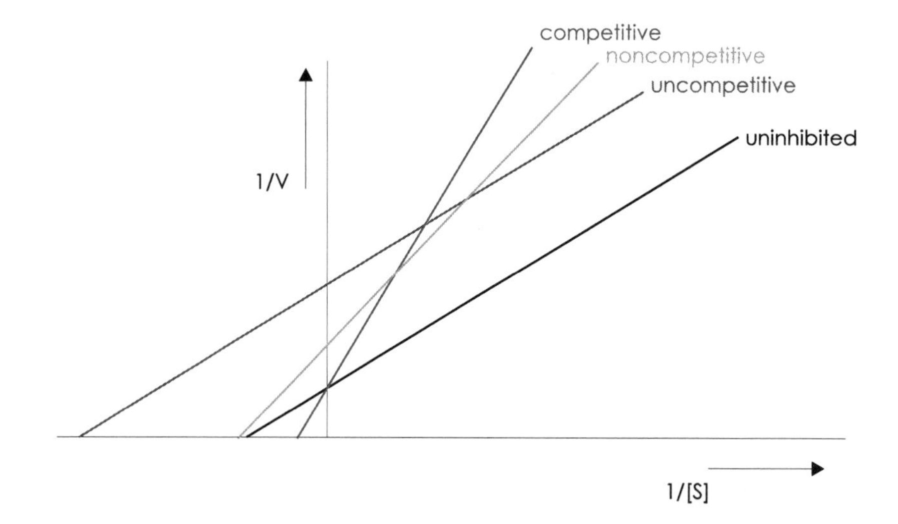

Figure 8.5.

Double reciprocal plots for an inhibited enzyme. The black line shows the double reciprocal plot for an enzyme in the absence of inhibitor. The other lines show the effect of an uncompetitive inhibitor (gray), a noncompetitive inhibitor (light purple), or a competitive inhibitor (purple).

dine ring binds to acetylcholinesterase and holds the hydroxylamine group in the right position to hydrolyze the enzyme–phosphoryl bond.

METABOLIC PATHWAYS

Many *inborn errors of metabolism* are due to the loss of a single enzyme activity in the body. To understand how this loss affects the body's function and learn how to treat these conditions, we need to see how individual enzymes fit into the overall picture of the body's metabolism using the concept of metabolic pathways.

A cell is rarely able to catalyze a desired chemical reaction in one step. There are various reasons for this, related to issues of energy conservation, control, and evolution. For example, the breakdown of glucose (glycolysis) takes place in many steps, including generating the largest possible number of ATP and other energy-rich molecules and giving the cell as much useful energy as possible from each glucose molecule. Thus, reactions occur through a sequence of steps, called a metabolic pathway. These pathways are controlled in complex ways in the body and understanding these control mechanisms is crucial for successful intervention in diseases involving metabolic processes or enzymes. Control refers both to the determination of which of the many possible reactions will actually be catalyzed (re-

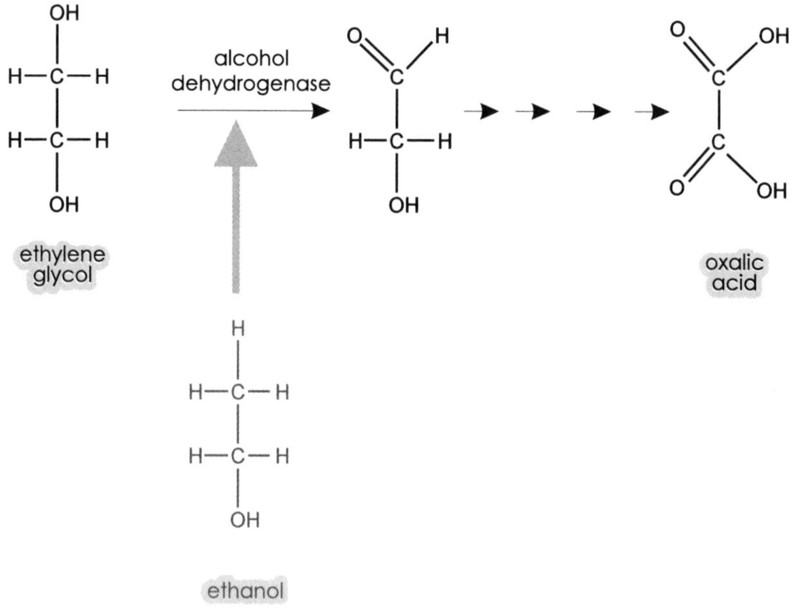

ethylene glycol
alcohol dehydrogenase
oxalic acid
ethanol

Figure 8.6.

Treatment of ethylene glycol poisoning. The metabolic pathway shown across the top of the figure illustrates the conversion of ethylene glycol to oxalic acid. This can be inhibited by the action of the competitive inhibitor, ethanol, shown in purple.

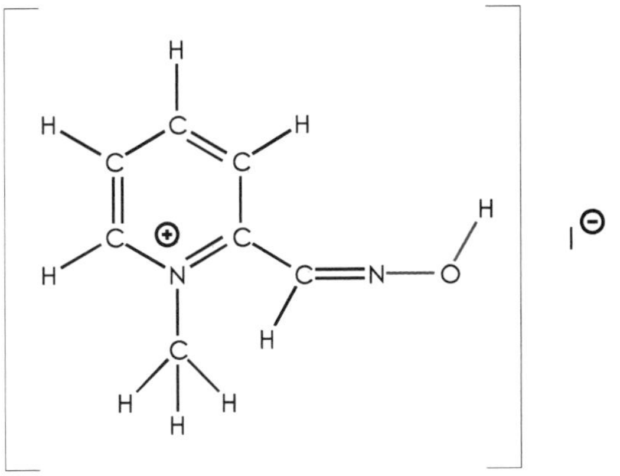

Figure 8.7.

Pyridine aldoxime methiodide. The active portion, shown with purple bonds, of this antidote is analogous to hydroxylamine.

action specificity) and the rate at which the reactions are catalyzed (reaction velocity).

Reaction specificity is limited by the enzymes available to the cell through encoding in the genes. Different cells can modify the spectrum of enzymes they use by controlling gene expression. Hence, some enzymes used by all cells (sometimes called "housekeeping" enzymes) are always available. Other, more specialized, enzymes are only available in the appropriate cells. Some of these latter enzymes can be synthesized or not in response to cellular needs; many are determined by the cell type and don't change much during the life of a fully *differentiated* cell.

However, a cell must respond to changing environments and changing requirements. This response is frequently achieved by changing the rate at which specific reactions are catalyzed, that is, by changing the activity of specific enzymes. Changes in enzyme activity are actually going on as a continuous process necessary to keep the body's metabolism in balance. *Diabetes* is an example of the problems that occur when the balance is lost.

The cell achieves its efficiency, stability, and responsiveness through the use of metabolic pathways. A *metabolic pathway* is a sequence of enzyme-catalyzed reactions in which the product of one reaction is the substrate for the next. Usually, a pathway ends with the production of a particular product, such as an amino acid needed for protein synthesis. A particular starting material, such as glucose, may be used for a variety of purposes. In this case, the metabolic pathways have an initial section that is common to each purpose; eventually, each pathway splits off into its own section. At this point, a key enzyme controls the flow of substrate into the specific section of the pathway. This enzyme is controlled allosterically, typically by a product of the pathway, in a negative feedback loop, as described below. This first step on a unique pathway is called the *committed step* or committed reaction.

How does the cell ensure that chemicals flow in the correct direction through the pathway, given that all enzyme reactions are theoretically reversible? Generally, the committed reaction is essentially irreversible under physiological conditions. Sometimes this is achieved by linking the reaction to the hydrolysis of ATP, a very favorable reaction.

Flow through the pathway can effectively be controlled by the committed reaction. The rate of each reaction after the committed reaction is determined by the concentration of substrate for that reaction. When the committed reaction speeds up, the substrate for the second reaction increases, increasing the rate of the second reaction, and so on down the pathway. Hence, the feedback mechanism only needs to work on the initial, committed step in the pathway.

Inhibition of TTP Synthesis in Cancer Chemotherapy—An Example of Metabolic Pathways

Cancer cells generally grow and proliferate faster than most cells in the body. Hence, one approach to treatment is to attack cells that are growing rapidly and slow them down. Proliferating cells need to replicate their DNA, while other cells do not. One compound that is required only for DNA synthesis is *thymidine triphosphate* (TTP). Two chemotherapy drugs work by inhibiting the synthesis of TTP.

To understand the process, we must look at the metabolic pathway that leads to TTP as the final product. The pathway is summarized in Figure 8.8, where rNDP represents any of the *ribonucleoside diphosphates*—ADP, CDP, GDP, or UDP

dNDP represents any of the *deoxyribonucleoside diphosphates*—dADP, dCDP, dGDP, or dUDP

dNTP represents any of the *deoxyribonucleoside triphosphates*—dATP, dCTP, dGTP, or dTTP

One drug, *5-fluorouracil,* is a precursor to *fluorodeoxyuridylate,* which is an inhibitor of *thymidylate synthetase*—it binds like *thymidylate* but the fluorine atom prevents addition of the methyl group.

The other drug, methotrexate, which is used to treat some forms of leukemia, is somewhat more indirect in its action. It is a competitive inhibitor of dihydrofolate reductase. This prevents the regeneration of *dihydrofolate* to *tetrahydrofolate,* thus blocking the

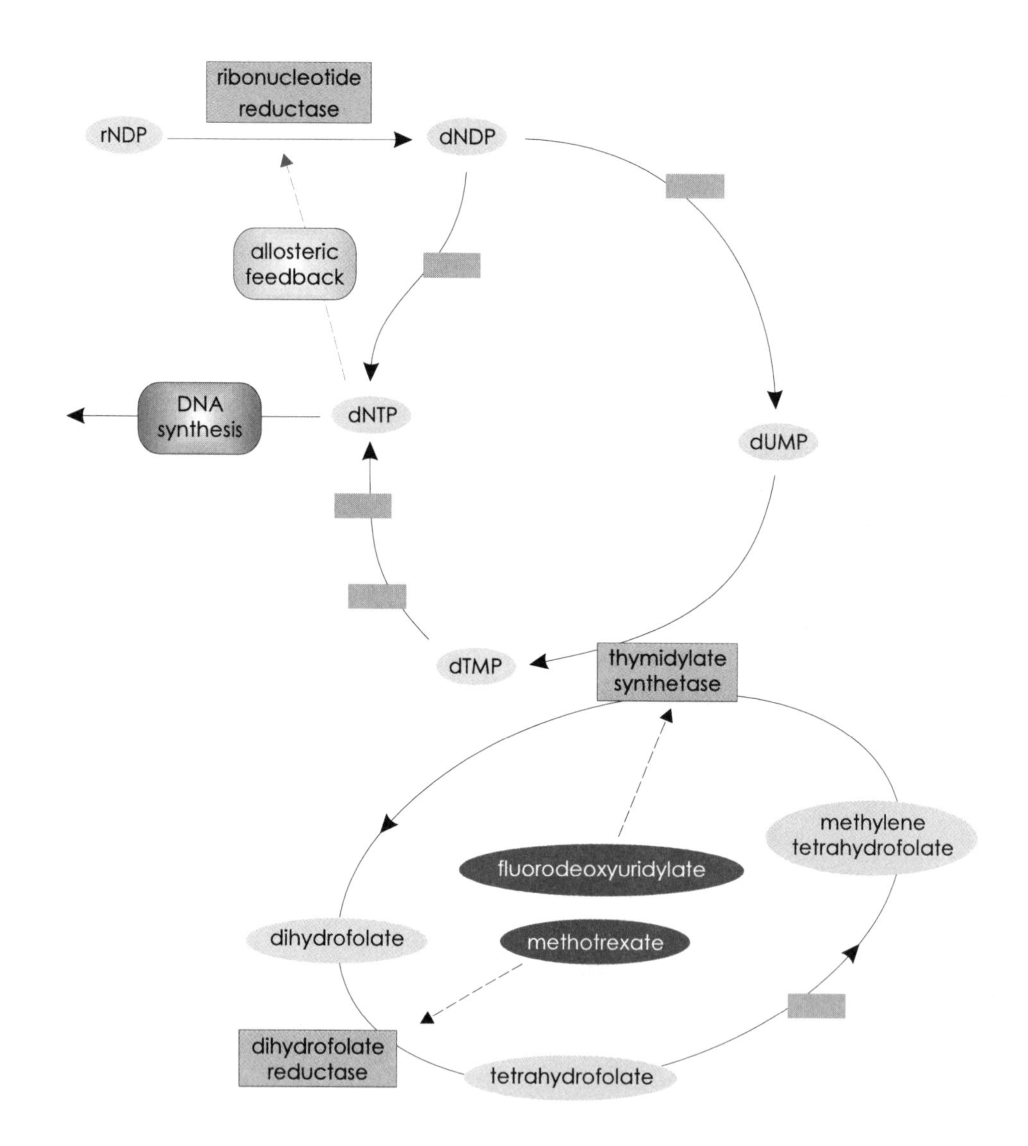

Figure 8.8.

TTP synthesis and chemotherapy. rNDP stands for the four ribonucleoside diphosphates—adenosine, cytosine, guanosine, and uridine diphosphate. dNDP stands for the corresponding four deoxyribonucleoside diphosphates. dNTP stands for the four deoxyribonucleoside triphosphates—deoxyadenosine, deoxycytosine, deoxyguanosine, and thymidine triphosphate. In the pathway, three of the dNDP are phosphorylated directly to form the corresponding dNTP. The fourth dNDP, deoxyuridine diphosphate, goes by a different route involving dUMP (deoxyuridylate or deoxyuridine monophosphate) and TMP (thymidylate or thymidine monophosphate) and is eventually converted to thymidine triphosphate, the fourth dNTP.

The gray rectangles indicate enzymes in the pathway. The light purple shaded ellipses represent substrates and products in the pathway. The dark purple shading indicates the two inhibitors. Allosteric feedback and the process of DNA synthesis have distinctive shading.

thymidylate synthetase reaction. Of course, both drugs affect normal cells, but at carefully controlled doses they are useful clinically until better chemotherapy agents are available. *Fluorouracil* is undergoing clinical trials with colon cancer patients, a very large group.

Allosteric Enzymes

Allosteric enzymes are used by the body to provide feedback control of metabolic pathways and maintain appropriate levels of product. The activity of an allosteric enzyme is modulated by binding substrate and/or nonsubstrate effectors. Allosteric means having an effect at a site different from the one immediately involved. This may be due to binding of a different molecule—an *allosteric effector*—at a nonsubstrate site.

For example, the rate of reaction of *ribonucleotide reductase* is regulated by *deoxyadenosine triphosphate* (dATP) which is a product of the pathway for which ribonucleotide reductase is the committing step (Fig. 8.9). Note that dATP is neither a substrate nor a product of ribonucleotide reductase itself; rather, it is an allosteric inhibitor. If the pathway (NDP→dNTP) is running at a rate too high for the rate at which dNTPs are being used (for DNA synthesis), the concentrations of the dNTPs will rise, including [dATP]. The increase will "feed back" to ribonucleotide reductase by the

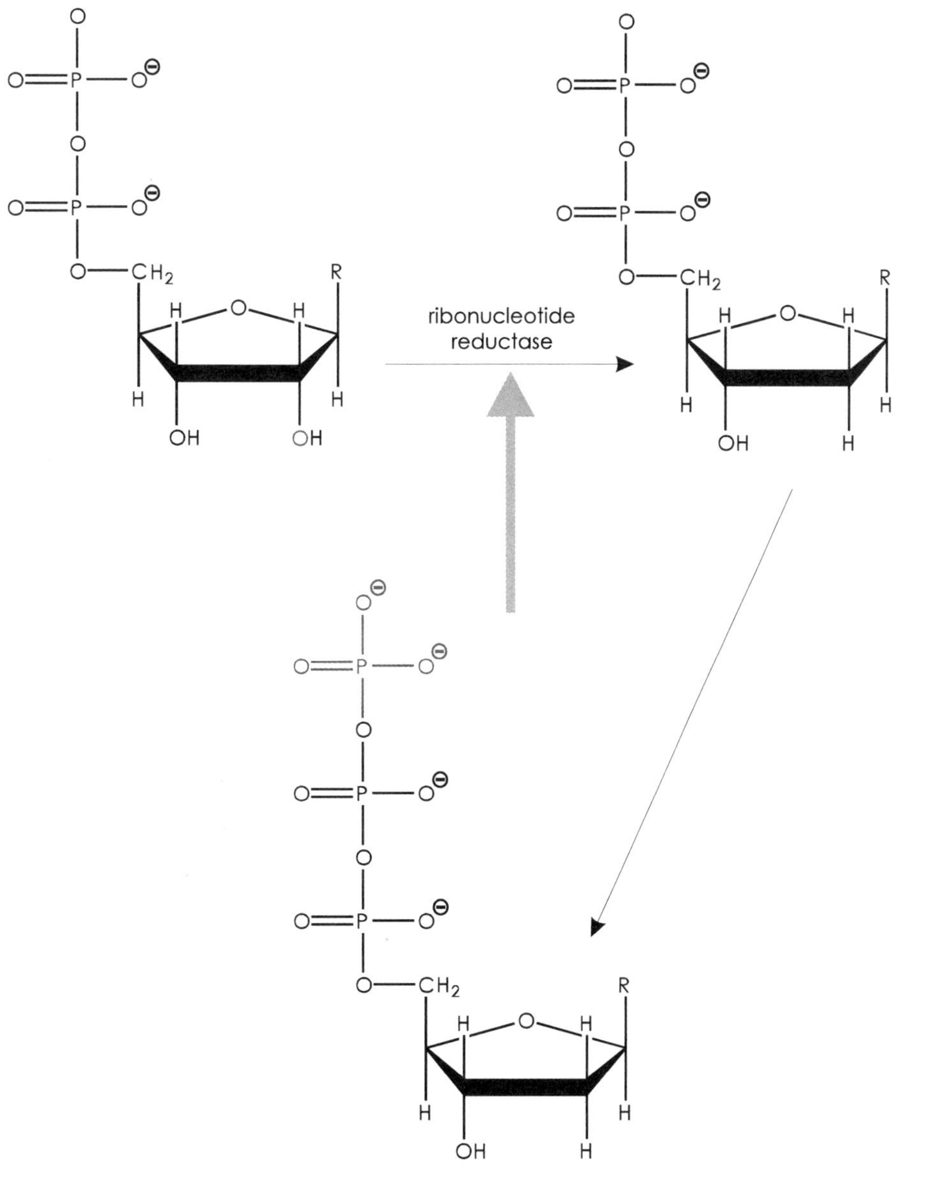

Figure 8.9.

Negative feedback. A ribonucleoside diphosphate, with 2′-OH shown in purple, is converted to the corresponding deoxynucleoside diphosphate in the reaction shown across the top. The deoxynucleoside diphosphate is then phosphorylated to the corresponding deoxynucleoside triphosphate with the γ-phosphate shown in purple. This is the final product of the pathway and it acts as a negative allosteric inhibitor of the initial enzyme, as shown by the large purple arrow.

binding of dATP to the allosteric site on ribonucleotide reductase, causing inhibition of ribonucleotide reductase and a slowing down of the pathway. Conversely, if the pathway is running too slowly, DNA synthesis will use up the existing dNTPs and their concentrations will fall. This will cause dATP to be released from ribonucleotide reductase, relieving the inhibition and allowing the reaction rate of ribonucleotide reductase to rise, increasing the production of dNTPs.

This type of allosteric regulation has several names: end-product inhibition, *feedback inhibition,* or retroinhibition. In this example, dATP is a negative modulator or negative effector. The reaction that is regulated directly, that is, the reaction catalyzed by ribonucleotide reductase, is the committing reaction or rate-limiting step, because all the following enzymes simply convert essentially all of their substrate to product as soon as substrate is available. Therefore, the rate of the initial reaction sets the rate for the whole pathway. This saves overproduction of dNTPs and of all the intermediate compounds through regulation of just one enzyme. Figure 8.10 shows a more generalized view.

A given allosteric enzyme may have both positive and negative modulators. Allosteric enzymes are generally large multi-subunit enzymes. The allosteric site(s) may be on the same subunit(s) as the catalytic site(s) or on different subunits. The number of allosteric sites may be equal to the number of catalytic sites or they may be different. Under physiological conditions, most allosteric enzymes do not show kinetics that fit the Michaelis–Menten equation, either in the presence or absence of effectors.

Allosteric enzymes show a *sigmoid curve* relating initial velocity to substrate concentration (Fig. 8.11). The sigmoid curve starts very slowly because the enzyme has low affinity for substrate. However, binding of one molecule of substrate increases the affinity of the enzyme for subsequent substrate molecules and the curve turns upward and the maximum velocity is soon reached. This is an example of ***positive cooperativity.***

There is some confusion in the literature about the use of the term "allosteric" for an enzyme. Many authors restrict this term to multi-subunit enzymes that show substrate cooperativity and *sigmoidal kinetics.* Other authors, however, are less specific and their definitions include enzymes that follow Michaelis– Menten kinetics and have non- or uncompetitive inhibitors. Fortunately, in metabolism, regulated enzymes are generally multi-subunit, cooperative, enzymes and fall into the more specific use of the term.

The $[S]_{0.5}$ for allosteric enzymes is defined as the substrate concentration [S] at which the reaction velocity is half its maximum (Fig. 8.12). The maximum velocity is just called V_{max}. The effect of positive cooperativity is to make the enzyme almost inactive at low substrate concentrations, but quickly activated when the substrate concentration gets into the region of $[S]_{0.5}$. This gives a much more sensitive response in this region, which is usually the physiological concentration range. For example, in a typical case, a noncooperative enzyme requires an 81-fold increase in [S] to bring the reaction rate from $0.1 \times V_{max}$ to $0.9 \times V_{max}$, whereas an enzyme showing positive cooperativity requires only a nine-fold increase in [S] for the same change in reaction rate.

A Hill plot, which shows log (V/V_{max}) as a function of log [S] (Chapter 6), is often used to characterize cooperative enzymes. The central portion of the Hill plot is a straight line whose slope is known as the Hill coefficient; it is 1 for noncooperative enzymes and greater than 1 for positive cooperativity. The Hill coefficient gives the minimum number of interacting sites on the enzyme.

Some positive regulators act by raising the affinity of all substrate binding sites (catalytic sites) to the maximum, independent of [S], thus destroying the cooperativity and raising the reaction rate at low [S]. An example of this behavior is the regulation of *phosphofructokinase* by *AMP (adenosine 5′-phosphate).*

Some negative regulators act in the opposite way. The free enzyme shows Michaelis–Menten kinetics and the regulator shifts it to positive cooperativity so that the $[S]_{0.5}$ is too high for much activity. An example of this behavior is given by *amidophosphoribosyltransferase.* This enzyme is the initial step in purine synthesis in which control by negative feedback occurs. The

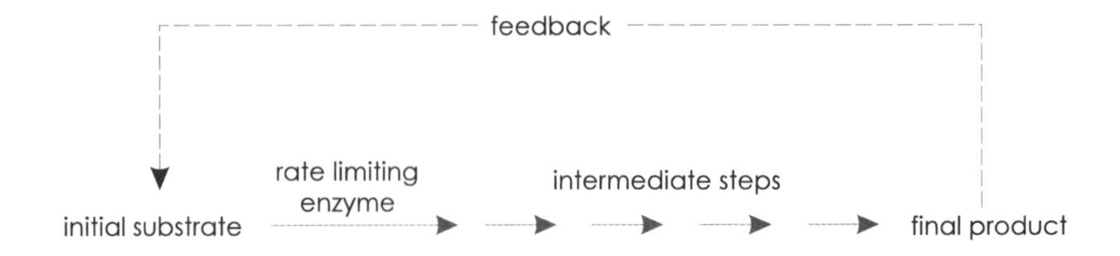

Figure 8.10.
Generalized feedback inhibition. Enzyme reactions in the metabolic pathway are shown as solid arrows. The allosteric effect of the final product on the reaction rate catalyzed by the initial enzyme is indicated by the broken line.

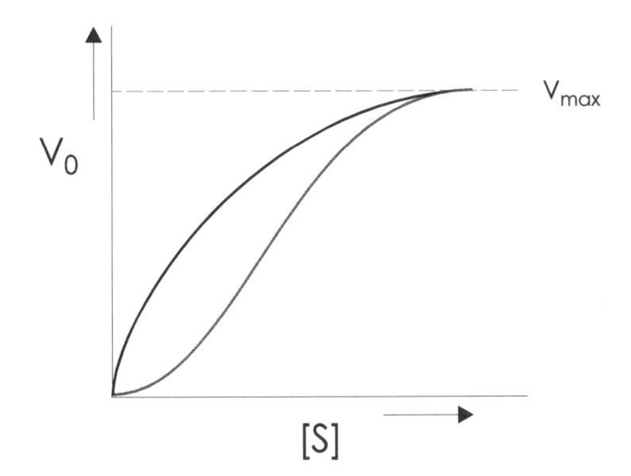

Figure 8.11.

Sigmoid rate curve. The purple line shows how the reaction rate, for a reaction catalyzed by a cooperative enzyme, varies with the substrate concentration. For comparison, the black line shows the shape of the hyperbolic curve seen with reactions catalyzed by enzymes that obey the Michaelis–Menten equation.

negative regulator is a purine nucleotide such as AMP or GMP. In the absence of regulator, purine synthesis is rapid and amidophosphoribosyltransferase works as a regular enzyme with a low $[S]_{0.5}$. When regulator concentration is high, this indicates purine synthesis is no longer needed and the regulator purine nucleotide binds to amidophosphoribosyltransferase and converts it to an enzyme with positive cooperativity, thereby pushing its $[S]_{0.5}$ well above the normal physiological range of substrate concentration.

Notice that AMP was the regulator in both cases. In one case it acted as a positive regulator, in the other, as a negative regulator. In the examples above, the regulator or effector changed the apparent $[S_{0.5}]$; in other cases, the apparent V_{max} or both the apparent V_{max} and the apparent $[S]_{0.5}$ may change.

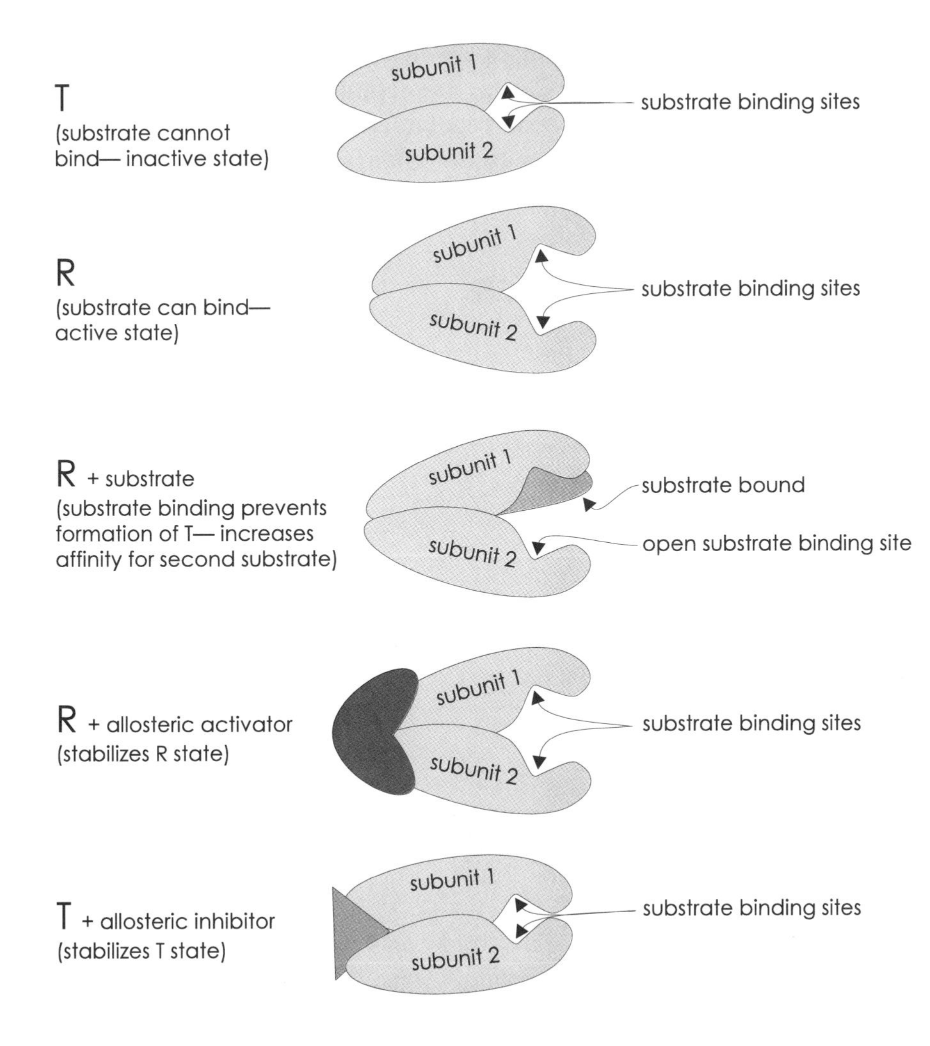

Figure 8.12.

Diagrammatic allosteric enzyme. An enzyme, comprised of two subunits, is shown in light purple. The substrate is shown in light gray; an allosteric activator is shown in dark purple and an allosteric inhibitor is shown in dark gray.

Allosteric effectors may modify the substrate specificity of an enzyme, as well as the reaction rate. Ribonucleotide reductase is a good example:

ribonucleoside diphosphate + NADPH + H^+
→ deoxyribonucleoside diphosphate
+ $NADP^+$ + H_2O

This reaction is the initial step in production of deoxynucleoside triphosphates (dNTPs) for DNA synthesis. Four dNTPs are needed for DNA synthesis—dATP, dCTP, dGTP, and TTP (TTP comes from dUDP). The proportions of these dNTPs need to be balanced for efficient synthesis. The feedback molecules, or effectors, are the final products of the pathway, the dNTPs, and they act on ribonucleotide reductase to modify its substrate specificity in order to balance the production of dNTPs. In addition to the specificity control site, an additional allosteric control site determines the overall rate of the reaction. At this site, ATP acts as a positive regulator and dATP as a negative regulator.

SUMMARY

1. The cell may regulate its activities by switching enzymes on or off, but more often it increases or decreases an activity, up-regulation or down-regulation, in a continuous manner.

2. Enzyme regulation may occur through a large number of processes; some are reversible and some are not. Many of these mechanisms operate on all enzymes. Some, however, operate with a class of enzymes that show cooperativity of substrate binding and thus do not follow the Michaelis–Menten equation.

3. Important reversible inhibitions occur by the binding of an inhibitor to the enzyme. If the inhibitor competes with the substrate for the active site, the inhibition is competitive and can be overcome by increasing the substrate concentration.

4. For example, ethylene glycol poisoning is treated with the competitive inhibitor of alcohol dehydrogenase—ethanol. Inhibitors of nerve-impulse transmission that act on the activation of receptors by acetylcholine esterase have a variety of uses in medicine and agriculture.

5. Many reactions in the body are broken down into steps, each of which is catalyzed by a specific enzyme. Regulation of such a metabolic pathway is achieved by feedback inhibition of the committing enzyme by the product of the pathway. Inhibition is by binding of the inhibitor to a site on the enzyme different from the active site. Typically, such enzymes obey cooperative substrate binding and the inhibition is referred to as allosteric inhibition.

6. Cooperative enzymes show sigmoidal curves of reaction rate as a function of substrate concentration. Allosteric regulators may change the V_{max} of such enzymes and/or change the cooperativity.

REFERENCES

Enzyme Kinetics, Irwin Segel, 1975, Wiley, New York.

Allosteric proteins and cellular control systems, J. Monod, J.-P. Changeux, and F. Jacob, 1963, *J. Mol. Biol.,* **6**:306–329.

Suicide enzyme inactivators, R. H. Abeles, 1983, *Chem. Eng. News,* **61**:48–56.

REVIEW QUESTIONS

1. In which way are cooperative enzymes similar to enzymes that do not display cooperativity? Enzymes
   a) Obey the equation: $V = V_{max} \cdot [S]/(K_M + [S])$
   b) Are affected by allosteric regulators
   c) Have a V_{max} that is unaffected by competitive inhibitors
   d) Only have one substrate
   e) May be found at the end of a metabolic pathway

2. Which of the following statements about Michaelis–Menten enzymes is false?

a) A graph of 1/initial velocity ($1/V_0$) as a function of 1/[substrate] is a straight line that intercepts the $1/V_0$ axis at $1/V_{max}$.
b) Only single-substrate enzymes obey the Michaelis–Menten equation.
c) The Michaelis–Menten equation is not appropriate for enzymes that exhibit cooperativity.
d) The K_M is independent of enzyme concentration.
e) The V_{max} depends on the enzyme concentration.

3. In any particular metabolic pathway in a steady state, the reaction carried out by the rate-limiting enzyme is slower than the nonrate-limiting enzyme reactions in that pathway.
   a) Always false
   b) Often false
   c) Equally likely to be true or false
   d) Often true
   e) Always true

ANSWERS TO REVIEW QUESTIONS

1. ***c*** For any enzyme, a competitive inhibitor is an inhibitor that competes with the substrate for the active site of the enzyme. Thus, at high enough substrate concentration, the effect of the competitive inhibitor will disappear and V_{max} will not be affected.

2. ***b*** Two substrate enzymes are common in the body. The Michaelis–Menten equation can be applied by holding one substrate concentration constant and using appropriate kinetic constants for the other substrate. Hence, ***b*** is false.

3. ***a*** Note carefully the language in the question: "... in a steady state ... is slower than ...." If a pathway is in steady state, all reactions must be occurring at the same rate. Hence, ***a*** is the correct answer.

CHAPTER

9

REGULATION OF ENZYME ACTIVITY BY COVALENT CHANGES

INTRODUCTION

The last chapter discussed noncovalent interactions that regulate enzyme activity. The body also uses covalent changes for enzyme regulation. In some cases, the changes are reversible, as in the case of protein phosphorylation; in other cases, the changes are irreversible, as in the use of proteolysis as an activation mechanism.

ZYMOGENS

Zymogens in Digestion

A *zymogen* is an inactive protein that is converted to an active protein by specific *proteolysis* (hydrolysis of peptide bonds). This mechanism is used when enzymes must be stored in an inactive form ready for rapid activation, as in blood clotting, or when enzymes must be transported in an inactive state from their site of synthesis to their site of action. The use of zymogens by the *digestive system* is an excellent example of the latter situation. Digestion involves breakdown of the proteins, and other molecules, in food without damaging the host tissues (Fig. 9.1).

Proteins in food are initially broken down into smaller polypeptides by *endopeptidases* that hydrolyze peptide bonds *within* the chain. Digestive enzymes have specificity or preference for particular side chains adjacent to the peptide to be cleaved, but there is sufficient variety to break the great majority of proteins into small polypeptides. The endopeptidases include the *pepsins, trypsin, chymotrypsin,* and *elastase.*

Pepsins are secreted by the *gastric mucosa.* They catalyze hydrolysis of internal peptide bonds under the acidic conditions found in the stomach.

Trypsin, chymotrypsin, and elastase are all serine proteases, which means that they use a serine residue as a key part of their reaction mechanism. They are made in the pancreas and then exported, or secreted, into the *pancreatic duct* leading to the duodenum. All serine proteases have the same catalytic mechanism but minor changes in the active site change the substrate specificity. Chymotrypsin has a deep hydrophobic pocket that fits an aromatic or large hydrophobic side chain from the substrate; trypsin has an aspartate residue in the bottom of the pocket, replacing serine. The charged aspartate encourages Lys and Arg sidechains and rejects others, so trypsin only cuts at Lys or Arg. Elastase has a small hydrophobic pocket because the chymotrypsin-like pocket is filled with Val and Thr, replacing the two Gly residues in trypsin and chymotrypsin.

Pancreatic carboxypeptidases are also secreted by the pancreas and act in the small bowel as *exopeptidases,* which trim amino acids one at a time from an end of a polypeptide. Carboxypeptidase A has a pref-

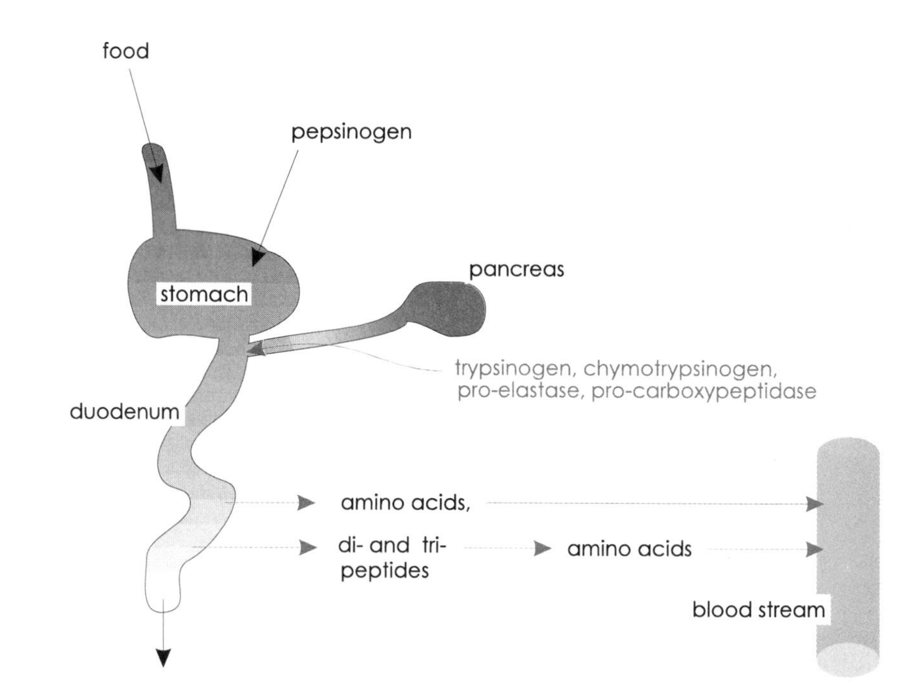

Figure 9.1.

Digestion of proteins. The stomach and intestines are shown diagrammatically, with the pancreas, shown in dark purple, connected by the pancreatic duct. Amino acids, peptides, and other nutrients are absorbed through the intestinal wall, as indicated by the straight purple arrows, and then passed into the bloodstream.

erence for carboxy-terminal neutral or aromatic residues; other carboxypeptidases prefer arginine or lysine or other terminal amino acids. These exopeptidases work well with short peptides and continue the digestion where the endopeptidases stop. Single amino acids, dipeptides, and tripeptides are absorbed into the *intestinal mucosa.*

The intestinal mucosa contain *aminopeptidases* that complete the digestion of di- and tripeptides to amino acids.

The pancreas is as susceptible to destruction by proteolysis as any other cell and damage that leads to release of active digestive enzymes in the pancreas produces acute *pancreatitis,* which is often fatal. Normally, the pancreas protects itself by three mechanisms:

1. The proteases are synthesized as inactive zymogens
2. The zymogens are packed into *lipoprotein granules*
3. A specific *inhibitor of trypsin* is synthesized by the pancreas

A zymogen (or a *proenzyme*) is longer than the active enzyme and the additional amino acids either block access to the active site or hold the protein in an inactive conformation. Activation is achieved by hydrolysis of one or more peptide bonds in the zymogen.

Pepsinogen is the zymogen of pepsin. Pepsinogen has 41 "extra" amino acids which make it completely inactive at neutral pH. At low pH, however, in the stomach, it has weak activity, which allows it to cleave another pepsinogen to produce pepsin. Pepsin cleaves pepsinogen to produce pepsin. Even if pepsinogen is split at neutral pH, some of the "extra" amino acids remain bound as an inhibitor until the pH is almost as low as pH 2. The inhibitory peptide prevents pepsinogen attacking the gastric mucosa but, as soon as pepsinogen reaches the acidic gastric juices in the stomach, the pepsin and low pH there rapidly produce full activity. (A further measure of protection is provided by the low pH optimum of pepsin itself.)

A peptide or part of a protein may inhibit an enzyme by binding to its active site. In the case of proteases, the amino-acid sequence of the inhibitor often resembles that of a substrate. Such an inhibitory sequence is called a *pseudo-substrate* sequence.

The pancreatic peptidases are also secreted as zymogens (*trypsinogen, chymotrypsinogen, proelastase,* and *procarboxypeptidase*). They are synthesized in cells in the pancreas, where they are packaged into granules with a lipid-protein "coat." The granules accumulate at the apex of the cell and are secreted into a duct leading to the intestine.

The stability of the zymogens is rather fragile; a very small amount of active trypsin can activate trypsinogen to create more trypsin and generate an *autocatalytic* reaction. Trypsin also activates the other zymogens. The normal pancreas protects itself from this catastrophe by a *trypsin inhibitor,* a small polypeptide that effectively inhibits trypsin activity. If more trypsin is released in the pancreas than the inhibitor can cope with, symptoms of pancreatitis occur. Pancreatitis is correlated with excessive consumption of ethanol, although it is not known why.

The normal activation of the zymogens takes place in the intestine after leaving the pancreatic duct. The intestinal mucosa secrete a highly specific protease called *enteropeptidase* (previously called *enterokinase*) which removes a particular sequence of amino acids from the amino terminus of trypsinogen. Trypsinogen begins with one to three nonpolar residues followed by -Asp-Asp-Asp-Asp-Lys-Ile-. The four negative charges on the aspartates hold trypsinogen in an inactive conformation. Enteropeptidase splits off the amino terminus up to and including the Lys in the sequence above. The highly charged peptide falls off and trypsin assumes its active conformation.

Similarly, chymotrypsinogen is activated by cleavage by trypsin after Arg-15. The active chymotrypsin (called π-chymotrypsin) can further digest other π-chymotrypsin molecules to release two small peptides and give α-chymotrypsin which is also active.

Proelastase and *procarboxypeptidase* are activated similarly.

Zymogens in Blood Clotting

Blood clotting is a situation in which the enzymes needed to form a clot must be present in the blood and ready to act rapidly when *trauma* to a blood vessel occurs. However, clots must not form in normal blood vessels or a *heart attack* or *stroke* will occur. The blood-clotting system is a complex and highly regulated mechanism which uses many steps of zymogen activation, with more examples of serine proteases.

Platelets

Platelets are cytoplasmic fragments of cells, *megakaryocytes,* found in *bone marrow.* Platelets circulate in the blood stream. They contain mitochondria and cytoplasmic enzymes and form a substantial part of the blood. Platelets adhere to each other and to foreign surfaces and this property is referred to as *platelet adhesion* or aggregation or similar terms. Platelet aggregation is the initial stage of formation of a blood clot and is accompanied by release of phospholipids and specific *clotting factors* that trigger *blood coagulation.*

Coagulation Pathway

A blood clot is formed mainly from a network of cross-linked fibrin molecules that traps platelets, erythrocytes, and other materials to form a solid clot. The aggregation and cross-linking of fibrin is the final stage of a *proteolytic cascade* or pathway which is triggered by one or both of two mechanisms: the *intrinsic pathway* and the *extrinsic pathway.*

The components of the intrinsic system are all present in the plasma or platelets. They are stimulated by contact with a foreign surface, especially *collagen.* The components of the extrinsic system require additional components for activation. In particular, a factor called *tissue factor* or *tissue thromboplastin* is released after damage to the walls of blood vessels and triggers the extrinsic system.

The two pathways converge and then lead to the aggregation and cross-linking of fibrin (Fig. 9.2). The pathways are called proteolytic cascades because the factors exist in the blood as inactive zymogens. Each factor becomes activated by the previous factor to become a specific proteolytic enzyme, which then activates the next factor by cleaving off the inhibitory part of the zymogen and exposing the active protease. Such a cascade allows a very small initial signal to be amplified to produce the macroscopic result seen in the blood clot.

Intrinsic Pathway

The intrinsic pathway begins with the activation of *Factor XII.* (The factors are identified by Roman numerals, with a subscript "a" for the active form.) Factor XII is activated by other proteins, including *kininogen* and *kallikrein.* Factor XII_a then activates *Factor XI* which activates *Factor IX.* Factor IX_a, together with another factor, Factor VIII_a, then activates *Factor X,* where the pathways converge (Fig. 9.3).

Recombinant DNA techniques have enabled the cloning of *Factor VIII,* which is missing from most patients with the inherited bleeding disorder *hemophilia.* Factor VIII is essential for proper blood clotting, and untreated hemophilia patients suffer from severe uncontrolled bleeding. Recombinant Factor VIII is used to restore clotting activity to the blood of hemophiliacs.

Extrinsic Pathway

In the extrinsic pathway, tissue factor activates *Factor VII* to produce Factor VII_a, which then activates Factor

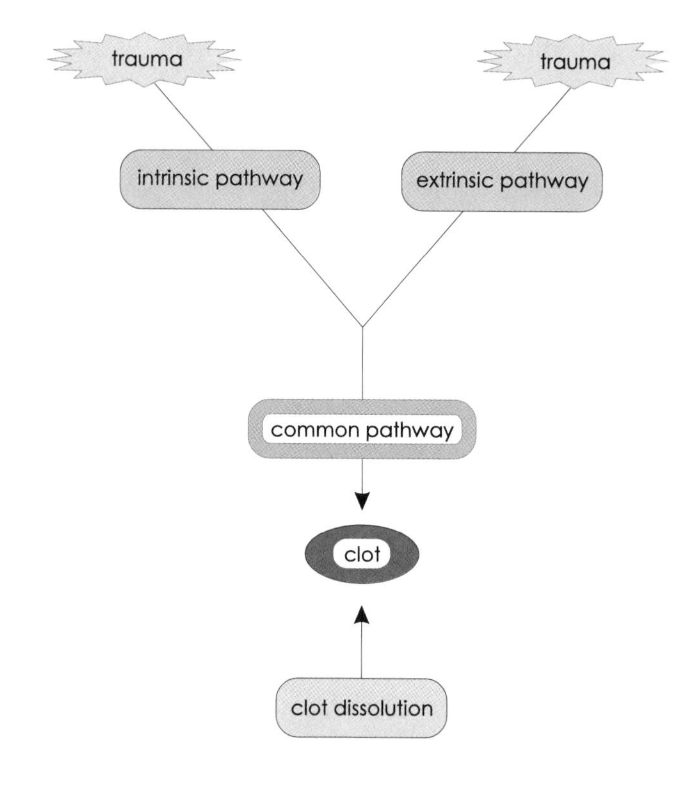

Figure 9.2.
Pathways of blood clotting—overall view.

X. In addition to the convergence of the pathways at the activation of Factor X, there are interactions between the pathways, because Factor VII_a can activate Factor IX and kallikrein can activate Factor VII (Fig. 9.4).

Common Pathway

The stages of the common pathway are better understood than the earlier stages because the concentrations of the factors are much higher and they are therefore easier to study. The first stage is activation of Factor X (Fig. 9.5).

Factor X

Factor X_a is a serine protease, like kallikrein, *thrombin,* and several other factors. A serine protease is a member of a family of proteases with a common catalytic mech-

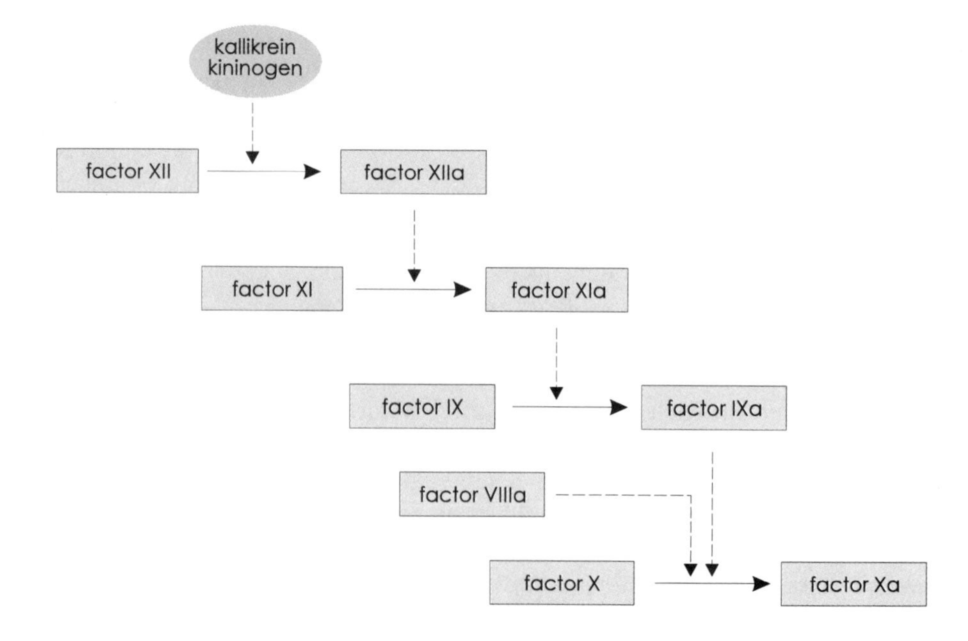

Figure 9.3.
Intrinsic pathway. The solid arrows indicate the enzymatic reaction in which part of the inactive protein factor, or zymogen, on the left is cleaved, leaving the active factor. The broken arrow indicates the enzyme, or active factor, that catalyzes the reaction.

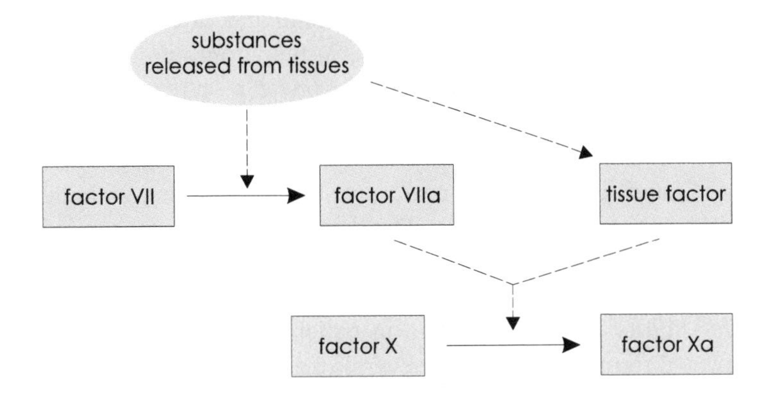

Figure 9.4.
Extrinsic pathway. The solid arrows indicate the enzymatic reaction in which part of the inactive protein factor or zymogen, on the left is cleaved, leaving the active factor. The broken arrow indicates the enzyme, or active factor, that catalyzes the reaction.

anism but varying specificity for the amino-acid sequence that is cleaved. Other examples of serine proteases include trypsin, chymotrypsin, and elastase. The common catalytic mechanism includes a serine in the enzyme active site. The –OH of this serine forms a covalent intermediate with the broken polypeptide chain.

Factor X_a binds to the phospholipid membranes released by *platelet aggregation.* Factor X_a cleaves the zymogen, *prothrombin,* at two specific sites to produce the active enzyme, thrombin.

Prothrombin and Thrombin

Prothrombin has two domains. The carboxy terminal domain is the catalytic domain, which is released by the

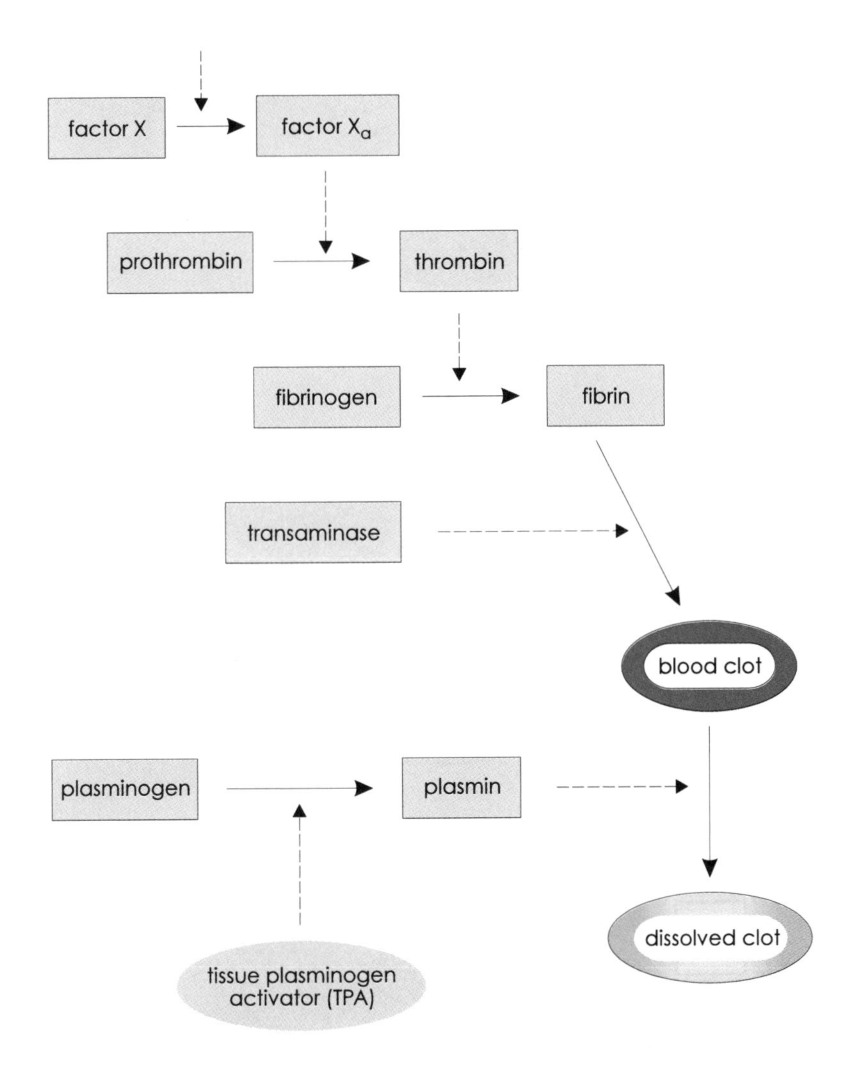

Figure 9.5.
Common pathway. The solid arrows indicate the enzymatic reaction in which part of the inactive protein factor or zymogen, on the left is cleaved, leaving the active factor. The broken arrow indicates the enzyme, or active factor, that catalyzes the reaction.

Both the formation of a blood clot and its dissolution are shown.

action of Factor X_a. The amino terminal domain has a dual function. First, it acts to inhibit the protease activity of the catalytic domain; second, it acts to bind prothrombin to the same phospholipid surfaces that contain Factor X_a. The resulting proximity of Factor X_a and prothrombin is essential for the efficient action of Factor X_a (Fig. 9.6). The binding of prothrombin to phospholipid surfaces is mediated through the binding of Ca^{2+}, which binds strongly to the phospholipid and to unusual residues in prothrombin, *γ-carboxyglutamates* (Fig. 9.7). Without γ-carboxyglutamate, prothrombin binds poorly to phospholipid and hence is not activated efficiently by Factor X_a. A similar effect is obtained by removing (*chelating*) Ca^{2+} from the blood with Ca^{2+}-binding compounds, such as *citrate* or *EDTA*; these reagents are used to inhibit blood clotting in samples drawn from patients.

Gamma-carboxyglutamate is synthesized post-translationally, that is, prothrombin is synthesized with glutamate in the amino-acid chain and, after synthesis, specific glutamate residues, in the amino-terminal domain, are modified by a *carboxylation* reaction that adds a carboxyl group. The carboxylation reaction requires *vitamin K* and hence it is sensitive to antagonists, such as *dicoumarol* and *warfarin,* that interfere with the action of vitamin K. Dicoumarol is found in spoiled sweet clover; warfarin is used as a rat poison. Similar antagonists may be used clinically to reduce the probability of blood clotting in patients with a tendency to thrombus formation. Several other clotting factors also contain γ-carboxyglutamate and they, too, will be affected by vitamin K antagonists.

The binding of prothrombin to phospholipids can be inhibited by chelating Ca^{2+} by EDTA (*ethylenediaminetetraacetate*), which binds both Mg^{2+} and Ca^{2+} and inhibits several stages of blood clotting; it can also be used to inhibit clotting in blood samples.

Thrombin can be inhibited by *heparin,* a complex

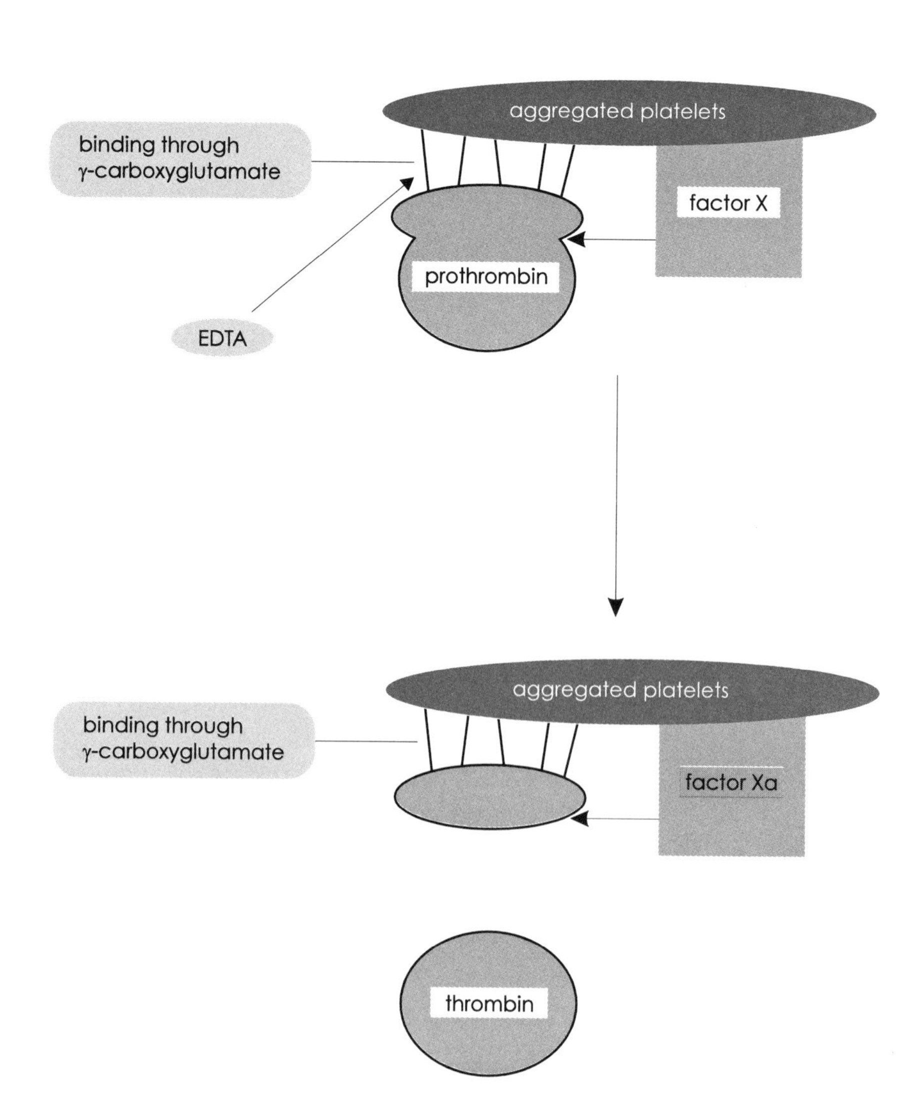

Figure 9.6.

Activation of prothrombin. The upper section shows the situation immediately before activation of the common pathway of blood clotting. Prothrombin is attached to platelets in an electrostatic linkage that can be broken by EDTA. The lower section shows the activation of the common pathway of blood clotting. Factor X has been activated to become Factor X_a, which immediately acts on adjacent prothrombin molecules and causes the release of thrombin into the plasma.

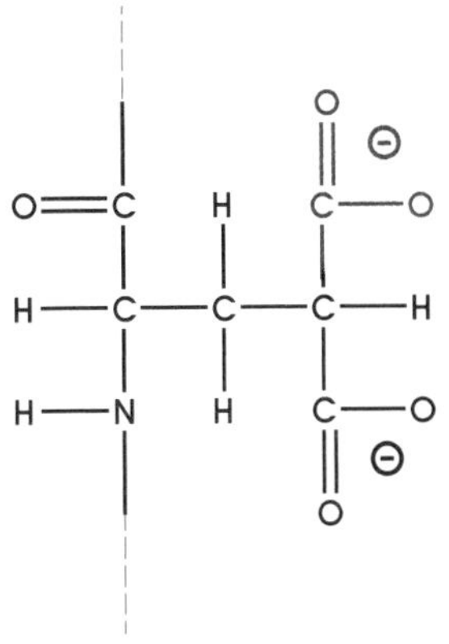

Figure 9.7.

γ-Carboxyglutamate. An extra carboxyl group, shown in purple, is added to glutamate residues in proteins. Part of the polypeptide backbone is shown on the left of the structure.

polysaccharide found in blood-vessel walls and other tissues. Heparin acts by enhancing the interaction of thrombin with *antithrombin III.* Antithrombin III is a protein found in the plasma and it binds to thrombin and prevents its action on other proteins. The balance of thrombin and antithrombin III is critical to proper operation of the coagulation system. Heparin is useful clinically for preventing blood coagulation.

Thrombin cleaves fibrinogen at several -Arg-Gly- sites and converts fibrinogen to fibrin. Cleavage of fibrinogen yields fibrin and *fibrinopeptides.* The function of the fibrinopeptides in fibrin is to keep fibrinogen molecules apart and thus prevent them from aggregating. This is achieved by the electrostatic repulsion of negative charges on glutamate, aspartate, and *tyrosine sulfate* residues on the fibrinopeptides. Removal of the fibrinopeptides allows the remaining fibrin to aggregate rapidly. Tyrosine sulfate is an unusual posttranslational modification of tyrosine (Fig. 9.8).

Aggregated fibrin molecules form a loose clot. It is strengthened into a hard clot by the action of a *transamidase* (*Factor* XIII_a), which forms an unusual covalent cross-link between strands of fibrin. The Lys and Gln residues are linked with the release of ammonia (Fig. 9.9).

Plasminogen and Plasmin

Blood clots are eventually dissolved. The mechanism involves the digestion of cross-linked fibrin by *plasmin,* another serine protease normally present as its zymogen, *plasminogen.* Plasminogen is activated by *tissue-type plasminogen activator* (*TPA*). TPA acts efficiently only on plasminogen adhering to blood clots. Hence, it does not affect plasminogen free in the blood stream.

TPA has recently become available for clinical use as *activase* owing to its production in mammalian cell cultures using recombinant DNA techniques. Activase is used clinically to dissolve blood clots.

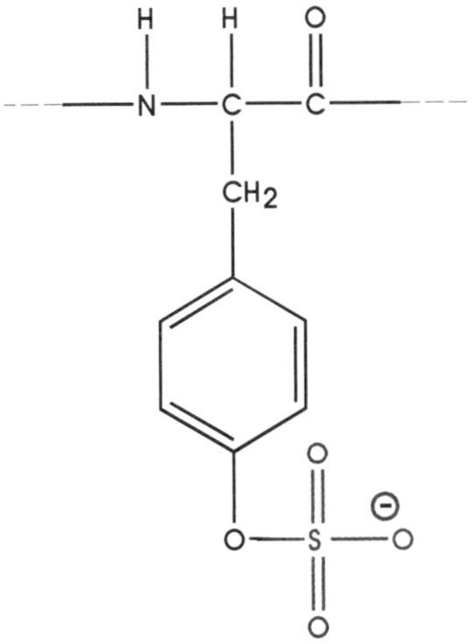

Figure 9.8.

Tyrosine sulfate. A sulfate group, shown in purple, is added to tyrosine residues in proteins. Part of the polypeptide backbone is shown at the top of the structure.

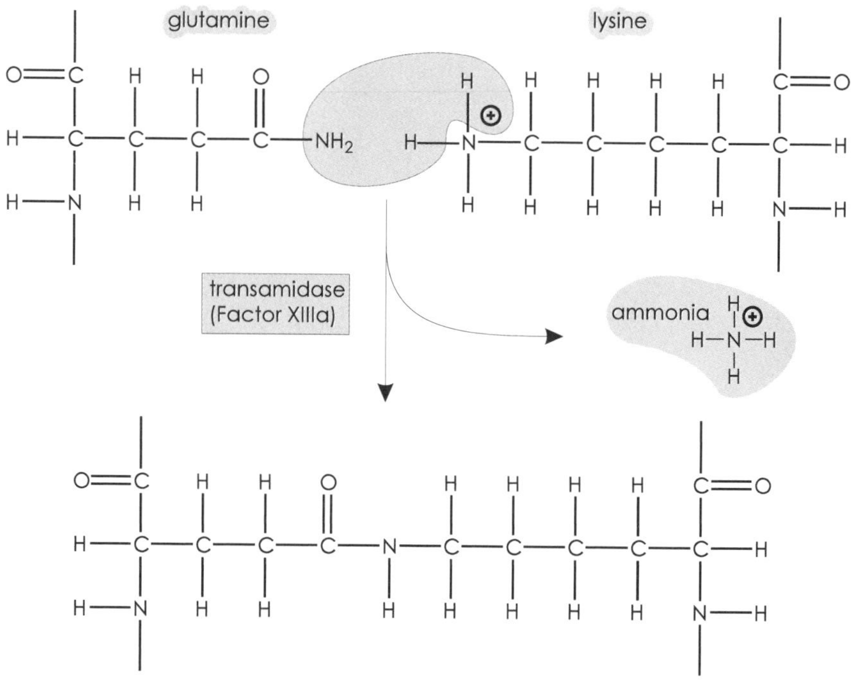

Figure 9.9.
Fibrin cross-links. The structures at the top are parts of two separate fibrin molecules. The polypeptide backbones are shown on the far left and the far right. The fibrin molecule on the left contains a glutamine side chain; the molecule on the right contains a lysine side chain. The structure at the bottom shows the same two fibrin molecules but now the two side chains have been joined by an isopeptide bond, making a covalent cross-link. The ammonia released in the reaction is shown in purple. The reaction is catalyzed by transamidase.

REVERSIBLE COVALENT MODIFICATION OF ENZYMES

We have seen how allosteric control can provide feedback regulation to maintain order in the body and respond to its needs. Allosteric control is reversible and the analogous type of regulation by covalent modification is the reversible modification of amino-acid side chains. Some enzymes are affected directly by both types of control. By far the most common modification of proteins is the phosphorylation of serine and threonine residues, and *tyrosine phosphorylation* is a key part of many control mechanisms. Protein phosphorylation is discussed below. Many other types of protein modification occur—some reversible and some irreversible.

The reversible covalent modification of enzymes is important in control of metabolism, cell growth and division, response to hormones, and other processes. Examples of the types of reversible side-chain modifications found in cells include:

1. Phosphorylation of amino-acid side chains, serine, threonine, histidine, and tyrosine
2. *ADP-ribosylation* (*adenosine diphosphoribosylation*) of amino-acid side chains
3. Acetylation of the amino group of a lysine residue

This list is not complete. In addition, many side-chain and *N*- or *C*-terminal modifications are irreversible.

In the cases of reversible modification, the protein exists in a minimum of two forms designated *o*- (for original, unmodified) and *m*- (modified). Two additional enzymes actually carry out the modification, one to add the modification—such as a *protein kinase*—and one to remove it—such as a *protein phosphatase.* The use of different enzymes to add or remove the modification provides a mechanism for controlling the degree of modification.

Generally, modification affects the activity of the enzyme either directly or by changing its response to allosteric effectors. Frequently, the active form of the enzyme is designated "a" and the inactive form "b."

Amplification by Covalent Modification

The main differences between direct allosteric control of an enzyme (allosteric regulator interacts directly with the enzyme to be regulated) and control through covalent modification are:

1. Covalent modification systems can amplify the effect of a regulator both by catalytic amplification and by cascade amplification.

2. Covalent modification systems have a greater capacity for biological integration.

Both systems allow for the controlled enzyme to be in a steady state with some molecules active and some inactive and for this steady state to be changed by changing levels of metabolites or by external stimuli. In rare cases, enzymes may be totally inactive (switched off) or totally active (switched on); normally, regulation controls the steady-state balance between active and inactive forms to give an overall change in activity somewhere between fully active and fully inactive.

Catalytic amplification occurs when one enzyme modifies a large number, say *N*, of the molecules being controlled. Hence, one molecule of regulator enzyme affects *N* molecules of the enzyme being controlled, an amplification factor of *N*. For example, *MAP kinase* activates the protein *ribosomal S6 kinase* (*Rsk*); each molecule of MAP kinase can phosphorylate many molecules of Rsk, giving catalytic amplification (Fig. 31.13).

Cascade amplification occurs when more than one covalent modification occurs in a sequence. For example, *Mek* activates MAP kinase by phosphorylation. One molecule of Mek activates many molecules of MAP kinase, each of which can activate many molecules of Rsk. If the amplification by Mek were *N*-fold and the amplification by MAP kinase were *M*-fold, the overall, cascade, amplification is $N \times M$-fold. Thus, if $N = M = 100$, the overall amplification is 10,000 times. Cascades can easily be extended so that a five-stage cascade with 100-fold amplification at each stage would have a potential overall amplification of 100^5 or 10 billion-fold.

If any one of the enzymes in a cascade becomes modified, such as through mutation of its gene, so that it is constitutively active (i.e., doesn't depend on activation by the previous enzyme in the cascade), the cascade will run out of control. This occurs in cancer, where a constitutively active component that leads to cancer is an example of an *oncogene* (Chapter 30).

Biological integration occurs when many different signals, such as external stimuli or levels of metabolites, all have an effect on enzyme activity and the system of regulation responds to the overall, integrated result of all the signals.

The body tends to use allosteric regulation for rapid intracellular control mechanisms that work on a time scale of a few seconds or less. Covalent modification tends to be used for both intra- and extracellular mechanisms and, even within cells, is often part of the signaling stimulated by extracellular signals. Although some systems based on covalent modifications respond very rapidly, others respond more slowly, with effects being seen in a matter of minutes rather than seconds.

SUMMARY

1. Covalent change in enzymes is an important regulatory mechanism complementing allosteric regulation, which is noncovalent.

2. Covalent changes may be irreversible—typically the activation of an enzyme by removal of an *N*-terminal sequence that holds the enzyme in an inactive state. The inactive form of such an enzyme is a zymogen.

3. Zymogens are critical to the production of digestive enzymes in the pancreas. Production of zymogens protects the normal pancreas from autodigestion. The zymogens are activated when they leave the pancreatic duct and are needed to digest proteins and other food molecules.

4. Zymogen activations may be linked to form a cascade, such as that involved in blood clotting.

5. A cascade multiplies the effects of linked catalytic amplifications to give very rapid massive amplification of small initial events.

6. Blood clotting begins with the aggregation of platelets and the activation of two pathways—intrinsic and extrinsic—which converge onto the common pathway. The common pathway begins with the activation of Factor X, a zymogen, to produce Factor X_a. Factor X_a, in turn, activates another zymogen, prothrombin, to produce thrombin. Thrombin then cleaves fibrinogen to produce fibrin, which forms the clot.

7. Blood clots are digested by plasmin, which is activated from its zymogen form, plasminogen, by tissue plasminogen activator, now available as a clot-dissolving drug.

8. Covalent changes may be reversible, the most important class being protein phosphorylation, which occurs on several side chains including serine, threonine, and tyrosine.

9. Protein phosphorylation cascades are involved in multiple intracellular signaling pathways.

REFERENCES

Introduction to the blood coagulation cascade and the cloning of blood coagulation factors, E. W. Davie, 1986, *J. Prot. Chem.*, **5**:247–253.

Growth factors and cancer, S. A. Aaronson, 1991, *Science*, **254**:1146–1153.

The role of protein phosphorylation in neural and hormonal control of cellular activity, P. Cohen, 1982, *Nature*, **296**:613–620.

Molecular and Cellular Basis of Digestion, P. Desnuell, H. Sjorstrom, and O. Noren, (Eds.), 1986, Elsevier, New York.

PH domains: diverse sequences with a common fold recruit signaling molecules to the cell surface, M. A. Lemmon, K. M. Ferguson and J. Schlessinger, 1996, *Cell,* **85**:621624.

Active and inactive protein kinases: structural basis for regulation, L. N. Johnson, M. E. Noble and D. J. Owen, 1996, *Cell,* **85**: 149–158.

REVIEW QUESTIONS

1. What is the effect of warfarin (a vitamin K antagonist)?
   a) Inhibits posttranslational modification of prothrombin
   b) Rapidly dissolves blood clots
   c) Only affects the intrinsic clotting pathway
   d) Inhibits the formation of tyrosine sulfate
   e) Promotes the cleavage of prothrombin by factor X

2. Which zymogen is not prematurely activated in pancreatitis?
   a) Trypsinogen
   b) Chymotrypsinogen
   c) Pepsinogen
   d) Procarboxypeptidase
   e) Proelastase

3. What is the main reason why activation of MAP kinase by the Raf–Mek pathway is known as a cascade?
   a) MAP kinase is activated by phosphorylation.
   b) The pathway is stimulated by Ras.
   c) Growth factor receptors with intrinsic tyrosine kinase activity trigger the activation of MAP kinase.
   d) There are two successive phosphorylation activation steps in the Raf–Mek–MAP kinase activation pathway.
   e) Mek is a protein kinase.

ANSWERS TO REVIEW QUESTIONS

1. ***a*** Vitamin K is required for the formation of γ-carboxyglutamate, a posttranslational modification that occurs on a number of proteins, including prothrombin. Hence, ***a*** is the correct answer.

2. ***c*** Pepsinogen is the zymogen form of pepsin. Pepsinogen is synthesized in the wall of the stomach and activated inside the stomach. This occurs quite separately from the pancreatic enzymes that are synthesized in the pancreas and exported to the digestive tract below the stomach. There is no effect on pepsinogen activation in pancreatitis. All the other zymogens mentioned in the question are made in the pancreas.

3. All these statements are true. However, the reason for using the term cascade is that there is *more than one* catalytic activation acting in sequence, in this case Raf phosphorylates Mek, activating it and allowing it to phosphorylate MAP kinase, thus activating it. This cascade arrangement has the capacity for massive amplification of the original signal.

CHAPTER 10

PROTEINS OF THE CYTOSKELETON AND THE EXTRACELLULAR MATRIX

INTRODUCTION

Proteins play a major role in the structures of cells by generating muscle contraction, controlling cell shape and movement, and providing the external environment.

THE CYTOSKELETON

Eukaryotic cells contain many *filamentous structures* that determine the overall shape of the cell, allow it to move, and provide a structure within the cytoplasm on which many cellular events take place. This structure is very complex, involving hundreds of different proteins, and is known as the *cytoskeleton.* The filaments of the cytoskeleton are classified into three groups: *microtubules, microfilaments,* and *intermediate filaments.*

Microtubules

Microtubules are made from the protein *tubulin.* They play a major role in organizing the cytoplasm and provide the substrate on which proteins and *vesicles* can be *actively transported* from one part of the cell to another. Microtubules are found in cells throughout the body and are especially prominent in nerve cells in the brain.

Each tubulin molecule is a heterodimer of *α-tubulin* and *β-tubulin.* The *tubulin genes* comprise a small family of α- and β-tubulins which probably have specific functions. However, the formation of microtubules is a common function for all tubulins. A microtubule is a long, thin, hollow cylinder, or tube, of 25 nm diameter. The walls of the cylinder are made from 13 long, thin *protofilaments,* aligned parallel to each other, that—viewed on end—make a 13-sided polygon that approximates the cylindrical structure. Each protofilament is a string of α-and β-tubulin subunits alternating along the protofilament. Thus, one end of a protofilament is an α-tubulin molecule and the other is a β-tubulin. The protofilaments align in a parallel fashion (as opposed to antiparallel), so that one end of a microtubule is the beginning of a long helix of α-subunits with 13 residues per turn—from an end-on view it looks like a ring of 13 subunits. Similarly, the other end is the beginning of a long helix of β-subunits that looks like a ring of 13 subunits when viewed end on. The two ends behave differently, as seen in Figure 10.1.

Except for the highly stable microtubules of *cilia*

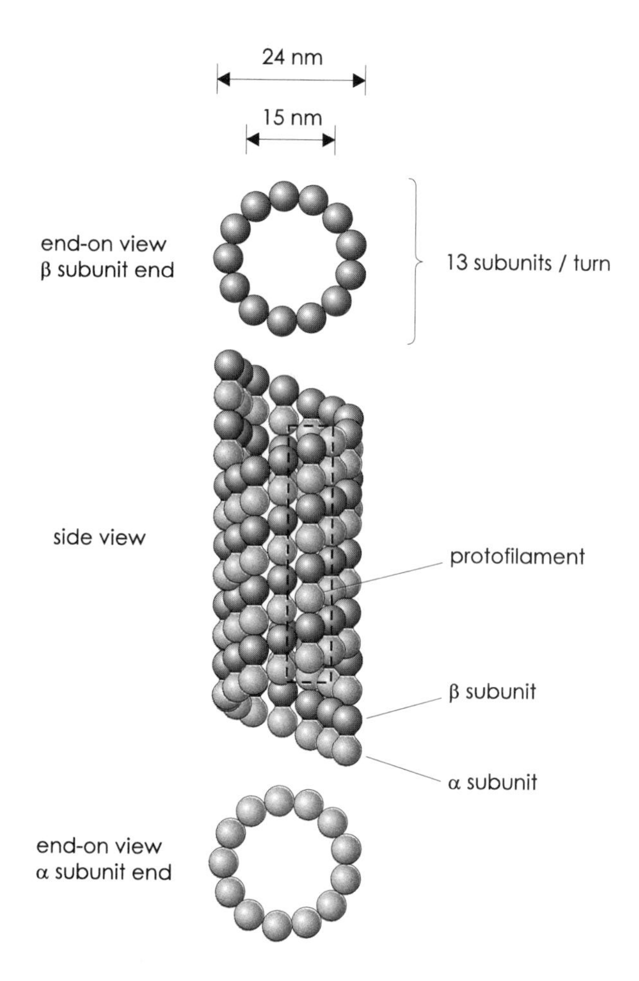

Figure 10.1.
Microtubules. A microtubule is shown as an array of tubulin molecules. Each tubulin is a dimer of an α subunit, shown in purple, and a β subunit, shown in gray.

and *flagella,* most microtubules are not stable structures; they display *dynamic instability.* They are continually elongating and collapsing in the cell. Thus, the life of a microtubule comprises initiation, elongation, and shortening (also called nucleation, polymerization, and depolymerization). There may be several cycles of polymerization and depolymerization, but depolymerization is often complete, making the total lifetime of any one microtubule relatively short. Tubulin molecules, on the other hand, have a long lifetime and participate in many microtubule structures. The lifetime of microtubules is dependent on cell type, being relatively long in *nerve cells,* which have a stable structure, and relatively short in ***fibroblasts,*** which are found in connective tissue and can respond rapidly to changes such as *tissue injury.*

Microtubules grow or shrink from both ends. However, one end, the "plus" end, changes much more rapidly than the other, "minus," end. Polymerized α-tubulin is slowly modified by removal of its *N*-terminal tyrosine and acetylation of a specific lysine residue. These modifications accumulate while the α-tubulin is polymerized and are rapidly removed when the α-tubulin is depolymerized. Addition of tubulin to a growing end requires a cofactor, GTP, bound to the tubulin. The GTP is not hydrolyzed as part of the addition process, but it is hydrolyzed soon after. The presence of GTP in tubulin stabilizes the microtubule; GDP destabilizes it. Thus, if the growth rate slows so that GDP is formed before the next tubulin is added, the microtubule will tend to begin shortening and the presence of GDP on the "older" tubulins in the microtubule will make this a rapid process, resulting in the collapse of the microtubule (Fig. 10.2). Microtubules can be stabilized by the binding of proteins to the ends.

Because microtubules are required for the separation of chromosomes along the spindle in mitosis, drugs that interfere with microtubule polymerization and depolymerization can be used to selectively attack cells that are in mitosis. This is one of the strategies used to treat cancer, and drugs such as *vinblastine* and *vincristine,* which disrupt *spindle microtubules,* or *taxol,* which stabilizes microtubules, are used as *anticancer drugs.*

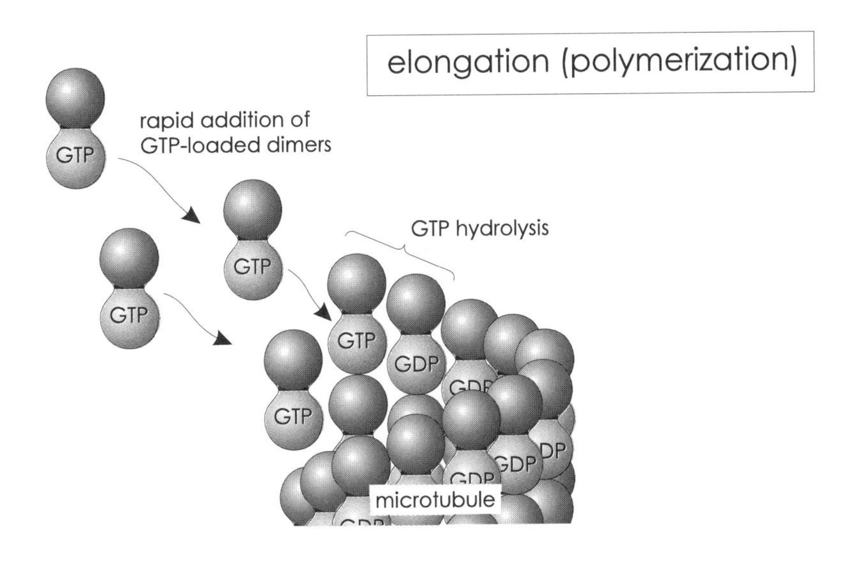

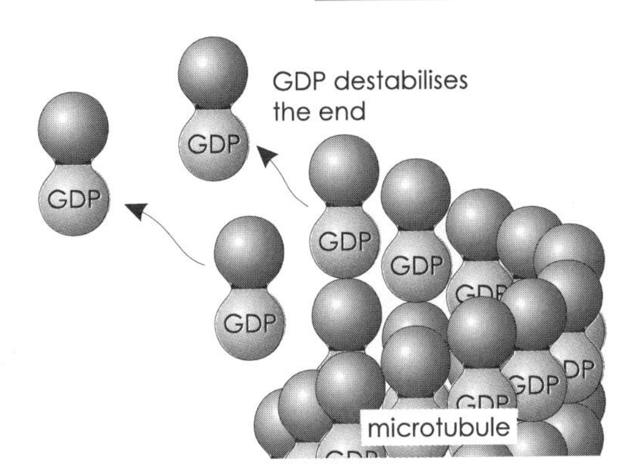

Figure 10.2.

GTP hydrolysis and microtubules. The top panel shows the elongation process for microtubules, with tubulin molecules, containing bound GTP, being added to the growing end. The lower panel shows shortening of a microtubule through dissociation of tubulin molecules containing bound GDP.

Microtubule-Associated Proteins

Many proteins other than tubulin bind to microtubules. These proteins are collectively known as *microtubule-associated proteins* or *MAPs.* (The same abbreviation, MAPs, is also used for *mitogen-associated proteins*—proteins that respond to *mitogens,* which are factors that cause increased *cell proliferation.*) The microtubule MAPs can be classified into a high-molecular-weight group which includes *MAP-1, MAP-2,* and others and a lower-molecular-weight group known as *tau proteins.* The MAPs bind all along the microtubules and help link them with other cell components. They also bind to several tubulin molecules, thus helping with initiation of a microtubule, and they tend to stabilize microtubules. There is currently great interest in the possible role of tau protein and its phosphorylation in *Alzheimer's disease.*

Microtubules provide routes for *active transport* of molecules and vesicles through the cell. The transport is carried out by *microtubule-dependent motor proteins,* the *cytoplasmid dyneins* and *kinesins.* Cytoplasmic dyneins are closely related to the *ciliary dyneins* found in cilia and flagella. Different kinesins and to, some extent, dyneins are involved in different cellular processes, such as the separation of chromosomes in mitosis, as well as transport through the cytoplasm. During cytoplasmic transport, one part of the motor protein is attached to the microtubule and another part is attached to a protein complex or organelle, the *cargo.* Conformational changes, driven by hydrolysis of ATP, cause the motor proteins to move along the microtubule, carrying their cargo with them. Kinesins transport their cargo toward the *plus ends* of microtubules; dyneins move toward the *minus end.* Since microtubules are often initiated at specific places in the cell, the directions of microtubules reflect the cell organization and microtubules thus provide directional as well as spatial routes for transport.

Actin Filaments

Actin is an abundant, ubiquitous, protein. It is found both as stable helical filaments and as unstable structures in cells, together with the monomer form. The monomer form is a globular protein of molecular weight 42,000; it is known as *G-actin.* Under physiological conditions, G-actin monomers spontaneously aggregate into a long, thin, helix, an *actin filament*; this *filamentous form* is known as *F-actin.* There is a small family of actin proteins with most of the sequence highly conserved. The family members show tissue specificity, with *α-actin* being found in muscle cells and *β- and γ-actin* being found in nonmuscle cells. In nonmuscle cells, actin filaments are found throughout cells, but are concentrated near the plasma membrane in the *cortex* of the cell.

Like microtubules, actin filaments either elongate or shorten at both ends, with one end changing faster than the other. The difference is larger for actin filaments and the plus and minus ends are often referred to as the *barbed end* and the *pointed end,* respectively, after the arrow-like appearance of actin filaments in the electron microscope. Actin filaments are dynamic, like microtubules, but unlike microtubules this often occurs as *treadmilling,* where actin monomers are alternately added at one end and removed at the other. In this way, the actin monomers in the filament are continuously being replaced, but the filament itself is in a *steady state* of constant length.

The polymerization and organization of actin filaments in the cell is controlled by a large variety of *actin-binding proteins.* For example, *profilin* is associated with the plasma membrane and may help with the initiation of actin filaments. Other actin-binding proteins, such as *α-actinin,* bind actin filaments to the plasma membrane.

The array of actin filaments, called the *actin cortex,* in the cell responds promptly to a number of signals that are transmitted through the plasma membrane. For example, *a T-cell* recognizes an *antigen* displayed on the surface of an infected cell. When a *cytotoxic, or killer, T-cell* binds to the antigen, the resulting signal—generated inside the T-cell—causes rearrangement of the actin cortex, followed by movement of the microtubules and eventually the *Golgi complex,* which can then efficiently direct the export of *toxic products* to the target (Fig. 10.3). Similarly, other white cells in the blood—*neutrophils*—have receptors that respond to a specific bacterial product—*N-formylated peptides.* When the receptors are triggered, the actin cortex is polarized and the cell moves by *chemotaxis* toward the region of higher concentration of the bacterial product. These changes are triggered through the activation of *G-proteins* related to *Ras* (see Chapter 30).

Intermediate Filaments

Intermediate filaments are assemblies of intermediate filament proteins that provide mechanical strength to animal cells. Intermediate filaments include *keratins, desmin filaments, vimentin filaments, nuclear lamins,* and *neurofilaments.* The diameter of these filaments (about 10 nm) is intermediate between the thin actin filaments (about 7 nm) and the thicker microtubules (about 25 nm). Networks of cytoplasmic intermediate filaments are found throughout the cytoplasm of most animal cells, with some concentration around the nucleus. The *lamin* proteins make a network of intermediate filaments, the *nuclear lamina,* that lies just inside the nuclear membrane.

Each intermediate filament protein has globular head and tail domains joined by a long α-helical domain. The α-helical domain contains many direct repeats of an amino-acid sequence motif called the *heptad repeat.* Since an α-helix makes just under two turns in seven residues, each position in a seven-residue repeat will slowly wind around the helix. Thus, if two such helices with heptad repeats interact through hydrophobic interactions of side chains, they will slowly wind around each other. This structure is known as an α-helical coiled coil and is stabilized by the presence of hydrophobic residues at positions 1 and 4 of the repeat. The intermediate filament is built from pairs of proteins in the coiled-coil configuration. Two of these pairs bind together, in an antiparallel, staggered, configuration. Further pairs can then add on where the overlap occurs (Fig. 10.4).

Intermediate filaments are relatively stable. They are regulated, however, and this is seen most dramatically in mitosis when phosphorylation of the lamin proteins causes the disassembly of the nuclear lamina at the beginning of mitosis and dephosphorylation of the lamin proteins causes reassembly in two daughter nuclear laminae at the end of mitosis.

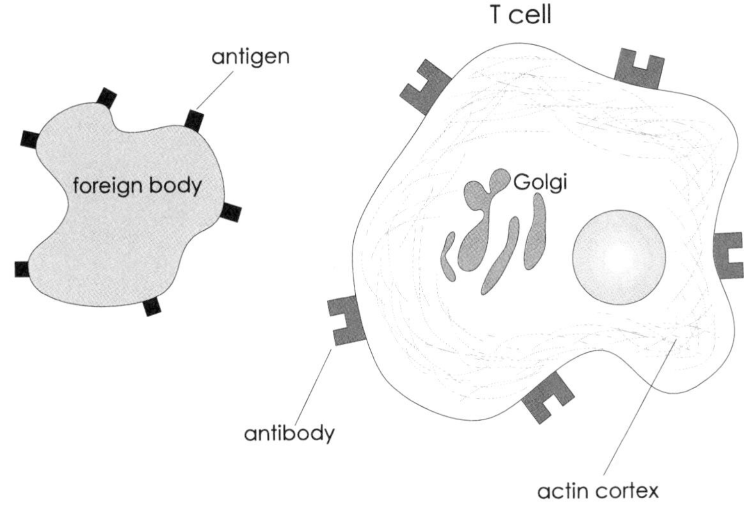

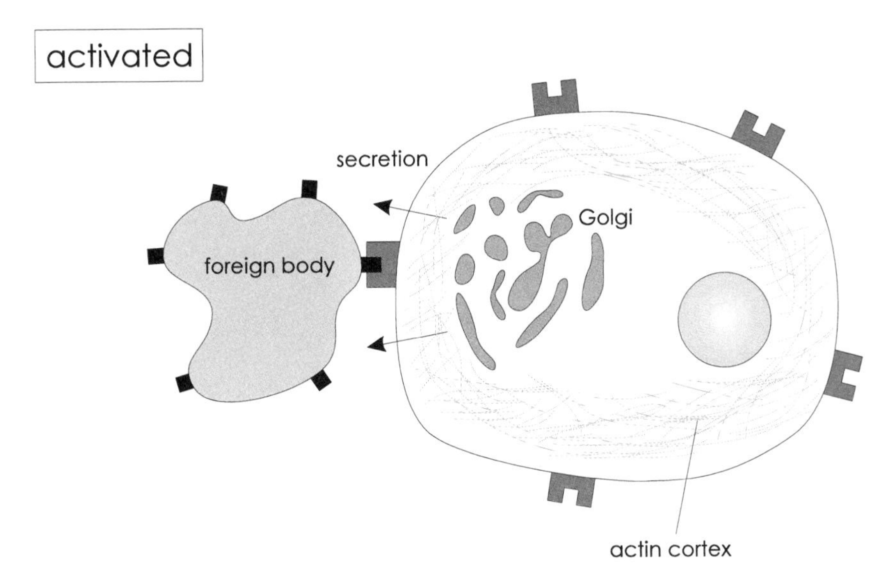

Figure 10.3.

Killer T cell activation. The cell in purple, possibly a cell infected with a virus or bacterium, is displaying antigen on its surface, indicated by the black rectangles. The T-cell has antibody-like molecules, shown in gray, on its surface that recognize and bind to the antigen. When binding takes place, the antibody-like molecule signals, inside the T-cell, that it has bound antigen and a variety of changes takes place. The figure shows the change in the actin cortex that brings the Golgi apparatus close to the bound antigen so that toxins can be exported to kill the infected cell.

MUSCLE

Actin, together with the proteins ***tropomyosin*** and ***troponin,*** forms the thin filaments of *muscle cells.* In muscle cells, thin filaments are interspersed longitudinally with *thick filaments* made from *myosin.* Myosin is a large protein containing two identical heavy chains. Each heavy chain has a long straight α-helical domain and one globular domain. In myosin, the two α-helical domains form a parallel coiled-coil structure with the two globular domains at one end (Fig. 10.5). Myosin additionally contains two each of two kinds of light chain which are bound to the globular domains of the heavy chains. These myosin molecules aggregate specifically to form the thick filaments. The actin in the thin filaments binds myosin to produce a complex of actin and myosin called *actomyosin.* Hydrolysis of ATP by myosin causes conformational changes in the globular domains of the heavy chains, leading to sliding of the thick filaments past the thin filaments, contracting the muscle.

The nature of the conformational changes is beginning to be understood with the determination of three-dimensional structures for actin and the globular domain of myosin. The globular domain of myosin, also called the "head" or *myosin S1,* forms two subdomains. One of these contains the ATP binding site and the actin binding site; the other is a long rod formed by an α-helix which is part of the heavy chain protected by the light chains. The rod subdomain acts as a lever,

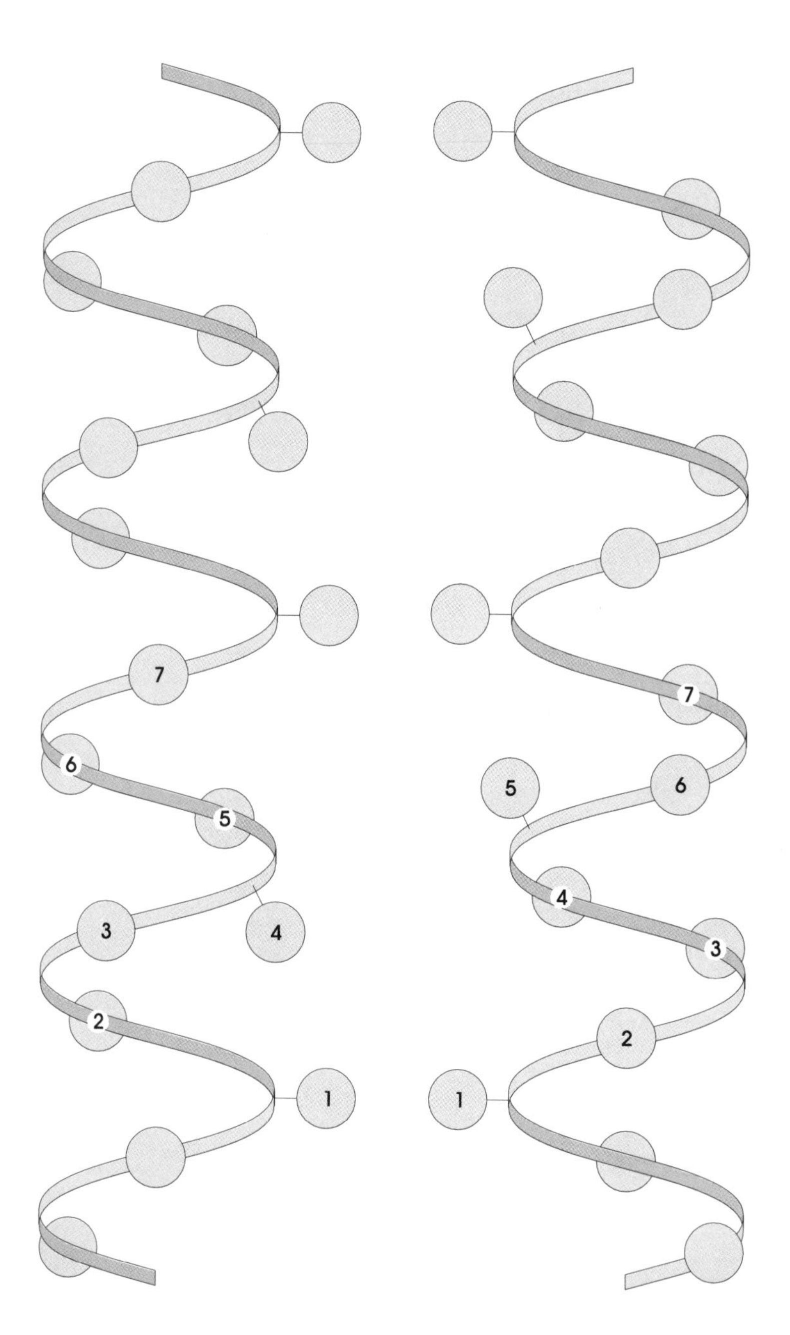

Figure 10.4.

Coiled cell. Each of the two ribbons represents an α helix. The purple spheres indicate the amino-acid side chains. The α helices run parallel to one another, and the numbering of the side chains shows that every seventh amino-acid side chain will approximately coincide, allowing a close interaction. In practice, the two chains are coiled around one another in a shallow double helix to match the seventh amino-acid side chains more precisely.

joining the coiled-coil thick filament domain to the subdomain with the actin and ATP binding sites. In the absence of ATP, the ATP binding site is a wide wedge shape. When ATP binds, the myosin is released from actin. The ATP binding site closes down around the ATP, eliminating the wedge-shaped opening and swinging the rod subdomain "lever" through a large distance. Hydrolysis of ATP occurs and the free phosphate is released while ADP remains bound and the binding site remains closed. However, phosphate release changes the conformation of the actin binding site, which then binds actin strongly. In the next step, the *power stroke,* the ADP binding site opens and returns to its wide-wedge shape, swinging the rod subdomain "lever" and causing the thick filament domain of myosin to move a considerable distance relative to the thin filament—the actin—and releasing the ADP. The cycle can then be repeated. Although this model must be confirmed, particularly in the details, it provides a convincing account of the way in which actomyosin converts the chemical energy of ATP into kinetic energy of motion.

Although the energy for muscle contraction comes from ATP, the stimulus to cause contraction is an increase in [Ca^{2+}]. The [Ca^{2+}] is regulated by the ***sarcoplasmic reticulum*** (the equivalent in muscle cells

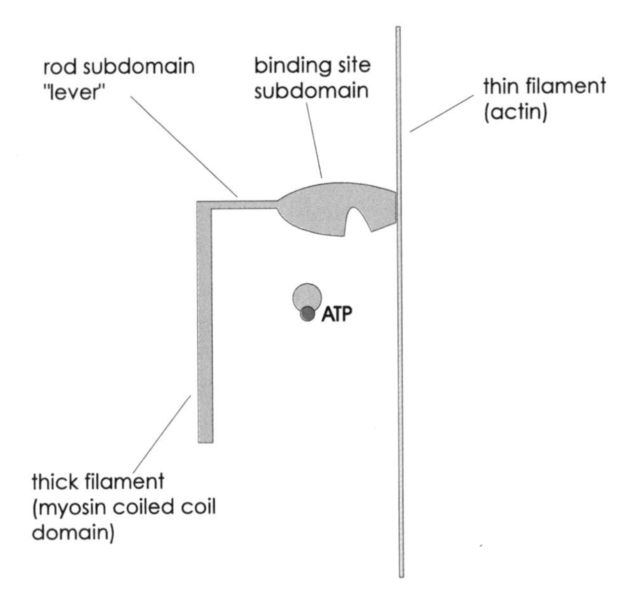

Figure 10.5.

Actomyosin structure. The actin thin filament is shown as the gray line on the right. One monomer of myosin is shown in purple, with the thick vertical part representing part of the thick filament. For clarity, the other monomer of the dimer, and additional myosin molecules in the thick filament, are not shown. The myosin head domain is shown interacting with the thin filament. The ATP binding site is shown as an indentation in the head domain.

to the endoplasmic reticulum), which normally sequesters Ca^{2+} but releases it when triggered by a *nerve impulse.*

In *skeletal muscle,* Ca^{2+} binds to troponin, leading to changes in tropomyosin and eventually to contraction. In contrast, smooth muscle does not use the Ca^{2+}–troponin–tropomyosin mechanism. Smooth muscle contracts when one of the myosin light chains is phosphorylated by myosin light-chain kinase (*MLCK*). MLCK is activated by Ca^{2+} bound to another protein, calmodulin. The muscle relaxes when the myosin light chain is dephosphorylated by a *protein phosphatase* (*PP2C*). Thus, continual rephosphorylation is required to maintain contraction (Fig. 10.6). The phosphorylation mechanism allows for additional control of smooth muscle contraction, through hormone action. *Cyclic AMP-dependent protein kinase*—activated by hormone binding at the cell surface—phosphorylates MLCK and reduces its affinity for Ca^{2+}, thereby tending to relax the muscle.

The usual cellular mechanisms for maintaining [ATP] are relatively slow, and vertebrate muscle cells

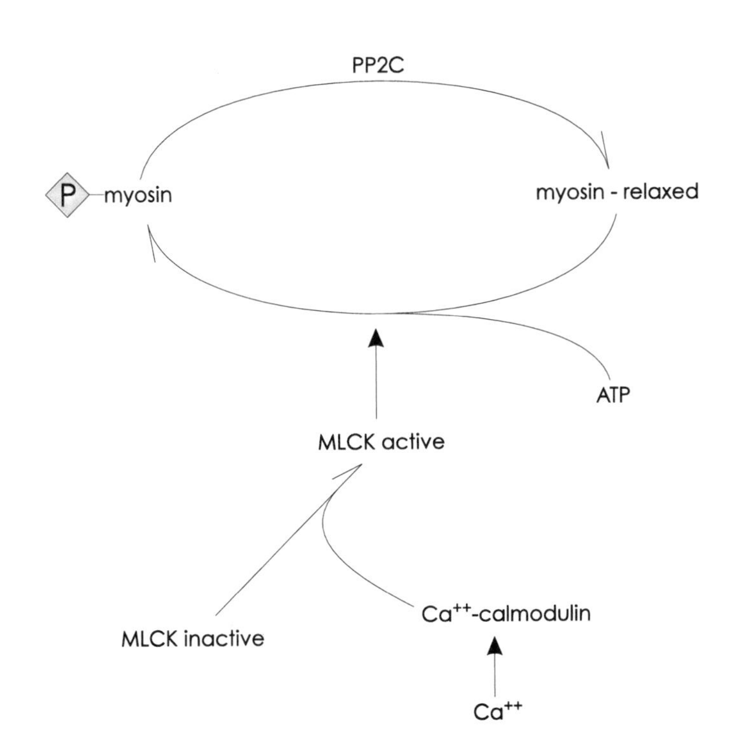

Figure 10.6.

Ca^{2+} regulation of smooth muscle contraction. Myosin light-chain kinase (MLCK) phosphorylates myosin and causes contraction, in response to calcium, as shown in this signaling pathway.

store energy in the form of *creatine phosphate,* which can rapidly donate its high-energy phosphate to ADP to maintain the [ATP].

CELL–CELL CONTACTS AND THE EXTRACELLULAR MATRIX

Although the bulk of the body is made from cells, there are extracellular components as well, such as the blood plasma. Tissues contain extracellular materials that help determine and maintain their structure. These materials are synthesized in the cell and exported to the extracellular space where they assemble into the *extracellular matrix.* Direct cell-to-cell contacts also provide tissue structure.

Cell Junctions

Chemically, the most intimate form of cell–cell contact is the gap junction. In this structure, a channel is formed, linking the cytoplasms of the two cells. The proteins of the channel regulate the flow of small molecules and ions between the two cells. The channel proteins keep the plasma membranes apart, hence the name gap junction.

Cell–cell contacts are made at intercellular junctions. Direct contacts between the plasma membranes of adjacent cells, particularly in *epithelial tissues,* are called tight junctions (Fig. 10.7). A tight junction prevents the passage of extracellular molecules and, to a varying extent, ions between the cells and even restricts the diffusion of transmembrane components within their own plasma membrane, polarizing the cells so that different membrane functions such as specific receptors, expressed on one side or another of the epithelium, are restricted to diffuse within their own section of the plasma membrane.

Anchoring junctions may link cellular *cytoskeleton proteins* through the plasma membrane to the extracellular matrix or an adjacent cell. Such an anchoring junction involves a transmembrane protein that binds the cytoskeleton on the cytoplasmic face of the plasma membrane. On the extracellular face of the plasma membrane, the anchoring junction binds either the extracellular matrix (*focal contacts, adhesion plaques,* or *hemidesmosomes*) or a receptor on another cell (*adhesion belt* or *desmosomes*).

Extracellular Matrix Proteins

Many of the extracellular materials are filaments made from *fibrous proteins,* mainly *collagen* and *elastin,* and *adhesion proteins* such as *fibronectin* and *laminin.* Collagens form a family of proteins with a tissue-specific distribution, including types I–III—found in *connective tissue* such as filaments—and types IV and V—found in *basal laminae*—forming sheets of tissue.

Collagens are synthesized as *propeptides* that form an intermolecular *triple-stranded helix, procollagen,* in which each strand is known as an *α-chain.* Procollagen is secreted by the cell and enzymatically cleaved to form *tropocollagen,* which then aggregates into collagen fibrils (Fig. 10.8). The tight packing in the procollagen triple helix requires a glycine every third residue in each α-chain. Procollagen is also characterized by specific posttranslational modifications of proline and lysine forming *hydroxyproline* and *hydroxylysine.* Hydroxyproline stabilizes the triple helix by hydrogen-bonding interactions. Hydroxylysine provides attachment points for *glycosylation* of procollagen and for fibril formation when tropocollagen aggregates into collagen fibrils. Proline hydroxylation requires *vitamin C*; deficiencies in vitamin C result in the disease *scurvy,* owing to reduced proline hydroxylation. Abnormal collagen production is also responsible for the symptoms of several human genetic diseases, including *Ehlers–Danlos syndrome* and *osteogenesis imperfecta,* as summarized in Table 10.1. A normal modification of collagen that accumulates during life is the oxidation of some lysine side chains, producing covalent cross-links between tropocollagen fibers. The cross-links reduce the flexibility of collagen structures, making bones more brittle and skin less elastic as we *age.*

Elastin is the other major protein of the extracellular matrix. Elastin molecules are cross-linked in the extracellular matrix, after synthesis and export by fibroblasts, to make networks of elastic fibers that provide *elasticity* to tissues.

Fibronectin and laminin are the principal *adhesive* glycoproteins in connective tissues and basal laminae (the structures underlying tissues that form sheets), respectively. Adhesive glycoproteins link the outer surfaces of cells to collagens and *proteoglycans* of the extracellular matrix. *Fibronectin* binds to cells through a *tripeptide* sequence (-Arg-Gly-Glu-) called *RGD* (from the one-letter amino-acid sequence). The *RGD motif* binds to the *fibronectin receptor,* one of a family of cel-

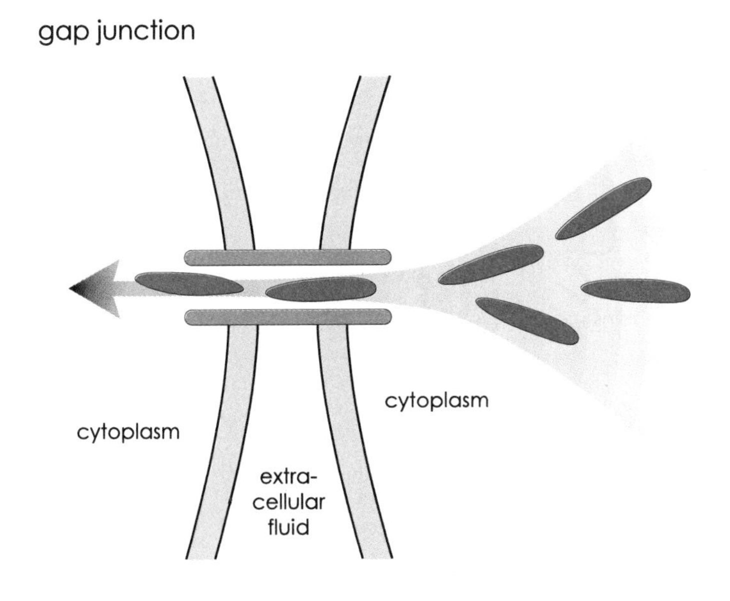

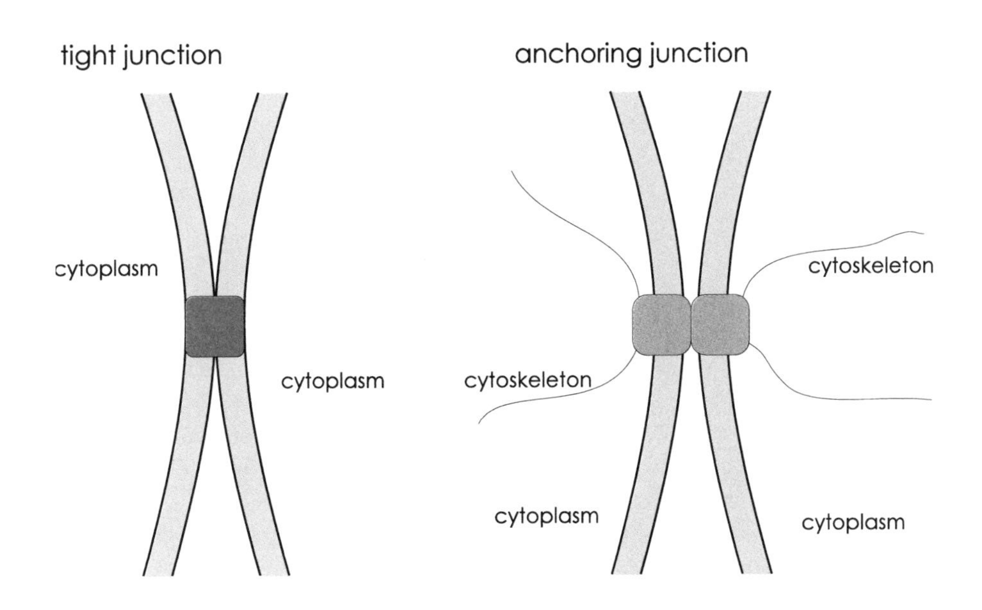

Figure 10.7.

Cell junctions. Each of the three panels shows parts of two plasma membranes, in gray, where they come together to form a junction between two cells. The gap junction shows purple objects, representing molecules, being exchanged through the junction.

lular transmembrane proteins called *integrins.* The fibronectin receptor domain of this integrin is on the outer cell surface and the cytoplasmic domain interacts with the actin cytoskeleton. Thus, integrins link the extracellular matrix to the cytoskeleton (Fig. 10.9).

Glycosaminoglycans and Proteoglycans

Other important extracellular materials include *glycosaminoglycans* (*GAGs*) and proteoglycans. The GAGs are *polysaccharides* with a repeating disaccharide structure. All GAGs, except *hyaluronic acid,* are covalently bound to proteins to form *proteoglycans.* The protein part forms a "backbone," with multiple GAGs attached along it during proteoglycan maturation, in the Golgi apparatus, before the proteoglycan is exported from the cell.

Hyaluronic acid is made from *glucuronic acid* and *N-acetylglucosamine*; It may have a molecular weight as high as 10 million and is much larger than the other GAGs. In the extracellular matrix, hyaluronic

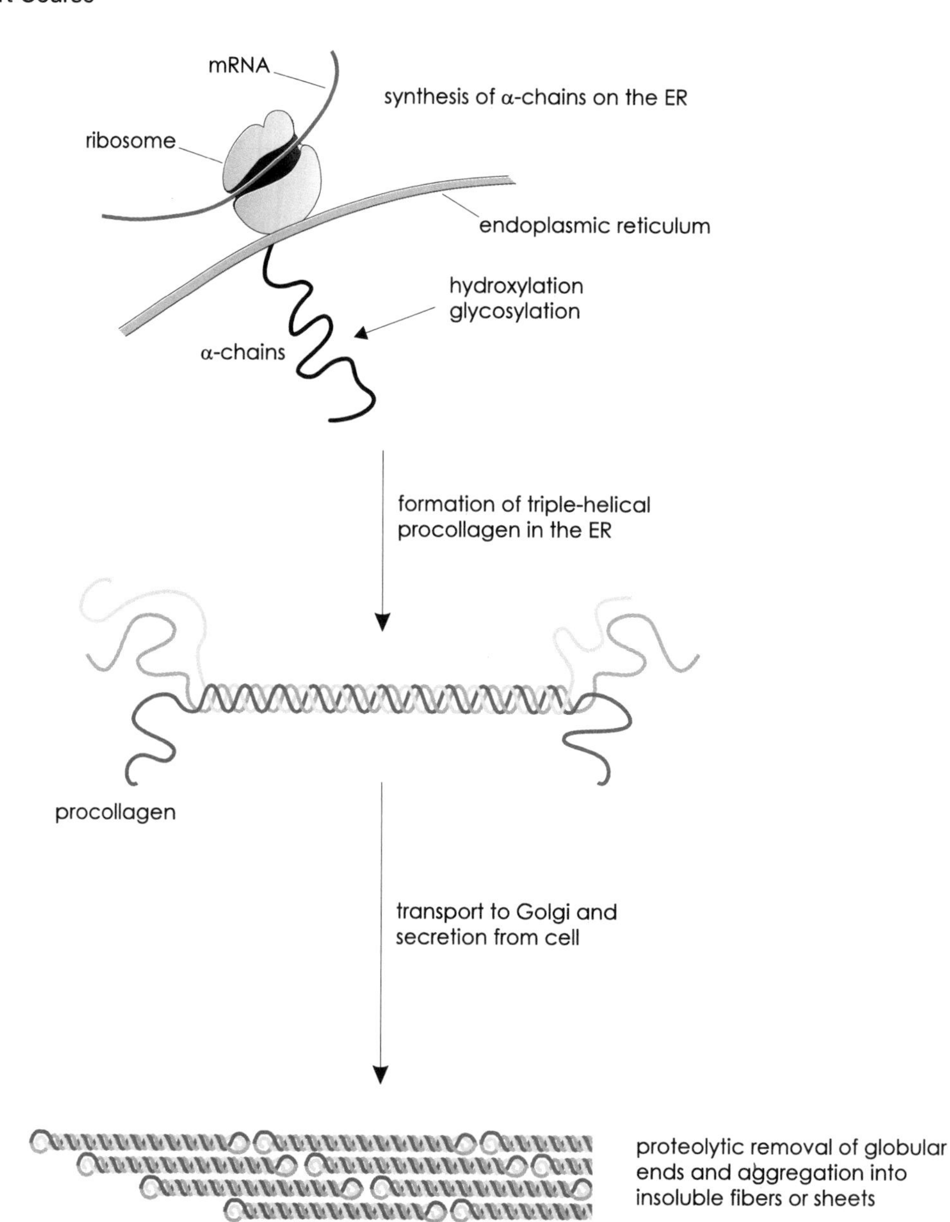

Figure 10.8.

Collagen maturation. The top part of the figure shows a ribosome, with a messenger RNA shown in purple and a nascent collagen chain shown in black. The ribosome is attached to the endoplasmic reticulum and the nascent polypeptide is passed directly into the lumen of the endoplasmic reticulum. In the next stage, three collagen monomers form a triple helix, with two head regions, known as procollagen. Finally, the head regions are removed by proteolysis, outside the cell, to form tropocollagen molecules which aggregate to form collagen fibrils.

Table 10.1. Some Disorders of Collagen Structure[a]

Disorder	Collagen defect	Clinical manifestations
Ehlers–Danlos IV	Decrease in Type III	Arterial, intestinal, or uterine rupture; thin, easily bruised skin
Osteogenesis imperfecta	Decrease in Type I	Blue sclerae; multiple fractures and bone deformities
Scurvy	Decreased hydroxyproline	Poor wound healing; deficient growth; fragile capillaries
Ehlers–Danlos VI	Decreased hydroxylysine	Hyperextensible skin and joints; poor wound healing; musculoskeletal deformities
Ehlers–Danlos VII	Amino terminal propeptide present	Hyperextensible; easily bruised skin; hip dislocations
Ehlers–Danlos V and *cutis laxa*	Decreased cross-linking	Skin and joint hyperextensibility

[a]Adapted from T. M. Devlin, *Biochemistry with Clinical Correlations,* Wiley, New York, 1992.

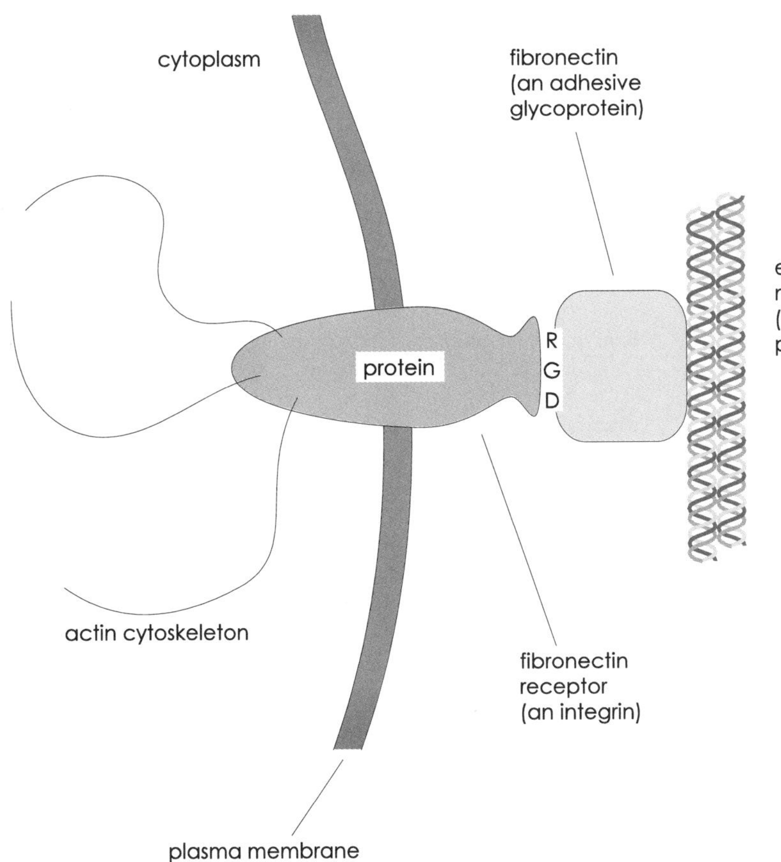

Figure 10.9.
Integrin function. The cytoplasm is on the left, separated from the extracellular space by the plasma membrane, shown in gray. The transmembrane protein, shown in purple, is an integrin. It is attached to the actin cytoskeleton inside the cell and to the extracellular matrix through an interaction with an RGD motif on fibronectin.

acid organizes proteoglycans into a large, water-rich structure that absorbs *compression stresses,* as in *knee joints* and other places.

SUMMARY

1. Proteins provide the structure of cells and extracellular spaces.

2. A microtubule is made from α- and β-tubulin molecules.

3. Microtubules are dynamically unstable in that they continuously grow and then collapse.

4. Microtubules provide routes for transport of other proteins by the motor proteins—cytoplasmic dyneins and kinesins.

5. Actin filaments are made from F-actin, the filamentous form of actin. F-actin is made from a long, thin, helical array of globular G-actin molecules.

6. Actin filaments are also dynamic, with actin monomers continuously being added at one end and removed at the other (treadmilling).

7. Much of the actin is found around the inside of the plasma membrane where it controls the shape of the cell.

8. Intermediate filaments are a family of structures intermediate in size between the thin actin filaments and the fat microtubules. Intermediate filaments are formed by coiled-coil interactions between intermediate filament proteins.

9. Muscle is made from interspersed thin (actin) and thick (myosin) filaments.

10. Muscle contraction is due to conformational changes in myosin caused by the hydrolysis of ATP. Structural studies are revealing how this conversion of chemical energy to force is accomplished.

11. Muscle contraction is regulated by calcium ion concentration.

12. Cells contact each other through tight junctions, gap junctions, and anchoring junctions.

13. Connective tissue between cells is made from the protein collagen. Collagen has a unique triple-helix structure requiring several modified amino acids, including hydroxyproline and hydroxylysine.

14. The elasticity of the extracellular matrix is provided by cross-linking collagen molecules with elastin.

15. The extracellular matrix is connected to cells by adhesive glycoproteins, fibronectin, and laminin, which bind to integrins in the plasma membrane.

16. Glycosaminoglycans and proteoglycans form a water-rich protein–carbohydrate structure that absorbs compressions stresses in the extracellular matrix.

REFERENCES

Kinesin II, a membrane traffic motor in axons, axonemes and spindles, J. M. Scholey, *J. Cell Biol.*, **133**:1–4.

The Collagens: Biochemistry and Pathophysiology, E. J. Kucharz, 1991, Springer Verlag, New York.

The molecular biology of intermediate filament proteins, K. Albers and E. Fuchs, 1992, *Int. Cytol.*, **134**:243–279.

Microfilament structure and function, A. Bretscher, 1991, *Ann. Rev. Cell Biol.*, **7**:337–374.

Integrins: Versatility, modulation and signaling in cell adhesion, R. O. Hynes, 1992, *Cell*, **69**:11–25.

Cell–matrix interactions and cell adhesion during development, P. Ekblom, D. Vestveber, and R. Kemler, 1986, *Ann. Rev. Cell Biol.*, **2**:28–47.

Kinesin-related proteins at mitotic spindle poles: function and regulation, C. E. Walczak and T. J. Mitchison, 1996, *Cell,* **85**:943–946.

Morphogenetic properties of microtubules and mitotic spindle assembly, A. A. Hyman and E. Karsenti, 1996, *Cell,* **84**: 401–410.

New clues to Alzheimer's disease: unraveling the roles of amyloid and tau, B. a. Yankner, 1996, *Nature Medicine,* **2**:850–852.

REVIEW QUESTIONS

1. Which of the following cell junctions allows exchange of cytoplasmic molecules between the two cells?
   a) Gap junction
   b) Tight junction
   c) Anchoring junction
   d) All of the above
   e) None of the above

2. Which statement about microtubules is false?
   a) Are smaller in diameter than intermediate filaments
   b) Grow by addition of tubulin dimers containing GTP
   c) Collapse by the loss of tubulin dimers containing bound GDP
   d) Undergo massive rearrangement during mitosis
   e) Are larger in diameter than microfilaments

3. Where is the collagen triple helix structure not found?
   a) Cytoplasm
   b) Lumen of the endoplasmic reticulum
   c) Golgi apparatus
   d) Intracellular vesicles
   e) Extracellular matrix

ANSWERS TO REVIEW QUESTIONS

1. ***a*** Only the gap junction provides a channel between the cells. No exchange of cytoplasmic molecules occurs through tight junctions or anchoring junctions.

2. ***a*** Microtubules are larger in diameter than intermediate filaments, whose diameter is intermediate between that of microfilaments and microtubules. Thus, ***a*** is false; the other statements are correct.

3. ***a*** The triple helix is formed in the lumen of the endoplasmic reticulum after synthesis of the individual α chains of collagen, forming procollagen. Thus, the collagen triple helix is never found in the cytoplasm.

CHAPTER

11

GLYCOLYSIS, CITRIC-ACID CYCLE, AND ELECTRON TRANSPORT SYSTEM

INTRODUCTION

This chapter introduces metabolism. We examine how glucose can be converted to lactate with the production of high-energy phosphates in the form of ATP. This occurs anaerobically. The lactate or pyruvate formed from glucose can be oxidized further to CO_2 and H_2O with production of more high-energy phosphates. This process requires O_2, because reduced coenzymes are produced which yield energy when oxidized in the cell. The major control of these systems is examined using skeleton outlines of metabolism. The critical potential control steps are presented first for understanding and integration. The complete pathway with structures and descriptions of the enzymes appear at the end of each section.

Metabolism is a conversion of one compound (i.e., substrate) to another compound either through one or a series of reactions. The normal function of metabolism is threefold: (1) energy production, (2) biosynthesis, and (3) excretion. In this section, we consider a number of metabolic pathways and particularly their interrelationship and control. When we examine a metabolic pathway, we look at an overview of the entire pathway, including all of the steps and intermediates. However, the major emphasis is on the steps that may be rate controlling or may help to determine net fluxes and directions.

THERMODYNAMICS

Most all of the steps in metabolism are enzyme catalyzed. The equilibrium of the reaction depends on the energy levels of the reactants and the products. If these differences are small (i.e., 1000 kcal/mole or less) the reactions are usually readily reversible. If the difference in energy $\triangle G^{\circ}$ is in the range of 3000 cal/mole or more, the reaction is usually physiologically unidirectional. For example, with a simple reaction of A to B and a $\triangle G^{\circ}$ of 1000 kcal/mole, the relative concentrations of the two reactants at equilibrium would be 5:1, respectively. The energies of compounds A and B are not so far apart that the reaction cannot easily be reversed. With $\triangle G^{\circ}$ of 3000 cal/mole, the relative concentrations at equilibrium would be 128:1 and much more difficult to reverse. The greater (positive or negative) the value of $\triangle G^{\circ}$ is, the greater this difference in ratio of reactants and products. With the $\triangle G^{\circ}$ of approximately 9000 cal/mole, the equilibrium ratio would be over 2,000,000:1.

The differences between $\triangle G^{\circ}$ and equilibria are not proportional because of the logarithmic nature of the equation $^{-}\Delta G^{\circ} = RT \ln K$ eq. This is analogous to jumping over a 3-ft high hurdle, which many can do, versus jumping over a 9-ft hurdle, which almost no one can do. Thus, it is not three times as hard to clear 9 ft as it is to clear 3 ft; the same relationship exists effectively for reversing reactions of various ΔG° values. ΔG is the true energetic consideration and this involves the concentration of reactants and products as well as ΔG° values, with the equation being $\Delta G = \Delta G^{\circ} + RT \ln$ ([Prod 1][Prod 2] etc./[Reactant 1][Reactant 2] etc.).

General Aspects of Metabolic Control

Since many reactions in metabolism are of the equilibrium type (i.e., high enzyme activity and attaining equilibrium in the tissue), these reactions cannot generate a flux or determine direction of flow, but can allow continuous flow in either direction, depending on the rate that substrates are either entering or leaving the pathway at some irreversible step. ***Since the irreversible steps*** (or the control and flux-generating steps) ***determine not only rate or flux but direction of flux, we examine these in greater detail.*** It will also become apparent that these are the steps at which metabolic control in the tissue is exerted by allosteric effectors, covalent modification, and so on. We examine a number of pathways individually and then attempt to integrate the controls with an emphasis on the rationale of the control as well as the details of these controls.

Another method of control is the use of isoenzymes. Isoenzymes are proteins with different genetic loci and polypeptide composition. They are enzymes that effectively carry out similar reactions, that is, phosphorylation of glucose by a family of different hexokinases. The physiological basis for isoenzymes is that, although they catalyze the same reactions, they may have different substrate specificities, kinetics, and control mechanisms. Since many of these isoenzymes are specific for various organs, they also afford the medical advantage of aiding in the identification of the injured organ when there is elevated enzyme leakage from a tissue into the blood. One example of this application is in the use of different isoenzymes of lactic dehydrogenase, since skeletal muscle and heart have different isoenzymes. If there is an increase in this enzyme in the blood, by examining the isoenzyme distribution using methods such as electrophoresis, one can determine whether a major source of the increased enzyme is skeletal muscle or heart.

Metabolism is carefully controlled in humans. This control is usually placed at key unidirectional steps and at branch points of metabolic flux. The enzymes involved in these key steps can be controlled by changing amounts of enzymes, altering the kinetics, or a combination of both methods. Changing the amount of enzymes is a slow process (from 10 min to several days) and is therefore a chronic, rather than an acute, change. Alterations in kinetics can affect either K_m, or V_m, alter the shape of the v vs. [s] curve, or a combination of all of these. The kinetic changes can be achieved by covalent modification (as phosphorylation) of the enzyme. Phosphorylation is rapid, but reversal (i.e., dephosphorylation) is usually slower. These changes usually occur in response to hormones and have effects that last from minutes to hours. The second-to-second control of metabolism is affected by allosteric effectors, whose concentration can change rapidly within the cell. These ***allosteric effectors are usually small molecules which are normal metabolites***. They have their effect at sites on the enzyme other than the active site, although with some enzymes with sigmoidal kinetics, the substrate reacting at the active site on one subunit can affect the kinetics of binding subsequent substrates on other subunits (e.g., hemoglobin).

Many of the important control enzymes exhibit sigmoid kinetics. This has the advantage of moving the steep part of the response curve (change in v with change in [S] or effector) into the physiological range. All substrates and effectors have a normal range of concentration in the cells, which may vary from n molar to m molar. In most of the sigmoidal reactions the K_s (0.5), K_a (0.5), and K_i (0.5) are near the middle of the physiological range of concentrations for substrate, activator, or inhibitor.

If a hyperbolic curve (Michaelis–Menten kinetics) and a sigmoidal curve are compared, where both K_m and K_s (0.5) are in the middle of the physiological range of the substrate, differences become apparent (Fig. 11.1). In the case of simple Michaelis–Menten kinetics, there is some change in v with changes in [S] in the physiological range, but it is not large. With Michaelis–Menten kinetics the greatest change in v with [S] is at zero substrate to well below K_m, a situation unlikely to arise physiologically. With sigmoidal kinetics with K_s (0.5) in the middle of the physiological range, the v changes rapidly with changes in [S], thus giving a strong response of activity to changes in [S], activators, or inhibitors. Also note the steepest part of the curve (response of v with changes in [S]) is in the physiological range. Sigmoidal reactions can also have

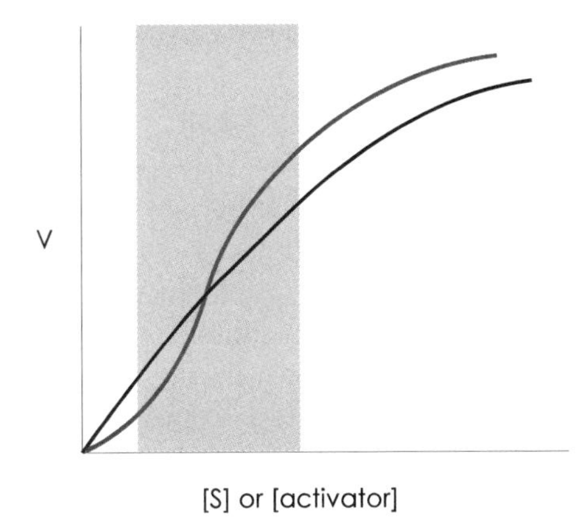

Figure 11.1.
Comparison of hyperbolic and sigmoidal enzyme kinetics. The plot of velocity, rate of reaction, versus concentration of substrate or activator shows the response of a hyperbolic (Michaelis–Menten) reaction in black. The response of a sigmoidal reaction is shown by the purple curve. The shaded area indicates the range of physiological concentrations of substrate or activator.

kinetics altered by effectors by changing the shape of the curve with a shift to the left (activation) or the right (inhibition). Inhibition, by shifting of sigmoidal curves to the right, can be overcome with excess substrate—but many times these overcoming concentrations cannot be achieved physiologically.

GLYCOLYSIS

To examine the rationale of control one must consider the biological function of the various pathways. We first consider glycolysis (i.e., the conversion of glucose to lactate which can be accomplished in the absence of oxygen). There are three major functions of glycolysis. The first is the production of energy (i.e., ATP, particularly in the absence of oxygen, anaerobic glycolysis). The second is to convert glucose to pyruvate (aerobic glycolysis) in preparation for final oxidation in the citric-acid cycle. The third is to form compounds that may be important in biosynthesis, such as α-glycerol phosphate required for phospholipid and triacylglycerol synthesis; also, intermediates for amino-acid formation (i.e., intermediates of glycolysis) can be converted to alanine and serine with appropriate nitrogen donors.

To understand thoroughly the controls and the logic therein, the function of each of these pathways must be studied. However, two other complications arise in examining controls. These are that ***different tissues have different metabolic priorities*** and, therefore, the control necessary to assure that these priorities are met may vary in different tissues. Further, some of these priorities may change in direction or magnitude in response to body needs—many of these are signaled by hormone responses. Therefore, at least a cursory examination of hormone messages to different tissues and how they relate to the control of metabolism is needed. First, the potential control step and rationale for the control, and then details of the control, will be examined. In glycolysis, there are three physiologically irreversible reactions—hexokinase, phosphofructokinase, and pyruvate kinase.

To understand the control of the metabolic pathways in a meaningful way, at least some general understanding of the metabolic message of some hormones is useful. The metabolic process of how these hormones effect their function and this relationship to the overall control of the function of a given tissue is also important. Thus, throughout this section three hormones are emphasized, although many more hormones interact in a complicated manner. The first is insulin, which is secreted by the beta cells of the pancreas, is the only truly and completely anabolic hormone in the body, and usually is an indication that the individual has been fed and that it is time to store the dietary materials that are present in excess of the energetic needs. Insulin stimulates storage of material by increasing glycogen, fat, and protein synthesis. The second, glucagon, which is secreted by the pancreas alpha cells, is a hormone that is indicative of low blood-glucose concentration. Glucagon's general message to the body is to increase blood glucose concentration by increasing rates of synthesis and decreasing rates of utilization of glucose. In some cases, such as diabetes, this may be an inappropriate message, because blood glucose concentration may already be high, and thus the severity of the disease is increased. The third is epinephrine (formerly known also as adrenaline), a catecholamine arising from the adrenal cortex. This hormone produces the classical fight or flight response. Increased epinephrine is an indication to the body that all energy should be made available for an episode of flight or a stand and fight. This hormone will affect many tissues, particularly the muscle where both force and energy for contraction are critical. Thus, the general systemic message presented to the body overall by each of these three hormones can be examined in a more rational manner. Indications of the metabolic advantages and, if the message is inappropriate, the metabolic disadvantages, are considered.

GLYCOLYTIC CONTROL

Transport

The flow of glucose can be controlled at four major steps—three enzymatic and one of transport. ***For glucose to be metabolized it must enter the cell.*** The transport phenomena in general is discussed in Chapter 4. Moving glucose from the blood into body tissues occurs primarily via a facilitative transport system on the cell membrane. There are several types of transporters. Some are insulin sensitive, as in skeletal muscle and adipose tissues, and others are noninsulin sensitive, such as brain, liver, and the red blood cell. Since these systems are all facilitative diffusion, glucose cannot be accumulated in the cell against a concentration gradient. This process can be rate limiting either in insulin-sensitive tissues with low insulin or in the brain when blood glucose concentration drops sufficiently well below the K_s (0.5) for blood glucose.

Hexokinase

The next step, once inside the cell, is hexokinase, catalyzing the phosphorylation of glucose to glucose-6-phosphate (G-6-P) and requiring ATP. Since the plasma membranes are predominantly nonpolar (see Chapter 4), the phosphorylation produces a charged molecule and traps the G-6-P in the cell. If hexokinase, which catalyzes a physiologically irreversible reaction, were allowed to proceed unabated, there could be a depletion of glucose in the blood due to the large mass of muscle and high hexokinase activity.

The hexokinase isoenzymes are, with the exception of glucokinase, inhibited by glucose-6-phosphate. Once in the cell, glucose-6-phosphate has a number of fates, one of which is glycolysis to produce energy. However, even in the presence of excess energy, glucose-6-phosphate can be used as a precursor of glycogen synthesis. Therefore, ATP inhibition would be inappropriate, because under certain high-energy conditions, glycogen synthesis would be advantageous (Fig. 11.2). However, if glucose-6-phosphate accumulates and is no longer required for energy production or glycogen storage, it would be disadvantageous for a cell to continue to remove blood glucose because this could potentially deplete the supply of blood glucose, Therefore, when glucose-6-phosphate is not further metabolized at a rapid rate, hexokinase is inhibited. Since this enzyme has a very low K_m for glucose, between 10^{-4} and 10^{-6} M, competitive inhibition would not be effective and it has been shown that glucose-6-phosphate inhibits in a noncompetitive manner. Under normal conditions, the hexokinase in the muscle is 95% inhibited. This low degree of steady-state activity, compared to potential maximal activity, is advantageous, since under extreme conditions the potential activity is about 20-fold that under steady-state conditions.

Glucokinase

One of the hexokinases, usually referred to as glucokinase, is found in the liver. Glucokinase has slightly simodial kinetics, in contrast to the hexokinases, where the kinetics are hyperbolic. Glucokinase also has a much higher K_s (0.5) for glucose—10 mM, as compared to 0.1–.01 mM for the other hexokinases. Thus, in the liver, the phosphorylation is sensitive to blood-glucose concentration (about 5 mM), especially those of the portal vein (1–20 mM, depending on dietary intake and time since eating). Glucokinase is not subject to G-6-P inhibition. The major purpose of glucokinase appears to be to allow the liver to take up glucose for glycogen synthesis and for potential formation of α-glycerol phosphate, which is important in triacylglycerol formation. The problem seen of removing blood glucose to a low concentration in muscle is not observed in the liver for two reasons. The first is the high K_m of glucokinase for glucose. Second, the liver contains the enzyme glucose-6-phosphatase, which can catalyze the conversion of glucose-6-phosphate to inorganic phosphate and glucose, thus raising blood–glucose concentration. Muscle is devoid of this enzyme.

Phosphofructokinase

The first irreversible step uniquely associated with glycolysis is catalyzed by phosphofructokinase (PFK), which phosphorylates F-6-P in the number one position using ATP and forms fructose-1,6-bisphosphate. Phosphofructokinase has been one of the most studied en-

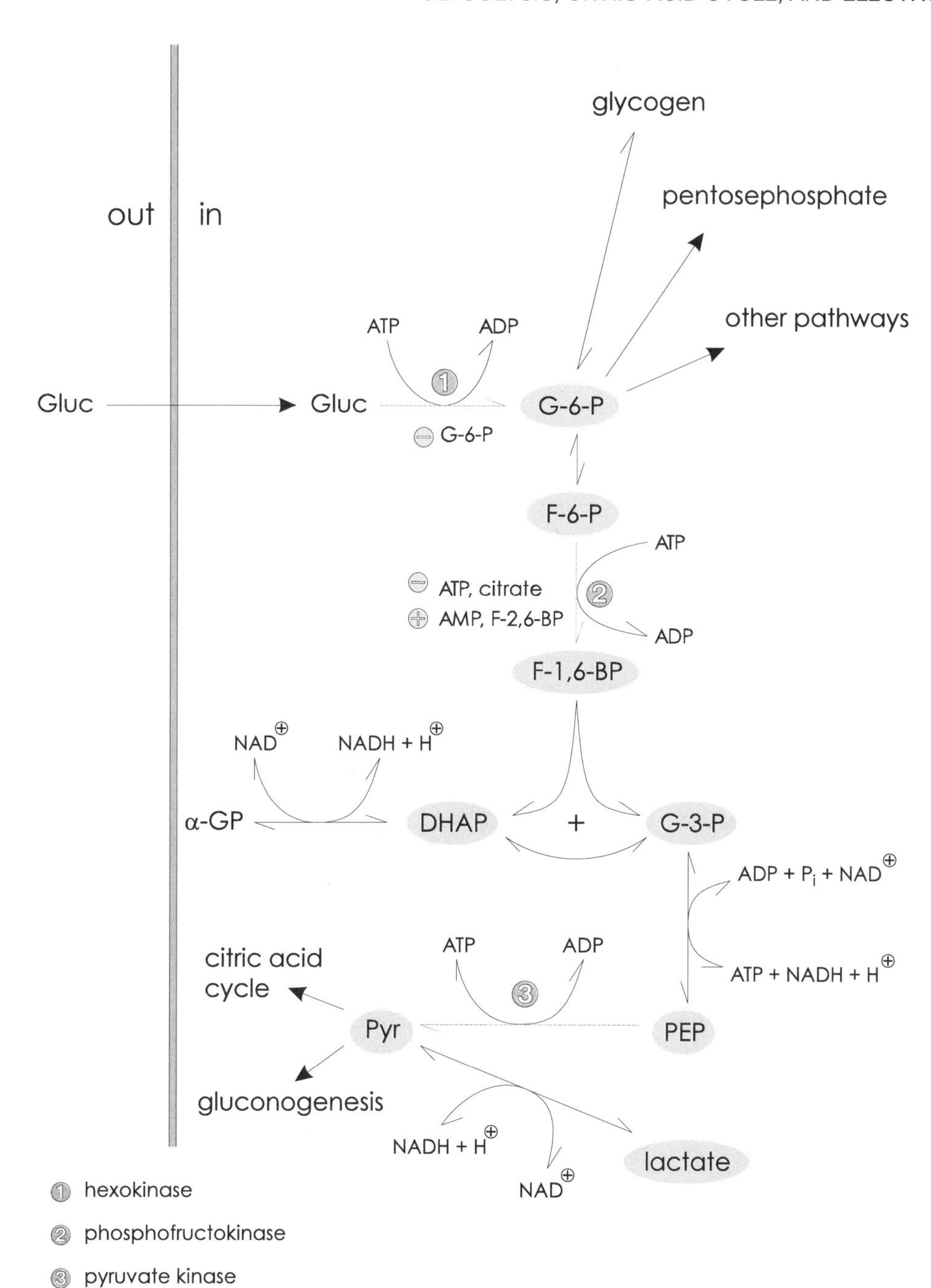

Figure 11.2.

Diagram of major steps in glycolysis and linking pathways. This illustration presents the major steps in glycolysis, including the control steps. Other fates of the glycolytic intermediate are indicated to present a more general aspect of glucose metabolism. Critical unidirectional reactions are shown with a purple arrow and intermediates of glycolysis, per se, are contained in a shaded circle. The circled numbers also indicate unidirectional reactions: (1) hexokinase, (2) phosphofructokinase, and (3) pyruvate kinase. The complete details of glycolysis and chemical structures are presented in Figure 11.5. The major positive and negative effectors for the first two irreversible reactions are also shown.

zymes for regulation and is highly regulated. It is the major control site of glycolysis. There are over a dozen effectors of PFK, both positive and negative; however, for ease of understanding, only the apparent major effectors are discussed in this chapter. Metabolism of glucose to the fructose-1,6-bisphosphate stage usually progresses further through glycolysis. Phosphofructokinase has sigmoidal kinetics in respect to F-6-P (Fig. 11.3). The shape of the PFK kinetic curve can vary, ranging from Michaelis–Menten kinetics to an extremely sigmoidal curve. Thus, at any given fructose-6-phosphate concentration, the activity of this enzyme in the cell will depend on the relative sigmoidal nature of the curve. The less sigmoidal the kinetics is, the greater the activity; the more sigmoidal, the less activity at any fixed F-6-P concentration. It would appear that even with an increased right shift of the curve, fructose-6-phosphate could continue to accumulate to a concentration where appreciable activity would occur. However, since fructose-6-phosphate and glucose-6-phosphate are in equilibrium, when fructose-6-phosphate accumulates, so will glucose-6-phosphate and, as indicated previously, the ele-

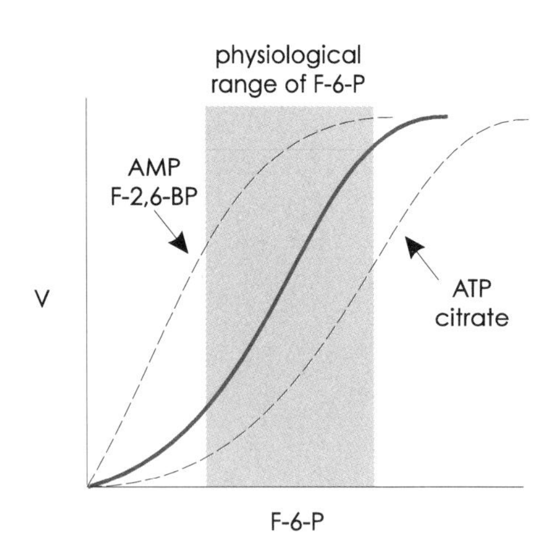

Figure 11.3.

Kinetics of phosphofructokinase. Phosphofructokinase has sigmoidal kinetics in respect to F-6-P concentration. The solid curve is the response at normal concentrations of effectors. The dotted lines indicate direction changes in the kinetics due to effectors. These changes are really an infinite number of curves with left or right shifts, with only one curve being shown for each. A right shift, as with increased ATP and/or citrate, will produce a decreased reaction rate with any physiological concentration of F-6-P. A left shift, as with AMP or F-2,6-BP, will produce an increased reaction rate with any physiological concentration of F-6-P. The physiological range of F-6-P concentrations (in peripheral tissues) is shown by the shaded area.

vated glucose-6-phosphate will inhibit hexokinase. This inhibition will prevent further increases of glucose-6-phosphate and fructose-6-phosphate from glucose as a carbon source.

Phosphofructokinase at any constant level of fructose-6-phosphate is inhibited by elevated ATP, which is further exacerbated by the presence of elevated citrate (Fig. 11.4). Citrate in the presence of noninhibitory levels of ATP will not in itself cause inhibition. The action in this case is one of synergism. Activators of PFK include 5′ AMP and fructose-2,6-bisphosphate (which is under hormonal control; see Chapter 13).

This control of PFK also illustrates some important aspects of allosteric regulation. For example, ATP reacts at one site as a substrate for this reaction and, when at a higher concentration, will also bind to an allosite, which will cause inhibition in the form of a right shift of the sigmoidal curve. When low, ATP behaves as one would expect from Michaelis–Menten kinetics. However, instead of remaining at V_{max} as ATP concentrations continue to increase, the reaction has a significant decrease of activity at a fixed F-6-P concentration, particularly at concentrations similar to those observed in intact cells. Most cells have ATP concentrations that are strongly inhibitory unless the cells, such as muscle cells, are exhausted from continuous contraction. Thus, a unique aspect in this reaction is that, although ATP is a substrate for PFK, the activity of this enzyme in the cell actually increases as ATP decreases in concentration owing to a partial deinhibition. Since the major function of glycolysis is the formation of ATP, the physiological advantage of having increased activity as ATP concentration declines is apparent. The inhibition by ATP is increased further in the presence of citrate, which is an indication from the mitochondrion that the potential energy generation system in this subcellular particle is maximized or near maximized. ATP is a signal of high energy and AMP is a signal of low energy. AMP, which causes an increased flux through PFK to help produce energy (as ATP) via glycolysis, will overcome some of the ATP inhibition.

The relative energy status of the cell can be esti-

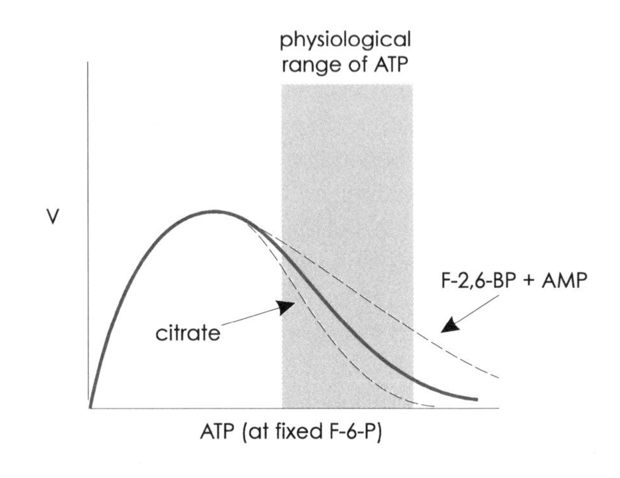

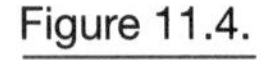

Figure 11.4.

Effect of ATP on activity or phosphofructokinase. The effect of varying ATP at a fixed F-6-P concentration is shown as a solid line. The physiological range of ATP is shown in the shaded area. In the physiological range of ATP, increases in ATP concentrations are further inhibitory. In this range, citrate causes further inhibition of PFK and either F-2, 6-BP, or AMP causes some deinhibition of the system, both shown with a dotted line.

mated by determining the adenylate charge. The adenylate charge is

$$\frac{2(\text{ATP}) + (\text{ADP})}{2(\text{ATP} + \text{ADP} + \text{AMP})}$$

Basically, this formula indicates the number of high-energy phosphates present (i.e., two high-energy phosphates in ATP and one in ADP) divided by the number of potential high-energy phosphates (i.e., each of the adenine nucleotides could potentially have two high-energy phosphates). This is an indication of a number of high-energy phosphates present compared to the number that could be present. This ratio in normal resting cells ranges between 0.85 and 0.9. Most of the nucleotides are in equilibrium with each other in regard to energy charge. Thus, if one determines a uridylate, cytodylate, or guanidylate charge, it would be similar to that observed in the adenylate charge. However, since the adenine nucleotides are in greater excess than the other nucleotides, they are usually used to determine energy charge. With a high adenylate charge, the ATP would be high and PFK would be inhibited. At a low adenylate charge, where ATP decreases and AMP increases, PFK activity would increase.

Fructose-2,6-bisphosphate (F-2,6-BP) formation is catalyzed by PFK 2. F-2,6-BP can be increased in response to hormones such as insulin in the liver and epinephrine in the muscle. F-2,6-BP in the liver is discussed in the section on gluconeogenesis; however, the aspect in the muscle is well related to the role of epinephrine. Epinephrine increases cAMP (3′5′-cyclic AMP) in the muscle, which activates a cAMP-dependent protein kinase, causing the phosphorylation of a number of enzymes. One of these enzymes, PFK 2 in the muscle, when phosphorylated, causes increased activity of this enzyme, producing more fructose-2,6-bisphosphate. This increased fructose-2,6-bisphosphate will stimulate phosphofructokinase 1 (F-6-P + ATP → F-1,6-BP + ADP) and increase the rate of glycolysis, enabling the muscle to obtain more energy under the fight-or-flight conditions signaled by epinephrine. Fructose-2,6-bisphosphate is a potent activator of PFK and is present in tissues at micromolar or nanomolar concentrations. Fructose-1,6-bisphosphate, which is an intermediate of glycolysis and is used in substrate amounts, is present in a millimolar range. Furthermore, fructose-2,6-bisphosphate is not on any major metabolic pathway and its only route of final metabolism is to be dephosphorylated back to fructose-6-phosphate by the enzyme fructose-2,6-bisphosphatase. Thus, the level of this activator is dependent on its relative rate of synthesis and degradation. Degradation is relatively constant in the muscle and the synthesis can be increased in the muscle owing to the phosphorylation of PFK 2.

Thus, we can see that PFK 1 is a carefully controlled enzyme with many effectors; the activity of this reaction depends on the concentration of several metabolites present in the cell. These include ATP, AMP, citrate, F-6-P, and F-2,6-BP.

Pyruvate Kinase

Pyruvate kinase is the next unidirectional reaction in glycolysis and can be inhibited by high ATP levels, although usually these are not high enough in the cell to decrease the activity of this enzyme significantly. Pyruvate kinase activity in most cells is considerably higher than that of hexokinase and phosphofructokinase. Thus, pyruvate kinase does not appear to play a major control role in the peripheral tissues. However, in the liver, a different isoenzyme of pyruvate kinase is under stringent control and plays an important role related to gluconeogenesis; therefore, this aspect is considered when gluconeogenesis is discussed.

Details of Glycolysis

Following is a detailed description of the steps involved in anaerobic glycolysis, especially those related to muscle, which is the major tissue of the body. The steps in this section correspond to the numbers used in Figure 11.5:

1. Glucose must be transported into the cell for glycolysis to occur. There are specific glucose transporters in most cells; some are insulin stimulated and others are refractory to insulin.
2. Hexokinase. This enzyme catalyzes the phosphorylation of glucose in the six position using ATP and Mg^{2+} and is physiologically irreversible. This requires an input of energy and accomplishes the trapping of glucose within the cell. We will see that throughout glucose metabolism, until there is a net generation of energy, all of the intermediates are phosphorylated and therefore retained within the cell. A number of different isoenzymes of hexoki-

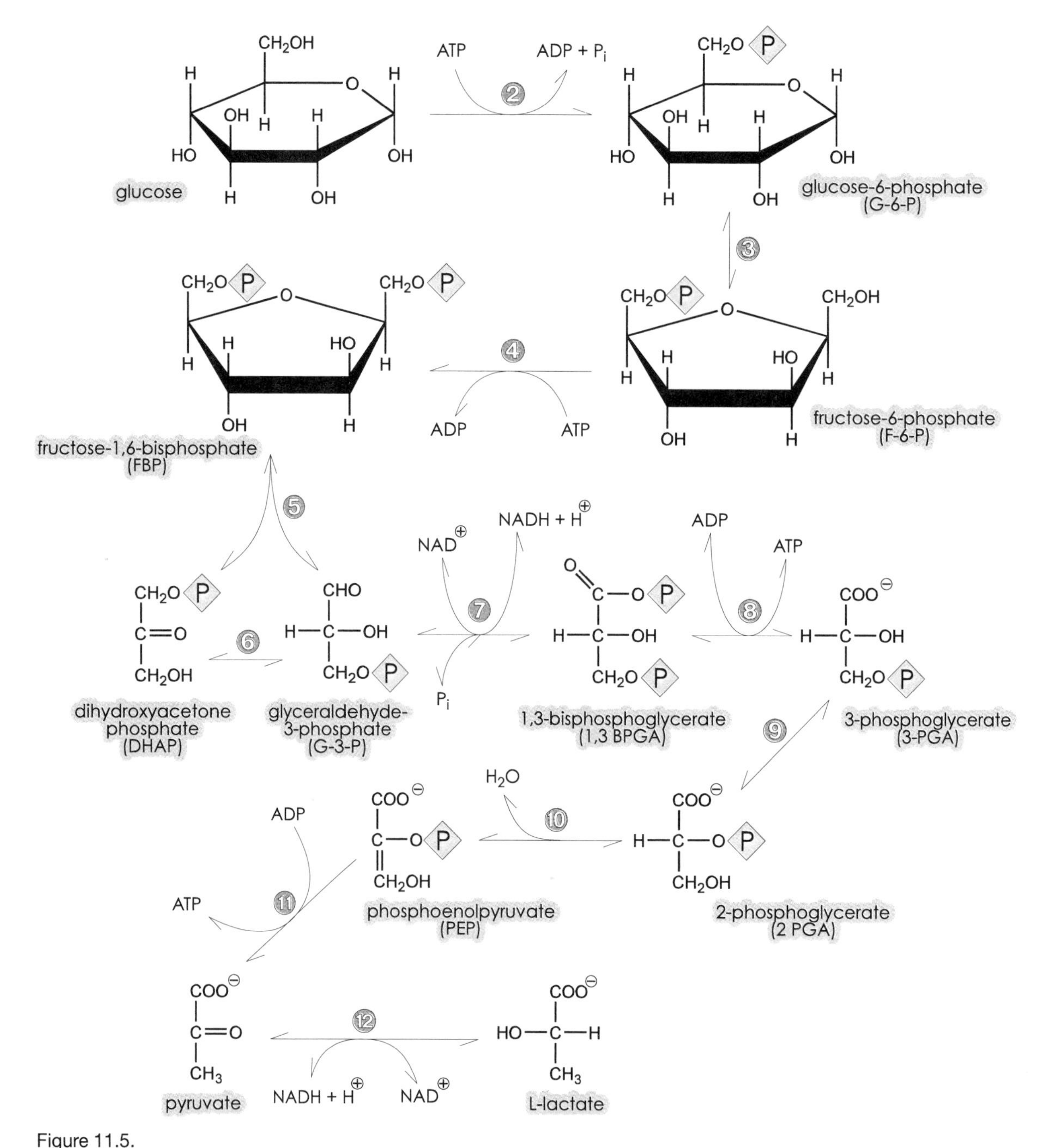

Figure 11.5.

Glycolysis. The detailed steps and structures of glycolysis are presented. The number of each reaction corresponds to the same number of the detailed discussion of glycolysis presented in the text.

nase occur in various cells, which vary in their K_m for glucose and their potential physiological inhibition.

3. Phosphoglucose isomerase catalyzes an equilibrium between glucose-6-phosphosphate and fructose-6-phosphate. This enzyme is usually in excess and produces an equilibrium mixture of the two hexophosphates.

4. Phosphofructokinase. This enzyme is traditionally thought of as the first enzyme that commits glucose metabolism to the glycolytic pathway. This enzyme requires ATP and Mg^{2+}, as was the case with hexokinase, and catalyzes the phosphorylation of fructose-6-phosphate in the number one position, forming fructose-1,6-bisphosphate.

5. Aldolase is an enzyme that achieves equilibrium or near equilibrium in the tissue. It catalyzes the conversion of fructose-1,6-bisphosphate to 2-triosephosphates. One is glyceraldehyde 3-phosphate, which will proceed further through glycolysis. Another is dihydroxyacetone phosphate, which can act as a precursor of α-glycerol phosphate for lipid synthesis or be converted to glyceraldehyde 3-phosphate to proceed through glycolysis.
6. Triphosphate isomerase. This enzyme achieves equilibrium. It should be noted that in most of these cases, equilibrium does not necessarily mean a one-to-one ratio. In this case, the ratio at equilibrium of the triophosphates is 24 parts of dihydroxyacetone phosphate to one part of glyceraldehyde 3-phosphate. This reaction is important because it allows the dihydroxyacetone phosphate formed from fructose-1,6-bisphosphate to be converted to glyceraldehyde 3-phosphate and to proceed further through glycolysis. In the absence of this enzyme, only half of the triosephosphates (i.e., glyceraldehyde phosphate) formed from fructose-1,6-phosphate would proceed through to glycolysis and dihydroxyacetone phosphate would continually accumulate.
7. Glyceraldehyde 3-phosphate dehydrogenase. This enzyme contains sulfhydryl groups (from cysteines in the polypeptide structure) and a tightly bound NAD^+. The mechanism of this enzyme is such that sulfhydryls form an adduct with the aldehyde group at G-3-P, thus forming an apparently new hydroxy group. This hydroxy group is then oxidized by the NAD^+ bound to the enzyme to form a thioester (i.e., an ester between the SH and a carboxyl group of the 3-carbon phosphorylated intermediate). Thioesters are high-energy compounds. NADH formed during this oxidation can be regenerated back to the NAD^+ by nonbound NAD^+ in the cytosol, thus producing a free (unbound) NADH (Fig. 11.6). The energy in the thioester link is conserved by a phosphorolytic cleavage which produces the enzyme with a free SH and 1,3-bisphosphoglycerate. The phosphate in the number 1 position is an anhydride between the carboxyl group and the phosphate; thus, there is a high-energy bond. These two concerted reactions—the dehydrogenase and phosphorolytic cleavage—are catalyzed by a single enzyme known as glyceraldehyde 3-phosphate dehydrogenase.
8. 3-Phosphoglycerate kinase will reversibly catalyze the conversion of 1,3-bisphoglycerate and ADP to 3-phosphoglyceric acid and ATP. The cell has invested two high-energy phosphates in glycolysis—one at hexokinase and one at phosphofructokinase. The hexose is converted to two triosephosphates and when both of the trioses proceed through the glycerate kinase step, two high-energy phosphates are formed (one for each triosephosphate produced). Thus, at this point in glycolysis, the cell is at the break-even point in terms of high-energy phosphates. Substrate-level phosphorylation, which occurs in several reactions, such as the 3-phosphoglycerate kinase reaction, is the formation of ATP from ADP with the required energy coming directly from a cleavage of a high-energy intermediate such as 1,3-bisphosphoglycerate or phosphoenolpyruvate. Substrate-level phosphorylations does not require oxygen and is the mechanism by which ATP is formed throughout glycolysis.
9. Phosphoglyceratemutase catalyzes the conversion of 3-phosphoglycerate to 2-phosphoglycerate and attains equilibrium. This reaction requires a cofactor of 2,3-bisphosphoglycerate. It is an unusual reaction where each product, in either direction, is converted temporarily to 2,3-bisphosphoglycerate and then to the final product.
10. Enolase catalyzes the removal of water from 2-phosphoglycerate, forming phosphoenolpyruvate. This reaction is freely reversible. The enol group is a weak acid and therefore the enolphosphate is effectively an anhydride and has the properties of a high-energy bond. This reaction is inhibited by fluoride ion, which can be added to red blood cells to prevent glucose consumption or to other preparations to prevent glycolysis.
11. Pyruvate kinase irreversibly catalyzes the formation of ATP and pyruvate from phosphoenolpyruvate (PEP) and ADP. This reaction is named for the direction that is opposite to the flow of the process. Since the high-energy

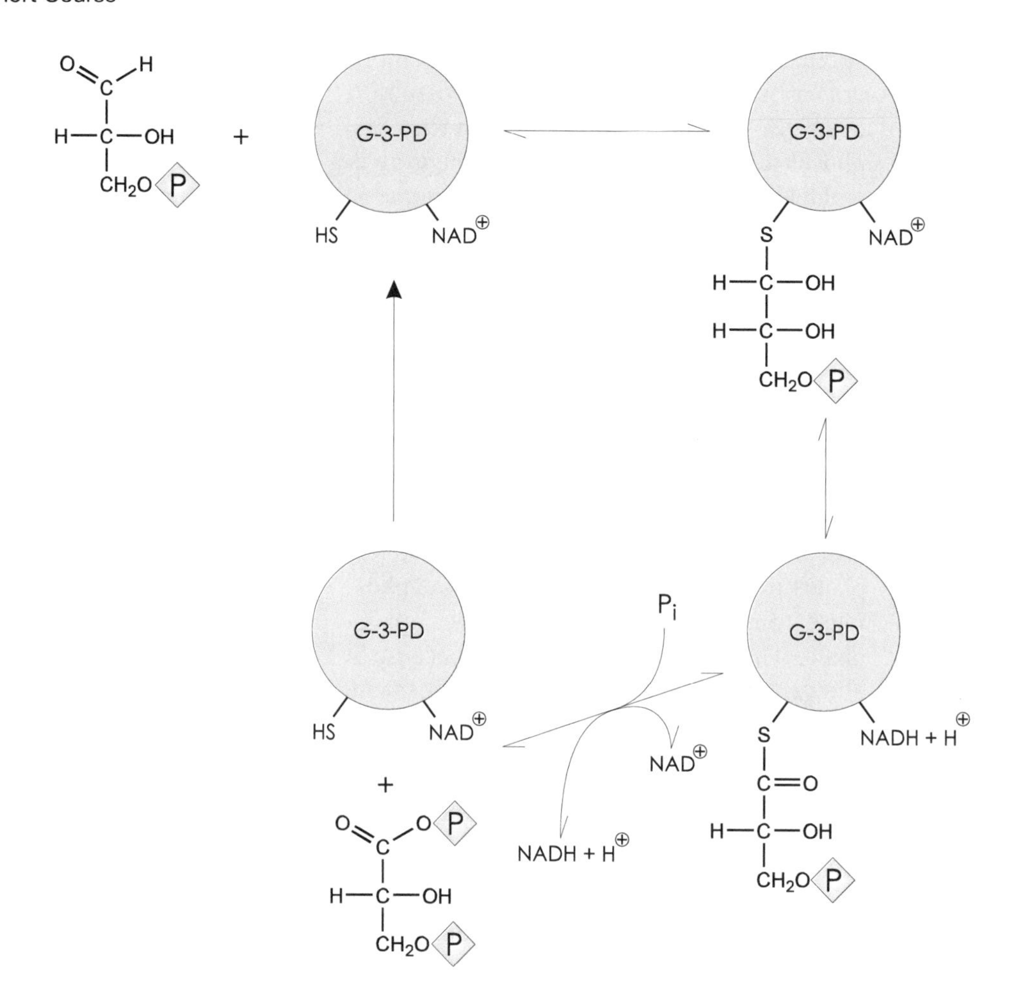

Figure 11.6.

Glyceraldehyde-3-phosphate dehydrogenase. A schematic presentation of the mechanism of glyceraldehyde-3-phosphate dehydrogenase with the conversion of glyceraldehyde-3-phosphate to 1,3 bisphoglycerate and NAD^+ to NADH.

phosphate break-even point occurred earlier, at the phosphoglycerate kinase step, this step produces a further gain of two high-energy phosphates. Thus anaerobic glycolysis yields a net of two high-energy phosphates as ATP per glucose consumed. Pyruvate is the first compound in the glycolytic pathway that is no longer phosphorylated. At this point, a net gain of ATP has been achieved and the intermediates (pyruvate and lactate) are no longer trapped in a cell.

12. Lactate Dehydrogenase. The glyceraldehyde 3-phosphate dehydrogenase step produced NADH from NAD^+. In the absence of oxygen, this NADH must be reoxidized to NAD^+ to allow glycolysis to proceed because there are limiting amounts of coenzymes available. This is accomplished by a reversible reaction catalyzed by lactate dehydrogenase, which in most tissues, reaches equilibrium. This enzyme catalyzes pyruvate plus NADH to form lactate and NAD^+. At usual cellular levels of NAD^+ and NADH, the ratio of lactate to pyruvate is approximately 10:1. During anaerobic exercise, for example, there is a formation and buildup of lactate, with some pyruvate, which is then released into the bloodstream; this accounts for the increased lactate found in the blood after exercise.

THE CITRIC-ACID CYCLE

Pyruvate produced in the glycolysis can be oxidized in the citric-acid cycle, producing CO_2, water, and primarily reduced coenzymes. Reoxidation of these coenzymes through the electron-transport system (the cytochromes) can transduce the energy to form ATP in a process known as oxidative phosphorylation.

The metabolism of glucose through glycolysis and the citric-acid cycle (including pyruvate dehydrogenase) produces 38 high-energy phosphates, if one includes the cytosolic NADHs formed during glycolysis. Actually, 40 high-energy phosphates were produced, but the process required the expenditure of two high-energy phosphates. Six of the high-energy phosphates are formed via substrate-level phosphorylations and 34 via oxidative phosphorylation. This indicates the importance of O_2 and oxidation in regard to high-energy phosphates during carbohydrate catabolism.

In the mitochondrial system, which includes pyruvate dehydrogenase and the citric-acid cycle, there are four physiologically irreversible steps (Fig. 11.7). These are pyruvate dehydrogenase, citrate synthase, isocitrate dehydrogenase (NAD-linked), and α-ketoglutarate dehydrogenase. These are all potential control points in metabolism. Each is examined in terms of (1) its role, (2) the advantage of using controls at each step, and (3) the potential controlls and the rationale for using each.

Pyruvate Dehydrogenase

The first enzyme, pyruvate dehydrogenase, is not truly part of the citric-acid cycle, but is effectively the entrance to the cycle for carbohydrates such as glucose.

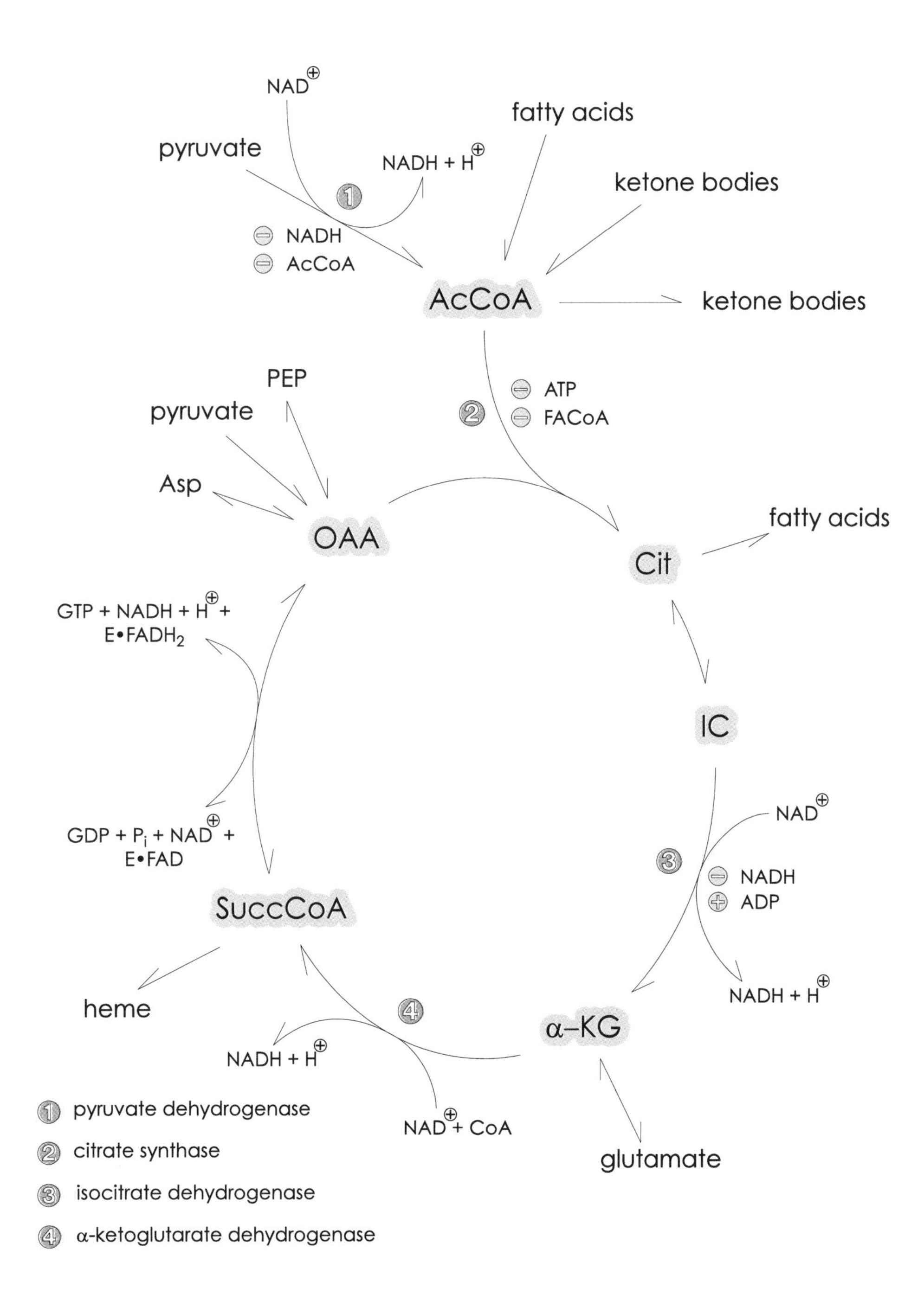

Figure 11.7.
The citric-acid cycle and pyruvate dehydrogenase in diagrammatic form. This presents the major steps in the citric-acid cycle and other metabolic paths associated with the citric-acid cycle. The intermediates of the citric-acid cycle are contained in shaded circles and the unidirectional steps are indicated with purple arrows. The three major control steps—pyruvate dehydrogenase, citrate synthase, and isocitrate dehydrogenase—are indicated along with major negative and/or positive effectors. A complete diagram of the citric-acid cycle with formulas is presented in Figure 11.9.

Pyruvate, one of the later products of glycosis, can either be reduced to lactate, regenerating NAD^+ to keep glycolysis flowing, or enter the cycle via pyruvate dehydrogenase. To appreciate fully the control of this enzyme, the body's glucose economy must be understood; therefore, before returning to pyruvate dehydrogenase, the glucose economy is considered.

Pyruvate and/or lactate are both potential precursors for glucose formation. However, ***after pyruvate is converted to acetyl CoA by pyruvate dehydrogenase, it is no longer a potential net precursor of glucose formation***. Thus, when glucose formation is required, any pyruvate converted to acetyl CoA is lost as a potential precursor of glucose, and any pyruvate saved by not proceeding via pyruvate dehydrogenase to acetyl CoA is potentially glucogenic. Although the liver and kidney cortex are the major organs for gluconeogenesis, even pyruvate saved in the muscle can be sent through the blood, as pyruvate or lactate, to the liver, where it can undergo conversion to glucose for the body's use, especially by those organs and cells that have an absolute requirement for glucose, such as the central nervous system and the erythrocytes.

Pyruvate dehydrogenase is the enzyme that interdigitates glycolysis and the citric-acid cycle and is carefully controlled. ***Pyruvate dehydrogenase has both allosteric and covalent modification as control mechanisms***. Included in the first of these is inhibition by the product acetyl CoA. It is important to recognize that acetyl CoA can arise not only from pyruvate, but also during fatty-acid, ketone-body, and amino-acid metabolism. Therefore, there are numerous sources of acetyl CoA. During times of starvation and potential hypoglycemia, one of the more important sources of acetyl CoA (see Chapter 14), is from fatty acids or their metabolic product, ketone bodies. When large amounts of fatty acids are metabolized, in an attempt to conserve glucose and glucose precursors, acetyl CoA will rise and inhibit pyruvate dehydrogenase. This reaction is of such importance that a single method of control is insufficient. This reaction is also inhibited by high NADH, which signals high-energy potential within the mitochondrion.

Pyruvate dehydrogenase can also be controlled by a covalent modification (phosphorylation/dephosphorylation). This phosphorylation, catalyzed by pyruvate dehydrogenase kinase, occurs more readily when there are high concentrations of acetyl CoA and/or NADH in the mitochondrion (both of which can arise from fatty-acid and ketone-body metabolism) and a high ATP to ADP ratio in the mitochondrion. Phosphorylation of pyruvate dehydrogenase gives a relatively inactive or less active pyruvate dehydrogenase. Under these conditions, the elevated acetyl CoA concentration will not only inhibit the enzyme allosterically, it will also stimulate the conversion into a less active phosphorylated form. This phosphorylated pyruvate dehydrogenase can be converted back to the active form by a phosphatase that requires Ca^{2+}. Thus, the activity of this enzyme depends on the rate of phosphorylation and dephosphorylation, as well as the allosteric effects. It appears that high pyruvate concentrations, which occur when there is rapid glycolysis, and are unlikely to occur during starvation, inhibit the pyruvate dehydrogenase kinase and keep the enzyme in an active form, whereas elevated acetyl CoA and NADH help stimulate the phosphorylation of this enzyme. The pyruvate dehydrogenase kinase and phosphatase are in the mitochondrion and not the cytosol, where cAMP-dependent protein kinase has its activity. Therefore, cAMP per se is not a major factor in the phosphorylation state of this enzyme. Most all of the organs in the body that have mitochondria will have effective control of pyruvate dehydrogenase via phosphorylation. One tissue that does not burn fatty acid regularly is the central nervous system. In the brain, under all conditions studied, pyruvate dehydrogenase remains in the nonphosphorylated (active) state. In other tissues, pyruvate dehydrogenase is primarily in a nonphosphorylated state during feeding of high-carbohydrate diets, and in a phosphorylated (inactive) state during starvation, diabetes, or high-fat feeding.

Citrate Synthase

The formation of citrate from oxaloacetate and acetyl CoA by citrate synthase is also a controlled unidirectional reaction. It is inhibited by ATP. This is a leaky control and is dissimilar to other ATP inhibition where AMP or ADP activate the reaction. In the case of citrate synthase, all three adenine nucleotides ATP, ADP, and AMP, appear to inhibit this enzyme, with ATP being somewhat more potent than ADP and AMP. The physiological significance of the inhibition of this enzyme by ATP may be questionable, because under high-energy conditions, it might be advantageous for citrate to be formed. When the energy of the cell is high, citrate can leak from the mitochondria to the cytosol, where it is an inhibitor of PFK; if ATP inhibited citrate synthase, the citrate could not accumulate and have this effect. There is a stronger inhibition by fatty acetyl CoA (an intermediate in fatty-acid catabolism), which appears to cause an inhibition of citrate synthase. This allows an accumulation of oxaloacetate and acetyl CoA, which may be important during gluconeogenesis (see Chapter 13).

Isocitrate Dehydrogenase

The NAD$^+$ linked isocitrate dehydrogenase in the mitochondrion is a major control in the citric-acid cycle. This enzyme has sigmoidal kinetics in relationship to isocitrate concentration. The kinetic curve shifts to the right (i.e., resulting in a greater relative inhibition) with increased levels of NADH. It is also activated by shifting the kinetic curve to the left; this is stimulated by ADP. These are reasonably appropriate effectors in a mitochondrion. Most of the potential energy generated in the citric-acid cycle and via pyruvate dehydrogenase is in the form of reduced coenzymes such as NADH. When there are increased NADH concentrations which cannot be converted readily to ATP (owing to a lack of either oxygen or available ADP), there is no advantage in accumulating more NADH. Thus, an increased NADH concentration, which is an indicator of potential for ATP formation in the mitochondrion, inhibits isocitrate dehydrogenase. The ADP, which is critical for the conversion of NADH to potential energy in the electron transport system, activates this enzyme and indicates that ADP is available for the formation of ATP. Thus, energy levels are the determinant of the control of isocitrate dehydrogenase. However, the high-energy indicator in this case is not ATP per se, but NADH, which could proceed through the electron transport system to form three high-energy phosphates (i.e., ATP from ADP) (Fig. 11.8). In this case, the low-energy indicator in the mitochondrion is ADP, which is similar to AMP as a low-energy indicator in the cytosol. One advantage of inhibition at this step, with potential high energy levels in the mitochondrion, is the accumulation of citrate, which can leak into the cytosol and increase the degree of inhibition of PFK in the presence of ATP. Thus, the energy levels in both the mitochondrion and the cytosol can work in unison to control the flow of glucose–carbon through glycolysis.

Details of the Citric-Acid Cycle

If oxygen is present, the pyruvate formed in glycolysis can undergo further oxidation to CO_2 and water, with a large production of useful energy in the final form of ATP or high-energy phosphate compounds. The following is a detailed description of the cycle and each of its steps. The step numbers correspond to the numbers used in Figure 11.9.

The process is initiated by the entry of pyruvate into the mitochondrial matrix, where the enzyme pyruvate dehydrogenase will irreversibly convert pyruvate to acetyl CoA. This enzyme is a complex of a large number of subunits of independent but cooperating individual enzymic activities. (1) The first subunit (pyruvate decarboxylase) causes the decarboxylation of pyruvate and requires thiamine pyrophosphate. A two-carbon thiamine pyrophosphate adduct is formed and then transferred from this subunit to a second subunit containing oxidized lipoic acid (i.e., S-S). (2) The two-carbon adduct is transferred to form an HS and

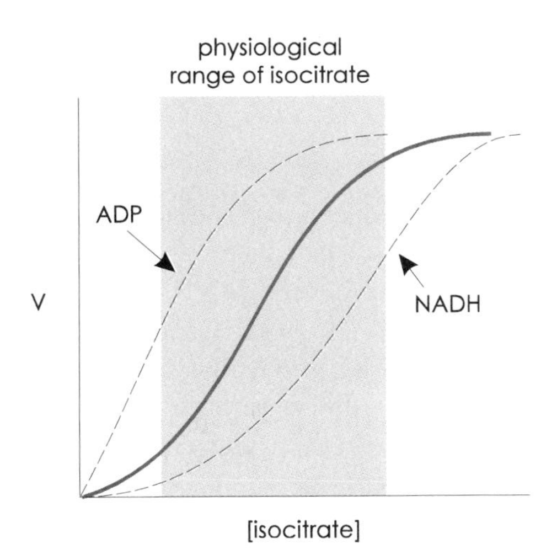

Figure 11.8.

Kinetics of isocitrate dehydrogenase. Isocitrate dehydrogenase has sigmoidal kinetics in respect to isocitrate concentrations. The solid curve is the response at normal concentrations of effectors. The dotted lines indicate direction changes in the kinetics due to effectors. These changes are really an infinite number of curves, with only one left and right shift shown. A right shift, as with increased NADH, will produce a decreased reaction rate with any physiological concentration of isocitrate. A left shift, as with increased ADP, will produce an increased reaction rate with any physiological concentration of isocitrate. The physiological range of isocitrate concentrations is shown by the shaded area.

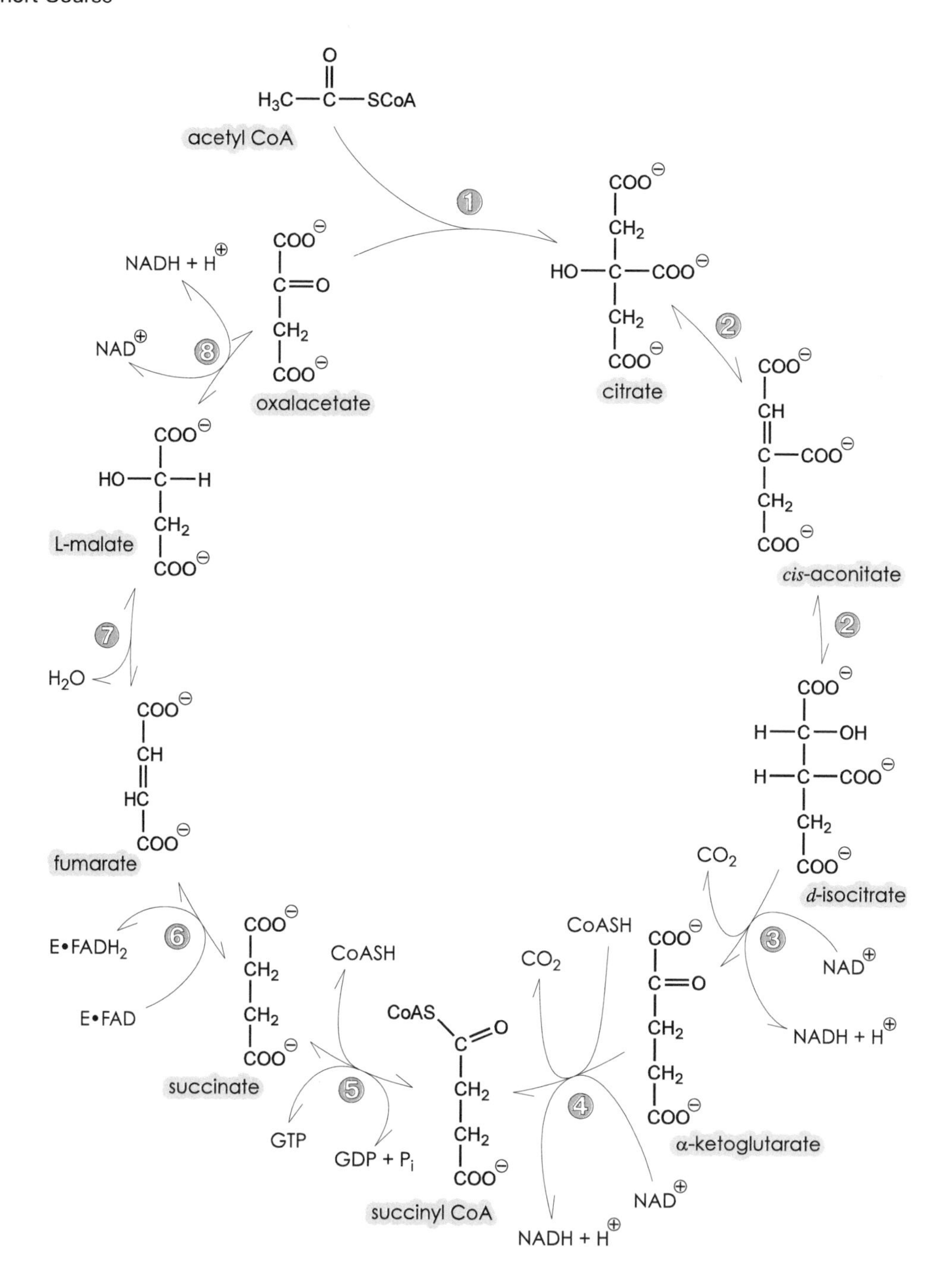

Figure 11.9.

The citric-acid cycle. The detailed steps and structure of the citric-acid cycle are presented. The number of each reaction corresponds to the same numbers of the detailed citric-acid cycle presented in the text. The carbons of the new entering acetyl CoA are in purple to indicate the asymmetric nature of citrate in the cycle. All carbons are equilibrated at succinate; therefore, this differentiation of carbons is discontinued after succinate is formed.

$S{-}\overset{\overset{O}{\|}}{C}{-}CH_3$ on the lipoic acid. (3) The acetyl group is then transferred by the same subunit to coenzyme A, which diffuses from this complex to the total mitochondrial matrix. (4) The lipoic acid containing the subunit now has lipoic acid with two SH groups and must be reoxidized back to the S-S form. This is catalyzed by another subunit lipoyl dehydrogenase, which contains a covalently bound prosthetic group of FAD. This enzyme catalyzes the reoxidization of lipoic acid and concurrently FAD is reduced to $FADH_2$. The subunit (i.e., lipodehydrogenase) can then reconvert the $FADH_2$ back to FAD simultaneously with NAD^+ being converted to NADH + H^+. The NADH and H^+ then leave the complex and diffuse throughout the matrix. Thus, the overall action of this pyruvate dehydrogenase, is to take pyruvate plus NAD^+ plus coenzyme A to NADH + H^+ + acetyl coenzyme A + CO_2. This step is irreversible and is not truly a part of the citric-acid cycle, but is an important interstep between glycolysis and the cycle.

The enzyme pyruvate dehydrogenase can be inhibited by arsenite, which complexes with the reduced form of lipoic acid, thus making the lipoic acid unavailable for continued reactivity. This enzyme is also decreased in amount and activity during a thiamine deficiency (Fig. 11.10).

1. The acetyl CoA formed can now enter the citric-acid cycle (also known as the tricarboxylic-acid cycle and the Krebs cycle, after the individual who discovered this pathway). The first step is a condensation between oxaloacetate and acetyl CoA to form citrate, catalyzed by citrate synthase. This reaction is also physiologically unidirectional. The coenzyme A released from acetyl CoA can then enter the general coenzyme A pool and participate once more in pyruvate dehydrogenase and other related reactions.
2. Aconitase catalyzes the removal of water from citrate to form *cis*-aconitate, and water is added to *cis*-aconitate to form either citrate again or d-isocitrate, depending on whether the hydroxyl is on the third or second carbon. This enzyme is fairly active and achieves an equilibrium distribution where 90% of the intermediates are citrate, 7% are D-isocitrate, and 3% are *cis*-aconitate. One interesting aspect of this reaction is that although citrate is a symmetrical molecule, it is acted upon by the enzyme aconitase as if it were asymmetric. This is due to a three-point attachment of the substrate to the enzyme with stereospecificity in the position at which water is added to *cis*-aconitate. The hydroxyl on isocitrate is always found distal to the newly added acetyl group and stereospecificity forms D-isocitrate.

   Aconitase can be inhibited by florocitrate, which can be formed from fluroacetate. Fluroacetate has been used as an animal poison and has occasionally been ingested by human infants.
3. NAD^+-linked isocitrate dehydrogenase converts isocitrate + NAD^+ to α-ketoglutarate + CO_2 + NADH + H^+ unidirectionally. Details of this reaction and its control have been discussed previously in this chapter.
4. α-Ketoglutarate dehydrogenase, a multienzyme with a mechanism similar to pyruvate dehydrogenase, but specific for α-ketoglu-

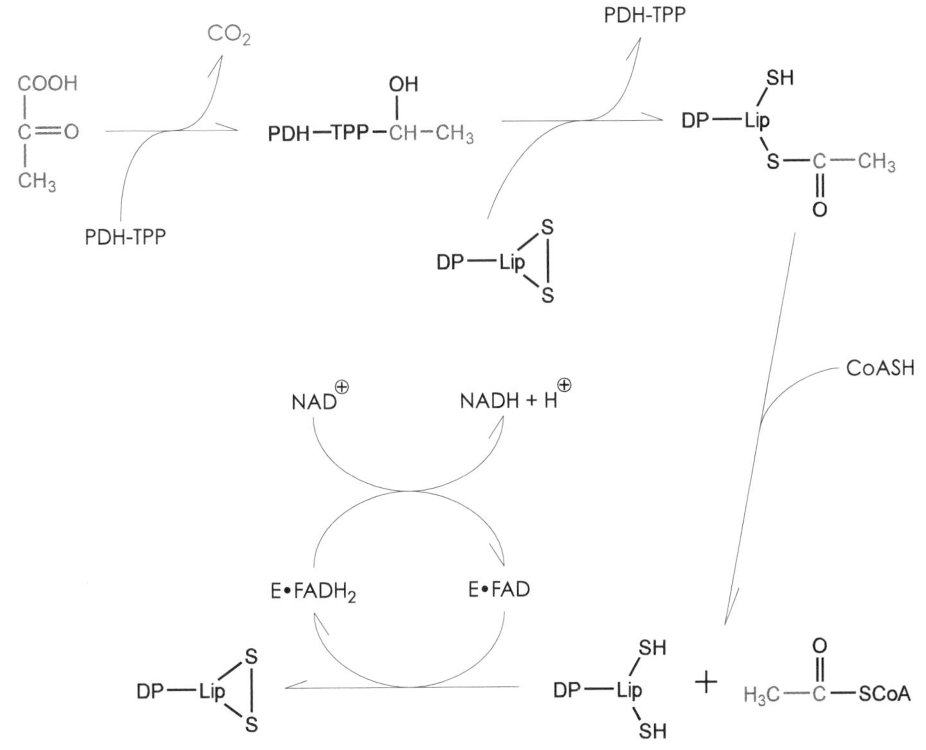

Figure 11.10.

Mechanism of pyruvate dehydrogenase. This presents the activities of the separate subunits of pyruvate dehydrogenase. The carbons of the pyruvate are followed throughout in red.

tarate, converts α-ketoglutarate + NAD^+ + coenzyme A to succinyl CoA + NADH + H^+ + CO_2. Acetyl CoA was the product of pyruvate dehydrogenase (i.e., an acyl CoA with one carbon less than the original pyruvate); in this case, succinyl CoA is the product, which is an acyl CoA with one carbon less than the original α-ketoglutarate.

5. Succinyl CoA synthetase, or succinate thiokinase, converts succinyl CoA + GDP + inorganic phosphate to succinate, coenzyme A, and GTP. This is the only reaction in the citric-acid cycle which has a substrate-level phosphorylation, as seen in glycolysis. This enables the system to conserve the energy in the thioester, succinyl CoA, for useful energy production. Note that energetically GTP, ATP, UTP, and CTP are identical in regard to the energy bond between the first and second phosphate, and the second and third phosphate. Thus, GTP is as effective a high-energy phosphate as ATP. Also, there are enzymes within the body that can easily and reversibly transfer phosphates from one nucleotide to another, such as GTP + ADP reversibly to ATP + GDP.
6. Succinate dehydrogenase is unique among the citric-acid enzymes because it is bound to the mitochondrial inner membrane. The enzyme contains FAD as a cofactor and catalyzes the removal of two hydrogens from succinate to form fumarate with a formation of an enzyme-bound $FADH_2$. Usually, NAD is associated with dehydrogenation of alcohols and aldehydes as well as α-keto acid dehydrogenases. When formed, the reduced coenzyme (NADH) is not autooxidizable and, therefore, is reasonably stable. FAD is usually associated with the formation of carbon–carbon double bonds. The coenzyme and reduced coenzyme are almost always attached as a prosthetic group, in contrast to NAD^+ and NADH which usually can easily dissociate from the enzyme. Reduced FAD (i.e., E-$FADH_2$) is autooxidizable, and if it comes into contact with O_2, can form E-FAD + H_2O_2 (peroxide). Thus $FADH_2$ is protected from this contact, except in peroxisome reactions, where the resulting H_2O_2 can be destroyed by catalase.
7. The fumarate formed can be converted reversibly by fumarase to L-malate. This is a rapid reaction, attaining equilibrium.
8. The final reaction in a citric-acid cycle is a regeneration of oxaloacetate to start the cycle again. This is accomplished by enzyme-malate dehydrogenase, which reversibly catalyzes the conversion of NAD^+ + L-malate to NADH + H^+ + oxaloacetate. As with a number of enzymes in the cycle, this enzyme, is apparently rapid enough to attain equilibrium within the mitochondrion. This appears to be the case for aconitase, fumarase, and the malate dehydrogenase. The citric-acid cycle also contains three effectively unidirectional reactions. Citrate synthetase, isocitric dehydrogenase, and α-ketoglutarate dehydrogenase. In addition, under most conditions, two of the other enzymes, succinyl CoA synthetase and succinate dehydrogenase, flow in the forward direction, although under some unusual conditions, these reactions can be reversed.

ELECTRON-TRANSPORT SYSTEM AND OXIDATIVE PHOSPHORYLATION

It is important for the cell to convert the potential energy in reduced coenzymes formed in glycolysis, and particularly the citric-acid cycle, into utilizable energy in the form of ATP. This is accomplished by a mitochondrial system that is part of the inner membrane and is referred to as the electron-transport system (Fig. 11.11).

The pathway for this transport is effectively a transfer of the hydrogens of NADH + H^+, reducing an enzyme-bound FMN to form E-$FMNH_2$. The E-$FMNH_2$ then transfers the two hydrogens to a quinone derivative, known as coenzyme Q, which is confined during this operation to the mitochondrial cristea. The reduced CoQ passes electrons to cytochrome b and releases H^+ into the mitochondrial matrix. The electrons are passed from cytochrome b to cytochrome c, then to cytochrome a, and finally to cytochrome a_3. From cytochrome a_3, the electron plus the H^+ from the mitochondrial matrix combine with molecular oxygen to form water. Thus, overall, the reduced coenzymes, either NADH or protein-bound $FMNH_2$, are reoxidized with the hydrogens, producing metabolic water. There are also other flavoprotein links with a cytochrome system that transfer the hydrogens directly to CoQ, such as α-glycerol phosphate dehydrogenase and succinate dehydrogenase. These two systems are distinct from the protein-bound FMN within the chain-oxidizing NADH.

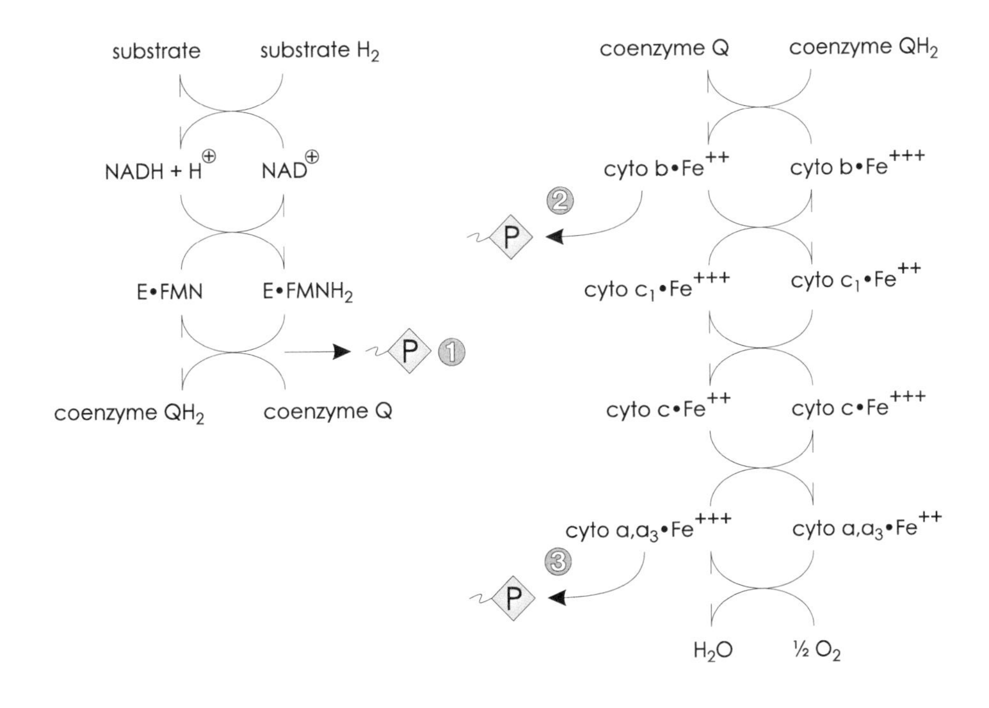

Figure 11.11.

Electron-transport system. Both the hydrogen- and the electron-transport portion are presented. The ~P indicates steps that have a sufficient energy drop to produce a high-energy phosphate. The circled numbers indicate where corresponding inhibitors of the electron-transport system have their effect.

Throughout this transport, there is a continual drop in energy level to proceed forward (Fig. 11.12). Three steps within the electron-transport system have a sufficient energy drop to account for potential formation of a high-energy phosphate. These occur between (1) NADH to protein bound FMN, (2) cytochrome b to c, and (3) cytochrome a/a_3 to O_2. The cytochrome system functions using iron containing heme proteins in which the valance of iron can be varied. Thus, when a cytochrome is reduced (i.e., accepts an electron), the iron changes from Fe^{+3} to Fe^{+2}. The cytochrome then transfers the electron down the system, with the transferring cytochrome being reoxidized to Fe^{+3} and the cytochrome accepting the electron being reduced to Fe^{+2}. The potential energy gained during these passages down the electron system are used to move hydrogen ions out of the mitochondrial matrix into the cytosol against a gradient. The estimated pH in the mitochondria is approximately 7.4, and that of a cytosol is approximately 7.0. Therefore, any movement of hydrogen ions out of the mitochondrion into the cytosol require energy supplied by the electron-transport system. The hydrogen ions can translocate with the concentration gradient back into the mitochondria from the cytosol, passing through a selective channel in the mitochondrial membrane known as the F_o channel (Fig. 11.13).

During the passage through the F_o channel, and interacting with the F_1 ATPase, the movement of the hydrogen ions into the mitochondrion are coupled with

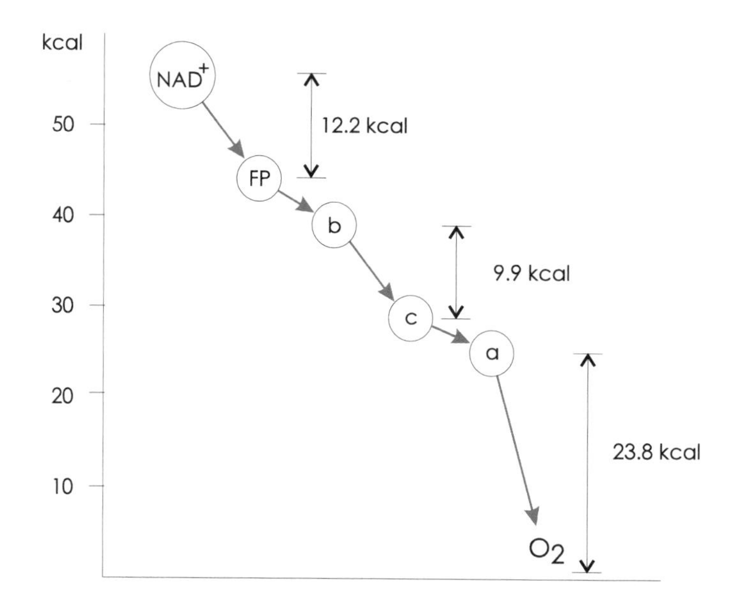

Figure 11.12.

Energy levels of electron-transport system. The energy levels and drops in energy along the electron-transport system are shown.

ATP formation from ADP and inorganic phosphate. The driving force of this phosphorylation is the pH gradient between mitochondrion and cytosol, with the accompanying movement of the hydrogen ions back into the mitochondria with the concentration gradient. This pH differential was due to the pumping of hydrogen ions out of the mitochondrion by the electron-transport system.

The electron-transport system and oxidative phosphorylation in most cells is closely coupled. That is, there is no movement of electrons down the electron-transport system without concomitant ATP formation. The ATP or high-energy phosphate yield from NADH is three high-energy phosphates; from a protein-bound $FADH_2$ reaction (such as glycerol phosphate dehydrogenase, succinic dehydrogenase, or other

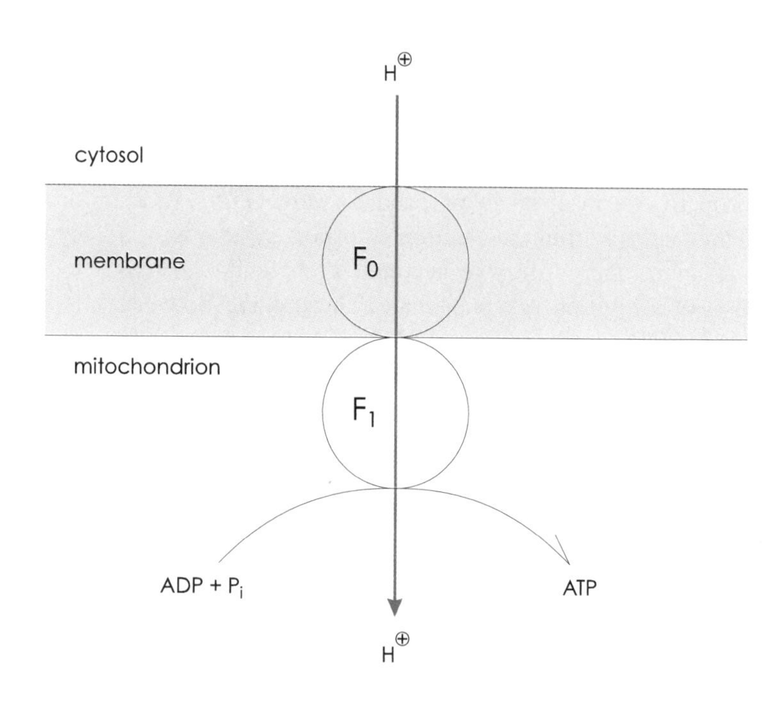

Figure 11.13.

ATP formation. A schematic of the F_o/F_1 action forming ATP is shown.

reactions seen in fatty-acid metabolism), the yield is two high-energy phosphates. Since the movement of two electrons down the electron-transport system finally reacts with one oxygen (i.e., 1/2 O_2) to water, there is accompanying oxygen uptake along with the formation of high-energy phosphates. This is referred to as oxidative phosphorylation, because the supply of cytochromes is limited, and unless they can be regenerated, with oxygen as the final hydrogen acceptor, this system cannot continue and thus produces no high-energy phosphates.

A useful measurement and concept is the P/O ratio, which measures how many high energy phosphates are obtained per atom of oxygen used. These values are 3/1 for NADH and 2/1 for $FADH_2$. Both the formation of high-energy phosphates and oxygen consumption can be measured using an oxygen electrode. In many cases, isolated mitochondria are used in conjunction with various substrates to measure P/O ratios. If the ratio obtained is close to the theoretical, the mitochondria are considered to be tightly coupled. If a much lower ratio of P/O is obtained, the mitochondria are said to be relatively uncoupled, which might be due to a physiological condition or damage of the mitochondria during preparation.

During the normal metabolic process, the rate-limiting factor for oxidative phosphorylation is usually the availability of ADP. In the presence of an oxidizable substrate with low ADP, a slight oxygen uptake is observed with mitochondria, owing to potential uncoupling of endogenous ADP, or ATP may be hydrolyzed to form low levels of ADP. This is referred to as resting respiration or state 4. Upon addition of ADP to these mitochondria, there is a rapid increase in oxygen utilization accompanied by ATP formation from the added ADP and inorganic phosphate; this is referred to as active respiration or state 3. When the added ADP is all converted to ATP, the mitochondria return to the slow-resting respiration state; thus, the importance of ATP formation concomitant with oxidation is clearly observed. If atractyloside, an ATP–ADP inner-membrane translocase inhibitor, which blocks ADP entry into the mitochondrial matrix, is added to the preparation, only resting respiration is observed, even in the presence of ADP. This shows the crucial nature of the intramitochondrial ADP for oxidative phosphorylation. It also illustrates that the ATP formed in the mitochondrion is exchanged for ADP in the cytosol by a specific translocator, assuring a continuous supply of cytosolic ATP in exchange for a continuous supply of mitochondrial ADP to keep respiration and energy-requiring functions proceeding. There are also both inhibitors and uncouplers of oxidative phosphorylation in the electron-transport system. Two examples of substances that inhibit between NADH and coenzyme Q (ubiquinone) are rotenone and amytal, a barbiturate. These effectively prevent the electron-transport system from flowing if the hydrogen donor is NADH; however, for reactions such as succinate dehydrogenase, which generate a protein-bound $FADH_2$, electron transport and oxidative phosphorylation can continue. Antimycin A inhibits between cytochrome b and c, and prevents electron-transport flow and oxidative phosphorylation from all normal substrates. Another inhibitor of this system is cyanide, which reacts with cytochrome a_3 (cytochrome oxidase); one may question why this would inhibit oxidative phosphorylation since many of the steps leading to cytochrome a_2 are those that generate sufficient energy to extrude hydrogen ions from the mitochondrion. The reason is that there are limited amounts of cytochromes, and when the flow is inhibited at cytochrome a_3, all the cytochromes become more reduced and cannot pass reducing power down the chain; thus, the final hydrogen ion and electron cannot be passed on to oxygen because of the inhibition of cytochrome oxidase. The amount of hemoglobin present in individuals is much greater than the amount of cytochromes; therefore, in certain cases of suspected cyanide poisoning, particularly where the release may be more protracted (e.g., eating seeds that contained complex cyanide derivatives), one treatment to use amylnitrite to oxidize some of the hemoglobin to the Fe^{+3} state (i.e., methehemoglobin), which can also react with cyanide. Cyanide only reacts with the Fe^{+3} heme. Thus, the methehemoglobin ties up cyanide before it reaches the cells and completely inhibits the cytochrome system.

An uncoupler of oxidative phosphorylation is dinitrophenol, which acts by carrying the hydrogen ions back into the mitochondrion without passing through the tightly coupled F_o/F_1 system. Although electrons traverse the electron-transport system, where oxidation per se is very rapid (similar to state 3), there is little or no ATP formed (depending on the concentration of dinitrophenol) because the hydrogen ions that leak back into the mitochondrion via pathways other than the F_o/F_1 channel are not accompanied by high-energy phosphate formation. An important difference is that inhibitors of an electron system prevent both oxidation of substrates as well as formation of high-energy phosphates, whereas an uncoupler allows rapid oxidation but is accompanied by less total high-energy phosphate production.

Malate and α-Glycerolphosphate Shuttle

If glucose undergoes glycolysis and the resulting pyruvate is oxidized in the citric-acid cycle, the pyruvate cannot also be reduced to lactate. Therefore, the NADH produced at glyceraldehyde-3-phosphate dehydrogenase remains in the cytosol. ***NADH cannot cross the mitochondrial membrane and therefore cannot directly enter the electron-transport system***. Glycolysis depends on NAD^+ and only limiting amounts of NAD^+ + NADH are present in the cytosol. Thus, the NADH formed in the cytosol must be regenerated back to NAD^+. This regeneration is accomplished by using small molecules to shuttle in the reducing power of NADH to regenerate cytosolic NAD^+. These smaller molecules are malate for the malate shuttle, and α-glycerolphosphate for the α-glycerolphosphate shuttle (Fig. 11.14).

The malate shuttle operates by reducing oxaloacetate in the cytosol to malate, regenerating NAD^+ from NADH. The malate is, via specific translocases, moved across the mitochondrial membrane in exchange for other ions such as α-ketoglutarate. After entering the mitochondrion, the malate can be reoxidized to oxaloacetate, generating a mitochondrial NADH, which can then be utilized by the mitochondrial electron-transport system. The oxaloacetate formed, which cannot transverse the mitochondrial membrane, is transaminated with glutamate via aspartate aminotransferase to form α-ketoglutarate and aspartate in the mitochondrial matrix. The aspartate can then exit the mitochondrion to the cytosol, where, again under the action of aspartate aminotransferase can form oxaloacetate and glutamate. The glutamate can exchange with aspartate to reenter the mitochondria for continual aminotransferase activity. The oxaloacetate formed in the cytosol can then be reduced

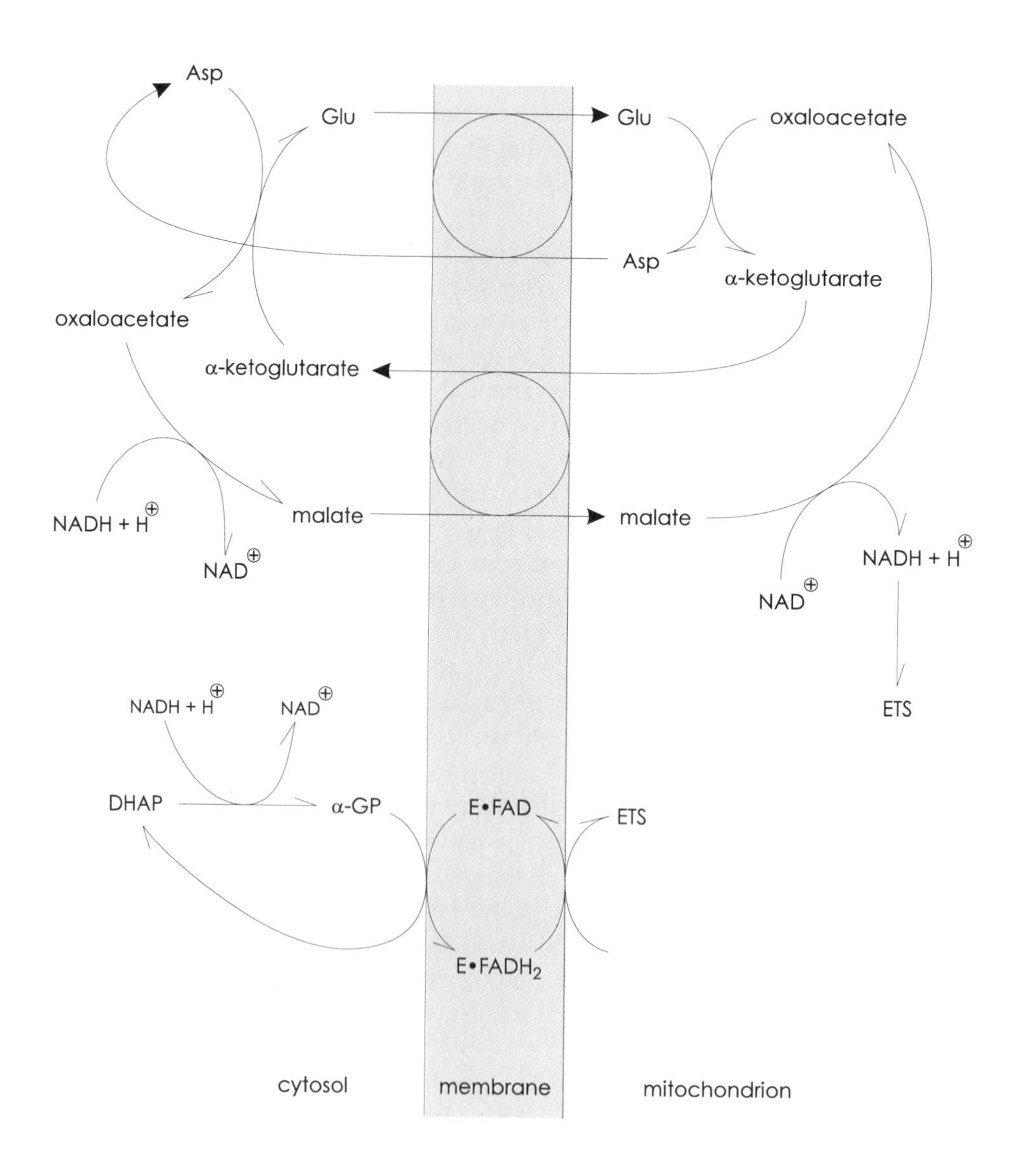

Figure 11.14.

Malate and α-glycerolphosphate shuttles. The shuttling of reducing power from cytosolic NADH into the mitochondrion. The malate shuttle (top portion) is more complex, but yields more high-energy phosphates. Malate can move reducing power into or out of the mitochondria. The α-glycerolphosphate shuttle only moves reducing power from the cytosol to the mitochondria.

to malate and the cycle begins again. Thus, small molecules such as malate and aspartate can, with specific translocases, transverse the mitochondrial membrane to allow a hydrogen shuttle to be effective in regenerating cystosolic NAD^+ from NADH. ***This cycle requires that both aspartate aminotransferase and malic dehydrogenase be present in both the cytosol and the mitochondria***. The transamination is required because the inner mitochondrial membrane is relatively impermeable to oxaloacetate and there are no specific translocases for this intermediate. However, there are translocases for aspartate, glutamate, and α-ketoglutarate. This makes transamination an essential part of the malate shuttle.

A second hydrogen shuttle system that is less active in mammals, including people, and more active in insects, is the α-glycerolphosphate shuttle. This shuttle consists of α-glycerolphosphate dehydrogenase, which converts dihydroxyacetone phosphate and NADH to α-glycerophosphate and regenerates NAD^+. The α-glycerolphosphate enters the mitochondrion via a translocase, where there is another α-glycerolphosphate dehydrogenase which is a flavoprotein. This mitochondrial enzyme contains FAD and oxidizes the α-glycerolphosphate to dihydroxyacetone phosphate and a protein-bound reduced $FADH_2$. The dihydroxyacetone phosphate formed can then exit the mitochondrion to start the cycle again. The reduced flavoprotein can be regenerated by the flavoprotein $FADH_2$ donating its hydrogens to the electron-transport system at the level of coenzyme Q within the mitochondrion.

Each of these shuttle systems has distinct advantages and disadvantages. The advantage of the malate shuttle is that the NADH generated in the mitochondrial matrix will yield three high-energy phosphates, whereas the $FADH_2$ in the α-glycerolphosphate shuttle will only yield two high-energy phosphates. The malate shuttle is, in part, reversible, thus reducing power can be moved either from the cytosol to the mitochondrion, as is the case for muscle and brain, or by malate formed in the mitochondrion translocating to the cytosol, as is the case in gluconeogenesis in the liver. The α-glycerolphosphate shuttle only moves reducing power from the cytosol to the mitochondrion.

SUMMARY

1. Sigmoidal kinetics produce a finer control of change in enzyme activity with change in substrate than Michaelis–Menten kinetics. Many, but not all, control enzymes exhibit sigmoidal kinetics.

2. Glucose can be converted to lactate with the production of two high-energy phosphates (ATP) in the absence of oxygen.

3. Transport of glucose into the cell can limit glycolysis.

4. Phosphofructokinase, the first step committed to glycolysis, appears to be the rate-determining enzyme of glycolysis.

5. ATP and citrate are major negative effectors (inhibition) of PFK, while AMP and F-2,6-BP are the major activators of PFK.

6. In the presence of oxygen, pyruvate formed during glycolysis can be metabolized to CO_2 and water in the citric-acid cycle.

7. The connection between glycolysis and the citric-acid cycle is pyruvate dehydrogenase. This mitochondrial enzyme is inhibited by acetyl CoA and NADH or by phosphorylation.

8. The citric-acid cycle, a mitochondrial process, starts with citrate synthase catalyzing the conversion of oxaloacetate and acetyl CoA to citrate. This enzyme can be inhibited weakly by ATP and more strongly by fatty acyl CoA.

9. A major control enzyme in the citric-acid cycle is isocitrate dehydrogenase. This enzyme is inhibited by NADH and activated by ADP.

10. Most of the potential energy from the citric-acid cycle is in the form of reduced coenzymes, NADH, and enzyme bound $FADH_2$.

11. These reduced coenzymes can be oxidized in the electron-transport system to produce metabolic water and ATP.

12. The cytochromes transfer electrons via heme iron and units of H^+ are pumped from mitochondria to cytosol.

13. The return of H^+ back to the mitochondrion is coupled with ATP formation.

14. Oxidation of NADH can produce three high-energy phosphates (ATPs), and enzyme-bound $FADH_2$ (as per succinate dehydrogenase) can produce two high-energy phosphates.

15. Reducing power (i.e., NADH) can be transported into the mitochondrion by the malate shuttle. Malate can transport reducing power out of the mi-

tochondrion. Reducing power can also be moved from the cytosol to the mitochondrion by the α-glycerolphosphate shuttle. In people, the malate shuttle plays a more important role than the α-glycerolphosphate shuttle.

REFERENCES

Facilitative glucose transporters: An expanding family, G. W. Gould and G. I. Bell, 1990, *Trends Biochem. Sci.* **15**:18–23.

Flux-generating and regulatory steps in metabolic control, E. A. Newsholme and B. Crabtree, 1981, *Trands Biochem. Sci.*, **6**:53–56.

Regulation of the pyruvate dehydrogenase multienzyme complex, R. H. Behal, D. B. Buxton, V. G. Robertson, and M. S. Olson, 1993, *Ann. Rev. Nutr.*, **13**:497–520.

Enzymatic regulation of liver acetyl-CoA metabolism in relation to ketogenesis, O. Wieland, L. Weiss, and I. Eger-Neufeldt, 1964, *Adv. Enzyme Reg.*, 85–99.

The glucose paradox: New perspectives on hepatic carbohydrate metabolism, J. Katz, M. Kawajima, D. W. Foster, and J. D. McGarry, 1986, *Trends Biochem. Sci.*, **11**:136–140.

Pyruvate kinase and other enzyme deficiency disorders of the erythrocyte, W. N. Valentine, K. R. Tanaka and D. E. Paglia, in *The Metabolic Basis of Inherited Disease*, Vol. 1, 6th ed., C. R. Scriver, A. L. Beaudet, W. S. Sly, and D. Valle, McGraw-Hill—New York, Eds. 1989, pp. 2341–2365.

REVIEW QUESTIONS

1. Which of these is not an effector of PFK?
   a) ATP
   b) ADP
   c) AMP
   d) F-2,6-BP
   e) Citrate

2. Which is the first enzyme that commits glucose to glycolysis?
   a) Hexokinase
   b) Glyceraldehyde-3-phosphate to dehydrogenase
   c) PFK
   d) Aldolase
   e) Lactate dehydrogenase

3. The first step in glycolysis that produces ATP is . . .
   a) PFK
   b) Pyruvate kinase
   c) Enolase
   d) Glyceraldehyde-3-phosphate dehydrogenase
   3) 3-Phosphoglycerate kinase

4. Which of these is not a cofactor for pyruvate dehydrogenase and α-ketoglutarate dehydrogenase?
   a) FAD
   b) Lipoic acid
   c) ATP
   d) NAD^+
   e) Coenzyme A

5. The only reaction in the citric-acid cycle that produces a substrate-level phosphorylation?
   a) Succinyl CoA synthetase
   b) Pyruvate dehydrogenase
   c) Citrate synthetase
   d) Succinate dehydrogenase
   e) α-Ketoglutarate dehydrogenase

6. The major system for transporting reducing power into the mitochondrion in people?
   a) Glyoxylate shuttle
   b) α-Glycerolphosphate shuttle
   c) The carnitine shuttle
   d) The malate shuttle
   e) The phosphate shuttle

7. Which of these is a major positive effector (activator) of isocitrate dehydrogenase?
   a) AMP
   b) ADP
   c) ATP
   d) NAD^+
   e) NADH

8. Hexokinase is controlled primarily by
   a) ATP concentrations
   b) NADH concentrations
   c) F-1,6-BP concentrations
   d) G-6-P concentrations
   e) AMP concentrations

9. Because of allosites, ATP can be both a substrate and an inhibitor in the physiological direction for

a) Hexokinase
b) PFK
c) Pyruvate kinase
d) Pyruvate dehydrogenase
e) Citrate synthase

10. Which of these is not a physiological controller of pyruvate dehydrogenase?
 a) Acetyl coenzyme A
 b) NADH
 c) covalent modification
 d) $FADH_2$
 e) ATP

11. Why is NADH the negative effector of isocitrate dehydrogenase instead of ATP, as an energy signal?
 a) NADH is higher in concentration than ATP.
 b) NADH is worth more potential high energy phosphates than ATP.
 c) The malate shuttle moves reducing power but not ATP.
 d) ATP, ADP, and AMP are in equilibrium via adenylate kinase.
 e) NADH and not ATP is the major potential energy compound produced in the citric-acid cycle.

12. Malic dehydrogenase is not a control reaction in the citric-acid cycle because
 a) It can occur both in the cytosol and mitochondria.
 b) It is the last step of the cycle.
 c) It is in equilibrium.
 d) It can form aspartate.
 e) Malate and oxaloacetate are both dicarboxcylic acids.

13. If hydrogen ions pass from the cytosol into the mitochondrion without going through the F_1 and F_o channels?
 a) Well coupled
 b) Oxidation
 c) Reduction
 d) Uncoupled
 e) Active transport

14. What is added to increase flow of reducing power, electrons, and oxygen uptake in intact isolated mitochondria in the presence of an oxidizable substrate?
 a) More citric-acid cycle enzymes
 b) ADP
 c) $FADH_2$
 d) NADH
 e) Cyanide

ANSWERS TO REVIEW QUESTIONS

---

1. ***b*** The answer is ADP. ATP and citrate are both negative effectors of PFK. Amp and F-2,6-BP are both positive effectors of PFK.

2. ***c*** The answer is PFK, which commits the hexose-6-phosphates to glycolysis. Hexokinase initiates glucose metabolism; however, the G-6-P formed can form glycogen, transverse the pentose phosphate pathway, or produce metabolites other than glycolytic intermediates. The other three enzymes catalyze glycolytic reactions, which occur after the action of PFK. These glycolytic enzymes catalyze physiologically reversible reactions, whereas PFK is physiologically unidirectional.

3. ***e*** The answer is 3-Phosphoglycerate kinase. This enzyme catalyzes the formation of ATP from ADP with the concurrent conversion of 1,3BP glycerate to 3-P glycerate. The pyruvate kinase reactions also forms ATP, but this occurs later in glycolysis. The reactions catalyzed by the other three enzymes do not produce ATP.

4. ***c*** The answer is ATP. All of the other cofactors participate in both the action of pyruvate dehydrogenase and the α-ketoglutarate dehydrogenase.

5. ***a*** The answer is succinyl CoA synthetase, which catalyzes the conversions of succinyl CoA and GDP + Pi to succinate, CoA, and GTP. None of the other reactions directly produces a high-energy phosphate.

6. ***d*** The answer is the malate shuttle, which occurs in almost all aerobic tissues. There is some transport by the α-glycerol phosphate shuttle, but considerably less than via the malate shuttle. None of the other systems transports reducing power (i.e., NADH).

7. ***b*** The answer is ADP. NADH is a negative effector of isocitrate dehydrogenase. The other three are not major effectors of isocitrate dehydrogenase.

8. ***d*** The answer is G-6-P concentrations. The concentrations of the other metabolites have other roles in metabolic controls.

9. ***b*** The answer is PFK. ATP is a substrate for, but not an inhibitor of, hexokinase. ATP is a substrate for pyruvate kinase, but only in the nonphysiological direction. ATP is not involved directly as a substrate for pyruvate dehydrogenase or citrate synthase.

10. ***d*** The answer is $FADH_2$. The other four are potential negative effectors of pyruvate dehydrogenase.

11. ***e*** The answer is e because only the succinyl CoA synthetase directly produces a high-energy phosphate. Other reactions of the citric-acid cycle catalyze molecular rearrangements and/or the formation of reduced coenzymes (i.e., NADH or enzyme-bound $FADH_2$).

12. ***c*** The answer is in equilibrium. An enzyme can do no better than achieve equilibrium; a reaction that achieves equilibrium cannot control flux.

13. ***d*** The answer is uncoupled. Hydrogen ions that do not flow through the F_1 and F_o channels reenter the mitochondrion without the formation of ATP and allow oxygen consumption to continue, which is referred to as uncoupled.

14. ***b*** The answer is ADP. ADP is required to form ATP, and in a well-coupled mitochondrion, there is no flow of electrons and oxygen uptake without concomitant ATP formation.

CHAPTER 12

GLYCOGEN METABOLISM, PENTOSE-PHOSPHATE PATHWAY, AND METABOLISM OF OTHER CARBOHYDRATES

GLYCOGEN METABOLISM

Introduction

The metabolic and physiological advantages of using glycogen to store glucose and glucose phosphates is examined in this chapter. The techniques used to control glycogen synthesis and degradation include metabolic effectors, covalent modification, and hormonal control. Two separate rate-limiting enzymes are involved: one for synthesis and, another for degradation. ***The physiological role of glycogen in muscle is different from that in the liver.*** Thus, as might be expected, controls in the two tissues may differ via the function of isoenzymes. The synthesis of glycogen requires more energy than is yielded by reconversion of glycogen to hexose phosphates; however, excessive energy loss is prevented because cycling of glycogen is kept to a minimum. To accomplish minimum recycling, the cell turns off synthesis when degradation is activated and vice versa. In this chapter, we also examine another pathway of glucose metabolism, known as the pentose-phosphate pathway, and the metabolism of other carbohydrates. The two major hexoses to be examined are fructose and galactose, both of which occur in human diets primarily as parts of disaccharides.

Glycogen is an animal storage form of glucose. The advantage of storing glycogen instead of free glucose is that glycogen, a very large polymer of glucose, will have very little osmotic effect, whereas storage of an equivalent amount of free glucose in the cell would have profound osmotic effects. Glycogen in most cells can be both synthesized and degraded. Despite the broad occurrence of glycogen, only two tissues contain large amounts of glycogen—muscle and liver.

Biosynthesis of Glycogen

The biosynthesis of glycogen occurs from glucose-6-phosphate. As indicated earlier, this is one of several alternative fates of glucose-6-phosphate in addition to glycolysis. Glucose-6-phosphate via phosphoglucomutase and glucose-1,6-bisphosphate can be reversibly converted to glucose-1-phosphate. This reaction mechanism is analogous to that for phosphoglycerate mutase discussed earlier. Glucose-1-phosphate can be activated further by a reversible reaction with UTP forming UDP glucose (UDPG) + pyrophosphate. Although this reaction is reversible, the pyrophosphate is rapidly degraded in the cell to two inorganic phosphates; thus, by removing one of the products of this reaction (i.e., pyrophosphate), the reaction can be pulled strongly in the direction of UDPG formation. This is an example of investing one more high-energy phosphate than necessary to assure that a reaction proceeds unidirectionally, despite unfavorable equilibrium considerations. For example, the body could have formed UTP glucose and inorganic phosphate; however, equilibrium constraints and the high concentration of inorganic phosphate would prevent this reaction from proceeding in a forward direction to a sufficient degree.

The substance UDPG is a reactant in two metabolic processes—glycogen synthesis and detoxification reactions. It can be oxidized with NAD^+ to UDP glucuronic acid, an important intermediate in conjugation and detoxification pathways, such as drug metabolism and excretion of bile pigments. ***Glycogen synthesis***, which is quantitatively more important, ***is catalyzed by glycogen synthase*** via the reaction of glycogen of n glucose units + UDPG forming glycogen of $n + 1$ glucose units + UDP (Fig. 12.1). This reaction is unidirectional and is under both allosteric and hormonal control. The glycogen links are primarily α-1,4 and α-1,6 linkages. Glycogen synthase forms only α-1,4 links. As the glycogen chain grows, a branching enzyme acts on a portion consisting of several glucoses linked α-1,4 and transfers this portion to an α-1,6 branch so that a highly branched polysaccharide is formed; this is in contrast to the synthesis of amylose (starch) where all the linkages remain in a straight α-1,4 chain. Since all of the branches in glycogen have the four hydroxyl groups of the terminal glucose moieties free, additional glycogen synthesis can continue by adding glucose from UDP glucose to the end of any of the branches. Once more, as the branches become longer, new α-1,6 branches are formed.

Breakdown

Glycogen breakdown occurs by a phosphorolysis at the α-1,4 link producing glucose-1-phosphate and a glycogen molecule with one less glucose moiety. The enzyme that catalyzes this reaction is phosphorylase. This reaction is readily reversible in a test tube, but in the cell, owing to the high level of inorganic phosphate and low level of glucose-1-phosphate, it proceeds unidirectionally. The major roles of glycogen are different in muscle and liver. ***In liver, glycogen is primarily a store of potential blood glucose; in muscle, it is primarily an energy source.*** Highly branched glycogen has more available terminal α-1,4 links, for which phosphorylase is specific, than does a linear polysaccharide such as amylose. This provides a greater availability of reactive ends and a more rapid degradation to form G-1-P for energy production. The contrasts between the functions of branched and linear polysaccharides are instructive. In plants, where mobility and rapid responses are not important, the storage of polysaccharide as an energy source occurs mainly as linear or only slightly branched polysaccharide. In contrast, in animals, where mobility and rapid responses are critical for success, the polysaccharide store is the highly branched glycogen. Glycogen can be degraded rapidly to G-1-P either for energy production or as a source of blood glucose.

Phosphorylase catalyzes degradation of glycogen to glucose-1-phosphate and a limit dextrin; that is, when the branches are within four glucose units of the α-1,6 link, phosphorylase cannot catalyze breakdown of this branch to glucose-1-phosphate and a branch of fewer glucose units. ***Further degradation of glycogen requires a debranching enzyme*** that will remove three of the α-1,4 linked four glucosyl units adjacent to the α-1,6 bond and move them to the end of another chain so that phosphorylase can continue its breakdown to form glucose-1-phosphate. The last glucosyl unit (i.e., the one with the α-1,6 link) is then removed in an α-1,6 glycosidase-catalyzed reaction, releasing free glucose. Glucose-1-phosphate formed by the action of phosphorylase can, through phosphoglucomutase, be converted to glucose-6-phosphate, which then can enter glycolysis in the muscle or, as we shall see, be converted to glucose in the liver.

Glycogen metabolism is of medical interest since there are a number of glycogen storage diseases where glycogen cannot be degraded in the appropriate manner. In all of these diseases, there are potentially significant consequences of excess glycogen accumulation leading to muscle weakness and inability of cells to use glycogen as a source of energy. If liver glycogen degradation is impaired, there may be not only excess storage, but also the possibility of hypoglycemia (and possible death) due to inadequate ability to mobilize liver glycogen to produce blood glucose.

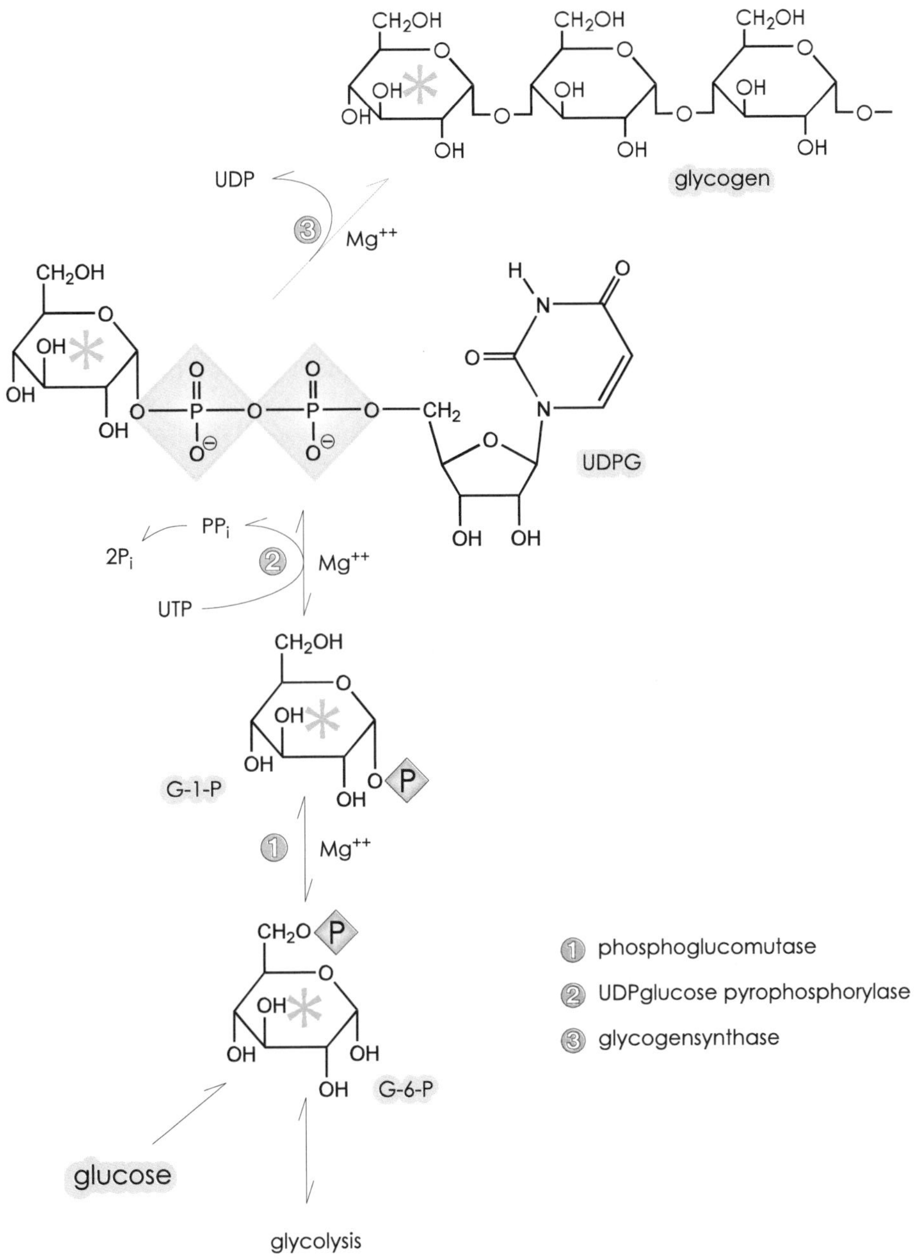

1. phosphoglucomutase
2. UDPglucose pyrophosphorylase
3. glycogensynthase

Figure 12.1.

Glycogen synthesis. The general pathway of glycogen synthesis is presented. The action of the key enzyme in synthesis, glycogen synthase, is shown by purple arrows.

Glycogen Storage Diseases

There are a number of glycogen storage diseases. Of those that are discussed, three involve deficiencies of systems directly related to glycogen degradation; another involves abnormal metabolism of the product of glycogen degradation—specifically, the hexose phosphate.

TYPE I (VON GIERKE'S DISEASE). This, the most common of the glycogen storage diseases, is due to a recessive gene defect and is one in which glycogen accumulates primarily in the liver and, to a lesser extent, in the kidney. Both organs are strongly involved in gluconeogenesis (see Chapter 13), a process that requires glucose-6-phosphatase for release of free glucose into the blood. Individuals with Von Gierke's disease are missing this enzyme; thus G-6-P accumulates in the liver and cannot be released as free glucose. The G-6-P from glycogen degradation or gluconeogenesis, which normally proceeds to glucose, now has two alternate fates: reformation of glycogen, leading to excessive glycogen accumulation, or glycolysis, leading to excessive blood lactate levels. The major problem, in addition to excess glycogen in the liver, is the inability to obtain glucose from gluconeogenesis and glycogenolysis, resulting in hypoglycemia. Some small amounts of glucose can be released from glycogen, but only at the α-1,6 branches, via the amylo-α-1,6-glucosidase catalyzed reaction.

TYPE II (POMPE'S DISEASE). This is due to the absence of a lysosomal amylo-α-1,6-glucosidase which catalyzes removal of the glucoses at the α-1,6 branches in a nonphosphorylitic cleavage producing free glucose instead of glucose phosphate. In this disease, glycogen accumulates in all tissues that store glycogen. The accumulation of glycogen particles in the lysosome without concomitant degradation causes a defect in lysosomal functions. This disease results in heart failure at an early age.

TYPE III (CORI'S DISEASE). This is due to the absence of the debranching enzyme. Glycogen accumulates because only the outer portion of the branches can be removed by phosphorylase. Thus, a highly branched glycogen accumulates and, because glycogen degradation is incomplete, there is an occasional mild hypoglycemia.

TYPE V (MCARDLE'S DISEASE). This is due to the absence of muscle phosphorylase, severely limiting glycogen degradation and the capacity for sustained strenuous exercise, because muscle glycogen is unavailable as an every source. There usually is muscle damage accompanying this disease.

Control of Glycogen Synthesis

There are both hormonal and allosteric controls on glycogen synthesis. The action of some allosteric effectors of glycogen metabolism can be observed in the test tube, but are not of physiological significance. Glycogen metabolism in the muscle is controlled by two opposing hormones: epinephrine, a fight-or-flight response, and insulin, the major anabolic hormone. Insulin can stimulate the synthesis of glycogen, whereas epinephrine can accelerate the breakdown of glycogen. The major control of glycogen synthesis in the muscle is the activity state of glycogen synthase. This enzyme can occur in either a phosphorylated or a nonphosphorylated form. Glycogen synthase, as with pyruvate dehydrogenase, is inactive in the phosphorylated form and is active in the nonphosphorylated form. Also, glucose-6-phosphate is a potential allosteric activator of this enzyme, particularly with the inactive form. The phosphorylated form of glycogen synthase has been referred to as the D form, which indicates that it is dependent on high levels of glucose-6-phosphate for activation. However, this appears to be of no significance physiologically in the muscle because, during glycogen synthesis from glucose, high amounts of glucose-6-phosphate do not accumulate owing to the glucose-6-phosphate inhibition of hexokinase. Epinephrine causes increases in the muscle content of cyclic AMP (cAMP), a second message for a number of hormones. The elevated cAMP can cause activation of a cAMP-dependent protein kinase in the following manner:

$$R_2C_2 + 4cAMP \rightarrow 2C + R_2\,(cAMP)_4$$

The C subunits are the catalytic subunits which are active in phosphorylating appropriate target proteins such as glycogen synthase. The R subunits are regulatory subunits which, when associated with catalytic subunits, renders them inactive in regard to catalyzing protein phosphorylation (see Chapter 16 for more detail). Insulin, apparently as one of its many effects, causes a decrease in cAMP and/or an increase in the rate of dephosphorylation of protein. The cAMP-dependent protein kinase can catalyze phosphorylation of certain proteins, and phosphoprotein phosphatases can remove this phosphate. ATP plus protein by catalysis of the cAMP-dependent protein kinase will yield phosphorylated protein and ADP. The phosphoprotein phosphatase can catalyze protein phosphate to protein + inorganic phosphate. In muscle, the rate of glycogen synthesis depends on the phosphorylation state of the glycogen synthase. The more this enzyme is in the dephosphorylated form, the faster glycogen synthesis occurs; the more it is in the phosphorylated state, the slower glycogen synthesis will proceed. In muscle, phosphorylase and glycogen synthase both seem to be associated with the large glycogen particle. As the particle gets larger, and as we approach approximately 2% of wet weight as glycogen, a greater proportion of the glycogen synthase occurs in the inactive or phosphorylated state regardless of the level of insulin and potential absence of epinephrine (Fig. 12.2). This then limits the muscle storage of glycogen to approximately 2% of wet weight. In some cases, long-distance runners go through dietary and exercise regimens to allow for a process known as "glycogen loading," where muscle glycogen can be elevated somewhat above the 2% level, but not as high as in the liver.

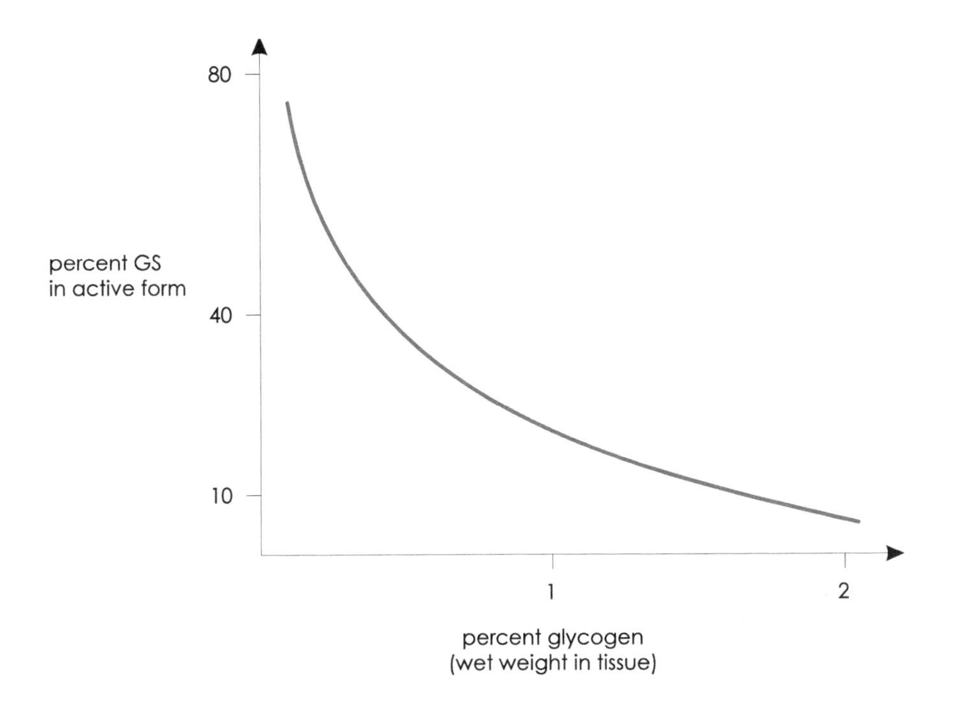

Figure 12.2.

Glycogen synthase as a function of glycogen concentration. The percent of glycogen synthase in the active form as a function of glycogen concentration is presented. The percent of glycogen synthase in the active form decreases as the concentration of glycogen rises.

These same controls occur in the liver cell; however, the ability to deposit glycogen in the liver is much greater. For example, under starve–refeed conditions, the liver can contain as much as 8% glycogen. The possibility exists that the glucose-6-phosphate allosteric activation of glycogen synthase may take place in the liver and allow for a larger deposit of glycogen. This may be possible because some glucokinase may not be inhibited, especially in the presence of F-1-P, and considerable glucose-6-phosphate can be formed during the process called gluconeogenesis (Chapter 13). An example of this phenomenon is Von Gierke's disease, where both glucose-6-phosphate and glycogen accumulate.

One major difference between liver and muscle is in hormonal sensitivity. In the interconversion of hepatic glycogen synthase, epinephrine favors a phosphorylated form in muscle; however, the major hormone causing increased cAMP in the liver is glucagon, which is released in response to low blood glucose. This is consistent with the role of liver glycogen as a storage site of potential blood glucose. In response to glucagon, glycogen synthase is phosphorylated and inactive. In both muscle and liver, the response of this enzyme is consistent with hormonal messages. That is, insulin signals that conditions are appropriate for storage, and epinephrine and glucagon signal the conditions to mobilize glycogen, either for energy (for fight or flight) or as a source of blood glucose to return blood glucose levels to normal.

Control of Glycogen Degradation

As with glycogen synthase, phosphorylase is under exquisite allosteric and hormonal control. It was indicated previously that an extra-high-energy phosphate was used for glycogen synthesis. Although this is not recovered directly, there are a number of advantages of glycogen utilization in the muscle. If glucose-1-phosphate formed from glycogen breakdown via phosphorylase is converted to lactate, the net gain is three high-energy phosphates instead of the two that can be obtained from glucose. The reason for this difference is that the cell is starting with a phosphorylated intermediate and does not require the first ATP for glucose phosphorylation. It was indicated that the only source of anaerobic energy is glucose and/or glycogen; therefore, this difference can be significant anaerobically—three high-energy phosphates via glycogen to lactate (per glycosyl unit), compared to two high-energy phosphates via glucose to lactate is actually an increased yield of 50% under anaerobic conditions. Another advantage of glycogen breakdown is the lack of strong allosteric feedback by hexose phosphates on phosphorylase; thus, a cell can build up high intracellular levels of glucose-6-phosphate and fructose-6-phosphate, allowing PFK to proceed under conditions where at lower levels of fructose-6-phosphate, this would not proceed at any reasonable rate. Such hexose phosphate accumulation cannot occur starting from free glucose because of the inhibition of hexokinase by G-6-P. Thus, in emergency conditions, large amounts of substrate for energy production can be mobilized within the cell, which is advantageous to the cell, the tissue, and the organism.

The control of glycogenolysis is more complex than that of glycogen synthesis, and it has a number of places where intervention can occur. Mobilization of energy sources is usually more rapid than storage, because an instant supply of energy from these sources may be critical, whereas storage can occur over a period of time and effectively at leisure. A major signal for glycogen mobilization is an increase in intracellular cAMP which activates cAMP-dependent protein kinase. In the case of phosphorylase, unlike many other enzyme systems, the AMP-dependent kinase does not catalyze the phosphorylation of the final affected enzyme, that is, phosphorylase (Fig. 12.3).

There is a cascade in which the cAMP-dependent protein kinase catalyzes a phosphorylation of phosphorylase kinase. This activates phosphorylase kinase, which catalyzes the phosphorylation of phosphorylase from the inactive to the active form, which can then rapidly catalyze the mobilization of glycogen. The advantage of the cascade system is a multiplicative effect. That is, each cAMP-dependent protein kinase can catalyze the phosphorylation of a large number of phosphorylase kinases, which in turn can catalyze the phosphorylation of a large number of inactive phosphorylase molecules (i.e., phosphorylase b) to the active phosphorylated form (i.e., phosphorylase a), and each of these active phosphorylases can accelerate production of hexose phosphates from glycogen. This amplifies the stimulatory effect of cAMP and causes a more rapid mobilization of glycogen than would result from catalysis of a direct phosphorylation of phosphorylase by the cAMP-dependent protein kinase. The increase in cAMP is a hormonal message which has an effect in two major target organs, liver and muscle, but different hormones affect increased cAMP in each. In the liver, the hormone is glucagon, whereas in the muscle, the hormone is epinephrine. In the liver, glycogen breakdown (i.e., glycogenolysis) can serve as a source of hexose phosphate, a precursor of blood glucose in the

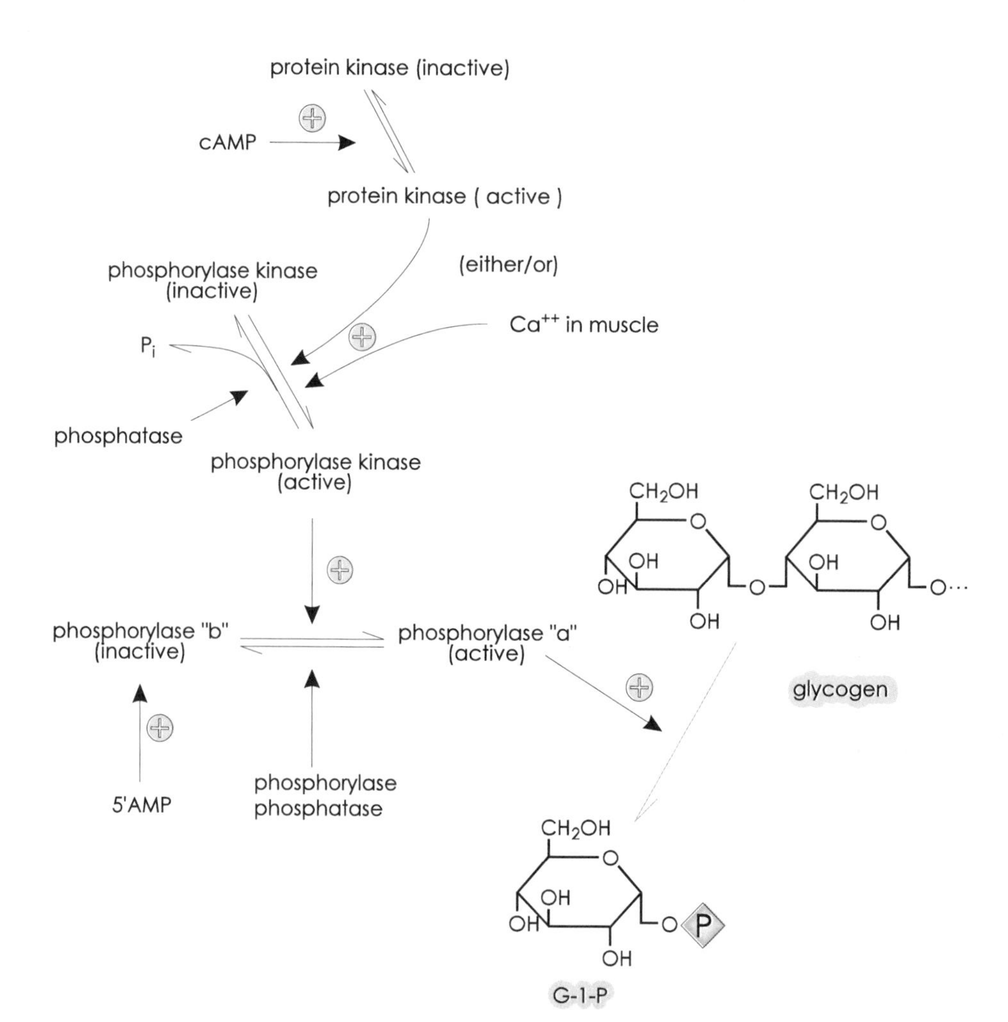

Figure 12.3.

Glycogen degradation. The activation of phosphorylase, the key enzyme in glycogen degradation, via covalent modifications and allosteric activation is presented. The action of phosphorylase per se is indicated in purple (see the text for more detail).

liver, thus increasing blood glucose levels at the expense of glycogen. In the muscle, this process can increase hexose phosphates from glycogen, which can be used for glycolysis; potentially, the resulting pyruvate can be used for oxidative metabolism. This would assure substantial energy to the muscle, both anaerobic and aerobic, during a flight-or-fight situation.

Another advantage of a cascade is the possibility of intervening anywhere along the cascade, not only at the initiation. In the case of the phosphorylase cascade, phosphorylase kinase in the inactive (i.e., nonphosphorylated) form can be activated allosterically by calcium, thus causing phosphorylase kinase to catalyze the phosphorylation of phosphorylase to the active form, causing increased glycogenolysis. One of the subunits of phosphorylase kinase is a polypeptide known as calmodulin. This polypeptide occurs in many proteins that require or have Ca^{2+} as an effector. This polypeptide in phosphorylase kinase results in the ca^{2+} activation of this enzyme. The binding of Ca^{2+} by the calmodulin subunit of phosphorylase kinase also facilitates a more rapid phosphorylation of this enzyme by the cAMP-dependent protein kinase.

This response in the muscle to calcium, which increases during contraction, allows for individual, active muscles to increase glycogenolysis without necessitating the hormonal message, epinephrine, which would affect all muscles. In the liver, epinephrine works via an alpha receptor and causes an increase in calcium, which would then activate phosphorylase kinase to activate phosphorylase and cause glycogenolysis. The metabolic advantage would be a higher level of blood glucose during a fight-or-flight situation, assuring available substrate to the central nervous system and the brain during an emergency.

It is important to distinguish clearly between a covalent modification for activation and an allosteric activation. A covalent modification remains with the enzyme until the modification is enzymatically reversed. For example, phosphorylase kinase and phosphorylase will remain active after phosphorylation until the enzymes are dephosphorylated as catalyzed by phosphoprotein phosphatase. In the case of allosteric activation, the enzyme will remain active as long as there is an elevated allosteric effector such as calcium. As soon as calcium is reconcentrated in subcellular fractions or subcellular organelles and/or extruded from the cell, the enzyme becomes relatively inactive because there is no longer an allosteric activator present. The same is true for allosteric inhibitors. Thus, it is advantageous that both of these mechanisms are available to the cell.

Usually covalent modifications are consequences of hormonal messages, whereas allosteric activation and inhibition are associated with localized changes in metabolism or activity. Another example of allosteric intervention in glycogenolysis is allosteric activation by 5′ AMP of phosphorylase b in muscle. The advantage of this activation is that a low adenylate or energy charge is related to an elevated AMP within a given cell, thus glycogen will be degraded to provide hexose phosphates to reestablish the energy charge. As the charge is reestablished, the elevated AMP will decrease by conversion to ADP and ATP, thus the allosteric signal will disappear and glycogenolysis will decrease. This is advantageous because various cells are at different energy stages. This allows each individual cell to maintain its energy charge, potentially at the expense of glycogen, without requiring neighboring cells to carry out glycogenolysis, as would be the case within a given muscle with increased calcium being the only message, or as would be the case with all muscles responding in the presence of epinephrine.

The liver contains an isoenzyme of phosphorylase which, unlike that in muscle, is not activated by 5′ AMP. Physiologically this is reasonable, since liver glycogen is a source of blood glucose which is not readily used by the liver for energy. Therefore, a low-energy charge in the liver would not be a reasonable signal to elevate blood glucose levels. The hepatic phosphorylated phosphorylase and/or phosphorylase kinase, as indicated earlier for muscle glycogenolysis, can be transformed back to the inactive enzyme by dephosphorylation. This process is stimulated by the hormone insulin, which is physiologically advantageous because insulin is elevated in the presence of high blood glucose and calls for a general state of anabolism instead of catabolism. Thus, the dephosphorylation of phosphorylase and phosphorylasekinase minimizes glycogenolysis and allows a rapid resynthesis of glycogen (Fig. 12.4).

Glycogen synthesis and glycogenolysis are controlled by hormones in tandem, that is, those hormones resulting in phosphorylation of both glycogen synthase and phosphorylase will activate phosphorylase, inactivate glycogen synthase, and allow maximum mobilization of glycogen, whereas hormones such as insulin, by stimulating dephosphorylation of both systems, will minimize glycogenolysis and allow maximum glycogen synthesis. Therefore, these two enzymes are controlled in a noncompeting manner for covalent modifications usually directed by hormonal messages, thereby assuming maximal unidirectional flux and a minimal of recycling.

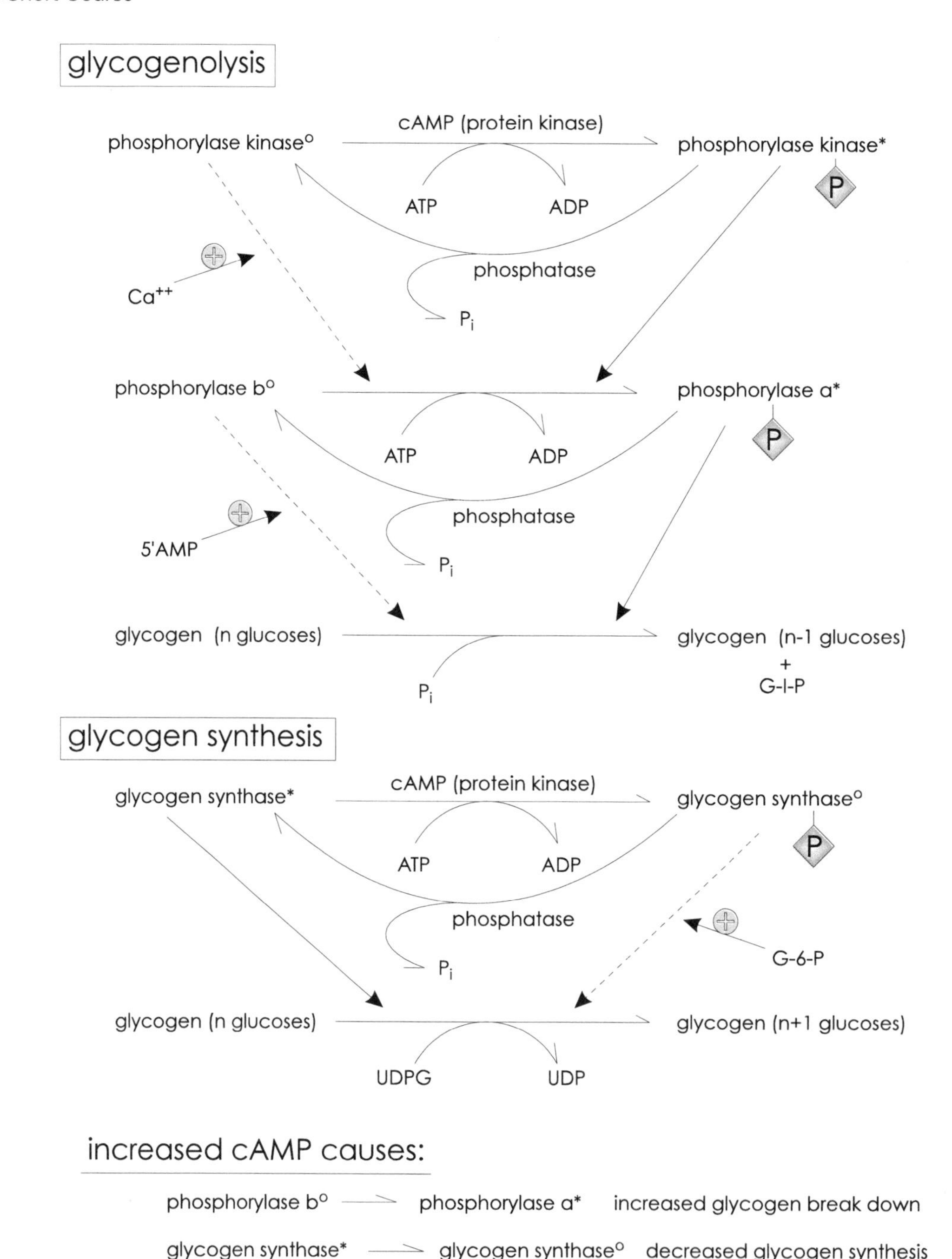

Figure 12.4.
Modifications of phosphorylase and glycogen synthase. Covalent and allosteric modifications of the enzymes associated with glycogen synthesis and degradation is presented. Note that hormonal responses increasing cAMP cause activation of phosphorylase, through a cascade, and inactivation of glycogen synthase, thus causing glycogenolysis.

THE PENTOSE-PHOSPHATE PATHWAY

The pentose-phosphate pathway, also known as the hexose monophosphate pathway, is a metabolic system with two important consequences. The first is production of NADPH for biosynthesis, and the second is production of ribose-5-phosphate for nucleotide and coenzyme biosynthesis (Fig. 12.5).

In the first step of this pathway, glucose-6-phosphate + $NADP^+$ is converted to 6-phosphogluconate + NADPH + H^+ and, in the second step, 6-phosogluconate + $NADP^+$ is converted to ribulose-5-phosphate + NADPH + H^+. These reactions are considered to be the oxidative arm of the pentose-phosphate pathway and are critical for producing much of the NADPH used in biosynthetic pathways. The remainder of the cycle (i.e., the nonoxidative portion) consists of converting 5-carbon phosphorylated sugars to 3-, 4-, 5-, and 7-carbon intermediates, finally achieving the resynthesis of hexose-6-phosphate and triose phosphate. Some of

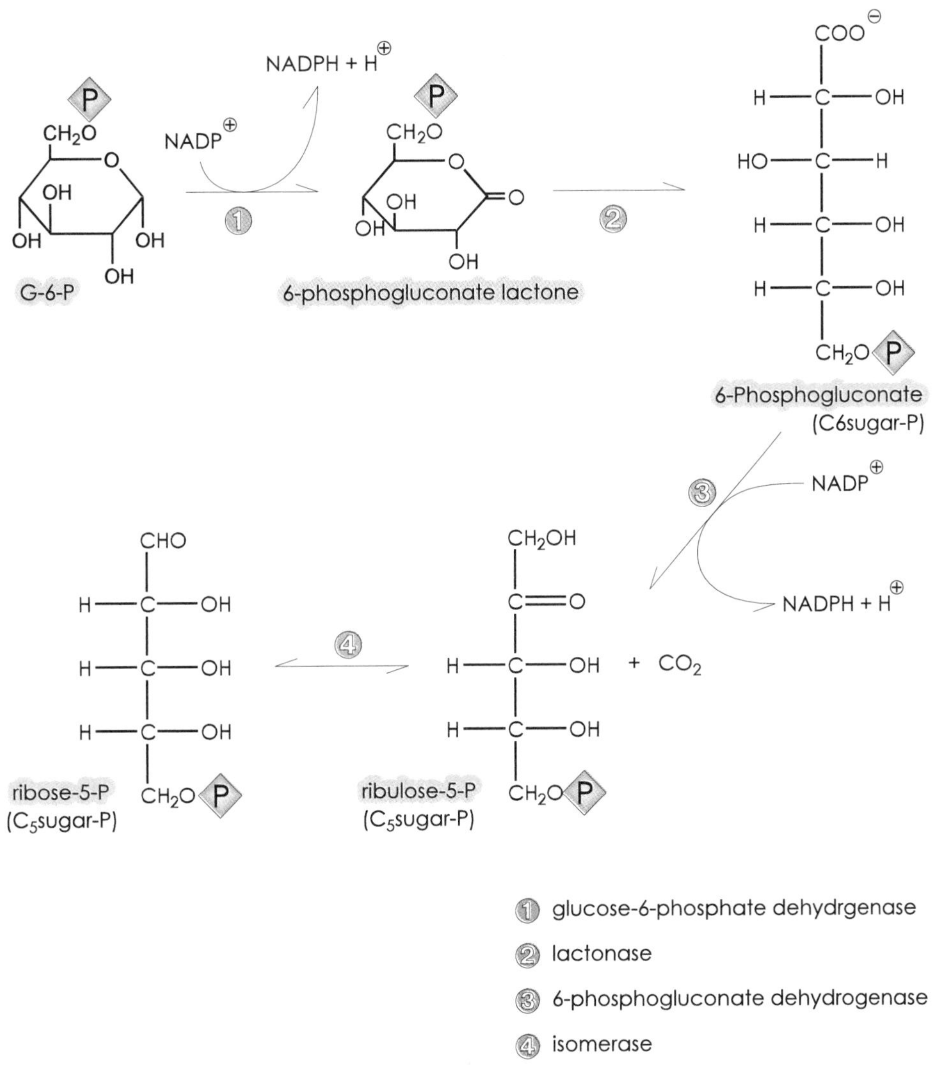

① glucose-6-phosphate dehydrgenase

② lactonase

③ 6-phosphogluconate dehydrogenase

④ isomerase

following is sugar rearrangements with only number of carbons shown:

2 C_5sugar-P → C_3sugar-P + C_7sugar-P

C_3sugar-P + C_7sugar-P → C_4sugar-P + F-6-P (C_6sugar-P)

C_4sugar-P + C_5sugar-P → F-6-P + glyceraldehyde-3-P (C_3sugar-P)

Figure 12.5.

The pentose-phosphate pathway. The early steps of the pentose-phosphate pathway are shown in detail as for a pentose-phosphate formation. This pathway produces two NADPHs and ribose-5-phosphate. The reformation of hexose phosphate involves C_3, C_4, C_5, and C_7 sugar phosphates and is not shown in detail.

these reactions utilize transketolase (which requires thiamine pyrophosphate) and transaldolase as well as isomerases and epimerases to achieve the final conversion of pentose phosphate to hexose and triose phosphates. Some tissues lack the oxidative arm of the pentose-phosphate pathway, but they have the nonoxidative portion and, thus, can utilize triose and hexose phosphates to produce pentose phosphates for nucleic acid and coenzyme production. The oxidative portion is most active in those tissues in which there is active biosynthesis, such as in the liver, where fatty-acid synthesis occurs, and some endocrine glands where there is rapid synthesis of steroid hormones. The major control of this system appears to be at the first step—the reaction catalyzed by glucose-6-phosphate dehydrogenase—which is increasingly inhibited as the $NADP^+$/NADPH ratio decreases. Normally, $NADP^+$/NADPH in most tissues is approximately 0.01, the 100-fold excess of NADPH provides the reducing equivalents needed for biosynthetic reactions. In contrast, NAD^+/NADH in most tissues is 400 or greater; thus, this nucleotide pair is poised to accept the reducing equivalents from biological oxidations. As we proceed, it will be observed that NADPH is used widely for reductive biosynthesis, whereas NAD^+ is used frequently in oxidative reactions, especially those associated with energy production.

A glucose-6-phosphate dehydrogenase deficiency in the red blood cell can result in hemolysis after treatment with sulfa antibiotics or antimalarials.

OTHER HEXOSES

In addition to glucose and glucose polymers, two other sugars are commonly encountered in the diet. These are galactose, from the milk sugar, lactose, and fructose from the common table sugar, sucrose. ***Both galactose and fructose are avidly and primarily taken up by the liver where they can be converted to glucose or glycogen.***

Galactose Metabolism

After absorption, galactose passes to the liver where it is phosphorylated to form galactose-1-phosphate. Galactose-1-phosphate is then converted to UDP-galactose by two mechanisms. The first, and primary, mechanism is a reaction between galactose-1-phosphate and UDP-glucose to form UDP-galactose + glucose-1-phosphate (Fig. 12.6).

The second mechanism, which is less active, is galactose-1-phosphate + UTP → UDP-galactose + pyrophosphate. Once more, the hydrolysis of pyrophosphate to two inorganic phosphates helps this reaction proceed in the forward direction. The UDP-galactose is then converted to UDP-glucose via an epimerase which has a tightly bound NAD. Note that glucose and galactose are epimers differing only by the orientation of the hydroxyl on a number 4 carbon. Therefore, UDP-galactose on dehydrogenation at the number 4 position, while still tightly bound to the enzyme, can be rereduced so that the hydroxyl forms in either orientation to reform UDP-galactose or to form UDP-glucose. This is a reversible reaction with an equilibrium ratio of approximately 3:1 in favor of UDP-glucose. Once UDP-glucose is formed, it can be used either to form glycogen and other products or to react with another galactose-1-phosphate, releasing the glucose-1-phosphate which originated from galactose. Glucose-1-phosphate then in the liver can be released as free glucose after several reactions, or UDP-glucose can proceed to glycogen synthesis. Galactose metabolism, as described for the liver, helps supply glucose to other tissues and organs in the body. Thus, many other cells do not have to express all of the enzymes needed to carry out galactose metabolism. Many nonhepatic cells do, however, have portions of this metabolic pathway, particularly the UDP-glucose–UDP-galactose epimerase. This enables these cells to use glucose to form UDP-glucose and UDP-galactose which is the precursor for essential galactose-containing compounds such as the galactolipids. Furthermore, UDP-galactose in the mammary gland is the precursor for the galactose moiety of the milk sugar lactose.

There is an inherited, recessive metabolic disease of galactose metabolism known as galactosemia. In this disease, the enzyme that catalyzes galactose-1-phosphate + UDPG → UDP-galactose + G-1-P is absent in homozygous individuals. Thus, galactose can be phosphorylated, but the ability for galactose-1-P to be further utilized is limited. In this disease, the liver accumulates large amounts of galactose-1-phosphate when lactose or galactose is ingested. Excess galactose may also be converted to galactitol in the eye, resulting in cataracts. This disease should not be confused with a lactose intolerance in which there is a small-intestine inability to hydrolyze lactose and, consequently, the ingested lactose is passed to the large intestine where it serves as a rich nutrient for the resident bacteria, leading to a gastrointestinal disorder.

Fructose

Fructose is phosphorylated primarily in the liver via fructokinase to form fructose-1-phosphate. Fructose can also be phosphorylated by the various hexokinases that catalyze formation of fructose-6-phosphate. However, the K_m values of these enzymes for fructose are extremely high, whereas the K_m values for glucose are orders of magnitude lower. Therefore, these hexokinases do not catalyze appreciable phosphorylation of fructose, especially in the presence of 5 mM glucose, the normal concentration in blood (Fig. 12.7).

Subsequent cleavage of fructose-1-phosphate, catalyzed by a liver-specific isoenzyme of aldolase, forms dihydroxyacetone phosphate + glyceraldehyde. The glyceraldehyde can then, either directly or through several steps, be converted to glyceraldehyde-3-phosphate. Glyceraldehyde-3-phosphate and dihydroxyace-

galactose, ATP, ADP, ①, Gal-1-P, UTP, ②, ③, G-1-P, 2P$_i$, PP, UDPGal, G-6-P, ④, UDPG

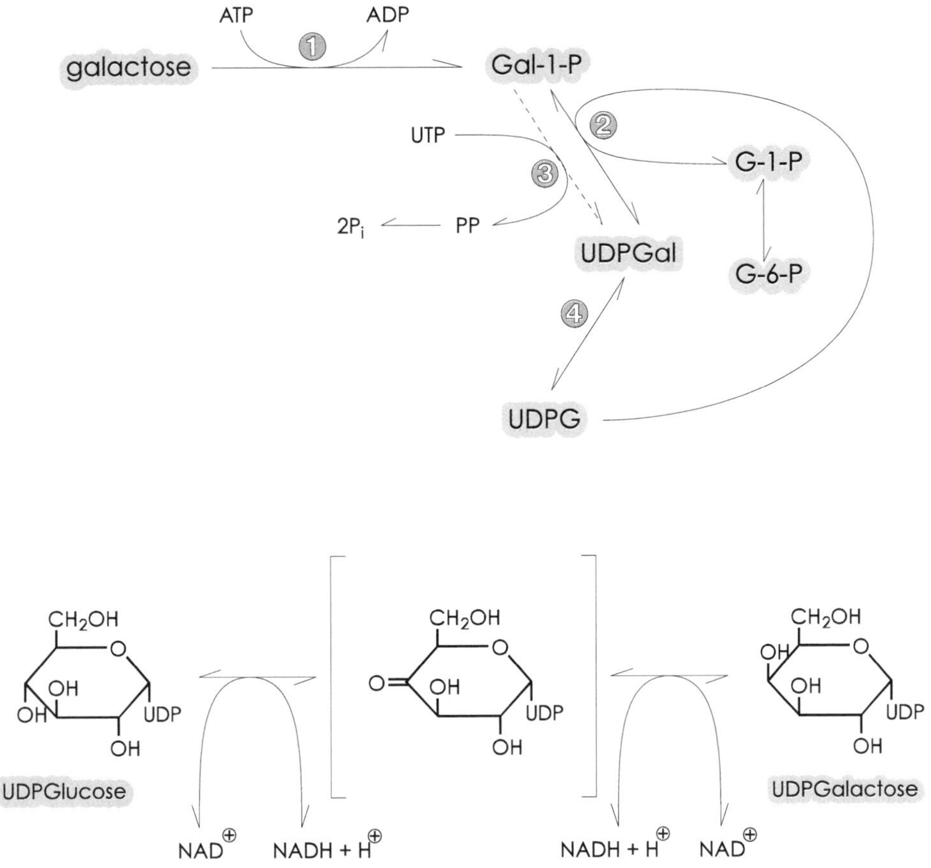

CH_2OH, OH, UDP, UDPGlucose, UDPGalactose, $NAD^{\oplus}$, $NADH + H^{\oplus}$

purposed mechanism of UDP galactose-4-epimerase

① galactokinase

② galactose-1-phosphate uridylyltransferase

③ UDPgalactose pyrophosphorylase

④ galactose-4-epimerase

Figure 12.6.

Galactose metabolism. The metabolism of galactose to form glucose phosphates is presented. The enzymes indicated with the circled numbers are listed in the figure as well as a suggested mechanism for UDP galactose-4-epimerase.

tone phosphate can proceed either through glycolysis or via aldolase to form fructose-1,6-bisphosphate and from there to glucose-6-phosphate, which can be released into the blood as glucose or stored as glycogen in the liver. The liver-specific aldolase catalyzes cleavage of fructose-1-phosphate and fructose-1,6-bisphosphate at approximately the same rate; thus, this enzyme is useful for both fructose metabolism and glycolysis or, as discussed in Chapter 13, gluconeogenesis. In contrast, the aldolase in muscle and other tissues cleaves fructose-1,6-bisphosphate at 50 times the rate with fructose-1-phosphate. The absence of fructokinase in peripheral tissues limits the use of fructose in these tissues.

There are two genetic diseases associated with fructose metabolism. One is fructosuria, in which the liver enzyme, fructokinase, is missing. Thus, fructose is not metabolized but, instead, accumulates in the blood and is passed into the urine. In this case, fructose is not useful as an energy or carbohydrate source, and sucrose and/or fructose should be avoided in a diet. The second disease, fructose intolerance, has more serious consequences. In this case, the liver-specific aldolase is missing and the muscle-type aldolase is expressed in the liv-

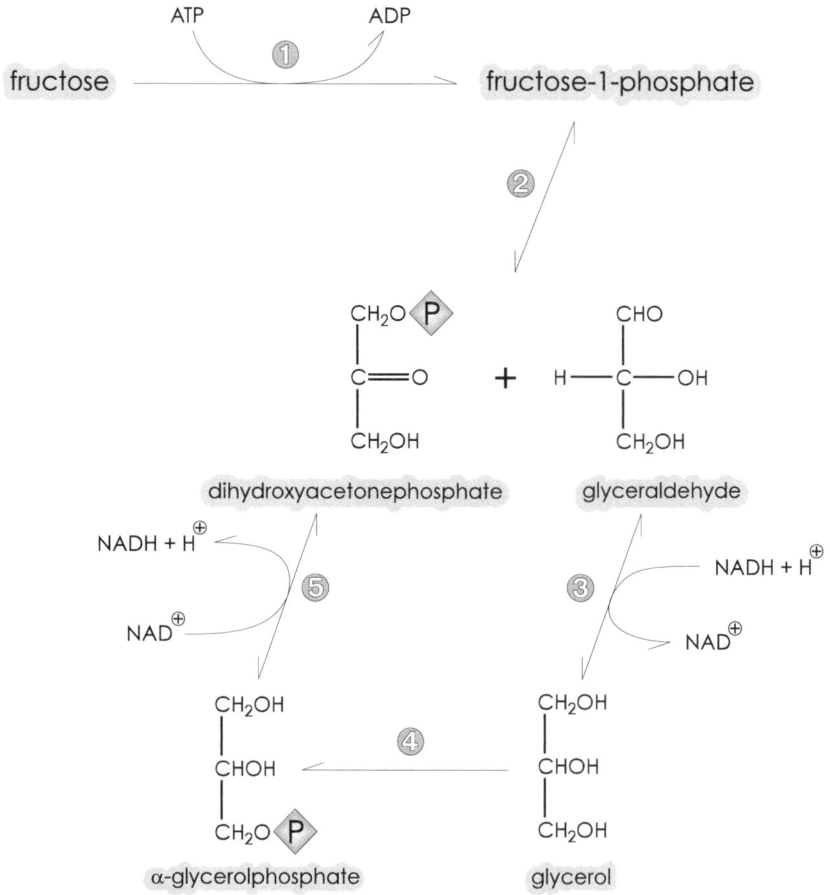

① fructokinase

② fructose-1-phosphate aldolase

③ glyceraldehyde reductase

④ glycerolkinase

⑤ α -glycerolphosphate dehydrogenase

Figure 12.7.

Fructose metabolism. The metabolism of fructose to the level of triose phosphates is presented. The enzymes indicated with the circled numbers are listed in the figure.

er. Muscle-type aldolase allows for normal glycolysis and gluconeogenesis, but not sufficient fructose metabolism. The fructokinase catalyzes the phosphorylation of fructose to fructose-1-phosphate, which cannot readily be cleaved to dihydroxyacetone phosphate and glyceraldehyde owing to the absence of the liver-type aldolase. In this disease, fructose-1-phosphate accumulates in the liver and interferes with normal hepatic metabolism, which can lead to serious consequences, including death. For this disease, a very strong effort must be made to eliminate fructose or fructose-containing compounds from the diet in a manner similar to avoiding galactose and galactose-containing carbohydrates in the diet of a galactosemic.

SUMMARY

1. Glycogen is a storage form of glucose in the cell and is a highly branched polymer of glucoses.

2. Glycogen can be synthesized from glucose phosphates via UDPG. The enzyme glycogen synthase can catalyze the transfer of glucose from UDPG and add it to glycogen.

3. Glycogen synthase has two major forms: a phosphorylated form, which is inactive, and a dephospho form, which is active.

4. Glycogen synthase is phosphorylated as cat-

alyzed by a cAMP-dependent protein kinase, which is stimulated by epinephrine and/or glucagon. Insulin helps stimulate the dephosphorylation of glycogen synthase.

5. Glycogen is degraded (glycogenolysis) by phosphorylase to form glucose phosphates.

6. Phosphorylase has two forms: a phosphorylated form (phosphorylase a), which is active, and a nonphosphorylated form (phosphorylase b), which is relatively inactive.

7. Phosphorylase phosphorylation is catalyzed by phosphorylase kinase. Phosphorylase kinase is inactive in a nonphosphorylated state and active when phosphorylated. Phosphorylase kinase phosphorylation is catalyzed by the cAMP-dependent protein kinase.

8. There are allosteric activators of phosphorylase kinase (Ca^{2+}) and muscle phosphorylase (AMP), which can cause activation of the appropriate enzyme without the covalent modification.

9. c-AMP will ultimately cause phosphorylation of both glycogen synthase and phosphorylase, producing net glycogenolysis (degradation). Insulin will stimulate dephosphorylation of both of these enzymes, leading to net glycogen synthesis.

10. The pentose-phosphate pathway produces NADPH, for reductive biosynthesis, and pentose phosphates, for nucleotide synthesis.

11. The first two steps of the pentose-phosphate pathway are oxidative; the other steps form 3-, 4-, 5-, 6-, or 7-carbon sugar phosphates to finally regenerate hexose phosphates.

12. When glycogen in the muscle reaches certain levels, it prevents dephosphorylation of glycogen synthase, producing a cessation of further glycogen synthesis.

13. Galactose is primarily metabolized in the liver, via galactose-1-phosphate and UDP-galactose. Galactose can be converted to glucose in the liver. In most tissues the enzyme, UDP-galactose epimerase, can interconvert UDPG and UDP-galactose, thus allowing for the formation in the body of galactose-containing compounds.

14. Fructose is primarily metabolized in the liver via fructose-1-phosphate. The F-1-P formed can rapidly be converted to glycolytic intermediates, as catalyzed by a liver-specific aldolase isoenzyme, and further to glucose, glycogen, or lactate.

REFERENCES

Short-term hormonal control of protein phosphatases involved in hepatic glycogen metabolism, W. Stalmans, M. Bollen, B. Toth, and P. Gergely, 1990, *Adv. Enzyme Reg.*, 305–327.

From dietary glucose to liver glycogen: The full circle round. J. D. MGarry, M. Kawajima, C. B. Newgard, D. W. Foster, and J. Katz, 1987, *Ann. Rev. Nutr.*, **7**:51–73.

Multifunctional Ca^{2+}/calmodulin-dependent protein kinase: Domain structure and regulation, H. Schulman and L. L. Lou, 1989, *Trends Biochem. Sci.*, **14**:52–66.

Glycogen storage diseases. H-G Hers, F. Van Hoof, and T. de Barsy, in *The Metabolic Basis of Inherited Disease*, Vol. 1, 6th ed., C. R. Scriver, A. L. Beaudet, W. S. Sly, and D. Valle (Eds.), 1989, McGraw-Hill: New York, pp. 425–452.

The regulation of the pentose phosphate cycle in rat liver, H. A. Krebs and L. V. Eggleston, 1974, *Ad. Enzyme Reg.*, 421–434.

Current issues in fructose metabolism, R. R. Henry, P. A. Crapo, and A. W. Thorburn, 1991, *Ann. Rev. Nutr.*, **11**:21–40.

Disorders of fructose metabolism, R. Gitzelmann, B. Steinmann and G. Van Den Berghe, in: *The Metabolic Basis of Inherited Disease*, Vol. 1, 6th ed., C. R. Scriver, A. L. Beaudet, W. S. Sly, and D. Valle, Eds., 1989, McGraw-Hill: New York, pp. 399–424.

Disorders of galactose metabolism, S. Segal, in *The Metabolic Basis of Inherited Disease*, Vol. 1, 6th ed., C. R. Scriver, A. L. Beaudet, W. S. Sly, and D. Valle, Eds., 1989, pp. 453–480.

The genetic polymorphism of lactase activity in adult humans, G. Flatz, in *The Metabolic Basis of Inherited Disease*, Vol. 1, 6th ed., C. R. Scriver, A. L. Beaudet, W. S. Sly, and D. Valle, Eds., 1989, pp. 2999–3006.

REVIEW QUESTIONS

1. It is a major function of the pentose-phosphate pathway to produce which of the following?

a) Ribose-5-phosphate and glucose
b) NADPH and glyceraldehyde-3-phosphate

c) NADH and ribose-5-phosphate
d) Ribose-5-phosphate and NADPH
e) Ribulose-5-phosphate and NADH

2. Through what does galactose conversion to glucose proceed?
   a) ADP glucose
   b) UDP glucose
   c) Galactose-6-phosphate
   d) Sorbitol
   e) Galactitol

3. How is the unfavorable equilibrium for UDP glucose formation overcome?
   a) Using two UTPs to form one UDPG
   b) Producing more enzymes
   c) Producing pyrophosphate which is hydrolyzed to two inorganic phosphates
   d) Using glucose-1,6-bisphosphate as a precursor which then will lose one of its phosphates
   e) Covalent modification of the enzyme

4. When both phosphorylase and glycogen synthase are phosphorylated via the cAMP-dependent protein kinase, which of these is true?
   a) One process is direct and one via a cascade and both enzymes are active.
   b) Both processes are direct and phosphorylase is active and glycogen synthesis inactive.
   c) Both processes are via a cascade and phosphorylase is inactive and glycogen synthase active.
   d) One process is direct and one via a cascade and phosphorylase is active and glycogen synthase is inactive.
   e) One process is direct and one via a cascade and both enzymes are inactive.

5. An activator of the phosphorylated form of glycogen synthase, although probably physiologically of no consequence, is
   a) UTP
   b) G-6-P
   c) F-1,6-BP
   d) 5′ AMP
   e) PEP

6. In an exercising muscle increased Ca^{2+} can cause increased glycogen breakdown by
   a) Directly activating phosphorylase
   b) Causing increases in cAMP
   c) Inhibiting glycogen synthase
   d) Activating phosphorylase kinase
   e) Reacting with the glycogen particle

7. In galactosemia, what enzyme is missing?
   a) Galactokinase (Gal+ATP ↔ Gal-1-P + ADP)
   b) UDPG-UDPGal epimerase (UDPG ↔ UDPGal)
   c) UTP, Gal-1-P pyrophosphorylase (UTP + Gal-1 ↔ P-UDPGal + PPi)
   d) UDPG, Gal-1-P transferase (UDPG + Gal-1-P ↔ UDPGal + G-1-P)
   e) UDPG kinase (UDP + ATP ↔ UTP + ADP)

8. In fructose intolerance, what enzyme is missing?
   a) Liver-specific aldolase
   b) Triose kinase
   c) Fructokinase
   d) Fructose-1,6-bisphosphatase
   e) PFK

9. In fructosuria, what enzyme is missing?
   a) Liver-specific aldolase
   b) Triose kinase
   c) Fructokinase
   d) Fructose-1,6-bisphosphatase
   e) PFK

ANSWERS TO REVIEW QUESTIONS

1. ***d*** The pentose-phosphate pathway produces NADPH for reductive biosynthesis and ribose-5-phosphate for nucleotide synthesis.

2. ***b*** UDP glucose is an intermediate in the conversion of galactose to glucose, glycogen, or other products. Galactose is phosphorylated to form galactose-1-P, not galactose-6-P.

3. ***c*** By removing one of the products, as pyrophosphate is removed by forming two inorganic phosphates, the reaction can be pulled in a thermodynamically, relatively unfavorable direction.

4. ***d*** Phosphorylase is phosphorylated in a cascade by phosphorylase kinase and is active in the phosphorylated form. Glycogen synthase is phospho-

rylated directly by the catalytic subunit of the cAMP-dependent protein kinase and is inactive in the phosphorylated form.

5. ***b*** G-6-P is an activator of the phosphorylated form of glycogen synthase, but physiologically does not reach a high enough concentration to activate this enzyme. The G-6-P inhibition of hexokinase prevents an excessive accumulation of G-6-P from glucose.

6. ***d*** Activating phosphorylase kinase allosterically stimulates the phosphorylation of phosphorylase to the active form. The active form of phosphorylase catalyzes glycogenolysis, the breakdown of glycogen.

7. ***d*** UDPG, Gal-1-P transferase, which causes increased concentrations of UDPGal and Gal-1-P.

8. ***a*** This cleaves F-1-P to DHAP and glyceraldehyde as fructose is phosphorylated in the liver to form F-1-P, not F-6-P. In this disease, F-1-P accumulates in the liver.

9. ***c*** This would have converted fructose to F-1-P in the liver. Thus fructose, after ingestion, accumulates in the blood and is excreted in the urine.

CHAPTER 13

GLUCONEOGENESIS

INTRODUCTION

Gluconeogenesis is the process by which compounds are converted into glucose. This process takes place primarily in the liver and kidney cortex. The major natural precursors for gluconeogenesis are other carbohydrates such as fructose and galactose, amino acids, lactate, pyruvate, and glycerol. Humans can survive a substantial period of food deprivation if water is provided (i.e., up to 30 days or more). Without the process of gluconeogenesis, the period of maximum starvation would be reduced to just several days. It is critical to maintain blood-glucose levels, particularly for those tissues dependent on glucose for energy, such as the central nervous system and red blood cells. Most individuals have a blood-glucose level, after an overnight fast, of about 80–120 mg/dL (4.4–6.7 mM). After prolonged starvation, this normally does not drop much below 70–80 mg/dL (3.9–4.4 mM) despite continual utilization of glucose for special metabolic requirements. Thus, gluconeogenesis is critical for maintaining blood-glucose levels over a variety of nutritional and physiological conditions. Many of the same enzymes that catalyze reversible reactions in glycolysis also participate in gluconeogenesis.

PATHWAY OF GLUCONEOGENESIS

Four enzymes are unique to the process of gluconeogenesis and are required to circumvent the unidirectional steps of glycolysis (Fig. 13.1). Two of these are pyruvate carboxylase, which converts pyruvate to oxaloacetate, and PEP carboxykinase, which converts oxaloacetate to phosphoenolpyruvate. These enzymes are required to effectively reverse the action of pyruvate kinase. The other two are fructose-1,6-bisphosphatase and glucose-6-phosphatase, which effectively reverse the actions of PFK and hexokinase or glucokinase.

Pyruvate carboxylase, which is a key, and potentially rate-limiting, enzyme in gluconeogenesis, from lactate and pyruvate, is located in the mitochondrion and not only requires ATP and CO_2, but also acetyl CoA as an essential activator. The acetyl CoA is normally formed in the mitochondrion from pyruvate, as indicated previously, and can be formed during fatty-acid oxidation. There is an investment of one high-energy phosphate in this reaction and the resulting oxaloacetate may proceed via various steps toward gluconeogenesis, depending on the species and the distribution of the enzyme PEP carboxykinase, which catalyzes oxaloacetate and GTP to form GDP, phosphoenolpyruvate, and CO_2.

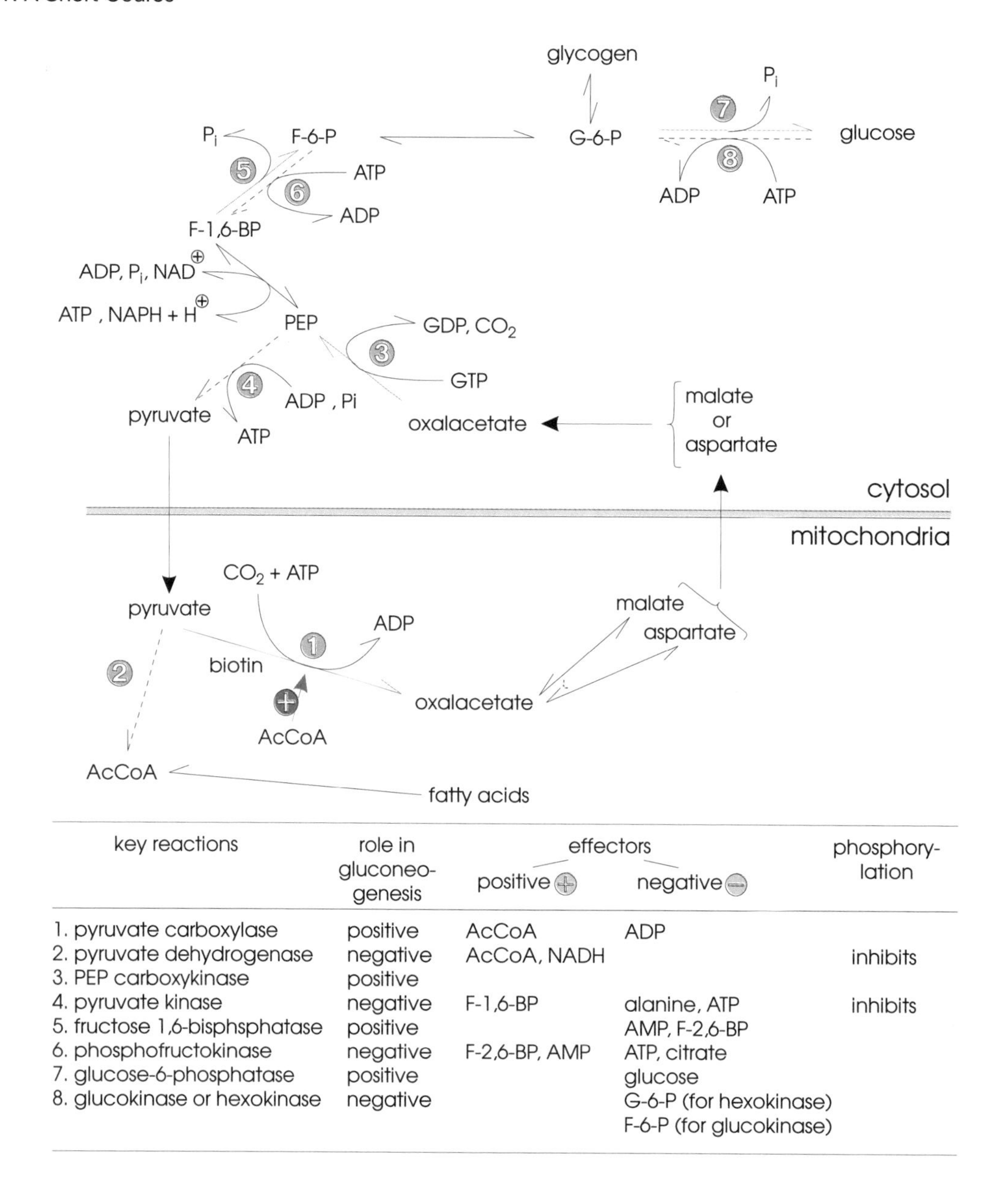

key reactions	role in gluconeo-genesis	effectors: positive ⊕	effectors: negative ⊖	phosphory-lation
1. pyruvate carboxylase	positive	AcCoA	ADP	
2. pyruvate dehydrogenase	negative	AcCoA, NADH		inhibits
3. PEP carboxykinase	positive			
4. pyruvate kinase	negative	F-1,6-BP	alanine, ATP	inhibits
5. fructose 1,6-bisphsphatase	positive		AMP, F-2,6-BP	
6. phosphofructokinase	negative	F-2,6-BP, AMP	ATP, citrate	
7. glucose-6-phosphatase	positive		glucose	
8. glucokinase or hexokinase	negative		G-6-P (for hexokinase) F-6-P (for glucokinase)	

Figure 13.1.

Gluconeogenesis. The pathways of gluconeogenesis and opposing pathways are presented in a diagrammatic form. The major steps associated with gluconeogenesis are indicated by purple arrows, and those opposing gluconeogenesis are indicated by dotted arrows. The key reactions are numbered and the enzymes are listed by number below, with an indication as to whether this reaction is positive (favors) or negative (decreases) for gluconeogenesis. The positive and negative effectors for each of these reactions are listed, as well as the effect of phosphorylation (if it occurs) on activity. The double line on the diagram separates the mitochondria and the cytosol.

If the PEP carboxykinase is located in the mitochondrion, the formation of PEP will take place in the mitochondrion and it will be translocated to the cytosol. If the PEP carboxykinase is located in the cytosol, oxaloacetate will be converted to either malate or aspartate and then transported to the cytosol where it will be reconverted to oxaloacetate, which, via the action of PEP carboxykinase, can be converted to PEP. In cases where the enzyme is located both in the mitochondrion and cytosol, as in humans, some of both of these processes take place. To maintain electroneutrality, the transport of these compounds via PEP, aspartate, or malate out of the mitochondrion requires cotransport of other ions into the mitochondrion by specific translocases. A large number of translocases are only capable of exchanging specific anions for one another, and potentially complex translocations are required for these processes.

Two high-energy phosphates are invested to convert a pyruvate into phosphoenolpyruvate. This two-step process overcomes the thermodynamic difficulty in reversing the pyruvate kinase reaction. If PEP reforms pyruvate as catalyzed by pyruvate kinase, only one high-energy phosphate is produced. Therefore, any recycling of the formed PEP back to pyruvate would be wasteful energetically, that is a loss of one high-energy phosphate. This recycling would also force the pyruvate to transverse the rate-limiting enzymes of pyruvate carboxylase and PEP carboxykinase to complete gluconeogenesis.

The PEP formed by these two reactions then can precede via the same pathway as glycolysis to the level of F-1,6-BP. Since the PFK reaction cannot be reversed, the enzyme that effectively reverses this is fructose-1,6-bisphosphatase, which catalyzes the hydrolysis of fructose-1,6-bisphosphate to fructose-6-phosphate and inorganic phosphate. In this reversal, there is no high-energy phosphate gain, although a high-energy phosphate was invested in a forward direction. The fructose-6-phosphate can then be isomerized to glucose-6-phosphate; this substance can be hydrolyzed by specific-enzyme glucose-6-phosphatase, which cleaves the glucose-6-phosphate to free glucose and inorganic phosphate. Glucose-6-phosphatase is located in the endoplasmic reticulum and requires more than one functional process for its action. There is a specific transport system for moving glucose-6-phosphate from the cytosol to the endoplasmic reticulum; the enzyme glucose-6-phosphatase can then catalyze the hydrolysis of G-6-P and release the free glucose into the circulation.

Gluconeogenesis is an energy-requiring process. Glucose formation from lactate requires six high-energy phosphates, three per triose incorporated. The steps requiring one high-energy each are pyruvate carboxylase, PEP carboxykinase, and 3-phosphoglycerate kinase. This high-energy phosphate cost is higher than that for the two high-energy phosphates gained during glycolysis. In addition, gluconeogenesis from pyruvate or many amino acids requires one cytosolic NADH per triose formed, for the glyceraldehyde-3-phosphate dehydrogenase step, or two NADHs per glucose formed.

GLYCEROL AND AMINO ACIDS

Glycerol can enter the gluconeogenic pathway by being phosphorylated to a glycerol phosphate and then oxidized to dihydroxyacetone phosphate, which then can proceed to fructose-1,6-bisphosphate and through gluconeogenesis as described previously. Any amino acid that can produce a glycolytic intermediate, which, in effect, could then produce pyruvate, or a net citric acid intermediate is potentially glucogenic. Any TCA cycle intermediate produced can proceed through the cycle until forming oxaloacetate, which can proceed to PEP and onto glucose.

LACK OF GLUCONEOGENESIS FROM ACETYL COA

Any compounds that are converted to acetyl CoA cannot provide net gluconeogenesis. This includes pyruvate becoming acetyl CoA or acetyl CoA arising from fatty-acid oxidation. This is an important aspect in considering gluconeogenesis and some of the protective controls. The reason for this lack of gluconeogenesis in higher animals from acetyl CoA is as follows: If the cell starts two oxaloacetates, and sufficient energy plus reducing power is supplied, one glucose can be formed. If two compounds also form two new net oxaloacetates, there are then four oxaloacetates which, with the proper energy and reducing power, can now form two glucoses. However, if the two oxaloacetates react with acetyl CoA and form citrate to enter gluconeogenesis, citrate must proceed around the citric-acid cycle to form oxaloacetate. During this time, each loses two CO_2's, which is equivalent to the acetyl unit of the acetyl CoA, and as the citrates spin around the cycle only two oxaloacetates are reformed (the same as at the start), and are still worth only one glucose. Thus, no matter how many times acetyl CoA spins around the citric-acid cycle, although it may provide reducing power and energy, it cannot produce net carbon for gluconeogenesis. However, owing to some peculiarities in the citric-acid cycle, asymmetry and randomization during metabolism of ^{14}C-labeled acetate via acetyl CoA can incorporate radioactivity into glucose formed, but still does not produce net glucose. Gluconeogenesis from fructose and galactose proceeds as discussed in the previous chapter.

CONTROLS OF GLUCONEOGENESIS

Gluconeogenesis is a critical process in the body and is closely controlled. Controlling gluconeogenesis, and protecting potential gluconeonic precursors from losing their gluconeogenic potential, occur via all of the mechanisms discussed previously (Chapter 11). These include allosteric control, covalent modification, and change in the rate of synthesis of enzymes. These aspects are best examined one at a time, with the broadest implications being considered. First, the conditions under which gluconeogenesis plays an extremely important role can be considered, for example, starvation. One major concern is to conserve gluconeogenic precursors—particularly those that are generated in the periphery, such as lactate and pyruvate, preventing them from proceeding to the formation of acetyl CoA.

As described earlier, control of pyruvate dehydrogenase is achieved via acetyl CoA inhibition of the enzyme and phosphorylation to the inactive form. Both of these mechanisms are stimulated by elevated acetyl CoA and NADH in the mitochondrion. Increased acetyl CoA may appear to arise only via pyruvate dehydrogenase, if only carbohydrate metabolism is considered; however, acetyl CoA in the peripheral tissue mitochondrion can arise from fatty acids and/or ketone-body metabolism, both of these compounds are elevated in the blood during starvation (Chapter 14). Thus, the increase in availability of these readily utilizable substrates (fatty acids and ketone bodies) will inhibit pyruvate dehydrogenase and protect the glucogenic precursors. The elevated citrate and ATP levels generated during lipid catabolism in the periphery will effectively work to inhibit PFK, which will cause a rise in F-6-P and G-6-P and thereby inhibit hexokinase. This is another method to help raise and maintain blood glucose levels (i.e., decreased utilization by many of the peripheral tissues).

In the brain, where fatty-acid catabolism is negligible, the utilization of glucose continues during starvation at a normal or relatively near-normal rate and glucose metabolized in the brain still is primarily converted to CO_2 and water through complete oxidation. The central nervous system (including the brain) is one of the few tissues in the body where pyruvate dehydrogenase remains in the active form during starvation. This again indicates continual glucose utilization in a central nervous system and the importance of maintaining blood glucose for the central nervous system.

Regardless of where it is saved, a glucose precursor such as pyruvate still has the potential to yield one-half glucose. If pyruvate is conserved in the muscle, it can pass from the muscle to the blood and then to the liver and be converted to glucose. If pyruvate is saved in the liver, it can proceed via the gluconeogenic pathway to glucose. It is important to appreciate that the glucose precursors must be spared not only in the liver and kidney cortex, where glucose is being formed, but in all tissues, because glucose metabolism and total metabolism is much greater in the total peripheral tissue than in the liver and kidney.

PYRUVATE CARBOXYLASE

The liver and kidney cortex have a unique role in gluconeogenesis—converting pyruvate to glucose. Some of the controls are bifunctional, for example, the increased acetyl CoA in the liver from fatty-acid metabolism will not only inhibit pyruvate dehydrogenase but also activate pyruvate carboxylase, where acetyl CoA is an essential activator (Fig. 13.2).

The activation of pyruvate carboxylase by acetyl CoA is a sigmoidal response, with the steep part of the curve between the normal low and high ranges of acetyl CoA. Thus, as acetyl CoA increases in concentration, there is a marked stimulation of pyruvate carboxylase (Fig. 13.3).

The interaction of pyruvate dehydrogenase and pyruvate carboxylase in the liver in response to acetyl CoA is analogous to the effects of phosphorylase and glycogen synthetase in tissues in response to increased cAMP, except that no covalent changes occur with pyruvate carboxylase. Nature uses similar general concepts regularly throughout metabolism; in many cases, when one pathway is activated, a competing pathway with an opposite effect is simultaneously decreased in activity. Pyruvate carboxylase appears to be a constitutive enzyme, that is one that normally does not change greatly in amount, and is primarily controlled by the concentration of acetyl CoA and, to a lesser degree, by pyruvate and ATP to ADP ratios. Increased fatty-acid metabolism, as would occur in starvation, would cause

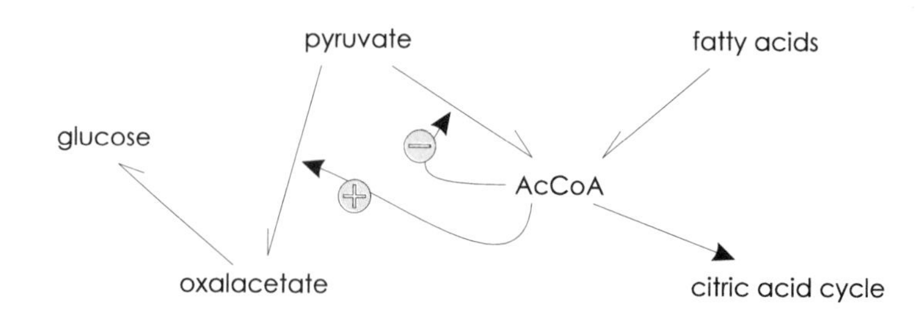

Figure 13.2.
Pyruvate metabolism. The control of pyruvate metabolism by acetyl CoA is shown in diagrammatic form. The two major routes examined are pyruvate dehydrogenase and pyruvate carboxylase.

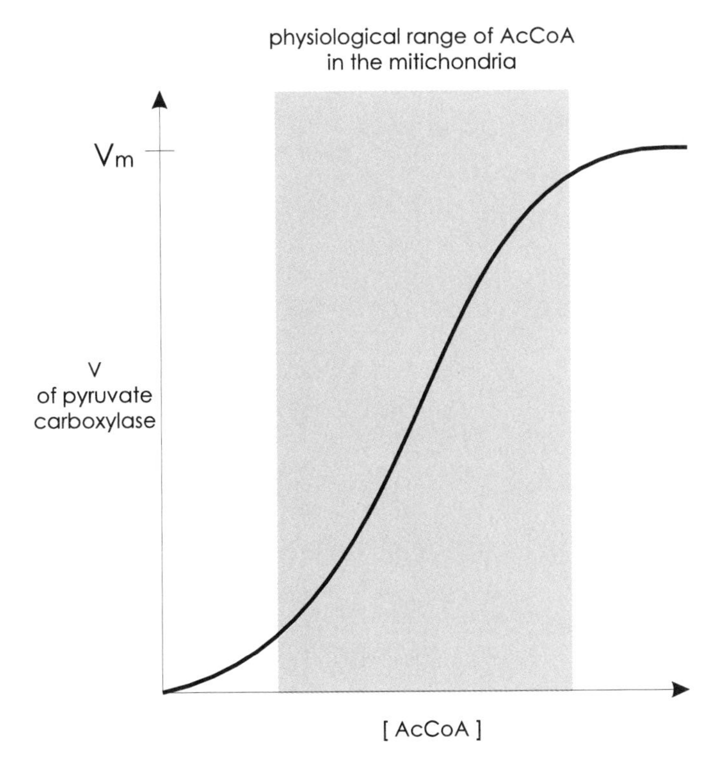

Figure 13.3.

Pyruvate carboxylase. Pyruvate carboxylase has hyperbolic kinetics in respect to its substrates pyruvate, CO_2, and ATP, but sigmoidal kinetics in respect to its essential allosteric activator acetyl CoA. The values of *v* for pyruvate carboxylase are plotted as a function of acetyl CoA concentration. The range of acetyl CoA concentrations in the mitochondria (subcellular localization of pyruvate carboxylase) is indicated by the shaded area.

a marked increase in mitochondrial-acetyl CoA in the liver and kidney. Usually the other metabolite, CO_2, is not a limiting factor.

PEP CARBOXYKINASE

The next step in gluconeogenesis is catalyzed by the enzyme PEP carboxykinase, which has neither normal physiological modification by allosteric effectors nor covalent modification. The major control of this enzyme during starvation and other conditions of increased gluconeogenesis appears to be an increased rate of enzyme synthesis; thus, it is an adaptive enzyme. PEP carboxykinase has a relatively short half-life; therefore, it can respond rapidly by both increasing and decreasing in amount. Hence, the major control at this point appears to be changes in enzyme amount. From a number of studies conducted with various animals, it appears that pyruvate carboxylase is more likely to be the rate-limiting or control enzyme for gluconeogenesis than is PEP carboxykinase.

PYRUVATE KINASE

A major enzyme that is indirectly associated with gluconeogenesis, but appears to be extremely important, is pyruvate kinase. Although it is unusual to see pyruvate kinase associated with gluconeogenesis, as indicated previously, the action of this enzyme would convert PEP back to pyruvate, which in effect would short circuit gluconeogenesis. This would cost a high-energy phosphate as well as forcing the pyruvate back through the potentially rate-limiting steps of pyruvate carboxylase and PEP carboxykinase. Unlike in muscle, where pyruvate kinase has little control and proceeds rapidly in a forward direction (Chapter 11), there is exquisite control of this enzyme in the liver. The major differences between the muscle and liver isoenzymes of pyruvate kinase result from an additional sequence of 28 amino acids in the liver isoenzyme. ***Pyruvate kinase in the liver is controlled by both allosteric effectors and covalent modification, with an interaction of these two.*** There are both positive and negative allosteric effectors of pyruvate kinase in the liver, in which the kinetics of pyruvate kinase relative to PEP concentration is sigmoidal (Fig. 13.4). The sigmoidal kinetics allows cellular PEP levels to rise before there is marked conversion of PEP to pyruvate. The elevated levels of PEP enable the other competing enzymes, particularly enolase and the phosphoglycerate mutase, to move the carbon fluxtoward glucose synthesis. A positive effector of this enzyme is fructose-1,6-bisphosphate, which tends to make the reaction kinetics less sigmoidal relative to PEP [causing a decrease in the K_s

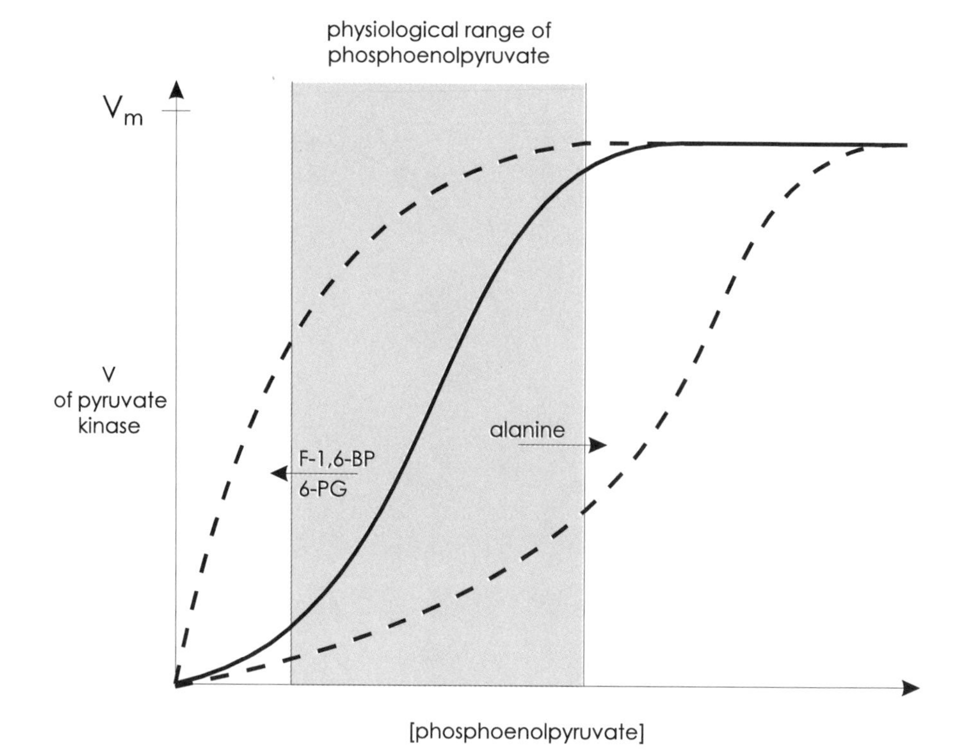

Figure 13.4.
Liver pyruvate kinase. A plot of pyruvate kinase activity versus PEP concentration is presented. The response of activity to PEP concentration is sigmoidal. Shifts in the curve (one of an infinite number, depending on the concentration of the effectors) in the positive (F-1,6-BP and 6-PG) and negative (alanine) direction are indicated by dashed lines. The physiological range of PEP in liver is shown by the shaded area.

(0.5) for PEP] and thus the flux through pyruvate kinase will increase. This is advantageous when the cell requires a larger flux through glycolysis. An elevated fructose-1,6-bisphosphate would potentially be the result of an increase in flux through PFK and would indicate that there was sufficient fructose-1,6-bisphosphate for gluconeogenesis. With a decreased flux through FBPase toward gluconeogenesis, and increased glycolysis, this would be appropriate. Another allosteric effector is alanine, which acts as a negative effector and causes the curve to become more sigmoidal [causing an increase in the K_s [(0.5) for PEP], which would cause a decrease flux through pyruvate kinase at any PEP level, other than saturating, and would allow a greater accumulation of PEP for gluconeogenesis. These controls are analogous to that seen earlier with PFK, shifting the sigmoidal curve by AMP and ATP. Alanine as an indicator of protein catabolism, which in many cases is occurring to supply carbon for gluconeogenesis, is an appropriate signal. The covalent modification is a ***phosphorylation stimulated by the cAMP-dependent protein kinase***. This ***will also make the reaction more sigmoidal relative to PEP***, and thus at any PEP concentration other than saturating, will slow down the flux through pyruvate kinase. This will increase the probability of the PEP proceeding to glucose. The hormone in the liver that causes increased cAMP is glucagon and, therefore, this response is consistent with the hormonal message to increase blood glucose levels. The increased blood glucose in this case would arise via gluconeogenesis, where, as was seen previously, the same message helped increase blood glucose through glycogenolysis.

There is an interaction between the cAMP-dependent proteinkinase and alanine. Phosphorylation makes the pyruvate kinase more responsive to inhibition by alanine; thus, the same amount of alanine will give a greater inhibition with the phosphorylated form of the enzyme than with the nonphosphorylated form. The presence of alanine also makes the pyruvate kinase a better substrate for the cAMP-dependent protein kinase; that is, the enzyme alanine complex is more readily and rapidly phosphorylated than the enzyme in the absence of alanine. Thus, there is an interaction between these two effectors. Both the hormonal message of glucagon plus the presence of alanine will be additive or synergistic in its effects on this enzyme.

Another activator of pyruvate kinase enzyme is 6-phosphogluconate, which is formed in the pentosphosphate pathway. The pentosphosphate pathway is particularly active in the liver during conditions of fatty-acid synthesis. The 6-phosphogluconate may well be an intervening signal between the pentosphosphate pathway and glycolysis because an increased activity of pyruvate kinase would produce more pyruvate, which potentially could proceed to acetyl-CoA and then to fatty-acid synthesis (Chapter 15). With higher insulin levels, the nonphosphorylated pyruvate kinase becomes predominant (Fig. 13.5). This signals higher glucose concentrations and is an optimal setting for lipogenesis to occur. The nonphosphorylated pyruvate kinase is more sensitive to 6-phosphogluconate and less sensitive to the inhibitor alanine. After phosphorylation via the hormonal message of glucagon to increase blood glucose, the pyruvate kinase becomes not only less active at any level of PEP, but more sensitive to inhibition by alanine and relatively refractive to activation by 6-phosphogluconate.

FRUCTOSE 1,6 BISPHOSPHATASE

The next unidirectional reaction of gluconeogenesis must be considered with the reverse reaction. The reaction for gluconeogenesis is FBPase; however, PFK is a reaction that reverses the result of FBPase. Therefore, if both mechanisms occur rapidly, we would have a fructose-6-phosphate + ATP $\rightarrow$ fructose-1,6-bisphosphate + ADP, followed by the FBPase reaction of F-1,6-BP $\rightarrow$ F-6-P + inorganic phosphate. The sum of these two reactions, ATP $\rightarrow$ ADP + Pi, has been referred to as a substrate or futile cycle. In liver, the activity of FBPase is considerably higher than that of PFK, which usually is also under some degree of ATP inhibition. However, if both reactions flow at maximal rate, the result is a considerable flux toward F-6-P and gluconeogenesis. There are controls for these reactions—the major allosteric effector is ***fructose-2,6-bisphosphate, which acts as an activator for PFK and an inhibitor of FBPase***. Fructose-2,6-bisphosphate is not part of any metabolic pathway, but is accumulated or destroyed in response to hormonal signals. In liver, there is an enzyme called PFK2, which in a non-phosphorylated form will catalyze the formation of fructose-2,6-bisphosphate (Fig. 13.6). PFK2 is likely to be nonphosphorylated or dephosphorylated by the presence of insulin, a hormone indicating a time for storage and such processes as lipid and glycogen synthesis. The nonphosphorylated PFK2 causes an accumulation of fructose-2,6- bisphosphate, which inhibits FBPase and stimulates PFK, allowing for a net flux in the glycolytic direction, even though the activity of FBPase is normally much higher than that of PFK. When phosphorylated by the cAMP-dependent protein kinase, in response to glucagon, PFK2 effectively becomes not a kinase but a phosphatase. When phosphorylated, PFK2, effectively is a fructose-2,6-bisphosphatase. Thus, it will carry out the reaction of fructose-2,6-bisphosphate $\rightarrow$ fructose-6-phosphate + inorganic phosphate. This will decrease the level of fructose-2,6-bisphosphate, removing the activation of PFK and the inhibition of FBPase, and al-

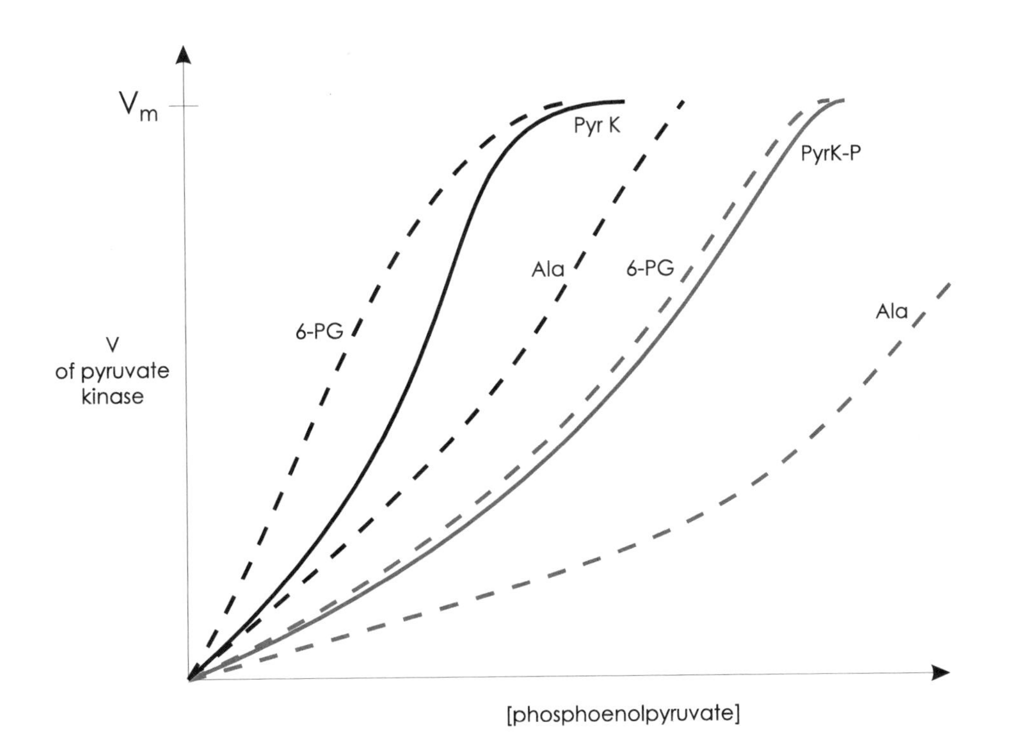

Figure 13.5.
Covalent modification of liver pyruvate kinase. A plot of pyruvate kinase versus PEP concentration is presented. The black solid curve indicates the response of velocity with change in PEP concentration for the nonphosphorylated enzyme in the absence of positive or negative effectors. The dashed black curves show the activity in the presence of a positive effector (6-PG) or a negative effector (alanine). The solid purple curve indicates the response of velocity with change in PEP concentration for the phosphorylated enzyme in the absence of positive or negative effectors. The dashed purple curve is the activity of the phosphorylated enzyme in the presence of a positive effector (6-PG) or a negative effector (alanine).

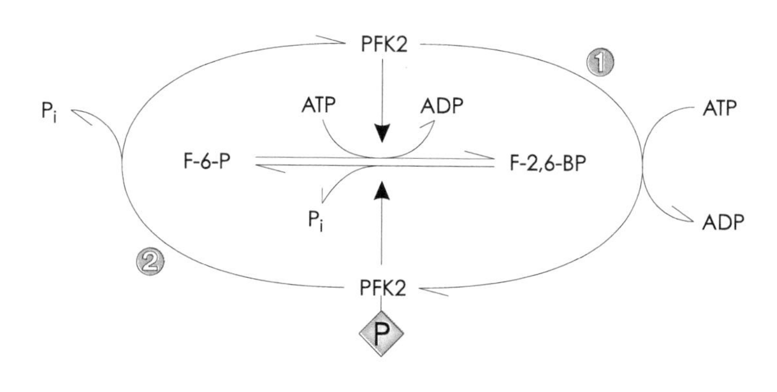

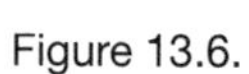
Figure 13.6.

Control of F-2,6-BP levels in the liver. This figure shows that PFK2 causes the formation of F-2,6-BP from F-6-P. When phosphorylated PFK2 becomes a F-2,6-BP phosphatase, converting F-2,6-BP to F-6-P, the phosphorylation of PFK2 is affected by the cAMP-dependent protein kinase. This phosphorylated PFK2 can be dephosphorylated by a specific phosphoprotein phosphatase.

lowing a net flux toward F-6-P and gluconeogenesis. FBPase is also inhibited by high level of 5′ AMP, but the physiological importance of this control in the liver is questionable.

GLUCOSE-6-PHOSPHATASE

The last enzyme of gluconeogenesis is G-6-Pase, which does not have strong allosteric controls but can have the appearance of inhibition with high glucose. During the hydrolysis of glucose-6-phosphate, there is an enzyme phosphoryl intermediate, which then can transfer the phosphate either to water, completing the hydrolysis, or back to glucose reforming glucose-6-phosphate. Thus when glucose is at a higher concentration, there will be a greater degree of retransfer back to glucose, reforming the glucose-6-phosphate and having the appearance and net effect of inhibition of glucose-6-phosphatase.

This apparent inhibition by high glucose has several potential advantages, such as causing an accumulation of G-6-P rather than releasing free glucose when glucose is already elevated. This increased G-6-P in the liver then can be used for glycogen synthesis. FBPase as well as G-6-Pase are both adaptive enzymes, similar to the PEP carboxykinase, which show increases in amount during conditions where chronic gluconeogenesis is occurring. These two enzymes increase in activity, although not to the same extent or rapidity as PEP carboxykinase. Thus it can be seen that gluconeogenesis is a critical process that is carefully controlled at several steps, some of which are very distant from the process per se, such as protecting pyruvate peripheral tissues as well as reactions closely associated with gluconeogenesis in the liver.

SUMMARY

1. Gluconeogenesis is the synthesis of glucose from nonglucose precursors and occurs primarily in the liver and kidney cortex.

2. There are four unique reactions associated with the conversion of lactate and/or pyruvate to glucose. These are pyruvate carboxylase, phosphoenolpyruvate carboxykinase, fructose-1,6-bisphosphatase, and glucose-6-phosphatase.

3. There are three reactions that are counterproductive relative to gluconeogenesis. These are pyruvate kinase, PFK I, and glucokinase (or hexokinase).

4. Acetyl CoA provides no carbon for net gluconeogenesis; thus, conversion of pyruvate to acetyl CoA eliminates the glucose potential of pyruvate.

5. Acetyl CoA, which can arise from fatty-acid metabolism as well as from pyruvate, is a required activator of pyruvate carboxylase and an inhibitor of pyruvate dehydrogenase.

6. Fructose-2,6-bisphosphate is an inhibitor of fructose-1,6-bisphosphatase and an activator of PFK I.

7. Fructose-2,6-bisphosphate can be formed by PFK2. When PFK is phosphorylated by the cAMP-

dependent protein kinase, it becomes a fructose-2,6-bisphosphatase and destroys fructose-2,6-bisphosphate.

8. Liver pyruvate kinase, which short circuits gluconeogenesis and favors glycolysis, can be phosphorylated by the cAMP-dependent protein kinase. Phosphorylation of this enzyme causes an increase in the K_s (0.5) for PEP and thus, effectively, inhibition.

9. There are also two activators of liver pyruvate kinase, fructose-1,6-bisphosphate and 6-phosphogluconate, and one inhibitor, alanine.

10. Phosphorylation of pyruvate kinase diminishes the activation by 6-phosphogluconate and increases the inhibition by alanine.

11. In the liver increased cAMP, via the cAMP-dependent protein kinase, increases gluconeogenesis and dephosphorylation of proteins causes a decrease in gluconeogenesis.

12. Any compound that produces a new net intermediate of glycolysis or of the citric-acid cycle is potentially glucogenic.

13. Glycerol can be converted to glucose via α-glycerol phosphate and dihydroxyacetone phosphate.

14. Gluconeogenesis from lactate requires six high-energy phosphates, whereas glycolysis only produces a net of two high-energy phosphates.

REFERENCES

Glucogenic substrate levels in fasting man, T. T. Aoki, C. J. Toews, A. A. Rossini, N. B. Ruderman, and G. F. Cahill Jr., 1975, *Adv. Enzyme Reg.*, 329–336.

Hormonal regulation of hepatic gluconeogenesis and glycolysis, S. J. Pilkis, M. R. El-Magkrabi, and T. H. Claus, 1988, *Ann. Rev. Biochem.*, **57**:755–831.

Hepatic gluconeogenesis/glycolysis: Regulation and structure/function relationships of substrate cycle enzymes, S. J. Pilkis and T. H. Claus, 1991, *Ann. Rev. Nutr.*, **11**:465–515.

Futile cycling in glucose metabolism, J. Katz and R. Rogustad, 1978, *Trends Biochem, Sci.*, **3**:171–174.

Substrate cycles: Their role in improving sensitivity in metabolic control. E. A. Newsholme, R. A. J. Challis, and B. Crabtree, 1984, *Trends Biochem. Sci.*, **9**:277–280.

REVIEW QUESTIONS

1. Levels of acetyl CoA in the liver and muscle mitochondrion are important because high levels of acetyl CoA in both these tissues will
   a) Activate pyruvate carboxylase
   b) Inhibit pyruvate dehydrogenase
   c) Activate pyruvate carboxylase and inhibit pyruvate dehydrogenase
   d) Inhibit pyruvate carboxylase and activate pyruvate dehydrogenase
   e) Activate ketone bodies

2. Phosphoenolpyruvate carboxykinase is important in metabolism in order to
   a) Utilize cellular GTP
   b) Form oxaloacetate from phosphoenolpyruvate
   c) Synthesize alanine in the liver
   d) Overcome the irreversibility of pyruvate kinase
   e) Stimulate the activity of pyruvate carboxylase

3. Covalent modification of pyruvate kinase in the liver due to increased cAMP alters the enzyme in the following way:
   a) Decreases the $K_s(0.5)$ for PEP and makes it more sensitive to alanine inhibition
   b) Increases the $K_s(0.5)$ for PEP and makes it less sensitive to alanine inhibition
   c) Increases the $K_s(0.5)$ for PEP and makes it more sensitive to alanine inhibition
   d) Decreases the $K_s(0.5)$ for PEP and makes it less sensitive to alanine inhibition
   e) Has no effect on $K_s(0.5)$ for PEP, but makes it more sensitive to alanine inhibition

4. The metabolic effector fructose-2,6-bisphosphate will do the following:
   a) Inhibit fructose-1,6-bisphosphatase and activate PFK
   b) Activate fructose-1,6-bisphosphatase and inhibit PFK

c) Activate pyruvate carboxylase and phosphoenolpyruvate carboxykinase
d) Inhibit both fructose-1,6-bisphosphatase and PFK
e) Activate phosphoenolpyruvate carboxykinase and inhibit pyruvate kinase

5. In the liver PFKII when phosphorylated in response to increased cAMP will
   a) Become inactive in all functions
   b) Become a more active kinase
   c) Become a G-6-P phosphatase
   d) Become an activator of phosphoenolpyruvate carboxykinase
   e) Become an F-2,6-BP phosphatase

6. An example of a futile or substrate cycle occurs if the following pair of enzymes are functioning rapidly:
   a) Glucokinase and PFK
   b) PFK and fructose-1,6-bisphosphatase
   c) PFK and pyruvate kinase
   d) Glucokinase and fructose-1,6-bisphosphatase
   e) Glucose-6-phosphatase and fructose-1,6-bisphosphatase

ANSWERS TO REVIEW QUESTIONS

1. ***b*** This will allow pyruvate to accumulate and prevent the irreversible catabolism of pyruvate. Pyruvate is a glucogenic precursor where the product of pyruvate dehydrogenase, acetyl CoA, cannot produce net glucose in people.

2. ***d*** Since pyruvate kinase is physiologically irreversible, another pathway of conversion of pyruvate to PEP for gluconeogenesis must be available. This pathway is pyruvate to oxaloacetate, via pyruvate carboxylase and oxaloacetate to PEP, via phosphenolpyruvate carboxykinase.

3. ***c*** Both the increase in K_S (0.5) for PEP and alanine inhibition decreases the rate of pyruvate kinase activity for physiological concentrations of PEP. Thus, the competing pathway of gluconeogenesis can occur more readily.

4. ***a*** which will slow gluconeogenesis and stimulate glycolysis. The concentration of this effector in the liver is increased by increased glucose and insulin concentrations and decreased by increased glucagon concentrations.

5. ***e*** This decreases the liver concentrations of F-2,6-BP allowing a more rapid gluconeogenesis (see question 4 explanation).

6. ***b*** Phosphofructokinase converts F-6-P to F-1,6-BP, utilizing an ATP. Fructose-1,6-bisphosphatase converts F-1,6-BP to F-6-P plus inorganic phosphate. Thus, when both enzymes are active, some ATP is wasted.

   (1) F-6-P + ATP $\rightarrow$ F-1,6-BP + ADP
   PFK

   (2) F-1,6-BP $\rightarrow$ F-6-P + Pi
   FBPase

   ATP $\rightarrow$ ADP + Pi
   Addition of reactions (1) + (2)

CHAPTER 14

FATTY-ACID METABOLISM: CATABOLISM

INTRODUCTION

In the next two chapters we examine aspects of lipid metabolism and its control. The major energy stores in humans are lipids in the form of triacylglycerol (three fatty acids esterified to glycerol). Their rates of synthesis and degradation are carefully controlled and loss of this control can lead to obesity. The fatty acids can arise from the diet or hepatic synthesis and, after conversion into triacylglycerol (triglycerides) in the liver, are repackaged as lipoproteins which contain phospholipids, cholesterol, cholesterol esters, and protein. The lipoproteins, as very low-density lipoproteins (VLDL), low-density lipoproteins (LDL), and high-density lipoproteins (HDL), are carried through the blood. This transport is to the adipose tissue for storage or to other tissues for other purposes. Lipids such as phospholipids and cholesterol are important in other cellular functions, such as synthesis of membranes, production of steroid hormones, and as intracellular signals. In this chapter, we examine the synthesis of fatty acids and cholesterol, as well as fatty-acid oxidation for energy, including ketone-body formation.

LIPID STORAGE AND TURNOVER

The major source of potential energy in the human body is in the form of a lipid, namely, triacylglycerol. Although triacylglycerol is the major form of lipid in the body, there are numerous other important lipids, including phospholipids, steroid hormones, prostaglandins, fat-soluble vitamins, and others. In addition to energy storage, lipids play important roles in membranes and lipoproteins, as discussed in earlier chapters. The triacylglycerols also play a role in mechanical and thermal insulation. In this chapter, the utilization of triacylglycerol and fatty acids as a source of energy is emphasized. Triacylglycerols are primarily stored in the adipose tissue. In addition to having large deposits in certain anatomic compartments, adipose cells can also be fairly diffuse throughout other areas of the body, such as muscle. Fatty acids for triacylglycerol synthesis arise from the diet or are synthesized in the body. Triacylglycerol in the storage cells are continually undergoing turnover (i.e., synthesis and breakdown). In many individuals with good weight control, this turnover is such that storage and breakdown of triacylglycerols (i.e., lipolysis) occur at approximately equal rates. However, this equality is over the long term; under shorter time conditions, there is usually net synthesis after eating and net lipolysis between meals or during starvation.

Triacylglycerol Hydrolysis

An increase in lipolysis can occur primarily in response to two hormones: epinephrine and glucagon, both of which cause increased cAMP in the adipose cell and result in a phosphorylation of a hormone-sensitive lipase, converting it from an inactive to an active form (i.e., the phosphorylated enzyme is the active form). Hormone-sensitive lipase is usually the rate-limiting enzyme of lipolysis, catalyzing hydrolysis of the first fatty acid from the triacylglycerol and leaving a 1,2 diacylglycerol that can readily be hydrolyzed by other lipases within the cell. The fatty acids released by this lipolysis leave the adipose cell and enter the bloodstream, where they are bound to circulating albumin. Nonesterified fatty acids are good emulsifying agents, and if not bound to the albumin, could cause emulsification and lysis of the red blood cell. Thus, the nonesterified fatty acids are carried throughout the body primarily bound to albumin with a slight dissociation. These fatty acids can enter many cells, including liver, kidney, muscle, and heart. As they enter these tissues and are taken up, the concentration in the plasma decreases and, owing to equilibrium, more of the albumin–fatty-acid complex dissociates to reestablish the circulating low levels of free, unbound fatty acids in the plasma.

Activation and Beta Oxidation

When fatty acids are taken up by cells capable of metabolizing them, they are activated to their coenzyme A derivatives by two reactions. The first is fatty acid + ATP ↔ fatty acid-AMP + pyrophosphate. The pyrophosphate is then hydrolyzed to two inorganic phosphates to pull the reaction in a forward direction, or equilibrium would favor the fatty acids and ATP. The fatty acid portion of the fatty acid–AMP complex, while still bound to the fatty-acid activating enzyme is then transferred to coenzyme A and AMP is released. This fatty-acyl CoA, also referred to as long-chain fatty-acyl CoA, is similar to the acetyl CoA discussed previously, except that in this case a long-chain fatty-acid is covalently bound via a thioester to the coenzyme A. This activation process occurs both on the endoplasmic reticulum and on the outer mitochondrial membrane. However, ***the oxidation of the fatty acids occurs primarily in the mitochondrial matrix***. Thus, the problem arises of how to transfer the large fatty-acyl CoA molecule into the mitochondria. As was the case for NADH and the malate shuttle, another shuttle mechanism is used here. It operates by transferring the fatty-acid moiety to carnitine via carnitine palmitoyltransferase 1, located on the outer membrane of the mitochondria (Fig. 14.1). The resulting smaller fatty-acyl carnitine is transferred into the mitochondrion, where carnitine palmitoyltransferase 2 transfers the fatty acid to coenzyme A in the mitochondrial matrix, forming fatty-acyl CoA. Short-chain fatty acids (C4–C8) are activated in the mitochondria and do not require carnitine transport for catabolism.

Fatty-acyl CoA then undergoes a process known as beta oxidation, so called because the beta carbon is continually oxidized in the cycle. This process results in reduced coenzymes and acetyl CoA (Fig. 14.2). The first enzyme of this process is fatty-acyl CoA dehydrogenase, which with FAD forms a double bond by removing two hydrogens between the alpha and beta carbon of the fatty acid and produces an enzyme-bound $FADH_2$, which can be reoxidized via the electron transport system, producing high-energy phosphates. The configuration of this double bond is trans which, as we will see, is different from that found in naturally occurring unsaturated fatty acids. The next step is enoyl CoA hydratase, which adds water across the double bond, similar to the reaction of fumarase, forming a beta (or 3) hydroxy, fatty-acyl CoA with a L configuration. 3-L-hydroxy acyl CoA dehydrogenase, in conjunction with NAD^+, removes the two hydrogens at the number 3 carbon, yielding a beta keto fatty-acyl CoA and NADH. The latter is also reoxidized via the electron-transport system, again producing high-energy phosphates. The beta keto (or 3 keto) fatty-acyl CoA then undergoes a cleavage catalyzed by the enzyme beta ketothiolase, with the participation of another coenzyme A, yielding a fatty-acyl CoA that is two carbons shorter than the original fatty-acyl CoA, plus an acetyl CoA. Once started, this process continues until all of the fatty acid is converted to acetyl CoA. This appears to be very rapid, at least down to the four-carbon intermediate stage, because when starting with palmitate (C16), one of the more predominant storage fatty acids, no significant concentrations of C6, C8, C10, or C12 fatty-acyl CoAs have been observed. ***In the process of beta oxidation, therefore, fatty acids are converted to acetyl CoA and produce reduced coenzymes*** which, via the electron-transport system and oxidative phosphorylation, are utilized for high-energy phosphate formation. With a C16 fatty acid, there will be seven cycles of beta oxidation, although eight acetyl CoAs are produced, since the last acetyl CoA formed is from the two terminal carbons.

The energy production from a C16 fatty acid is 35 high-energy phosphates through beta oxidation from the formation of seven $FADH_2$'s and seven NADH's. There are also eight acetyl CoA's, each of which can, on

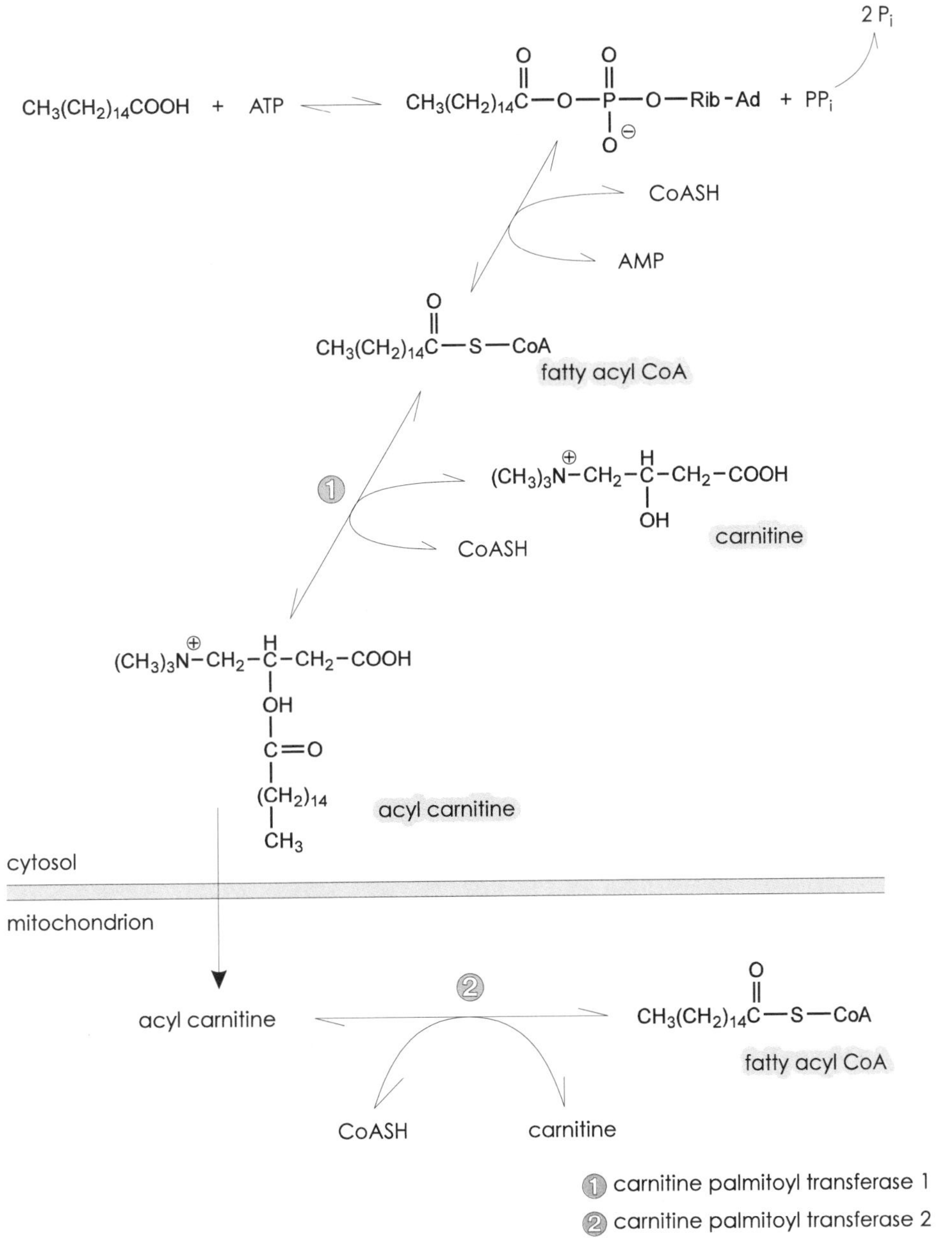

$CH_3(CH_2)_{14}COOH$ + ATP ⇌ $CH_3(CH_2)_{14}C(=O)-O-P(=O)(O^{\ominus})-O-$Rib-Ad + PP_i → 2 P_i

CoASH

AMP

$CH_3(CH_2)_{14}C(=O)-S-CoA$

fatty acyl CoA

① $(CH_3)_3\overset{\oplus}{N}-CH_2-CH(OH)-CH_2-COOH$ carnitine

CoASH

$(CH_3)_3\overset{\oplus}{N}-CH_2-CH(O-C(=O)-(CH_2)_{14}-CH_3)-CH_2-COOH$ acyl carnitine

cytosol

mitochondrion

acyl carnitine ② $CH_3(CH_2)_{14}C(=O)-S-CoA$ fatty acyl CoA

CoASH carnitine

① carnitine palmitoyl transferase 1

② carnitine palmitoyl transferase 2

Figure 14.1.

Activation and transport of long-chain fatty acids. The activation of long chain fatty acids to their CoA derivatives and subsequent transport, via carnitine, into the mitochondria is depicted in this figure.

complete oxidation, produce the equivalent of 12 high-energy phosphates per acetate unit, forming 96 high-energy phosphates. This totals 131 high-energy phosphates, but from this number must be subtracted the two high-energy phosphates required for the initial activation step. (Note that only one ATP was used; however, an AMP was produced and this requires two high-energy phosphates to return to the original ATP state). Thus, the net energy production from complete oxidation of a C16 fatty acid is 129 high-energy phosphates. To further emphasize the advantage of storing excess energy in the form of triacylglycerol rather than carbohydrate for a mobile animal (such as a person), is a marked increase in the caloric density of triacylglycerols (9.5 cal/gm) versus a carbohydrate like glycogen (4.0 cal/gm), which yields over twice the energy potential per unit weight for a triacylglycerol compared to glycogen. In addition, polysaccharides are very hydrophilic compounds, and for every gram of glycogen stored there is at least one gram or more of associated water. With triacylglycerols, which are extremely hydrophobic, storage occurs without any additional necessity for water. Thus, the high-energy yield from 1g of stored triacylglycerol is over five times as great as for 1g of stored glycogen. This is particularly important where mobility is critical. For example, plants, which

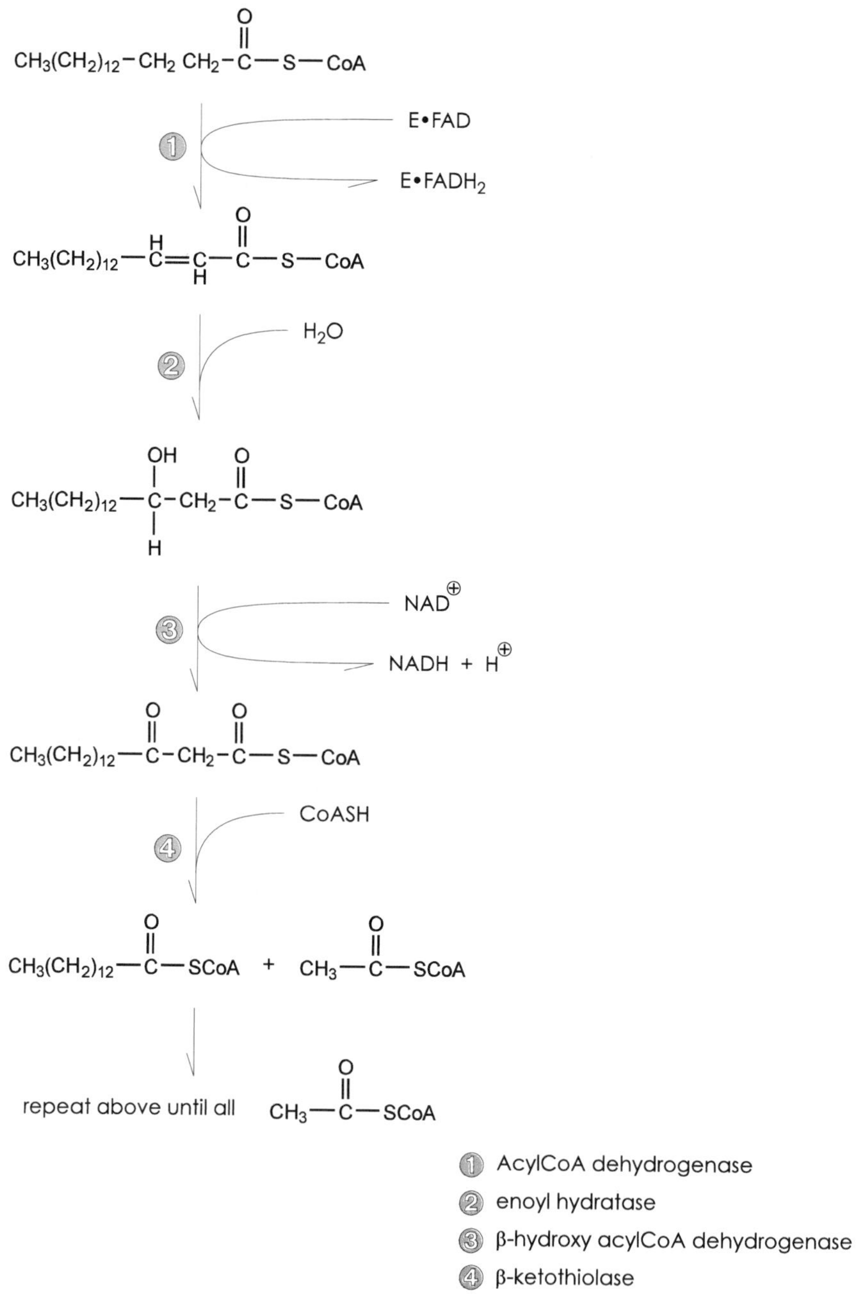

$CH_3(CH_2)_{12}-CH_2\,CH_2-C(=O)-S-CoA$

① E•FAD → E•$FADH_2$

$CH_3(CH_2)_{12}-CH=CH-C(=O)-S-CoA$

② H_2O

$CH_3(CH_2)_{12}-CH(OH)-CH_2-C(=O)-S-CoA$

③ $NAD^{\oplus}$ → NADH + $H^{\oplus}$

$CH_3(CH_2)_{12}-C(=O)-CH_2-C(=O)-S-CoA$

④ CoASH

$CH_3(CH_2)_{12}-C(=O)-SCoA$ + $CH_3-C(=O)-SCoA$

repeat above until all $CH_3-C(=O)-SCoA$

① AcylCoA dehydrogenase
② enoyl hydratase
③ β-hydroxy acylCoA dehydrogenase
④ β-ketothiolase

Figure 14.2.

Beta oxidation. The steps of beta oxidation occurring in the mitochondria are shown in this figure. These steps are repeated until all of the fatty acid is converted to acetyl CoA.

are not very mobile, store most of their excess energy as carbohydrate. In contrast, the seeds of plants, which require much greater mobility, store a high percentage of their energy as lipid.

With all of the apparent advantages of storing energy in the form of lipid, it might be asked why any energy is stored in the form of glycogen. However, there are some very important differences between the metabolism of glucose and fatty acids. First and foremost, ***fatty acids require oxygen to be utilized for energy and cannot provide energy under anaerobic conditions.*** Carbohydrate is the only substrate that can produce energy under anaerobic conditions. Because of peculiarities of the membranes, the central nervous system does not take up significant quantities of fatty acids and thus does not utilize these substrates for energy to any appreciable extent. Under most circumstances, therefore, the central nervous system is dependent on glucose, although it is an aerobic and not an anaerobic tissue.

From even-chain fatty acids (i.e., those containing an even number of carbons and which represent the major, if not sole, store of fatty acids and triacylglycerols in people), ***there is no net glucose formation***. Even-chain fatty acids produce exclusively acetyl CoA, which cannot produce any net gluconeogenesis. In the presence of oxygen, fatty acids are an excellent source of energy and, as indicated in previous chapters, they can play an important role in conserving glucose. This occurs via the formation of acetyl CoA, which is an important inhibitor of pyruvate dehydrogenase and stimulates the phosphorylation and inactivation of this enzyme. In muscle, the high ATP and citrate levels produced during fatty-acid degradation can inhibit PFK, causing elevated F-6-P and G-6-P inhibition of hexokinase, and thus a decrease in glucose utilization. In liver, the increased level of acetyl CoA not only inhibits pyruvate dehydrogenase, but also acts as an allosteric activator for pyruvate carboxylase, a key enzyme in gluconeogenesis. Thus, as will be seen throughout metabolism, a single pathway or control system cannot be considered in isolation; there are very strong interactions, as illustrated here in the case of carbohydrate and lipid metabolism in the body.

Ketone Bodies

In muscle, most of the fatty acids undergoing beta oxidation are completely oxidized to CO_2 and water. In liver, however, there is another major fate for fatty acids; this is the formation of ketone bodies, namely acetoacetate and β-hydroxybutyrate. ***The fatty acids must be transported into the mitochondrion for normal beta oxidation. This may be a limiting factor for beta oxidation in many tissues and ketone-body formation in the liver.*** The extramitochondrial fatty-acyl portion of fatty-acyl CoA can be transferred across the outer mitochondrial membrane to carnitine by carnitine palmitoyltransferase I (CPTI). This enzyme is located on the inner side of the outer mitochondrial membrane. The acylcarnitine is now located in mitochondrial intermembrane space. The fatty-acid portion of acylcarnitine is then transported across the inner mitochondrial membrane to coenzyme A to form fatty-acyl CoA in the mitochondrial matrix. This translocation is catalyzed by carnitine palmitoyltransferase II (CPTII; Fig. 14.1), located on the inner side of the inner membrane. This later translocation is also facilitated by carnitine–acylcarnitine translocase, located in the inner mitochondrial membrane. The CPTI is inhibited by malonyl CoA, an intermediate of fatty-acid synthesis (see Chapter 15). This inhibition occurs in all tissues that oxidize fatty acids. The level of malonyl CoA varies among tissues and with various nutritional and hormonal conditions. The sensitivity of CPTI to malonyl CoA also varies among tissues and with nutritional and hormonal conditions, even within a given tissue. ***Thus, fatty-acid oxidation may be controlled by the activity and relative inhibition of CPTI.***

Fatty acids undergo beta oxidation in the liver and form acetyl CoA. If acetyl CoA accumulates instead of traversing the citric-acid cycle, the liver becomes more dependent on beta oxidation than on the citric-acid cycle for its source of high-energy phosphate. To continue beta oxidation, coenzyme A is required. If all the CoA is tied up as fatty-acyl CoA and acetyl CoA, no beta oxidation occurs. However, there is a metabolic pathway, unique to the liver, through which to overcome this apparent dilemma—ketone-body formation (Fig. 14.3). It occurs as follows in the mitochondrial matrix.

Two acetyl CoAs can combine to form acetoacetyl CoA by the reverse of β-ketothiolase. The acetoacetyl CoA then combines with another acetyl CoA to make hydroxymethyl glutaryl CoA (HMG CoA) by the enzyme hydroxymethyl glutaryl CoA synthase. The HMG CoA in the mitochondrion can be cleaved by HMG CoA lyase in the mitochondrion to form acetoacetate and acetyl CoA. In this conversion, the formation of acetoacetyl CoA from two acetyl CoAs releases a free CoA and formation of HMG CoA from acetyl CoA and acetoacetyl CoA also releases a free coenzyme A. Thus, the release of free coenzyme A allows beta oxidation to continue with the production of acetoacetate. During diabetes and starvation, almost 90% of carbon from a fatty acid such as oleate can be accounted for in the form of ketone bodies during experiments with perfused livers. At this time, it would be worth noting that this process occurs in the mitochondrion; later it will be seen that HMG CoA in the cytosol is a major precursor for cholesterol synthesis.

Within the mitochondria of most cells, including the liver, the enzyme β-hydroxybutyrate dehydrogenase can reversibly catalyze a reaction NADH + H^+ + acetoacetate ↔ β-hydroxybutyrate + NAD^+. Through this reaction, much of the acetoacetate formed in liver is converted to β-hydroxybutyrate. The ratio of β-hydroxybutyrate to acetoacetate is determined by the NAD^+/NADH ratio in the mitochondrion. The ratio of β-hydroxybutyrate to acetoacetate circulating in humans is approximately 3:1 to 4:1. Since ketone bodies are an important source of energy to tissues other than the liver, the advantage of the higher level of β-hydroxybutyrate is that it is a stable compound; whereas acetoacetate, being a beta keto acid, is relatively less stable and

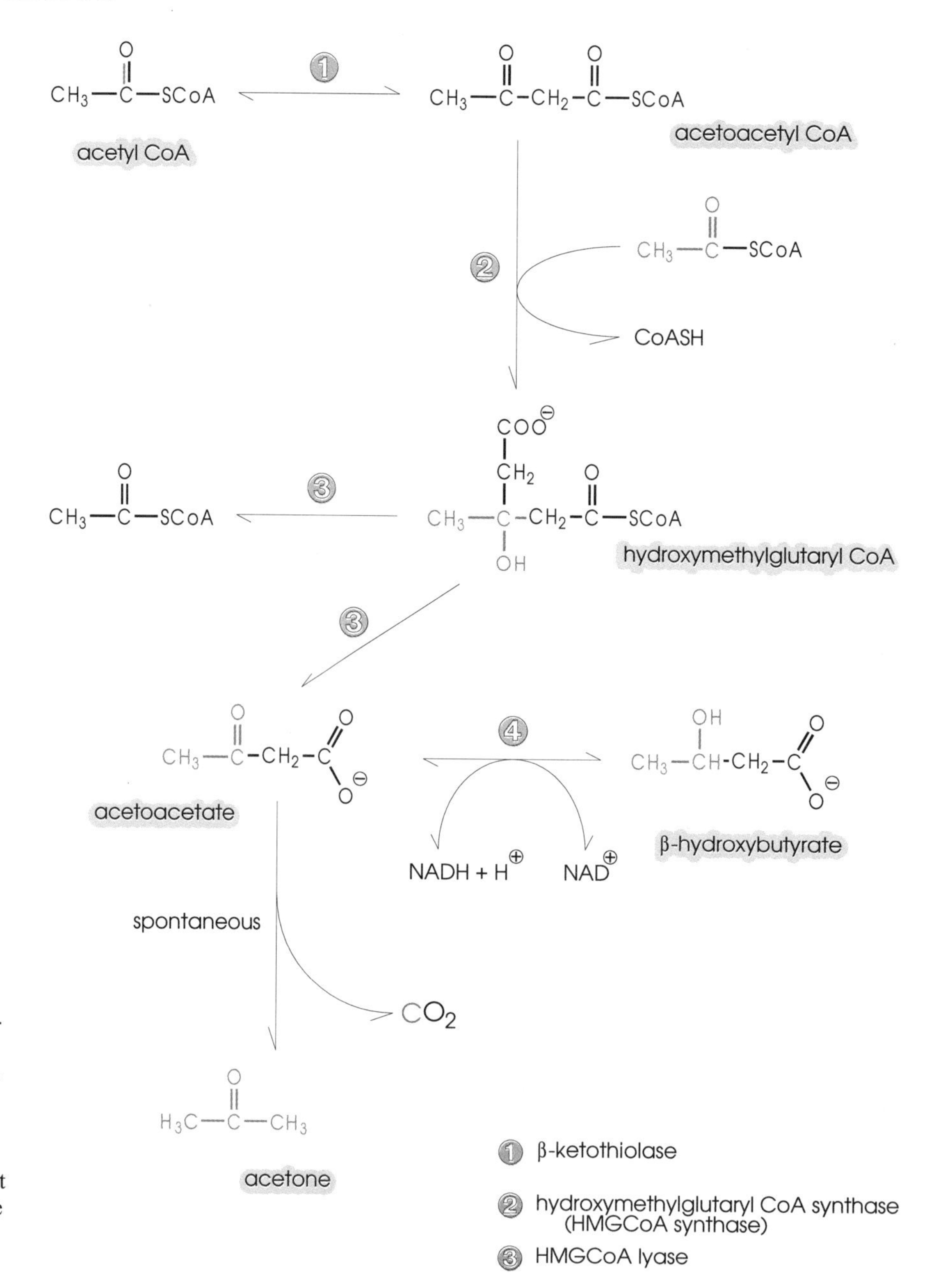

Figure 14.3.

Ketone-body formation. This diagram shows the steps in ketone-body formation. Acetoacetate and β-hydroxybutyrate are both ketone bodies, as is acetone, a decarboxylation product of acetoacetate. The last acetyl CoA added is shown in purple. Note that since the methyl groups on acetone are indistinguishable, both are shown in purple.

can spontaneously decarboxylate to form CO_2 and acetone. This is why acetone is often smelled on the breath of individuals with ketosis, whether it is due to prolonged starvation or untreated diabetes. The ketone bodies then circulate in the blood and can reach high levels, up to 5mM or more, whereas fatty acids always remain at about 1mM or lower. One advantage of ketone bodies over fatty acids in circulation is that they are small molecules, are very water soluble, and will not cause emulsification and lysis of red cells, as would high levels of fatty acids. Also, the ketone bodies can be used by almost all cells, including the central nervous system. Two cell types that cannot use ketone bodies are the red blood cell, owing to the lack of mitochondria, and the liver, particularly parenchymal cells, owing to the absence of the major enzyme for activating these compounds.

The major process for activating ketone bodies occurs in the mitochondrion, catalyzed by succinyl CoA–acetoacetate thiotransferase, which, starting with succinyl CoA + acetoacetate, forms succinate + acetoacetyl CoA; the latter, via beta keto thiolase, can then react with CoA to produce two acetyl CoAs (Fig. 14.4). Although the GTP that would have been obtained from

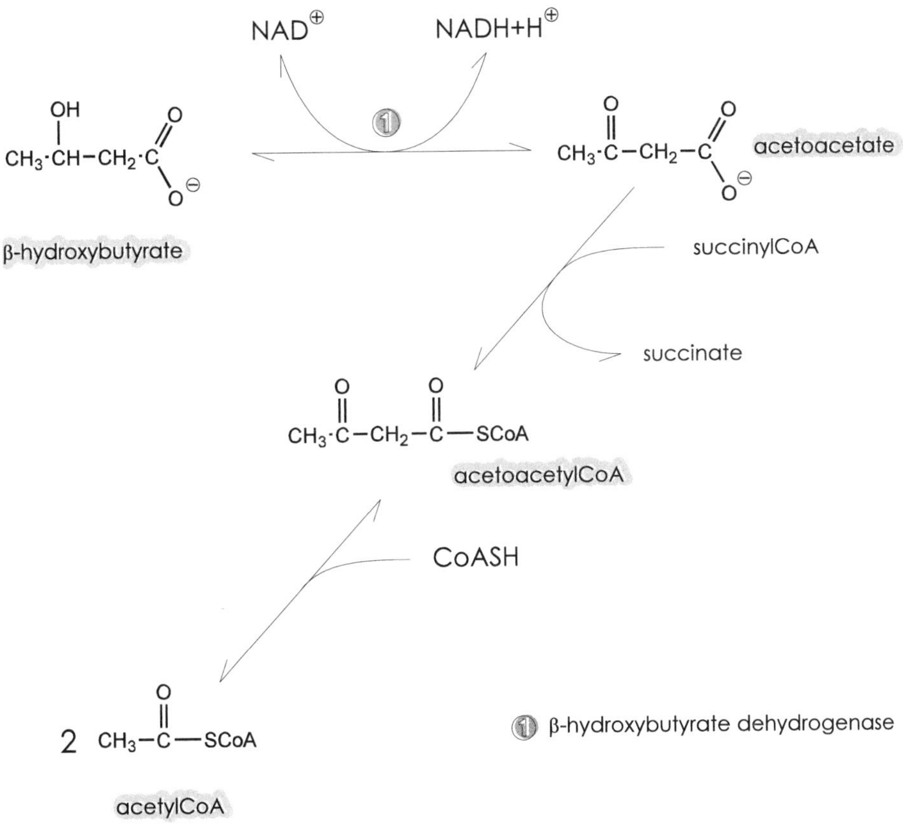

NAD$^{\oplus}$ NADH+H$^{\oplus}$

$CH_3 \cdot CH(OH)-CH_2 \cdot COO^{\ominus}$ (1) $\rightleftharpoons$ $CH_3 \cdot CO-CH_2-COO^{\ominus}$

β-hydroxybutyrate acetoacetate

succinylCoA → succinate

$CH_3 \cdot CO-CH_2-CO-SCoA$

acetoacetylCoA

CoASH

$2\ CH_3-CO-SCoA$

acetylCoA

(1) β-hydroxybutyrate dehydrogenase

Figure 14.4.

Activation of ketone bodies. This diagram shows the steps in the activation of ketone bodies and their conversion to acetyl CoA. This occurs in the mitochondria of many types of cells.

succinyl CoA will not be formed, the loss of this one high-energy phosphate saves two high-energy phosphates that normally would be required if acetoacetate were activated in a manner similar to that used for activating a fatty acid. After activation and conversion to acetyl CoA in the mitochondrion, the acetyl CoAs from ketone bodies can be metabolized similarly to any other acetyl CoA.

In certain cells, there is some activation of ketone bodies in a manner similar to that of fatty acids in the cytosol, forming acetoacetyl CoA, which then may be utilized as a carbon source for either fatty-acid synthesis or cholesterol synthesis.

Ketone-body utilization has priority over both fatty-acid and carbohydrate utilization. The ketone bodies are activated within the mitochondrion from a critical intermediate of the citric-acid cycle. Therefore, even with lower availability of CoA, the ketone bodies have priority for activation; fatty acids activated outside the mitochondrial matrix must be carried into the mitochondrial matrix (via the carnitine shuttle) where CoA must be available for reformation of mitochondrial fatty-acyl CoA. In most tissues, the lowest priority for use when all three substrates and oxygen are available is that of carbohydrate, because of the large number of controls involved in carbohydrate utilization (which include hexokinase, PFK, and pyruvate dehydrogenase). Ketone-body utilization in the brain can account for as must as 60% of the energy requirements, particularly during prolonged periods of starvation. However, there is still a critical need for glucose to supply the remaining 40% of the energy needs.

Ketone bodies are normal and useful metabolites but, like many other metabolites, when their concentration is excessive, they can produce problems. The major problems with ketone bodies are (1) loss of energy when these are at high levels and are excreted into the urine, and (2) the loss of cations that accompanies excretion of the ketone bodies. These can be sodium, potassium, and, in later stages, even calcium. As will be seen, the body attempts to compensate by increasing the output of ammonium ion as the cation to accompany ketone bodies. The pK of the ketone bodies is approximately 4, and at blood pH, and even at most acidic urine pH's, they exist primarily in the ionized form. Therefore, if excreted, they require an accompanying cation. The pH problems associated with the ketone bodies are reversed when these compounds are

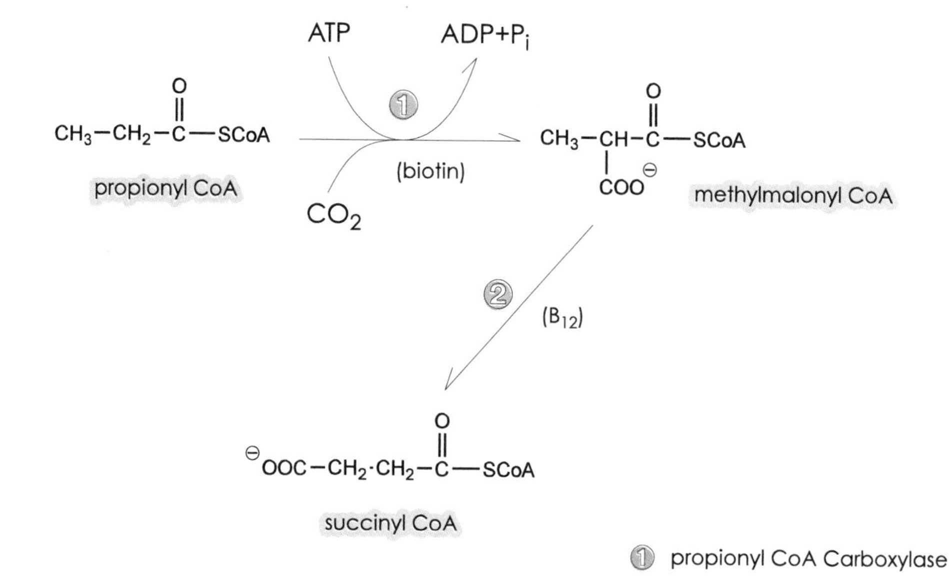

Figure 14.5.
Propionate metabolism. The conversion of propionyl CoA to succinyl CoA is shown stepwise in this figure. This occurs in the mitochondria. The production of succinate from propionate makes propionate glucogenic.

metabolized to CO_2 and water. However, that cannot happen if they are excreted from the body in the urine. Therefore, although ketone bodies have a negative connotation, they are normal and useful metabolites if not in excess. Only when they are in excess and there is ketonemia (excessive elevated ketone bodies in the blood) and particularly ketonuria (excess ketone bodies in the urine) are there problems. The problems are primarily those of acid–base balance. Ketonemia and ketonuria are most likely to occur in prolonged starvation or diabetes.

Odd-Chain Fatty Acids

Odd-chain fatty acids (those containing an odd number of carbon atoms) also undergo beta oxidation. However, in this case, the final product is one molecule of proponyl CoA. This cannot be oxidized directly in the citric-acid cycle, as can acetyl CoA. Proponyl CoA is metabolized, in the mitochondrion, through several steps, beginning with the formation of methylmalonyl CoA (MMCoA) catalyzed by a biotin-containing enzyme requiring CO_2 and ATP (see Fig. 14.5). Then, catalyzed by an isomerase requiring vitamin B_{12}, MMCoA is converted to succinyl CoA, a citric-acid cycle intermediate. The succinyl CoA can form oxaloacetate via the citric-acid cycle, and then net glucose. Therefore, because two molecules of ***odd-chain fatty acids*** can produce one molecule of glucose, these compounds ***are gluconeogenic in character***.

SUMMARY

1. Lipids are the major energy storage in people. They contain over five times the energy of glycogen per weight of stored reserve.

2. Triacylglycerols are stored primarily in the adipose. Triacylglycerols can be hydrolyzed by lipases, one of which can be activated by glucagon or epinephrine. These hormones cause activation of hormone-sensitive lipase, which results in release of nonesterified fatty acids into the blood.

3. Many tissues can activate fatty acids to fatty-acyl CoA in the cytosol and transfer them to the mitochondria via a carnitine-linked system. The transfer of fatty acids to the mitochondrial matrix may limit the rate of fatty-acid oxidation and ketone-body production. This transfer can be controlled by the activity of carnitine palmitoyltransferase I (CPTI). CPTI is inhibited by malonyl CoA.

4. The mitochondrial fatty-acyl CoA can undergo β-oxidation, producing acetyl CoA's, NADH's, and enzyme-bound $FADH_2$'s. The NADH and $FADH_2$ can be used to form ATP's via the electron-transport system. Fatty-acid catabolism requires oxygen.

5. Fatty-acid metabolism in the muscle can increase ATP and citrate, which can inhibit PFK raising G-6-P levels, which inhibits hexokinase and decreased use of glucose.

6. Fatty-acid metabolism in the liver can increase gluconeogenesis, owing to increased mitochondrial acetyl CoA levels.

7. Ketone bodies are formed from fatty acids in the liver. This releases free CoA to continue beta oxidation and supplies supplemental energy to other tissues, including the brain. Ketone bodies can be activated in the mitochondria of many tissues and produce acetyl CoA.

8. The priority of usage by cells which can use glucose, fatty acids, and ketone bodies is, in the presence of O_2, ketone bodies > fatty acids > glucose.

9. Fatty acids containing an odd number of carbon atoms can produce net glucose.

REFERENCES

Regulation of hepatic fatty acid oxidation and ketone body production, J. D. McGarry and W. Foster, 1980, *Ann. Rev. Biochem.*, **49**:395–420.

Regulation of ketogenesis and the renaissance of carnitine palmitoyltransferase, J. D. McGarry, K. F. Woeltje, M. Kawajima, and D. W. Foster, 1989, *Diabetes Met. Rev.*, **5**:271–284.

Regulation of mitochondrial carnitine palmitoyltransferases from liver and extrahepatic tissues, D. Saggerson, I. Ghadiminejad, and M. Awan, 1992, *Adv. Enzyme Reg.*, 205–306.

Enzymatic regulation of liver acetyl-CoA metabolism in relation to ketogenesis, O. Wieland, L. Weiss, and I. Eger-Neufeldt, 1964, *Adv. Enzyme Reg.*, 85–99.

The AMP-activated proteinkinase: A multisubstrate regulator of lipid metabolism, D. G. Hardie, D. Carling, and A. T. R. Sim, 1989, *Trends Biochem. Sci.*, **14**:20–23.

REVIEW QUESTIONS

1. Fatty acids are transported into the mitochondrion via
   a) Coenzyme A
   b) Carnitine
   c) α-Glycerolphosphate
   d) Lipoic acid
   e) Phospholipids

2. Fatty acids are released from adipose triglyceride stores through the initial action of
   a) Hormone-sensitive lipase
   b) Lipoprotein lipase
   c) Lecithin acyl transferase
   d) Fatty-acyl CoA hydrolase
   e) Hydroxymethylglutaryl CoA lyase

3. A natural inhibitor of fatty acid oxidation in the intact cell is
   a) Citrate
   b) G-6-P
   c) Malonyl CoA
   d) HMGCoA
   e) NADPH

4. Fatty acids affect carbohydrate oxidation via
   a) Direct effect on hexokinase
   b) Indirect effect on pyruvate kinase
   c) Indirect effect on glycogen synthesis
   d) Indirect effect on oxaloacetate dehydrogenase
   e) Indirect effect on pyruvate dehydrogenase

5. Ketone bodies (BHB and acetoacetate) cannot be used for energy by
   a) Muscle
   b) Kidney
   c) Brain
   d) Red blood cells
   e) Heart muscle

ANSWERS TO REVIEW QUESTIONS

1. ***b*** The answer is via palmitoyl acylcarnitine transferase 1 and 2, which forms fatty-acyl CoA in the mitochondrial matrix.

2. ***a*** This enzyme is activated by phosphorylation.

3. ***c*** Malonyl CoA, an intermediate in fatty-acid synthesis, is an inhibitor of palmitoyl acylcarnitine transferase 1, and the transfer of fatty acids to the mitochondrial matrix, the major site of fatty-acid oxidation.

4. *e* The increases in mitochondrial acetyl CoA concentration, due to fatty-acid oxidation, inhibits pyruvate dehydrogenase and stimulates phosphorylation of this enzyme to a relatively inactive enzyme.

5. *d* Red blood cells cannot use ketone bodies for energy, because they have no mitochondria. Ketone-body utilization for energy is an oxidative process and requires the presence of mitochondria.

CHAPTER 15

LIPID METABOLISM: SYNTHESIS AND TRANSPORT

INTRODUCTION

There are several types of lipids, including triacylglycerols (also known as triglycerides), phospholipids, sphingolipids, glycolipids, and cholesterol. This chapter describes the synthesis and transport of these lipids. In many cases, lipid synthesis or lipogenesis refers to the de novo formations of fatty acids and cholesterol, which then may be followed up by the formation of more complex lipids, such as triacylglycerol and phospholipids. We examine the synthesis of fatty acids from nonfatty-acid precursors (e.g., glucose and pyruvate). ***The major site of de novo fatty-acid synthesis in humans appears to be the liver.*** There can be some degree of fatty-acid synthesis in the adipose tissue; however, this appears to be very small, unless the diet contains 5% fat or less—an unlikely situation in most modern societies. Since relatively little glycolysis occurs in the liver, the major carbon source for fatty acid synthesis appears to be lactate and pyruvate returning to the liver from other tissues via the hepatic circulation.

FATTY-ACID SYNTHESIS FROM ACETYL CoA

This process is basically taking acetyl units from acetyl CoA to form a long-chain fatty acid. Fatty-acid synthesis occurs in the cytosol. The first committed step of fatty-acid synthesis is the carboxylation of acetyl CoA to form malonyl CoA, catalyzed by acetyl CoA carboxylase, a biotin-containing enzyme. The malonyl moiety of malonyl CoA can be transferred to fatty-acid synthase (FAS), which contains pantothenic acids (i.e., similar to coenzyme A without the adenine nucleotide ribose moiety). An acetyl group from acetyl CoA is transferred to another pantotyl group of the FAS. The attachments to the pantotyl groups are as thioesters. The acetyl moiety is then transferred to the malonyl moiety with the release of CO_2. The loss of CO_2 helps assure unidirectionality of the reactions of fatty-acid synthesis. This forms acetoacetyl FAS, which is reduced to β-hydroxybutyryl-FAS. The β-hydroxybutyryl-FAS is dehydrated and reduced; NADPH provides the reducing power for both reductions (Fig. 15.1). Another malonyl moiety, from malonyl CoA, is added to the FAS. The butyryl moiety is then added to the malonyl moiety of the FAS with the release of CO_2. The reduction, dehydration, and reduction are repeated as with acetoacetyl-FAS. The product is a six-carbon fatty acid attached to the FAS. This is repeated until a C14 to C18fatty acid, usually a 16-carbon fatty acid, is formed attached to FAS. These fatty-acid FASs can then be cleaved to form free FAS and a free fatty acid. During this process the growing, fatty-acid chain is always added to the new malonyl moiety so that the acetyl CoA first added is always at the methyl end of the

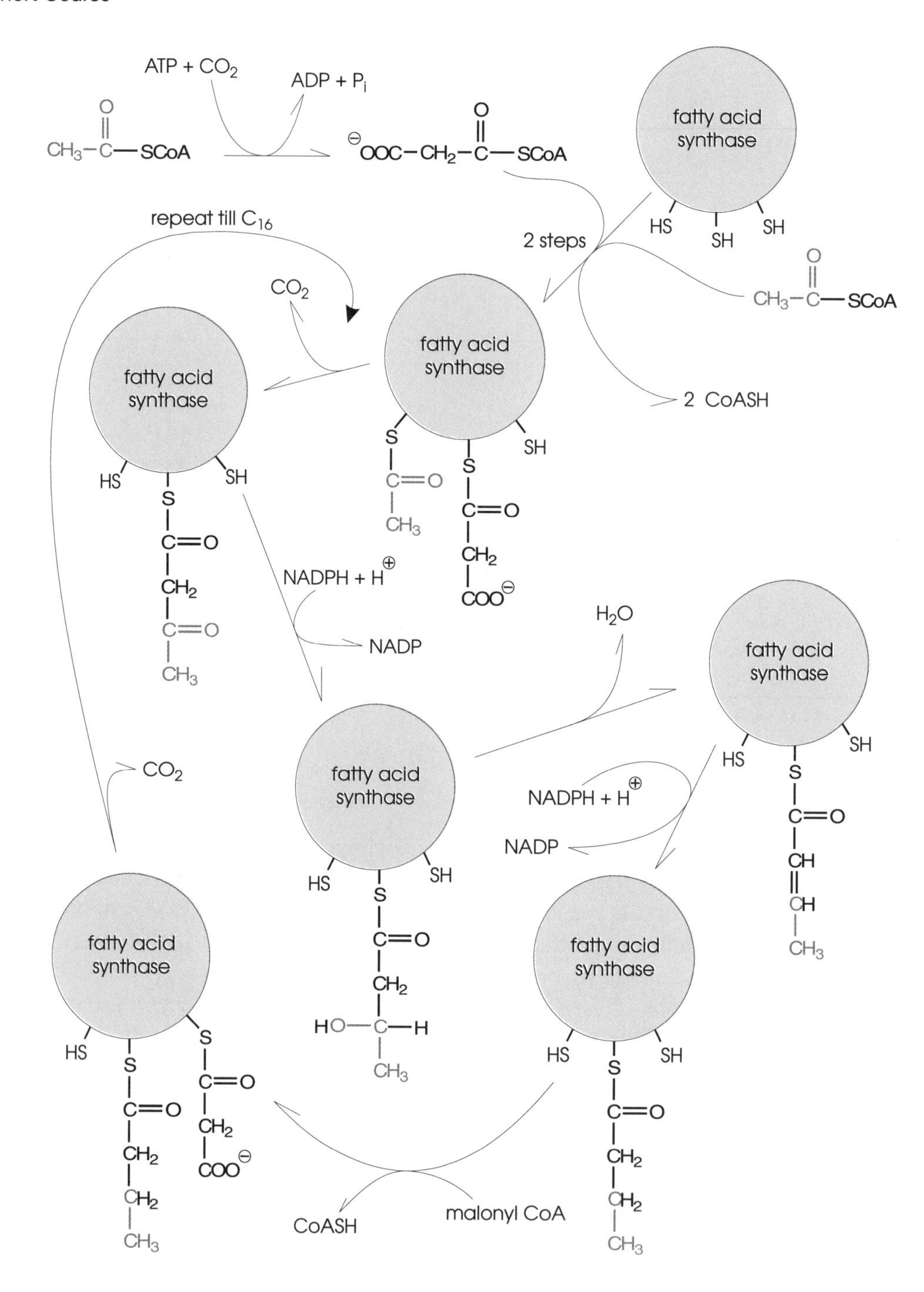

Figure 15.1.

Fatty-acid synthesis. The pathway and mechanism of fatty-acid synthesis is depicted in this figure. The fatty-acid synthase is a large enzyme complex which contains a number of enzymic activities and acyl carrier proteins. The synthesis is followed only to the formation of butyryl acyl carrier proteins; however this, in fact, is repeated until palmytyl acyl carrier protein is formed by the addition of two carbons at a time. The portion added as acetyl CoA, instead of the malonyl CoAs, is indicated in purple to show that the growing chain is added to new malonyl CoA at each addition step.

fatty acid. ***The reducing power for fatty-acid synthesis is supplied by NADPH.*** The advantage of having the entire process on a single multifunctional protein is that the intermediates can readily be transferred without being a free-floating compound.

Moving Acetyl Units from the Mitochondrion to the Cytosol

The acetyl units needed for fatty-acid synthesis are generated in the mitochondrion, but utilized in the cytosol. The units originate in the pyruvate-dehydrogenase reaction forming acetyl CoA, which cannot cross the mitochondrial membrane. To overcome this problem, a unique shuttle system exists. In times of sufficient or excess energy, citrate can accumulate in the liver owing to inhibition of isocitrate dehydrogenase by NADH and depletion of ADP. Citrate is a much smaller molecule than acetyl CoA and can exit the mitochondria via a specific citrate translocase (Fig. 15.2). However, in the liver, unlike in the muscle, the exiting citrate is not a signal to slow glycolysis, because very little glycolysis occurs in the liver, but a signal to stimu-

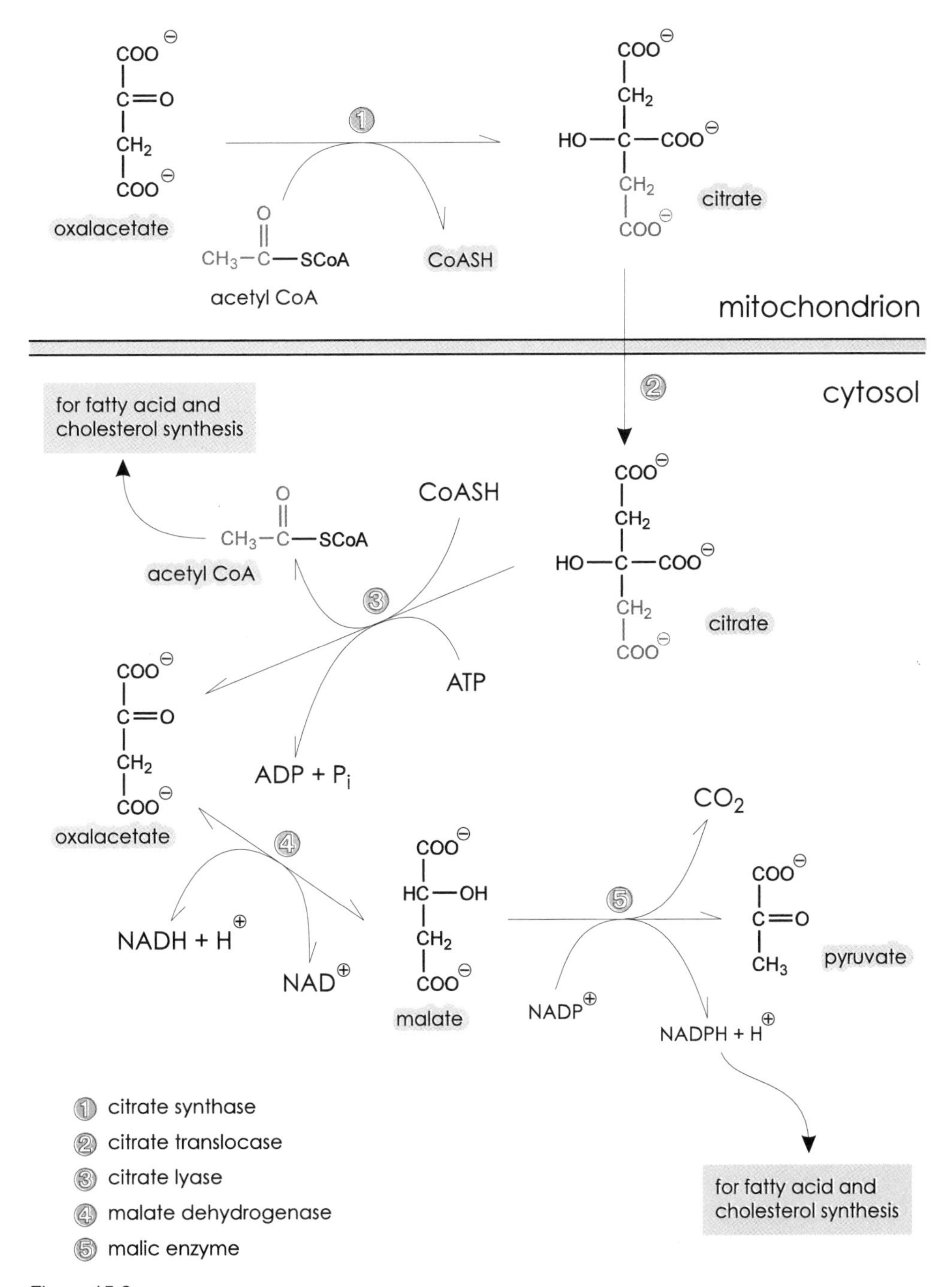

Figure 15.2.

Transport of acetyl CoA from mitochondria to cytosol. The transport of the acetyl unit of acetyl CoA is moved out of the mitochondrion as citrate and reformed to acetyl CoA in the cytosol. The resulting oxaloacetate can participate in effective conversion of NADH to NADPH, the latter participating in cholesterol and fatty-acid synthesis. The acetyl carbons of acetyl CoA are depicted in purple so as to follow their fate from the mitochondrion to the cytosol.

late fatty-acid synthesis. As will be seen, citrate plays a multifaceted role in fatty-acid synthesis, both as supplier of substrate and an effector molecule.

In the cytosol, citrate lyase catalyzes the conversion of citrate + CoA + ATP to acetyl CoA + oxaloacetate + ADP + inorganic phosphate. This reaction requires ATP, as the citrate synthase reaction is effectively irreversible. The resulting acetyl CoA in the cytosol can then act as a direct precursor of fatty-acid synthesis via the action of acetyl CoA carboxylase and FAS, as previously described.

Control of Fatty-Acid Synthesis

Lipogenesis is controlled by a number of mechanisms, including allosteric effectors, covalent modification, and availability of substrate. Pyruvate is an excellent potential precursor for fatty acids, particularly in the liver. One of the difficulties encountered is that pyruvate can proceed to acetyl CoA in the mitochondrion; however, acetyl CoA in the mitochondrion will not directly produce fatty-acid synthesis because this process occurs in the cytosol.

Thus, there are important controls that allow acetyl CoA in the mitochondria formed from pyruvate to be transported to the cytosol (Fig. 15.3). Fatty-acid synthesis occurs primarily when there is sufficient or excess energy coming from sources other than fatty acids themselves, because it would be energetically wasteful to rapidly synthesize and degrade fatty acids. When fatty acids are not elevated and in the presence of high insulin, as with a high-carbohydrate diet, pyruvate dehydrogenase is in the active form. Pyruvate could proceed to rapidly form citrate, because there would be little or no inhibition of pyruvate dehydrogenase or citrate synthetase. As the energy charge in the cell built up, there would be higher ATP and lower ADP, which would potentially slow the electron transport system and allow NADH to accumulate in the mitochondrion. The elevated NADH and lower than normal ADP would cause inhibition of the NAD^+-linked isocitric dehydrogenase. Thus, citrate would accumulate in the mitochondrion; this elevated citrate could then be translocated to the cytosol and be acted upon by citrate lyase. Citrate lyase appears to be present in greater amounts when high-glucose or potentially lipogenic diets are fed, and in lesser amounts under conditions where lipogenesis is minimal. Citrate lyase catalyzes the formation of acetyl CoA and oxaloacetate. The oxaloacetate is reduced by malate dehydrogenase using NADH to malate, and the malate, via malic enzyme, can form CO_2, pyruvate, and NADPH, which provides reducing power during lipogenesis. Thus, the net effect of coupling the NAD^+ malate dehydrogenase and the $NADP^+$ decarboxylating malic dehydrogenase, also known as malic enzyme, is effectively to transfer reducing power from NADH to NADPH where it can be used for lipogenesis (see Fig. 15.2).

Acetyl CoA Carboxylase

Acetyl CoA is converted to malonyl CoA and into fatty acids as described previously. ***The enzyme that carries out the first committed step for fatty-acid synthesis, acetyl CoA carboxylase, is finely controlled both allosterically and covalently.*** This enzyme can occur in a monomeric inactive form or a polymeric active form. One factor that affects this is citrate, which stimulates the polymeric or active form of acetyl CoA carboxylase. Thus, citrate plays an important role in lipogenesis as (1) a source of cytosolic acetyl CoA, (2) an allosteric positive effector of acetyl CoA carboxylase, and (3) a provider of oxaloacetate in the cytosol, which can allow transhydrogenation from NADH to NADPH. An allosteric inhibitor of acetyl CoA carboxylase that causes dissociation to the monomeric form is fatty-acyl CoA. Thus, if exogenous fatty acids are available, there is little reason to synthesize more fatty acids. ***Fatty-acyl CoA in the cytosol decreases malonyl CoA formation by inhibiting acetyl CoA carboxylase.***

A phosphorylation, catalyzed by the cAMP-dependent protein kinase, will cause the enzyme to favor an inactive form, and more likely form monomers. Increased cAMP-dependent protein kinase activity is stimulated by glucagon. Glucagon is a signal for increased gluconeogenesis. At this time, it is inappropriate to use pyruvate as a substrate for lipogenesis and is time to use pyruvate for gluconeogenesis. Thus, phosphorylation causes the enzyme to take on an inactive confirmation. Removal of this phosphate group (as would be stimulated by insulin) would convert the enzyme into a more active confirmation. Thus, a major control of fatty-acid synthesis appears to lie with the first committed step of fatty-acid synthesis, formation of malonyl CoA and the enzyme acetyl CoA carboxylase. If sufficient reducing power as NADPH is available, and there is sufficient cytosolic malonyl CoA and acetyl CoA, lipogenesis will proceed rapidly.

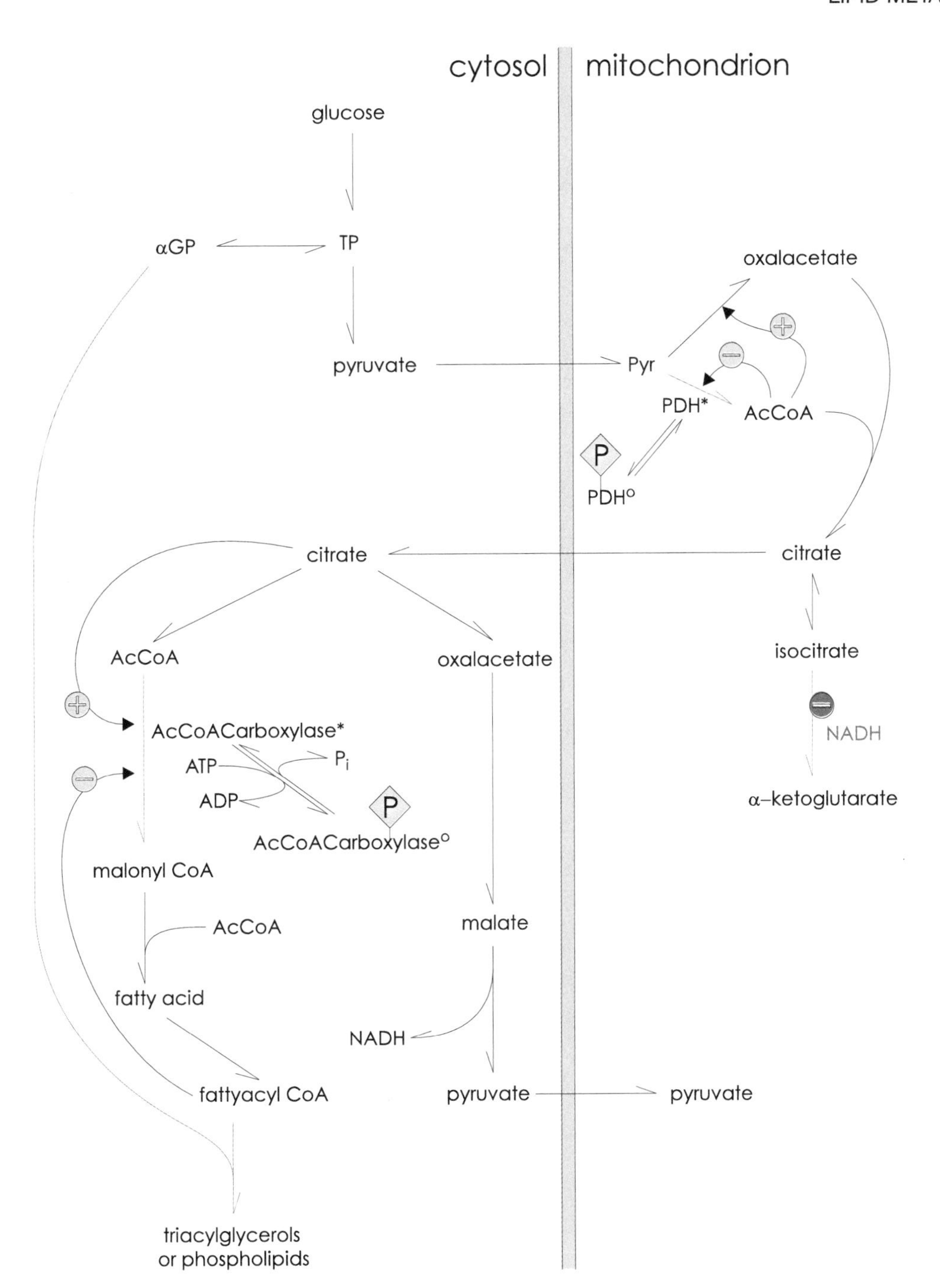

Figure 15.3.
Control of fatty-acid synthesis. A schematic of the control of fatty-acid synthesis is presented with the major control steps shown in purple with effectors (both positive and negative) indicated.

Oxidation versus Synthesis

The control of fatty-acid oxidation is related to the availability of circulating fatty acids and the activity of palmitoyl carnitine transferase 1. When circulating fatty acids are elevated, considerable fatty-acyl CoA is formed in a number of tissues, including the liver, which is sufficient to inhibit both acetyl CoA carboxylase in the cytosol and, indirectly, pyruvate dehydrogenase in the mitochondrion. Under this condition, neither malonyl CoA nor citrate would accumulate; thus, there would be a diminution of fatty-acid synthesis. When large amounts of fatty acids reach the liver, they are sufficient to allow both rapid rates of esterification to α-glycerophosphate and beta oxidation. Why are fatty acids that are being synthesized in the cytosol not transported into the mitochondrion and rapidly degraded? This wasteful substrate cycle is prevented by malonyl CoA inhibition of palmitoyl carnitine transferase 1, thus preventing fatty acids that are newly synthesized from entering the mitochondrion and undergoing beta oxidation (Fig. 15.4). This allows newly synthesized fatty-acyl CoAs to be incorporated into triacylglycerol and phospholipids. When rapid fatty-acid synthesis occurs, the formation of malonyl CoA may be more rapid than utilization by fatty-acid synthase, allowing malonyl Co to accumulate and inhibit acyl carnitine transferase 1.

When large amounts of exogenous fatty acid enter the liver, how can they be transferred into the mitochondrion, for beta oxidation, with malonyl CoA inhibition? This dichotomy is overcome because fatty-acyl CoAs inhibit acetyl CoA carboxylase and the malonyl CoA present proceeds onto fatty acids. When the malonyl CoA is converted to fatty acid, the level of malonyl CoA drops and is not restored. Thus, the inhibition of acyl carnitine transferase 1 is removed and fatty-acid oxidation can proceed.

Desaturation and Elongation of Fatty Acids

The body has the ability to desaturate fatty acids. This desaturation produces a cis double bond, in contrast to the trans double bond produced during synthesis and degradation. The desaturation process in humans can occur at the C9 position or positions closer to the carboxyl group. Fatty acids with polyunsaturation at either C9, C12, and/or C15, or C9 and C12, cannot be formed by humans and are needed in the diet. These particular polyunsaturated fatty acids comprise the pool of essential fatty acids. The body can elongate palmitate primarily to C18 (sterate) fatty acids and longer C20–C24. Elongation occurs at the endoplasmic reticulum using

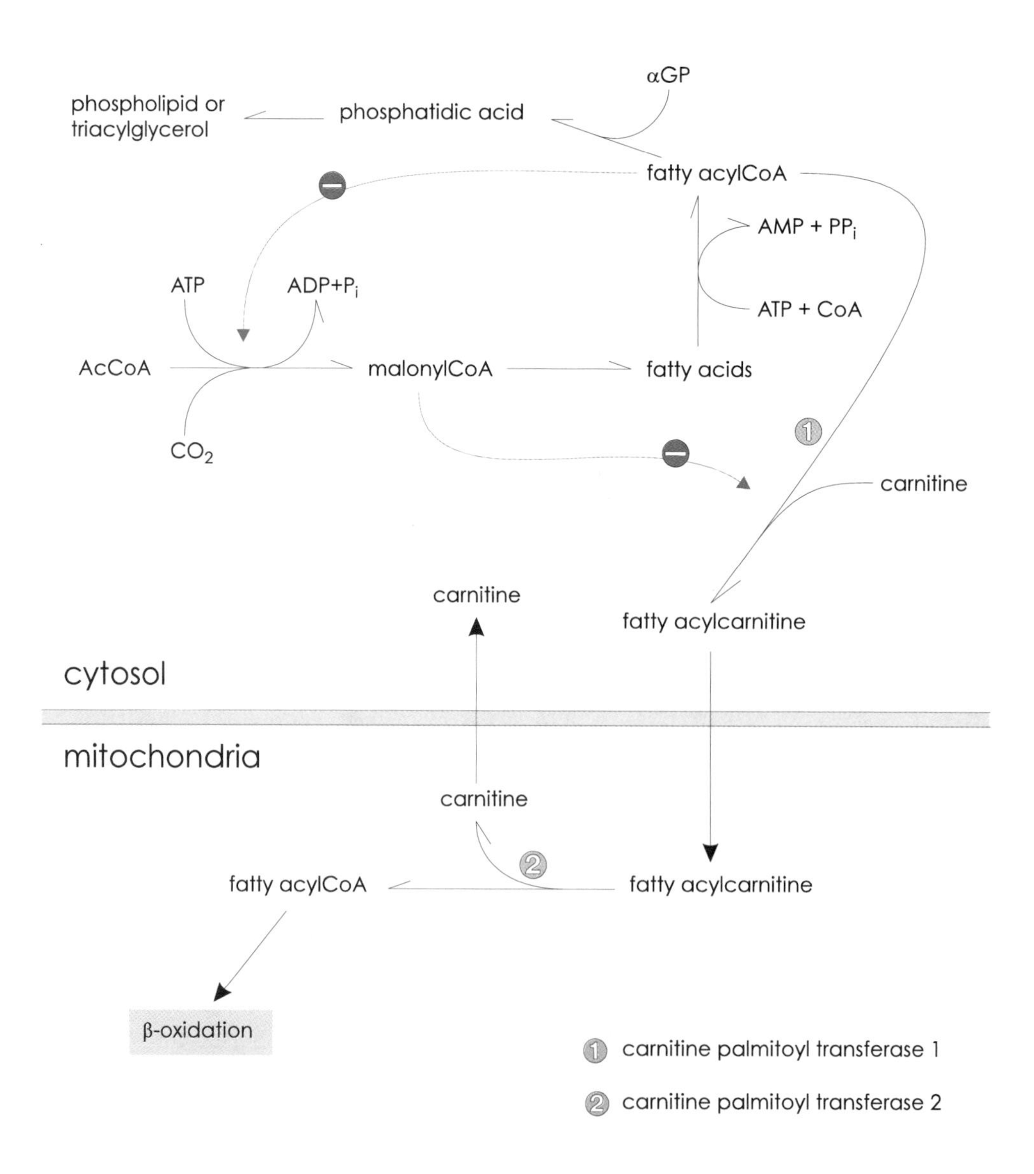

Figure 15.4.

Relationship between fatty-acid synthesis and degradation. This diagram shows that malonyl CoA inhibits CPTI and thereby β-oxidation, avoiding the oxidation of newly synthesized fatty acids. Fatty-acyl CoA inhibits acetyl CoA carboxylase and thereby fatty-acid synthesis. When fatty-acyl CoAs are converted to triacylglycerols or phospholipids, they are effectively removed and will no longer be inhibitory.

malonyl CoA and NADPH, but differs from fatty-acid synthesis in that elongation occurs with CoA derivatives instead of fatty-acid synthase.

Linoleic acid (18:2), an essential polyunsaturated fatty acid, can be elongated and further desaturated to form arachadonic acid. Arachadonic acid is a normal constituent of phosphatidylinositols, occurring in the S_n2 position. Arachadonic acid can be converted to prostaglandin catalyzed by the prostaglandin synthase complex. The first step is catalyzed by the cyclooxygenase component of the prostaglandin synthase complex. This and closely related systems can form prostaglandins and thromboxanes.

SYNTHESIS OF PHOSPHOLIPIDS AND TRIACYLGLYCEROLS (TRIGLYCERIDES)

Fatty acids released from the acyl carrier protein moiety of FAS can be activated in the cytosol and esterified to alpha glycerol phosphate to form phosphatidic acid. Phosphatidic acid can then proceed in two major directions. The first is the cleavage of phosphate by a phosphatase, which forms diacylglycerol, and then another fatty acyl CoA transfers the fatty acid to the diacylglycerol to form a triacylglycerol, also referred to as a triglyceride. Triacylglycerol formed in the liver must be transported via the blood for storage in the adipose tissue, which is discussed later in this chapter. Another major direction for phosphatidic acid, whether formed from newly synthesized fatty acids or dietary fatty acids, is the formation of phospholipids. Phospholipid synthesis also proceeds via phosphatidic acid, which can react with a series of CDP derivatives. These CDP derivatives can include choline and ethanolamine. Phosphatidyl serine can be formed from phosphatidyl ethanolamine by a reaction exchanging serine for ethanolamine.

The major pathway of phosphatidylcholine (lecithin) synthesis is via preformed choline (Fig. 15.5). Phosphotidylethanolamine can be converted to phosphatidylcholine in a minor pathway by the addition of $3CH_3$ groups (from methionine). Thus, phosphatidylcholine can be synthesized *de novo* if choline is not available and there is a source of "CH_3" groups. In a major pathway, phosphatidylcholine can also be synthesized more directly starting with choline. Choline can be phosphorylated with ATP to form phosphocholine. Phosphocholine can react, via phosphocholine cytidyltransferase, in the presence of CTP, to form CDP choline + pyrophosphate, which is pulled in the forward direction by hydrolysis of the pyrophosphate. The CDP choline then can react with diacylglycerol to form phosphatidylcholine and CMP. This reaction is catalyzed by the enzyme phosphocholinetransferase. The major role of phospholipids in cell membranes is discussed in Chapter 4.

There are numerous other phospholipids. One group includes phosphatidylinositol and its phosphorylated derivatives (e.g., phosphatidylinositol 4,5-bisphosphate), which are important in transduction of hormone messages (see Chapter 16). Phosphatidylinositol is important as an anchor for many membrane proteins. Cardiolipin, composed of two phosphatidic acids linked together through a glycerol, is in high concentration in the inner mitochondrial membrane. One plasmologin differs from phosphatidylethanolamine by having an O-(1-alkenyl) substituent in place of acyl moiety. These ethanolamine plasmologins occur primarily in myelin. A special phosphatidylcholine, namely dipalmitoylphosphatidylcholine, is a surfactant and is needed for normal lung function.

Lipoprotein and Lipid Transport

Fatty acids, whether synthesized by the liver or taken in through the diet, must be transported through the blood, usually for storage in the adipose tissue. Triacylglycerols have relatively low solubility in the plasma, and therefore are transported as lipoproteins (Fig. 15.6).

The classes of lipoproteins include chylomicrons, which transport dietary fat, and very-low-density lipoprotein (VLDL), low-density lipoprotein, (LDL) and high-density lipoprotein (HDL), which transport endogenous fats. These lipoproteins vary in size, density, relative composition of triglycerides, phospholipids, cholesterol, cholesterol ester, and proteins. In addition, each of these classes of lipoproteins may contain different apoproteins. These specific apoproteins are important for transformations within the lipid particle, and are signals for cellular receptors allowing uptake and/or endocytosis.

Dietary triacylglycerols are hydrolyzed in the intestine by lipases to form fatty acids, glycerol, and monoacylglycerols. These are then absorbed and resynthesized back into triacylglycerols in the intestine. The intestine also synthesizes phospholipids and proteins, which are important in forming the chylomicrons. ***Chylomicrons, large lipoproteins, are formed in the intestine and are absorbed via the lymphatic system***, thus avoiding a direct flow to the liver via the portal vein. Chylomicrons contain a large percent of triacylglycerol,

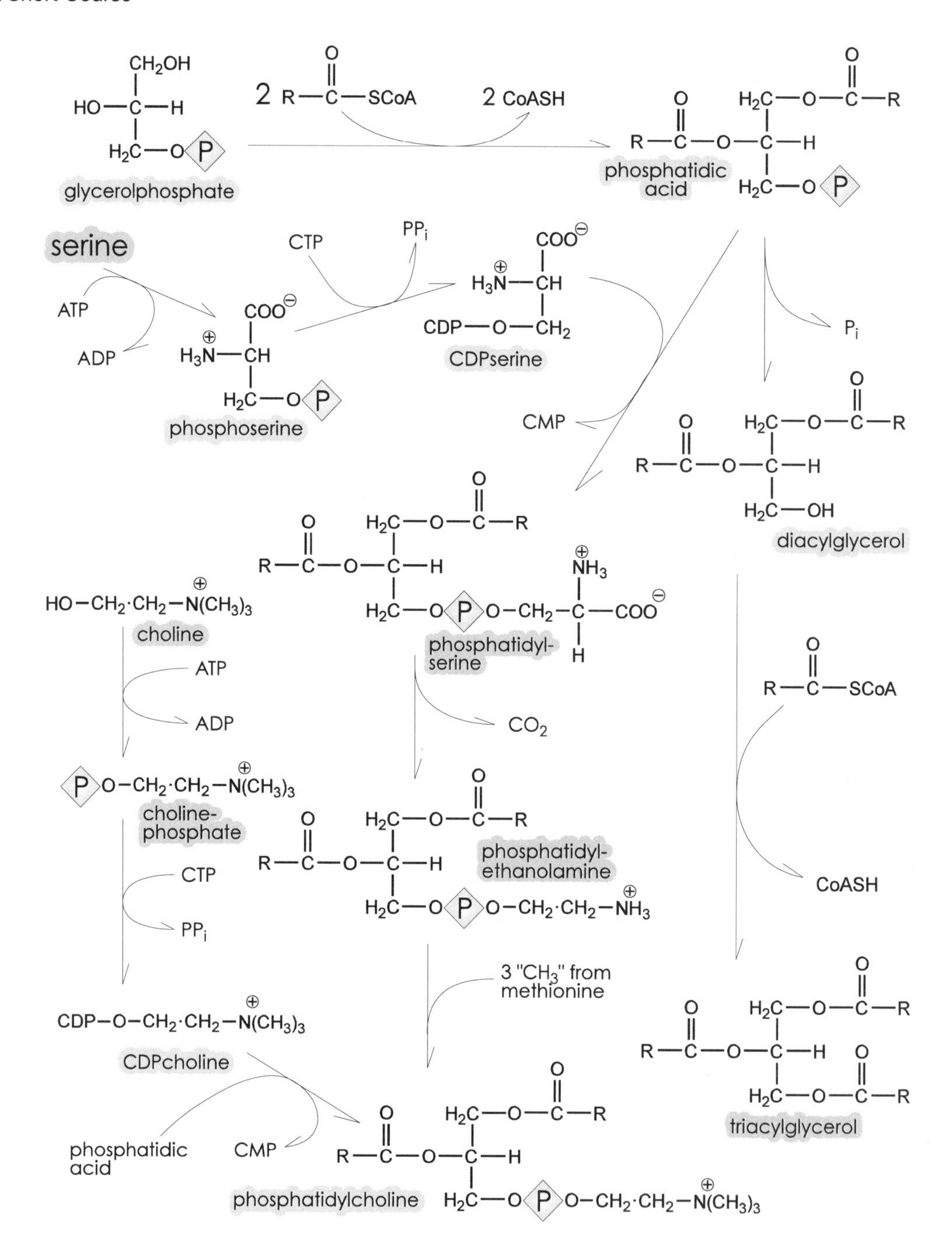

Figure 15.5.

Triglyceride and phospholipid formation. This figure depicts the formation of triacylglycerol from α-glycerolphosphate and fatty-acyl CoA. The formation of phosphatidylethanolamine and phosphatidylcholine from scratch (i.e., from serine and methionine methyl groups) is also shown. The formation of phosphatidylcholine, starting with choline, is also depicted and is the major pathway for phosphatidylcholine

with considerably lesser percentages of phospholipids, cholesterol, cholesterol esters, and protein. Because of their large size, an abundance of chylomicrons in the circulation gives the plasma or serum a milky-white appearance. Circulating chylomicrons are processed by the action of lipoprotein lipase. This enzyme catalyzes the hydrolysis of fatty acids from triacylglycerols contained within lipoproteins. Lipoprotein lipase is synthesized in the cell and extruded to the interior cell walls of blood vessels of various tissues, such as adipose, muscle, and liver. When fatty acids are cleaved from the triacylglycerol, they can then be taken up by the neighboring cells, either as a source of energy through beta oxidation (Chapter 14), as for the muscle, or storage, as in the case of the adipose tissue. The glycerol moiety continues in the circulation, where it can then be metabolized by the kidney or the liver, and utilized for energy or glucose formation or, in the liver, for resynthesis of triacylglycerol and phospholipids. Repeated action of lipoprotein lipase converts chylomicrons to chylomicron remnants, which contain a specific apoprotein that is "recognized" by receptors in the

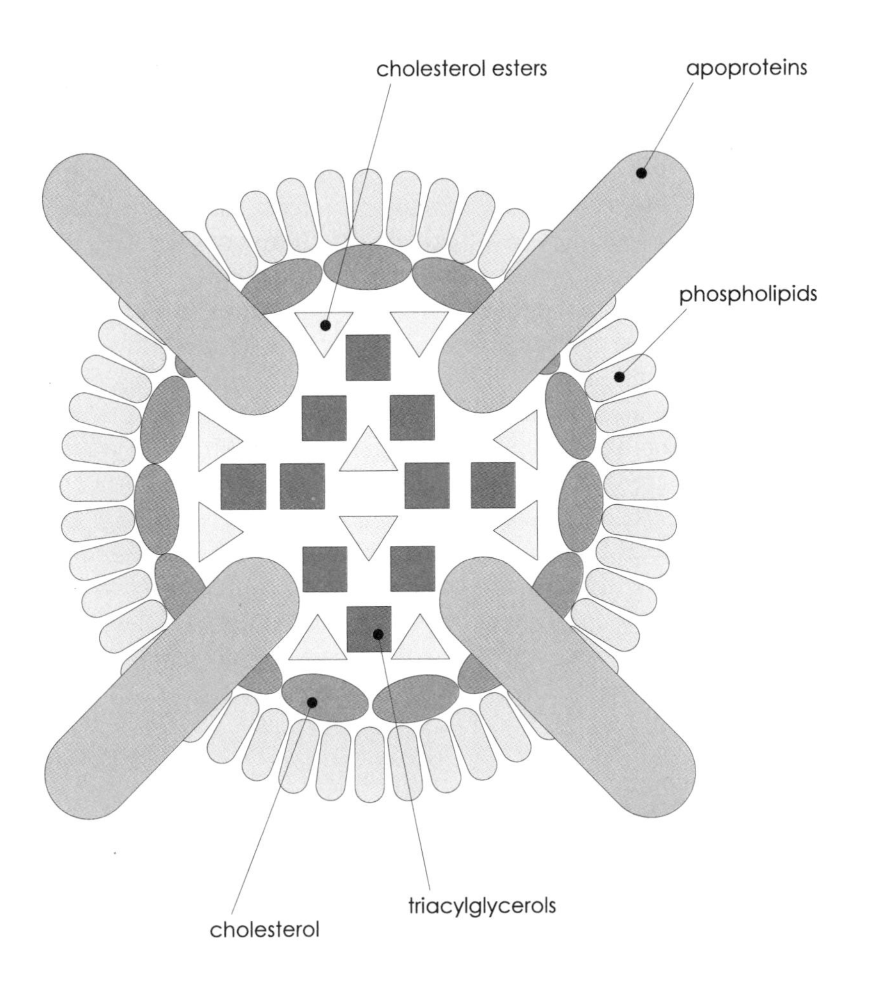

Figure 15.6.

Lipoprotein structure. The figure depicts the structure of lipoproteins, cut through the middle with a side view of the cut surface, in a general way. Notice the more polar compounds, such as the apoproteins and phospholipids, are on the outside surface, in contact with the plasma. The more nonpolar compounds, such as triacylglycerols and cholesterol, are near the center of the lipoprotein.

liver. ***The liver can take up the chylomicron remnants***, repackage the lipid portion, and reutilize the amino acids in the protein portion.

Two major lipoproteins are produced by the liver for the purpose of transporting endogenous cholesterol and triacylglycerol. These are VLDLs and HDLs (Table 15.1). As indicated by the names, these two differ in density and can be separated by gradient-density centrifugation. As with all lipoproteins, these contain triacylglycerol, cholesterol, cholesterol esters, phospholipids, and proteins. The chylomicrons, which transport dietary fat, contain approximately 90–95% triacylglycerols, with relatively small percentages of protein, phospholipid, and cholesterol. In contrast, the lipoproteins, which transport endogenous triacylglycerol and cholesterol packaged by the liver, such as VLDLs, contain only 60% triacylglycerol and larger proportions of total cholesterol and phospholipids as well as 10% protein.

As the VLDLs circulate and the triacylglycerols are hydrolyzed by lipoprotein lipase, particularly at the adipose for storage, the density of this particle and its

TABLE 15.1 Approximate Composition of Plasma Lipoproteins

	Composition in percent of total (by weight)				
Lipoprotein	Triacylglycerol	Phospholipid	Free Cholesterol	Cholesterol Ester	Protein
Chylomicrons	86	8	1	3	2
VLDL	52	18	7	14	9
IDL	27	23	8	31	11
LDL	10	23	8	38	21
HDL	8	25	4	18	45

percentage of triacylglycerols decreases to where there is 25% protein, 10% triacylglycerols, 20% phospholipids, and a threefold increase in the percentage of cholesterol content (compared to the original VLDL) up to 45%. The depletion of triacylglycerols from VLDL produces IDL (intermediate-density lipoprotein) and then LDL. During these conversions, the cholesterol percentage not only increases, but the proportion of cholesterol in the esterified form increases. The circulating LDL, which is cholesterol rich, seems to be most associated with CHD (chronic heart disease) and atherosclerosis. The LDL, especially after oxidative damage, has a higher propensity for depositing cholesterol in areas of the blood vessel where some damage has started. In most cases the ***LDL is removed by the liver or other organs, such as the adrenal gland via specific LDL receptors***. In these tissues, the cholesterol can be used for the formation of steroid hormones or bile acids, or in the liver for repackaging as VLDL. In individuals who are deficient in this receptor, chronic heart disease occurs almost without exception at an early age, as early as 20 or 30 years of age. Some individuals who are heterozygous for this recessive defect also are more prone to heart disease. In both of these conditions, there is an unusually high level of circulating total cholesterol in the blood, and unless intervention is used, fatality occurs at an early age. Despite the intervening therapy, the probability of an early heart attack is still extremely high in these individuals.

In addition to VLDL delivering triacylglycerol to tissues, ***LDL can deliver cholesterol to peripheral tissues, thereby delivering cholesterol to tissue that cannot make cholesterol***.

The fourth major type of lipoprotein is the HDL, which does not arise from VLDL or LDL and is synthesized by the liver. This lipoprotein contains a very high percentage of protein, as much as 50%, very low triacylglycerides, high phospholipids, and moderate cholesterol levels. Transport of triacylglycerols from the liver, where they are synthesized to the adipose for storage, is not a major purpose of this lipoprotein. The HDL contains a different, specific set of apoproteins from chylomicrons, VLDLs, or LDLs. Although some of the apoproteins (i.e., polypeptide chains) are similar among the four different lipoproteins, there are some unique combinations of apoproteins associated with each of these particles. The HDL particle contains an apoprotein that will activate lecithin cholesterol acyltransferase (LCAT), which transfers a fatty acid from the phosphatidylcholine (lecithin) molecule to the three-hydroxyl group on the cholesterol to form the esterified form of cholesterol. The esterified form of cholesterol is even less polar than the free cholesterol and apparently is moved to the center of the circulating HDL. The HDL is an effective scavenger of cholesterol molecules from other locations, possibly even from some early plaque formations and, therefore, has been considered to be a "good lipoprotein." High-density lipoprotein cholesterol can also be transferred to VLDL and LDL. The cholesterol associated with HDL has been referred to as "good cholesterol." The total circulating lipoproteins in the plasma do not tell a complete story unless the relative concentrations of VLDL, LDL, and HDL are considered. For example, at the same total lipoprotein (or cholesterol) concentration, an individual with very high LDL and low HDL could be a good candidate for chronic heart disease or atherosclerosis; with the same level of total lipoproteins (or cholesterol), an individual with low LDL and high HDL would have a positive prognosis of not being affected by CHD or atherosclerosis.

CHOLESTEROL

Cholesterol is an important metabolic compound occurring in membranes and lipoproteins. It is also a precursor to bile acids and steroid hormones. The steroid hormones have only small structural differences which cause major differences in functions. This group includes progesterone, testosterone, mineral corticoids, glucorticoids (cortisol), and others. ***The body has the ability to synthesize and redistribute cholesterol.*** The main organ that synthesizes cholesterol is the liver. The amount of cholesterol synthesized by the body can be two to three times or more the amount ingested. Cholesterol is not an essential nutrient and can be made in the body from simple compounds via acetyl CoA.

Effectively, three acetyl CoAs are combined to form HMGCoA, which is reduced by NADPH catalyzed by HMGCoA reductase. This enzyme, the rate-limiting enzyme of cholesterol synthesis, is located on the endoplasmic reticulum. The subcellular localization of cholesterol synthesis, cytosol, and endoplasmic reticulum, compared to ketone-body formation in the mitochondria, assures that the fate of HMGCoA is determined by subcellular localization of HMGCoA formation. The reduction of HMGCoA by HMGCoA reductase, using 2NADPHs as the source of reducing power, forms mevalonate (Fig. 15.7). Mevalonate, with the input of two ATPs, forms 5-pyrophosphomevalonate. Decarboxylation and intramolecular rearrangements of 5-pyrophosphomevalonate produce 3,3-dimethyallyl py-

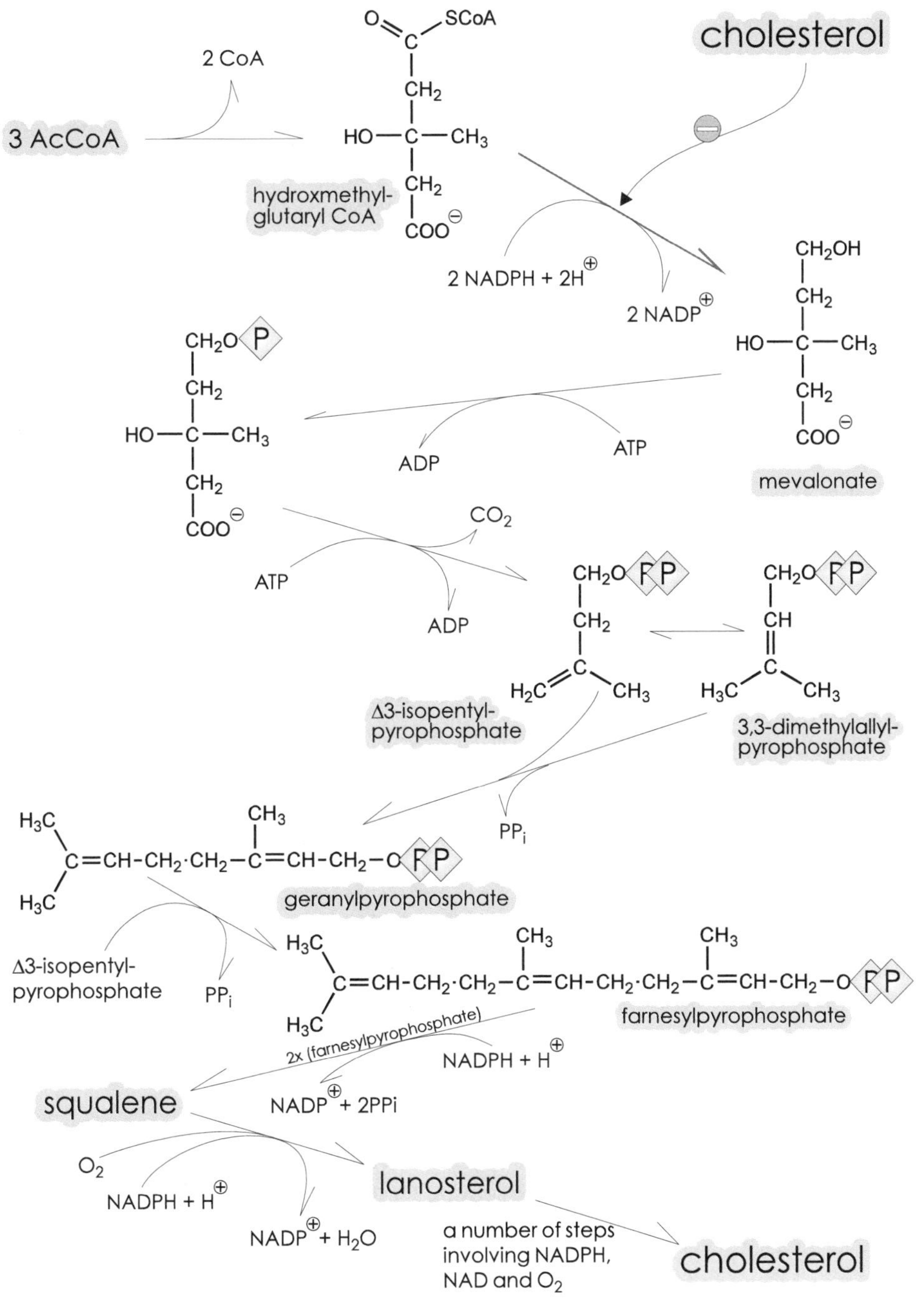

Figure 15.7.

Cholesterol synthesis. The early steps of cholesterol synthesis are presented, including the rate-limiting reaction HMGCoA reductase (in purple), with a general outline of the latter steps. Note, as with fatty-acid synthesis, NADPH is an important source of reducing power.

rophosphate and Δ3-isopentenyl pyrophosphate, both isopentenyl units (C5). Three C5 units combine to make a C15 pyrophosphate unit, farnesyl pyrophosphate. Two C15 pyrophosphate units plus NADPH produce squalene, a C30 unit. After enzymatic rearrangements, oxidation to form the 3-OH, and further reductions utilizing NADPH, the squalene forms cholesterol.

Cholesterol synthesis, storage, and delivery is controlled. ***The major control of cholesterol synthesis is at the HMGCoA reductase step***, which can be inhibited either by cholesterol or phosphorylation of the enzyme, as mediated indirectly by increases in intracellular cAMP. With severe hypercholesterolemia, people may be treated with an HMGCoA reductase inhibitor, such as lovastatin. Cholesterol storage and delivery is controlled by a number of LDL receptors. Low-density lipoprotein is a cholesterol-rich lipoprotein which can deliver cholesterol to many tissues via LDL receptors. As tissue cholesterol levels rise, the importance of cholesterol delivery is less important. When cellular cholesterol levels are high, there is a down-regulation of LDL receptors (a decrease in the number of receptors), thus decreasing cholesterol delivery to that tissue. The tissues rich in LDL receptors are liver, adrenal gland, ovaries, testes, and others that convert cholesterol to important metabolic products.

Quantitatively, the greatest loss of cholesterol is via bile acids and bile salts formed by the liver. These bile acids and bile salts are important in digestion for emulsification of lipids, including fat-soluble vitamins. The bile acids and bile salts are secreted from the liver to the intestines via the bile. A considerable portion of bile acids return to the liver by the enterohepatic circulation. This decreases the need to convert as much cholesterol to bile acids. Therefore, another attempt to decrease cholesterol levels is by increasing conversion of cholesterol to bile acids, by slowing recirculation of the bile acids and bile salts, has been used. This has been suggested as one of the benefits of dietary fiber, which binds bile acids and causes greater excretion in the feces instead of recirculation back to the liver. Several bile-salt binding drugs have been used to accomplish the same result, one of the more popular of which is cholestyramine.

ETHANOL

Ethanol is metabolized primarily in the liver. There are two major pathways, one for excess ethanol over extended periods. With extended alcohol intake a microsomal (ER) catalase-linked system becomes important. The second pathway functions under all conditions where ethanol is present. This is alcohol dehydrogenase catalyzing $CH_3CH_2OH + NAD^+ \rightarrow CH_3CHO + NADH + H^+$ present in the cytosol. The resulting acetaldehyde moves into the mitochondrion, where aldehyde dehydrogenase converts it to acetate. One of the problems during ethanol metabolism is a marked increase in $NADH/NAD^+$. This in turn causes an increase in αGP (DHAP + NADH $\rightarrow$ αGP + NAD). The increase in αGP makes it easier to form triacylglycerols from fatty acids entering the liver. This may be one of the factors associated with fatty livers observed during alcoholism.

SUMMARY

1. The liver is the major organ for fatty-acid synthesis in humans.

2. Acetyl CoA is the precursor of fatty acids. For palmitate synthesis (C16), seven acetyl CoAs are converted to malonyl CoA by acetyl CoA carboxylase. These seven malonyl CoAs, plus one acetyl CoA, are combined stepwise on FAS, with addition of one malonyl CoA at a time.

3. The FAS reduces the keto group of the growing chain on the FAS, utilizing two NADPHs per "acetyl unit" added, as the reducing power to form the fatty acids.

4. Acetyl CoA formed in the mitochondrion from pyruvate can be converted to citrate. When energy stores (ATP and NADH) are high in the liver from nonlipid sources, citrate can accumulate. The citrate can exit the mitochondria to the cytosol.

5. In the liver cytosol, the site of fatty-acid synthesis, the citrate is cleaved to produce acetyl CoA and oxaloacetate.

6. The oxaloacetate can be reduced to malate by NADH and then oxidized by malic enzyme to form pyruvate and NADPH. The latter is needed for fatty-acid synthesis.

7. Fatty-acid synthesis appears to be controlled primarily by the activity of acetyl CoA carboxylase. This enzyme is activated by citrate and inhibited by fatty-acyl CoA and/or phosphorylation.

8. Citrate plays a threefold role in fatty-acid synthesis. It brings the acetyl CoA out of the mitochondria, produces oxaloacetate, which affects transhydrogenation of NADH to NADPH, and activates acetyl CoA carboxylase.

9. Fatty acids can be desaturated from C9 forward and can be elongated to C24.

10. Fatty acids can be activated and form triacylglycerols or phospholipids, such as phosphatidylethanolamine and phosphatidylcholine. The formation of phospholipids includes the use of cytodine nucleotides combined with ethanolamine or choline.

11. Lipids formed in the liver and from the diet must be transported through the blood for storage or use. This transport involves chylomicrons, from the intestine, or VLDLs from the liver. These lipoproteins contain triacylglycerols, phospholipids, cholesterol, cholesterol esters, and proteins. Some proteins help determine action or target tissues for the lipoproteins.

12. When VLDLs lose some triacylglycerols, they become LDLs, which are cholesterol rich and important in delivering cholesterol to other tissues that cannot make cholesterol. The LDLs appear to be detrimental to health if they are too high in the blood or remain circulating for too long.

13. The HDLs like VLDs, are synthesized in the liver, but appear to have a beneficial effect. The HDLs can act as a cholesterol scavenger and reduce the risk of CHD.

14. Cholesterol can be synthesized, primarily in the liver, from HMGCoA. The control of enzyme for cholesterol synthesis is HMGCoA reductase, which is inhibited by cholesterol and/or phosphorylation. The major quantitative fate of cholesterol in the human is the formation of bile acids or bile salts.

REFERENCES

The activation of acetyl CoA carboxylase by tricarboxylic acids, M. D. Lane and J. Moss, 1971, *Adv. Enzyme Reg,.* 237–251.

Regulation of plasma LDL-cholesterol levels by dietary cholesterol and fatty acids, D. K. Spady, L. A. Woollett, and J. M. Dietschy, 1993, *Ann. Rev. Nutr.*, **13**:355–381.

Phosphorylation and degradation of HMG-CoA reductase, S. J. Miller, R. A. Parker, and D. M. Gibson, 1989, *Adv. Enzyme Reg.*, 65–77.

Familial hypercholesterolemia. J. L. Goldstein and M. S. Brown, in *The Metabolic Basis of Inherited Disease*, Vol. 1, 6th ed., C. R. Scriver, A. L. Beaudet, W. S. Sly, and D. Valle, Eds., 1989, McGraw-Hill, New York, pp. 1215–1250.

Familial lecithin: Cholesterol acyltransferase deficiency, including fish eye disease, K. R. Norum, E. Gjone and J. A. Glomset, in *The Metabolic Basis of Inherited Disease*, Vol. 1, 6th ed., C. R. Scriver, A. L. Beaudet, W. S. Sly, and D. Valle, Eds., 1989, pp. 1181–1194.

Biosynthesis of prostaglandins, W. E. M. Lands, 1991, *Ann. Rev. Nutr.*, **11**:41–60.

REVIEW QUESTIONS

1. The major control enzyme for fatty acid synthesis is
   1) Fatty-acid synthetase
   b) Citrate lyase
   c) HMGCoA lyase
   d) Acetyl CoA carboxylase
   e) Hormone-sensitive lipase

2. The major control enzyme in sterol synthesis is
   a) HMGCoA synthetase
   b) HMGCoA reductase
   c) HMGCoA lyase
   d) Acetylacetyl CoA synthetase
   e) NADPH production

3. The acetyl units used for fatty-acid synthesis are transported from the mitochondrion as
   a) A CoA derivative
   b) A carnitine derivative
   c) Citrate
   d) A glycerolphosphate derivative
   e) Free acetate

4. A positive effector of acetyl CoA carboxylase is
   a) cAMP
   b) Fatty-acyl CoA
   c) α-Glycerolphosphate
   d) Citrate
   e) Malonyl CoA

5. Fatty acids are moved from the liver to the adipose via
   a) High-density lipoproteins
   b) Very-low-density lipoproteins
   c) Low-density lipoproteins
   d) Chylomicrons
   e) Free fatty acids

6. Which of these is not required for fatty-acid synthesis?
   a) ATP
   b) CoA
   c) NADPH
   d) CO_2
   e) Choline

ANSWERS TO REVIEW QUESTIONS

1. ***d*** Although both A and B are required for fatty-acid synthesis, acetyl CoA carboxylase is the major (note: not necessarily sole) control enzyme for fatty-acid synthesis.

2. ***b*** This enzyme is considered to be the major control enzyme in sterol synthesis. In drug treatment of hypercholesterolemia, inhibitors of this enzyme are used.

3. ***c*** Citrate formed from oxaloacetate and acetyl CoA is translocated to the cytosol. Citrate can then be cleaved by citrate lyase to form oxaloacetate and acetyl CoA. The latter can be used for fatty-acid synthesis.

4. ***d*** Citrate is not only the carrier of acetyl units out of the mitochondria, but also is an allosteric activator of acetyl CoA carboxylase. Both A and B can cause inactivation or inhibition, respectively, of acetyl CoA carboxylase.

5. ***b*** However, (a) are also produced by the liver, but not primarily for fatty-acid transport; (c) are formed from (b) after the action of lipoprotein lipase; (d) are formed by the intestine, primarily for transport of ingested fatty acids.

6. ***e*** Although choline is needed for the synthesis of many phospholipids, it is not required for fatty-acid synthesis.

CHAPTER 16

HORMONES

HORMONE INTERACTIONS

Introduction

We will examine the mechanisms and metabolic effects of hormones. Hormones can either work at the membrane or enter the cell for their functions. Three mechanisms are primary for hormones acting at the membrane, which require a transduction of the message into the cell. These transductions include (1) formation of cAMP, (2) formation of inositol trisphosphate and diacylglycerol, and (3) direct covalent modifications. The hormones that enter the cell have their effects in the nucleus. To show the physiological effects of the hormones and review the integration of carbohydrate and lipid metabolism, these hormones are examined in detail. The effects we examine include not only metabolism and its control, but the roles of individual tissues.

Metabolism and interactions of metabolic pathways are strongly influenced by hormones and hormonal balance. Therefore, it is worthwhile to examine the effects of hormones on metabolism as well as the potential mechanism of action of the hormones. The mechanism of three major hormones (i.e., insulin, glucagon, and epinephrine) discussed previously are examined as well as others, such as steroid hormones. Many hormones, including those discussed earlier, which cause acute effects on metabolism, do not enter the cell, but must have a transduction to cause intracellular effects, although the hormone itself is bound to a receptor on the membrane. Others, such as steroid and thyroid hormones, enter the cell and have effects in the nucleus.

Transduction of Signals

Hormones affect only cells in which there are specific receptors. For many hormones, these receptors are on the outside of the plasma membrane and may transverse the membrane, although the receptor portion for those hormones that do not enter the cell is always on the extracellular side. The receptors act as discriminators to respond only to specific hormones in specific tissues. For example, there are specific β-receptors in the muscle and adipose that will recognize epinephrine, there are α-receptors in the muscle and liver that will recognize epinephrine, and there is an α-2 receptor in the β-cell of the pancreas that will recognize epinephrine. Thus, one hormone may effect different cells in different ways, depending on the specific receptor and the transduced message. The major transducer between the hormone message external to the cell, and the activities internal to the cell, is a protein known as a G protein. The G protein consists of three subunits—α, β, and γ. The α subunit, when bound to the receptor, binds a GDP molecule (Fig. 16.1). Upon the signal by the hormone, the GDP is exchanged for a GTP and the α subunit migrates to the internally oriented effector and activates the system. ***This system acts not only as a transducer, but also as a multiplicative step*** because a single hormone molecule can produce a large response owing to the enzymatic-like activity of the transduction. The α subunit also contains a GTPase which will split GTP to GDP and inorganic phosphate. When this occurs and the α subunit now binds a GDP, and it no longer activates the adenylcyclase, it then migrates back to the receptor and recombines with the β and γ subunit. If hormone is still bound, the GDP can be exchanged for GTP, and the activity can start once more. If the hormonal message is not present, the α subunit will remain with the β and γ subunit at the receptor area until a new message (i.e., hormone) is received by the receptor.

The specificity for the effects are due to the specificity of the α subunit, which can interact in various

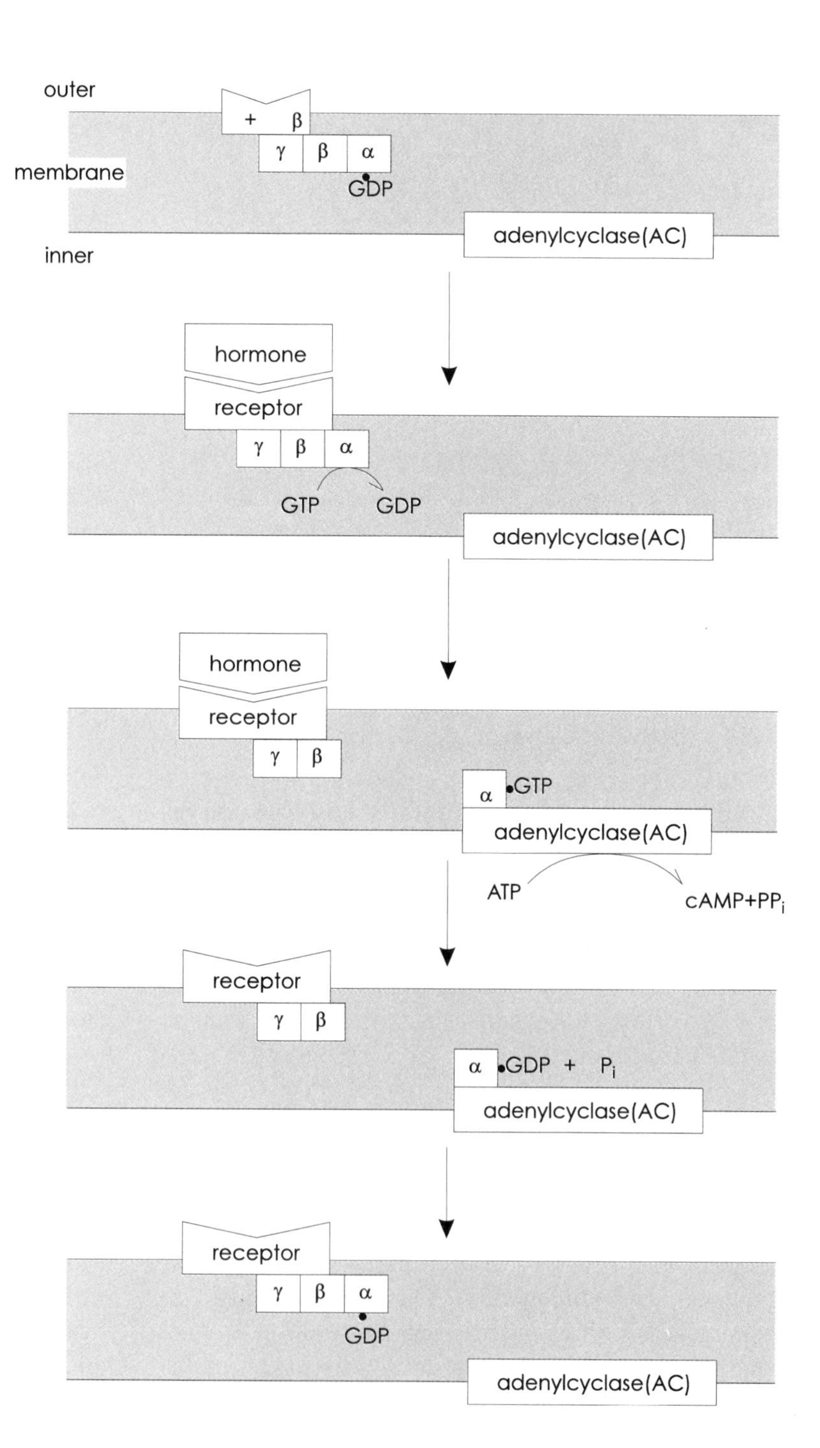

Figure 16.1.

Transduction through the G protein. This figure shows stepwise the action of α1, α2, and β receptors for hormone transduction. Although the diagram emphasizes the action at a β receptor with resulting increased cAMP, it is similar for the other two receptors. With α2 receptors there is inhibition of adenylcyclase instead of the activation shown for the β receptor. The action of the α1 receptor is similar to that of the β receptor, except that there is an activation of phospholipase C (see Fig. 16.3). See the text for further discussion of this figure.

ways depending on the nature of the α subunit and receptor. Some interactions are:

1. Glucagon and many polypeptide hormone receptors, as well as β receptors, as activated by epinephrine and norepinephrine, cause the α subunit to migrate, activate adenylate cyclase, and form cAMP (see Fig. 16.1).
2. At the α-2 receptor, a different α subunit will react with GTP and migrate to the adenylate cyclase, where inhibition of adenylate cyclase is observed.
3. The α-1 receptors bind certain agonists as epinephrine in the liver and muscle; in this case, the α subunit of the G protein with GTP interacts to increase phospholipase C (see Fig. 16.3).

cAMP

An attempt will be made to examine the results of interactions with each of these receptors and their potential implications. The intracellular aspects and mechanisms of these various hormones and products of the transduction will be examined. ***The first, and probably best known, second message is that of cAMP, which activates a protein kinase (R_2C_2).*** cAMP works by reacting with the R subunit of R_2C_2 to produce 2R $(cAMP)_4$ + 2C. In the R_2C_2 complex, the R protein is a regulatory subunit and C is the catalytic subunit. When the catalytic subunit is bound to the regulatory subunit, it remains inactive. When the catalytic subunit is freed, such as by reaction of cAMP with a regulatory subunit, the C subunit then has the catalytic activity of a protein kinase. In many cases, this system is referred to as a cAMP-dependent protein kinase, which is one of many protein kinases. The catalytic subunit will phosphorylate a large number of specific proteins and enzymes on serine or threonine groups. These proteins can range in diversity from glycogen synthetase and pyruvate kinase, which are decreased in activity upon phosphorylation, to hormone-sensitive lipase and phosphorylase kinase, which are activated upon phosphorylation. In the latter, for the final effect, glycogenolysis, to occur, the phosphorylase kinase must catalyze the phosphorylation of phosphorylase to convert it from the inactive to the active form. Another example of the effects of cAMP-dependent protein kinase is the phosphorylation of troponin, a protein involved in muscle contraction.

The catalytic subunit primarily phosphorylates proteins in the cytosol. However, ***the catalytic subunit,*** but not R_2C_2, ***can enter the nucleus and cause phosphorylation of nuclear proteins. The cAMP-dependent protein kinase does not directly cause the phosphorylation of mitochrondrial proteins*** (Fig. 16.2).

The cAMP can be degraded to 5′ AMP by the enzyme phosphodiesterase. Upon the removal of cAMP, the R subunits will recombine with the C subunits to reform R_2C_2, which is inactive. Phosphodiesterase can be

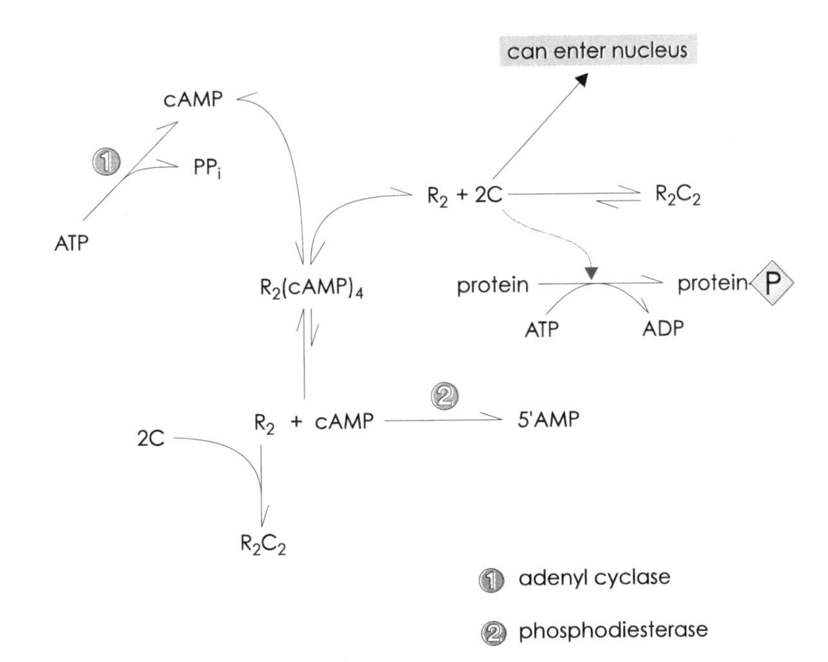

Figure 16.2.

cAMP action. The action and mechanism of cAMP is presented. The C subunit released by cAMP action is the active cAMP-dependent protein kinase. Although the equilibrium favors R_2 $(cAMP)_4$, there is some dissociation, so that when cAMP is destroyed by phosphodiesterase, more free R_2 units are released, which can recombine with the C unit and form the inactive R_2C_2.

inhibited by methylxanthines, such as caffeine and theophylline, thus producing larger increases in cAMP due to a given hormone stimulation. Most cells have specific hormone messages that will increase cAMP and bring about various metabolic changes. In many cases, although all cells respond similarly to cAMP, that is, causing catalytic subunits to phosphorylate specific proteins, the results may vary in different tissues owing to the metabolic machinery within that tissue; for example, the adrenal cortex responds to ACTH by a rise in cAMP to cause a greater synthesis of glucocorticords, whereas the polypeptide hormone TSH will cause increased cAMP in the thyroid gland, resulting in an increased synthesis and release of thyroid hormone (i.e., thyroxine and triiodothyronine). Activation of the α-2 receptor inhibits adenylcyclase and the production of cAMP. Thus, for example, in the β-cell of the pancreas, epinephrine interacts with an α-2 receptor, inhibits adenylate cyclase, and inhibits the release of insulin; glucagon, at the β-cell of the pancreas, reacts with a receptor, causing an increase in both cAMP and insulin secretion. Thus, it can be seen, when cAMP is an important factor, both the rapid formation and inhibition of formation of cAMP can play an important role, as can the rate of destruction by phosphodiesterase.

Inositol Trisphosphate and Diacylglycerol

The α-1 receptors can be activated by compounds such as epinephrine in the muscle, heart, and liver, which ***upon transduction activates phospholipase C, which catalyzes the hydrolysis of phosphatidylinositol 4,5-bisphosphate*** (diacylinositol trisphosphate) ***found in the membrane*** (see Fig. 16.3).

This hydrolysis produces two important second messages. The first is inositol 1,4,5-trisphosphate (IP_3), which increases cytosolic calcium levels by releasing calcium from storage vesicles, such as the T tubules and sarcoplasmic reticulum, as well as potentially opening calcium channels to allow extracellular calcium to enter the cell. Diacylglycerol (DAG), the other portion of the molecule after hydrolysis, is an activator of protein kinase C, which appears to play an important role in cell division and proliferation. Another group of compounds that activate protein kinase C and stimulate proliferation are phorbal esters, which have been identified as promoters, but not necessarily initiators, of carcinogenesis. It appears that a number of growth factors may work via protein phosphorylation and protein kinase C. The DAG and inositol phosphate, formed after phosphatases have removed two of the phosphates from IP_3, can recombine to make phosphatidylinositol, which can reform phosphatidylinositol trisphosphate. In addition to these effects, the DAG formed is particularly rich in arachidonic acid, which can be hydrolyzed via action of a lipase. Arachidonic acid is an effective precursor of prostaglandins, which may modify hormone action.

The insulin receptor system, which functions independently of the G protein, causes proteins to be phosphorylated on tyrosine groups in contrast with the serine and threonine phosphorylations due to cAMP-dependent protein kinase (Fig. 16.4). A number of tyrosine phosphorylations occur, including autophosphorylation of the insulin receptor, per se, which may have some important relationship to down regulation of the insulin receptors during continual high levels of insulin. The tyrosine kinase system has not been studied in the same detail as have cAMP, IP_3, or DAG systems. A number of proteins appear to be phosphorylated on tyrosine residues but there is no clear indication of one critical enzyme that may be activated or inhibited. There are strong suggestions that there are effects on phosphoprotein phosphatase, which would cause dephosphorylation of proteins and work in opposition to cAMP effects. One of the difficulties in elucidating the action on insulin and proteins effected by tyrosine phosphorylation may be related to protein interactions. For example, it is possible that in the purification not only may a protein that is phosphorylated, change its activity, but a protein which may be phosphorylated on the tyrosine group may work in conjunction with the critical enzyme. Thus, on purification, the two interacting proteins may be separated from one another. ***Insulin also has major effects on certain tissues by increasing glucose transporters and the rate of glucose transport into the cell***, both in regard to maximum rate as well as sensitivity or K_m for glucose.

A number of other factors, including some growth factors, may act in a manner similar to insulin, causing phosphorylation of proteins on both tyrosine and serine moieties, and having protein tyrosine phosphate phosphatase activity.

A number of hormones act by entering the cell and interacting with intracellular receptors. These include steroid hormones, triiodothyronine, and other compounds such as retonic acid and 1,25-dihydroxy vitamin D. In most cases, the receptors are located in the nucleus, where the active effector–receptor complexes have nuclear action (see Chapter 29) (Fig. 16.5).

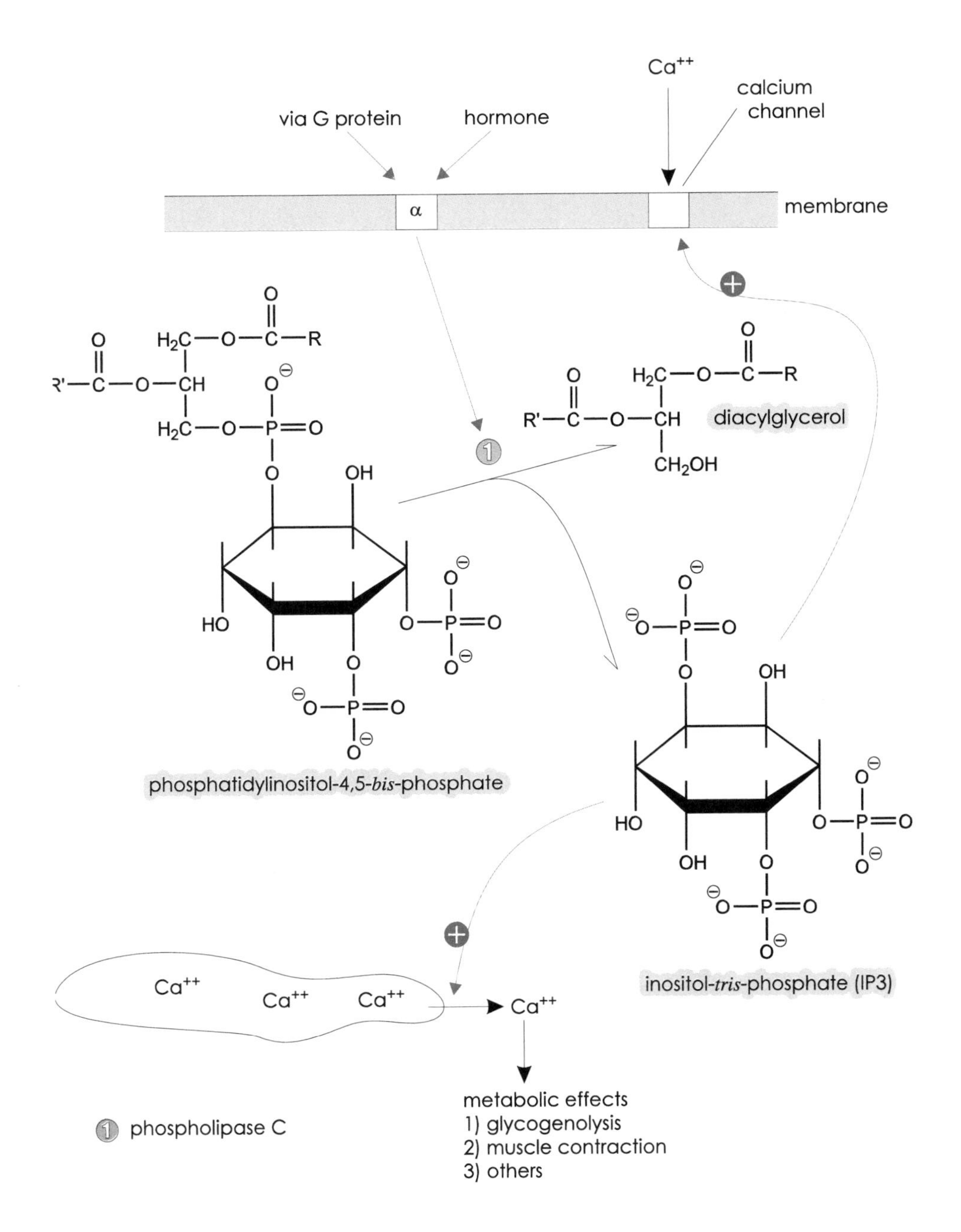

Figure 16.3.

Effects at α_1 receptors. When hormones acting at α_1 receptors interact with the receptors, they cause an activation of phospholipase C via transduction, utilizing the G protein (as seen in Fig. 16.1). The products of phospholipase C are inositol trisphosphate (IP_3), which increases cytosolic Ca^{2+}, and diacylglycerol (DAG), which activates protein kinase C.

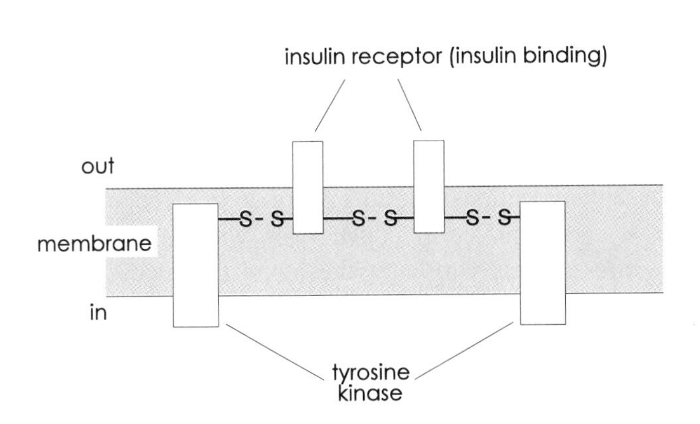

Figure 16.4.

Insulin receptor. A schematic of the insulin receptor is shown. Note that this works independently of the G protein. The transduction of the insulin message results in the activation of a tyrosine kinase, which results in the phosphorylations of cytosolic proteins on the tyrosine moieties.

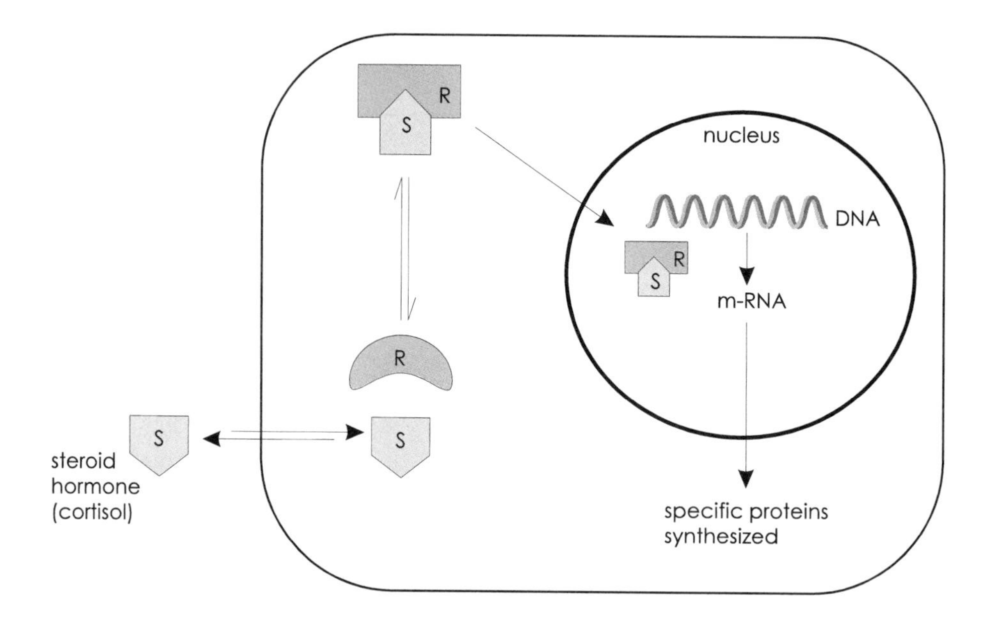

Figure 16.5.
Steroid hormones. A schematic for cortisol and its receptor is presented. In this case, the receptor is intracellular and has effects in the nucleus. For other steroid hormones, the receptors are usually in the nucleus—unlike those for cortisol, which are in the cytosol and, when combined with cortisol, migrate to the nucleus.

In the case of glucocorticoids, the receptors are located in the cytosol. Upon formation of the glucocorticoid receptor complex, this complex can migrate into the nucleus, where it has its metabolic effect (see Chapter 29). The interaction and ratios of circulating hormones are as important as absolute levels. Therefore, absolute levels of glucagon and insulin, for example, cannot be interpreted without knowing the concentration of the other hormone (i.e., glucagon/insulin ratios instead of just concentrations of glucagon and/or insulin). In many cases, hormones give opposing messages and normally, when one is elevated, there is a general suppression of the other hormone, such as the case with glucagon and insulin.

Glucagon

One way of reviewing integration of metabolism is to examine hormonal control. It would be well to consider the basic systemic messages causing response of elevated hormones and the various metabolic perturbations and ramifications brought about by this given hormone to effect the final result. The first example to be examined is glucagon. Glucagon is secreted by the α-cells of the pancreas, usually in response to a low blood-glucose level or a perceived low blood-glucose level. Since the α-cell requires insulin for entry of glucose, in the case of the diabetic, despite the elevated blood glucose, the perception of the α-cell is a deficit of blood glucose and therefore, an increased secretion of glucagon, which exacerbates the problems observed in diabetes. The release of glucagon by the α-cells in response to low blood glucose will have major effects on two tissues, liver and adipose tissue. In both cases, ***the glucagon receptor will cause an increase in cAMP***. An interesting aspect is that although glucagon has major metabolic effects on only these two tissues, the ramifications of these changes will be observed throughout the body.

In the adipose tissue, the increased cAMP will activate hormone-sensitive lipase, through a cAMP-dependent phosphorylation reaction. This will cause an increased release of fatty acids into the blood for utilization by other tissues, including the liver and muscle (Fig. 16.6). ***The effects of glucagon in the liver due to increased cAMP are numerous***, including a phosphorylation of glycogen synthase and phosphorylase kinase, which cause phosphorylation of phosphorylase. These two reactions lead to a decrease in glycogen synthesis and a marked glycogenolysis, where the glucose moieties formed can enter the blood as glucose via a number of enzymatic transformations starting with phosphorylase and ending with glucose-6-phosphatase. The increased cAMP also has other effects favoring gluconeogenesis. These include, in the liver, the phos-

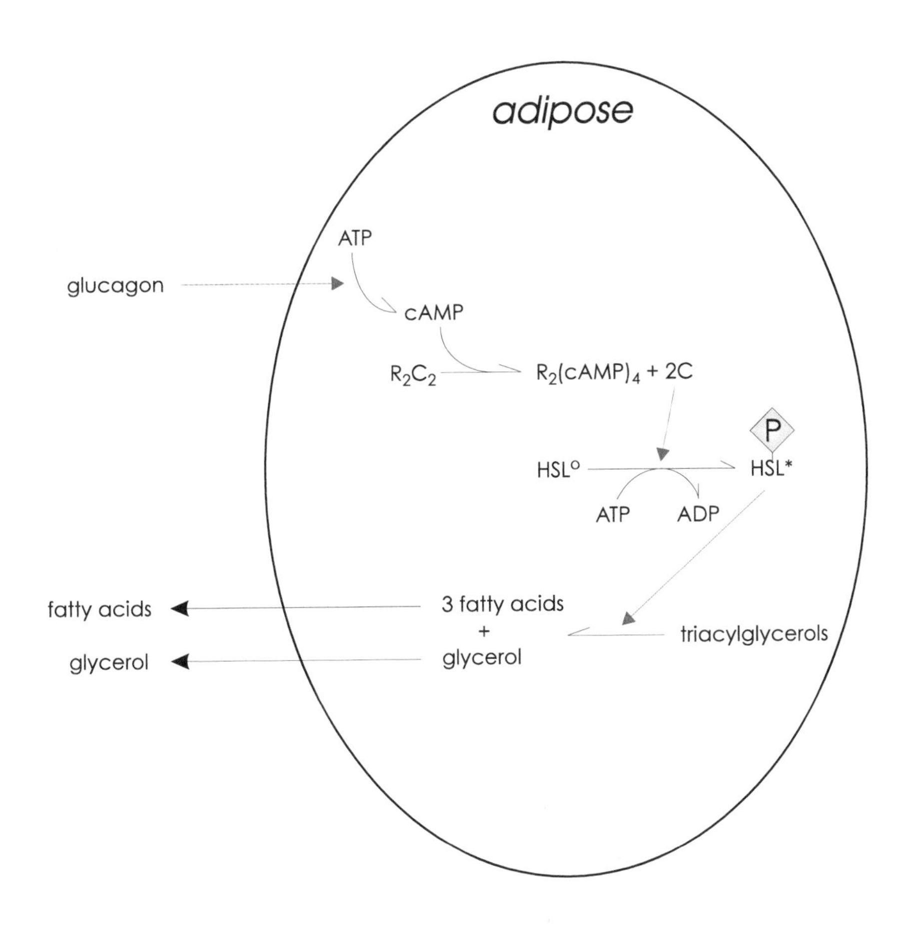

Figure 16.6.
Glucagon and adipose. The action of glucagon on the adipose is shown above. The major overall effect is the release of fatty acids into the bloodstream (see text for more detail).

phorylation of PFK2, turning PFK2 into a fructose-2,6-bisphosphatase, and lowering the level of fructose-2,6-bisphosphate. This decreases the concentration of F-2,6-BP, an activator of PFK and an inhibitor of fructose-1,6-bisphosphatase. Thus, the basic activity of fructose-1,6-bisphosphatase, which is considerably greater than that of PFK, will predominate and allow substrate to pass in the gluconeogenic direction toward fructose-6-phosphate, glucose-6-phosphate, and glucose.

The elevated cAMP will also cause a phosphorylation of pyruvate kinase, which will cause an increase of the K_s (0.5) for PEP and, at physiological levels of PEP, the pyruvate kinase activity will slow markedly. This allows enolase and other multifunctional enzymes, which can be used for gluconeogenesis, to increase the gluconeogenic flux, which would be pulled by the increased FBPase activity. There are reports that the contribution of phosphorylation of pyruvate kinase by cAMP is one of the most important covalent modifications to allow a marked increase in gluconeogenesis, since the normally high activity of pyruvate kinase would short-circuit the PEP formed from pyruvate. This pyruvate from PEP would again have to transverse pyruvate carboxylase, be transported out of the mitochondrion, and be acted upon by PEP carboxykinase before proceeding back to PEP and glucose. This would also cost a high-energy phosphate.

Important controls in gluconeogenesis in addition of covalent modifications are allosteric effects. Many of these may be brought about by the elevated availability of fatty acids from lipolysis in the adipose. After entering the liver and being activated, the fatty acids can be moved into the mitochondrion where, via β-oxidation, they can provide both high-energy phosphates, in the form of ATP, and reducing power, in the form of NADH; both may be important for gluconeogenesis. Reducing power is critical in the case of gluconeogenesis from amino acids and pyruvate. The catabolism of fatty acids will cause an accumulation of acetyl CoA in the mitochondrion which, in conjunction with high ATP concentrations, leads to certain other major effects. Of particular importance is the inhibition of pyruvate dehydrogenase and the activation of pyruvate carboxylase. This assures that the gluconeogenic precursor, pyruvate, is not converted to the nongluconeogenic precursor, acetyl CoA, but is strongly converted to oxaloacetate, a potential glucose precursor.

During fatty-acid metabolism, the high NADH, acetyl CoA, and ATP lead to the phosphorylation of pyruvate dehydrogenase, further inactivating the system and maintaining pyruvate levels (Fig. 16.7).

This conversion of pyruvate dehydrogenase to the less-active phosphorylated form is not due to the elevated cAMP directly, but appears to be strongly related to the high rate of fatty-acid catabolism. The high level of fatty-acyl CoA will cause inhibition of citrate synthetase, which allows the acetyl CoA to accumulate, as well as forcing much of the oxaloacetate to alternate pathways such as gluconeogenesis. If there was no inhibition of citrate synthetase, but isocitrate dehydrogenase were inhibited, the high levels of oxaloacetate and acetyl CoA would combine to cause an accumulation of citrate and isocitrate. This is not observed during starvation nor glucagon treatment of intact animals or isolated cells.

Since the liver is obtaining considerable energy from β-oxidation as compared to the full cycling of the citric-acid cycle, it is important to have free CoA available for β-oxidation. This ***free CoA can be regenerated***

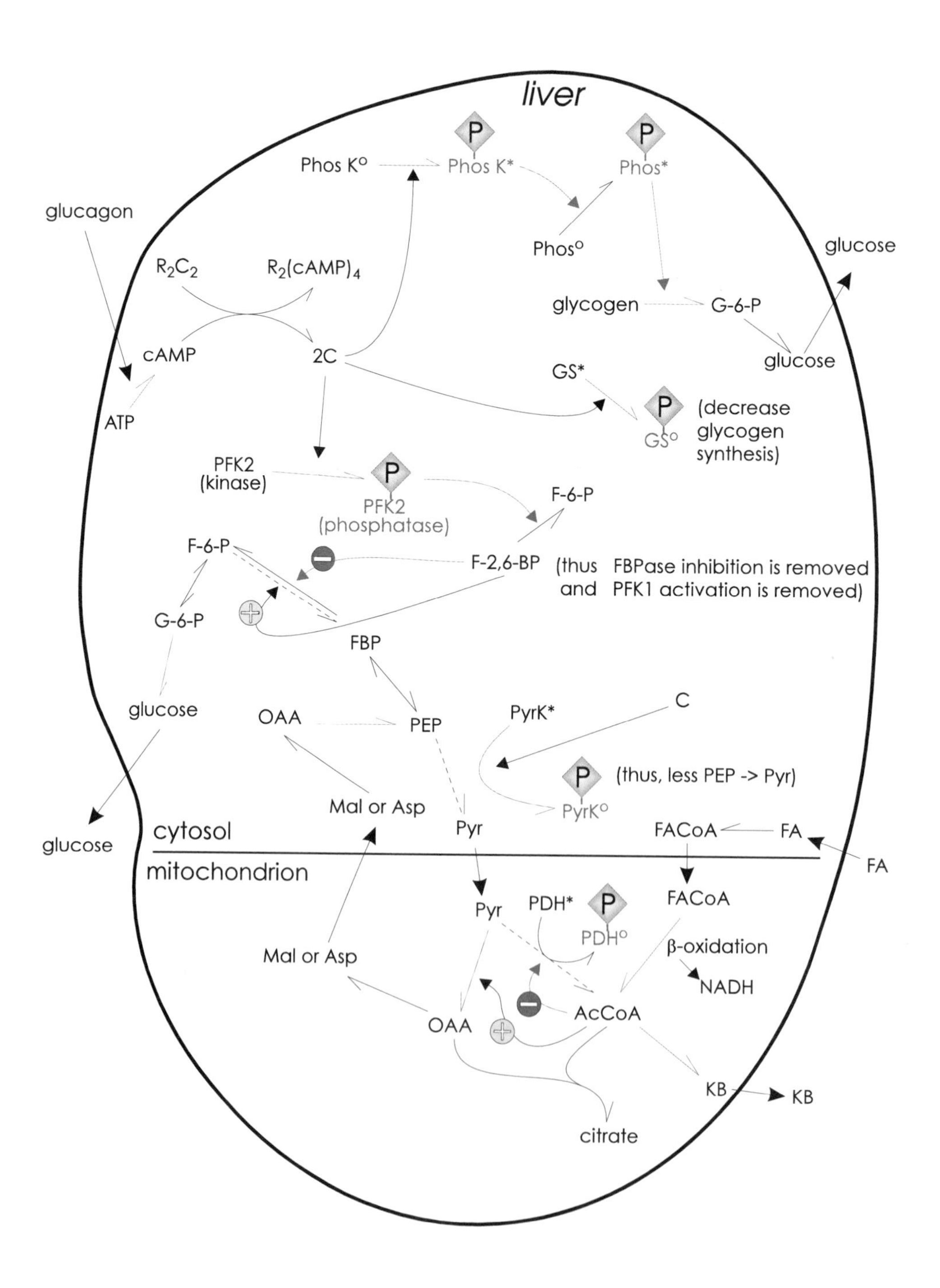

Figure 16.7.

Glucagon and the liver. The action of glucagon on the liver is shown above. The major overall effect is an increase in blood glucose, from glycogen and gluconeogenesis, and an increase in ketone bodies (see text for more detail). Major allosteric and covalent effects are shown, with the final form of covalently converted enzymes shown in purple, and reactions increased by glucagon shown with purple arrows and those slowed shown with dashed lines (see text for more detail).

by the conversion of acetyl CoA, through a number of steps described earlier, to ketone bodies. These ketone bodies provide potential energy for muscle, heart, kidney, and also even the brain. The increased fatty-acyl CoA and increased cAMP also cause an inhibition of fatty-acid synthesis, which would be inappropriate at a time of low blood glucose. This occurs via two independent mechanisms. The first is the phosphorylation of acetyl CoA carboxylase, yielding a relatively inactive enzyme, and the second is the inhibition of acetyl CoA carboxylase by fatty-acyl CoA, the concentration of which is markedly elevated at this time in the liver owing to the entrance of fatty acids released from the adipose.

One major organ in the body that does not have glucagon receptors and is not directly affected by glucagon, but is strongly indirectly affected, is the muscle. There are two potential aspects to raising blood glucose: the first is increasing glucose production through glycogenolysis and gluconeogenesis; the other would be to decrease the utilization of glucose. Since the muscle does not carry on gluconeogenesis, the only method for raising blood glucose from this tissue, including the heart, would be to decrease glucose utilization. Muscle is the largest tissue in the body and potentially can have the greatest glucose utilization in the body. Diminishing the use of blood glucose by the muscle would be an extremely important contribution toward raising blood-glucose levels. This diminution occurs mainly as a result of the effects at the adipose tissue, which causes the release of fatty acids, which are also the precursors of ketone bodies in the liver. The muscles can utilize glucose, fatty acids, or ketone bodies for energy. Each of these has some control, although the control on glucose utilization is considerably more exquisite than that on fatty-acid and ketone-body metabolism. Ketone bodies are small, soluble molecules that can move across both plasma and mitochondrial membranes and are activated in the muscle mitochondria catalyzed by acetoacetate:succinyl CoA–CoA-transferase, to form acetoacetyl CoA, which can be converted to acetyl CoA and metabolized in the citric-acid cycle. This metabolism of ketone bodies produces reducing power, ATP, and acetyl CoA, which will lead to a phosphorylation of pyruvate dehydrogenase in the muscle, thus protecting pyruvate from becoming a nongluconeogenic precursor, namely, acetyl CoA.

Saving pyruvate in the periphery is probably as important or more important than saving this compound in the liver, since potentially much more pyruvate can be converted to acetyl CoA in the muscle than the liver. Any time pyruvate is protected from proceeding to acetyl CoA, it can then leave the tissue as pyruvate or lactate and return to the liver to undergo gluconeogenesis. Wherever you lose a pyruvate as acetyl CoA, you have effectively lost a precursor for net gluconeogenesis. Fatty acids can also be activated in the muscle to fatty-acyl CoA, which can be transported into the mitochondria by the acyl carnitine transferase system. Normally, this is a very effective process as long as free CoA is available in the mitochondrion and malonyl CoA is below strongly inhibitory concentrations. Ketone bodies will have some preference over fatty acids for catabolism because they can enter the mitochondria directly and are activated via the normal intermediate of the citric-acid cycle (i.e., succinyl CoA). α-Ketoglutarate dehydrogenase has a lower K_m for CoA than the acyl carnitine transferase 2, which moves the fatty acids from acyl carnitine to acyl CoA in the mitochondria. In the absence of high concentrations of ketone bodies, fatty acids can rapidly be catabolized through β-oxidation to acetyl CoA, which can be metabolized in the citric-acid cycle to form reducing power (NADH) and ATP, conditions that favor phosphorylation of pyruvate dehydrogenase. When fatty acids or ketone bodies are rapidly being metabolized, elevated concentrations of ATP and NADH can decrease the flux through isocitric dehydrogenase, allowing citrate to accumulate and exit the mitochondrion (Fig. 16.8). The elevated concentrations of citrate and ATP can further inhibit PFK, which will cause an accumulation of F-6-P and G-6-P, with the latter inhibiting the activity of hexokinase. This will decrease the phosphorylation of glucose, thus allowing a more rapid rise in blood glucose from gluconeogenesis and glycogenolysis than would occur if the muscle were using the increased glucose available. ***The ketone bodies can also be used by the brain to relieve some of the glucose requirement*** so that potentially, at maximum effectiveness, ketone bodies can supply 60% of the energy required by the brain, but glucose is still required for at least 40% of the central nervous system energy. Thus, glucose utilization can never be reduced to nil in the central nervous system, although this can come close to being achieved in the muscles. There are also tissues, such as erythrocytes, that metabolically function anaerobically and require glucose, but only for conversion to lactate.

Glucagon can work in multifaceted mechanisms to bring about increases in blood glucose despite directly affecting only two tissues, liver and adipose tissue. ***The indirect effects of glucagon on peripheral tissues and even the central nervous system have profound repercussions on glucose utilization and metabolism.***

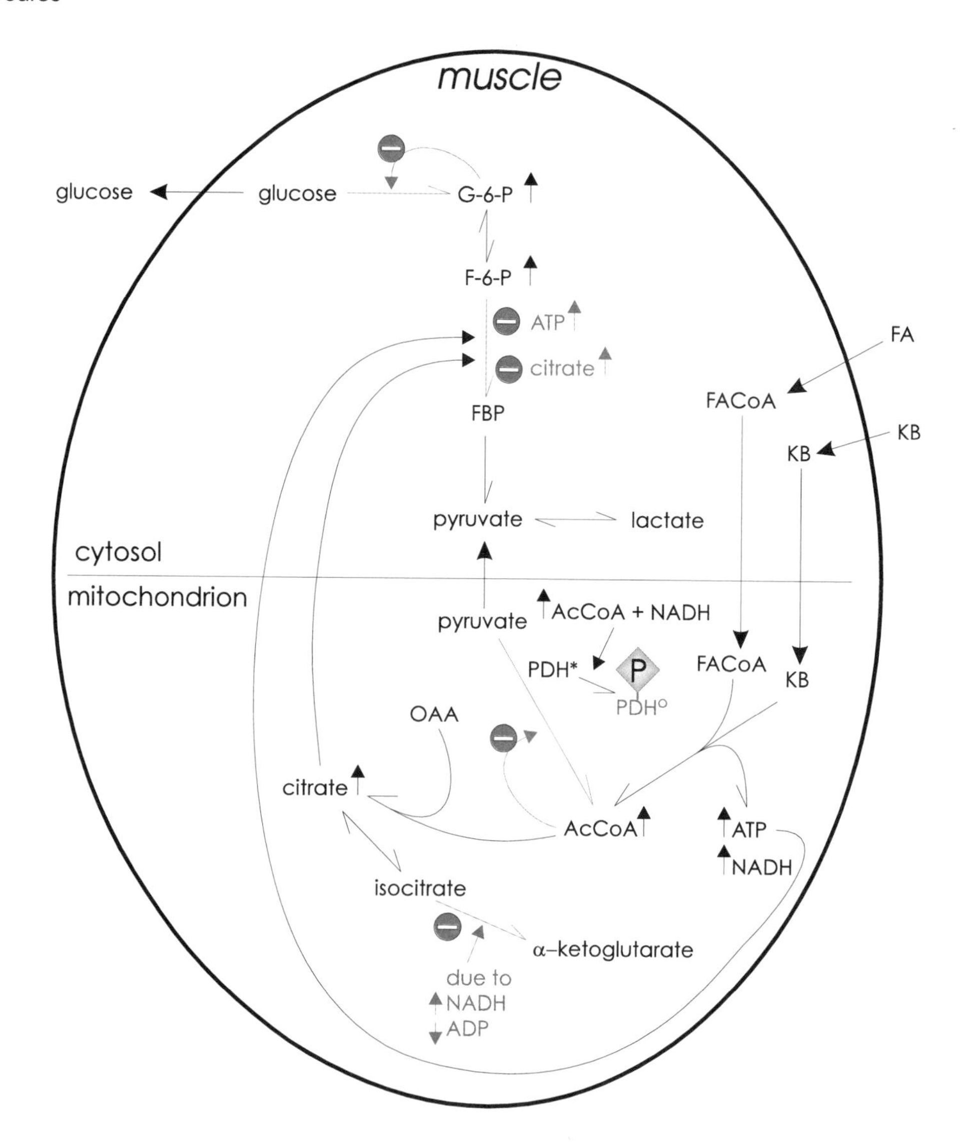

Figure 16.8.

Effect of increased fatty-acid availability on muscle metabolism. The effect of fatty acids and ketone-body metabolism on glucose metabolism in the muscle is shown above. The major effect is a decrease of glucose utilization by the muscle. Negative aspects on this metabolism are shown in purple (for more detail, see the text).

EPINEPHRINE

Epinephrine is a hormone released by the adrenal medulla in response to a fight-or-flight situation. Today it would more readily be associated with anxiety in either a physical or mental situation. In the past, this hormone readily prepared people and animals for a fight-or-flight situation. This response is still probably appropriate in animals, but less so in people at the present time. This hormone elicits a series of responses related to affecting rapid muscle contraction, as well as protecting blood-glucose levels for the central nervous system. This hormone can affect α-1, α-2, and β-receptors in various tissues. Therefore, a mechanism and rationale for epinephrine action clarify the integration of metabolic controls in the whole body, interacting with many of the tissues in an attempt to elicit the fight-or-flight response.

Epinephrine in both skeletal muscle and heart affects both α-1 and β-receptors. The major effect on the α-1 receptors is an ***increase of inositol trisphosphate (IP_3), causing an increase in cytosolic calcium levels*** both from intracellular sources, such as sarcoplasmic reticulum, T-tubules, and so on, and from opening some calcium channels to increase cytosolic calcium from extracellular sources. The increased calcium has several major effects, the first of which is to initiate and strengthen muscle contraction both in the skeletal muscles for the flight-and-fight response and in the heart for a more rapid and stronger contracting system, thus increasing delivery of blood to the tissues of the body, with the concomitant increased supply of substrates and oxygen (Fig. 16.9).

The second effect of calcium is to increase the responses of glycogen degradation by further activating phosphorylase kinase, an allosteric activator of even the

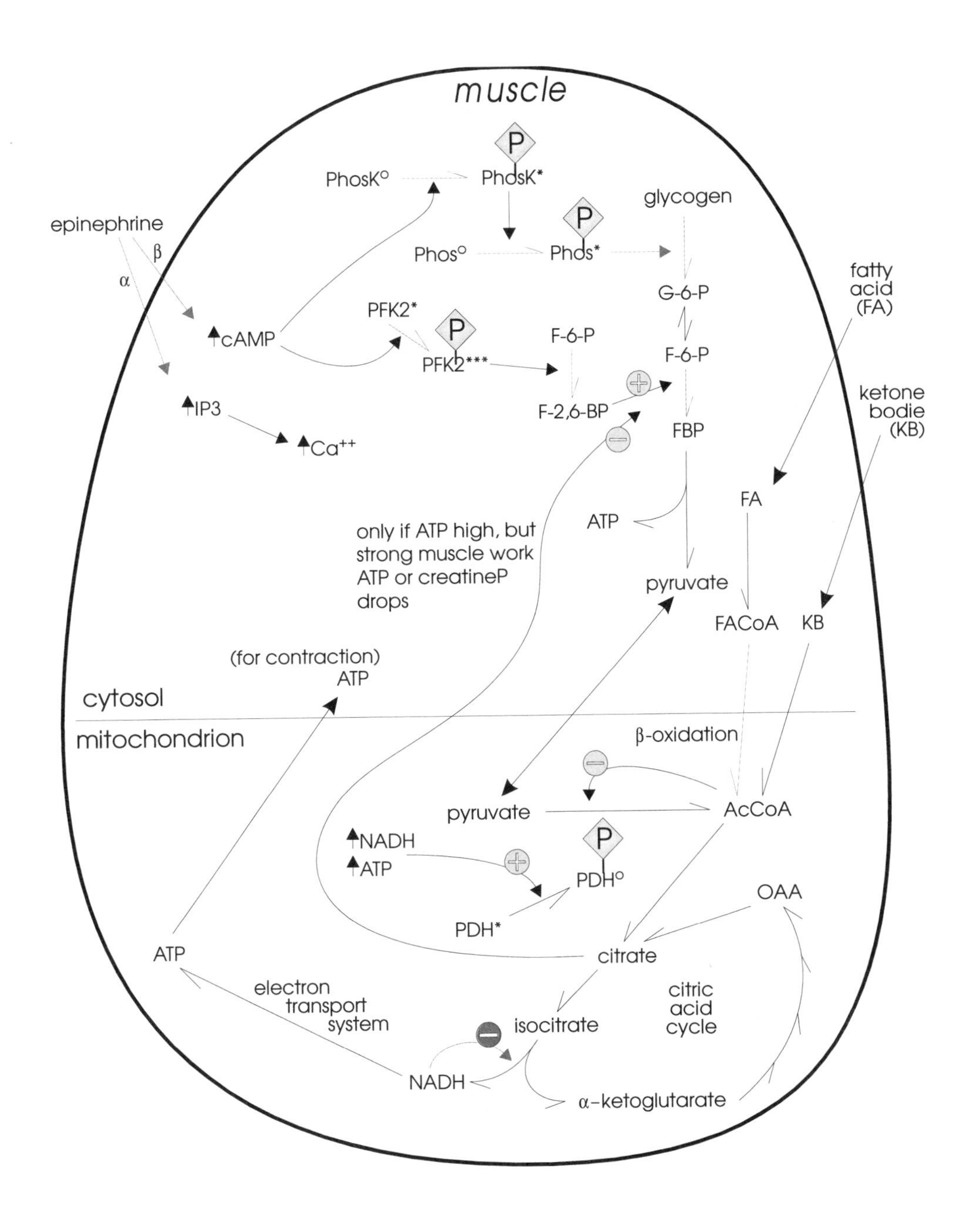

Figure 16.9.
Epinephrine and the muscle. The action of epinephrine and increased fatty-acid availability on muscle metabolism is shown above. Epinephrine works at both α_1 receptors to increase cytosolic Ca^{2+}, which is needed for contractions, and the β receptor to increase cAMP. The increased cAMP leads to glycogenolysis and a more rapid glycolysis for anaerobic energy. The reactions increased by epinephrine are indicated by purple arrows (see text for more detail).

nonphosphorylated form of phosphorylase kinase, as well as facilitating the phosphorylation of phosphorylase kinase by the cAMP-dependent protein kinase. This interaction is related to the calmodulin subunit of phosphorylase kinase. These activations increase glycogenolysis and the supply of glucose-6-phosphate intracellularly for glycolysis and anaerobic energy. The pyruvate formed can further be degraded via pyruvate dehydrogenase and the citric-acid cycle. This glycogenolysis is of greater significance in the skeletal muscle than in the heart because skeletal muscle can contain approximately 2% of its wet weight as glycogen, whereas the heart contains only a few tenths of a percent of glycogen. Furthermore, the heart is very efficient at using other sources of energy aerobically; some skeletal muscles are particularly efficient at anaerobic glycolysis, producing high-energy phosphates and lactate from glucose-6-phosphate.

The effect of epinephrine on the β-receptors of the muscle causes an increase in cAMP, which will cause the phosphorylation of a number of proteins, in particular, phosphorylase kinase, which will then phosphorylate phosphorylase from the inactive to the active form, causing an increased breakdown of glycogen. Another major protein phosphorylated in the muscle is PFK 2, which unlike the liver PFK 2, becomes more active when phosphorylated and increases the levels of fructose-2,6-bisphosphate. Fructose-2,6-bisphosphate

(F-2,6-BP) is an important activator of PFK, a potential rate-limiting enzyme of glycolysis, particularly when starting with intracellular hexosphosphates, as is the case with glycogenolysis. The increase in F-2,6-BP helps increase the PFK flux by decreasing the K_s (0.5) for F-6-P. Glycogenolysis can increase F-6-P (in equilibrium with G-6-P) from a source other than glucose and helps increase the PFK flux. This increase in F-6-P cannot be accomplished from glucose, owing to G-6-P inhibition of hexokinase. One further example of a protein phosphorylated is troponin, one of the contractile elements that may have major effects on effectiveness and efficiency of muscle contraction.

The liver is also affected by epinephrine; in this case, the α-1 receptors are activated by epinephrine. This will cause an increase in intracellular liver Ca^{2+}, via IP_3, an allosteric activator of phosphorylase kinase which will catalyze the phosphorylation of phosphorylase and thereby activation. This increase in active phosphorylase will rapidly increase glycogenolysis and provide hexosphosphates, which in the case of liver are not used primarily for glycolysis but, via the action of glucose-6-phosphatase, augment the blood glucose levels (Fig. 16.10).

The adipose tissue is also affected by epinephrine, primarily at the β-receptors, causing an increase in

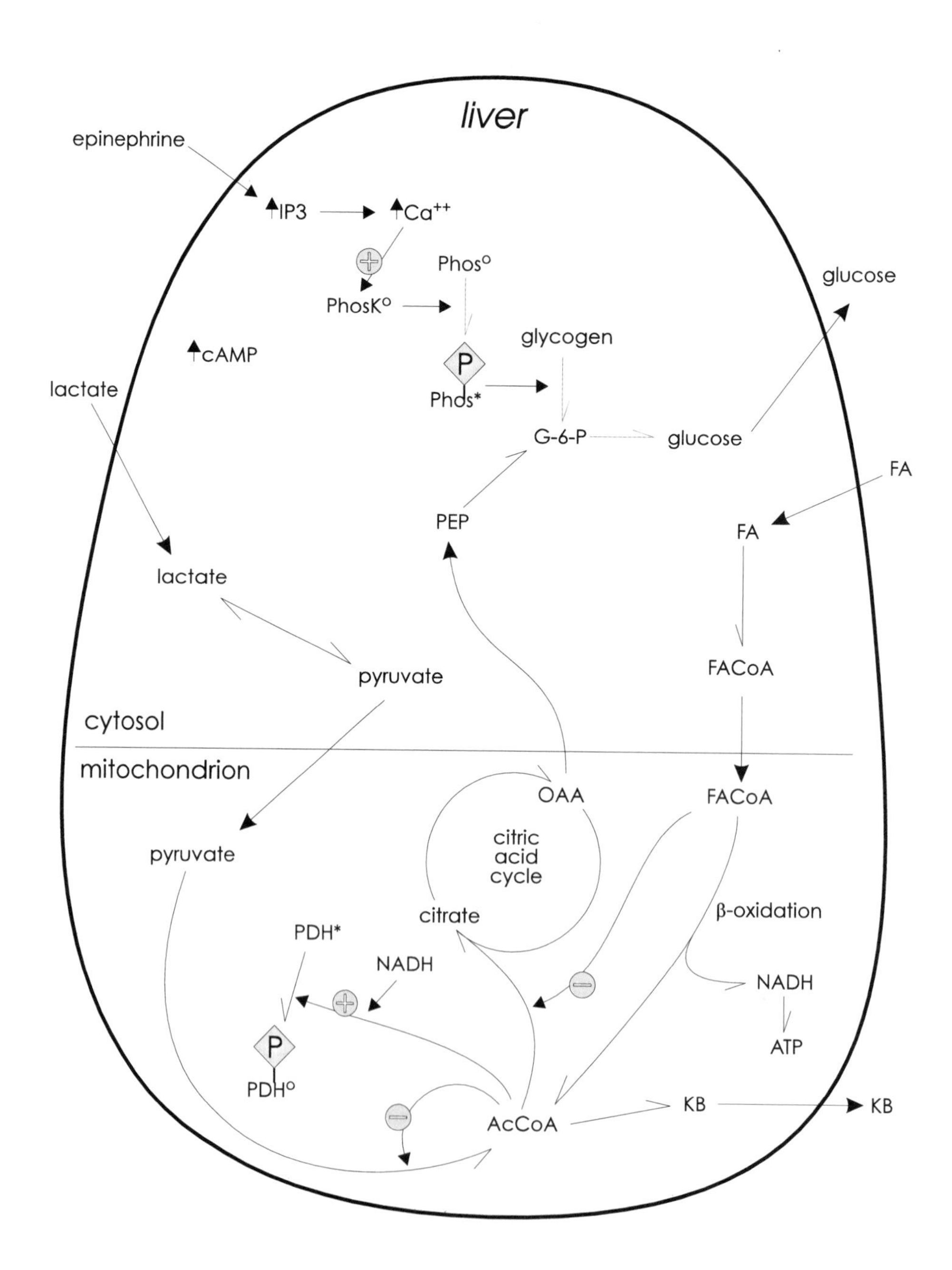

Figure 16.10.

Epinephrine and the liver. The action of epinephrine and increased fatty-acid availability on the liver is shown above. Epinephrine works at α_1 receptors in the liver, increasing IP_3, which increases cytosolic Ca^{2+}. The increased Ca^{2+} causes increased glycogenolysis and release of glucose into the blood. The reactions affected by Ca^{2+} are shown with a purple arrow (see text for more detail).

cAMP which will cause phosphorylation of hormone-sensitive lipase catalyzed by the cAMP-dependent protein kinase. The phosphorylated form of hormone-sensitive lipase (the active form) will stimulate lipolysis from the adipose and release of nonesterified fatty acids into the blood (same mechanism as glucagon; see Fig. 16.6). These nonesterified fatty acids can act as a major energy source for the heart and aerobic skeletal muscle, thus providing extra energy for the fight-or-flight situation. These increased fatty acids will also help spare glucose utilization, particularly by aerobic muscles, and help stimulate gluconeogenesis in the liver by increased levels of fatty-acyl CoA and acetyl CoA, as described in detail in the section on glucagon. This effect on the adipose helps release some of the plentiful stores of energy within the body.

Epinephrine acts on the β cells of the pancreas, where it interacts with an α-2 receptor, inhibiting cAMP formation and release of insulin. This is advantageous under these conditions because the muscles become more sensitive to insulin when exercising. Therefore, increased insulin release under these conditions could cause an increased glucose influx into the muscles and a marked decrease in blood glucose levels, thus potentially impairing the central nervous system. Muscles will utilize their own glycogen supply and fatty acids will be released from the adipose tissue. Therefore, this decrease in insulin release is important. This is the underlying reason why type I diabetics may suffer from hypoglycemia during strenuous exercise. Type I diabetics take exogenous insulin, which is not under control of pancreatic release. The level of insulin is relatively fixed and, since exercise makes the muscle more sensitive to insulin, the "normal level" becomes an overdose (relative hyperinsulinemia) and causes hypoglycemia. Exercise and insulin administration must be coordinated since type I diabetics have no effective feedback on insulin release, as normally occurs. Epinephrine prepares the body for an all-out effort of physical exertion.

The old advantages of epinephrine may be a disadvantage in modern western culture, and be a link between stress or coronary heart disease. In modern-day stress, when epinephrine is released, it is usually not a case of fight-or-flight, but more a problem of meeting deadlines or mental pressures. Thus, the released epinephrine causes the metabolic aspects previously described. The release of fatty acids from the adipose may be particularly detrimental. If there is little utilization of ATP to ADP, substrate catabolism will be minimal, as the limiting factor in catabolism is usually the availability of ADP. The breakdown of glycogen in the muscles may provide more energy than is required and the oxidation of the released fatty acids may be minimal. The circulating nonesterified fatty acids are taken up by the liver and repackaged primarily as VLDL, which then can be returned to the circulation. There is some utilization of the VLDLs by the adipose tissue for reforming triacylglycerol with the formation of LDL. At an inappropriate time (i.e., not after fatty-acid synthesis nor after the intake of a fat-containing meal) the increased circulating LDLs may lead to cholesterol deposition and plaque formation within the vessel, resulting in the formation of atherosclerosis and coronary heart disease. Thus, a prehistoric mechanism of nature to allow people to respond to life-threatening situations, where the release of epinephrine was usually followed by strong physical exertion, has turned into a present-day potential disadvantage, causing the inappropriate release of fatty acids at a time when physical exertion is neither planned nor undertaken.

INSULIN

The next hormone to be considered within this chapter is insulin. Insulin is synthesized in the β-cell's islets of Langerhans in the pancreas. It is synthesized as a long single peptide with several disulfide bonds. Before release there are two proteolytic cleavages yielding a C peptide, and the α- and β-chains of insulin are held together by disulfide bonds. The synthesis as a single peptide allows for the proper folding and disulfide bond formation. Insulin is secreted in response to a number of stimuli; among the most potent is an increase in glucose, with a rapid increase in glucose being more potent than a slow increase in glucose. A number of amino acids cause a release of insulin, the most potent of these are leucine and arginine. Leucine is an indication of protein intake and is a ubiquitous amino acid with a reasonably high molar percentage in most all proteins. Arginine appears to cause the release of insulin as a result of the formation of nitric oxide (NO), a potent neuro transmitter, which works at a number of sites in addition to the central nervous system. The release of insulin is also increased by factors that increase cAMP, including glucagon. In general, insulin release is a response to eating and food intake mediated by blood glucose and certain amino-acid concentrations. These stimuli cause the release of insulin, which is a truly anabolic hormone stimulating protein, glycogen, and lipid synthesis.

The systemic action of insulin is at least twofold. The first is to increase the number of glucose transporters and the kinetics of glucose uptake into insulin-sensitive cells, which may be related to mobilization of glucose transporters from the intracellular location to the membrane. ***The second set of major effects are metabolic transformations, which appear to be initiated by the insulin receptor via tyrosine kinases in insulin-sensitive cells.*** Not all cells are sensitive to insulin; most in the central nervous system are not sensitive, with the exception of the hypothalamus, which contains the feeding and satiety centers. A number of tissues, such as the liver, are sensitive to insulin in regard to metabolic responses but apparently not in regard to glucose transport. Other cells, such as muscles and adipose, are sensitive to insulin, both in regard to glucose transport and metabolic alterations. ***Insulin, similar to other hormones which do not apparently enter the cell, attaches to a specific membrane receptor which activates a tyrosine kinase on the cellular side of the membrane.*** Tyrosine kinase phosphorylates proteins primarily at tyrosine residues. The tyrosine kinase will also cause self-phosphorylation and potential destruction of insulin receptors, a phenomenon known

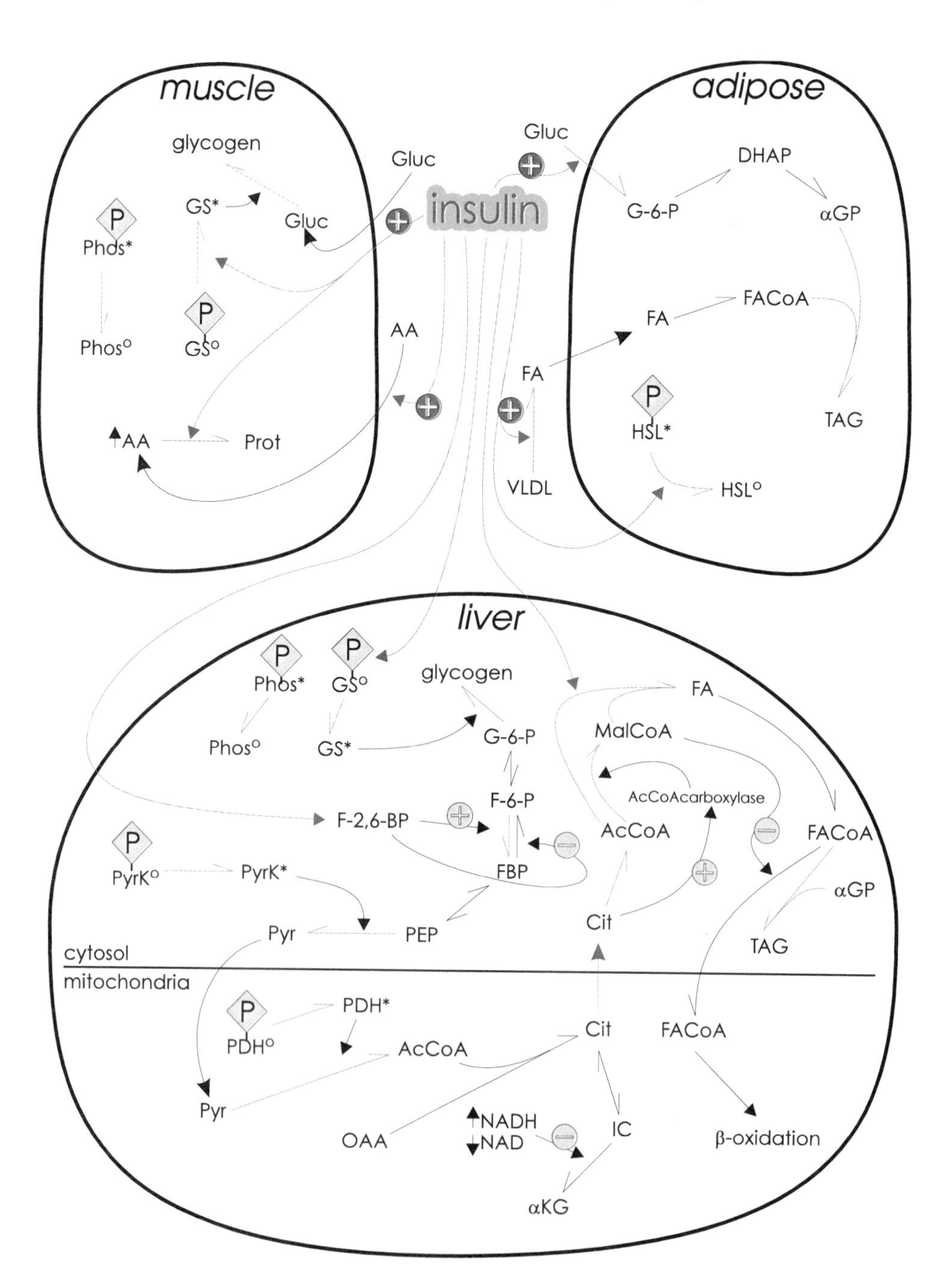

Figure 16.11.

The effect of insulin on muscle, adipose, and liver. The effect of insulin on these three tissues is shown above. The reactions stimulated by insulin are indicated with purple arrows. Insulin stimulates the dephosphorylation of many proteins phosphorylated by the cAMP-dependent protein kinase. Insulin stimulates glucose transport, glycogen synthesis, fat synthesis and storage, and protein synthesis—the latter especially in the muscle (see text for more detail).

as down regulation. This is a condition where excess insulin causes a decreased number and/or response of insulin receptors to insulin and the body becomes refractory to the insulin action. This can occur naturally in type II diabetes (i.e., noninsulin-dependent, nonketotic diabetes).

One of the major overall effects of insulin is to cause a dephosphorylation of many of the proteins phosphorylated by the cAMP-dependent protein kinase both in the liver and peripheral tissue. The exact mechanism is not completely clear at this time, although there are strong suggestions of a potential activation of the phosphoprotein phosphatase by insulin action. This dephosphorylation of proteins and increased permeability and transport of glucose and amino acids into the cell causes a number of important metabolic consequences (Fig. 16.11). These are examined on a tissue-by-tissue basis, realizing there is an integration and interplay among the various tissues. In the muscle, a major consequence of insulin is decreased proteolysis and increased protein synthesis, as well as increased availability of intracellular glucose. A conversion, by dephosphorylation, of glycogen synthetase to the active form and phosphorylase to the inactive form allows synthesis of glycogen. It has been noted that pyruvate dehydrogenase after insulin treatment reverts from the inactive to the active form. However, whether this is a direct effect of a second message of insulin or is the result of decreased circulating fatty acids and ketone bodies plus increased intracellular glucose is still a point of debate.

A major effect of insulin is on the adipose tissue, both in regard to glucose uptake and intracellular metabolic effects. The major metabolic conversion in response to insulin is a dephosphorylation of hormone-sensitive lipase to convert this enzyme into the inactive form. A second important metabolic aspect is an increase in synthesis and secretion to the blood-vessel surface of lipoprotein lipase. This enzyme is responsible for breaking down circulating triacylglycerols, particularly from VLDL as well as from chylomicrons, into fatty acids and glycerol. The fatty acids can then be taken up by the adipose. These fatty acids can be both those produced by the lipoprotein lipase, as well as those that may be circulating as nonesterified fatty acids. They can then be activated to fatty-acyl CoA. One difficulty in storing the acyl CoA as triacylglycerol may be the availability of α-glycerol phosphate. The adipose tissue has negligible activity of glycerol kinase, and cannot form α-glycerol phosphate from glycerol released either from within the tissue or from circulating triacylglycerol. ***Therefore, the important role of glucose entering the cell becomes paramount and is stimulated by insulin.*** The intracellular glucose in the adipose tissue can be phosphorylated to glucose-6-phosphate and via glycolysis to dihydroxyacetone phosphate, a precursor of α-glycerol phosphate. The α-glycerol phosphate and fatty-acyl CoAs can then combine to form triacylglycerol, the storage form of fatty acids. Thus, insulin has an important role in storing triglycerides in the adipose through several metabolic effects within the cell, as well as allowing glucose to enter the adipose cell. The general anabolic effects of insulin in the body, and in this case, in lipid storage, in the adipose tissue are apparent.

SUMMARY

1. Hormones can have their effects at cell membranes or enter the cells.

2. Hormones acting at cell membranes must interact with specific receptors and transduce the messages to form various second messages or actions within the cell.

3. This transduction usually, but not always, involves a G protein to carry information from the outer-cell membrane to the inner-cell membrane.

4. The second messages include cAMP, inositol trisphosphate (IP_3), and diacylglycerol.

5. cAMP frees C subunits from R_2C_2. The free C subunit, also known as the cAMP-dependent proteinkinase, will catalyze the phosphorylation of specific proteins on serine and threonine residues.

6. IP_3 increases cytosolic Ca^{2+} levels. The Ca^{2+} comes from intracellular organelles and/or from extracellular Ca^{2+}.

7. Diacylglycerol activates protein kinase C, which may be associated with cell proliferation.

8. Insulin receptors are independent of the G protein and transduction activates a tyrosine protein kinase, which phosphorylates specific proteins on tyrosine residues.

9. Steroid hormones, thyroid hormones, and some other hormone-like compounds enter the cell and interact with specific receptors, if present, and have their effect in the nucleus.

10. Glucagon causes increases in cAMP in the adipose and liver. In the adipose this causes an increase in lipolysis and release of nonesterified fatty acids to the blood. These fatty acids stimulate gluconeogenesis in the liver and decreased use of glucose in the muscle. The increased cAMP in the liver results in increased glycogenolysis and increased gluconeogenesis, both of which help increase blood glucose levels.

11. Epinephrine causes increases in cAMP in the adipose and muscle and increases in cytosolic Ca^{2+}, via IP_3, in the muscle and liver. The results in the adipose are similar to that of glucagon. The increased Ca^{2+} in the liver leads to increased glycogenolysis and release of glucose into the blood. The increased cAMP and Ca^{2+} in the muscle leads to increased glycogenolysis and G-6-P formation. The increased Ca^{2+} may aid in more effective contraction. The effects of increased fatty acids from the adipose has similar effects on liver and muscle, as in the case of glucagon.

12. Insulin causes activation of a protein tyrosine kinase in many cells. Insulin also causes increased glucose transport in selected tissues. Insulin will stimulate dephosphorylation of many proteins phosphorylated in response to cAMP. Insulin is an anabolic hormone and in many tissues causes increases in protein and glycogen synthesis. Insulin also causes an increase in adipose lipoprotein lipase and triacylglycerol storage in the adipose tissue.

REFERENCES

The G protein connection: Molecular bases of membrane association, A. M. Speigel, P. S. Backlund, Jr., J. E. Butrynski, T. L. Z. Jones, and W. F. Simonds, 1991, *Trends Biochem. Sci.*, **16**:338–341.

cAMP-dependent protein kinase: Framework for a diverse family of regulatory enzymes, S. S. Taylor, J. A. Buechler and W. Yonemoto, 1990, *Ann. Rev. Biochem.*, **59**:971–1005.

Consensus sequences as substrate specificity determinants for protein kinases and protein phosphatases, P. J. Kennelly and E. G. Krebs, 1991, *J. Biol. Chem.*, **266**:15555–15558.

Regulation of phospholipase C by G proteins, P. C. Sternweis and A. C. Simcka, 1992, *Trends Biochem. Sci.*, **17**:502–506.

Regulation of inositol phospholipid-specific phospholipase C isoenzyme, S. G. Rhee and K. D. Choi, 1992, *J. Biol. Chem.*, **267**:12393–12396.

Structure of mammalian steroid receptors: Evolving concepts and methods, M. R. Sherman and J. Stevens, 1984, *Ann. Rev. Physiol.*, **46**:83–105.

The insulin signaling system. M. F. White and C. R. Kahn, 1994, *J. Biol. Chem.*, **269**:1–4.

Signal transduction and protein phosphorylation in the regulation of cellular metabolism by insulin, J. C. Lawrence, Jr., 1992, *Ann. Rev. Physiol.*, **54**:177–193.

The nature and regulation of the insulin receptor: Structure and function, M. P. Czech, 1985, *Ann. Rev. Physiol.*, **47**:357–382.

Insulin, ketone bodies and mitochondrial energy transduction, K. Sato, Y. Kashiwaya, C. A. Keon, N. Tsuchiya, M. T. King, G. K. Radda, B. Chance, K. Clarke, and R. L. Veech, 1995, *FASEB J.*, **9**:651–658.

Insulin-sensitive phospholipid signaling systems and glucose transport: An update. R. V. Farese, 1996, *Proc. Soc. Exp. Biol. NY*, **213**:1–12.

Protein–tyrosine phosphatases, R. L. Stone and J. E. Dixon, 1994, *J. Biol. Chem.*, **269**:31323– 31326.

Molecular mechanisms of action of steroid/thyroid receptor superfamily members. M. J. Tsai and B. W. O'Malley, 1994, *Ann. Rev. Biochem.*, **63**:451–486.

REVIEW QUESTIONS

1. The action of hormones on specific tissues and not other is due to
   A. Enzyme content of the cell
   B. Metabolite levels in the cell
   C. Mitochondrial content of the cell
   D. Specific receptors on or in the cell
   E. The endocrine gland releasing the hormone

2. Beta receptors have action through
   A. A G protein and phosphodiesterase
   B. cAMP and phosphoinositides
   C. A G protein and cAMP
   D. An R protein and cAMP
   E. An R protein and phosphoinositides

3. Various cells respond differently to a second given message (as increased cAMP) because they have
   A. Different receptors
   B. Different enzymatic composition
   C. Different levels of phosphodiesterase
   D. Different nuclei
   E. Different membrane lipids

4. The number of potential intracellular messages from an α_1 agonist is

A. Two (both directly formed)
B. Three (two directly formed; one indirectly formed)
C. Three (all directly formed)
D. Two (one directly formed; one indirectly formed)
E. Two (both indirectly formed)

5. The hormone indicating low blood glucose concentration is
   A. Glucagon
   B. Epinephrine
   C. Insulin
   D. Thyroxine
   E. Progesterone

6. Which of these processes is not stimulated by glucagon?
   A. Gluconeogenesis
   B. Lipolysis
   C. Glycogenolysis
   D. Ketogenesis
   E. Lipogenesis

ANSWERS TO REVIEW QUESTIONS

1. ***d***. Since most hormones are carried in the blood, the discrimination among tissues is due to specific receptors and the sensitivity of various receptors toward specific hormones.

2. ***c***. Transduction of the hormone signal at beta receptors is through the G protein to activate adenylcyclase. The activated adenylcylase increases the cellular cAMP.

3. ***b***. Although the cAMP-dependent protein kinase phosphorylates many proteins, which affects the activity of many enzymes, the cell must contain these enzymes for a particular function to be altered.

4. ***b***. The two compounds formed by the action of phospholipase C are inositol trisphosphate (IP_3) and diacylglyceride (DAG). The hydrolysis of DAG produces a precursor of prostaglandins.

5. ***a***. In response to hypoglycemia, there is an increased release of glucagon from the α-cells of the pancreas. This hormone causes increased glycogenolysis and gluconeogenesis in the liver and increased lipolysis in the adipose.

6. ***e***. Lipogenesis. The other four processes are activated by cAMP arising from glucagon action.

CHAPTER

17

AMINO-ACID METABOLISM: UREA, GLUTAMATE, GLUTAMINE, AND ASPARTATE

INTRODUCTION

In this chapter we examine maintenance of nitrogen balance. This includes pathways and controls of nitrogen excretion as urea or ammonia. The method the body has for transferring nitrogen from one carbon skeleton to another is described. The metabolism of several amino acids, their physiological role, and their metabolic products are considered. Specific tissue metabolism and tissue–tissue interactions related to several amino acids are discussed.

Nitrogen has a major role in the metabolic and structural aspects of all species. Nitrogen can be found, not only in amino acids, but in proteins, nucleic acids, nucleotides, hormones, and many other critical biological compounds. In the animal system, including humans, metabolic nitrogen is obtained from the diet, unlike some microorganisms which can fix nitrogen. During growth, more nitrogen is taken in the diet than is excreted via feces, urine, loss of skin, loss of hair, and so on—a positive nitrogen balance. In cases of starvation or protein deficiency, there is more nitrogen excreted than ingested—a negative nitrogen balance. Most adults are in what is referred to as neutral nitrogen balance. That is, nitrogen intake and excretion are in balance. For most adults, if there is a major gain in weight, it is in terms of lipid gains, and not gain in protein, unless they are undergoing a specific exercise regime, or an equivalent training to increase protein. However, the amount of increased nitrogen usually is not as great as the potential increase in body lipid. Thus, in adult animals, the equivalent of all the nitrogen taken in must be excreted from the body. Some of this occurs during sloughing of skin and via the feces; however, most of the excess nitrogen is excreted via the urine.

NITROGEN EXCRETION

There are three major ways that nitrogen is excreted, with the primary form being dependent on either species or metabolic conditions. The easiest and least expensive method in terms of ATP requirement is as ammonia. However, ammonia per se is extremely toxic and permeable across many membranes. This poses few difficulties with marine animals, where the ammonia can be diluted in the surrounding water and utilized by other equatic life forms such as plants. However, in mammals, including people, excess ammonia is toxic. Except in the case of acidic urine, such as is seen in starvation and diabetes, this method of excretion is not widely utilized in people. The movement of ammonia from the tubules, back into the blood, is too great if there is not a large pH difference between the urine and the blood. Most of the ammonia at pH 7.4 is in the form of ammonium ion, at pH 7.4, the ratio of NH_4^+ to NH_3 is approximately 400:1. The permeability of the free ammonia with rapid equilibrium would make it impossible to concentrate the nitrogen in the urine, except as salts at lower pHs. Therefore, to remove the nitrogen as ammonia would require copious amounts of urine, amounting to several hundred liters per day. This is obviously impractical for land-dwelling animals; thus, an alternative excretion product of nitrogen is required. This is urea.

UREA

There are several advantages to utilizing urea as an end product of nitrogen metabolism. First, it is relatively nontoxic and extremely soluble, and its formation leads to an acidifying process ($2NH_4 + CO_2 \rightarrow$ urea $+ H_2O + 2H^+$). If ammonia were excreted as the ammonia ion, even without toxic effects, there would be an alkalization of the body fluids due to the removal of hydrogen ion by the ammonia as ammonium ion. This is not a difficulty under acidoic conditions, where there is an excess of hydrogen ions and decreased body pH. In this case, there are advantages to excreting ammonium ion. A third major way of excreting nitrogen, which occurs to a greater degree in avian and reptilian species, is as uric acid. This is more expensive metabolically and energetically than forming urea, but is relatively nontoxic and insoluble and may have advantages for egg-laying animals. It should be noted that people also excrete nitrogen as uric acid, not as an end product of nitrogen metabolism, but as an end product of purine metabolism. Excess of uric acid can lead to a disease known as gout, which is discussed in the chapter on purine metabolism (Chapter 20).

AMMONIA GENERATION

The major emphasis in this chapter related to excretion is on ammonia and urea, particularly on the control of the latter as the normal excretion product. ***There are two major nitrogenous precursors of urea: ammonia and aspartate.*** Before delving into the mechanism and control of urea synthesis, it would be wise to look at the different ways that these precursors can arise and the important tissues concerned with urea synthesis. In people, the predominant synthesis of urea occurs in the liver, with very small amounts of urea formed from arginine in other tissues. The complete synthesis of urea, from amino acids and other nitrogen sources, occurs via a process known as the urea cycle.

There are two important enzymes for generating ammonia, one of the precursors of urea synthesis. The first is glutamate dehydrogenase, a mitochondrial enzyme that occurs in a number of tissues and is particularly active in the liver mitochondria. This enzyme catalyzes the reversible reaction of glutamate + $NAD(P)^+$ to α-ketoglutarate + NAD(P)H + H^+ + ammonia. The equilibrium of this enzyme is strongly toward glutamate synthesis. However, the predominant flux in the mammalian system is toward ammonia production. One way that flux through a reaction against the equilibrium can be achieved is by the removal of end products. An end product of this reaction is α-ketoglutarate, which can be metabolized in the citric-acid cycle in the mitochondrion. Another product is NADH which, via the electron-transport system in the mitochondrion, can be reoxidized with the formation of water and high-energy phosphates. The third product, which is probably the most crucial to remove for pulling the reaction from glutamate toward ammonia, is ammonia itself. ***Ammonia can be activated in the liver mitochondrion to begin the formation of urea.*** Thus, the advantage of glutamate dehydrogenase occurring in the mitochondrion is apparent.

In other tissues, when there is excess ammonia, glutamate dehydrogenase will catalyze the reductive amination of α-ketoglutarate to form glutamate and potentially deplete α-ketoglutarate. This has been suggested as one of the possible mechanisms of ammonia toxicity, particularly if it occurs in the central nervous system, where maintaining citric-acid-cycle intermediates and continuing the flux through the citric-acid cycle is critical to maintain normal function and energy levels (see Chapter 11).

A second major method of generating ammonia in the liver and kidney is via glutaminase, which catalyzes the hydrolysis of the amino-acid glutamine to form glutamate and ammonia. Glutamine is one of the amino acids that is in very high concentration in the circulation and is a carrier of ammonia in a nontoxic form from other tissues to the liver for urea synthesis and to the kidney for ammonia excretion. This enzyme is unidirectional, is associated with the mitochondrion, and is activated by inorganic phosphate. Glutaminase is adaptive and can increase or decrease in amount in the liver and/or kidney, leading to preferential synthesis of urea or ammonia excretion respectively. In addition to these two major producers of ammonia, a number of amino acids directly produce ammonia during their degradation. Most are degraded selectively by the liver via unidirectional reactions. These amino acids include threonine, histidine, glycine, asparagine, and others. Since most of the amino acids, which are the major source of nitrogen for excretion, neither produce ammonia directly nor are glutamate or glutamine, there needs to be other methods of transferring the nitrogen from these acids to a potential ammonia producer.

The process of transferring nitrogen from one carbon skeleton to another is known as transamination. One of the acceptors of the nitrogen in this reversible reaction is usually α-ketoglutarate, and when transferring nitrogen to other carbon skeletons, the donor is usually glutamate via the following reaction: α-ketoglutarate + amino acid ↔ glutamic acid + α-keto acid (i.e., the α-keto acid corresponding to the original amino acid) (Fig. 17.1). All amino acids except threonine, proline, hydroxyproline, and lysine will transaminate, although this need not be a normal first step in the catabolism of all the other amino acids. Thus, one can affect the transfer of nitrogen from one amino acid to another and effectively convert most of the nitrogen to glutamate as a potential precursor of ammonia and urea synthesis. Two transaminations are widely spread and are of great importance. The first is aspartate aminotransferase, which reversibly catalyzes glutamate + oxaloacetate to aspartate + α-ketoglutarate. This is active in most tissues and present in both the mitochondrion and cytosol. It is by far the most active of any of the aminotransferases and is very active in the liver. This enzyme is important not only in transferring nitrogen between glutamate and oxaloacetate for urea synthesis, but is also an important participant in the malate shuttle. This shuttle is important wherever oxidative metabolism of carbohydrate proceeds (Chapter 11). The second active transamination is catalyzed by alanine aminotransferase, an enzyme widely distributed, but particularly active in the liver, primarily in the cytosol. This will reversibly catalyze the transamination of glutamate and pyruvate to form alanine and α-ketoglutarate. The importance of this aminotransferase is that some of the amino acids that have primarily extra hepatic aminotransferases appear to transfer nitrogen to the liver primarily in the form of alanine, as part of a process known as the alanine cycle. ***In the liver, after α-ketoglutarate accepts nitrogen from many amino acids to form glutamate, the glutamate can either form ammonia or transaminate to form aspartate, both precursors of urea.*** Thus, it can be seen the glutamate has an important central role in nitrogen metabolism.

The first step in the urea cycle is catalyzed by carbamoyl phosphate synthetase (CPSI). This requires ammonia, CO_2, and 2 ATP. The product is carbamoyl phosphate, 2 ADP, and inorganic phosphate. The requirement for 2 ATP assures the physiological irreversibility of this reaction and helps drive ammonia toward carbamoyl phosphate (Fig. 17.2). *N*-acetylglutamate is a required allosteric activator for CPSI. The next enzyme, ornithine transcarbamoylase, is also mitochondrial and catalyzes the reaction of ornithine with carbamoyl phosphate to form citrulline + inorganic phosphate. Citrulline then exits the mitochondria, where it is a substrate for a reaction catalyzed by argininosuccinate synthetase. This reaction combines aspartic acid + citrulline to form argininosuccinate and requires ATP, which is broken down to AMP + pyrophosphate. The hydrolysis of the pyrophosphate by the active pyrophosphatase in the cell assures the physiological irreversibility of this reaction. Argininosuccinate lyase catalyzes a hydrolytic cleavage of the argininosuccinate to form fumarate + arginine. The fumarate thus formed can react with fumarase to form malate, via malate dehydrogenase to form oxaloacetate, which via aspartate aminotransferase can reform aspartate. This is a potential subcycle of the major urea cycle. The major urea cycle now has formed the compound arginine. This can be hydrolytically cleaved, catalyzed by the enzyme arginase, to form urea and ornithine. The urea can then enter the bloodstream and be excreted via the kidney, and the ornithine can reenter the mitochondrion to restart the urea cycle.

The major control of the urea cycle, other than the availability of nitrogen, ***is exerted at the first potentially irreversible step of urea synthesis, that is, carbamoyl phosphate synthetase***. Carbamoyl phosphate

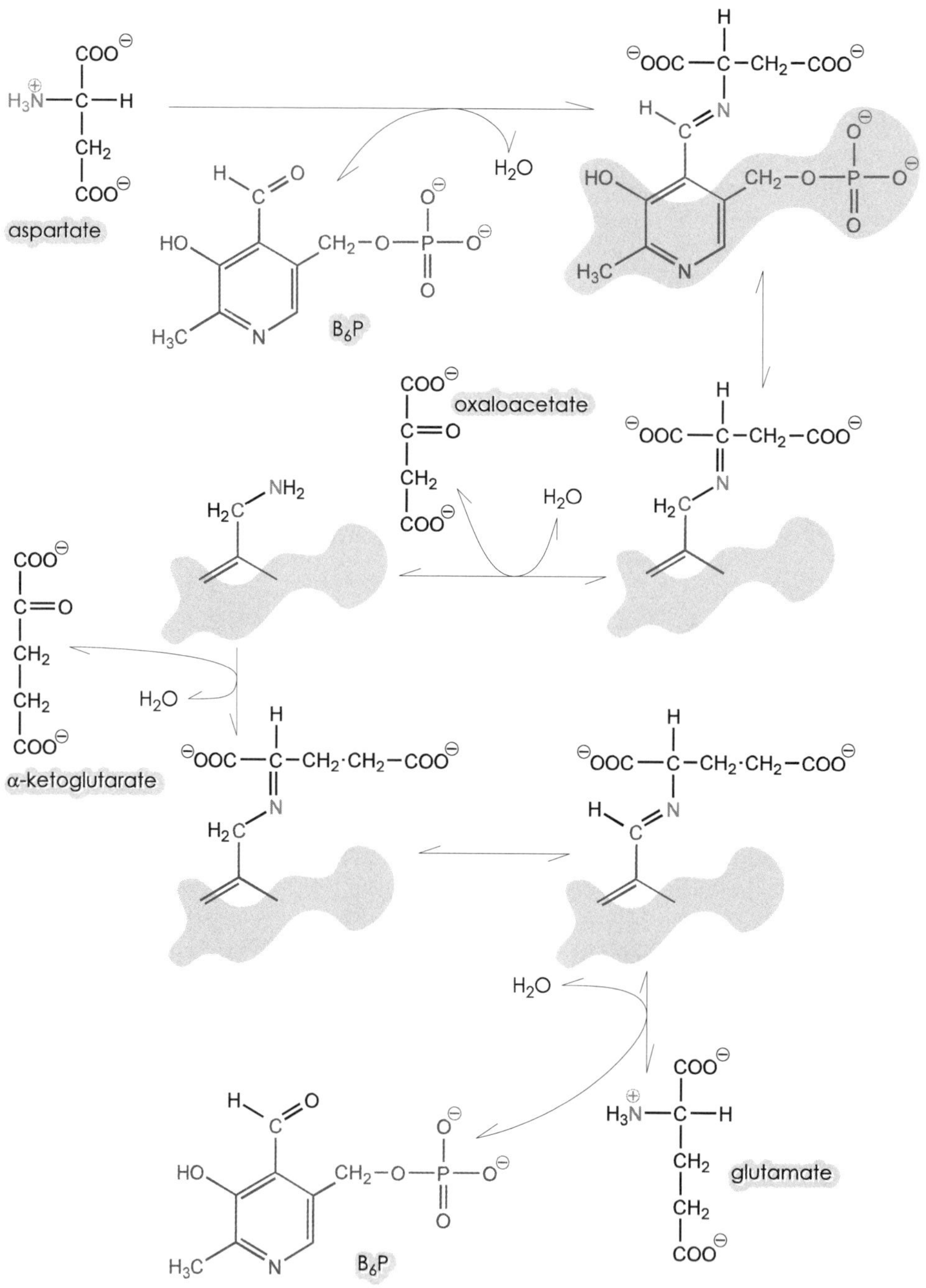

Figure 17.1.

Transamination. The mechanism and intermediates of transamination are shown above. Transamination is a reversible process which transfers amino groups from one carbon skeleton to another carbon skeleton. The nitrogen being transaminated is shown in purple.

synthetase (CPSI) has an absolute requirement for the allosteric effector *N*-acetylglutamate. This activator is formed by the reaction of acetyl CoA + glutamic acid to yield *N*-acetylglutamate. It is quite clear that the level of *N*-acetylglutamate is critical in determining carbamoyl phosphate synthetase activity and, if nonexcessive levels of ammonia are available, it is quite likely that this will be a rate-limiting system. Physiologically, this may be advantageous, because the liver is likely to have a higher level of *N*-acetylglutamate with higher glutamate concentrations. This is likely to occur when there are excess amino acids which, via aminotransferases, push the equilibrium of α-ketoglutarate and glutamate toward glutamate formation, with the potential formation and accumulation of *N*-acetylglutamate. This would lead to an increased activity of carbamoyl phosphate synthetase during a time of excess nitrogen, which would be appropriate for the excretion of excess nitrogen.

After the removal of nitrogen from amino acids,

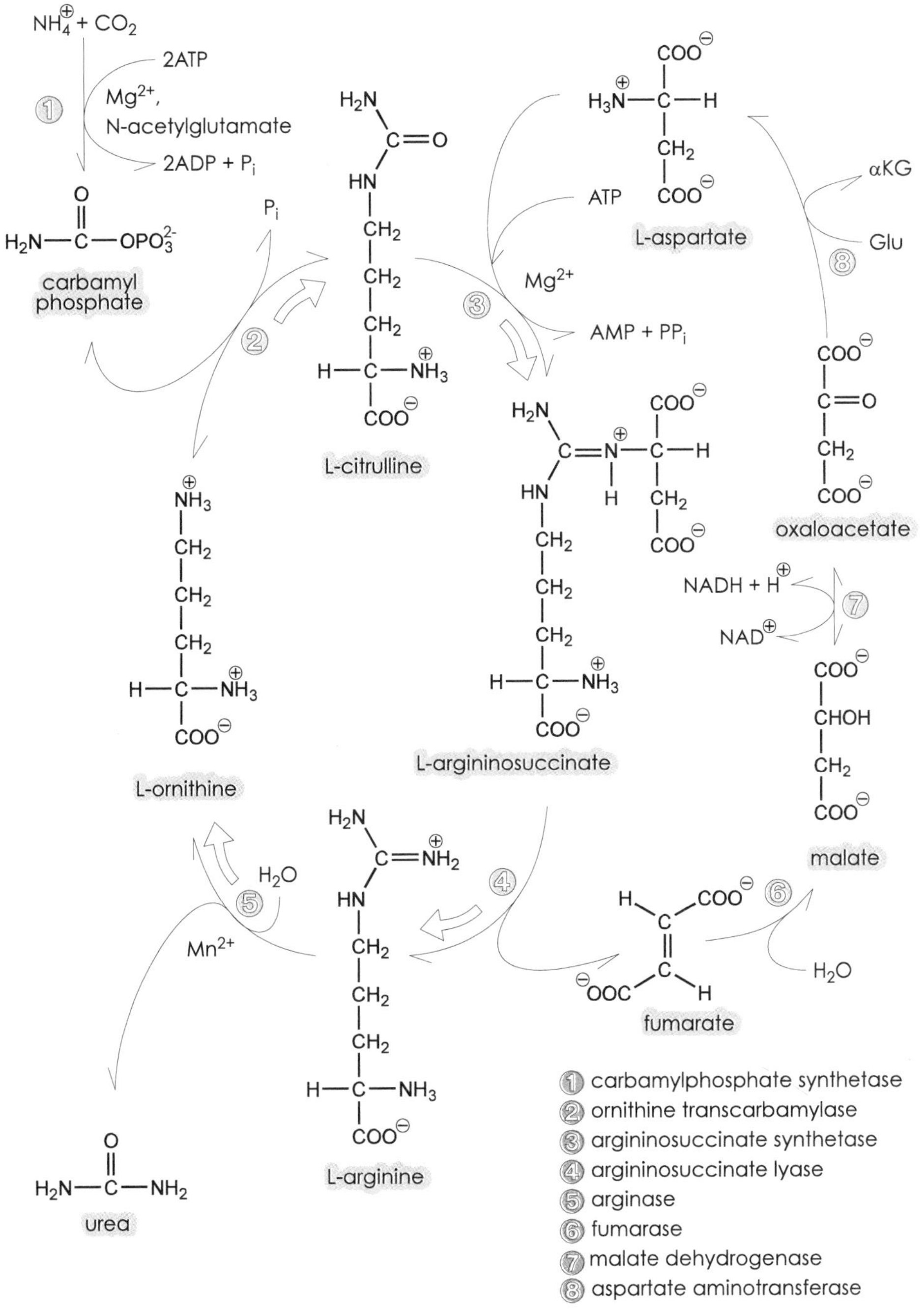

1. carbamylphosphate synthetase
2. ornithine transcarbamylase
3. argininosuccinate synthetase
4. argininosuccinate lyase
5. arginase
6. fumarase
7. malate dehydrogenase
8. aspartate aminotransferase

Figure 17.2.

Urea cycle. The steps and structures of the urea cycle, as well as the ancillary cycle for regenerating aspartate, is shown above.

the carbon portion of the molecule can be used for other metabolic purposes such as energy, gluconeogenesis, or even lipogenesis. During periods of low availability of nitrogen, where other keto acids may be in excess, there will be a tendency to decrease the level of glutamate in favor of α-ketoglutarate, thus decreasing the availability for *N*-acetylglutamate formation. When nitrogen availability is low, the lower *N*-acetylglutamate would decrease the activity of CPSI, thus less nitrogen would be converted to urea. This would facilitate a potential preservation of nitrogen for the synthesis of nonessential amino acids. Indeed, when protein intake is high, there is a marked increase in urea synthesis and excretion; on a low-protein diet, there is a marked decrease of urea synthesized and excreted.

The activity of ornithine transcarbamoylase, the next enzyme in the urea cycle, appears to be extremely high. In the case of excess ammonia, almost all of the ornithine present in the liver is converted to citrulline, indicating a sufficient activity of carbamoyl phosphate synthetase and ornithine transcarbamoylase when ammonia levels are elevated. The next step in the urea cycle, argininosuccinate synthetase, appears to be the enzyme with the lowest activity in the urea cycle, which would be consistent with the fact that citrulline accumulates during high levels of ammonia. In experimental conditions with perfused liver or isolated hepatocytes, it has been shown that when ammonia, ornithine, and an energy source such as lactate is present, there is a marked accumulation of citrulline and the rate of urea synthesis is very similar to the activity of argininosuccinate synthetase. It should be noted that although this enzyme has the lowest activity in the urea cycle, it is not under allosteric control and probably is not rate limiting or flux controlling under normal conditions since most individuals do not have very high levels of ammonia and ornithine circulating.

Arginase, which catalyzes the hydrolysis of arginine to ornithine and urea, is extremely active in the liver; however, it has a very high K_m value, thus allowing hepatic arginine to be used preferentially for protein synthesis rather than urea synthesis. However, the high activity of this enzyme does deplete arginine to levels in the liver insufficient to supply the body with net arginine synthesis for extra hepatic protein synthesis. In fact, there is sufficient arginase activity in the liver to produce an individual's body weight in urea each day. The only aspect that prevents this is the limited activities of the other enzymes in the urea cycle and the high K_m values for arginine. The enzymes of the urea cycle are adaptive, that is, they increase in amounts in response to high-protein diets, cortisol treatment, or diabetes. These increases occur over long periods (hours to days) in contrast with the acute changes in *N*-acetylglutamate concentrations, which affect the activity, but not amount, of carbamoyl phosphate synthetase.

Metabolic Diseases of the Urea Cycle

A number of inherited diseases are associated with the urea cycle. The mutations result in changes in either V_m or K_m as defective proteins are produced. These include disruptions of *N*-acetylglutamate synthase, carbamoyl phosphate synthetase, ornithine transcarbamoylase (the most prevalent of the urea cycle deficiencies), argininosuccinate synthetase, argininosuccinate lyase, and arginase. In these diseases, when applicable, treatments are low-protein diets, to put less strain on urea cycle flux and, when appropriate, addition of amino acids as required, such as ornithine and/or arginine.

ARGININE

Arginine can be synthesized de novo in the human, but not rapidly enough for maximum growth. Therefore, it is considered an essential dietary amino acid for growing children, but not adults. One might think that as an intermediate in the urea cycle, which occurs primarily in the liver, arginine for use in body protein synthesis would be synthesized in the liver. However, owing to the very high arginase activity in the liver, little arginine escapes from the liver to the bloodstream. The urea cycle is a cyclic process and does not form net arginine. ***Most of the arginine formed in the liver goes to urea or liver protein*** (Fig. 17.3). The protein synthesis has priority, with a low K_m for amino-acid activation and a high K_m for arginase. ***The formation of arginine for body-protein synthesis is a cooperative venture of two tissues: intestine and kidney.*** The intestine converts glutamine or glutamate to citrulline; the intestine contains CPSI, but almost no argininosuccinate synthetase. The citrulline formed in the intestine flows through the blood to the kidney, where it is converted to arginine and released into the blood stream for use by other tissues. The kidney has a much higher ratio of argininosuccinate synthetase to arginase than the liver, which is useful in net arginine production and accumulation. This role of the kidney in arginine synthesis explains why individuals with nonfunctional kidneys have high levels of circulating citrulline. In treating individuals with kidney failure by dialysis, arginine must be included in the amino-acid mix, because these individuals can no longer synthesize sufficient arginine for the body's need.

The brain and nerve cells can synthesize arginine efficiently from circulating citrulline. In these tissues, nitric oxide synthase can convert arginine to citrulline and nitric oxide (NO), an important neurotransmitter, which is thought to play a role in short-term memory, as well as many other functions. Interestingly, these cells have argininosuccinate synthetase and argininosuccinate lyase, but no arginase.

The major pathway for catabolism of arginine is arginase to form ornithine, followed by ornithine transamination and oxidation of the resulting glutamic semialdehyde to form glutamate. For the most part, this catabolism occurs in the liver. The resulting glutamate can be converted to glucose; thus arginine, is glucogenic.

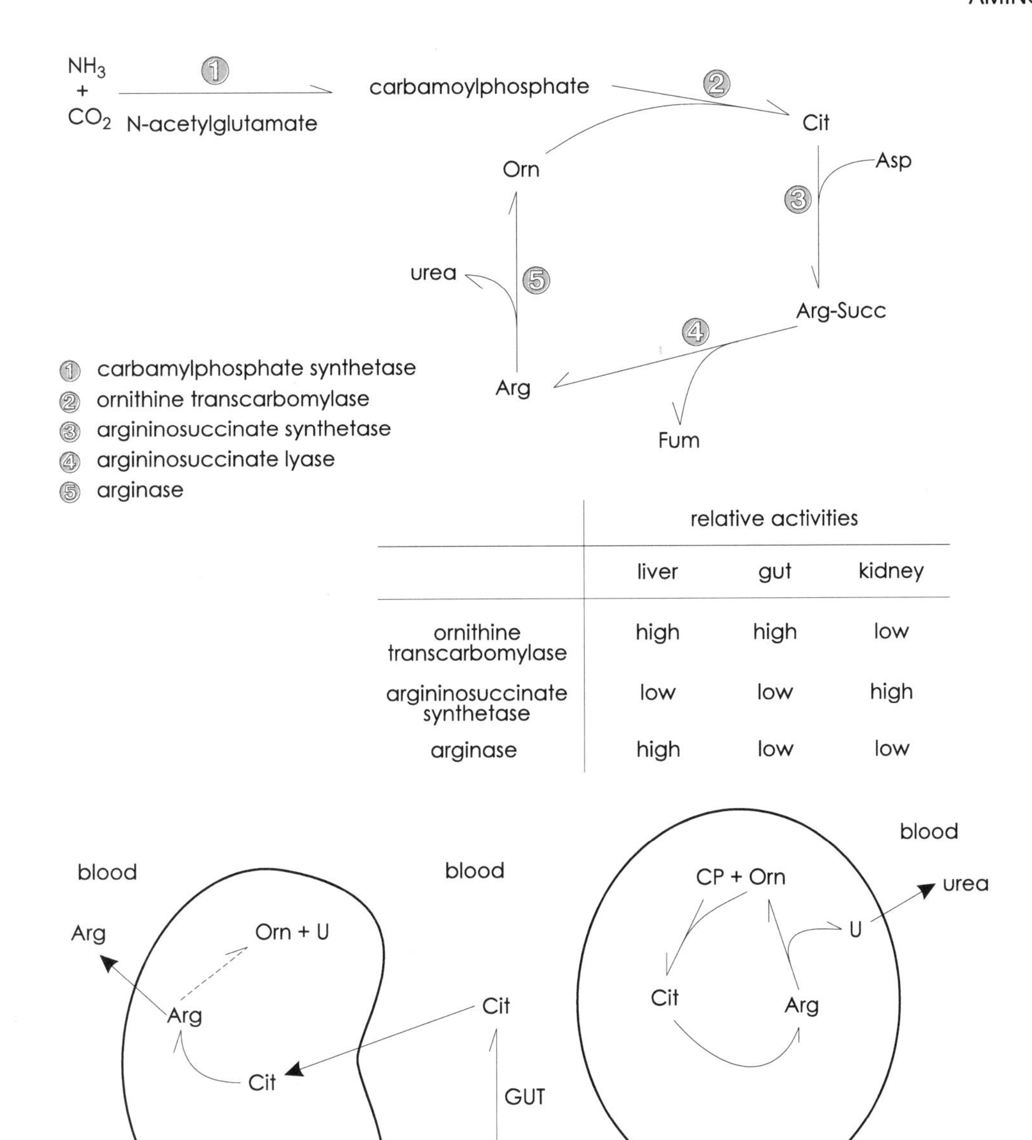

	relative activities		
	liver	gut	kidney
ornithine transcarbomylase	high	high	low
argininosuccinate synthetase	low	low	high
arginase	high	low	low

Figure 17.3.
Urea cycle and arginine synthesis. The general aspect and control of the urea synthesis and the role of the intestine and kidney in arginine synthesis are shown above.

GLUTAMIC ACID (GLUTAMATE)

Some of the major aspects of each of the individual amino acids will be examined. Before starting this section, it should be realized that ***the most important single function of any amino acid is that of protein synthesis.***

Glutamic acid and glutamate are synonymous. In the case of the two acidic amino acids (glutamate and aspartate), this depends on the ionization state. In the body, they exist predominantly as the fully charged zwitterions and therefore glutamate is an appropriate name for this amino acid. Glutamate, as indicated earlier, is at the center of amino acid catabolism. It can undergo transamination with many α-keto acids to form the amino acids and α-ketoglutarate (Fig. 17.4). In a reverse reaction, α-ketoglutarate can transaminate with many amino acids to produce the corresponding α-keto acid and glutamate. Glutamate, as previously indicated, is an ideal center point for nitrogen metabolism because glutamate can form ammonia, NADH, and α-ketoglutarate by the action of glutamate dehydrogenase. The ammonia can be used for urea synthesis, excretion as NH_4^+, or formation of glutamine and asparagine. The keto acid formed from glutamate is α-ketoglutarate, an intermediate in the citric-acid cycle, which could help continue the flow of the

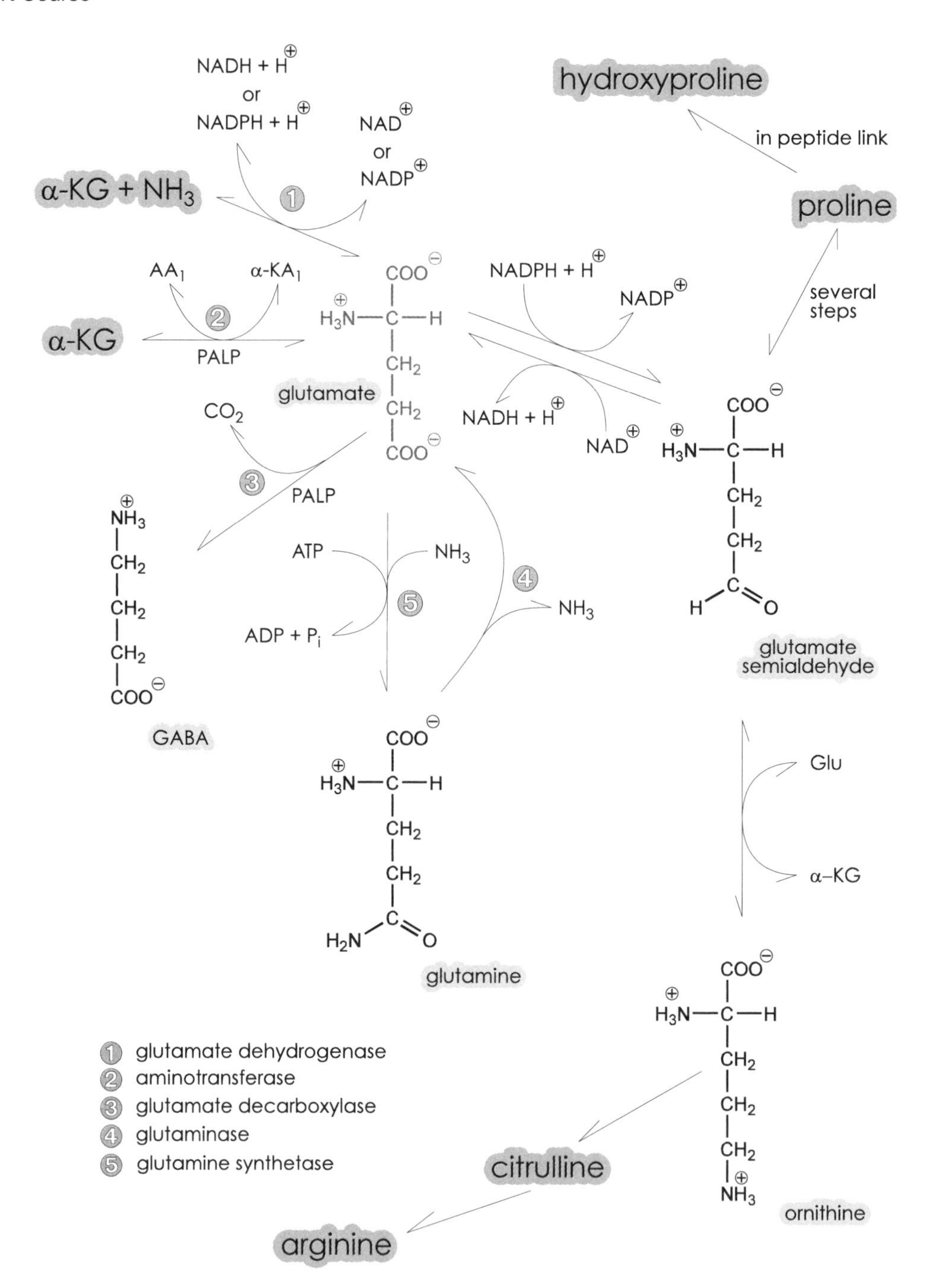

Figure 17.4.

Glutamate metabolism. The metabolism of glutamate, including fates and sources of glutamate, is shown.

citric-acid cycle. In liver and kidney, α-ketoglutarate can be converted to glucose. Any amino acids that yield net glutamate in metabolism are also potentially gluconeogenic in nature. Glutamate is also the precursor of glutamine, which is considered later in this section, via glutamine synthetase, which catalyzes ammonia, ATP, and glutamate to glutamine, ADP, and inorganic phosphate in a physiologically irreversible reaction. Glutamate is also the product of the enzyme glutaminase, which catalyzes the hydrolysis of glutamine to ammonia and glutamate. Glutamine synthesis and degradation occur by two separate irreversible enzymatic processes. Thus to control the levels and fluxes, toward and away from glutamine, having two separate enzymes are used, each of which can be controlled separately (i.e., similar to glycolysis and gluconeogenesis with PFK and fructose-1,6-bisphosphatase as well as hexokinase versus glucose-6-phosphatase).

ORNITHINE

Glutamate can be the precursor of ornithine as well as the product of ornithine catabolism. In this case, not only do separate enzymes work on this process, but they do so in two diverse tissues. The degradation of ornithine occurs primarily in the liver, where ornithine aminotransferase catalyzes ornithine plus α-ketoglutarate to form glutamate and glutamate semialdehyde. This involves a transamination of the distal, not the α-amino group, of ornithine. Glutamate semialdehyde oxidation is catalyzed by glutamate semialdehyde dehydrogenase, forming glutamate and NADH from NAD^+ and glutamate semialdehyde. This latter reaction is effectively unidirectional in the liver owing to the redox state of the NAD^+/NADH. Thus, ornithine is primarily catabolized in the liver and there is little net synthesis of ornithine other than from arginine. Ornithine can be formed in the liver from arginine by the action of arginase.

In contrast, in the intestine, glutamate can be reduced to glutamate semialdehyde using NADPH, which is favorable to reductive biosynthesis because of the reduced nature of the $NADP^+$/NADPH nucleotide pair as indicated previously (Chapter 12). The glutamate semialdehyde can transaminate with glutamate to form α-ketoglutarate and ornithine in the intestine. This ornithine can proceed further in the intestine, undergoing a reaction with carbamoyl phosphate to form citrulline. The resulting citrulline can then move to the general circulation and can be converted to arginine in the kidney.

The glutamic semialdehyde formed in the intestine can also cyclize and be reduced to form proline. Therefore, proline is not an essential amino acid in people. Any amino acid that can form glutamate can be converted to glucose. Among the amino acids that can produce glutamate, which will be examined in more detail when they are discussed individually, are glutamine, arginine via ornithine formation, proline, and histidine.

Glutamate in the brain and central nervous system can be converted to gamma amino butyric acid (GABA), which is catalyzed by glutamate decarboxylase. GABA is a neurotransmitter inhibitor compound that can be metabolized by transamination followed by oxidation.

GLUTAMINE

Glutamine, a metabolically important amino acid, occurs naturally in proteins, can be synthesized in the body, and is one of the two amino acids found highest in concentration in plasma. Glutamine can be formed from the reaction of glutamate, ammonia, and ATP catalyzed by glutamine synthetase; the other products are ADP and inorganic phosphate. This enzyme is found in high activity in the liver and moderate activity in muscle, the central nervous system, and most other tissues. Glutamine synthesis is one mechanism of detoxifying ammonia, carrying potential ammonia through the bloodstream in a nontoxic form from one tissue to another. For example, glutamine synthesized in the muscle can be transported through the blood to the liver where ammonia, for urea synthesis, can be produced by the action of glutaminase. If the kidney is the destination of the glutamine, ammonia can be released by glutaminase and excreted into the urine as the ammonium ion. In addition to being converted to glutamate, glutamine has many other important metabolic functions. These include first and foremost the synthesis of protein, as well as the synthesis of the tripeptide glutathione, purines, and pyrimidines (Chapter 20). Glutamine is a glucogenic amino acid. Glutamine is a nonessential amino acid, meaning that it is not required exogenously in a diet, since it can be synthesized from simple precursors, such as ammonia and glucose.

ASPARTATE

Aspartate can be transaminated to form oxaloacetate, an intermediate of the citric-acid cycle. As with most transaminations, this is a reversible reaction, and aspartate can also be synthesized by a transamination reaction with glutamate and oxaloacetate to form aspartate and α-ketoglutarate. Therefore, aspartate is a nonessential amino acid. The aminotransferase with aspartate and α-ketoglutarate is particularly active in most tissues and occurs both in the mitochondria and the cytosol. The importance of this reaction is greater than simply forming the oxaloacetate or aspartate. Aspartate aminotransferase is an important reaction in the malate shuttle (see Chapter 11) wherein, reducing power can be transferred from the cytosol to the mitochondrion. Aspartate also plays a role in purine and pyrimidine synthesis and is particularly important in pyrimidine synthesis, where it donates both carbon and nitrogen for the formation of these compounds (Chapter 20).

Aspartate is an important nitrogen source for urea synthesis. The transamination with glutamate is important in achieving urea synthesis from nitrogenous precursors. A large portion of the nitrogen is funneled into glutamate, which can then transaminate with oxaloacetate for urea formation, because half of the nitrogen in urea must come from aspartate. Aspartate can also be aminated by asparagine synthetase using aspartate, glutamine, and ATP to yield glutamate, AMP, pyrophosphate, and asparagine (Asp + Gln + ATP → Asn + Glu + AMP + PPi). Asparagine is an important constituent of proteins. The formation of asparagine is considerably slower than that of glutamine. Asparagine can be destroyed by the enzyme asparaginase, which is contained in numerous tissues. If asparaginase is added to the blood to destroy circulating asparagine, an internal asparagine deficiency may result. The addition of asparaginase to blood has been attempted as one mode of chemotherapy. The lack of sufficient asparagine may slow the rapid growth of many tumors, which depend on exogenous asparagine for rapid growth, such as in certain leukemias. Aspartate can also be decarboxylated to form β-alanine, which is found in coenzymes such as coenzyme A as part of pantothenic acid. However, this is normally accomplished in microorganisms because people need pantothenic acid as such and cannot add the β-alanine portion to the pantetheine. Thus, under most normal conditions, both aspartate and asparagine are not essential amino acids and are not required as such in the diet.

SUMMARY

1. The major excretion of nitrogen in the normal human is as urea, formed from ammonia and aspartate in the liver.

2. Ammonia can be generated from glutamate, via glutamate dehydrogenase, or from glutamine, via glutaminase.

3. Nitrogen as α-amino nitrogen can be moved between carbon skeletons by a process known as transamination, requiring pyridoxal phosphate. Two important active enzymes for transamination are aspartate aminotransferase and alanine aminotransferase. In most transaminations, glutamate is either the donor of the amino group or the product of the transamination.

4. Urea is synthesized in the liver in the urea cycle. The first step is formation of carbamoyl phosphate from ammonia, CO_2, and ATP. This is followed by a number of other steps, including formation of citrulline, argininosuccinate, and arginine, which is split to urea plus ornithine. The second nitrogen of urea is donated by aspartate in the formation of argininosuccinate.

5. The control of urea synthesis is availability of nitrogen and activity of carbamoyl phosphate synthetase (CPSI). CPSI is activated by *N*-acetylglutamate as a required allosteric effector.

6. Glutamate is a central amino acid in general amino-acid metabolism. It plays a major role in transamination, ammonia production, formation of ornithine, proline, glutamine, and γ-amino butyric acid (GABA).

7. Glutamine is a transporter of ammonia in a nontoxic form from peripheral tissues to the liver, kidney, and intestine. It is formed from glutamate and ammonia by glutamine synthetase and can be degraded back to glutamate and ammonia catalyzed by glutaminase. If glutaminase acts in the liver, the most probable fate of the ammonia is urea. If the glutaminase acts in the kidney, the most probable fate of the ammonia is as urinary ammonium. In cases of acidosis, the amount of nitrogen in the urine as ammonia can far exceed that in urea.

8. Aspartate can be formed from oxaloacetate and glutamate, via transamination. This is important in urea synthesis, the malate shuttle, purine, and pyrimidine synthesis.

9. Aspartate can be converted to asparagine, an important component of proteins. Asparagine can be destroyed by asparaginase.

REFERENCES

Regulation of enzymes of urea and arginine synthesis, S. M. Morris, 1992, *Ann. Rev. Nutr.* **12**:81–101.

Regulation of carbamoyl-phosphate synthase (ammonia) in liver in relation to urea cycle activity, A. J. Meijer, 1979, *Trends Biochem. Sci*, **4**:83–86.

Regulation of hepatic ammonia metabolism: The intercellular glutamine cycle, D. Häussinger, 1986, *Adv. Enzyme. Reg.*, 159–180.

Glutamine: A key substrate for the splanchnic bed, W. W. Souba, 1991, *Ann. Rev. Nutr.*, **11**:285–308.

Regulation of biosynthesis of nitric oxide, C. Nathan and Q. Xie, 1994, *J. Biol. Chem.*, **269**:13725–13728.

Urea cycle enzymes, S. W. Brusilow and A. L. Horwich, in *The Metabolic Basis of Inherited Disease*, Vol. 1, 6th ed., C. R. Scriver, A. L. Beaudet, W. S. Sly, and D. Valle, Eds., 1989, McGraw-Hill, New York, pp. 629–6630.

REVIEW QUESTIONS

1. Ammonia (NH_3) is not produced by which of the following reactions?
   A. Glutamate dehydrogenase
   B. Alanine amino transferase
   C. Threonine dehydratase
   D. Glutaminase
   E. Histidase

2. A major allosteric effector of the urea cycle is
   A. Glutamine
   B. cAMP
   C. Carbamoylphosphate
   D. *N*-Acetylglutamate
   E. Arginine

3. Glutamate cannot be formed totally in net amounts from
   A. α-ketoglutarate
   B. Oleate
   C. Histidine
   D. Glutamine
   E. Arginine

4. The major tissues involved in net arginine synthesis from glutamate for body protein synthesis are
   A. Liver and brain
   B. Intestine and liver
   C. Intestine and kidney
   D. Liver and kidney
   E. Kidney and brain

ANSWERS TO REVIEW QUESTIONS

1. **b.** This reaction transfers α-amino nitrogen from one carbon skeleton to another without the release of any free ammonia.

2. ***d.*** *N*-Acetylglutamate is an essential activator of carbamoylphosphate synthetase, the first reaction of the urea cycle.

3. ***b.*** Oleate can produce acetyl CoA, which can form glutamate, but only starting with oxalacetate. If oleate could form net amounts of glutamate, a potential precursor of glucose, it would be glucogenic.

4. ***c.*** Intestine converts glutamate to citrulline, which is released into the blood. The circulating citrulline is converted to arginine by the kidney and released into the blood.

CHAPTER

18

METABOLISM OF OTHER ALIPHATIC AMINO ACIDS

INTRODUCTION

In this chapter we examine the metabolism of aliphatic amino acids, including glycine. All of the amino acids, except lysine and leucine, are at least partially glucogenic and can provide carbon atoms for the formation of one half of a glucose molecule. Some of these amino acids are simple in nature, containing only hydrocarbon side chains. Others in addition contain sulfur, hydroxyl groups, or distal amino groups. All but lysine and threonine can transaminate. We will examine both the synthesis of dietary nonessential amino acids and the catabolism to common metabolic intermediates. Many of these amino acids are precursors of special metabolic products such as carnitine, choline, creatinine, and taurine. Interrelationships of various amino acids, as well as intertissue relationships, are discussed.

ALANINE

Alanine is the simplest L-amino acid found in protein. It has a simple metabolism, but complex physiological roles and functions. Alanine can transaminate reversibly with α-ketoglutarate, forming pyruvate and glutamate. This transamination occurs in many tissues, including liver and muscle. This is the only metabolic fate of pyruvate, other than protein synthesis. Therefore, alanine is glucogenic and is not required in the diet.

Several amino acids, other than alanine, are transaminated in the periphery to form glutamate, with the carbon skeleton being metabolized. The problem then exists of how to move the nitrogen from the periphery, in a nontoxic form, to the liver for urea synthesis and removal from the body. One possibility would be to use glutamate for this purpose, however, there are two problems with using glutamate. The first problem is that high glutamate levels in the blood are toxic, as shown in studies with total parental nutrition. Amino acid mixtures used for intravenous administration are devoid of glutamate (and aspartate, which is also toxic at higher levels) to alleviate the potential toxic effects. The second problem is a depletion of citric-acid cycle intermediates (αKG →Glu), which would slow the flux through the citric-acid cycle at times where this may be critical (e.g., NH_3 toxicity in the central nervous system). If one intermediate is pulled out of the cycle, all intermediates are depleted, owing to the cyclic nature of the process. To help counteract this to some degree, the brain contains pyruvate carboxylase, to replenish citric-acid cycle intermediates from pyruvate. In the muscle, there is no pyruvate carboxylase; therefore, the only way of replenishing citric-acid cycle intermediates in this tissue is from amino acids, which would not occur if glutamate were the carrier of nitrogen from the muscle.

Some amino acids are primarily transaminated in the muscle (see branched-chain amino acids). In these cases, the nitrogen is transported as alanine to the liver for urea synthesis, or as glutamine to the liver for urea synthesis or the kidney for urinary ammonia production. An advantage of alanine over transporting the glutamate formed during transamination is the prevention of toxicity caused by high glutamate and/or aspartate in the bloodstream. A question may arise as to what happens to dietary aspartate and glutamate released during proteolysis. The answer appears to be that they are converted to alanine and/or glutamine, but primarily alanine. When portal-vein levels of amino acids were measured after protein ingestion, there was very little glutamate or aspartate in the portal blood and markedly higher levels of alanine than expected. Since then, it has been shown that the nitrogen from aspartate and glutamate is transferred to pyruvate to form alanine, a nontoxic nitrogen carrier.

In the muscle, unusual amounts of alanine are formed as a transport system to the liver via the alanine cycle. The alanine cycle, as shown in Figure 18.1, consists of pyruvate conversion to alanine in the muscle, followed by alanine transport to the liver. The alanine nitrogen in the liver can be converted to urea, and the carbon of alanine (pyruvate) converted to glucose, which can return to the muscle to form pyruvate and start the cycle over. The importance of alanine is not only as a nitrogen carrier, but also a carrier of carbon for gluconeogenesis. This probably explains why alanine is the amino acid that is the negative effector of liver pyruvate kinase, an enzyme that affects gluconeogenesis negatively (Chapter 13). When there is a marked degradation of muscle protein as a carbon source for gluconeogenesis, there would also be a larger than expected rise in plasma alanine, not only supplying a source of carbon but also aiding the process by allosterically inhibiting liver pyruvate kinase.

The enzyme catalyzing the reversible transamination of alanine and α-ketoglutarate is alanine aminotransferase (ALT), and isoenzymes occur in both the cytosol and mitochondria. This enzyme occurs in many tissues and is particularly active in the liver. An increase of this enzyme in the serum (sometimes referred to as SGPT, serum glutamic pyruvic transaminase) is indicative of hepatic damage.

SERINE AND THREONINE

Serine is one of the two hydroxyamino acids, the other being threonine. ***Serine has two major pathways of catabolism.*** The first, and apparently predominant, direction in many mammals is catalyzed by serine dehydratase, where water is removed between the alpha and beta carbons of serine. A rearrangement of the double bond forms an amino acid with spontaneous hydrolysis to form pyruvate and ammonia. Pyruvate then can be metabolized as discussed in previous chapters. This enzyme is primarily active in the liver, where the ammo-

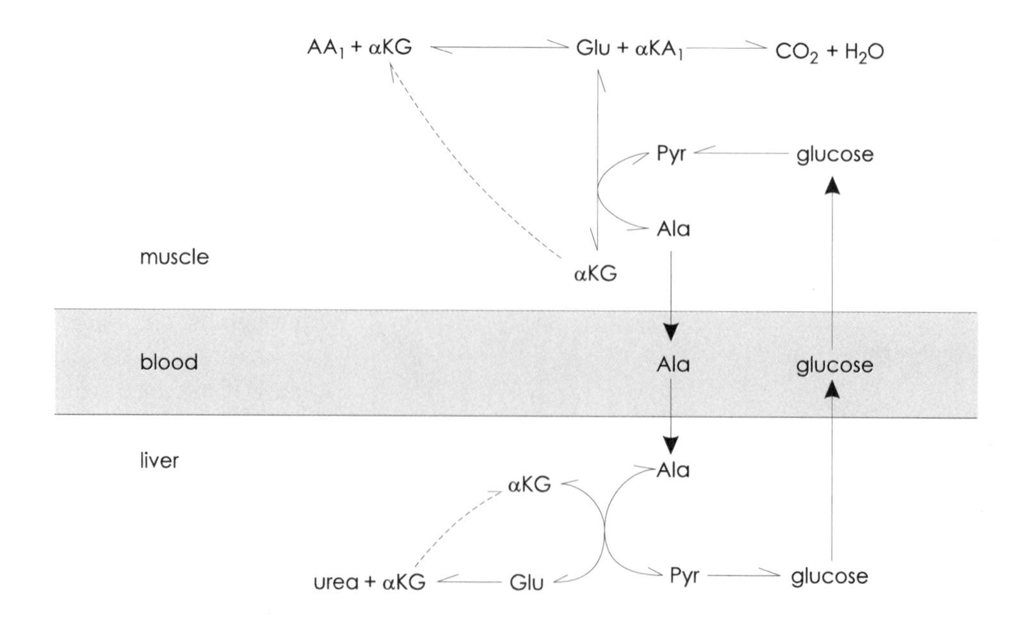

Figure 18.1.
Alanine cycle. The essentials of the alanine cycle are depicted in the above figure. Note that three tissues considered in this figure are muscle, blood, and liver.

nia can be converted to urea and the carbon to glucose. Serine dehydratase is adaptive in nature, with increasing amounts of enzyme as protein content in the diet gluconeonic needs rise. This assures that excess serine can be catabolized so that the carbon structures can be used for biosynthesis and energy. Serine can also be transaminated to form hydroxypyruvate, which can then be reduced and phosphorylated to form 3-phosphoglycerate, which can be converted either to glucose or proceed to glycolysis. This transaminase is in the liver, and it can play a significant role in many species, including humans, in serine catabolism (Fig. 18.2).

One advantage of this pathway over serine dehydratase is that in transamination free ammonia is not released and the potential difficulty of ammonia toxicity is minimized. Serine in people is not an essential amino acid and can be synthesized in the liver from the glycolytic intermediate 3-phosphoglycerate, which can be oxidized to phosphohydroxypyruvate and then transaminated with glutamate to form phosphoserine. The

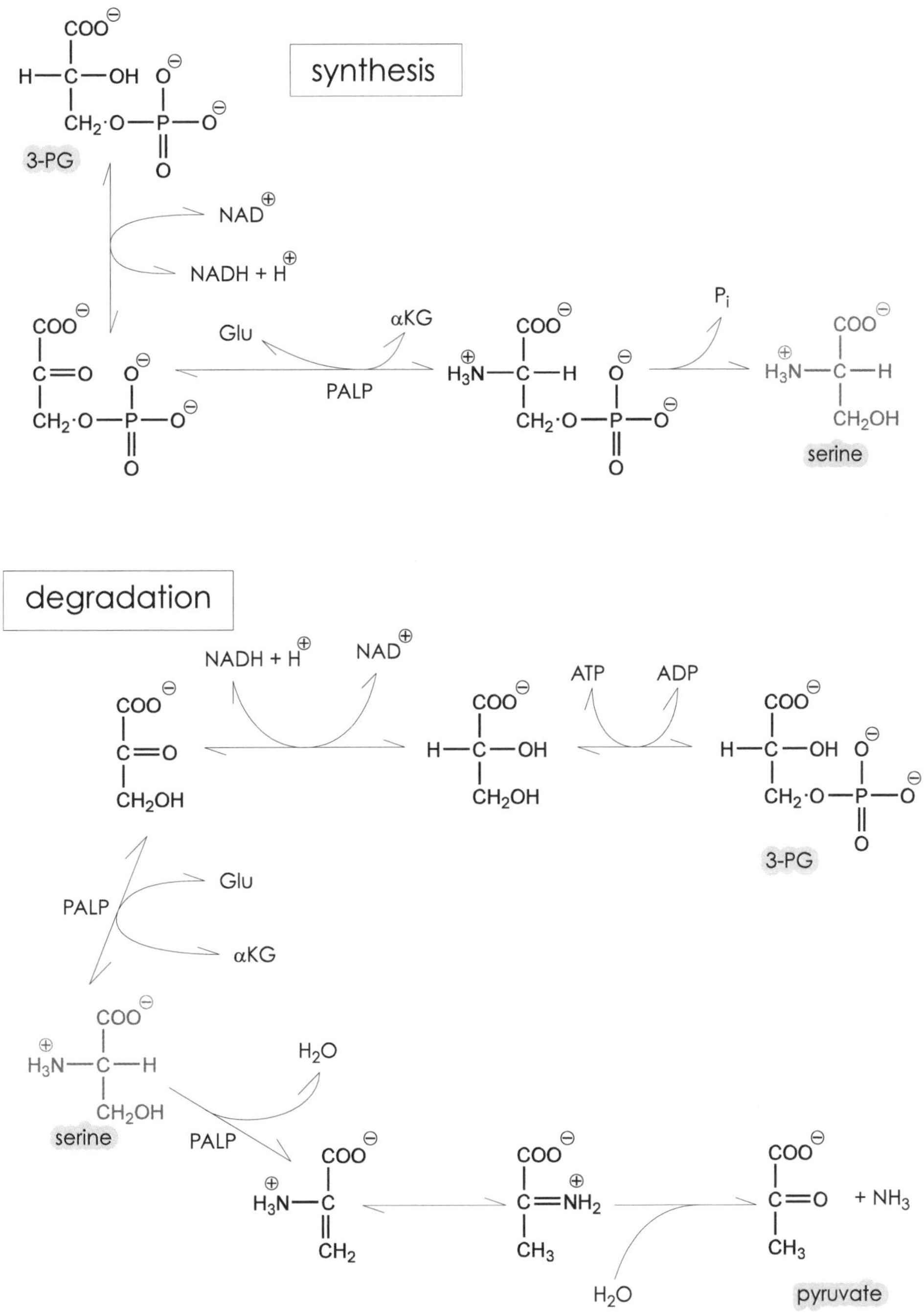

Figure 18.2.

Serine metabolism. The pathways and structures of serine synthesis and degradation are shown. The major pathway of synthesis is presented as well as two pathways for degradation.

phosphoserine is then hydrolyzed to form serine, which then proceeds to meet the many needs of serine, including protein synthesis. Phosphoserine is found in protein, but only serine is incorporated into the protein, not the phosphoserine per se. Specific serine residues present in protein are phosphorylated by protein kinases, such as the cAMP-dependent protein kinase.

Serine has a number of other important metabolic fates in addition to simple catabolism. Decarboxylation produces ethanolamine, which can be incorporated into a phospholipid, phosphatidylethanolamine. This then can be converted to phosphatidylcholine by the addition of three methyl groups. Hydrolysis of phosphatidylcholine can produce choline, where serine and the methyl groups of methionine were the precursors (see Fig. 15.3). This choline in the phospholipids is the same choline as in acetyl choline. Compounds that act to help move fat from liver to adipose, which is the appropriate storage site, are known as lipotrophic factors. Although it is a choline precursor, serine does not fall into this category because it can be made rapidly. Among those compounds that fall into the category of lipotrophic factors are choline, for phosphatidylcholine synthesis, and methionine, the donor of the methyl groups needed to convert serine to phosphatidylcholine.

Another fate of serine is the loss of the hydroxymethyl group, which is catalyzed by serine hydroxymethyl transferase, utilizing tetrahydrofolate as a cofactor, ***forming glycine*** and methylene tetrahydrofolate. Because of this reaction, glycine is not an essential amino acid in people. In some species, where the major excretion product is uric acid, such as birds and reptiles, glycine and/or serine can be an essential amino acid, but it is not so in people. ***Serine is also the precursor for the production of cysteine,*** providing the entire carbon chain structure with the SH group being donated by a metabolite of methionine (homocysteine).

Threonine, the other hydroxy amino acid, occurs in protein and can also be phosphorylated, when present in a protein chain, by several protein kinases. Threonine is an essential amino acid, with a high stringency of essentiality in people, because threonine will not transaminate nor can it be synthesized by higher animals. D-Threonine or the α-keto derivative is not a useful precursor for threonine for protein synthesis. There are two pathways for threonine catabolism, both occurring in the liver. The first is a threonine dehydratase, which is analogous to the serine dehydratase and produces ammonia and α-ketobutyrate. α-Ketobutyrate via a dehydrogenase in the mitochondrion can form CO_2 and propionyl CoA; the latter can be converted to succinyl CoA and then to glucose. Threonine metabolized via this pathway will yield potential glucose. The other suggested pathway is a threonine aldolase, which catalyzes the production of glycine and acetaldehyde. The acetaldehyde can be oxidized to acetate and be used for energy or fat formation but not gluconeogenesis. The resulting glycine has the potential for gluconeogenesis, although it is substantially less effective and slower than the α-ketobutyrate or propionyl CoA.

CYSTEINE

Cysteine is similar in structure to serine, but has an SH group on the β-carbon, instead of an OH group as in serine. This amino acid is extremely important in protein structure with regard to forming disulfide bonds and potential chelation. Cysteine is not considered an essential amino acid in people because it can be formed from serine and methionine. However, since methionine is an essential amino acid, in some cases improved growth can be obtained by adding cysteine to the diet, because it will spare the amount of methionine required to form the cysteine. Thus, on low-methionine diets, cysteine can be beneficial toward growth. A number of proteins are low in sulfur amino acids and, therefore, this methionine–cysteine relationship may become important.

Cysteine has a number of potential metabolic pathways. With low cysteine intake, cysteine desulfhydrase may be of major importance. Cysteine desulfhydrase requires pyridoxal phosphate as a coenzyme and catalyzes the loss of H_2S, similar to the loss of H_2O with serine dehydratase. Thus, the final products are pyruvate, ammonia, and H_2S. This pathway will produce pyruvate, a glucogenic precursor. However, the pathway, particularly with excess cysteine, must be limited, owing to the production of H_2S, which is extremely toxic (Fig. 18.3).

Cysteine, via cysteine decarboxylase, a pyridoxal phosphate requiring enzyme, ***is decarboxylated to form mercaptoethanolamine.*** Mercaptoethanolamine can combine with pantothenate to form coenzyme A or become part of the pantoyl moiety of fatty-acid synthase. The mercaptoethanolamine is the active portion of the coenzyme A, with the SH group from the decarboxylated cysteine combining with acyl groups to form thioesters. Cysteine can also be oxidized, requiring oxygen to give cysteinesulfinate with an oxidation of the SH group. This then can proceed in two pathways. One is an oxidation and decarboxylation, to produce

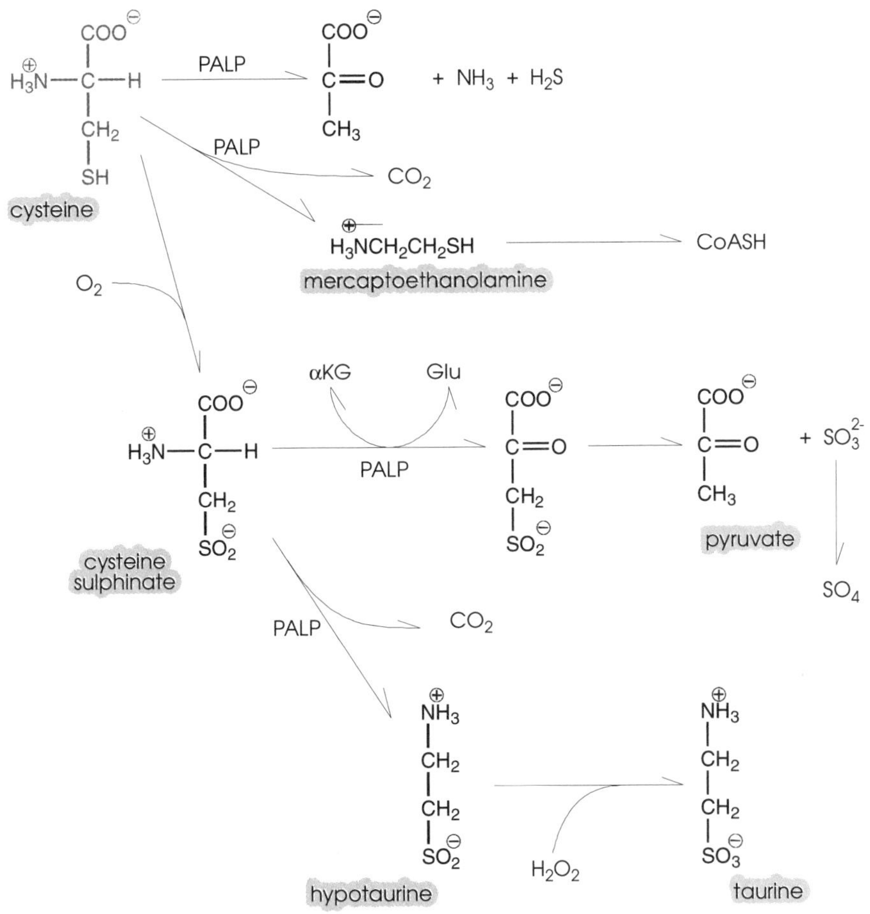

Figure 18.3.
Cysteine catabolism. Several pathways of cysteine catabolism are presented, including the formation

taurine, which has a number of important functions ranging from eye structure and heart function to formation of bile salts, such as taurocholate. Cysteinesulfinate can also form alanine and sulfate. Sulfate formed in this reaction is an important source of sulfate, particularly important in diets that are low in sulfate. Sulfate is needed in the formation of certain sulfated carbohydrate polymers such as mucopolysaccharides. ***Cysteine is also a key component of glutathione,*** which is important in methemoglobin reduction to hemoglobin, detoxification, reduction of disulfides in proteins to sulfhydryls, and amino acid transport.

CYSTINE

Free cysteine occurs only in very low amounts in the blood. Most of the potential free cysteine occurs as cystine where the two SH groups are oxidized to a disulfide bond. This compound can be reduced to cysteine where and when needed. Normally, the kidney reabsorbs both cysteine and cystine. In the disease cystinuria, the transport of cystine into many cells is defective. This causes an increase in plasma cystine, resulting in a spillage of cystine in the urine. Cystine is rather insoluble and forms stones in the kidney, bladder, and ureter, which can be extremely painful. Cystine has a lower solubility as pH decreases.

BRANCHED-CHAIN AMINO ACIDS

There are three branched-chain amino acids: leucine, isoleucine, and valine. These are all essential for higher animals, including humans, since they cannot be formed *de novo*. They can transaminate and, therefore, both the D- (after the action of D-amino acid oxidase) and keto amino acid derivatives can be utilized to form the L-amino acids needed for growth and protein synthesis. In adult humans, protein is in neutral balance; therefore, the equivalent of all of the branched-chain amino acids ingested each day must be catabolized, although they are not necessarily the same molecules ingested. Many of the ingested branched-chain amino acids will be incorporated into protein; however, an equivalent amount of branched-chain amino acids will be released during protein degradation.

Degradation

All three of these amino acids are metabolized reasonably similarly and will be discussed as a group with the small differences indicated. The degradation of these amino acids begins with a transamination where α-ketoglutarate is the acceptor. The product of this reaction will be glutamate and the α-keto derivative of the branched-chain amino acids; this reaction is reversible. ***A unique aspect of branched-chain amino-acid metabolism is that they are the only essential amino acids where the first step of degradation is a reversible reaction.*** This aspect of degradation is more difficult to control than in the case of other amino acids where there are unidirectional reactions. The second step of degradation occurs in the mitochondrion with an irreversible oxidative decarboxylation of the branched-chain keto acid, catalyzed by branched-chain α-keto acid dehydrogenase, to form a branched-chain fatty-acyl CoA. The mechanism of this reaction is identical to that of pyruvate dehydrogenase.

An aspect of branched-chain amino-acid metabolism which may help in the control of fluxes is that the aminotransferases are very high in muscle, almost missing in liver, and have intermediate activity in the kidney, with the branched-chain α-keto acid dehydrogenase being very high in liver and kidney and less active in muscle. The nitrogen from the branched-chain amino acids catabolized in the muscle is transported to the liver and kidney in the form of either alanine, as described in the alanine cycle, or glutamine. The second step of the pathway, the oxidative decarboxylation, is irreversible and is the rate-determining step in the degradation of the branched-chain amino acids. Thus, a considerable proportion of the branched-chain amino acids may be completely metabolized in the muscle, which due to the large abundance of muscle tissue may play an important role despite the low activity of the branched-chain ketoacid dehydrogenase. It has also been shown that branched-chain keto acids are at circulating levels in the blood; these may be catabolized by liver and kidney. The relative role of muscle, kidney, liver, and other tissues in the final catabolism of the branched-chain amino acids is still controversial and unclear. However, it is apparent that the muscle does play an appreciable role in this degradation. The separation of major activities in different tissues may be an important potential control. If transamination occurred in the liver, there would be rapid formation of α-keto acids, which would be rapidly catabolized. The rapid removal of the α-keto acids would push the flux of the transamination further in the direction of forming branched-chain α-keto acids. This would cause a deficiency of branched-chain amino acids. This deficiency is avoided by having transamination occur in the muscle, where the oxidative decarboxylation is less active.

CONTROL OF BRANCHED-CHAIN KETO-ACID DEHYDROGENASE

The control of branched-chain amino acid catabolism lies within the activity of branched-chain α-keto acid dehydrogenase. This enzyme-like pyruvate dehydrogenase can occur in an active nonphosphorylated or an inactive phosphorylated form. The enzyme is phosphorylated by a specific kinase in the mitochondrion, the location of both the kinase and the branched-chain α-keto acid dehydrogenase. The kinase is inhibited by branched-chain α-keto acids; thus, when these are in excess, the enzyme will be nonphosphorylated and active, allowing catabolism of the excess keto acids and, therefore, catabolism of excess branched-chain amino acids. The mitochondrion also contains a branched-chain α-keto acid dehydrogenase phosphatase, which returns the phosphorylated enzyme back to the active form. Thus, the major control of branched-chain amino-acid catabolism is the activity of the branched-chain α-keto-acid dehydrogenase, which is controlled by phosphorylation, primarily by the specific kinase, and dephosphorylation.

ISOLEUCINE

Isoleucine is a good example of branched-chain amino acids for a semi-in-depth examination. Unique aspects of the metabolism of valine and leucine are highlighted. After transamination and oxidative decarboxylation to form the branched-chain fatty-acyl CoA, a double bond is formed between α and β carbons utilizing FAD; then water is added to form a β hydroxy derivative (Fig. 18.4). Then a NAD^+-dependent dehydrogenase produces a keto derivative of the branched-chain fatty-acyl CoA. The similarity to straight-chain fatty-acid oxidation should be noted. This keto fragment is cleaved with participation of coenzyme A to form acetyl CoA, which ei-

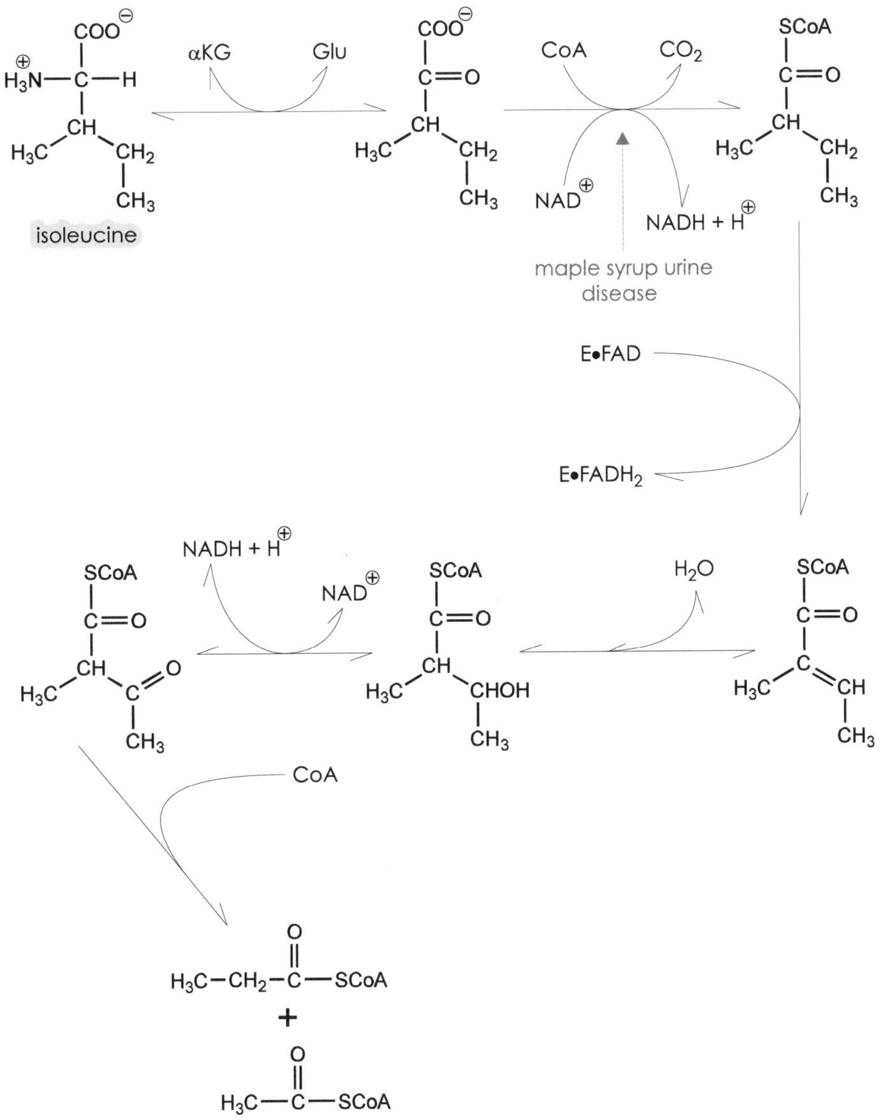

Figure 18.4.

Isoluecine catabolism. The catabolism of leucine is presented as a representative of branched-chain amino acid (leucine, isoleucine, and valine) metabolism. This can occur in more than one tissue (see text for more detail). The purple lettering under an enzyme indicates the resulting disease when this enzyme is missing.

ther enters the citric-acid cycle or is converted to ketone bodies, and propionyl CoA, which is gluconeogenic. Although the muscle is not able to carry out gluconeogenesis, the propionyl CoA can be converted to citric-acid cycle intermediates and on to form glutamine, which can be transported to liver and kidney. The liver and kidney then use the nitrogen to make urea or urinary ammonia and the carbon for gluconeogenesis. The propionyl CoA, when it is transformed to succinate, can also be used to replenish citric-acid cycle intermediates in the muscle. Therefore, isoleucine is both glucogenic and ketogenic. The end product for valine is primarily methylmalonyl CoA, which can be converted to succinyl CoA, a citric-acid-cycle intermediate. Methylmalonyl CoA is an intermediate between propionyl CoA and succinate. Therefore, valine is strictly a glucogenic amino acid, because no ketogenic precursors are formed. In its metabolism, leucine produces β-hydroxy β-methylglutaryl CoA (HMGCoA), which can be split to acetoacetate and acetyl CoA. Therefore this amino acid is strictly ketogenic.

MAPLE-SYRUP URINE DISEASE

In this disease, the α-keto derivatives for all three branched-chain amino acids are found in the urine. The transamination of the amino acids is normal, but ***the enzyme related to the oxidation of the α-keto acid derivatives (branched-chain α-keto acid dehydrogenase) is genetically defective or missing***. Thus, there is an accumulation of the branched-chain amino acids and keto acids since the two are in equilibrium. This causes difficulty with central nervous system function, which may be related to the transport of neutral amino acids into the brain. The incidence of this disease is about 1:200,000. Maple-syrup urine disease can be treated nutritionally, albeit poorly, by using a diet which is low in branched-chain amino acids and, therefore, prevents the large accumulation of these amino acids and α-keto acids in the blood. It should be noted that, in this disease, the requirement for these amino acids is commensurate with growth and, upon reaching the adult stage, the requirement is very low.

The large accumulation in the urine of the α-keto acid compared to the amino acid occurs because the kidney is adapted to effectively reabsorb amino acids, but not to reabsorb the branched-chain α-keto acids as effectively. Therefore, there is a greater spillage of the α-keto acids than the amino acids. We will see the same occurrence of large amounts of α-keto acid compared to amino acid in the urine when we look at another metabolic disease associated with aromatic amino-acid metabolism.

LYSINE

As with threonine, lysine is an absolutely essential amino acid in which no substitutions, such as the *D*-amino acid or keto acid, are useful for the formation of *L*-lysine for protein synthesis. ***Lysine degradation does not proceed via an α-amino transamination.*** The first step is unique. ***Lysine reacts directly with α-ketoglutarate to form a reductive complex at the epsilon amino group, utilizing NADPH*** (Fig. 18.5). This then splits oxidatively, utilizing NAD^+, to form glutamate and an aldehyde on the epsilon group of the former lysine (i.e., α-aminoadipic semialdehyde). The α-aminoadipic semialdehyde then proceeds via a number of metabolic transformations to form acetoacetyl CoA, and ketone bodies; thus, lysine is strictly ketogenic. The degradation of lysine illustrates one of the interesting aspects of amino-acid metabolism, where the first step in the degradation of all of the essential amino acids is a unidirectional step, with the exception of the branched-chain amino acids, where the second step is a unidirectional reaction. Having a unidirectional reaction at the first step makes it easier to control the level and degradation of the essential amino acids. The enzymes for degradation have relatively high K_m values, which assure that excess amino acids will be catabolized at a rapid rate, relative to enzymatic capacities, and a slower rate when the amino-acid concentrations are low. However, there is always a requirement for the essential amino acids, because even at rather low concentrations, there is always some catabolism proceeding. ***Lysine is also one of precursors of carnitine, along with methionine.***

SULFUR AMINO ACIDS

There are two sulfur amino acids: cysteine and methionine. The metabolism of cysteine was discussed earlier. References have been made to methionine for methylation previously, but it was not examined in detail. Methionine is an essential amino acid in the diet, but interestingly enough is toxic in excess. In normal foods this may not be a problem; however, in some fortified foods, methionine is added because many of the plant proteins are low in sulfur amino acids. This may also be a problem for vegetarians who wish to increase their sulfur amino-acid intake and overdose with methionine tablets available at health food stores.

CATABOLISM OF METHIONINE

Methionine can be transaminated; however, the transamination is normally a dead end for methionine. The major pathway of methionine metabolism is reaction with ATP to give *S*-adenosylmethionine, inorganic phosphate, and pyrophosphate (Fig. 18.6). The pyrophosphate is split and assures unidirectional mechanism of this reaction. The *S*-adenosylmethionine has the adenosyl attached to the sulfur atom on the methionine. This *S*-adenosylmethionine (i.e., SAM) is effective in donating methyl groups. ***Most methyl groups donated in the biological system come from SAM.*** These methyl groups are used for the formation of creatine, choline, melatonin, epinephrine, and so on. After the methyl group is transferred from SAM, there is a dissociation to produce adenosine and homocysteine. Homocysteine is a close relative to cysteine, but contains one extra methylene group. Many of the enzymes effecting cysteine will not react with this homologue. Catabolism of homocysteine proceeds through reaction with serine to form cystathionine with the elimination of water. The cystathionine is then split, catalyzed by cystathionase, with water, at the opposite side of the sulfur group from where the initial formation occurred, producing cysteine and homoserine, a homologue of serine with one extra methylene group. The carbon

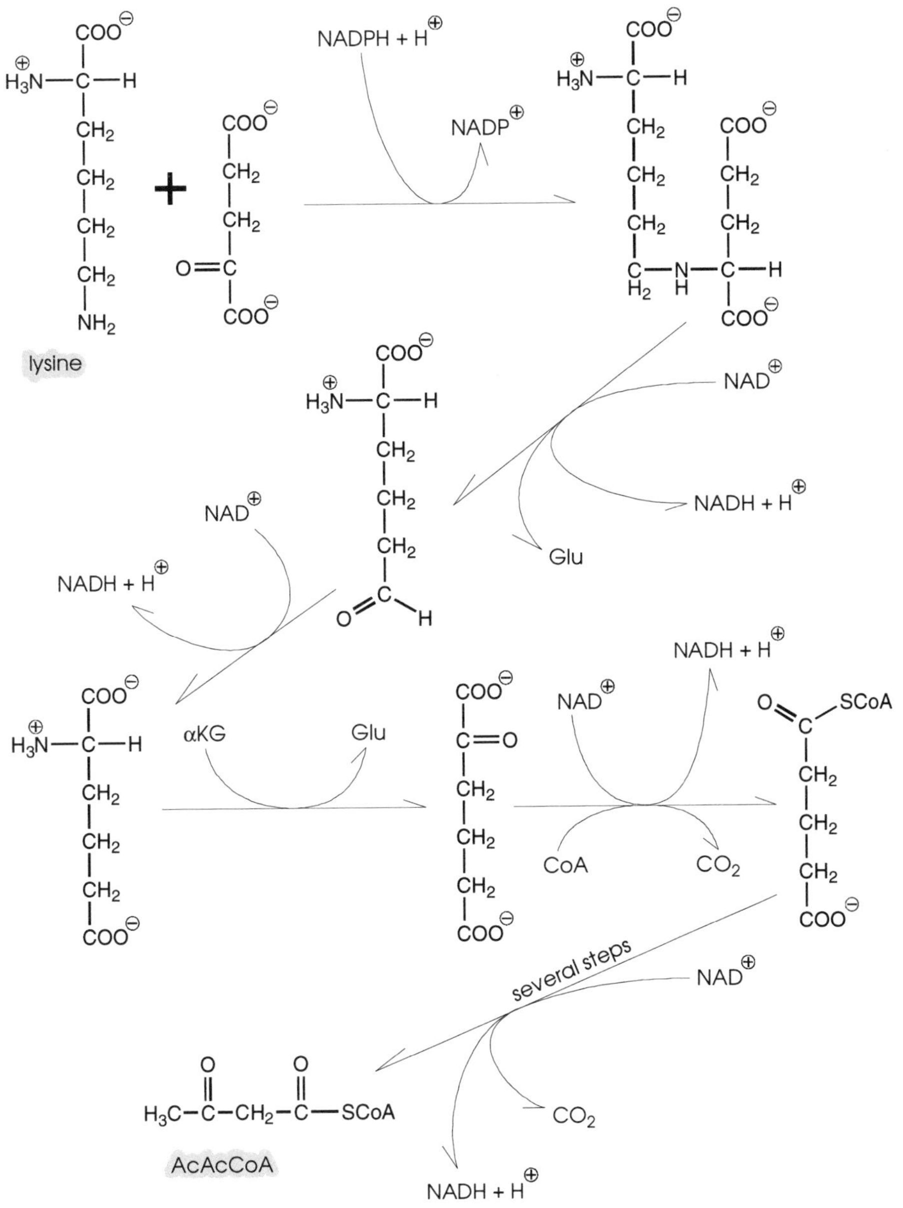

Figure 18.5.

Lysine metabolism. The pathway for the conversion of lysine to acetyl CoA is outlined in the figure above.

atoms of cysteine that are synthesized by the body come from serine, and the sulfur from methionine, whereas the carbons of the homoserine arise from methionine. This is the reason that cysteine is considered a nonessential amino acid—it can be formed from methionine. ***Since many of these reactions are physiologically irreversible, cysteine cannot give rise to methionine.*** Thus, methionine is the essential amino acid and cysteine is not. When an essential amino acid is able to produce a nonessential amino acid, there is a sparing effect. Thus, the requirement for methionine is less in the presence of cysteine than in the absence. In the absence of cysteine, there must be sufficient methionine to carry out all the functions of methionine, as well as producing all of the cysteine that is required.

The homoserine goes through a dehydratase reaction similar to serine, producing intermediates that rearrange to give an amino acid which spontaneously hydrolyzes to produce ammonia and α-ketobutyrate (a homologue of pyruvate with one extra methylene group). The α-ketobutyrate in the mitochondrion can react with an α-keto acid dehydrogenase, requiring coenzyme A, thiamin pyrophosphate, and NAD^+ to form propionyl CoA. Propionyl CoA can be rearranged to form succinate and then oxaloacetate, which is gluconeogenic. Most of the catabolism of methionine via this

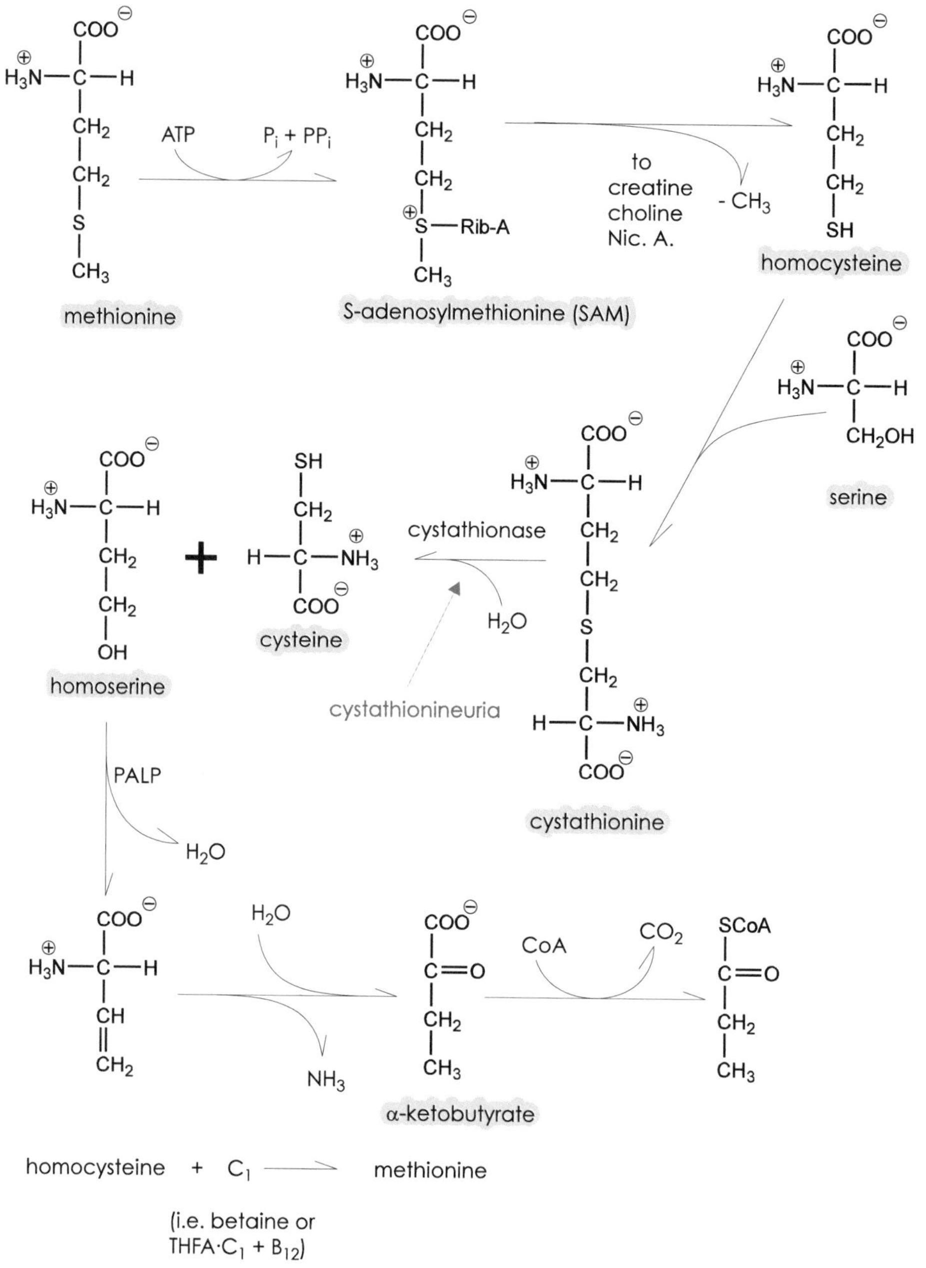

Figure 18.6.

Methionine metabolism. The major pathway of methionine metabolism is presented. This includes the donation of methyl groups, cysteine formation, and potential remethylation of homocysteine. The purple lettering under an enzyme indicates the resulting disease when this enzyme is missing.

pathway occurs in the liver, where the gluconeogenic potential of methionine can be realized. However, some of the metabolism occurs in a number of other tissues, because the formation of compounds such as melatonin and epinephrine occurs in extrahepatic tissues.

Cystathioninuria

In the genetic disease cystathioninuria, the enzyme cystathionase is defective or missing. In this case, cystathionine accumulates and is excreted in the urine. Thus, almost all of the equivalent of methionine intake is put out as cystathionine. If cystathionase is missing, an individual cannot make cysteine from methionine or homocysteine. Therefore, individuals with this disease have a requirement for cysteine and it becomes an essential amino acid for those people.

Remethylation of Homocysteine

An important factor related to the cysteine sparing effect on methionine is the fact that homocysteine has two potential metabolic fates. The first, as was discussed previously, is a formation of cystathionine which can then lead to cysteine formation. The second is a transmethylation with a C-1 fragment, either from betaine (a metabolite of choline), or from an N^5-methyl tetrahydrofolate (also requiring vitamin B_{12} in this action) to reform methionine. The methionine thus formed can either go on to form protein or can again become a methyl group donor to methylate compounds as previously discussed. Thus, ***it is possible for the homocysteine formed, after methylation of a compound by S-adenosylmethionine, to be reconverted to methionine***, which can then participate in further methylations or proceed to cystathionine, which would end the ability to reform methionine. The relative proportion passing on to cystathionine and then to cysteine would be increased in the absence of dietary cysteine and be somewhat decreased in the presence of dietary cysteine, thus accounting for the sparing action of the cysteine. However, in no case is this remethylation efficient enough to continually form methionine without loss as cystathionine. There is always a given amount of homocysteine destroyed via cystathionine. Thus, homocysteine is not as effective in recycling as is the case for certain coenzymes such as tetrahydrofolate and NAD^+.

LIPOTROPHIC FACTORS

Lipotrophic factors are those required for transportation of triacylglycerol from the liver to the adipose tissue for storage. These factors are those that cannot be synthesized from nonlipotrophic components of the diet. The major role of lipotrophic factors is the formation of phosphatidylcholine, which is critical in VLDL formation. One of the lipotrophic factors obviously would be choline, which can be incorporated into phosphatidylcholine. Two other lipotrophic factors are related to the potential de novo synthesis of choline. The first and foremost is methionine, which can be used to donate the methyl groups for choline formation in the absence of dietary choline, thus allowing lipids to be moved from liver to adipose tissue (Fig. 18.7).

A second lipothrophic factor is betaine, which is effective because the transfer of at least one of its methyl groups to homocysteine is very efficient and can replenish methionine for choline formation. In the absence of sufficient lipotrophic factors, a fatty liver develops, and there is insufficient movement of fats either ingested or synthesized in the liver to the adipose tissue. As fats enter or are synthesized in the liver, they are repackaged or packaged as VLDLs to be moved out for transport from the blood to adipose tissue. The VLDLs contain protein, triacylglycerol, cholesterol, cholesterol esters, and phospholipids, especially phosphatidylcholine (lecithin). ***If one has either a protein deficiency or a lipotrophic factor deficiency,*** the movement of triacylglycerols from the liver to adipose is ineffective and a ***fatty liver can develop***. Choline can be present in the diet and need not be synthesized de novo. Phospholipid synthesis has been discussed previously (Chapter 15).

Choline is also catabolized and, therefore, needs constant replacement. The catabolism of choline proceeds primarily in the liver mitochondrion where the alcohol portion is oxidized with two molecules of NAD^+ to produce betaine. Betaine usually donates one of its methyl groups to homocysteine to form methionine, as indicated earlier. The other two methyl groups can be released either as CO_2 or C-1 fragments.

GLYCINE

Glycine has a very simple structure; there is no D- or L-form. Glycine is not a required dietary amino acid for humans. It can be synthesized enzymically from serine by the removal of the C-1 unit (i.e., hydroxymethyl group), requiring tetrahydrofolate. This is rapid enough to meet all the requirements for glycine of most mammals, including humans. The reverse pathway can also occur, that is, the addition of a hydroxymethyl group to glycine, to form serine; however, this is not rapid or critical. If glycine proceeds through serine, it then becomes glucogenic. However, since this pathway is rather slow, glycine provides carbon for glucose synthesis much slower than other glucogenic amino acids. The major quantitative pathway for glycine catabolism is a reaction catalyzed by a kidney enzyme, glycine oxidase (an FAD-containing enzyme), to produce ammonia directly and glyoxylate. This enzyme is adaptive, and if an individual is starving or on a high-protein diet, it is very active. Therefore, glycine is, potentially, a high-ammonia producer, if fed in large amounts to a starving animal.

Glyoxalate can be removed via a number of pathways. The most beneficial for people is to form CO_2 and a tetrahydrofolate C1 fragment, which then can be used in the C1 pool. Glyoxalate can also be oxidized with NAD^+ to produce oxalate. Oxalate is a chelating agent and can react with calcium to form an insoluble chelate. If in sufficient concentration, the insoluble chelate can form stones in the kidney, bladder, or ureter.

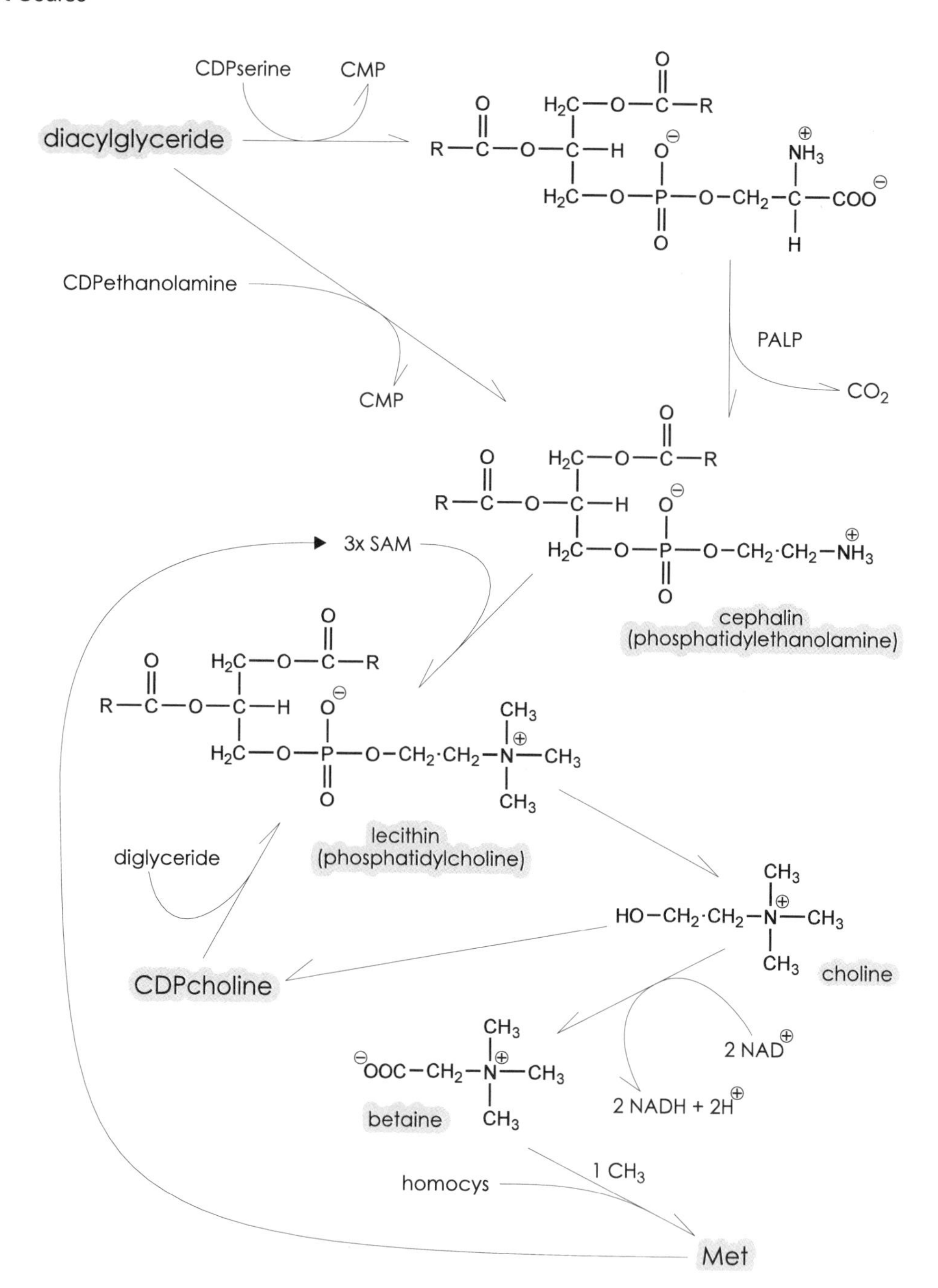

Figure 18.7.

Lipotrophic factors. This figure outlines various pathways for lecithin (phosphatidylcholine) formation. Lecithin is essential for transport of triglycerides from liver to adipose (or muscle). These include complete de novo synthesis, formation from choline, and remethylation of homocysteine from betaine (a catabolic product of choline).

Conjugation and Other Products Utilizing Glycine

Glycine is also important for certain detoxifications. For example, benzoic acid, which is found in numerous foods as a preservative, is detoxified by conjugation with glycine in an amide link to produce hippuric acid. The hippuric acid is then rapidly removed via the kidney. This amide bond to an acid group is not unique to benzoic acid. It also occurs in glycocholic acid (a combination of glycine and cholic acid) and taurocholic acid (a combination of taurine and cholic acid), both important bile salts.

Glycine is also important in the synthesis of purines. In mammals, this is a qualitatively important process, but is not a major flux for either glycine or nitrogen. This is in contrast to observations in birds and reptiles, where purines are the end product of nitrogen metabolism. Glycine is also an important precursor of porphyrins, which can then proceed to form hemes such as those found in hemoglobin and the cytochromes. In Chapter 20, there is a more detailed discussion of purine and porphyrin synthesis.

CREATINE

Glycine is also important in creatine synthesis. This compound is particularly important in muscle as an energy store in the form of creatine phosphate (Fig. 18.8). The first step in creatine biosynthesis is catalyzed by a transamidinase where the guanido group of arginine is transferred to glycine, forming guanidoacetate. This step occurs in the kidney. The advantage of this happening in the kidney is that there is a much lower arginase activity and arginine can accumulate to reasonable amounts in the kidney, as was discussed with arginine synthesis. If this were to take place in the liver, it would be difficult for this reaction to compete with the very active arginase. The guanidoacetate is transported through the blood to the liver, where a reaction with *S*-adenosyl methionine transfers a methyl group to the guanidoacetate to form creatine. The advantage of this occurring in the liver is that this organ is the major site of methionine catabolism and *S*-adenosyl methionine formation. Thus, there is cooperation between two organs in the formation

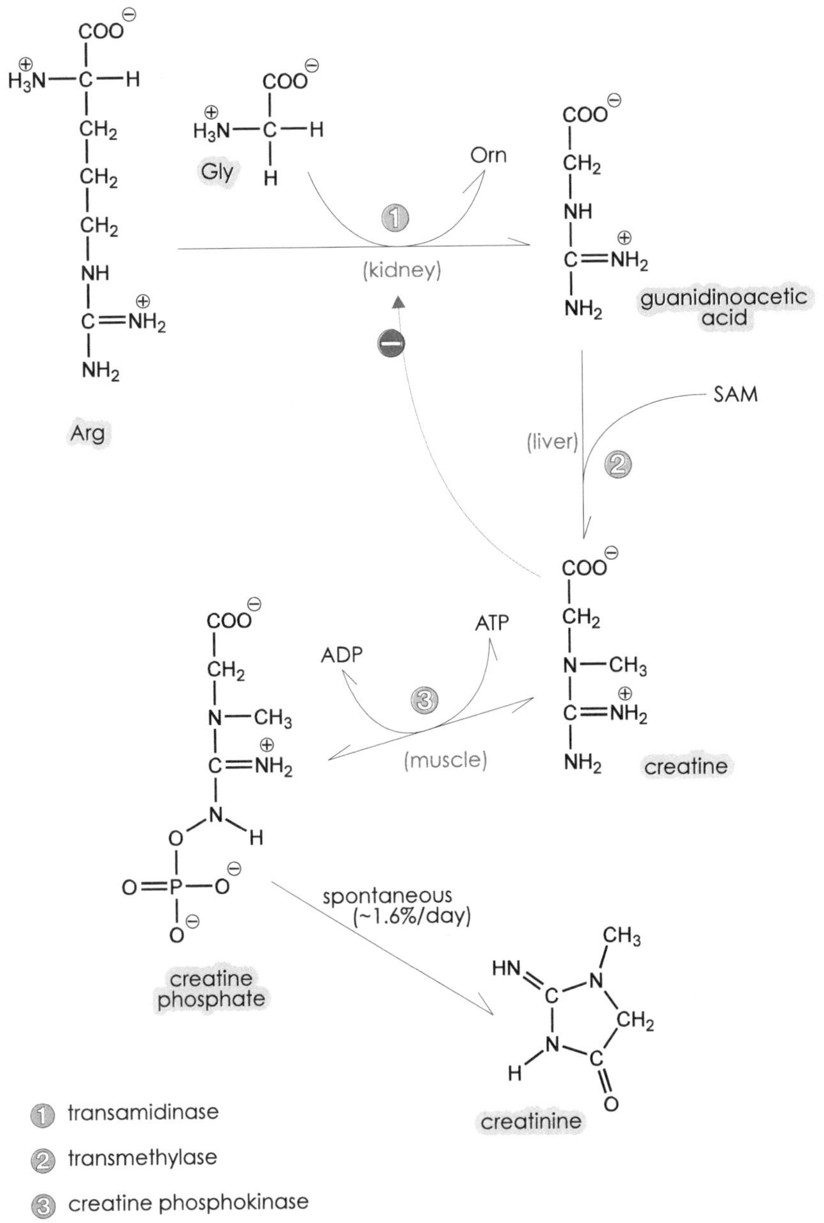

① transamidinase

② transmethylase

③ creatine phosphokinase

Figure 18.8.

Creatine synthesis. The synthesis of creatine, including the major control, is depicted above. The tissue where the reaction occurs is indicated in parenthesis in purple. The formation of creatinine from creatine phosphate is also shown. (See text for more detail.)

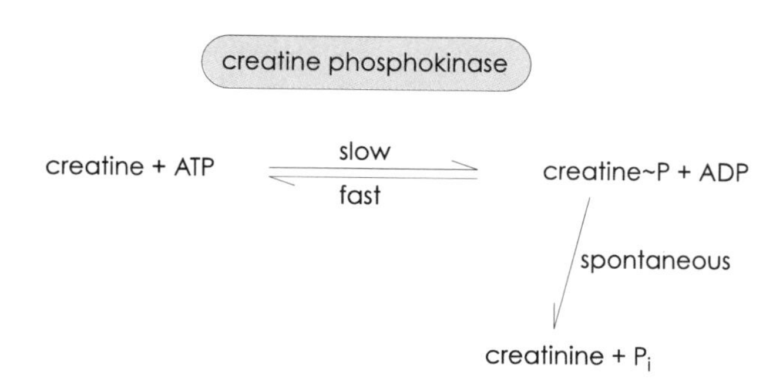

Figure 18.9.
Creatine phosphokinase. The formation of creatine phosphate and its use to reform ATP is shown above.

of creatine and a third organ, the muscle, in the storage of energy. This is similar to branched-chain amino-acid metabolism and arginine biosynthesis; thus, in many cases, there is a collaborative effect of several organs to fulfill a given biosynthetic or biodegradative pathway. Creatine then moves out of the liver through the blood to the muscle, where it is phosphorylated to form creatine phosphate, which is catalyzed by creatine kinase. Creatine phosphate is a rapid reserve of potential high-energy phosphate, which can regenerate ATP. The phosphorylation of creatine takes place continually and at a moderate pace, with creatine plus ATP forming creatine phosphate plus ADP (Fig. 18.9). In most tissues, this reaction is at or near equilibrium. The reverse reaction of creatine phosphate plus ADP to form ATP and creatine is much more rapid, which is advantageous in recouping the high-energy phosphates from creatine phosphate as ATP when needed. In many cases, the muscle may contain four to five times as much creatine phosphate as ATP. Creatine phosphate will spontaneously cyclize to form creatinine because the high-energy bond of the guanido phosphate is not completely stable. About 1.5% of the creatine phosphate cyclizes per day. The only fate of creatinine is to be excreted in the urine.

Control of Creatine Synthesis and Creatinine Excretion

Creatine synthesis is well controlled. The control occurs at the transamidinase step in the kidney. If creatine is fed, a marked decrease in the amount of the transamidinase occurs. Thus, excess production of creatine is avoided. If creatinine is fed, there is no effect on the transamidinase, because creatinine is only an end product and will be excreted.

The output of creatinine remains relatively constant, because muscle creatine phosphate concentration is relatively constant and a given portion is converted to creatinine, which is then excreted. Thus, ***the amount of creatine excreted over a 24-hour period is fairly constant***; this is of clinical advantage when studying single collections of urine. In clinical laboratories, clinicians measure the amount of material being examined in urine per gram of creatinine in the urine. This partially corrects for dilution of the urine when there is a large fluid intake, or concentration of the urine sample if fluid intake is markedly decreased. Since most of the creatinine is excreted in the urine, a marked increase in blood creatinine is usually indicative of kidney disorder.

SUMMARY

1. All amino acids except lysine and threonine can reversibly transaminate. All amino acids except lysine and leucine are at least partially glucogenic (i.e., provide the carbon atoms for the formation of glucose).

2. Alanine can form pyruvate and vice versa via the action of alanine aminotransferase.

3. Alanine carries nitrogen from the periphery to the liver in a nontoxic form for urea synthesis; this is part of the alanine cycle.

4. Alanine is a negative effector of liver pyruvate kinase, allowing more rapid gluconeogenesis.

5. Serine can be formed from a glycolytic intermediate and nitrogen from glutamate in the liver. Serine can be degraded by either transamination or a dehydratase in the liver.

6. Serine is a precursor of phospholipids, glycine, and cysteine.

7. L-Threonine is an absolute dietary essential amino acid. The degradation of the threonine can be via a dehydratase or a specific aldolase in the liver.

8. Cysteine is a sulfur-containing amino acid. Cysteine is a precursor of mercaptoethanolamine, taurine, and sulfate. Cysteine can be reversibly oxidized to form cystine.

9. The branched-chain amino acids, leucine, isoleucine, and valine, are metabolized similarly. They are transaminated, primarily in the muscle, then oxidatively decarboxylated to a branched-chain fatty-acyl CoA in many tissues.

10. These branched-chain fatty-acyl CoAs are metabolized similarly to fatty acids.

11. Branch-chain amino acids are the only essential amino acids where the first step of normal catabolism is reversible and the second step is irreversible.

12. The control of branched-chain amino-acid metabolism lies with the branched-chain α-keto-acid dehydrogenase. This enzyme can be phosphorylated to produce an inactive form and, in turn, that enzyme can be dephosphorylated to produce the active form.

13. The absence of sufficient branched-chain α-keto acid dehydrogenase causes maple-syrup urine disease, which leads to the excretion of branched-chain α-keto acids. There is a nutritional treatment for this disease.

14. L-Lysine is an absolutely essential amino acid. Lysine is ketogenic and a precursor or carnitine.

15. Methionine can reversibly transaminate to form a primarily dead-end product.

16. Methionine in its normal catabolism is converted to *S*-adenosylmethione. This is an important methyl donor in many methylations. After methylation homocysteine is formed.

17. The homocysteine can accept a methyl group from betaine or tetrahydrofolate-C_1 to reform methionine. However, much of the homocysteine combines with serine to form cystathionine. The cleavage of cystathionine forms cysteine and homoserine. Thus, cysteine is not an essential amino acid, but it does have a sparing effect on methionine.

18. Cystathioninuria occurs when cystathionase is genetically missing. In this disease, cystathionine is excreted in the urine and cysteine becomes an essential amino acid.

19. Methionine, betaine, and choline are lipotrophic factors. They can help form phosphatidylcholine needed for moving fat from the liver to other tissues. The key is the formation of choline. Choline can be catabolized in the liver via betaine.

20. Glycine can be catabolized to CO_2, C-1 fragments, or oxalate. Glycine can also form serine, be conjugated to bile acid, or serve as a precursor for synthesis of purines and hemes.

21. Creatine is formed from glycine, arginine, and methionine. The first step, which is carefully controlled, is in the kidney. The second step is in the liver. Creatine's major action is in the muscle as creatine phosphate, a reserve of high-energy phosphates. Creatine phosphate can spontaneously cyclize to form creatinine, which is excreted in the urine.

REFERENCES

The regulation of the degradation of methionine and of the one-carbon units derived from histidine, serine, and glycine, H. A. Krebs and R. Hems, 1976, *Adv. Enzyme Reg.*, 493–513.

Metabolism of sulfur-containing amino acids, M. H. Stipanuk, 1986, *Ann. Rev. Nutr.*, **6**:179–209.

Regulation of the branched-chain α-ketoacid dehydrogenase and elucidation of a molecular basis for maple syrup urine disease, R. A. Harris, B. Zhang, G. W. Goodwin, M. J. Kuntz, Y. Shimomura, P. Rougraff, P. Dexter, Y. Zhao, R. Gibson, and D. W. Crabb, 1991, *Adv. Enzyme Reg.*, 1991, 245–263.

Disorders of branched chain amino acid and keto acid metabolism, D. J. Danner and L. J. Elsas, II, in *The Metabolic Basis of Inherited Disease*, Vol. 1, 6th ed., C. R. Scriver, A. L. Beaudet, W. S. Sly, and D. Valle (Eds.), 1989, 671–692.

Disorders of transsulfuration, S. H. Mudd, H. L. Levy, and F. Skovby, in *The Metabolic Basis of Inherited Disease*, Vol. 1, 6th ed., C. R. Scriver, A. L. Beaudet, W. S. Sly, and D. Valle (Eds.), 1989, 693–734.

REVIEW QUESTIONS

1. Glycine can be formed in the body from
   a) Aspartate
   b) Serine
   c) Threonine
   d) Histidine
   e) Tryptophan

2. A vitamin which, as a coenzyme, plays the greatest role in amino acid metabolism is
   a) Folic acid
   b) Thiamine
   c) Riboflavin
   d) Pyridoxine
   e) Biotin

3. Cysteine is not a precursor of
   a) Sulfate
   b) Taurine
   c) Methionine
   d) Glutathione
   e) Mercaptoethanolamine

4. One dietarily essential amino acid which must be present as the L-amino acid (free or in protein) for continued existence is
   a) Lysine
   b) Tryptophan
   c) Methionine
   d) Phenylalanine
   e) Isoleucine

5. The alanine shuttle moves
   A. Carbon and nitrogen from muscle to kidney
   B. Carbon and nitrogen from muscle to liver
   C. Carbon and nitrogen from liver to kidney
   D. Carbon and nitrogen from aromatic amino acids
   E. Carbon and nitrogen from brain to liver

6. Creatine synthesis requires which of the following?
   a) Methionine, glycine, and serine
   b) Arginine, glycine, and serine
   c) Tryptophan, arginine, and glycine
   d) Methionine, tryptophan, and glycine
   e) Methionine, arginine, and glycine

ANSWERS TO REVIEW QUESTIONS

1. ***b*** Serine can be converted to glycine by the action serine hydroxymethyl transferase, producing glycine and methylene THFA.

2. ***d*** Pyridoxine coenzymes play a role in transaminations, dehydratases, cystathionine synthesis, and other reactions.

3. ***c*** Although methionine can form cysteine, methionine is a dietary essential amino acid and cannot be formed from cysteine or any other amino acid found in protein.

4. ***a*** Lysine and threonine are the only essential α-amino acids that cannot be α-transaminated. These two are required as the L-amino acids; the D-amino acids or ketoacids will not substitute for the L-amino acid.

5. ***b.*** The nitrogen can be used for urea synthesis and the carbon for gluconeogenesis, both occurring in the liver.

6. ***e*** Glycine plus arginine form guanidoacetate. The addition of the methyl group from methionine then forms creatine.

CHAPTER

19

METABOLISM OF AROMATIC AMINO ACIDS AND PROTEIN ECONOMY

INTRODUCTION

In this chapter, the metabolism of aromatic amino acids, including histidine, is examined. The aromatic amino acids are dietary essentials, with the exception of tyrosine, which can be formed from phenylalanine. Each of the amino acids has a unique metabolism, including the formation of important compounds other than proteins as neurotransmitters, pigments, and a vitamin. A number of metabolic diseases are also associated with aromatic amino-acid metabolism. An overall view of nitrogen economy and protein metabolism is discussed at the end of the chapter.

HISTIDINE

Histidine appears to be essential for both growing animals and adult animals. The difficulty in establishing its importance occurred because hemoglobin is rich in histidine but people require only small amounts; therefore, it took a long time for histidine depletion to show up in adult, nongrowing animals. Although histidine can be transaminated, this reaction is not on the pathway to catabolism. The α-keto acid cannot be catabolized further and must be reaminated back to the L-histidine for further degradation. ***The first step in the degradation of L-histidine is a physiologically irreversible step catalyzed by the enzyme histidase.*** This produces ammonia with a double bond in the side chain of the histidine (i.e., urocanic acid). Water is then added across the conjugated double bonds (Fig. 19.1).

The ring portion of the histidine is then split, with the addition of water giving an intermediate of formiminoglutamate. This is converted to glutamate and catalyzed by an enzyme that transfers the formimino group to tetrahydrofolate (THFA), producing glutamate and N^5-formimino THFA. The N^5-formimino THFA then loses NH_3 and is converted to N^5, N^{10}-methylene THFA. The inclusion of THFA in the conversion of formimino glutamate (FIGlu) can be used to detect potential folate deficiencies. If an individual is given a large amount of histidine, it is normally catabolized to C-1 fragments, ammonia, and glutamate, all of which can be handled readily by the body. If folate is deficient, the FIGlu cannot be converted to glutamate at a rate commensurate with the metabolism of histidine and, therefore, the FIGlu will accumulate in the blood and be excreted in the urine. Thus, increased urinary FIGlu is an indication of folate deficiency. This is quantitatively the major pathway of histidine catabolism. Therefore, histidine is fundamentally a glucogenic amino acid because glutamate can give net glucose synthesis.

It is important here, and when considering aromat-

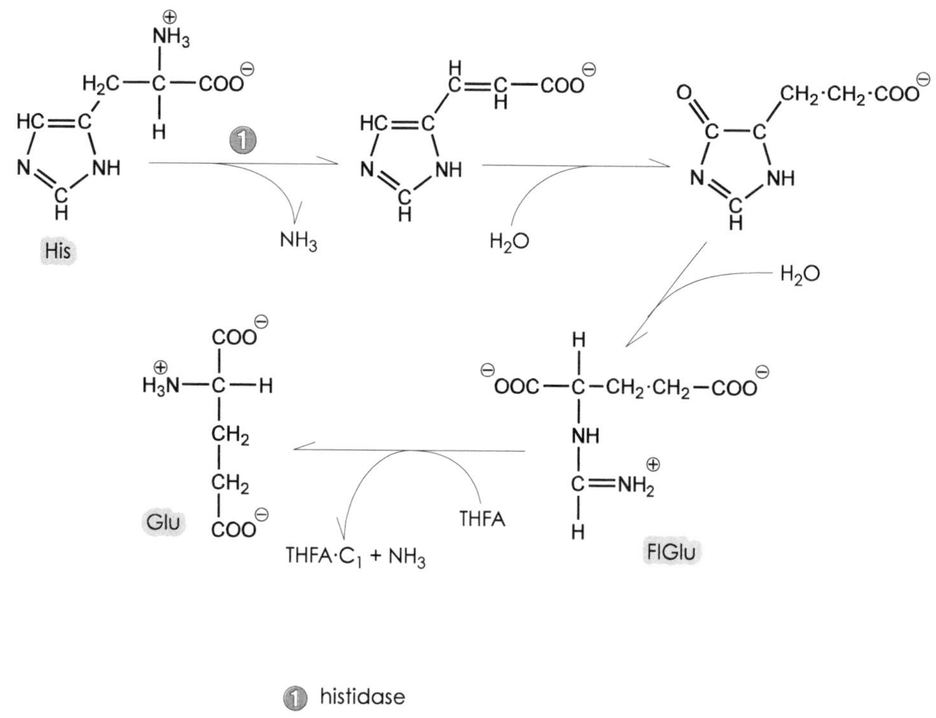

His → (1) → urocanate → FIGlu → Glu

① histidase

Figure 19.1.
Histidine catabolism. The major pathway of histidine catabolism to form ammonia and glutamate is shown above.

ic amino-acid metabolism, to differentiate between quantitatively important fluxes (e.g., histidine to glutamate) and qualitatively important fluxes—which are extremely important—even though only a small proportion of amino acid proceeds down the pathway. ***Another pathway of histidine that is qualitatively important but quantitatively minor is the decarboxylation of histidine, catalyzed by histidine decarboxylase, to form histamine.*** Histamine has physiological and pharmacological importance in response to a number of cell and body injuries. The histamine formed has two amino functions, one in the carbon chain, and one within the ring. Histamine is a good substrate for the enzyme diamine oxidase, which helps destroy the histamine. Diamine oxidase, a flavin enzyme, catalyzes the oxidation of the primary amine, producing an aldehyde and ammonia. The aldehyde is oxidized to an acid, with concomitant reduction of NAD^+. This acid can be further metabolized or excreted and allows for the destruction of the histamine so that the response to the histamine does not continue forever. Thus, in many allergic reactions, histamine is formed, has its action, and then is destroyed.

PHENYLALANINE AND TYROSINE

Two closely related aromatic amino acids are phenylalanine and tyrosine. The metabolism of these two amino acids is of medical interest for two reasons. First, a large number of metabolic diseases is associated with the metabolism of these two amino acids; second, a large number of important biological compounds other than protein are formed from these amino acids. ***Phenylalanine can be converted to tyrosine in a unidirectional, physiologically irreversible reaction.*** Phenylalanine is an essential amino acid that must be preformed in the diet, whereas tyrosine is not considered an essential amino acid because it can be formed from L-phenylalanine. However, the relationship is analogous to that previously indicated for cysteine and methionine; the amount of phenylalanine required in the diet depends on the tyrosine content of the diet, that is, the lower the tyrosine content, the more phenylalanine required. This is referred to as a sparing effect that tyrosine has on the phenylalanine requirement.

Conversion of Phenylalanine to Tyrosine

Phenylalanine catabolism occurs primarily in the liver, where the first step is a hydroxylation in the para position to form L-tyrosine, catalyzed by phenylalanine hydroxylase. This reaction requires oxygen, NADH, and a pteridine derivative. This reaction sequence follows a general rule of essential amino acids—the first step of catabolism is an irreversible reaction. Phenylalanine can also undergo transamination to form phenylpyruvic acid, where either pyruvate or α-ketoglutarate can be the amino acceptor to form alanine or glutamate respectively. However, this is a dead-end reaction, because there is no metabolic pathway for phenylpyruvate. The only way phenylpyruvate can be utilized by the body is through reamination back to L-phenylalanine and via phenylalanine hydroxylase for catabolism. Therefore, as with methionine, the keto derivative would be a good precursor of the L-amino acid in the diet.

Phenylketonuria

A major disease associated with the absence of phenylalanine hydroxylase is phenylketonuria (PKU). This disease, if untreated, results in mental retardation, but does not appear to be fatal at an early stage such as some other metabolic diseases. This disease appears to be caused by a simple autosomal recessive gene, where two carriers can produce a phenylketonuric offspring, as can a carrier and an individual with the disease. Two individuals with this disease always produce phenylketonuric offspring. Phenylketonuria, like Maple Syrup Urine disease, can be treated nutritionally, in this case, by feeding a low phenylalanine diet from infancy, where the dietary phenylalanine is commensurate with that needed for growth. In an adult with phenylketonuria, marginal phenylalanine is required for maintenance, because the catabolism of this amino acid is minimal. The absence of phenylalanine hydroxylase in individuals with PKU is not necessarily always 100%. Some activity may be left, but it is minimal and insufficient to convert phenylalanine to tyrosine at a rate commensurate with dietary intake. It appears that in those with the disease, the proteins for phenylalanine hydroxylase is formed; however, owing to mutations, it is enzymatically ineffective. In this disease, phenylalanine cannot be converted to tyrosine. Thus, phenylalanine rises in the blood. Since both phenylalanine and its keto derivative are in equilibrium through transamination, there is also a concomitant increase in the keto derivative, phenylpyruvic acid, in the blood. Both phenylalanine and its keto derivative are excreted into the urine, with higher amounts of the keto derivative, because the kidney lacks the ability to reabsorb much of it. The phenylpyruvate (a phenylketone) is found in the urine, hence the name phenylketonuria. The problem in PKU results from excess phenylalanine and its metabolites, not the inability to make tyrosine (if sufficient tyrosine is present in the diet). ***In individuals with PKU, tyrosine becomes an essential amino acid, because these individuals can no longer synthesize tyrosine from phenylalanine.***

TYROSINE CATABOLISM

Tyrosine can arise from two major sources—one is dietary tyrosine and the other is from phenylalanine that is converted to tyrosine. Tyrosine is further metabolized via a number of different pathways; the major quantitative pathway of metabolism occurs in the liver. ***Almost all essential amino-acid metabolism, with the exception of the branched-chain amino acids, occurs primarily in the liver.*** Tyrosine is transaminated, with α-ketoglutarate being the acceptor, to form *p*-hydroxyphenylpyruvate and glutamate (Fig. 19.2). This reaction is freely reversible. Tyrosine is not an essential amino acid, therefore, the reversibility of its first step does not violate the usual dogma of the first step of catabolism of essential amino acids being physiologically irreversible. *p*-Hydroxyphenylpyruvate is formed by a complex reaction in which the keto derivative loses a carboxyl CO_2, the ring is hydroxylated, and the remaining carbon chain is relocated on the phenyl ring. This reaction requires ascorbate (vitamin C). Most animals make their own ascorbate, but people, higher primates, and guinea pigs do not, and therefore can become scorbutic. One of the ways to test whether an individual is scorbutic is by giving a tyrosine load. If there is not sufficient ascorbate, the first intermediate of the pathway of *p*-hydroxyphenylpyruvate cannot be further converted and will accumulate in the blood and be excreted into the urine.

Normally, *p*-hydroxyphenylpyruvate is converted to homogentisate. The homogentisate is further metabolized by the opening of the ring by catalyzed homogentisate oxidase and is isomerized to form fumarylacetoacetate, which is cleaved to form acetoacetate (i.e., ketogenic) and fumarate (glucogenic). Fumarate can proceed via the citric-acid cycle to oxaloacetate and then through gluconeogenesis.

There is another disease associated with tyrosine catabolism called alcaptonuria. In this disease, homogentisate oxidase is missing and, therefore, homogentisate accumulates and is excreted in the urine. Homogentisate has two hydroxyls on the aromatic ring; thus, it is a hydroquinone. These hydroxy groups may oxidize to keto groups. The resulting compound is colored, so you have a dark urine (alcaptonuria). This disease is not fatal, but usually results in arthritis in the later years. The metabolic effect is minor; one only fails to obtain the normal products from phenylalanine and ty-

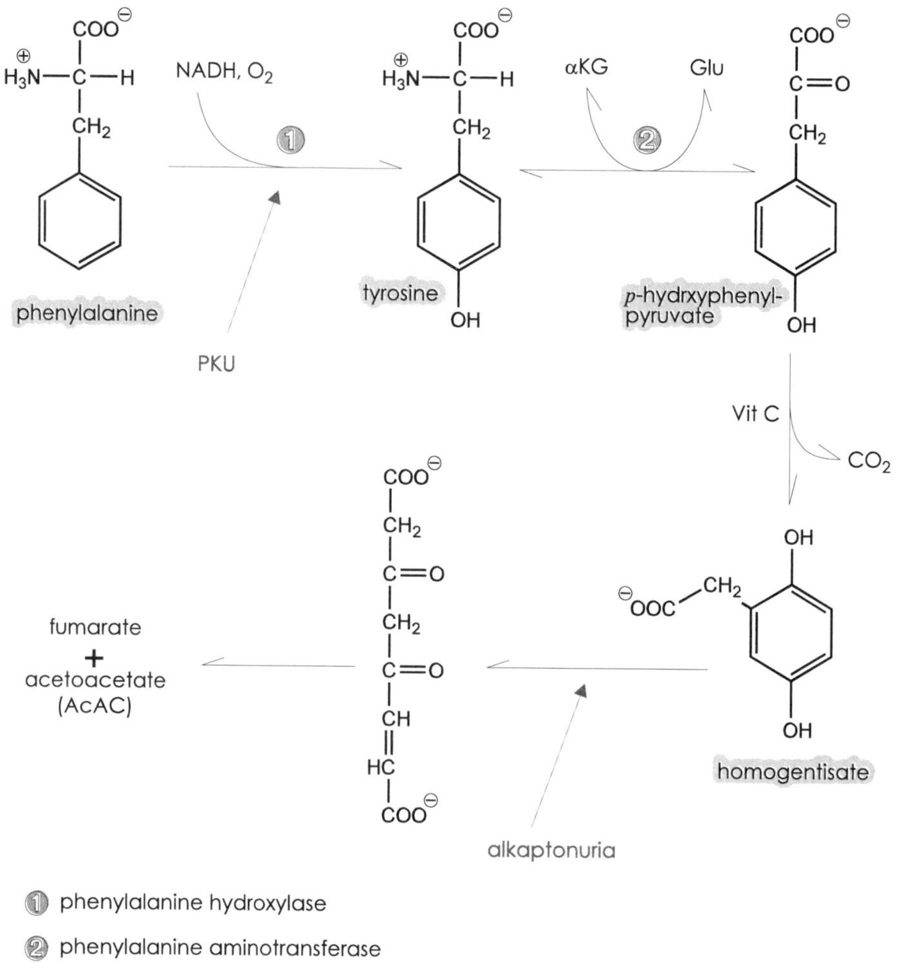

Figure 19.2.

Phenylalanine and tyrosine catabolism. The major pathway of phenylalanine and tyrosine catabolism is presented above. The fate of fumarate and acetoacetate is familiar to readers of this text and is not followed further. The purple lettering under an enzyme indicates the resulting disease when this enzyme is missing.

1 phenylalanine hydroxylase

2 phenylalanine aminotransferase

rosine metabolism and the resultant energy thereof, as well as losing potential glucose formation from these two aromatic amino acids.

Phenylalanine and tyrosine are glucogenic and ketogenic. Even though these amino acids contained nine carbons, they still only produced half of a glucose molecule, that resulting from fumarate. Therefore, it appears that in most cases, those amino acids that are glucogenic can produce half of a glucose, regardless of the number of carbons. This ranges from alanine, which can produce the equivalent of half of a glucose with only three carbons, to amino acids such as phenylalanine and tyrosine, with nine carbons, which still only produce effectively half of a glucose molecule. Even tryptophan, with 11 carbon atoms, only forms half of an equivalent of glucose during its catabolism.

Other Products from Tyrosine

A number of important products are formed from tyrosine which, although quantitatively not of the same significance as those discussed above, are extremely important qualitatively. Among these are the hormone thyroxine, which is formed from the iodination of tyrosine in the peptide link, migration of one of the aromatic chains to form thyroxine, and triiodothyronine in the peptide link, which after hydrolysis in the thyroid gland, releases thyroxine and triiodothyronine.

Tyrosine also has an important role in the central nervous system and melanocyte and is the precursor of both melanins and catecholamines (epinephrine and norepinephrine). The conversion to these products takes place in the appropriate tissues, usually melanocyte, the central nervous system, or the adrenal gland. In each of these tissues, the enzyme tyrosinase catalyzes the conversion of tyrosine to dihydroxyphenylalanine (DOPA) by hydroxylating the ring adjacent to the parahydroxy group. This is a catechol ring. If this were an amine instead of an amino acid, it would be a catecholamine. The DOPA is a precursor of catecholamines in the adrenal gland and central nervous system. In melanocyte, the DOPA is converted to melanine. In the disease albinism, the tyrosinase in the melanocyte is missing, but the isoenzyme in other tissues is still present, so the formation of catecholamines for hormonal formation is normal (Fig. 19.3). It has clearly been observed that albinos can still make sufficient epinephrine and norepinephrine (catecholamines), similar to individuals without albinism, despite being unable to form normal pigmentation.

Catecholamine Formation

Dihydroxyphenylalanine is an important precursor of biologically active amines that have action throughout the body, including the central nervous system. In the central nervous system and adrenal glands, as well as other selected tissues, DOPA is decarboxylated (requiring pyridoxalphospate) to form dopamine. The dopamine can then be converted to norepinephrine by the hydroxylation of the side chain in the number 2 position. This hydroxylation may be the rate-determining reaction in the formation of norepinephrine and epinephrine. Norepinephrine is a major messenger at nerve endings and within tissues. The advantage of this being a monoamine is that when a message goes from one nerve ending to another, shortly after its action, there is a potential for monoamine oxidase to destroy this compound and, therefore, the action can be of short duration. It is thus more effective as a local message rather than a major systemic message. In the adrenal gland, a methyl group from *S*-adenosylmethionine is used to methylate norepinephrine, producing epinephrine. Epinephrine is not a primary, but a secondary, amine and is not a substrate for monoamine oxidase activity. Thus, it can be released into the blood as a circulating hormone without the risk

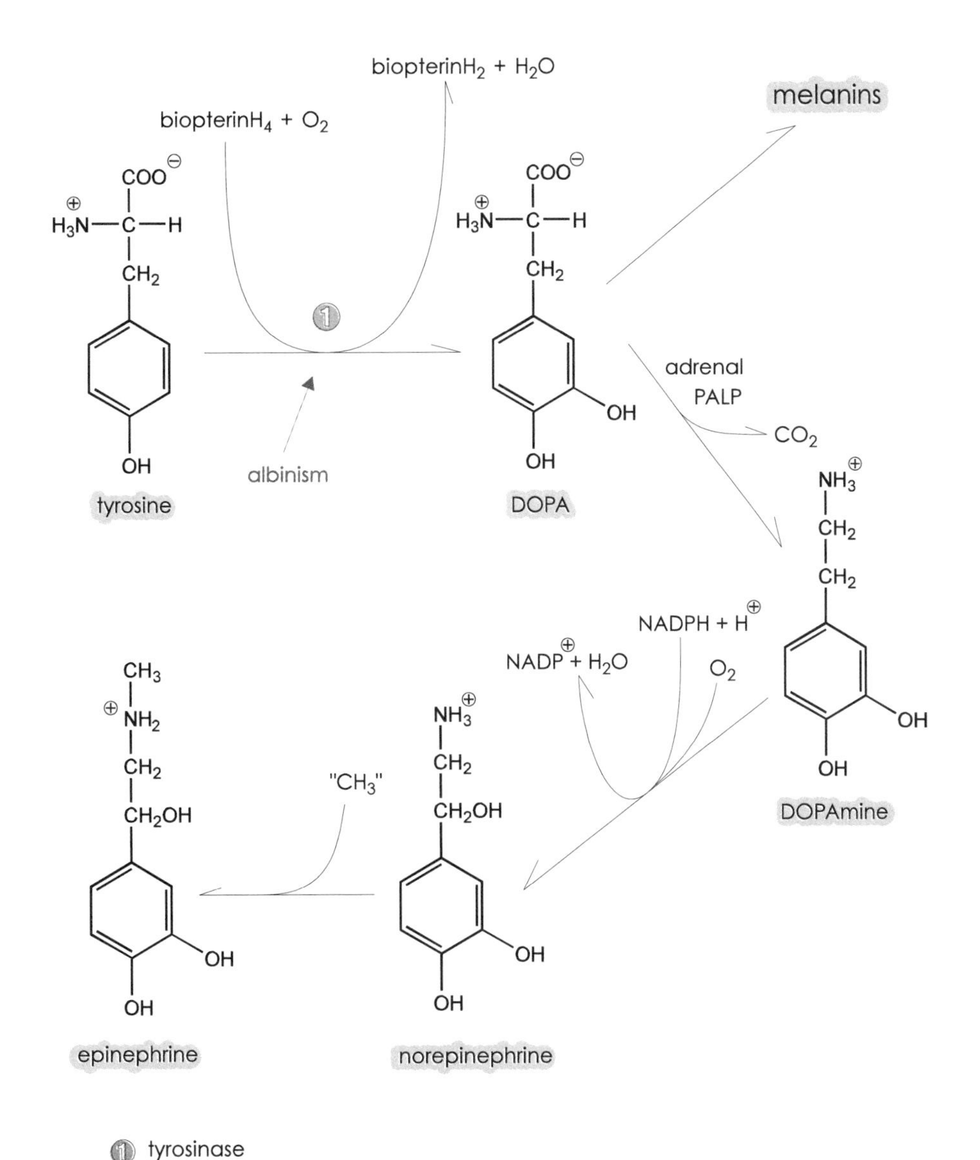

Figure 19.3.

Catecholamine formation. The conversion of tyrosine to catecholamines and melanins is outlined above. The purple lettering under an enzyme indicates the resulting disease when that enzyme is missing. In the case of albinism, an isoenzyme of tyrosinase is missing in the melanocyte, but not in other tissues.

of being destroyed by the ubiquitous monoamine oxidase. In the case of epinephrine, it is usually metabolized for excretion from the body by methylation of one of the hydroxyl groups on the ring and not by oxidation of the amine.

Parkinson's Disease

An interesting relationship between anatomy and biochemistry occurred in the treatment of Parkinson disease. An anatomist noticed that the substantia nigra in the brains of people with Parkinson's disease lacked the normal pigmentation. This suggested a possible lack of sufficient DOPA and dopamine and, therefore, a possible lack of norepinephrine. The biochemists recognized the relationship in these pathways, and in clinical studies individuals were treated with L-DOPA, which did enter the central nervous system. This caused more normalization of the pigmentation and relieved many of the symptoms of Parkinson's disease. Presently, other derivatives related to this compound are used for treatment. However, the first successful treatment was with L-DOPA. Even though dopamine would appear to have been a more effective agent, there were two difficulties with its use. Dopamine is a potent neurotransmitter and can be destroyed by monoamine oxidase. A greater problem is that most amines do not cross the blood–brain barrier and enter the brain. However, L-DOPA has all the appearances of an aromatic amino acid, not an amine; therefore, it can enter the brain and central nervous system, be decarboxylated there, and be converted to both pigmentation and norepinephrine.

TRYPTOPHAN

Tryptophan is a dietary essential amino acid. Tryptophan can transaminate; however, as with phenylalanine and histidine, this transamination is a dead end, and for major catabolism, the keto derivative must be transaminated to L-tryptophan for further normal degradation. Similar to other aromatic amino acids, there is a major quantitative pathway and some very important qualitative pathways related to tryptophan metabolism. The major quantitative pathway is a splitting of the five-membered ring, catalyzed by the enzyme tryptophan oxygenase, to form *N*-formylkynurenine, which is then hydrolyzed to produce formate plus kynurenine (Fig. 19.4). This is followed by a number of other steps, including the cleavage of the side chain to produce alanine. As indicated earlier, alanine is gluconeogenic amino acid, and thus one can achieve glucose formation from tryptophan. Metabolism proceeds to form 3-hydroxy anthranilate. From this point there is a branch; the major direction yields acetyl CoAs which are not glucogenic, and the minor pathway can produce niacin. Apparently, no more than one-sixtieth of the tryptophan can be converted to niacin. However, in normal protein-containing diets, this may be sufficient to meet most of the niacin requirement. The disease pellagra, a niacin deficiency, is usually only observed with low-protein diets, particularly those that derive protein from poor-quality sources, such as processed corn, that are low in tryptophan and niacin. During the 1930s, this was a major nutritional deficiency; today, because of our understanding of the disease and its treatment, it has become much less prevalent.

5-Hydroxytryptophan and Derivatives

Another qualitatively important, but quantitatively minor, pathway for tryptophan metabolism is the hydroxylation in the number 5 position. This can occur in a number of tissues, but it particularly important in the brain and the pineal gland. This enzyme appears to be a particulate enzyme with a reasonably high K_m for tryptophan. Thus, if insufficient tryptophan is present, protein synthesis may continue normally, owing to the low K_m for activation of tryptophan to the amino acyl *t*-RNA; however, there may be a problem with hydroxylation of tryptophan (Fig. 19.5). The hydroxylation of tryptophan produces 5-hydroxytryptophan, which can then be decarboxylated, catalyzed by tryptophan decarboxylase, a PALP-requiring enzyme, to 5-hydroxy tryptamine, also known as serotonin. Serotonin is an important compound in normal brain function and tranquility. Therefore, any disturbance of tryptophan metabolism via this pathway can lead to mental disturbances. Serotonin can be destroyed by the enzyme monoamine oxidase (a flavo protein), which catalyzes the formation of ammonia and 5-hydroxyindole acetaldehyde in an irreversible reaction. The aldehyde is rapidly oxidized enzymatically, utilizing NAD^+ to form 5-hydroxy indoleacetate, which is then usually excreted. The formation and turnover of serotonin can be estimated by 5-hydroxy indoleacetate output in the urine.

In untreated individuals with phenylketonuria, there is a significant drop in the 5-hydroxy indoleacetate in the urine, indicating an interference with the system by the excess phenylalanine. This may be due to

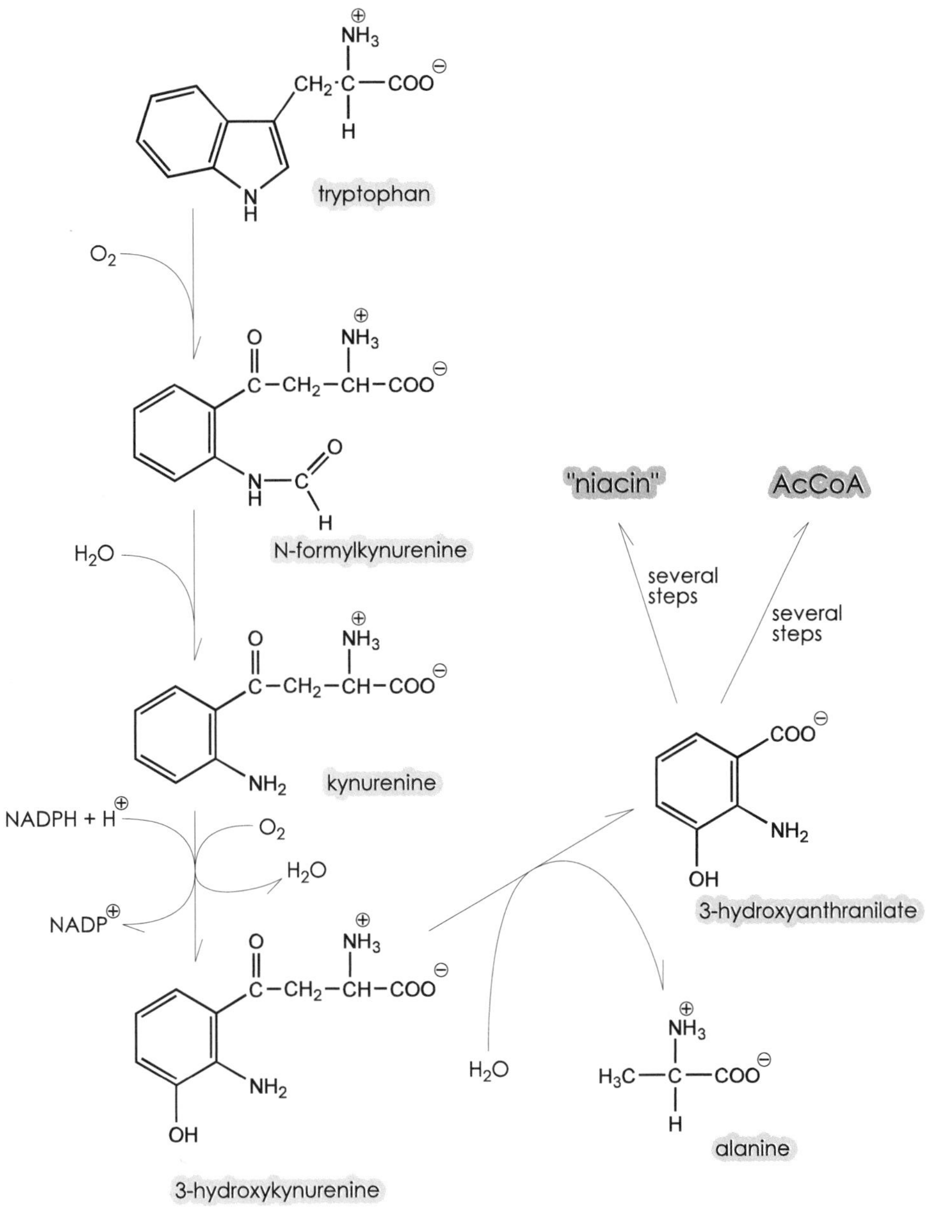

Figure 19.4.
Tryptophan catabolism. The major pathway of tryptophan catabolism is presented above. The early steps are shown in detail, with the formation of niacin and acetyl CoA from 3-hydroxyanthranilate shown with no detail.

competition for aromatic amino-acid uptake in the brain and/or inhibition of the hydroxylase in the brain or a linked enzymatic alteration. If individuals with this disease are given a low-phenylalanine diet, the excretion of 5-hydroxy indoleacetate returns to normal. This indicates that there is no basic defect in the uptake and hydroxylation of tryptophan in the absence of excess phenylalanine in these individuals.

MONOAMINE OXIDASE

Monoamine oxidase, which can destroy a number of amines, including norepinephrine, serotonin, and others, can affect the relative level of these amines in the brain. Therefore, in some cases where these are apparently at a low level, individuals may be treated with a drug, or a family of drugs, that are monoamine oxidase inhibitors. In a number of cases, these drugs bring about a restitution of normal mental functions. However, individuals on these drugs must avoid certain foods, particularly those that are heavily fermented, such as Swiss cheese. These foods contain high amounts of tyramine which, unless destroyed, can prove fatal. Normally, they are readily destroyed by monoamine oxidase before reaching the central nervous system and having a deleterious effect. However, in the presence of monoamine oxidase inhibitors, this destruction is mini-

tryptophan

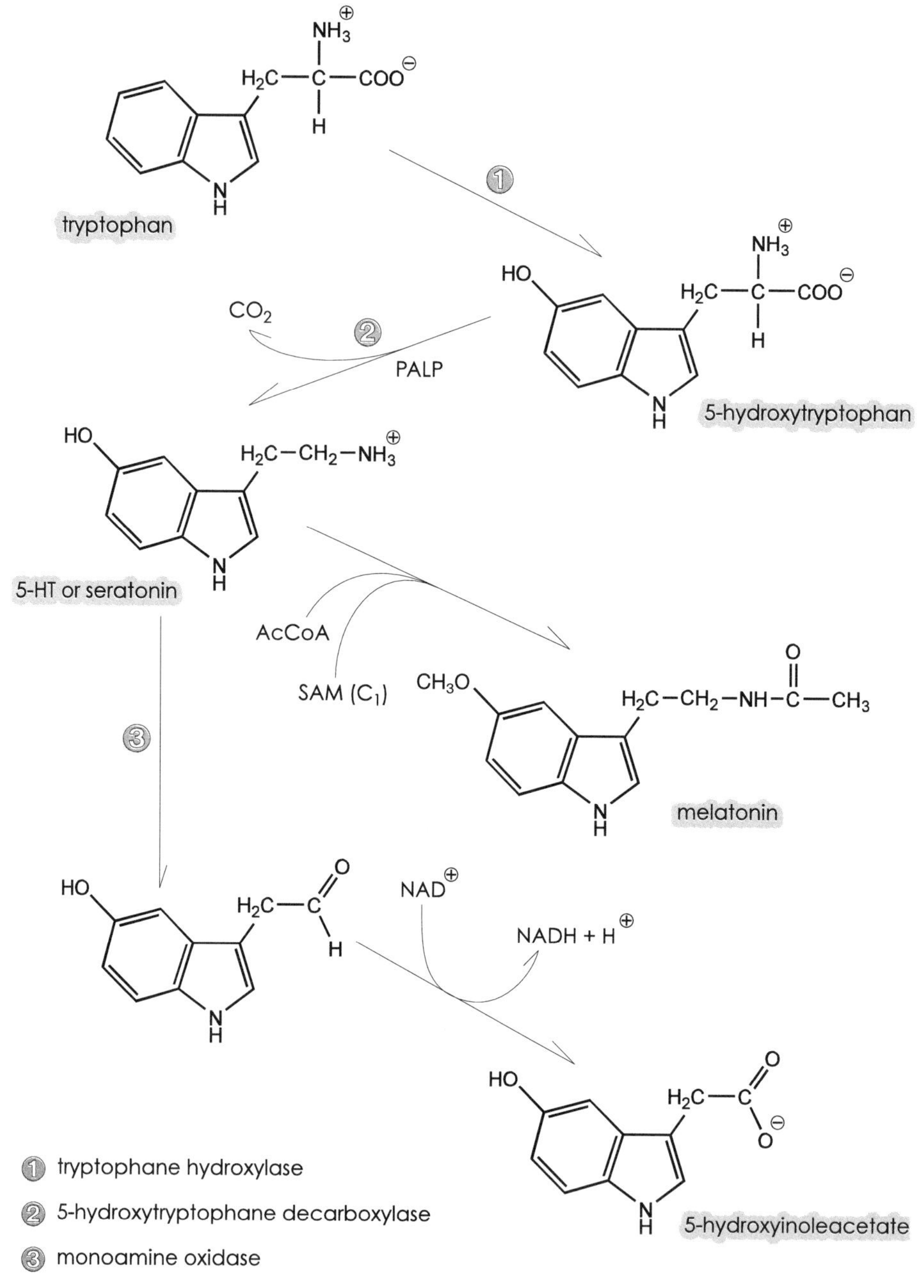

5-hydroxytryptophan

5-HT or seratonin

melatonin

5-hydroxyinoleacetate

① tryptophane hydroxylase

② 5-hydroxytryptophane decarboxylase

③ monoamine oxidase

Figure 19.5.
The hydroxylation pathway of tryptophan. The pathway of tryptophan hydroxylation in the 5 position and its products are shown above. These include the formation of serotonin and its degradation product, 5-hydroxyindole acetate, as well as melatonin formation. (See text for more detail.)

mized and tyramine can have extremely deleterious effects.

MELATONIN

Another fate of 5-hydroxytryptamine in the pineal gland is the acetylation of the amino group and an addition of a methyl group via *S*-adenosylmethionine, to the hydroxy group ***to form melatonin***. This compound plays a role in seasonal habits observed by animals due to changes in length of days. This compound causes birds to fly north or south at the appropriate times and Arctic rabbits to change the pigmentation of their fur to dark in the summer and white in the winter (hence the name melatonin, because it affected the melanines in these animals). In people, it may also affect behaviors that vary with the season and the amount of light obtained during the day, and reregulation may require some time for adaptation, particularly when traveling from northern to southern hemispheres.

GENERAL

Although diverse in structure, most amino acids lead to a few central compounds that flow into the major metabolic pathways. All of them effectively produce pyruvate, α-ketoglutarate, oxaloacetate, succinate, fumarate, acetoacetate, or acetyl CoA (Fig. 19.6). Therefore, a large number of unusual compounds are not formed, which would require a new set of enzymatic machinery for metabolism. This is also advantageous because the carbons from degraded amino acids can be funneled into the major pathways of metabolism with normal metabolic controls.

Protein and amino acids are always being turned over. ***There is no storage protein as such***; all proteins are functional. This is in contrast to carbohydrates, which have glycogen as a storage carbohydrate and lipid, which has triacylglycerol, as a storage lipid. To enable the cell and animal to remodel the metabolic machinery, proteins are in constant flux—being broken down to amino acids and resynthesized—allowing for a redistribution of proteins without necessarily a net change in total protein or amino acids. As proteins are degraded and produce amino acids, some of those amino acids will always be degraded by the body, thus the requirement for protein in the diet. This dietary protein is needed to replenish the amino acids that are catabolized. If there was a condition in people where amino-acid catabolism could be decreased to zero, there would be no (or a very low) dietary protein re-

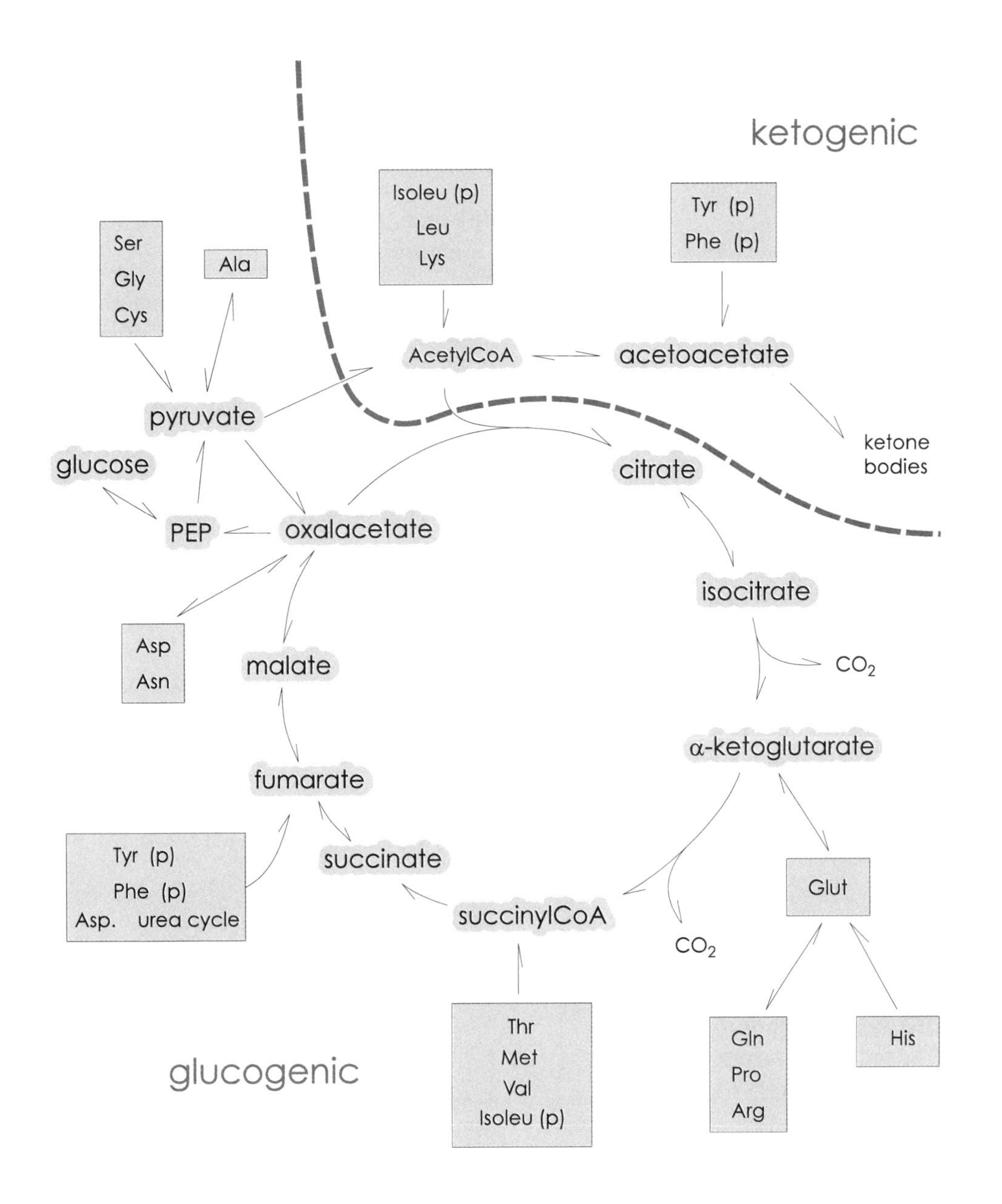

Figure 19.6.
Metabolism of amino acids to form common metabolic intermediates. This figure depicts the major fate of the amino acids that form common metabolic intermediates. Those formed above the dotted line are ketogenic and those formed below are glucogenic. Any amino acid followed by (p) is found both above and below the dotted line. As the (p) indicates, this amino acid produces both ketogenic and glucogenic products during metabolism.

quirement. All amino acids are needed for the synthesis of protein. ***If there is a deficiency of one amino acid, insufficient protein will be synthesized.*** All the other amino acids that would have formed the normal protein, had the deficient amino acid been present in normal amounts, will be degraded and no longer available for use as a precursor for protein. The deficiency of one amino acid is metabolically a more severe condition than a low-protein diet or starvation, because the higher protein levels will increase the rate of amino-acid degradation. This will produce a more rapid amino-acid depletion, due to protein turnover, than occurs in a low-protein diet or during starvation.

In starvation, it is critical to maintain both body-protein levels and blood-glucose levels. Since the large store of lipids can produce only a small amount of glucose (i.e., that coming from glycerol), it is incumbent on glycogen and amino acids to produce the remainder for glucose maintenance. During starvation, the nitrogen content of the urine, after several days, decreases and continues to decrease to a constant level. The reason for this decrease, is the availability of fatty acids and ketone bodies for energy, which decreases glucose utilization and, through hormonal mechanisms, decreases the rate of protein catabolism. The percent of the nitrogen excreted as ammonia in the urine is low at the start of starvation; after prolonged starvation, almost 80% of the nitrogen in urine can be excreted as ammonia. The advantage of this change is not only that it is energetically more favorable for ammonia to be produced compared to urea, but ammonia, as the ammonium ion, can provide an anion to help the body conserve sodium, potassium, and other minerals. This anion is particularly important during starvation because not only is the urine acidic, but there is an excretion of ketone bodies which requires a cation to accompany them. Thus, the ammonium ion can substitute for sodium and potassium. Furthermore, if the ammonium ion were converted to the urea, this would not only cause a loss of cations, but also produce two hydrogen ions, exacerbating the acidic burden. The same problem occurs in diabetes as in fasting or starvation.

SUMMARY

1. Histidine can be degraded to form ammonia and formiminoglutamate. The latter, with tetrahydrofolate, can form ammonia, N^5, N^{10} methylene tetrahydrofolate and glutamate.

2. Decarboxylation of histidine produces histamine.

3. Phenylalanine, an essential amino acid, can be converted to tyrosine in the liver, catalyzed by the enzyme phenylalanine hydroxylase. If phenylalanine hydroxylase is genetically insufficient, phenylketonuria results, with the excretion of phenylpyruvate. In this disease, tyrosine becomes an essential amino acid. This disease can be treated by feeding low-phenylanine diets.

4. The major quantitative pathway of tyrosine catabolism produces acetoacetate and fumarate. If homogentisate oxidase is missing, the result is alcaptonuria—dark urine.

5. Tyrosine is also the precursor of thyroxine, norepinephrine, epinephrine, DOPA, and the melanines.

6. Tryptophan can be converted to alanine and acetyl CoAs through a complex pathway.

7. Tryptophan is also the precursor of niacin, serotonin, and melatonin.

8. Most amino-acid metabolism produces simple metabolic products that readily flow into the metabolic flow. These products include acetyl CoA, glycolytic, and citric-acid cycle intermediates.

9. Protein continually turns over. In starvation, protein catabolism is decreased with time and the urinary nitrogen changes from primarily urea to primarily ammonia.

REFERENCES

Regulation of the activity of hepatic phenylalanine hydroxylase, S. Kaufman, 1986, *Adv. Enzyme Reg.*, 37–64.

Phenylketonuria, R. Koch and E. Wenz, 1987, *Ann. Rev. Nutr.*, **7**:115–117.

Albinism, C. J. Witkop, Jr., W. C. Quevedo, Jr., T. B. Fitzpatrick, and R. A. King, in *The Metabolic Basis of Inherited Disease*, Vol. 1, 6th ed., C. R. Scriver, A. L. Beaudet, W. S. Sly, and D. Valle, Eds., 1989, pp. 2905–2947.

Alcaptonuria, B. N. LaDu, in *The Metabolic Basis of Inherited Disease*, Vol. 1, 6th ed., C. R. Scriver, A. L. Beaudet, W. S. Sly, and D. Valle, (Eds.), 1989, McGraw-Hill, New York, pp. 775–790.

The hyperphenylalaninemias, C. R. Scriver, S. Kaufman, and S. L. C. Woo, in *The Metabolic Basis of Inherited Disease*, Vol. 1, 6th ed., C. R. Scriver, A. L. Beaudet, W. S. Sly, and D. Valle, (Eds.), 1989, pp. 495–546.

REVIEW QUESTIONS

1. The precursor of the catecholamines is
   a) Serine
   b) Tyrosine
   c) Tryptophan
   d) Histidine
   e) Glutamine

2. Which of the following directly produces ammonia during its normal metabolism?
   a) Phenylalanine
   b) Tyrosine
   c) Alanine
   d) Histidine
   e) Valine

3. Which of the following diseases can be treated by lowering dietary phenylalanine?
   a) Alhatonuria
   b) Phenylketonuria
   c) Albinisim
   d) Parkinson's disease
   e) Hypertension

4. The precursor of serotonin is
   a) Phenylalanine
   b) Tyrosine
   c) Histidine
   d) Tryptophan
   e) N^5,N^{10} methylene THFA

5. During starvation (prolonged), the following happens:
   a) Nitrogen and urea excretion rises
   b) Nitrogen and the proportion found as ammonia rises
   c) Nitrogen and the proportion as ammonia decreases
   d) Nitrogen excretion remains constant, but the proportion as ammonia rises
   e) Nitrogen excretion decreases and the proportion as ammonia rises

6. Glutamine is not the precursor of
   a) Arginine
   b) Proline
   c) Protein
   d) Purines
   e) Histidine

ANSWERS TO REVIEW QUESTIONS

1. ***b*** Tyrosine is the precursor of catecholamines, which include DOPA, dopamine, epinephrine, and norepinephrine.

2. ***d*** Histidase action upon histidine directly produces ammonia.

3. ***b*** Phenylketonuria is due to the inability to convert phenylalanine to tyrosine rapidly enough to maintain normal blood concentrations of phenylalanine. Lower and/or normal blood concentrations of phenylalanine can be maintained by feeding a low-phenylalanine diet.

4. ***d*** Tryptophan, after hydroxylation in the 5 position and decarboxylation, forms seratonin (5-hydroxytryptamine).

5. ***e*** Nitrogen excretion decreases to conserve body protein; also, lipids provide a greater percentage of energy use. The proportion of nitrogen excreted as ammonia (in the NH_4^+ form) helps with acid–base balance, especially with increased ketone-body excretion. normal blood concentrations of phenylalanine can be maintained by feeding a low-phenylalanine diet.

6. ***e*** Histidine is a dietary essential amino acid and cannot be formed from other amino acids.

CHAPTER

20

PURINES, PYRIMIDINES, AND HEME METABOLISM

INTRODUCTION

In this chapter we examine the synthesis and degradation of purines, pyrimidines, and hemes. These have complex structures, but are formed from simple precursors. All three can be synthesized in the body and have roles ranging from nucleic acids to hemoglobin. In addition to synthesis control of all three classes of compounds, a number of metabolic diseases associated particularly with purine and heme metabolism are discussed. The use of antimetabolites, as in chemotherapy, and the rationale for their use is presented.

Synthesis of these substances is carefully controlled, because it is metabolically expensive in terms of both energy and precursor molecules. Many, but not all, purines and pyrimidines can be interconverted. The catabolism of purines and hemes produces unique products, whereas pyrimidine catabolism produces common products. The catabolism of purines produces uric acid, which in excess can produce gout. The catabolism of hemes produces bile pigments which, when in excess in the circulation, are referred to as jaundice, and when excreted normally, add pigmentation to the feces.

PURINES

Purines can be found in nucleotides such as ATP, NAD^+, FAD, and coenzyme A, and play an important role in both RNA and DNA. There are numerous steps in the complex synthesis of purines, as outlined in Figure 20.1. However, it would be counterproductive to simply memorize all of the steps. It is important to visualize how a simple compound can be converted into purines, some important precursors, and particularly some of the steps where antimetabolites might enter the picture.

Purine synthesis starts with ribose-5-phosphate, which can be formed from the pentosephosphate pathway either via the oxidative arm or, in the reverse fashion, via the nonoxidative arm. Many tissues have the ability to form ribose-5-phosphate and their own purines. The first reaction is with ATP, where pyrophosphate is added to ribose-5-phosphate to produce phosphoribosylpyrophosphate (PRPP).

Phosphoribosylpyrophosphate can be used in a number of reactions. It is important in the synthesis of purines, pyrimidine nucleotides, and coenzymes such as NAD^+. Therefore, it would not be prudent for strict

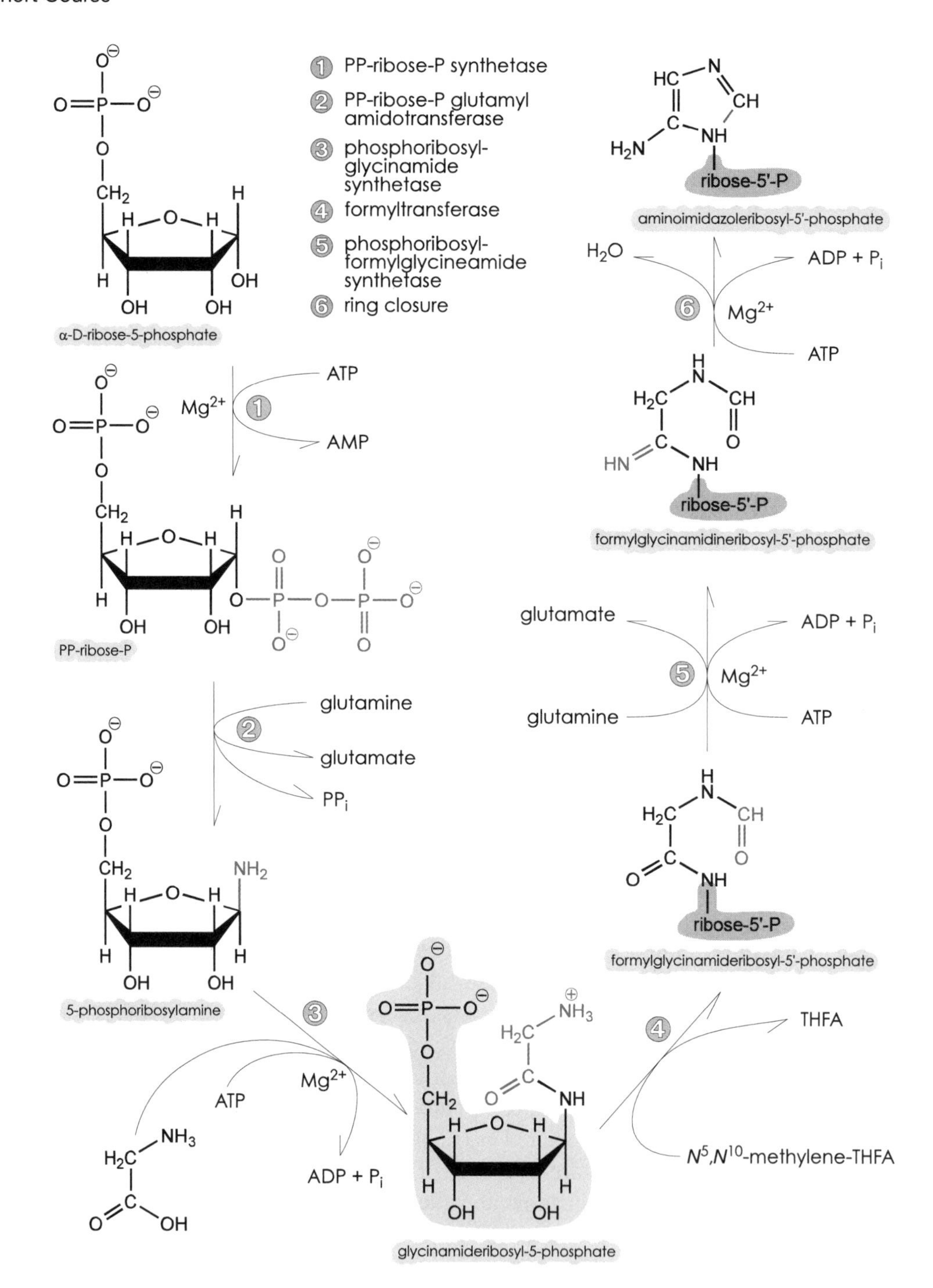

Figure 20.1.(a)

Purine synthesis. The synthesis of IMP is presented with structures and cofactors. The last constituent added in each reaction is shown in purple. THFA is tetrahydrofolate.

control of this system to be exerted by purines themselves. ***In purine synthesis, bases are formed around the ribosephosphate.*** This assures that the intermediates in the synthesis will not be lost from the cells. If the purine bases were to be made first, they would easily be lost from the cells, because purine bases without sugar phosphates are readily permeable out of cells into the bloodstream.

IMP Formation

In the next step, which is the first step uniquely related to purine synthesis, the amide nitrogen from glutamine is added to the PRPP to form 5-phosphoribosylamine, catalyzed by PRPP amidotransferase. This step can be inhibited by azaserine, an antimetabolite of glutamine. Glycine is then added, forming an amide bond. This re-

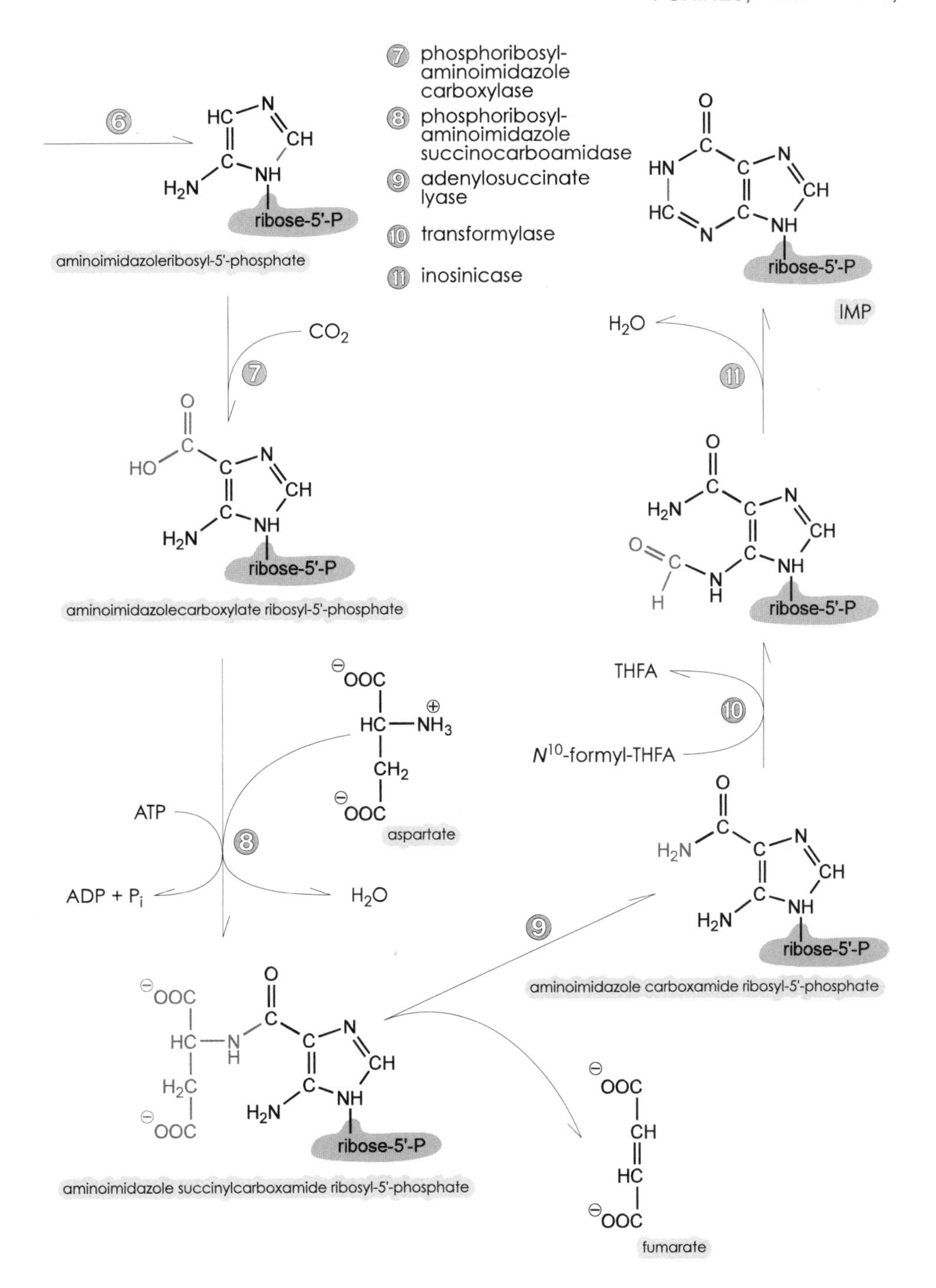

Figure 20.1.(b)

quires ATP. Glycine is the only amino acid in purine synthesis added intact. Next, a C1 fragment is transferred from N^{10}-formyl tetrahydrofolate onto the amino group of glycine to produce an aldehyde derivative. This is similar to tryptophan degradation, where a C1 aldehyde was removed from an amino group with tetrahydrofolate. A folate deficiency or an ineffective amount of tetrahydrofolate can slow the formation of nucleic acid synthesis at this and other steps. All cells suffer, but obviously those growing the fastest with the highest requirement of purines for nucleic-acid synthesis suffer the most. Therefore, drugs or antimetabolites that might effect these reactions can be used for chemotherapy to affect the most rapidly growing cells. Two examples of antimetabolites of tetrahydrofolate are methotrexate and aminopterin, which prevent the formation of tetrahydrofolate from dihydrofolate. Thus, even in the presence of sufficient dietary folate, there is a deficiency of tetrahydrofolate. These drugs and other related drugs have been used to treat leukemias as well as other oncological diseases. These drugs have a major effect on TMP synthesis (see page 280), as well as a lesser effect on purine synthesis.

Next, glutamine adds an amide nitrogen to the aldehyde group to form an amino derivative; once more azaserine could affect this step. The next step is a cyclization of the five-member ring of the purine requiring ATP. This is followed by a CO_2 fixation, a carboxylation of the five-member ring to start the six-member ring. This reaction is an unusual carboxylation, requiring neither biotin nor ATP. This intermediate then reacts with aspartate, requiring ATP, to give an aspartyl intermediate (this is similar to synthesis of arginine in the urea cycle). The aspartyl intermediate is then cleaved, with fumarate splitting off and leaving the nitrogen behind as part of the intermediate. A C1 unit from N^{10}-formyl tetrahydrofolate is then transferred to the amino group, once more to produce an aldehyde-type compound. Antimetabolites of tetrahydrofolate can inhibit this step as well as the earlier step. The aldehyde and the amino NH_2 cyclize and form a Schiff-base-like compound with the closing of the second ring. This produces the first nucleotide in the synthesis of inosine monophosphate (IMP). Note that this is not merely a base, but is also a nucleotide; IMP at the crossroads is an important precursor of other purine nucleotides. There are separate pathways to form guanosine monophosphates (GMP) and adenosine monophosphate (AMP).

Formation of Adenine and Guanine Nucleotides

Inosine monophosphate is oxidized with NAD^+ in a complex reaction, where FAD is a cofactor of the enzyme, which then passes the reducing power to NAD^+ to form NADH. This produces a keto group on the number 2 carbon of the ring. This is followed by donation of the amide nitrogen of glutamine to form an amino group at carbon number 2 and product GMP; this step requires ATP and produces pyrophosphate (Fig. 20.2). This last step could be inhibited by azaserine. The conversion of IMP to GMP is an important decision point. Once GMP is formed, the purine can no longer be used efficiently to form either IMP or AMP. Also, the energy for GMP formation comes from ATP; as will be seen later, the energy to form AMP will come from GTP. Thus, the lack of a specific nucleotide will not prevent its own formation owing to lack of energy.

For the conversion of IMP to AMP, some reactions are similar to those seen in the urea cycle. Aspartate interacts with the keto group of the IMP to give adenylosuccinate. The energy for this reaction comes from GTP. There is then a cleavage which produces fumarate, leaving the nitrogen on the ring to produce AMP. Adenosine monophosphate and GMP are not directly interconvertible; therefore, the cell must make a metabolic decision of which nucleotide to form from any given IMP. Adenine via a simple deamination can form IMP. The IMP could be reutilized to form GMP. However, once GMP is formed, the reconversion to IMP is slower and more complex.

Control of Purine Synthesis

One of the major controls of purine synthesis is the availability of PRPP. The enzyme forming this compound is of relatively low activity and is controlled by a number of intermediates, including purine nucleotide Fig. 20.3). However, since the fate of the PRPP can be to form purine nucleotides, pyrimidine nucleotides, or coenzymes, the control must be complex and not limited to only purine nucleotides, pyrimidine nucleotides, or coenzymes. ***The next enzyme under control by purine nucleotides is PRPP transamidinase, catalyzing the formation of the phosphoribosyl amide (PRPP + glutamine → $PRNH_2$ + glutamate + pyrophosphate), the first unique precursor of purines.*** Since the synthesis of PRPP is slow, there must be controls to allow its utilization in other processes when purine nucleotide pools are satisfied. The control of this system is interesting and shows the phenomena of synergism. That is, AMP, the product of one branch of the pathway, can inhibit the enzyme activity approximately 10%; GMP, the product of the other branch of the pathway, can also cause approximately a 10% inhibition. However, when both are sufficient, there is as much as a 90% inhibition of the PRPP transamidinase. Thus, there is a synergistic action where the two effectors together work much more effectively than each would by itself. Both at this step and at the branch, the K_I for GMP is lower, reflecting the normally higher adenylnucleotide to guanylnucleotide ratio in the cell.

There are no major apparent controls for $PRNH_2$ to inosine monophosphate. ***For GMP or AMP, each of the purine nucleotides inhibits the pathway for the formation of more of the same nucleotide.*** For example, GMP will inhibit the oxidation of IMP, which is the first step toward GMP formation, whereas AMP will

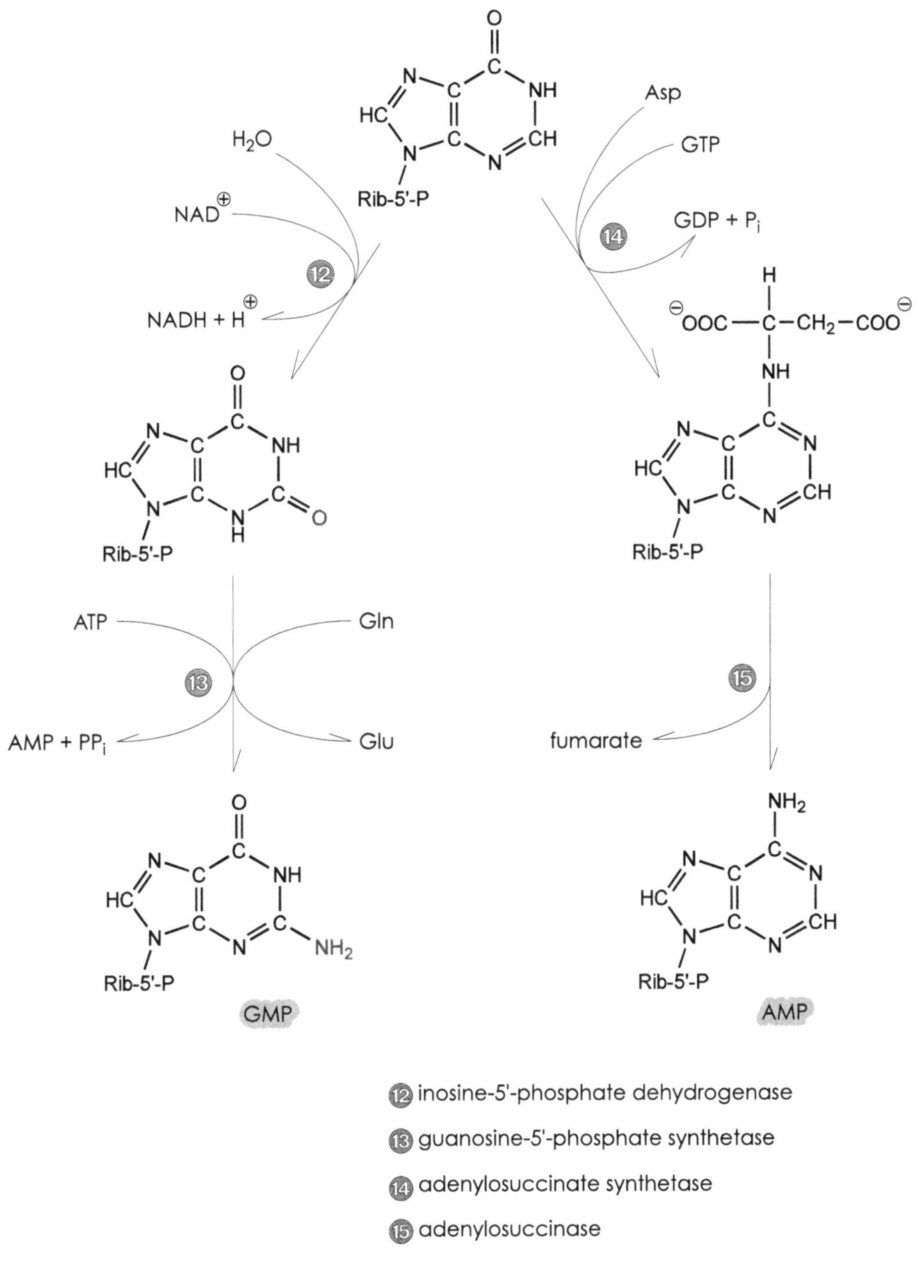

⑫ inosine-5'-phosphate dehydrogenase

⑬ guanosine-5'-phosphate synthetase

⑭ adenylosuccinate synthetase

⑮ adenylosuccinase

Figure 20.2.
Synthesis of AMP and GMP. This figure shows the pathways of IMP conversion to AMP or GMP. The number of each reaction corresponds to enzymes listed in the figure.

inhibit the formation of adenylosuccinate, the first step in AMP formation. Both GMP and AMP can form the di- and triphosphates, which will be in equilibrium, depending on the energy charge. The deoxynucleotides can be formed both from purines and pyrimidines, basically via a similar mechanism. There is a major difference between the ribonucleotides and the deoxyribonucleotides; the former plays roles in coenzymes and RNA, whereas the latter plays a role in DNA. ***The formation of deoxynucleotides starts with the diphosonucleotides.*** The cofactor to form the deoxynucleotides is NADPH via thioredoxin. The control of deoxynucleotides is discussed in Chapter 8.

Purine Degradation

The body is continually turning over and catabolizing nucleic acid and nucleotides. Therefore, the nucleotides and the bases are continually being destroyed. The first enzyme in AMP degradation is AMP deaminase, which catalyzes the conversion of AMP to IMP and ammonia. This occurs in most tissues and, in many cases, the ammonia generated leaves the tissues, such as the muscle, as glutamine. The IMP formed can either be further degraded or reutilized to reform AMP or GMP. If there is excess IMP and sufficient adenine and guanine nucleotides, the reactions to form further AMP

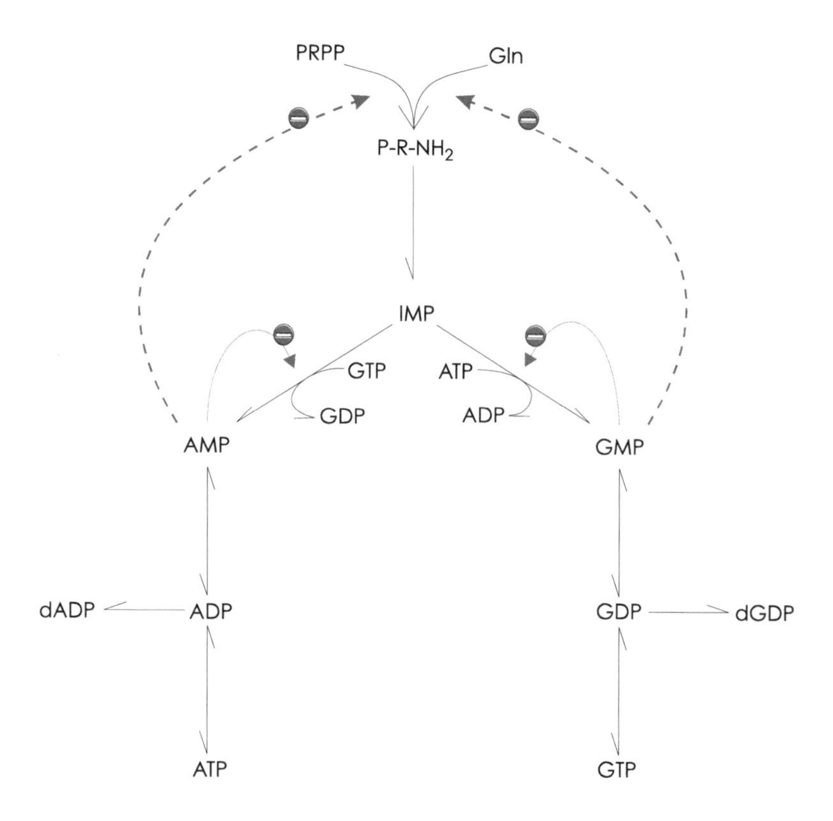

Figure 20.3.

Control of purine synthesis. A schematic of the control of purine synthesis, with only the control steps shown, is presented in this figure. The inhibition of the first reaction is synergistic and only partial for AMP or GMP alone.

or GMP from IMP will be inhibited and IMP will be degraded. First, inorganic phosphate is removed by a phosphatase and the ribose is removed by phosphorolytic cleavage to produce hypoxanthine and R-1-P. Hypoxanthine can also be salvaged to produce IMP via the reaction hypoxanthine + PRPP to form IMP + pyrophosphate. However, the availability of PRPP is low and, therefore, much of the hypoxanthine will be further degraded. ***The central nervous system and many other nonhepatic cells and tissues depend heavily on a salvage pathway to maintain purine nucleotide levels.*** If the enzymes of the salvage pathway are deficient, especially in the central nervous system, where they are normally high, so that there is a lack of ability to salvage purine bases back to nucleotides, the result is a Lesch–Nyhan syndrome (Fig. 20.4). In this syndrome, some people will self-mutilate, biting off fingers, toes, and other extremities. There is a correlation between this disease and an insufficiency of the salvage pathway of purine bases plus PRPP back to form nucleotides.

Guanine nucleotides follows a similar pathway where the amino group is oxidized to a keto group and there is removal of phosphate and then ribose to form xanthine. Thus, the two products of purine metabolism are hypoxanthine and xanthine. These are water soluble and can exit the cell and be carried to the liver for further oxidation. An enzyme called xanthine dehydrogenase (which contains molybidium) can oxidize hypoxanthine to xanthine and xanthine to uric acid.

Guanosine monophosphate can be converted to IMP by a reductive deamination and the resulting IMP can produce AMP. The salvage pathway is of great importance to extrahepatic cells, which cannot either make purines *de novo* or synthesize purines in sufficient quantities. The salvage pathway also allows for the use of purine derivatives such as 6-mercaptopurine and 6-thioguanine as antimetabolites in chemotherapy.

Gout

Uric acid is relatively insoluble and can form crystals, resulting in a painful condition. If uric acid accumulates in excess, it can precipitate in certain bone joints, causing a disease known as gout. For people with a propensity toward this problem, eating foods high in nucleic acid, like meat and yeast, may exacerbate the situation. ***One of the treatments of gout is the use of an analogue of uric acid known as allopurinol*** (Fig. 20.5). Allopurinol has a structure somewhat similar to hypoxanthine and acts as a competitive inhibitor of xanthine dehydrogenase, thus slowing the conversion of hypoxanthine and xanthine to uric acid. In that situation, xanthine and hypoxanthine rise in the blood and

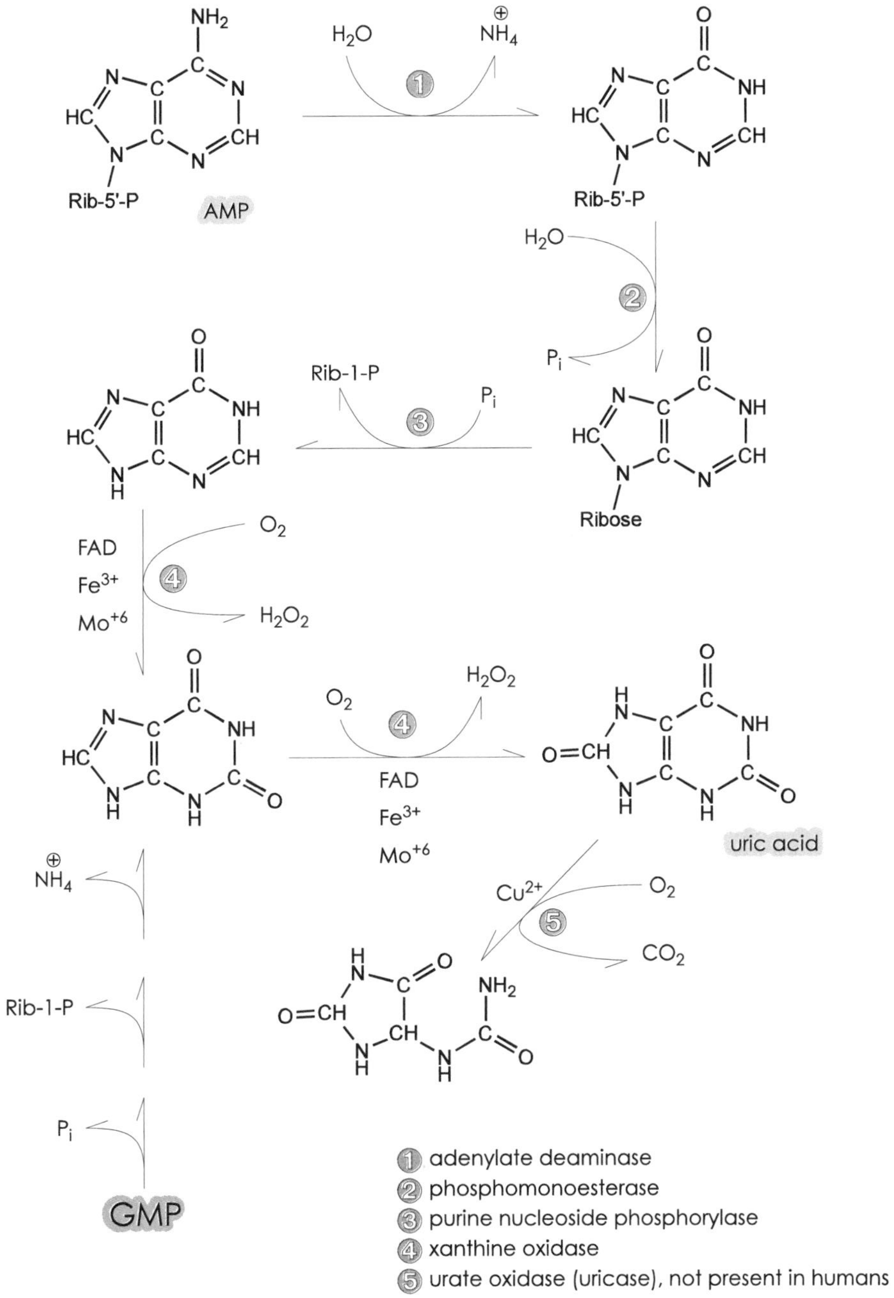
AMP
GMP
uric acid
Rib-5'-P
Ribose
Rib-1-P
FAD
Fe^{3+}
Mo^{+6}
Cu^{2+}
① adenylate deaminase
② phosphomonoesterase
③ purine nucleoside phosphorylase
④ xanthine oxidase
⑤ urate oxidase (uricase), not present in humans

Figure 20.4.

Purine catabolism. The pathways to uric acid from guanine and adenine nucleotides are presented above.

hypoxanthine allopurinol

Figure 20.5.

Allopurinol. The structure of allopurinol is shown with that of hypoxanthine to depict the structural similarities.

are spilled into the urine, avoiding the problem of high uric acid, since both xanthine and hypoxanthine are soluble and will not cause harm when excreted. This problem is due to the relative insolubility of uric acid. The end product of purine catabolism in people is uric acid and therefore, all of the precursors used for purine synthesis, such as C1 fragments and glycine, are lost to the body when the uric acid is excreted and are not available for potential reutilization.

PYRIMIDINES

Synthesis

Pyrimidines synthesis occurs in almost all cells. ***The first step of pyrimidine synthesis is the formation of carbamoyl phosphate.*** In the case of pyrimidine synthesis, carbamoyl phosphate is formed using the amide nitrogen of glutamine, CO_2, and two ATPs (the use of two ATPs assures that the reaction is driven in a forward direction). These substrates are different than those with the carbamoyl phosphate used for urea synthesis. The enzyme for the glutamine-dependent carbamoyl phosphate synthetase (CPS II) is in the cytosol, whereas that for urea synthesis (CPS I) is in the mitochondrion. The glutamine-dependent carbamoyl phosphate synthetase is present in most cells, whereas the mitochondrial carbamoyl phosphate synthetase is present primarily in the liver, kidney, and intestines (Fig. 20.6).

Carbamoyl phosphate is combined with aspartic acid catalyzed by aspartate carbamoyl transferase, in which the carbamoyl group is added to the amino group of aspartic acid. This reaction is similar to the formation of citrulline from ornithine and carbamoyl phosphate. Carbamoyl aspartate cyclizes to form a six-membered ring. A difference can immediately be seen between purine and pyrimidine synthesis. Purine synthesis started with the building of a base around a sugar phosphate; pyrimidine synthesis starts with the synthesis of the pyrimidine base. Then, catalyzed by a FAD-linked enzyme, two hydrogens are removed from the ring to form orotic acid. These hydrogens are then transferred to NAD^+. The carboxyl group on the orotic acid helps to retain the intermediates in the cell until final synthesis is achieved. The next step is the addition of PRPP to the orotic acid to form orotidine monophosphate (OMP). The OMP is then decarboxylated to form UMP. The carboxyl group is now not needed to keep the intermediate in the cell because the sugar phosphate will prevent leakage from the cell. Depending on the energy charge, UMP can form UDP and UTP reversibly. ***The UTP can be converted directly to CTP,*** with the only structural difference being that CTP has an amino group in place of the keto group on carbon number 4. This conversion utilizes the amide nitrogen of glutamine, requires ATP, and can be allosterically activated by GTP. Cytidine nucleotides can also be deaminated, as was the case for AMP and adenylate deaminase, and produce uridine nucleotides. In contrast to the purine synthesis, there are no branches in pyrimidine synthesis, and the two final products can readily be interconverted. Thus, it is unlikely that there will be an imbalance of CTP compared to UTP, and if this did occur, CTP could be converted to UTP to reestablish normal balance. The reverse could be true also of excess UTP being converted to CTP to reform the balance.

Control of Pyrimidine Synthesis

There are two potential control steps; one is used primarily by bacteria, the other in animal systems. Bacteria use inhibition of aspartate carbamoyl transferase by CTP—this is the second step of pyrimidine synthesis. This functions as a control in bacteria because the same carbamoyl phosphate used for pyrimidine synthesis can be used for arginine synthesis (Fig. 20.7); thus, the carbamoyl phosphate is not used solely for pyrimidine synthesis. In the animal system, the carbamoyl phosphate used for arginine and urea synthesis is formed in the mitochondrion via CPS I and the cytosolic enzyme (CPS II) functions only for pyrimidine synthesis; hence, the control is effectively different. ***In people, UTP feeds back and inhibits the cytosolic carbamoyl phosphate synthetase (CPS II)***; UTP has no effect on the mitochondrial carbamoyl phosphate synthetase (CPS I). One major control system in pyrimidine synthesis does not require the synergism seen in purine synthesis. This is so because uridine nucleotide and cytidine nucleotide synthesis is a linear pathway with no branches and the end products are readily interconverted. This is in strong contrast to the synthesis of purine nucleotides. Another control, as in the case of purine nucleotides, is the availability of PRPP. Since the normal pathway of pyrimidine synthesis is to form the base first and then add the sugar phosphate, salvage in this pathway is more efficient than the salvage of purine nucleotides, which are made by building the bases around the sugar phosphate. Uracil, orotate, and thymine, but not cytosine, are efficiently salvaged. An interesting ex-

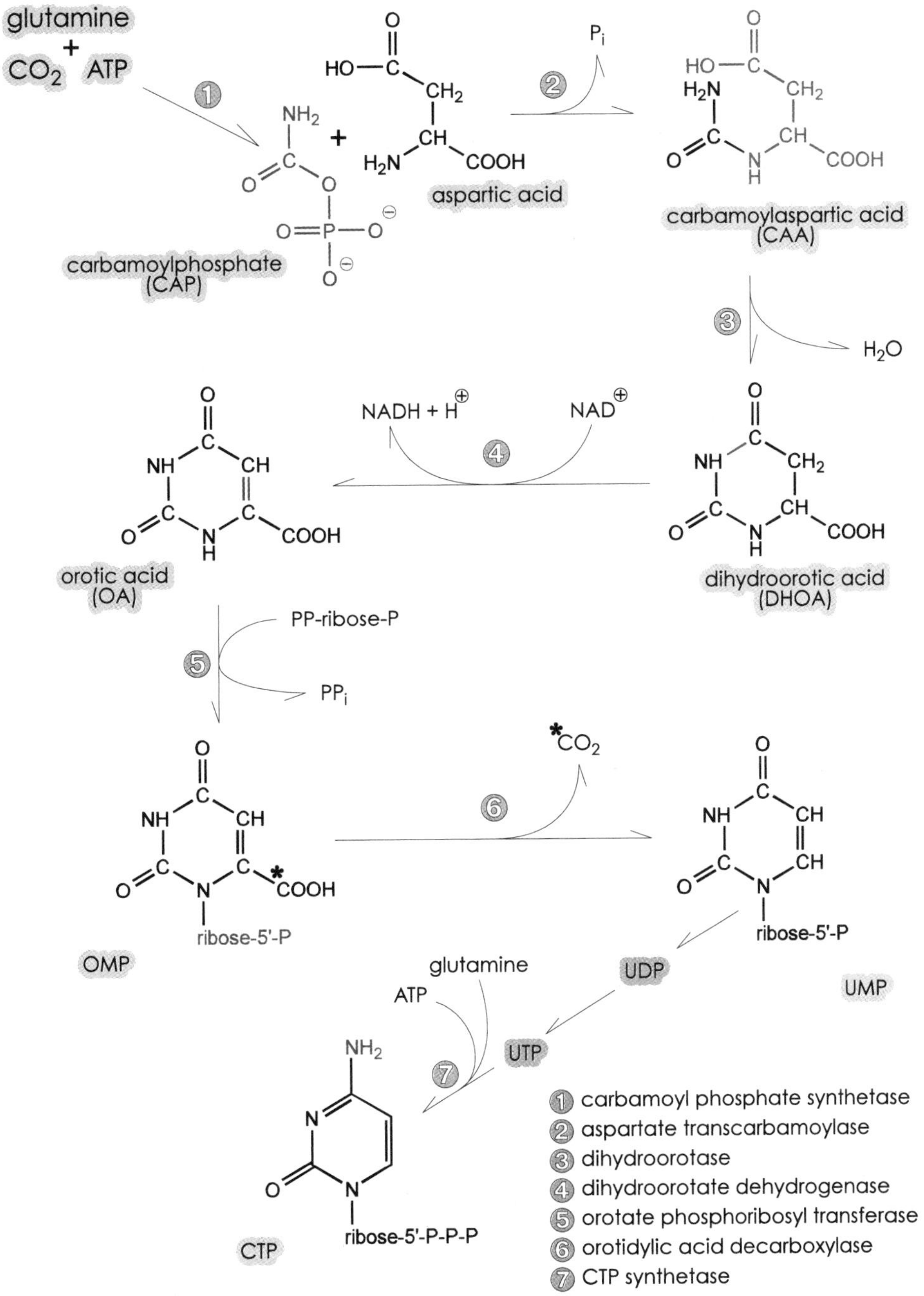

Figure 20.6.
Pyrimidine synthesis. The synthesis of pyrimidines, UTP and CTP, is presented with structures and cofactors. The last constituent added in each reaction is shown in purple.

ample related to the limitation of PRPP occurs with pyrimidine synthesis in cases of a severe deficiency of arginine or ornithine carbamoyl transferase. In these cases, the liver mitochondrion forms carbamoyl phosphate, but there is insufficient enzyme activity or ornithine, which normally arises from arginine. The carbamoyl phosphate will accumulate and leak out of the mitochondrion. Since this is past the control step for pyrimidine synthesis normally found in animals, the carbamoyl phosphate will combine with aspartic acid to form carbamoyl aspartate and orotic acid. However, owing to the limitation of PRPP, the orotic acid may accumulate and be excreted in the urine during these conditions.

Degradation

The degradation of pyrimidines has some similarities and dissimilarities with that of purines. They are similar

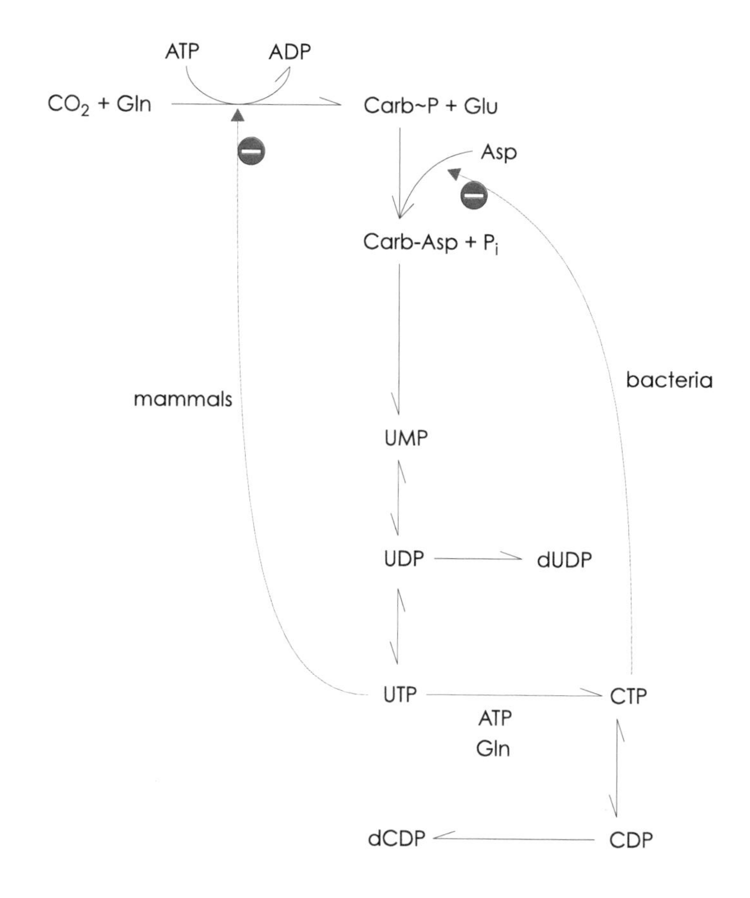

Figure 20.7.

Control of pyrimidine synthesis. The control of pyrimidine synthesis is presented in outline form. Control for both bacteria and mammals (including humans) is shown.

in that the phosphate is removed and then the ribose to give the bases cytosine or uracil. There is also a salvage pathway, as mentioned previously, for some of the bases. The major degradative pathway for cytosine is a loss of ammonia to produce uracil, which can be salvaged (Fig. 20.8). If not salvaged, uracil can be converted to dihydrouracil; the ring is cleaved open, followed by the loss of ammonia and CO_2, forming β-alanine. Beta-alanine can give rise to malonate and be activated to malonyl CoA. Thus, the catabolism of pyrimidines produces no unique end products in the urine, such as uric acid for the purines, and all of the components are potentially reutilizable.

Thymine Formation

All the pyrimidine nucleotides can proceed to the deoxy compounds as diphosphates, similar to the purines. However, ***uridine does not occur in DNA; therefore, it must be converted to a thymine derivative.*** This occurs by the following basic pathway: CDP is converted to dCDP. This is converted to dCMP, which can produce dUMP. The dUMP is then methylated in the number 5 position, via N^5, N^{10}-methylene tetrahydrofolate. This produces deoxythymidine monophosphate (dTMP) and dihydrofolate. The dTMP then can be converted to dTDP and dTTP, which can be incorporated into DNA (Fig. 20.9).

Antimetabolites

The unique aspects of thymine synthesis allow the design of antimetabolites that may have greater specificity in this process. Aminopterin and methotrexate have been used with some success. These are inhibitors of tetrahydrofolate formation where, at the correct dosage, the conversion of dUMP to dTMP appears to be more sensitive than many other reactions and is preferentially inhibited. This inhibits DNA formation and cell proliferation. Other antimetabolites, 5-bromouracil and 5-flurouracil, have been used in cancer therapy. These are activated to the nucleotides in the same fashion as orotic acid. The 5-bromouracil (which is less popular now) is incorporated into the DNA and causes misreading of the genetic code. This results in the synthesis of ineffective protein, but can also produce mutations in normal cells as well as carrying out its action in malignant cells. Bromine is large enough to look like

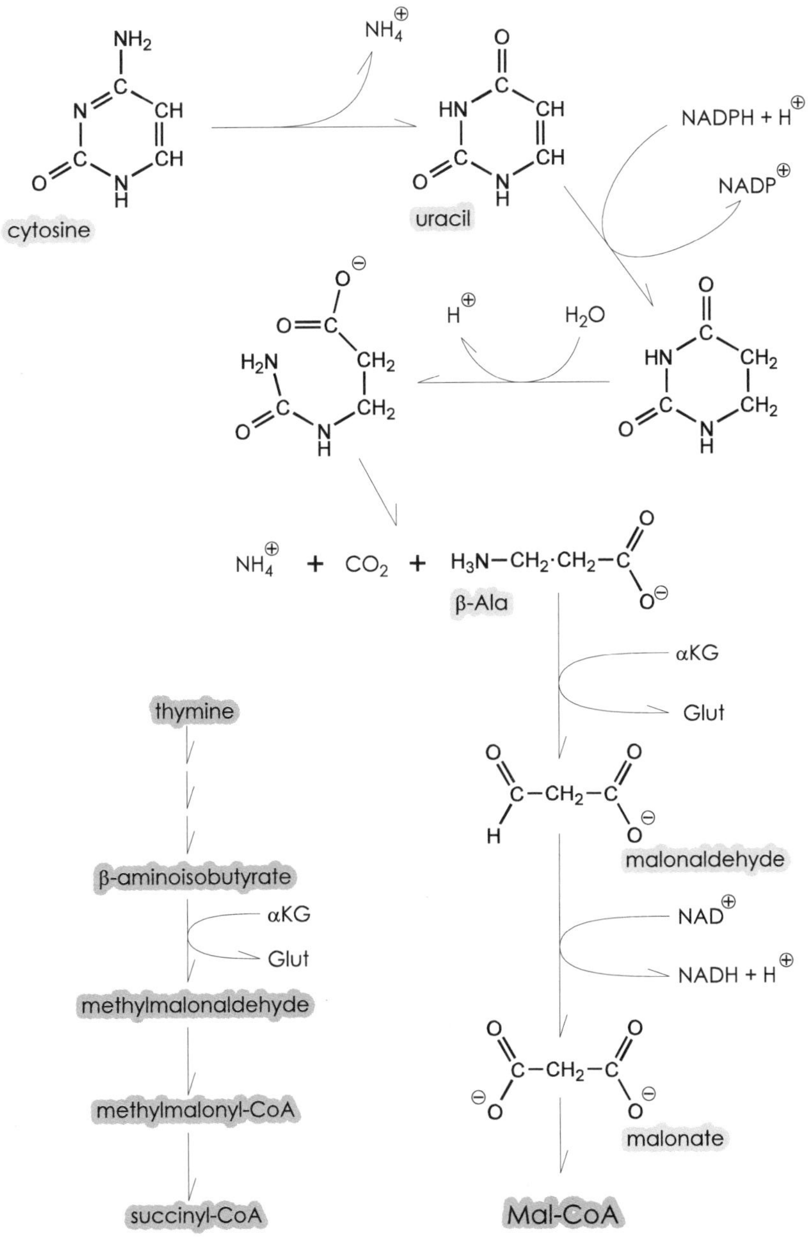

Figure 20.8.

Pyrimidine catabolism. The steps in the catabolism of cytosine and uracil are shown with corresponding structures; the steps for thymine are shown only in outline form.

the methyl group so that the compound appears like dTTP, but is really five-bromo UTP, and is incorporated as thymine, but does not have as much fidelity in reproduction of daughter molecules. 5-Flurouracil is activated and can be converted to dUMP with a fluoro group on a five position. The fluorine is small, but binds much tighter to the carbon than hydrogen would, so the fluorine cannot be exchanged for the methyl group. The 5-FdUMP and the tetrahydrofolate form a complex with thymidylate synthase which is inactive and thus prevents thymidine synthesis. This has the advantage of affecting DNA synthesis almost solely and, therefore, the most rapidly growing cells, without affecting other major metabolic pathways, as could be the case with methotrexate and azaserine.

There are numerous other chemotherapeutic agents, including 6-mercaptopurine, cytosine arabinoside, and 3′-azido-3′-deoxythymidine (AZT). 6-Mercaptopurine is converted to 6-mercaptopurine ribonucleoside 5′-monophosphate, which accumulates and inhibits PRPP-amidotransferase and the conversion of IMP to adenine and guanine nucleotides. Cytosine arabinoside is converted to cytosine arabinoside 5′-triphosphate, which competes with dCTP in the DNA polymerase reaction. The cytosine arabinoside 5′-triphosphate is incorporated into DNA, which halts the

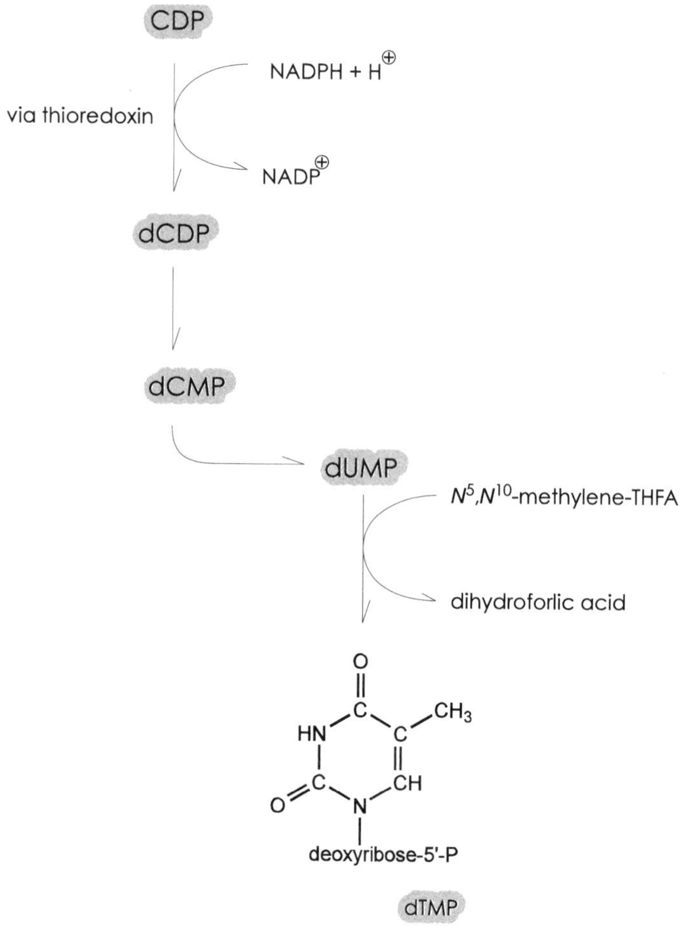

CDP
$NADPH + H^{\oplus}$
via thioredoxin
$NADP^{\oplus}$
dCDP
dCMP
dUMP
N^5,N^{10}-methylene-THFA
dihydroforlic acid
HN
C
O
C
CH_3
C
CH
N
O
deoxyribose-5'-P
dTMP

Figure 20.9.

Thymine formation. The pathway for CDP to dTMP is presented. Only the structure of dTMP (*d*-5-methyluridine monophosphate) is shown.

further synthesis of the DNA strand. The AZT is phosphorylated by cellular kinases to form AZT triphosphate. This compound inhibits a HIV–DNA polymerase, a RNA-dependent polymerase.

HEME

Heme is a tetrapyrrole consisting of a porphyrin ring with a metal constituent. A major amount of heme in the human body is found in hemoglobin. There are other important heme-containing proteins, such as the cytochromes and microsomal oxidizing enzymes; however, hemoglobin accounts for the largest portion of heme found in the human body. Heme combines with a protein to produce specific functions. For example, heme can be found in hemoglobin in combination with the protein globin used in O_2 transport. Various cytochromes that contain heme are involved in electron transfer. Heme can be combined with the protein by ionic and hydrophobic bonding, as with hemoglobin or with covalent bonding, as found with cytochrome C.

Despite the complexity of the structure, the porphyrin portion of heme can be synthesized from simple molecules and the metals arise from dietary sources.

A major site of heme synthesis is developing red blood cells, called reticulocytes. Much of the synthesis and control of heme synthesis has been studied in reticulocytes. In contrast to the mature red blood cells, reticulocytes contain a nucleus, mitochondria, and other subcellular organelles which are lost on maturation. All cells make some porphyrins for their own enzymic machinery, such as the cytochromes and endoplasmic reticulum oxidases; however, the reticulocytes are unusually active in heme synthesis.

Synthesis of the porphyrin involves the integration between the mitochondrion and cytosol, as in a number of metabolic processes previously examined such as urea synthesis, gluconeogenesis, and some amino-acid catabolism. The first step is mitochondrial, where glycine and succinyl CoA combine, catalyzed by the pyridoxal phosphate-dependent enzyme, amino levulinate synthase (Fig. 20.10).

It should be recognized that the advantage of initi-

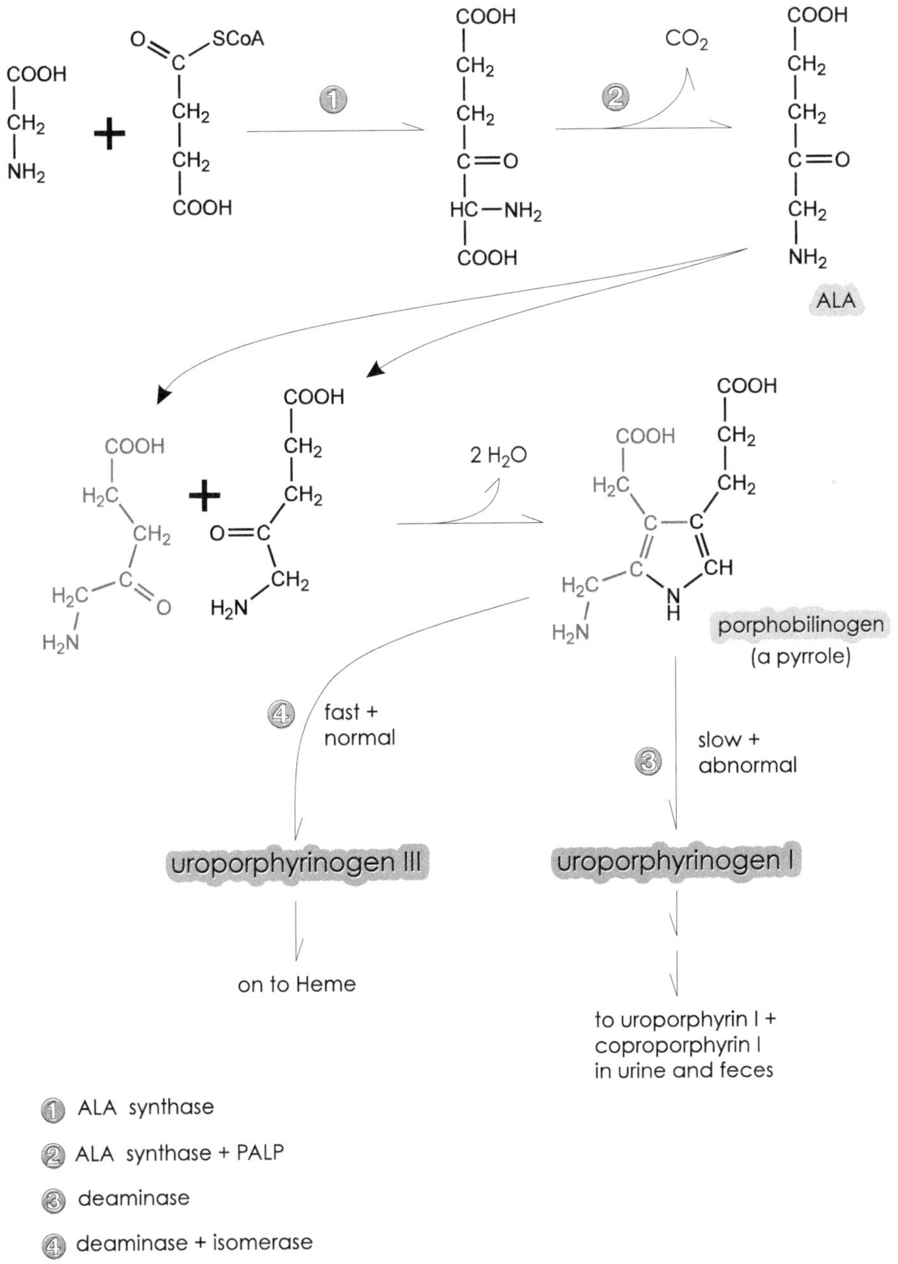

ALA

$2\ H_2O$

porphobilinogen
(a pyrrole)

④ fast + normal

③ slow + abnormal

uroporphyrinogen III

uroporphyrinogen I

on to Heme

to uroporphyrin I + coproporphyrin I in urine and feces

① ALA synthase
② ALA synthase + PALP
③ deaminase
④ deaminase + isomerase

Figure 20.10.
Porphyrin synthesis. The synthesis of porphyrins is shown, with the structures of metabolites up to porphobilinogen (a pyrrole) presented. After that point, only an outline is shown.

ating this step in the mitochondrion is the availability of succinyl CoA, which would not permeate across the mitochondrial membrane to the cytosol. The first step is the slow step in heme synthesis and, as will be seen, a major control. A glycine–succinate intermediate is formed which, while bound to the enzyme, is decarboxylated and released as delta amino levulinate (δ ALA). The δ ALA then crosses the mitochondrial membrane to the cytosol, where all but the last three reactions take place. The first reaction occurs when two molecules of δ ALA combine with the loss of two waters to form a five-member ring called porphobilinogen. This is a pyrrole ring, which is the building block for porphyrin synthesis. Porphobilinogen is converted in a series of steps to a compound with four pyrrole rings, namely uroporphyrinogen 3. This requires a number of enzymes and forms a normal compound that will form heme. ***One of the key enzymes in this formation is an isomerase, which incorporates one of the porphobilinogens in the tetrapyrrole in a "reverse direction."*** Thus, although three of the four porphobilinogens are incorporated in the expected arrangement, one is in an unexpected backward position (see Fig. 20.11). The tetrapyrrole, when formed, has two side groups per pyrrole—an acetyl and propionyl side

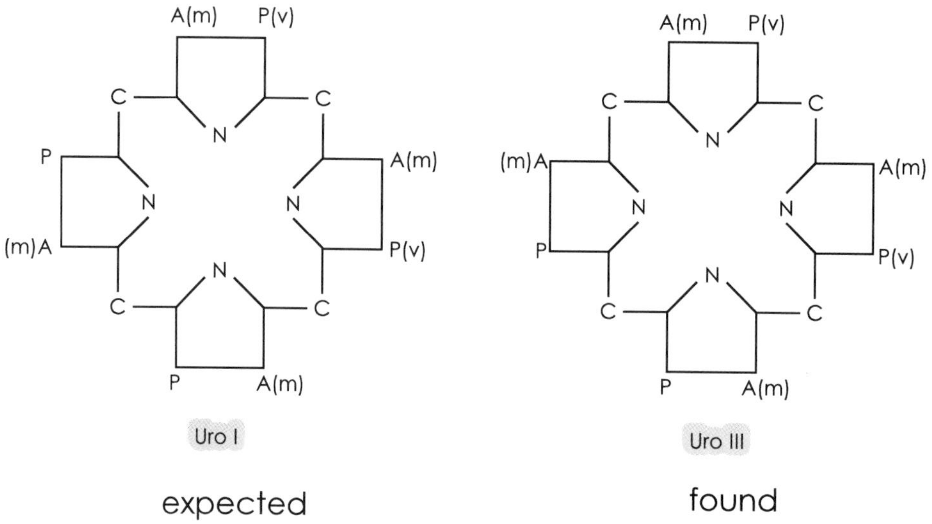

Figure 20.11.

Uroporphyrinogens. The structures of uroporphyrinogen I, the expected structure, and uroporphyrinogen III, the normal structure observed in heme, are shown for comparison. The side groups are propyl and acetyl groups in the original tetrapyrrole formed. These are converted from acetyl to methyl [A(m)]; two of the propyl group are converted to vinyl groups [P(v)] and two propyl groups remain as propionates.

Uro I

expected

Uro III

found

group. In the case of heme, all of the acetyl groups are converted to methyls and two of the propionyl groups are converted to vinyl groups in a formation of coporphyrinogen. One would expect a final product where the side chains considering the original source would be up as a-p-a-p-a-p-a-p.

However, it was found that the lineup was a-p-a-p-a-p-p-a; thus, one of the pyrroles was apparently incorporated in a backward configuration in the heme. This is required for normal heme synthesis. Occasionally, the coenzyme (uroporphrinogen III cosynthase) that aids in catalyzing the normal formation of uroporphrinogen 3 is present in low amounts. If no coenzyme is present, death will result, probably in the embryonic state. If it is low, abnormal porphyrins will be formed that have no physiological function (i.e., uropophyrinogen I). This is a nonfunctional end product and cannot be used to make heme.

Uroporphyrinogen decarboxylase catalyzes the conversion of uroporphyrinogen to coproporphyrinogen. The coporphyrinogen III can be oxidized to protoporphyrinogen IX, which is then oxidized to protoporphyrin IX. Ferrochelatase then inserts ferrous iron to form heme. The last three steps take place in the mitochondrion.

Control of Heme Synthesis

Normally, uroporphyrinogen III is converted to protoporphyrin IX and Fe^{2+} is inserted to form heme. Heme can combine with globin to form hemoglobin. To form the hemoglobin, we need not only the heme, but also the synthesis of the protein globin. If the reticulocyte has sufficient hemoglobin, feedback controls will control globin synthesis (Fig. 20.12). If globin is not synthesized, heme then accumulates in the presence of oxygen, and Fe^{2+} is converted to Fe^{3+} to give hematin. ***Both heme and hematin are signals that feed back and cause a decrease in the synthesis of δ ALA synthase and inhibit this enzyme.*** This allows for coordination between heme formation and globin formation to synthesize sufficient hemoglobin. When the heme synthesis exceeds that of globin, this feedback system is in effect.

Hemoglobin Maintenance

One important consideration is that only the immature red blood cell can synthesize the hemoglobin and, once mature, the mount of hemoglobin cannot be increased by new synthesis. Thus, during the process of the red blood cell carrying oxygen with the hemoglobin, there is some oxidation of the hemoglobin to methemoglobin (where the iron goes from 2+ to 3+), a form that will no longer carry oxygen. If this were allowed to continue, the red cell would soon be ineffective in carrying oxygen. Therefore, the red blood cell contains an enzyme, methemoglobin reductase, which catalyzes the oxidation of glutathione and reduces the iron in methemoglobin from 3+ to 2+, reforming hemoglobin and oxygen-carrying capacity.

Abnormal Synthesis and Porphyria

There are numerous types of porphyrias of genetic origin. One of the more common is acute intermittent

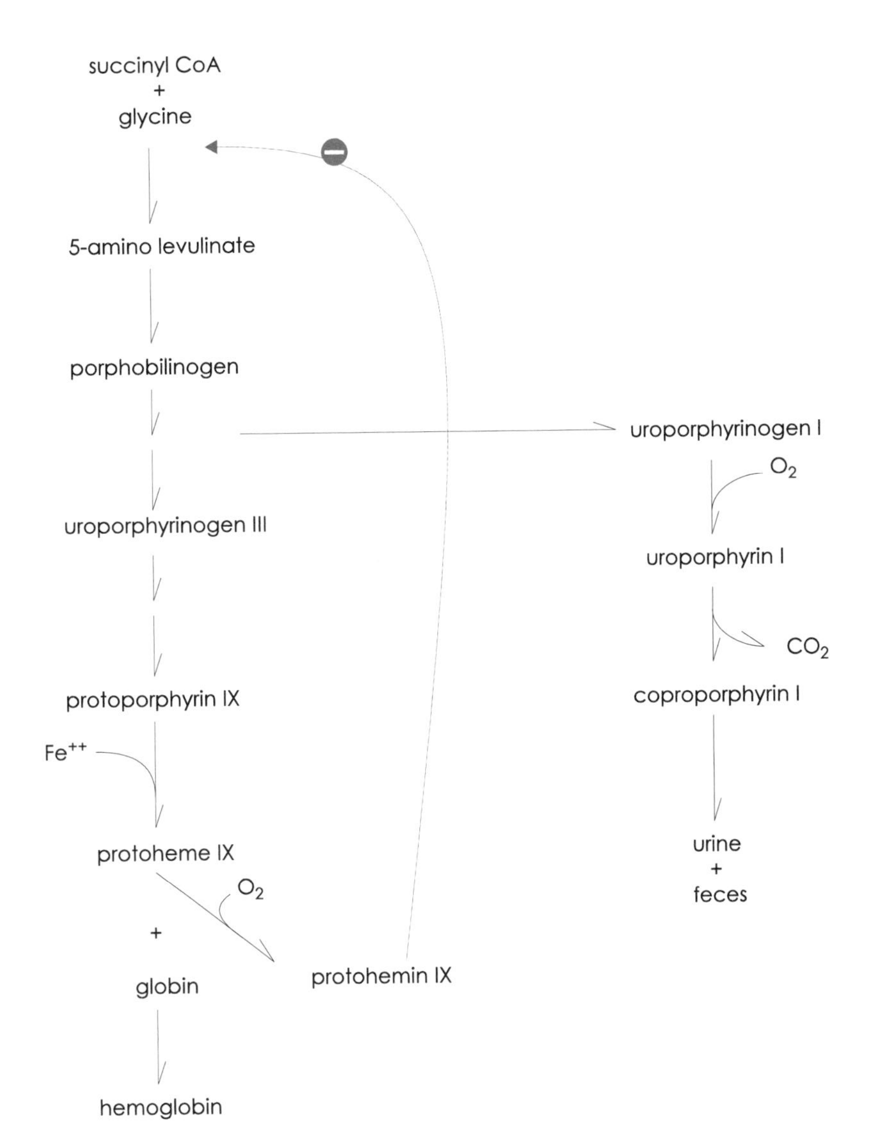

Figure 20.12.

Control of porphyrin synthesis. The control of porphyrin synthesis is shown in outline form, including inhibition of the first step, as well as a decreasing amount of the enzyme by protohemin 1X (where Fe^{2+} is oxidized to Fe^{3+}). The pathway of the synthesis of abnormal porphyrins is also depicted.

porphyria, a genetic disorder in which there is an increase in δ ALA synthase and a decrease in urophyrinogen I synthase (i.e., the synthesis of the tetrapyrrole), causing an elevation of porphobilinogen, which is excreted in the urine. Sufferers of porphyria have been associated with legends of vampires, owing to their sensitivity to light and unusually red lips. It is thought that King George III of England suffered from this disease, which led to intermittent fits of madness. Lead poisoning leads to an inhibition of two enzymes of heme synthesis including δ ALA dehydrase (i.e., the formation of prophobilinogen) and ferrochelatase (formation of heme); therefore, δ ALA accumulates and is excreted in the urine.

Degradation of Hemes

Hemoglobin will be used as the heme example, although similar aspects follow for all the body's heme. Red blood cells have a lifetime in humans of about 120 days. Once the red blood cell is destroyed, the hemoglobin is degraded. This degradation can occur in the spleen and other reticuloendothelial cells. The heme of hemoglobin, no longer functional, is converted to biliverdin by a microsomal (endoplasmic reticulum) enzyme called heme oxygenase, which releases the iron and products CO, from one of the methene bridge carbons. This is a unique reaction and the only one that has been established as producing carbon monoxide (CO) in the human body. In fact, a measurement of CO production by people can be used as a measurement of heme degradation. When the ring opens, we have a linear tetrapyrrole known as biliverdin. This compound has a greenish color and accounts for the greenish color of a bruise, because blood cells die at the site of the bruise and produce biliverdin. In the spleen and liver, biliverdin can be reduced to bilirubin (which has a reddish color), utilizing NADPH. This bilirubin can circulate in the blood bound to albumin. Bilirubin would normally be excreted; however, it is relatively insoluble and highly nonpolar (Fig. 20.13). This nonpolar nature can be a problem if there is elevated bilirubin, where it can cross the lipid membranes and cause interference with normal function, particularly in the central nervous system. In the liver, two molecules of glucuronic acid from UDP glucuronic acid are conjugated to the bilirubin to form a bilirubindiglucuronide; the conjugated bilirubin is excreted through the bile to the intestine and then excreted. The color of the feces is due in a large part to the excreted conjugated bilirubin, also known as bile pigments. Both biliverdin and bilirubin are referred to as bile pigments.

One of the problems with premature infants that causes an excess of bile pigment accumulation (known as jaundice) may be related to the fact that the enzymes for conjugation of these bile pigments are developmental in nature. That is, they are low to absent in the fetus

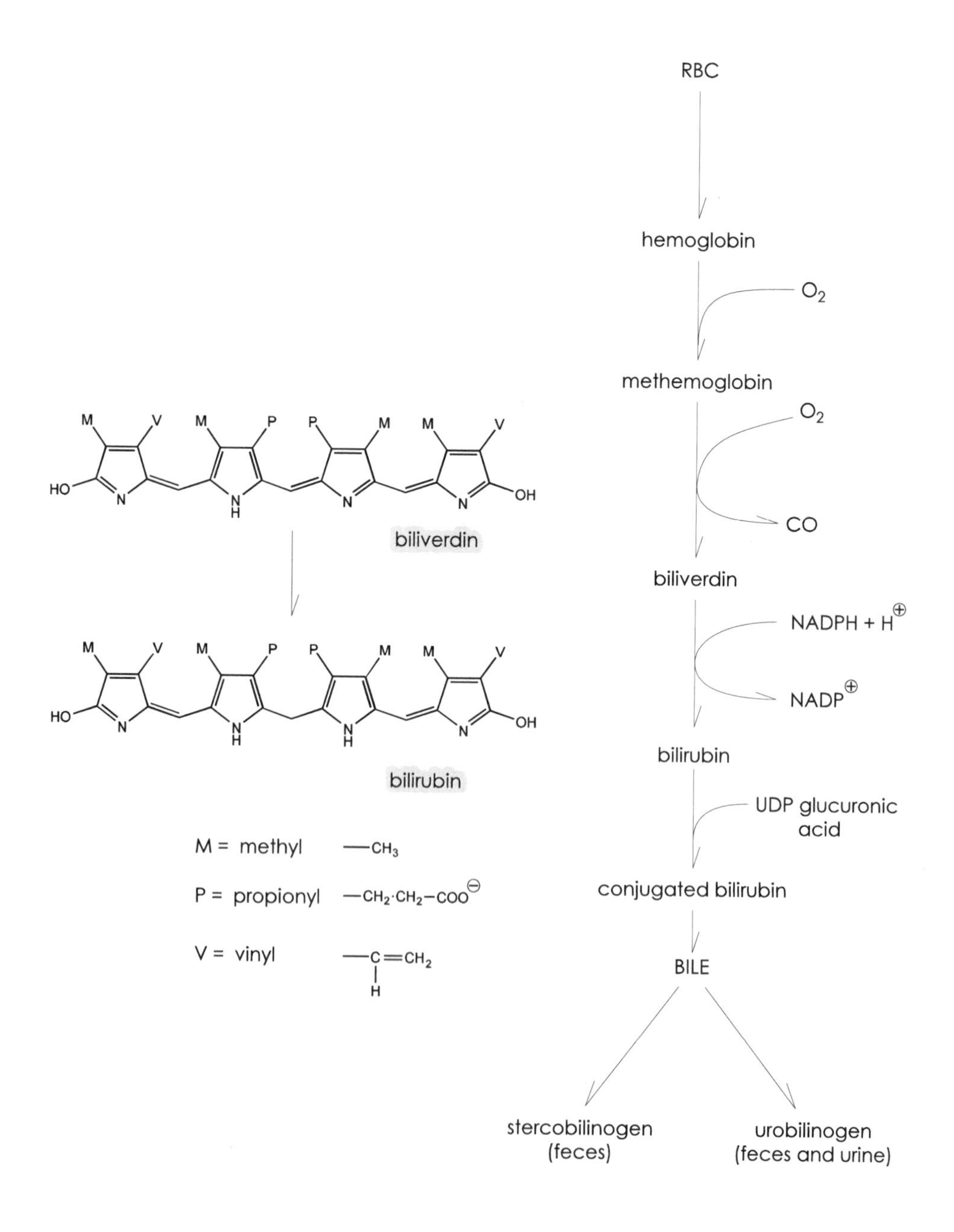

Figure 20.13.

Hemoglobin degradation. The pathway of hemoglobin degradation is shown in outline form. The structures of biliverdin and bilirubin are also illustrated.

and increase prior to term. Prior to term, the maternal system can handle detoxification of these compounds; however, after birth, the newborn must depend on its own system. In some cases, if the infants are premature, they do not have sufficient conjugating enzymes, causing problems related to bilirubin removal and jaundice.

Jaundice

Other than some inborn errors of metabolism, three conditions can lead to jaundice in adults. The first is excess destruction and fragility of the red blood cells, thus producing more bile pigments than can be handled by the liver for conjugation and excretion. The second is the problem of liver damage due to hepatitis or cirrhosis, as in the case of alcoholics. In these conditions, the liver is unable to take up and/or conjugate and excrete the bile pigments at a rate commensurate with their formation. The third is a blockage of the bile duct, for example, by a stone, where the liver is taking up and conjugating the bile pigments but cannot move the conjugated bilirubin into the bile and thus into the intestine. Therefore, these compounds accumulate in the liver cells and move back into the blood, causing jaundice. Jaundice causes a discoloration of the skin and a strong yellowing of the whites of the eyes. Jaundice can be a severe problem and requires appropriate treatment.

With liver cirrhosis and excess red-cell breakdown, many of these bile pigments may be in the unconjugated form. In the case of bile blockage, most of the bile pigments may be in the conjugated form, because the liver is working normally, but cannot put the bile pigments into the bile and gall bladder for excretion into the intestine. Another indication of jaundice is that the urine usually takes on a very strong, orangy red color because many of the bile pigments accumulate and appear in the urine.

SUMMARY

1. Purines are synthesized from simple compounds such as glycine, glutamine, aspartate, CO_2, and tetrahydrofolate–C1 compounds.

2. The control of purine synthesis is with an early step—formation of phosphoribosylamine by PRPP amidotransferase. This enzyme is partially inhibited by either AMP or GMP and strongly inhibited by AMP and GMP together.

3. IMP is a branch and can be converted to AMP or GMP with sufficient amounts of either nucleotide inhibiting its own formation.

4. Purines can be degraded to uric acid. They can also be salvaged prior to uric acid formation, with resynthesis of purine nucleotides. Excess uric acid formation can lead to gout.

5. Pyrimidines are formed from aspartate, CO_2, and glutamine.

6. Unlike purines, which are formed around the ribose phosphate, pyrimidine bases are formed first, then the ribose phosphate is added.

7. In humans, the control of pyrimidine synthesis lies with UTP inhibition of carbamoyl phosphate synthase II, the first step of pyrimidine synthesis.

8. Purine synthesis is a branched pathway; pyrimidine synthesis is a linear pathway with interconversion of products UTP and CTP.

9. Pyrimidine degradation produces no complex unique end products, but only common products.

10. Deoxynucleotides are formed from the diphosphonucleotides.

11. dUMP is converted to dTMP (with the additions of a methyl group requiring tetrahydrofolate) because thymine, but not uracil, is found in DNA.

12. A number of antimetabolites can be used to inhibit purine and pyrimidine synthesis and degradation. This is one of the bases of chemotherapy.

13. Heme is synthesized from glycine and succinyl CoA, plus a metal (usually Fe). Heme is a tetrapyrrole compound.

14. If the heme is not utilized in heme proteins, heme and hematin both inhibit and decrease the synthesis of δ-aminolevulinate, the first unique intermediate in heme synthesis.

15. One of the pyrroles is inserted in a reverse position by uroporphrinogen III cosynthase. If this process is deficient, a type of porphyria occurs.

16. There are a number of diseases associated with heme synthesis resulting in porphyrias.

17. Hemes can be degraded to linear compounds as biliverdin or bilirubin plus CO. One of these, bilirubin, is conjugated with glucuronic acid

and excreted into the bile. These are referred to as bile pigments.

18. If bile pigments are not excreted normally, jaundice can result.

REFERENCES

Hypoxanthine phosphoribosyltransferase deficiency: The Lesch–Nyhan syndrome and gouty arthritis, J. T. Stout and C. T. Caskey, in *The Metabolic Basis of Inherited Disease,* Vol. 1, 6th ed., C. R. Scriver, A. L. Beaudet, W. S. Sly, and D. Valle (Eds.), 1989, McGraw-Hill, New York, pp. 1007–1028.

Hyperuricemia and gout, T. D. Palella and I. H. Fox, in *The Metabolic Basis of Inherited Disease,* Vol. 1, 6th ed., C. R. Scriver, A. L. Beaudet, W. S. Sly, and D. Valle (Eds.), 1989, McGraw-Hill, New York, pp. 965–1006.

Thymidylate synthase: A target for anticancer drug design, K. R. Harrap, A. L. Jackman, D. R. Newell, G. A. Taylor, L. R. Hughes, and A. H. Calvert, 1990, *Adv. Enzyme Reg.,* 161–179.

Methotrexate 5-aminoallyl-2′-deoxyuridine 5′-monophosphate: A potential bifunctional inhibitor of thymidylate synthase, L. M. Stuhmiller, R. Mazarbaghi, S. Webber, and J. M. Whiteley, 1990, *Adv. Enzyme Reg.,* 141–157.

Properties of purine and pyrimidine analogs, G. H. Hitchings, 1991, *Adv. Enzyme Reg.,* 433–443.

Hereditary orotic aciduria and other disorders of pyrimidine metabolism, D. P. Suttle, D. M. O. Becroft, and D. R. Webster, in *The Metabolic Basis of Inherited Disease,* Vol. 1, 6th ed., C. R. Scriver, A. L. Beaudet, W. S. Sly, and D. Valle (Eds.), 1989, McGraw-Hill, New York, 1095–1126.

The porphyrias, A. Kappas, S. Sassa, R. A. Galbraith, and Y. Nordmann, in *The Metabolic Basis of Inherited Disease,* Vol. 1, 6th ed., C. R. Scriver, A. L. Beaudet, W. S. Sly, and D. Valle (Eds.), 1989, McGraw-Hill, New York, 1305–1365.

Hereditary jaundice and disorders of bilirubin metabolism, J. R. Chowdhury, A. W. Wolkoff, and I. M. Arias, in *The Metabolic Basis of Inherited Disease,* Vol. 1, 6th ed., C. R. Scriver, A. L. Beaudet, W. S. Sly, and D. Valle (Eds.), 1989, McGraw-Hill, New York, 1367–1408.

REVIEW QUESTIONS

1. The step in purine synthesis that is controlled synergistically is
   a) Phosphoribosylpyrophosphate formation
   b) Phosphoribosylamide formation
   c) IMP formation
   d) AMP formation
   e) GMP formation

2. A drug used to control gout (or production of uric acid) is
   a) 6-Amino purine
   b) Azaserine
   c) Aminopterin
   d) Allopurinol
   e) 6-Mercaptopurine

3. The major control step in pyrimidine synthesis in people is the formation of
   a) Carbamoylphosphate
   b) Carbamoylaspartate
   c) Orotic acid
   d) UMP
   e) CMP

4. Heme synthesis requires which two precursors?
   a) Alanine and succinyl CoA
   b) Glycine and acetyl CoA
   c) Alanine and acetyl CoA
   d) Glycine and succinyl CoA
   e) Glycine and arginine

5. The direct (immediate) precursor of dTMP is
   a) dCMP
   b) dAMP
   c) dUMP
   d) dGMP
   e) dCDP

ANSWERS TO REVIEW QUESTIONS

1. ***b*** This reaction, which is the first committed step in purine synthesis, can be inhibited weakly by AMP or GMP. However, the two together in a synergistic manner can strongly inhibit this reaction.

2. ***d*** Allopurinol is a competitive inhibitor of xanthine dehydrogenase, thus inhibiting the formation of uric acid.

3. ***a*** The cytosolic formation of carbamoylphosphate from glutamine, CO_2, and ATP is inhibited by UTP.

4. ***d*** Glycine plus succinyl CoA are combined, and the combination is decarboxylated, catalyzed by amino levulinate synthase, to form δ amino levulinate in the mitochondrion as the first step in heme synthesis.

5. ***c*** dUMP plus N^5, N^{10}-methylene THFA produces dTMP. The major direct precursor of dUMP, but not dTMP, is dCMP (a).

CHAPTER

21

GENOME ORGANIZATION

INTRODUCTION

Individual cells are capable of performing a large array of biochemical reactions and of responding to multiple stimuli. Together, cells in a multicellular organism, can coordinate and subspecialize their efforts to achieve a high degree of control over the environment and to efficiently utilize limited resources. Every cell in an organism contains essentially the exact same genetic information in what is collectively called a **genome.** This information needs to be accessed in a highly ordered and efficient way, similar to how a software program is needed to search a computer database. Before we can examine the details of the necessary biochemical processes required to manage and maintain a genomic "database," this chapter will first present an overview of how DNA is organized within the genome. Figure 21.1 shows the general flow of information from DNA to RNA through a process called **transcription,** and from RNA to protein by a series of biochemical synthesis reactions collectively called **translation.** The schematic drawing in Figure 21.1 also emphasizes that **genes** are the functional unit of information stored in DNA.

This chapter describes how genomes differ between organisms and presents approaches being used to identify and characterize the complete nucleotide sequence of several genomes. Because of the tremendous potential for understanding human development and disease, there is now a worldwide effort to obtain the entire sequence of the human genome. However, there is still a lot we need to understand about how the human genome is organized and there is great debate over how we will decipher the large amount of human DNA sequence once it is available.

GENOME SIZES

Prokaryotic Genomes

Evolution has heavily influenced the size of genomes which can be tolerated with any given cell type. For example, a typical bacterial virus (also called a **bacteriophage**), consists of a protein coat surrounding a **nucleocapsid core** containing a relatively small DNA genome. The genome of **bacteriophage ϕX174** is a 5386-nucleotide long, single-strand circle, while the **bacteriophage T7** genome is 39,936 nucleotides long and exists as a double-stranded linear molecule. Genetic and biochemical analysis of the T7 genome revealed that there are a minimum of 45 gene products made from the T7 genome. Figure 21.2 shows how the T7 genome is organized and it can be seen that many of the T7 genes encode proteins needed for DNA replication and RNA synthesis. The genes are arranged along the

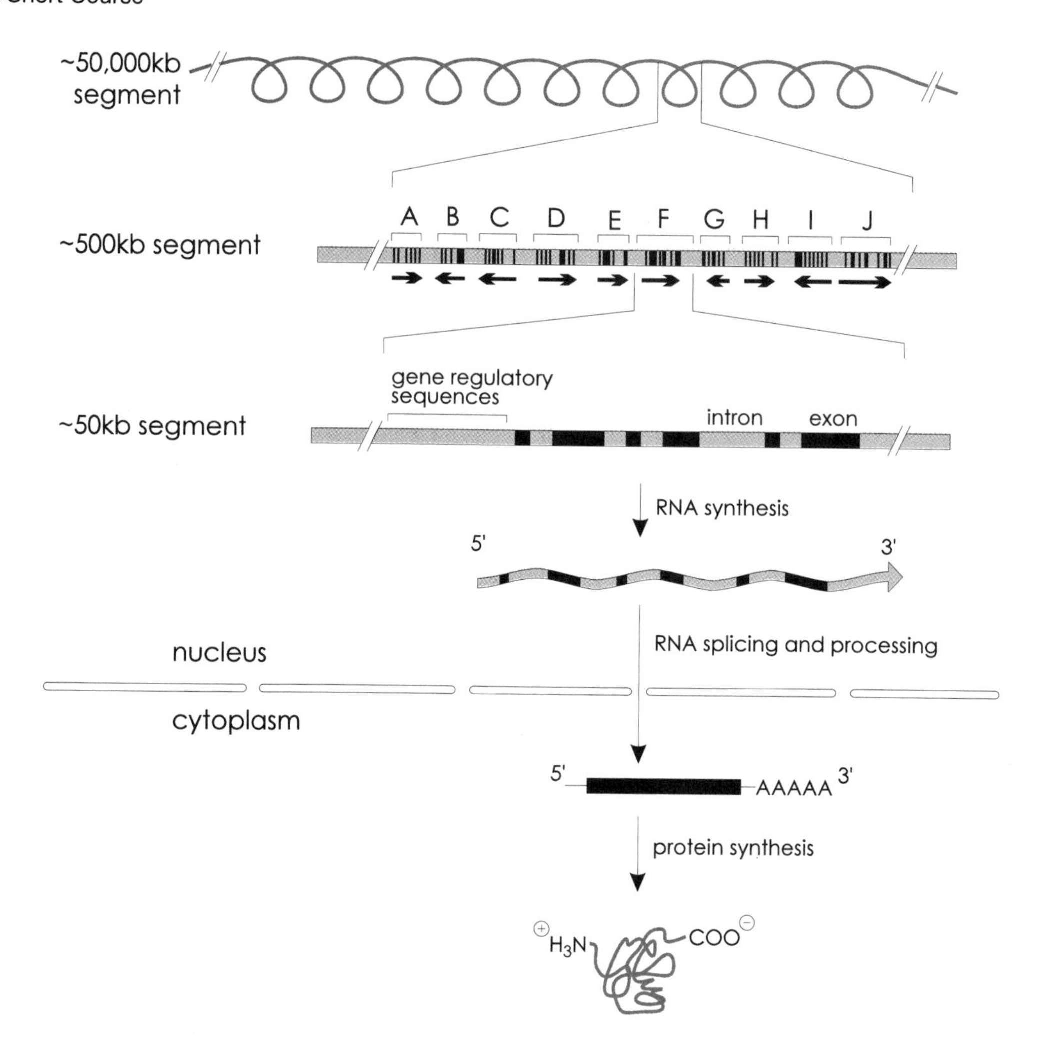

Figure 21.1.
The molecular flow of information from DNA to RNA and proteins. The example shown is for vertebrates, where it is common for the coding sequence of genes to be broken into exons and introns. Genes A-J are distributed along a chromosomal segment and the expanded view of gene F shows that it consists of both regulatory and coding sequences. The processes of transcription, RNA processing, and translation are discussed in detail in later chapters.

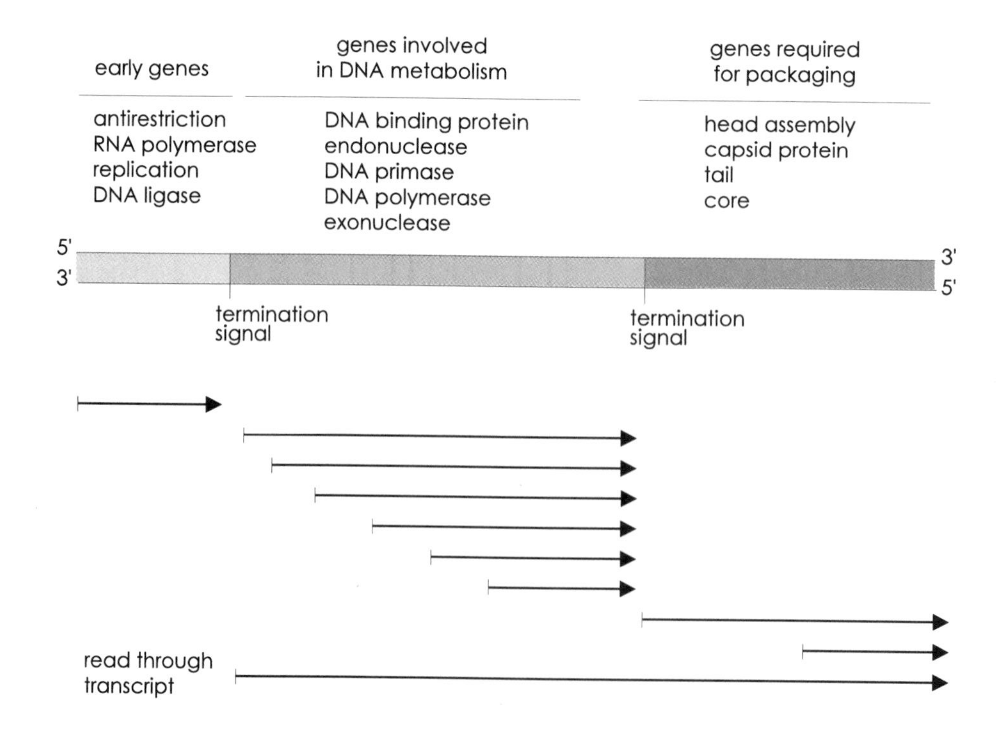

Figure 21.2.
Map of the bacteriophage T7 genome showing the ordered transcription of genes from various portions of the genome. Gene products required for bacteriophage replication are synthesized prior to those required for viral packaging.

genome in a 5′ to 3′ orientation such that their temporal expression optimizes a productive infection by T7.

The *E. coli* bacterial genome is a large double-stranded circle that is tightly associated with the cell membrane. Although it has not been entirely sequenced, physical measurements have estimated that it contains 5×10^6 base pairs of DNA. So far, close to 1000 genes have been mapped on the *E. coli* chromosome, and based on genetic analyses and the DNA sequencing information currently available, it is predicted that there are ~5000 *E. coli* genes in total.

The Human Genome

How much DNA is required to specify the life cycle of humans, perhaps the most complex organism on this planet? The human genome consist of $\sim 3.3 \times 10^9$ nucleotide base pairs of DNA. This large amount of genetic material is distributed among 23 human chromosomes found in sperm and egg cells. **Somatic** human cells (all other cell types besides **gametes**) contain twice this amount of DNA because they have 23 chromosomes from each parent, making 46 chromosomes in all. Current estimates are that there are about 100,000 genes encoded in the human genome.

The C-Value Paradox

The human genome contains perhaps only 20 times the number of genes as the *E. coli* genome, even though the human genome is about 1000 times as large. The lack of direct correlation between the number of genes (or complexity of the genetic information), and the total mass of DNA per cell, as compared across species, is called the **C-value paradox.** For example, some forms of algae contain 100 times more DNA than humans ($\sim 10^{11}$ base pairs), yet this lower plant form has a smaller number of unique gene sequences. The reason for this difference in genome size is that many types of algae contain large amounts of repetitive DNA sequences. Similarly, human DNA contains much more repetitive DNA than *E. coli,* as well as "spacer" DNA that is interspersed between gene-coding sequences. The nature of repetitive DNA and spacer DNA is described later in this chapter.

GENOMES IN CELLULAR ORGANELLES

Some cell organelles also contain DNA and it is referred to as **extrachromosomal DNA.** The complete sequence of the human **mitochondrial** genome is known and it has been found to be a 16,569-nucleotide double-stranded circle. Each mitochondrion contains 5–10 copies of this circular genome. Although most mitochondrial proteins required to perform the complex process of cellular respiration are encoded by the nuclear genome, there are 13 proteins encoded by the mitochondrial genome and seven of these specify the subunits of **NADH-Q reductase.** There are also 22 mitochondrial tRNA and 2 rRNA genes in the human mitochondrial genome which direct protein synthesis within a mitochondrion. The map of human mitochondrial DNA is shown in Figure 21.3 and the location of mutations in the mitochondrial genome which have been shown to cause the human disease **Leber's hereditary optic neuropathy (LHON)** are shown. LHON is a maternally inherited genetic disorder characterized by poor neural development in some tissues such as the optic nerve. Since the egg cell provides the mitochondria to the zygote, mitochondrial DNA mutations are maternally inherited.

The genomes of plant **chloroplasts** have also been studied and in some cases completely sequenced. For example, the chloroplast genome of the algae ***Euglena gracilis*** was found to contain 143,172 base pairs and it exists as a double-strand circle. As with mitochondria, chloroplasts contain multiple copies of this extrachromosomal DNA genome with about 50 copies per chloroplast. Based on what is known about bacterial genomes, and the function of mitochondria and chloroplasts in cellular processes, it has been proposed that these cell organelles may have arisen by a mutual symbiotic relationship between early eukaryotic cells and prokaryotic ancestors.

EUKARYOTIC CHROMOSOMES

Chromosome Organization

How can all that DNA fit into a cell and how can it be duplicated and divided efficiently during cell division? The answer in eukaryotes is to distribute the total amount of genomic DNA among a number of large linear molecules called **chromosomes.** These highly ordered chromosome structures contain proteins, such as histones (see Chapter 5), in addition to DNA. The important processes of DNA replication, and chromosome segregation, need to be accommodated by these large DNA–protein molecules. **Telomeres** are the structures at the two ends of each chromosome which facilitate DNA replication, and **centromeres** are the structures

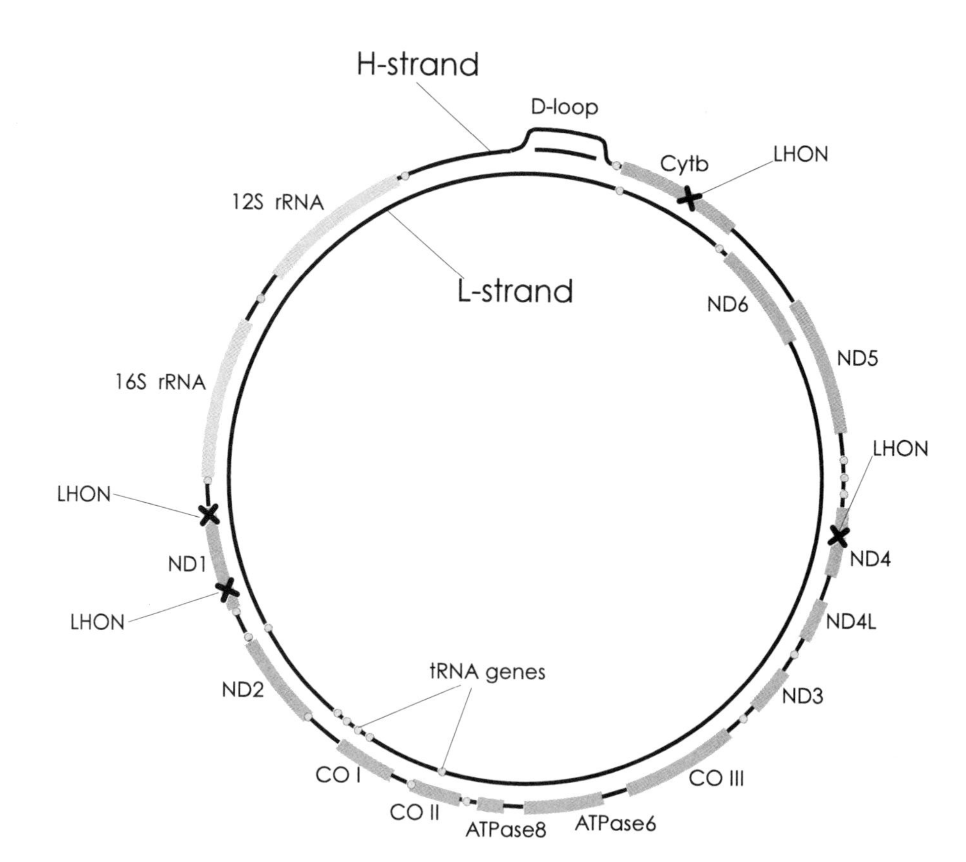

Figure 21.3.
Map of the human mitochondrial genome. ND refers to NADH dehydrogenase subunits, CO denotes cytochrome oxidase subunits, and cytochrome B is shown as cyt b. The D-loop is required for mitochondrial genome replication. Most of the genes are encoded by the H-strand (outer circle). Locations of several LHON mutations known are shown by an "X."

near the middle of each chromosome which are required for proper chromosome segregation during mitosis (see Chapter 5). Figure 21.4 shows the general layout of a typical eukaryotic chromosome identifying the types of DNA sequences known to be important for telomere and centromere function.

Telomeres

Telomeres are made up of short, highly repetitive, DNA sequences consisting of the repeating sequence (5′) C_xA/TG_y (3′), where x and y represent 1–3 bases in yeast telomeres. The sequence of the telomere repeat in humans is (5′) TTAGGG (3′). In yeast, there are about 100 of these repeats, while in humans the repeat length is variable and can include up to several kilobases of DNA. In addition to facilitating DNA synthesis at the end of linear chromosomes, telomeres seem to also function by protecting chromosomal termini from degradation. The G residues on the most distal telomere sequence repeat, at the end of the chromosome, are capable of forming a hairpin loop through nonstandard G-G base pairing. This structure would eliminate any free single-strand ends which could serve as substrates for cellular exonucleases. Interestingly, it has been found that the length of human telomeres are greatest in gamete cells and are progressively shorter in somatic cells of older individuals, suggesting that aging may be associated with shortening of telomeres. In support of this idea, some tissue culture cells have been shown to lose telomeric sequences as they age in culture through a process known as senescence. An enzyme called **telomerase** is required for telomere maintenance. Telomerase is an RNA-dependent DNA polymerase which adds T_xG_y repeats to the ends of chromosomes. The telomerase ribonucleoprotein complex contains an RNA template which is an integral component of the enzyme-active site.

Centromeres

The other important functional element in a chromosome is the centromere. Yeast centromeres have been shown to contain an extremely A-T rich region of about 130 base pairs, which is tightly associated with centromere-binding proteins. This protein–DNA complex on the chromosome is called the **kinetochore** and it interacts with the microtubules of the **mitotic spindle.**

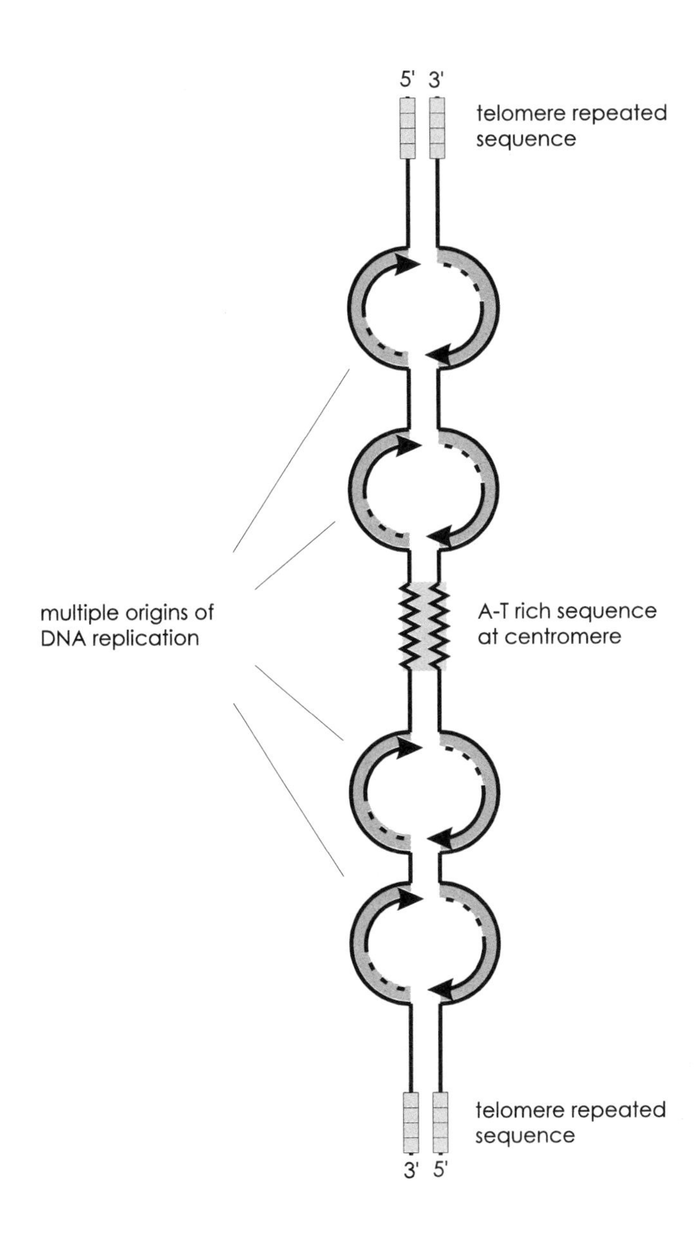

Figure 21.4.
Structure of a eukaryotic chromosome showing the relative locations of telomeres and the centromere. Multiple origins of DNA replication are also shown.

During mitosis, mitotic spindles attach to centromeres and provide the critical force needed to separate chromosomes prior to segregation into the two daughter cells. Human centromeres have also been shown to contain long stretches of A-T rich sequences, including a highly conserved repetitive sequence $(GGAAT)_n$.

EUKARYOTIC GENE ORGANIZATION

Genes are the fundamental unit of genetic information. Single-copy genes, most of which encode proteins, are scattered throughout the genome with very little recognizable organization. Therefore, it is not possible to predict where functionally related genes will be located among the 22 autosomal and 2 sex-linked (X and Y) chromosomes found in humans. There are also a large number of RNA-encoding genes, for example, **ribosomal RNA (rRNA)** and **transfer RNA (tRNA)** genes. In bacteria and viruses where genome sizes are small, related functional genes are often grouped physically into DNA segments called operons, which allow for coordinate gene expression (Chapter 28). In eukaryotic cells, the story is quite different. Unless there is a strong evolutionary pressure to preserve gene organization along a chromosome, it appears that genes can be "shuffled," most likely as a result of gene rearrangements. Mice and humans contain many of the same genes that are common to mammals; however, the organization of these genes is not always colinear among the various mouse and human chromosomes. This observation suggests that since the time that mice and men have diverged (~70 million years), large DNA segments have been moved between chromosomes by DNA recombination (see Chapter 23) and chromosomal translocation (Chapter 31). In eukaryotes, genes with similar functions are sometimes physically linked to each other in **gene clusters.** Gene clusters are analogous to bacterial operons except that each gene is independently regulated at the level of transcription (Chapter 29). The arrangement of genes within a eukaryotic gene cluster appears to have biological significance since the pattern of gene expression is often temporally and spatially regulated.

Through an evolutionary process known as **gene duplication** and **sequence divergence**, some related genes are found tightly associated with each other in the same chromosomal location. The globin gene cluster in humans appears to be a good example of this evolutionary process. The **α-globin genes** are located on human chromosome 16 and the **β-globin gene** cluster is found on chromosome 11. As shown in Figure 21.5, the five β-globin genes (ϵ, γ^G, γ^A, δ, and β) are arranged along the chromosome in the developmental order in which they are expressed. The same general feature is found in the α-globin gene cluster. The mechanistic basis for **β-globin gene "switching"** (the embryonic ϵ gene is not expressed in the fetus and the γ genes are not expressed in the adult) is not completely known. One model to explain the mechanism of β-globin gene switching is that a **locus control region** (LCR) directs the transcription of appropriate β-globin genes through the activity of specific DNA binding proteins (see Chapter 29). Note also that there is a β-globin **pseudogene** (a nearly identical but nonfunctional gene) which is presumably the result of a gene duplication and DNA mutations.

The vertebrate **Hox genes** are a second example of clustered genes that have a developmentally restricted pattern of expression. This large family of genes is required for a number of developmental processes such as tissue modeling during embryogenesis. The 5′ to 3′ arrangement of Hox genes along the chromosome reflects the relative spatial expression of the genes in cells throughout the body axis during very early development. Hox genes expressed in a region of the embryo destined to become the head and neck are on the same chromosome and physically adjacent to Hox genes that are expressed in the limbs. Thus, similar to the β-globin gene cluster, the Hox gene cluster is likely to have arisen from gene duplication of an ancestral gene required for embryonic development.

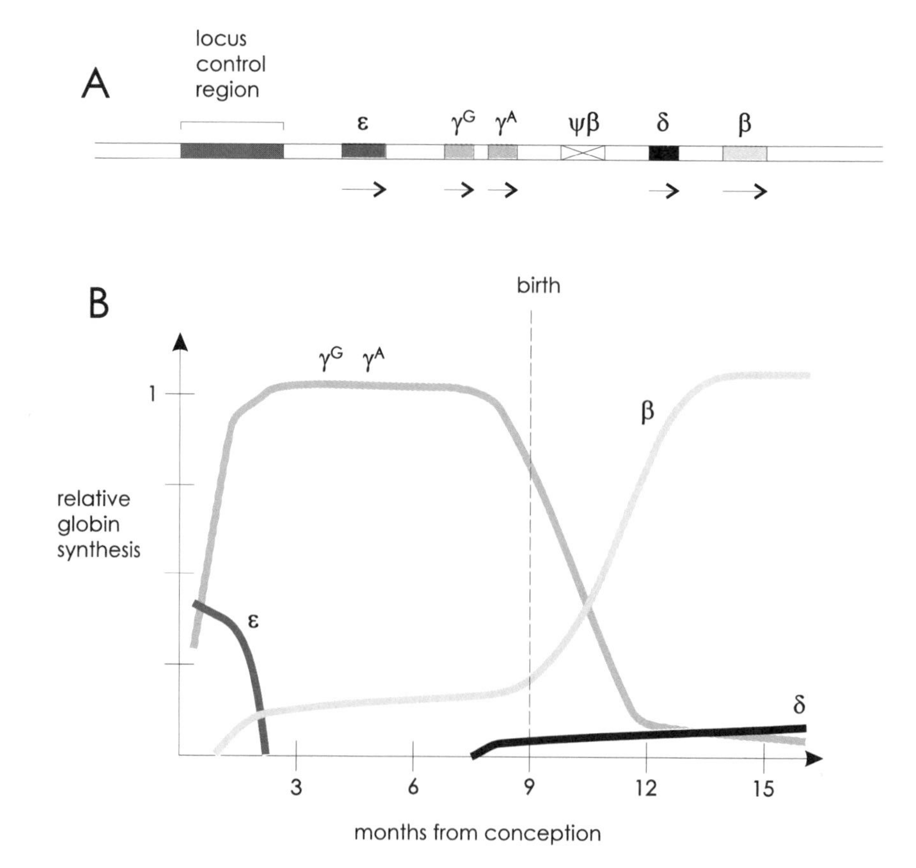

Figure 21.5.

Genomic arrangement and developmental expression of human β-globin genes. (A) Schematic map of the human β-globin gene cluster, including the β-globin pseudogene and the upstream locus control region (LCR), which is required for proper control of globin gene expression. (B) Expression pattern of the β-globin gene cluster in humans showing the relationship between age and transcription of specific globin-gene subclasses.

Another type of **multigene family** is one in which there are a large number of nearly identical copies of the same gene sequence. An example of this is the **tandem array** of rRNA genes. In this case, a long repeating unit encoding the large and small subunit rRNAs (**18S rRNA and 28S rRNA** in mammals) and the **5.8s RNA** is found in multiple head-to-tail copies on the chromosome (see Chapter 24). There exists a nontranscribed spacer region between adjacent rRNA gene clusters which may play an important role controlling rRNA gene expression. Humans have about 200 copies of the 18S and 28S rRNA genes per haploid genome, which are organized into tandem arrays on five different chromosomes. It is likely that long tandem arrays of rRNA genes greatly enhances the efficiency of rRNA synthesis at times in the cell cycle when large numbers of ribosomes need to be assembled.

REPETITIVE DNA SEQUENCES

It is estimated that less than 10% of the human genome is required to encode the nearly 100,000 genes thought to exist. What about all the rest of the DNA—does it have a function? The answer is not yet clear, but it seems likely that there is less "junk" DNA than previously thought, based on the finding that some human genetic diseases are caused by alterations in regions of noncoding DNA.

Di- and Trinucleotide Repeats

A large component of the noncoding DNA is made up of highly repetitive DNA sequences. The shortest of these are dinucleotide repeats including CG, which may be involved in controlling gene expression. These repeats are called **CpG islands** because they are clustered over ~1 kb and are often found near the 5′ ends of gene regulatory regions. Another **dinucleotide repeat** is characterized by long stretches of alternating A and T residues. It has recently been found that deletions in AT repeat lengths is associated with colorectal cancer. The most common dinucleotide repeat in the human genome is CA. The repeat length of specific CA dinucleotides in the genome has been found to be variable among different individuals. Geneticists have been able to take advantage of this observation to aid in the derivation of a **genetic linkage map** of the human genome. Other short repeated sequences are the satellite repeats found in telomeres and the AT-rich sequences located in and around centromeres. **Trinucleotide repeats** have also been found in the human genome and several human diseases have now been shown to be due to expansions in the number of trinucleotide repeating units at specific chromosomal locations. Table 21.1 lists the five **genetic diseases** that have been associated with trinucleotide repeat expansion, but how they cause disease is not yet known.

Short and Long Interspersed Elements

The next largest class of repetitive element, in both size and abundance, are the short interspersed sequences known as **SINEs (short interspersed elements).** Human DNA contains up to 500,000 copies of the SINE commonly referred to as the **Alu repeat,** which is ~300 base pairs in length. SINEs appear to have the ability to "hop" around the genome, by a process involving an RNA intermediate and DNA integration. This proposed mechanism is similar to the life cycle of a retrovirus, which also utilizes an RNA intermediate (Chapter 30). SINE "hopping" can be detrimental because it can lead to **insertional mutations,** an example of which is the disruption of the **neurofibromatosis-1** gene leading to elephant's man disease. **LINEs (long interspersed elements),** are similar to SINEs, but are about 7000 base pairs in length. Although it is not yet clear why these different types of repetitive sequences have accumulated in the human genome, they may serve an important evolutionary function, such as promoting gene shuf-

Table 21.1. Genetic Diseases of Trinucleotide Reiteration

Disease name	Repeat sequence	Normal number	Disease number	Location of repeat
Kennedy's	CAG	11–33	40–62	Protein coding
Huntington's	CAG	11–34	42–100	Protein coding
Spinocerebellar ataxia type 1	CAG	29–36	43–60	Protein coding
Fragile X	CGG	6–54	250–4000	5′ Untranslated
Myotonic dystrophy	CTG	5–30	>50	3′ Untranslated

fling through homologous recombination mechanisms (Chapter 23).

GENOME MAPPING

With advances in molecular biological techniques, it became apparent that it should be possible to obtain the nucleotide sequence of the entire human genome. This idea was further promoted by the identification of genes which contained mutations that caused several human genetic diseases. For example, the gene responsible for **cystic fibrosis** (CF) was identified by combining standard genetic analysis of afflicted family members with molecular cloning techniques. This approach eventually led to the isolation of a region of DNA which was different between normal individuals and those with the disease. The **CF gene** was found on human chromosome 7 and it encompasses over 250,000 nucleotides. The gene product is thought to encode a transmembrane protein involved in ion transport. About 70% of the CF gene mutations found so far are due to the deletion of a phenylalanine residue at amino acid 508 in the CF gene protein-coding sequence. This mutation may cause a defect in the ATP binding activity of the CF protein.

Restriction Fragment-Length Polymorphisms (RFLPs)

One of the techniques used to identify regions of DNA that may contain a gene mutation is called **restriction fragment length polymorphism (RFLP).** This strategy is based on the finding that no two individuals have the exact same DNA sequence in all 3.3×10^9 base pairs of their DNA. These differences are referred to as **DNA polymorphisms** and are the result of nucleotide alterations, DNA deletions, or DNA rearrangements. Since many regions of the human genome contain noncoding sequences, most all DNA polymorphisms have negligible effects on the genetic information. However, by using these polymorphisms as genomic "landmarks," it is possible to determine if a particular trait is inherited in the same pattern as the DNA polymorphism. If the occurrence of an inherited disease is nearly always found to be associated with a specific DNA polymorphism, then it is often possible to isolate DNA in the vicinity of the **RFLP marker** as a means to begin searching for the defective gene.

Figure 21.6 illustrates how **RFLP mapping** works. Note that the RFLP does not have to be located within the disease gene, but just close enough to serve as a marker of nearby sequences that are said to be "linked." RFLP mapping is an effective tool to clone disease genes because closely **linked sequences** in the genome are infrequently separated by genetic recombination (see Chapter 23). Therefore, the closer the DNA polymorphism is to the defective gene, the more often it will be associated with the disease. If the DNA polymorphism is the genetic *cause* of the disease, then there would be a 100% correlation between its presence and the disease, however, this is rarely the case.

The Human Genome Project

Molecular geneticists began to realize that if the DNA sequence of the entire human genome were available, it may be possible to more rapidly apply the genetic information obtained from family histories to the direct identification of specific disease genes. The **Human Genome Project** was officially born in 1988 as a government-sponsored effort to meet the goal of sequencing the human genome. This has come to involve large teams of American, European, and Japanese scientists. Individual pieces of DNA have been isolated and their relative positions in the human genome have begun to be identified. This is a huge endeavor when one considers the size of the human genome and the difficulty of data management. The basis of constructing such a "library" of DNA fragments comes from the very essence of recombinant DNA methodology, the ability to clone pieces of DNA using restriction endonucleases, as described in Box 21.1.

One of the obstacles that had to be overcome before DNA sequencing of whole genomes could be done efficiently, was to devise strategies to manipulate large pieces of DNA. Methods were developed to subdivide the human genome into segments of a **megabase** (10^6 base pairs) or more, such that all $\sim 3 \times 10^9$ base pairs could be contained on ~30,000 fragments. Smaller regions could then be analyzed by further dividing the megabase-sized pieces into another 1000 fragments, which are of a manageable size for standard DNA sequencing (see Chapter 2). Such an approach has been applied to the genomes of yeast and the nematode ***C. elegans.*** By assembling the smaller fragments into overlapping clones, and then sequencing them, researchers have so far been able to obtain the continu-

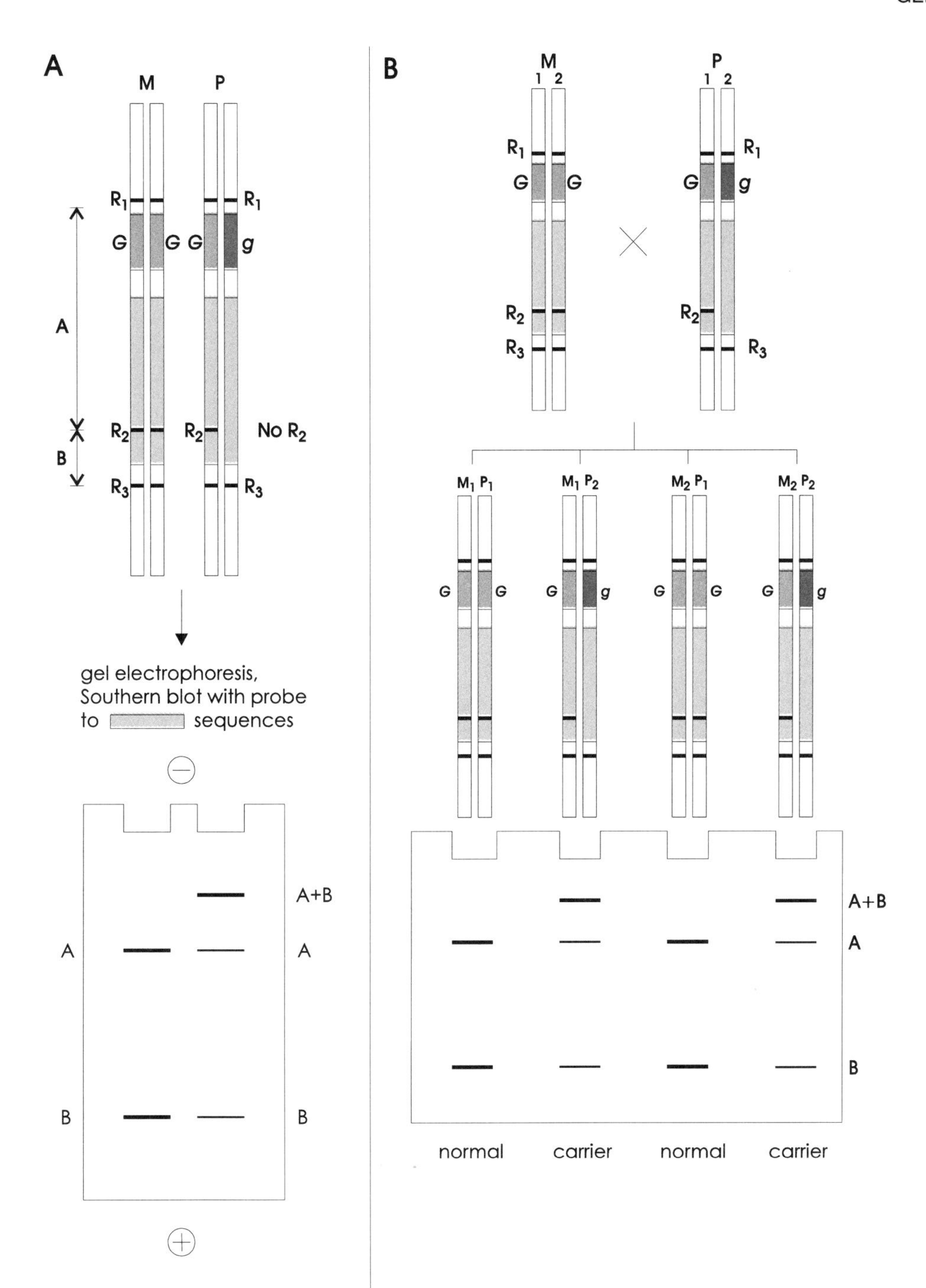

Figure 21.6.
RFLP mapping of a gene that is **autosomally inherited** (not located on the X or Y chromosome). (A) The maternal genome (M) is homozygous for both the R2 restriction site and the linked gene (G), whereas the paternal genome (P) is heterozygous because it contains one normal (G) and one defective (g) copy of the gene. In this example, the paternal chromosome lacking the R2 site also contains the tightly linked defective gene (g). Following gel electrophoresis and Southern Blotting, it can be seen that the paternal genome is heterozygous for the RFLP marker resulting from loss of the R2 site. (B) A pedigree chart showing an association between the presence of the R2 RFLP marker as identified in the Southern blot, and the genetic phenotype of disease carriers (G/g heterozygotes).

ous sequence of 2,181,032 nucleotides of the *C. elegans* chromosome 3.

Figure 21.7 shows how **yeast artificial chromosomes (YACs)** have been developed for use as the primary tool in manipulating large segments of DNA. By utilizing what was known about telomeres to protect the ends of chromosomes, centromeres for proper chromosome segregation at mitosis, and origins of DNA replication, **YAC cloning vectors** were developed that could accommodate large pieces of DNA inserted into the YAC vectors by standard cloning.

Collections of human YACs representing the entire genome are called "libraries" and are maintained as individual yeast strains, each of which contains a single replicating YAC with a segment of human DNA that is usually between 0.5–2 megabases in length. To

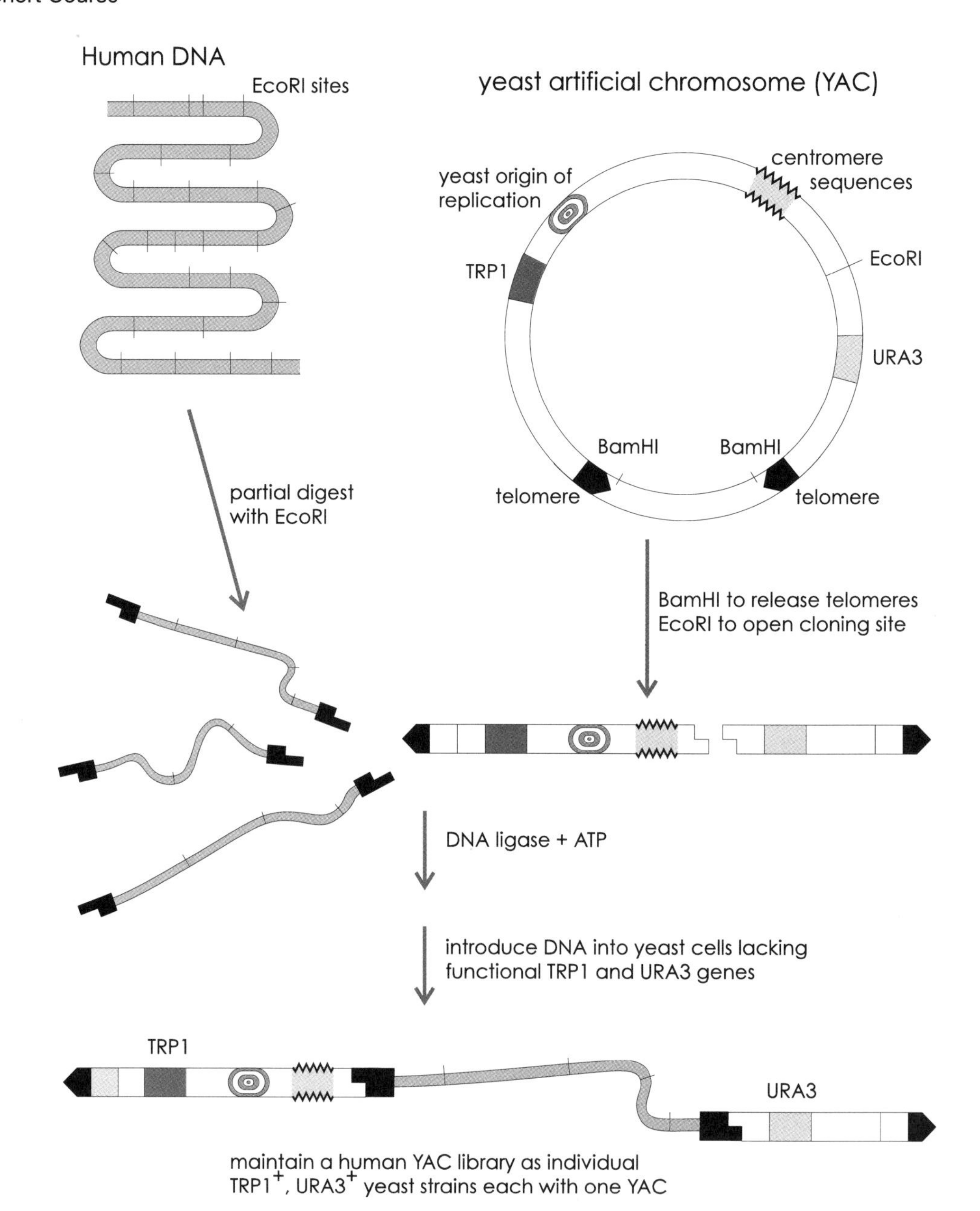

Figure 21.7.
Strategy of how a YAC vector is used to clone DNA. TRP1 and URA3 are genes which encode yeast biosynthetic enzymes that allow for selection of the YAC vector in yeast strains containing defects in their ability to synthesize the amino acids tryptophan and uracil, respectively. Bam HI and Eco RI are site-specific restriction endonucleases.

ensure that a complete **YAC library** contains at least one copy of the entire human genome, the library is constructed so that it has an ~8-fold redundancy (the collection of yeast strains together contain $8 \times 3 \times 10^9$ base pairs of human DNA). A typical human YAC library requires approximately 500 microtiter plates, each containing 96 wells for individual yeast strains. The YAC library can be screened with a radioactive DNA probe containing sequences from the region of interest using the technique of DNA hybridization. Once the appropriate YAC is identified, it can be subdivided by additional cloning steps using bacterial plasmids (see Box 21.1).

Physical Mapping of the Human Y Chromosome

One of the goals of the Human Genome Project is to construct a physical map made of overlapping YAC clones encompassing the entire human genome. Such a map of the human **Y chromosome** was recently completed using 160 **sequence tagged sites (STSs),** which are short unique DNA sequences that serve as genomic landmarks similar to RFLPs. Using the approaches described above, 196 Y chromosome YACs have been sequentially ordered relative to each other to cover approximately 60 megabases, as depicted in Figure 21.8.

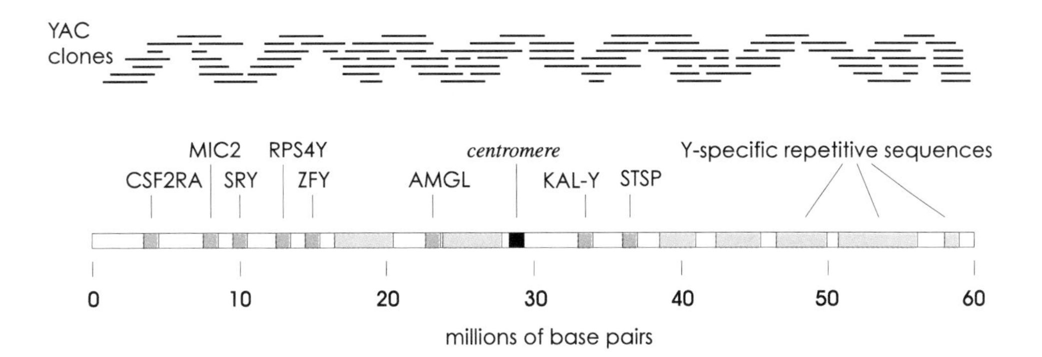

Figure 21.8.
Schematic drawing showing the human Y chromosome physical map. In all, 196 individual YAC clones were assembled into 10 contiguous segments separated by nine short gaps. The location of several known genes and the putative centromere region are shown.

This same type of YAC physical map is already available for the entire *C. elegans* genome, which contains a total of 100 megabases (only 3% of the human genome!).

Within the next decade, it is anticipated that physical maps, like the one shown in Figure 21.8 for the human Y chromosome, will be assembled for all of the human chromosomes and, moreover, that the DNA sequence of a substantial percentage of the human genome will be determined. However, a large number of technical difficulties still need to be overcome before this goal is realized, not the least of which is how to manage and access all this information. Nevertheless, most researchers agree that the insights gained by developing methodologies to achieve the goals of the human Genome Project, as well as the information itself, will undoubtedly provide a large number of ancillary discoveries that will benefit mankind as a whole.

Box 21.1. The use of specific DNA modifying enzymes for manipulating DNA.
Like many key discoveries in science, the age of genetic engineering was born by applying information gained by purely basic research to a new area. In this case, it was the observation that bacteria carry endonucleases that are able to cleave double-stranded DNA at specific sites to yield one or more fragments. These enzymes, known as **restriction endonucleases,** do not attack the bacterial DNA because it is protected by base modification by site-specific methylases. It is thought that bacteria evolved the ability to specifically degrade DNA to protect against infecting bacteriophage or other foreign DNA. In the 1970s, Paul Berg, Herbert Boyer, and Stanley Cohen shared a Nobel Prize for their discovery that purified restriction enzymes could be used as "molecular scissors" to cut DNA in vitro. They showed that the resulting DNA fragments could be rejoined with foreign DNA using a second enzyme, **DNA ligase,** which acts like "molecular tape" to covalently reattach the DNA fragments. This method has become known as **recombinant DNA technology.**

The key to using recombinant DNA as a tool is based on three important steps. First, the restriction enzymes need to cleave the DNA target in a specific location, for example, at the 5′ and 3′ ends of a gene. Second, the DNA must be recombined into a second piece of DNA, which will allow isolation of the gene fragment. This step is most often done using a modified bacterial DNA molecule called a **plasmid,** which retains the ability to replicate autonomously in bacterial cells, and importantly, can be retained in cells because it contains an **antibiotic resistance gene** for **ampicillin** (Amp^r). The use of bacteria as a host for DNA amplification is called cloning since all of the plasmid DNA obtained from a single strain of bacteria carrying the recombined plasmid DNA is identical. The third important step is the ability to identify plasmid clones containing the DNA of interest from a library of random clones. This latter step is done by DNA hybridization as described in Chapter 2.

Figure 21.9 shows the basic steps in recombinant DNA methodology. Note that nearly 500 different restriction enzymes and other DNA modifying enzymes have been discovered and many of these are commercially available. Since most restriction enzymes cleave DNA asymmetrically, overlapping complementary ends are the result. Base pairing between two separate DNA fragments that have been treated with the same restriction enzyme, such as EcoRI, is possible through the four bases present in the single-stranded overhangs. Subsequent treatment with a DNA ligase in the presence of ATP results in covalent bonds at the two junctions which joins 5′ and 3′ ends of the DNA fragments. This recombined DNA is then introduced into a bacterial strain that does not contain endogenous restriction enzymes and is also sensitive to ampicillin. One aspect of the biotechnology industry is to clone genes of commercial value for use in producing large amounts of protein in a way that is economically feasible.

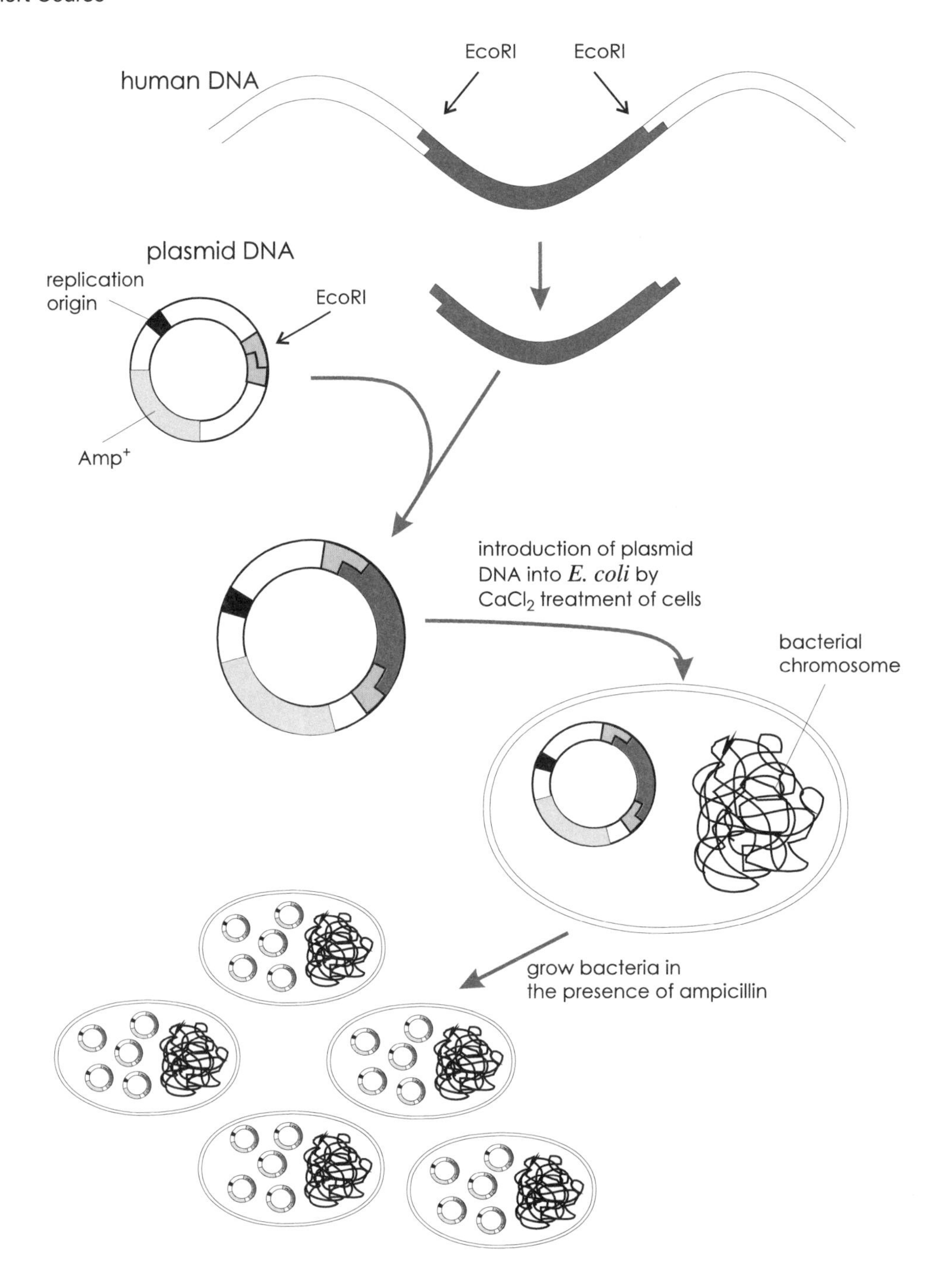

Figure 21.9.
The general-flow scheme of recombinant DNA. Eco RI is a specific endonuclease that recognizes the sequence (5′) GAATTC (3′) and makes a double strand cleavage between the G and A residues on both strands, leaving a "**sticky end.**" A multiple cloning site is a segment of DNA which has been synthesized in vitro and is designed to contain the recognition sequences of many common restriction enzymes.

SUMMARY

1. Genome sizes vary from very small in bacteriophage (~10^3–10^4 base pairs), to much larger in bacteria (~10^6 base pairs), and even larger in humans (~10^9 base pairs). However, the absolute size of a eukaryotic genome does not directly correlate to the number of genes, because many organisms contain large amounts of repetitive and spacer DNA. This discrepancy between genome size and genome complexity is called the C-value paradox.

2. Mitochondria and chloroplasts contain multiple copies of double-stranded circular genomes that encode genes required for organelle-specific functions. The complete DNA sequence of representative mitochondrial and chloroplast genomes is known and has been used to characterize genetic mutations.

3. Eukaryotic chromosomes contain three key functional elements: telomeres, centromeres, and multiple origins of DNA replication. Telomeres are

located at the ends of chromosomes and protect chromosome ends from nucleolytic degradation. Centromeres are found near the middle of chromosomes and function during mitosis to assist in chromosome segregation. Both telomeres and centromeres contain multiple copies of short, repeated DNA sequences.

4. Single-copy genes are found scattered throughout eukaryotic genomes and are interspersed between repetitive and spacer DNA. Some eukaryotic genes are found in clusters such as the β-globin and Hox genes; rRNA genes are in long tandem arrays. The temporal and spatial expression of genes within gene clusters can correspond to their arrangement along the chromosome.

5. Eukaryotic genomes contain large amounts of repetitive DNA. Short di- and trinucleotide repeats can vary in repeat length between individuals, and although their function is unknown, expansion of these short repeats has been associated with several human genetic diseases. Larger repeat sequences called SINEs and LINEs are also distributed throughout the human genome.

6. Several approaches have been developed to aid in the physical mapping of the human genome. RFLP mapping can be used to locate segments of DNA that contain putative disease genes and can also be useful in genetic diagnoses. The Human Genome Project has made use of yeast artificial chromosomes (YACs) to physically isolate large pieces of DNA that can then be subdivided by recombinant DNA methods and characterized by DNA sequencing.

REFERENCES

Antequera, F., Bird, A. (1993): Number of CpG islands and genes in human and mouse. *Proc Natl Acad Sci USA* 90:11995.

Charlesworth, B., Sniegowski, P., Stephan, W. (1994): The evolutionary dynamics of repetitive DNA in eukaryotes. *Nature* **371**:215.

Collins, F. S. (1992): Cystic fibrosis: Molecular biology and therapeutic implications. *Science* **256**:774.

De Lange, T. (1994): Activation of telomerase in a human tumor. *Proc Natl Acad Sci* **91**:2882.

Dib, C., Fauré, S., Fizames, et al. (1996): A comprehensive genetic map of the human genome based on 5,264 microsatellites. *Nature* **380**:152.

Foote, S., Vollrath, D., Hilton, A., Page, D. C. (1992): The human Y chromosome: Overlapping DNA clones spanning the euchromatic region. *Science* **258**:60.

Grady, D. L., Ratliff, R. L., Robinson, D. L., McCanlies, E. C., Meyne, I., Moyzis, R. K. (1992): Highly conserved repetitive DNA sequences are present at human centromeres. *Proc Natl Acad Sci USA* **89**:1695.

Greider, C.W. and Blackburn, E.H. (1996): Telomeres, telomerase and cancer. *Sci.Am.* **274**:92.

Grivell, L. A. (1994): Intron mobility. *Current Biol* **4**:161.

Huntington's Disease Collab Res Grp (1993): A novel gene containing a trinucleotide repeat that is expanded and unstable on Huntington's disease chromosomes. *Cell* **72**:971.

Jakobovits, A. (1994): YAC vectors: Humanizing the mouse genome. *Curr Biol* **4**:761.

Jonsson, J. J., Weissman, S. M. (1995): From mutation mapping to phenotype cloning. *Proc Natl Acad Sci USA* **92**:83.

Miklos, G.L., Rubin, G.M. (1996): The role of the genome project in determining gene function: Insights from model organisms. *Cell* **86**:521.

Newman, M., Strzelecka, T., Dorner, L. F., et al. (1994): Structure of restriction endonuclease BamHI and its relationship to EcoRI. *Nature* **368**:660.

Nowak, R. (1994): Mining treasures from "junk DNA." *Science* 263:608.

Oliver, S.G. (1996): From DNA sequence to biological function. *Nature* **379**:597.

Singer, M. S., Gottschling, D. E. (1994): *TLC1:* Template RNA component of *Saccharomyces cerevisiae* telomerase. *Science* **266**:404.

Sulston, J., Du Z., Thomas, K., et al. (1992): The *C. elegans* genome sequencing project: A beginning. *Nature* **356**:37.

Wagner, A. (1994): Evolution of gene networks by gene duplications: A mathematical model and its implications on genome organization. *Proc Natl Acad Sci USA* **91**:4387.

Warren, S. T. (1996): The expanding world of trinucleotide repeats. *Science* **271**:1374.

Wilson, R., Ainscough, R., Anderson, K., et al. (1994): 2.2 Mb of contiguous nucleotide sequence from chromosome III of *C. elegans*. *Nature* **368**:32

REVIEW QUESTIONS

1. A gene is
   a) a segment of DNA 50 kilobases in length
   b) the product of protein synthesis
   c) The smallest functional genetic unit as defined by mutations
   d) the number of nucleotides in virus
   e) always a protein-coding segment of DNA

2. Which statement is NOT TRUE regarding the cor-

relation between genome complexity (amount of unique sequence information) and genome size?
a) The more complex a viral life cycle is, the larger the viral genome.
b) Bacteria have a smaller and less complex genome than humans.
c) In eukaryotes, there is a 1:1 correlation between genome size and complexity.
d) The explanation for the C-value paradox is that some organisms contain larger amounts of repetitive DNA sequences in their genomes than other organisms.
e) The human mitochondrial genome encodes only a subset of genes required for mitochondrial function.

3. Evidence that telomeres are important for chromosome function includes
a) origins of replication contain telomere sequences
b) telomeres contain repetitive sequences capable of forming a hairpin loop that protects chromosomal ends from exonuclealytic digestion
c) telomere length is inversely correlated with the age of a cell
d) linear eukaryotic chromosomes all have telomeric sequences at both ends
e) b, c, and d are all correct

4. A large number of mouse and human genes appear to have been "shuffled" over evolutionary time, this observation suggests that
a) mice and humans share a common ancestor
b) the mammalian genome is too small to contain all necessary information
c) individual genes cannot function well when dispersed throughout the genome
d) genome arrangements are tolerated evolutionarily as long as individual genes are not physically disrupted or altered
e) most mouse and human genes are in fact colinear and have not been rearranged in the last 70 million years

5. Some types of repetitive DNA sequences can be used as molecular markers to screen for some types of human diseases. Choose the answer that best explains why this is true.
a) Large numbers of CpG islands are associated with some diseases.
b) Individuals with too few repetitive sequences have shorter lifespans.
c) Expansion of trinucleotide repeats have been associated with specific genetic disorders.
d) High levels of SINEs and LINEs indicate a viral infection.
e) Repetitive sequences are actually examples of genomic junk DNA and are not associated with any known genetic disorders.

6. RFLP mapping can be a useful tool to isolate DNA sequences associated with genetic defects because
a) it identifies the precise location of point mutations within a mutated gene
b) RFLPs that are tightly linked to a genetic phenotype represent molecular landmarks and can be used as a starting point to begin a gene search
c) males and females have different RFLP maps in the area of interest
d) loss of a restriction site in the genome is usually fatal
e) Southern blots using RFLP marker probes can be used to identify mRNA transcripts arising from the defective gene

ANSWERS TO REVIEW QUESTIONS

1. ***c*** A gene consists of both coding sequences and regulatory sequences that control gene expression.

2. ***c*** Many species of plants have more DNA than human cells even though their genomes are less complex. This is the C-value paradox, which is explained by genome differences in the amount of repetitive DNA.

3. ***e*** Telomeres are necessary for DNA synthesis of chromosomal termini and for protection from degradation (b), which explains why their loss is detrimental (c and d).

4. ***d*** Interspersed repetitive DNA and unique sequences most likely facilitate gene "shuffling" over evolutionary time, which is tolerated in most cases as long as genes themselves are not disrupted.

5. ***c*** At least five genetic diseases are indicated by measuring the length of trinucleotide repeats (Table 21.1).

6. ***b*** RFLP markers are most often serendipitous DNA polymorphisms that reside within the vicinity of a mutant allele and can therefore be used as entry points to begin cloning nearby DNA segments in search of the causative genetic mutation.

CHAPTER

22

DNA REPLICATION

INTRODUCTION

DNA must be faithfully duplicated before it can be passed on to the two daughter cells. **DNA replication** is the process whereby an entire genome is duplicated by DNA synthesizing enzymes called **DNA polymerases.** The time required for **DNA synthesis,** and the accuracy with which it must be performed, are both critical elements in assuring that the process of DNA replication is error-free. In humans, all 3×10^9 base pairs are replicated in about 8 hours, while in bacteria it takes about 30 minutes to replicate the 5×10^6 base pairs in the *E. coli* genome. What accounts for the speed at which DNA is replicated and how can mammalian DNA be replicated in proportionately less time than bacterial DNA? Another important question is, how are mistakes in DNA replication (incorporation of the wrong nucleotide) kept to less than 1 in 10^9 nucleotides added? Most of what is known about DNA replication at the molecular level has been learned by studying bacterial and viral replication. However, more recently, there has been progress in understanding the regulation of eukaryotic DNA synthesis in yeast and an in vitro DNA replication system using human replication proteins has been developed. In this chapter, we first describe what is known about the biochemistry of DNA replication using the prokaryotic systems as examples, and then discuss the control of DNA replication initiation in eukaryotes, because of its pivotal role in controlling eukaryotic cell division.

OVERVIEW OF DNA REPLICATION

As described in Chapter 2, base-pairing between G-C and A-T residues along the backbone of double-stranded DNA, provides the fundamental rule for copying the information stored in the chemical bonds of DNA. Since the two strands of DNA are complementary (a G on one strand is always opposite to a C on the other strand), it doesn't matter which strand is being read, because you can always determine the sequence of the other strand. Similarly, during DNA replication, each single strand of DNA is copied by enzymes that use the base sequence on one strand, to polymerize the complementary sequence on a new single strand. The two complementary strands of DNA are noncovalently associated with each other such that the 5′ to 3′ polarity of one strand is **antiparallel** to the 3′ to 5′ polarity of the other strand. This relationship is important to keep in mind when trying to understand DNA synthesis, because all known DNA polymerases can only synthesize DNA in the 5′ to 3′ direction, as will be described in more detail later in this chapter.

DNA Replication Is Semiconservative

Two models were proposed to explain how an individual DNA molecule could be duplicated to produce two identical "daughter" molecules. The first suggested that the DNA is replicated **conservatively,** meaning that both strands of the parent DNA are copied such that a single daughter molecule with two newly synthesized strands is produced as shown in Figure 22.1. Alternatively, as Watson and Crick proposed, based on their structural model of DNA, the parental DNA molecule is replicated **semiconservatively,** so that two daughter molecules are produced, each containing one original strand from the parent and one newly synthesized complementary daughter strand (Fig. 22.1).

In one of the classic experiments in early molecular biology, Matthew Meselson and Frank Stahl designed a test to distinguish between these two models of DNA replication. By metabolically labeling bacterial DNA with a "heavy" isotope called **nitrogen ^{15}N**, and then allowing the cells to grow for one cell division in normal "light" **nitrogen ^{14}N**, they could analyze the daughter strand of DNA and determine if it was made up of all light nitrogen (all ^{14}N) or half light and half heavy nitrogen (a mixture of ^{14}N and N^{15}). As shown in Figure 22.1, the results of their experiments supported the model that DNA is replicated semiconservatively. In fact, if they let the bacterial cells grow for a second cell division, they found that no heavy DNA molecules remained, only intermediate (^{14}N and ^{15}N) and light

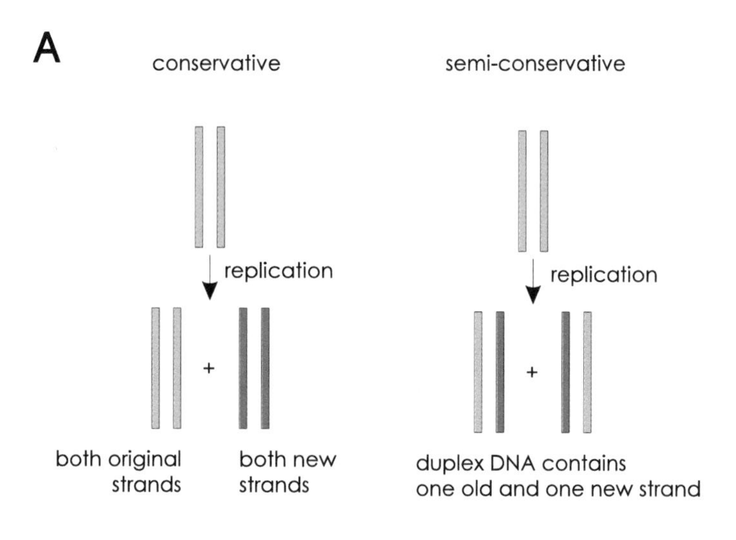

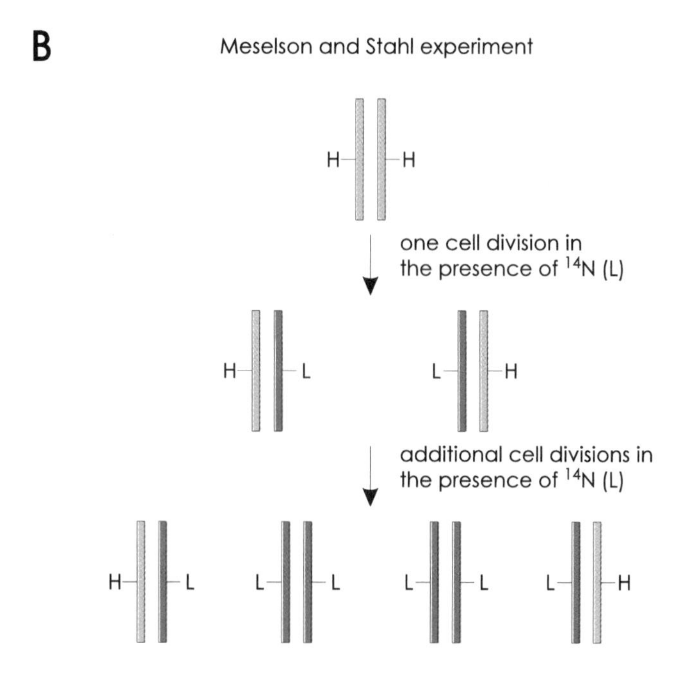

Figure 22.1.

DNA is replicated semiconservatively. (A) Schematic drawing showing the difference between conservative and semiconservative replication. (B) The results of the Meselson and Stahl experiment demonstrating that DNA is replicated semiconservatively. H represents ^{15}N (heavy), and L refers to ^{14}N (light), DNA.

molecules (^{14}N), as would be predicted by the **semiconservative replication model.**

DNA Replication Proceeds Bidirectionally

The finding that replication is semiconservative raised several new questions. For example, where does replication begin in a genome and does it proceed in only one direction at a time (**unidirectional replication**) or both directions simultaneously (**bidirectional replication**). Experiments using radioactive thymidine to mark newly synthesized DNA in actively dividing cells, proved that replication is bidirectional, as shown in Figure 22.2. High-magnification electron microscopy revealed that the first step in DNA replication is the formation of a "bubble" which is now known to correspond to the local unwinding of the DNA helix at the **origin of DNA replication.** The points at which the parental DNA is being unwound, and the new DNA strands are being synthesized, is called the **replication fork** (Figure 22.2). Subsequent genetic and biochemical studies confirmed that replication began at a specific site and proceeded bidirectionally. This was most easily done using *E. coli,* because bacteria contain only a single circular chromosome that has one unique origin of replication.

By studying the structures of replicating DNA using electron microscopy, and then localizing the origin of replication physically, it was found that ~30 minutes after replication began at the origin of replication, the two replication forks met half way around the chromosome. This was the predicted result for bidirectional replication of a circular genome (Fig. 22.2).

Replication Forks Have Leading and Lagging Strands

The question of how the growing fork could accommodate both 5′ to 3′ and 3′ to 5′ polymerization was a more difficult problem to solve because DNA polymerases were known to synthesize DNA only in the 5′ to 3′ direction. Using ^{3}H-thymidine, Reiji Okazaki determined the size of newly synthesized bacterial DNA by isolating DNA from cells that had been grown for a very short period of time (seconds) in media containing this radioactive nucleotide. He found that a large portion of the radioactive DNA existed as relatively small fragments (~1000 nucleotides) which became known as **Okazaki fragments.** Based on these observations, it was determined that one strand of the replication fork proceeds continuously in the 5′ to 3′ direction (the **leading strand**), while the other strand requires discontinuous 5′ to 3′ synthesis through the formation of short Okazaki fragments (**lagging strand**). As shown in Figure 22.3, this combination of continuous and discontinuous syn-

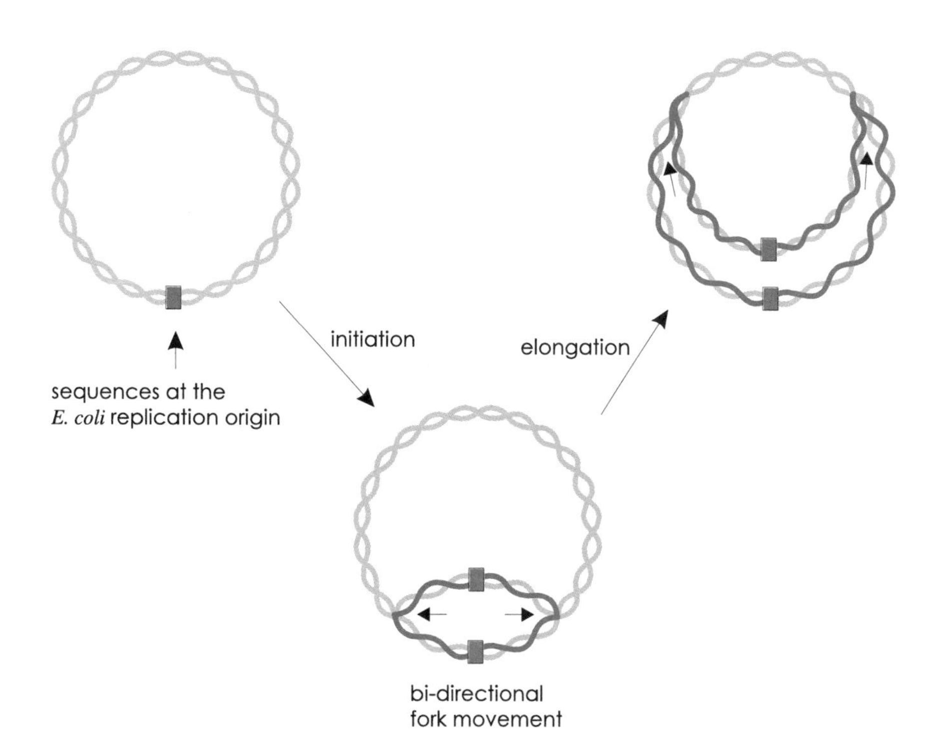

Figure 22.2.

DNA replication initiates at a unique location on the *E. coli* origin known as the replication origin. Bidirectional fork movement leads to the complete replication of the chromosome as the two replication forks meet approximately half way around the chromosome.

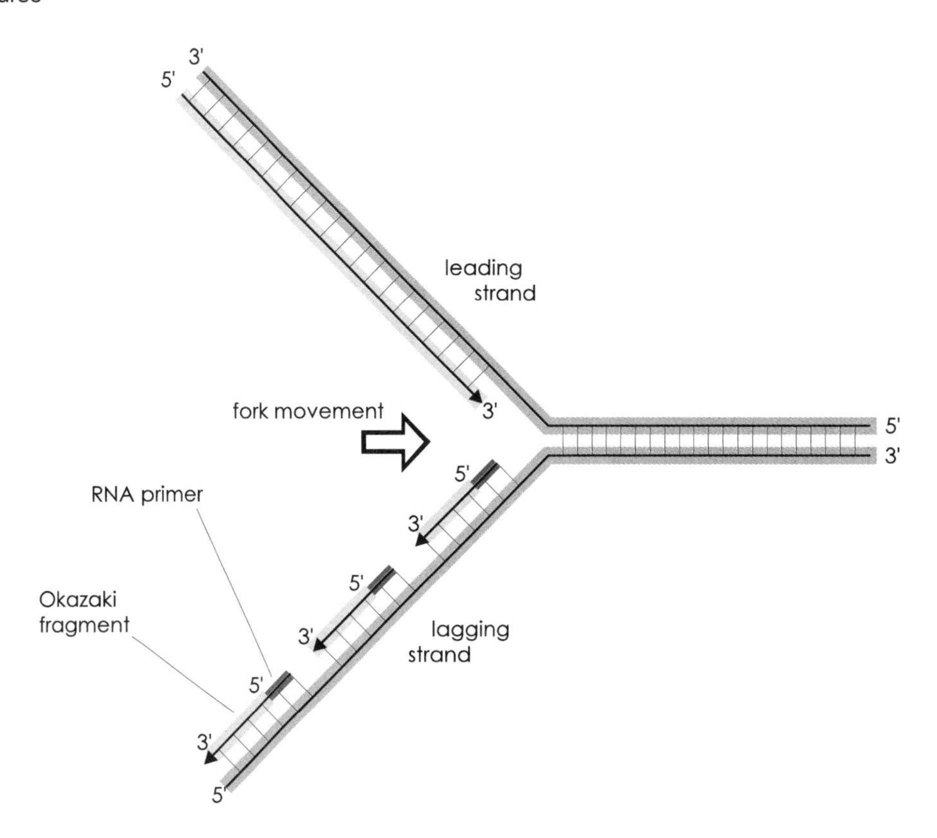

Figure 22.3.
Schematic drawing of the replication fork showing continuous 5′ to 3′ synthesis on the leading strand and discontinuous 5′ to 3′ DNA synthesis on the lagging strand through the formation of Okazaki fragments.

thesis results in the net movement of the fork in one direction. Also note that DNA synthesis on the lagging strand requires multiple **RNA primers** which are required before DNA polymerase can initiate synthesis of the Okazaki fragments. As discussed later in the chapter, all known DNA polymerases require a free 3′ OH end provided by an RNA or DNA primer annealed to the template strand.

DNA POLYMERASES

In the decade following the description of Okazaki fragments, a combination of biochemical and genetic approaches led to the isolation and characterization of a large set of proteins required for DNA replication. The first class of enzymes to be described were the DNA polymerases. Table 22.1 lists the properties of the sev-

Table 22.1. Comparative Characteristics of *E. coli* and Mammalian DNA Polymerases

DNA polymerase	Major function	Mol. wt. (Daltons)	Has a 3′-5′ exonuclease	Comment
E. coli enzymes				
pol I	Gap filling and removal of RNA primers on the lagging strand	109,000	Yes	Also has 5′-3′ exonuclease activity
pol II	DNA repair	90,000	Yes	Mutants are viable
pol III	Processive-chain elongation in DNA synthesis	~900,000	Yes	Absolutely required or E. coli DNA replication
Mammalian enzymes				
α (I)	Lagging-strand synthesis	300,000	No	Also includes priming activity
δ (III)	Leading-strand synthesis	~200,000	Yes	
ϵ (II)	DNA repair	250,000	Yes	
β	DNA repair	40,000	No	
γ	Mitochondrial DNA replication	~250,000	Yes	Nuclear-encoded gene

eral DNA polymerases isolated from both *E. coli* and mammalian cells. DNA polymerases I and III in *E. coli* are functionally analogous to DNA polymerases α and δ in mammalian cells, respectively.

All DNA polymerases have the ability to add nucleotides to the 3′ end of an existing single strand, provided there is a template strand to specify the base composition of the nascent chain. The study of DNA replication in *E. coli* has so far been more informative because of the combined approaches of biochemistry and genetics. It is generally assumed that the basic mechanism of DNA replication, specifically fork movement, is similar between prokaryotes and eukaryotes, based on the finding that both prokaryotic and eukaryotic DNA polymerases are large multisubunit complexes and that fork movement in both systems required leading and lagging strand synthesis.

Enzymology of DNA Replication

The basic enzymatic reaction of DNA synthesis involves the nucleophilic attack by the 3′-hydroxyl at the end of the growing chain on the 5′ α phosphate of the incoming dNTP, as illustrated in Figure 22.4. This cleavage results in the formation of one pyrophosphate (PP_i) per nucleotide added to the nascent chain. Pyrophosphate is subsequently cleaved by pyrophosphatase in a highly favorable reaction to yield inorganic phosphate. However, pyrophosphate cleavage is not ab-

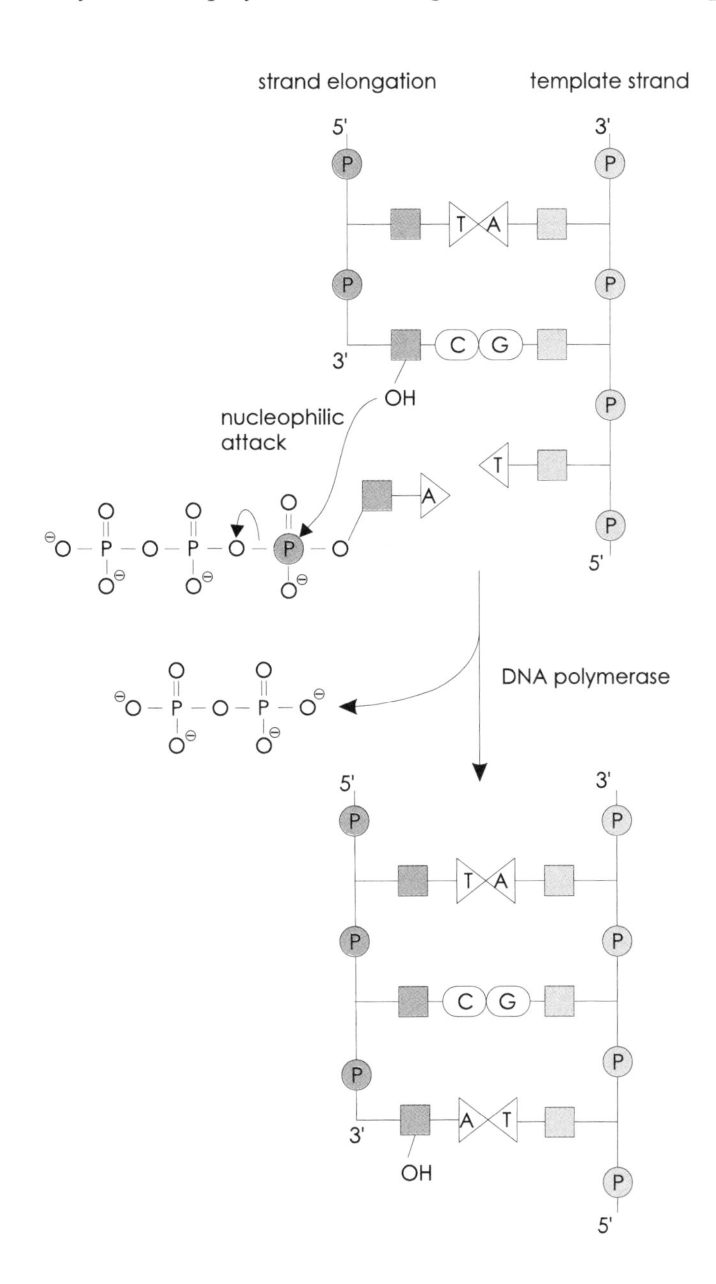

Figure 22.4.

Enzymology of the polymerization reaction. A nucleophilic attack by the 3′ hydroxyl on the growing strand results in the formation of a covalent bond with the α phosphate of the incoming dNTP.

solutely required for the DNA polymerization reaction to proceed and, therefore, it is thought that the newly formed noncovalent bonds between the added base and the complementary base on the other DNA strand, and base-stacking interactions within the same DNA strand, must also contribute to the favorability of this reaction.

Although a large fraction of the DNA polymerase activity in a bacterial cell is due to **DNA polymerase I** (DNA pol I), the multisubunit **DNA polymerase III** (DNA pol III) enzyme is responsible for DNA replication at the growing fork. DNA pol I is required for completing replication on the lagging strand between Okazaki fragments, and DNA polymerase II (DNA pol II) is involved in DNA repair. In addition to the ability to catalyze the polymerization reaction shown in Figure 22.4, three other important activities are shared between the three DNA polymerases; however, there are quantitative differences between the enzymes. The first is the rate of **chain elongation,** which is found to be highest for DNA pol III. This makes sense because DNA pol III needs to be able to complete replication in a relatively short time.

The second functional difference between the DNA polymerases is the level of **processivity,** which refers to the number of nucleotides added before the polymerase dissociates from the template. Again, DNA pol III is the clear winner in this category, as might be expected based on its central role in DNA replication. The **β subunit** of DNA pol III is designated as the processivity subunit. Two β subunits form a "clamp" around the template strand, which then can slide along the DNA (Fig. 22.5). This arrangement ensures that an active holoenzyme (all the DNA pol III subunits combined) remains tightly associated with DNA to enhance processivity.

DNA Polymerases Have Proofreading Activity

The third, and probably the most important, property shared by all three polymerases, is the ability to **proofread** the accuracy of the polymerization reaction. This requires an exonuclease activity which can remove misincorporated nucleotides in the 3′ to 5′ direction. This is analogous to using a correcting ribbon on a typewriter to remove typos. Figure 22.6 illustrates how this 3′ to 5′ exonuclease activity functions during polymerization. If an incorrect nucleotide is inserted, it does not properly base pair and the DNA polymerase will stop, back up, and remove the nucleotide by cleaving the phosphodiester bond. A significant number of these corrections are due to the insertion of a tautomeric form of cytosine which can hydrogen bond with adenine. When the cytosine undergoes a tautomeric shift to the normal configuration, the hydrogen bonds are broken with adenine and the mispaired nucleotide is then removed by the proofreading activity. The proofreading activity of the DNA pol III complex is contained on a separate subunit, whereas this activity is encoded within the single DNA pol I polypeptide chain.

Not all misincorporated nucleotides are removed by the DNA polymerase proofreading activities. A second mechanism, called DNA **mismatch repair,** is also involved in removing misincorporated nucleotides. As described in Chapter 23, DNA mismatch repair proteins are capable of recognizing unpaired bases in the nascent DNA strand. A segment of DNA including the mismatched nucleotide is removed by nucleotide excision and then DNA pol I and DNA ligase repair the strand. Together, the mechanisms of proofreading by DNA polymerases, and DNA mismatch repair, contribute to the extremely low error rate ($<1 \times 10^{-9}$ errors per base pair replicated) observed in DNA replication.

Similarities Between Bacterial and Mammalian DNA Polymerases

Eukaryotic DNA polymerases have also been isolated and characterized as listed in Table 22.1. Based on studies of SV40 DNA replication in vitro, it has been found that **DNA polymerase δ** has high processivity and is required for leading-strand synthesis, making it analogous to *E. coli* DNA pol III. DNA polymerase δ requires ATP and is stimulated by two additional DNA replication proteins, RF-C and PCNA. **DNA polymerase α** serves the same role as *E. coli* DNA pol I in that DNA polymerase α is necessary for lagging-strand synthesis. In addition to DNA polymerase α and δ, three other DNA polymerizing activities have been identified. **DNA polymerase ϵ** is involved in DNA repair and is most similar to *E. coli* DNA pol II. **DNA polymerase β** is also a repair enzyme, and **DNA polymerase γ** is required for mitochondrial DNA synthesis.

FORK MOVEMENT REQUIRES A MULTIPROTEIN COMPLEX

The simplistic model of a growing replication fork, as shown in Figure 22.3, cannot explain all the topological and enzymatic problems that must be solved to achieve

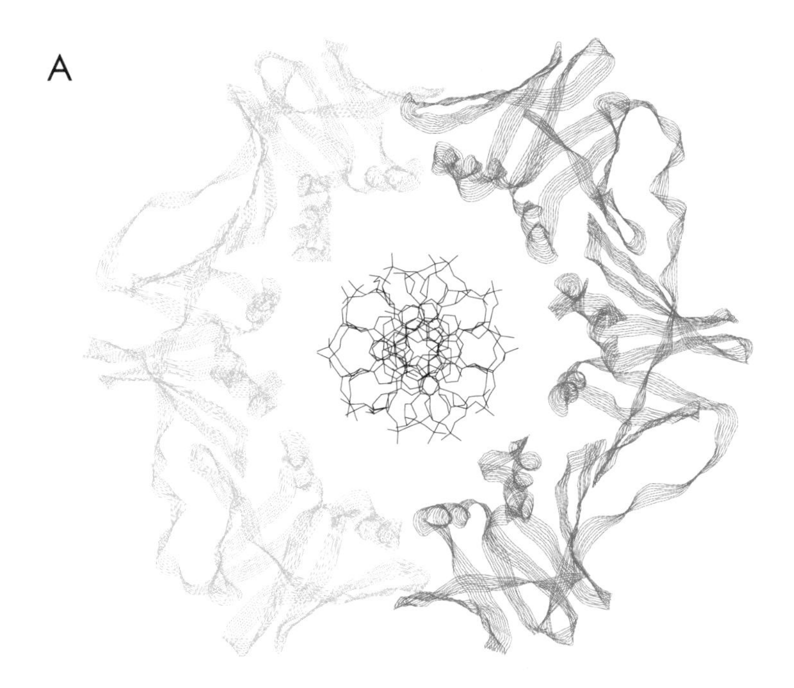

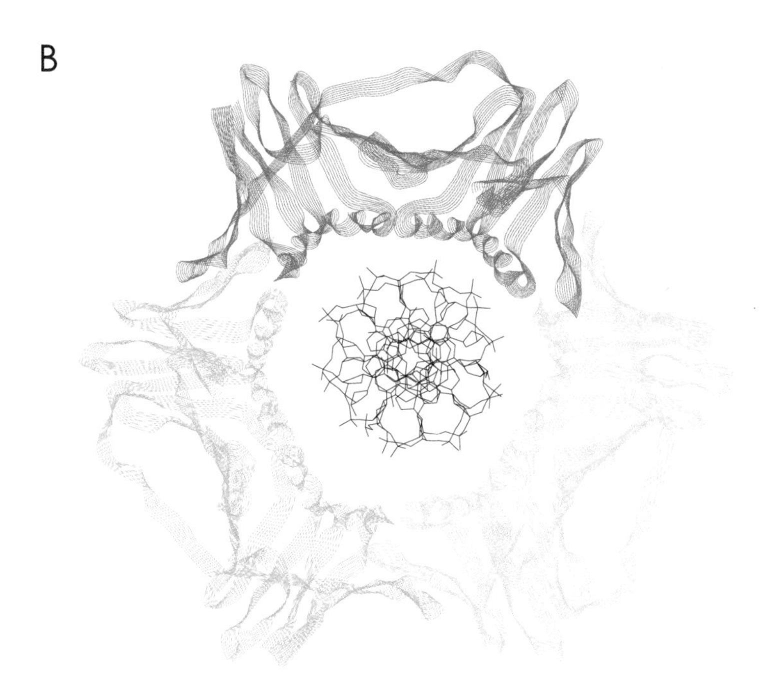

Figure 22.5.
Structure of the DNA pol III β subunit clamp based on X-ray crystallography data. (A) The two β subunits form a protein "clamp" which appears to wrap around the DNA helix. This arrangement explains its role as the processivity factor because it prevents the catalytic subunit of DNA pol III from dissociating with the DNA. (B) The structure of the eukaryotic processivity factor, PCNA, also forms a clamp around the DNA, but it is made up of three subunits.

the high rate of DNA synthesis observed in vivo. For example, how does the fork proceed without causing torsional stress on the double helix in front of the replication fork? What proteins are required to maintain the replication fork in an "open" configuration while DNA synthesis is in progress, and what enzyme provides the RNA primer on the lagging strand? The answers to most of these questions came from work on both *E. coli* and bacteriophage T4 DNA replication. In vitro DNA replication assays were developed to biochemically characterize the proteins required for DNA synthesis, and genetic studies were performed to test the effect of mutations in the genes encoding these same proteins. Together, these in vitro and in vivo studies, led to the theory that a DNA **replisome protein complex** coordinates DNA synthesis at a replication fork as described below.

In addition to the requirement for DNA pol III for the polymerization reaction, at least four other classes of proteins are needed for fork movement. These key

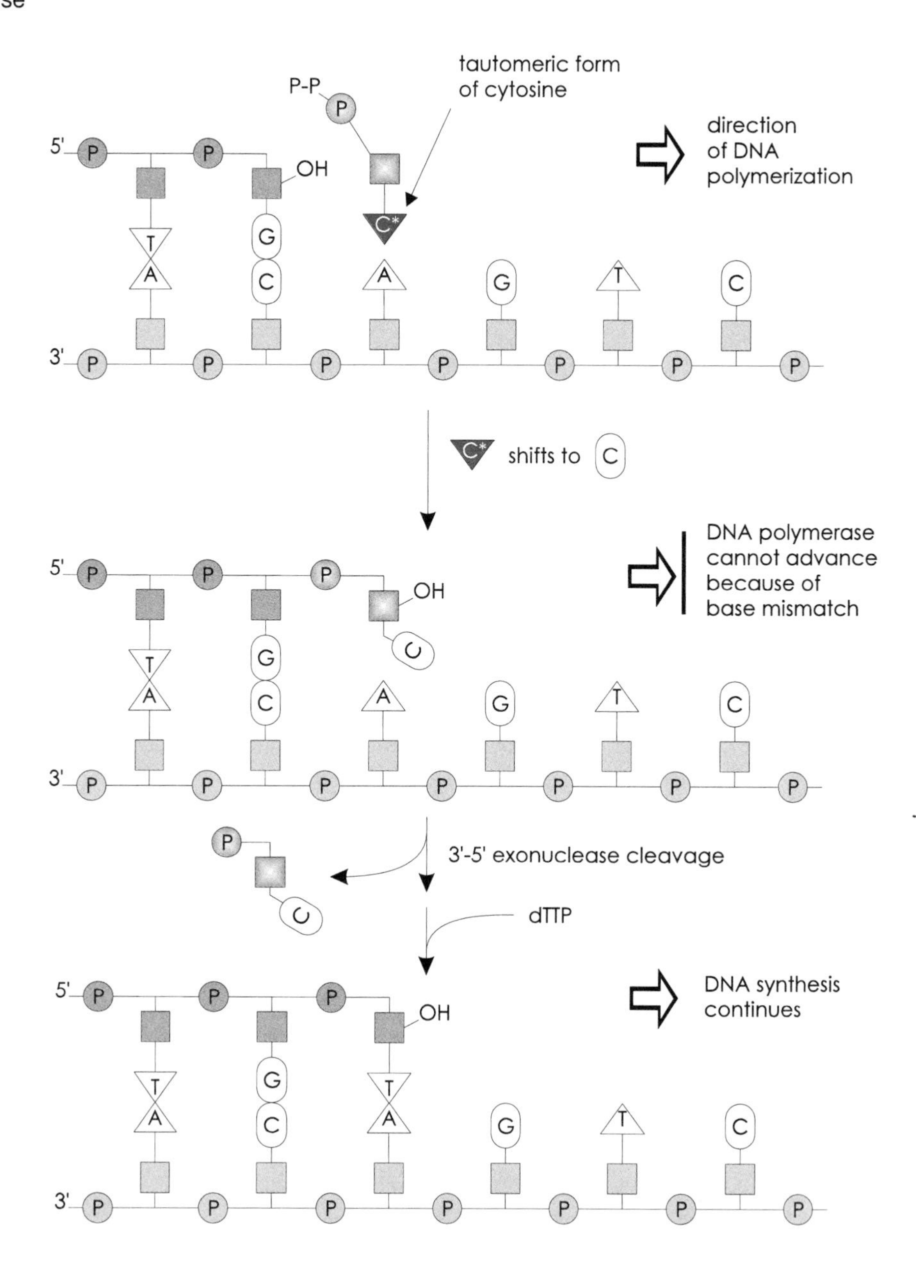

Figure 22.6.
The proofreading activity of the 3′ to 5′ exonuclease activity associated with DNA polymerases.

proteins are helicases, topoisomerases, single-strand binding protein, and DNA primase. The functional role of these proteins in fork movement and DNA replication is shown in Figure 22.7.

Classes of Proteins at the Replication Fork

Helicases are ATP-requiring enzymes that slide along single-stranded DNA just ahead of the moving replication fork. The double-stranded DNA in front of the helicases is unwound, creating the single-strand template needed for DNA pol III and DNA primase. Two other enzymes, **topoisomerase I and II,** are required to relieve the torsional stress that builds up as a result of the unwinding action of the helicases. Topoisomerase I cleaves DNA on one strand, which allows the DNA strands to swivel, while topoisomerase II cleaves both strands of DNA.

Single-strand binding protein (SSB) prevents the helix from simply zipping back up once the helicase has passed. As shown in Figure 22.7, SSB, which has an extremely high affinity for single-strand DNA, is primarily needed on the lagging strand because the leading strand is being replicated continuously as the helix is being unwound. The fourth critical enzyme in the replication machine is **DNA primase**. This enzyme synthesizes short RNA primers at regular intervals

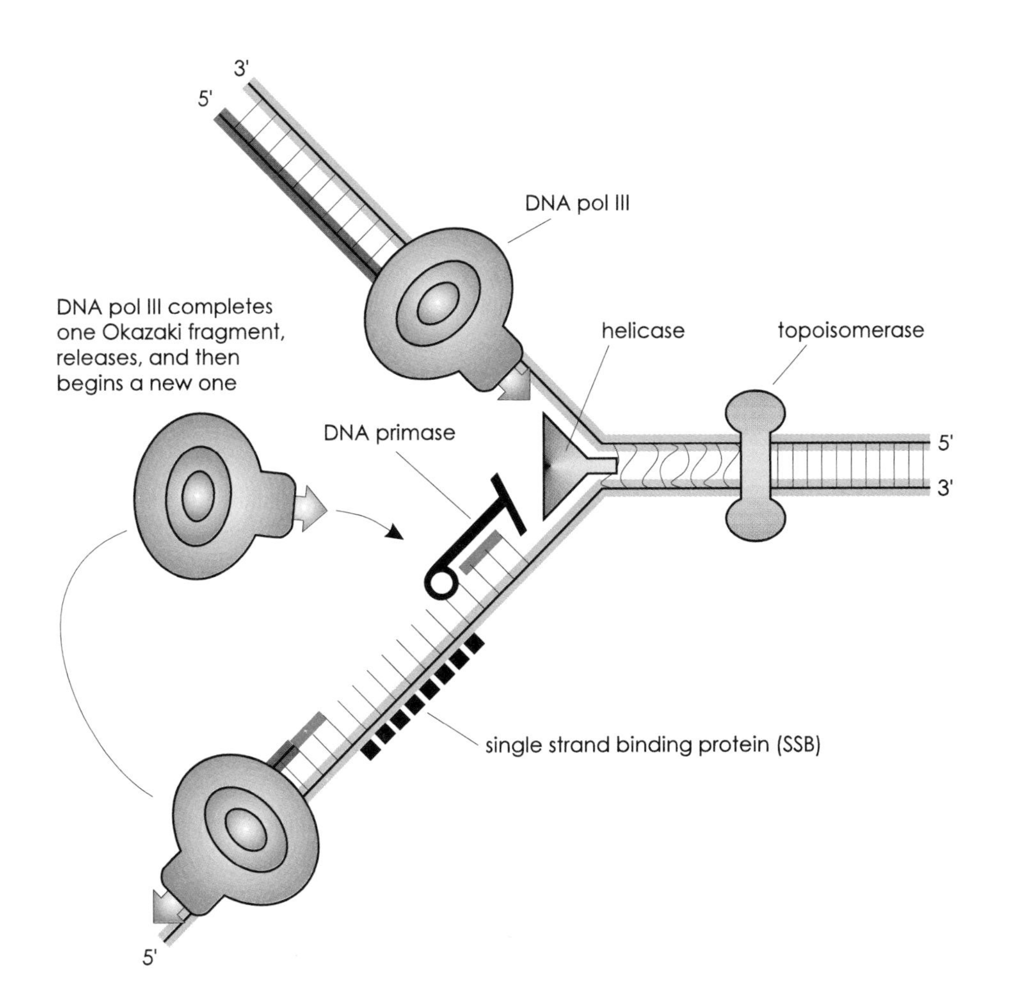

Figure 22.7.
Proteins involved in DNA replication fork movement. DNA pol III is responsible for de novo DNA synthesis on both the leading and lagging strands. Helicases unwind the DNA in front of the replication complex, and topoisomerases relieve the torsional stress on the DNA double helix that results from fork movement. Single-strand binding proteins (SSBs) prevent the single-strand regions from reannealing, and DNA primase synthesizes the necessary primers for lagging-strand synthesis of Okazaki fragments.

along the lagging strand. Extension of these primers in the 5′ to 3′ direction by DNA pol III creates Okazaki fragments, as shown in Figure 22.3. The RNA primers are removed by the 5′ to 3′ exonuclease activity of DNA pol I (see Table 22.1) and replaced by the corresponding deoxynucleotides. Note that DNA primase is also required for initiation of leading-strand synthesis at the origin of replication.

DNA ligase is required to seal the single-strand nick between the growing chain on the lagging strand and the adjacent Okazaki fragment. DNA ligase from *E. coli* requires NAD^+ to provide the phosphate from AMP, whereas T4 DNA ligase utilizes ATP for this reaction. The functions of DNA pol I and DNA ligase in lagging-strand DNA synthesis are illustrated in Figure 22.8.

The Multiprotein Complex Is Called a "Replisome"

Biochemical analysis of the large multiprotein complex used in T4 DNA replication revealed that it had a molecular weight in excess of 10^6 Daltons. In addition, it was determined that many of the proteins required for fork movement (Fig. 22.7), are actually tightly associated with each other in what has been called a **replisome.** One way to explain how the replisome can perform multitask functions such as lagging-strand synthesis, and still attain high rates of elongation (500 nucleotides per second in bacteria), is to propose that the lagging strand is looped back through the replisome to bring the 3′ end of the RNA primer into the same orientation as the 3′ end of the growing end of the newly synthesized DNA on the leading strand. This DNA looping model is shown in Figure 22.9.

Two key elements in this version of replisome-driven fork movement are that the DNA primase and helicase are tightly associated with each other, and that two molecules of the DNA pol III holoenzyme coordinately synthesize the new strands. The lagging-strand synthesis would still be discontinuous because the DNA pol III molecule on that strand would have to intermittently release and reattach as one Okazaki fragment was completed and before a second one was started. The DNA

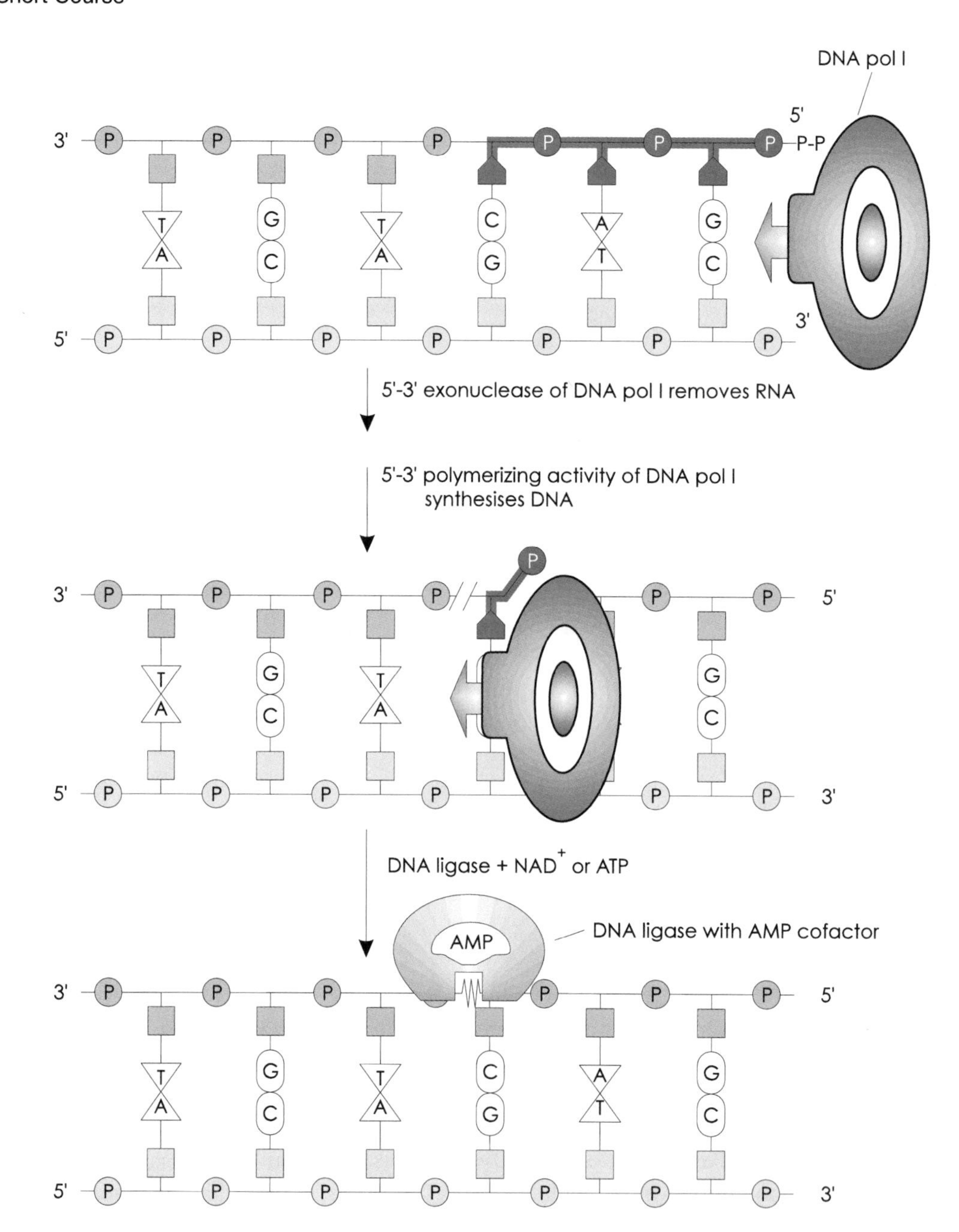

Figure 22.8.
The enzymatic activities of DNA polymerase I and DNA ligase are required to complete DNA synthesis of the lagging strand. *E. coli* DNA ligase requires NAD^+ as a cofactor, whereas T4 DNA ligase uses ATP.

replisome has been compared to a sewing machine where individual parts function together to provide a smooth and continuous sewing motion.

ORIGINS OF DNA REPLICATION

Although little is known about the steps involved in the termination of DNA replication, we do understand some of the control points needed for initiation of DNA synthesis at replication origins. The timing of bacterial DNA replication, for example, is tightly linked to cell division, such that each daughter cell receives only a single copy of the genome. Similarly, eukaryotic cells replicate their DNA only during the S phase of the cell cycle in a tightly controlled series of steps which requires about one third of the cell cycle time. In contrast, most viruses replicate their genomes in an opportunistic way so as to maximize viral production. The key determinants in specifying when DNA replication is initiated, regardless of the genome, are the replication proteins and their binding to specific DNA sequences at replication origins. In this section we describe what is known about prokaryotic and eukaryotic replication origins, and the corresponding proteins that initiate DNA synthesis at these sequences.

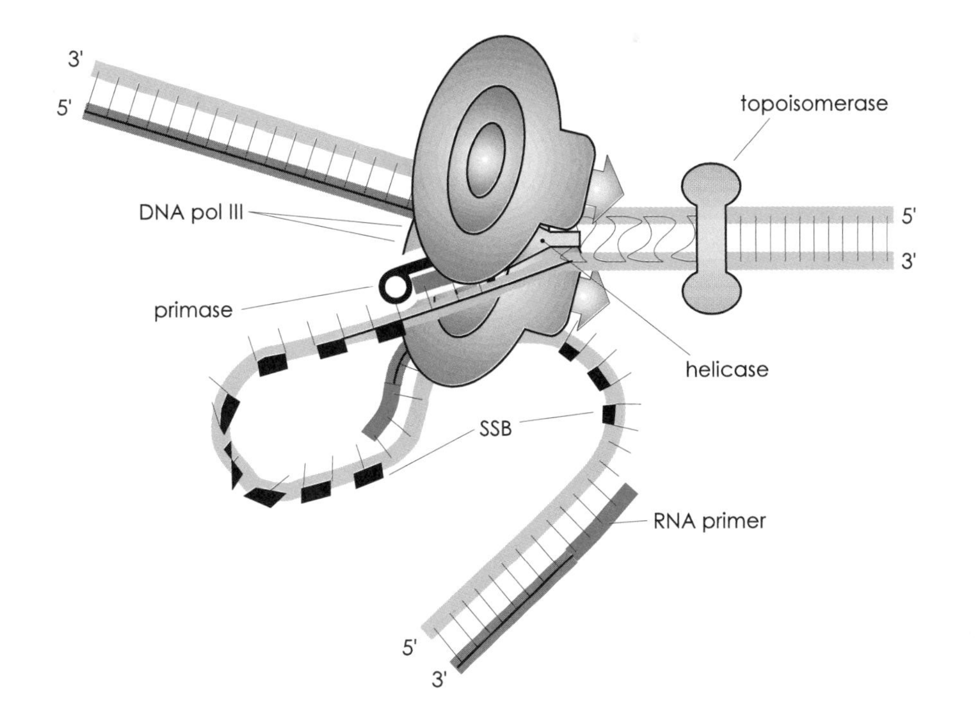

Figure 22.9.
A schematic drawing of the replisome model of fork movement showing how lagging-strand DNA synthesis could be accomplished by looping one strand of the replication fork back through the replisome complex.

The *E. coli* Origin of Replication

In bacteria and most viruses, a specific DNA sequence determines where replication begins, and in the case of *E. coli,* this is at a single origin located at the genetic locus called **OriC.** Biochemical analyses of initiation events at OriC have identified many of the same auxiliary proteins that are involved in fork movement (besides DNA pol III), such as a helicase, a DNA primase, SSB, and topoisomerases. However, these studies also revealed an origin binding protein called **DnaA protein.** Such a specific DNA binding protein was predicted based on the finding that mutations in OriC caused defects in origin function.

It was found that the 245-base-pair OriC region contained multiple copies of two types of repeated sequences, as shown in Figure 22.10. The DnaA binding site has a specific **9-base-pair repeat** sequence which is present four times. A second element is the **A-T rich repeat** which is present three times. The model of initiation at OriC is illustrated in Figure 22.10B.

It is thought that about 20 molecules of DnaA protein bind to the 9-base-pair repeats and cause a change in the DNA structure such that the double helix is denatured in the region of the 13-base-pair A-T rich repeats. Since A-T base pairs are much weaker, because they have only two hydrogen bonds, as compared to G-C base pairs, which contain three hydrogen bonds (Chapter 2), the location of the A-T rich repeats function to position the center of the **replication bubble.** Once this region is accessible to the other DNA replication proteins, the replication bubble is extended by the helicases and SSB in both directions to create the bidirectional replication forks. In the next step, DNA primase synthesizes RNA primers and DNA pol III initiates leading- and lagging-strand synthesis. The close proximity of the DnaA protein-binding sites to the A-T rich region results in a physical overlap between the genetic location of OriC and the actual replication origin.

Origins of Replication in Eukaryotes

Initiation of DNA replication in eukaryotes appears to be more complex than what has been found in bacteria. First, since there is much more DNA, and it resides on separate chromosomes, many more origins of replication are required. Based on the size of the human genome, and the elongation rate of eukaryotic DNA polymerases (~2000 nucleotides/minute), bidirectional origins would need to be spaced at a minimum of every 2×10^6 base pairs to finish replication during the S phase (~8 hours). However, studies have shown that replication origins occur more frequently (one about every 1×10^5 base pairs), and that some origins are utilized early in the S phase, while others are functional late in the S phase. Second, genetic and biochemical

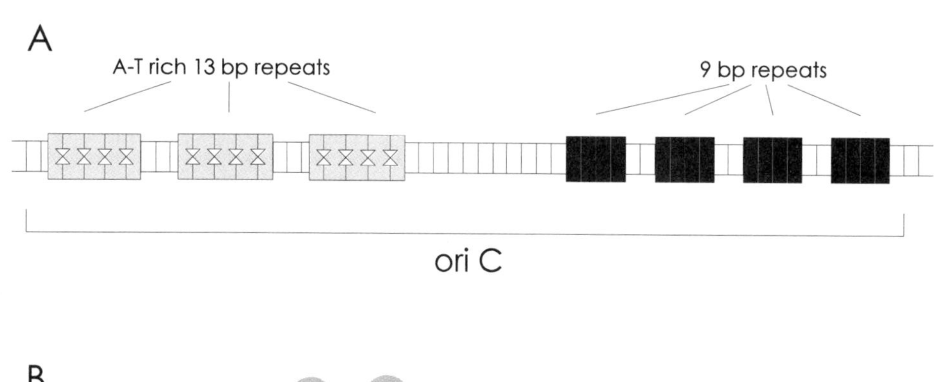

Figure 22.10.
Organization of functional sequence elements in the *E. coli* origin of DNA replication. (A) Location of A-T rich sequences and the four 9-base-pair repeats within OriC. (B). Model showing how DnaA protein binding to the 9-base-pair repeats leads to melting of the A-T rich sequence and formation of the two bidirectional replication forks.

analyses of origin sequences indicate that a fairly large region may be necessary for origin function. Physical origins of bidirectional replication, mapped by two-dimensional electrophoresis, are found in large replication zones that contain clusters of individual replication origins.

Taken together, these findings indicate that there may be less sequence specificity in eukaryotic origins than what has been found in bacteria and viruses. One explanation may be that functional origins in eukaryotes could be determined more by chromatin structure and auxiliary DNA binding proteins that interact at sequences distant from the physical bidirectional origin, rather than by specific sequences with the origin itself.

Autonomously Replicating Sequences in Yeast

Because of the difficulties in studying DNA replication in higher eukaryotes, yeast has been exploited as the organism of choice because of the opportunity to use genetic analysis. It was found almost 20 years ago that yeast genomic sequences could function as **autonomously replicating sequences (ARS)** when cloned into yeast plasmids that contained no endogenous origins of replication. Subsequent studies have shown that a subset of these ARS sequences do in fact correspond to yeast origins of replication in their normal chromosomal location. Molecular genetic analyses of ARSs have revealed several functional similarities between ARSs and the *E. coli* OriC. As shown in Figure 22.11, ARSs contain a DNA sequence called the **origin recognition element (ORE)** that is specifically bound by a set of proteins referred to as the **origin recognition complex (ORC).** The yeast ORC contains six protein subunits and requires ATP for DNA binding.

Binding of the yeast ORC proteins to the ORE DNA sequence is similar to the binding of *E. coli* DnaA proteins to the 9-base-pair repeat of OriC. A second similarity between yeast and *E. coli* is the presence of an A-T rich sequence in the yeast ARS which is called the **DNA unwinding element (DUE).** This sequence can easily be unwound and has been identified as the actual origin of DNA replication within the ARS segment.

A third class of DNA sequences also stimulate

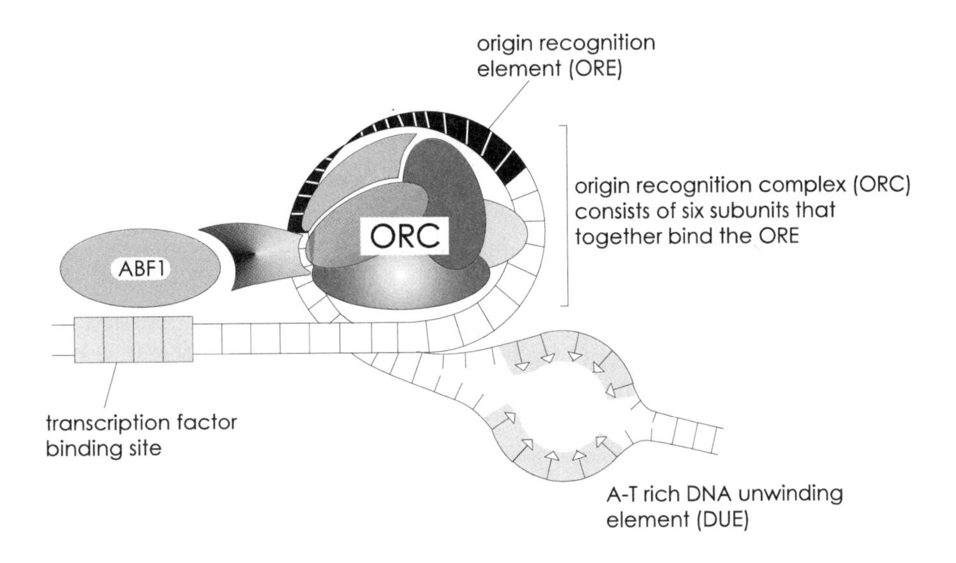

Figure 22.11.
Schematic representation of a yeast replication origin.

DNA replication at the ARS origin in yeast. This sequence corresponds to a binding site for the yeast **transcription factor ABF1.** Studies with eukaryotic viral origins of replication have suggested that some of the proteins involved in RNA synthesis (also called transcription factors, see Chapter 24) have dual functions in transcription and initiation of DNA replication. Current models suggest that transcription factors may be necessary to recruit replication proteins to the origin.

The regulatory mechanisms controlling replication origin function in eukaryotes are not at all clear. The accessibility of the replication proteins to single-strand DNA in a denatured A-T rich region would seem to be a critical step. Linking activation of origin-binding proteins to cell signaling, perhaps through phosphorylation (many of the cell cycle-control points are dependent on phosphorylation–dephosphorylation events), may be an important initiation point. By understanding more about the regulatory steps that control initiation of DNA replication, we may be able to identify possible gene defects that contribute to the onset of tumor growth in human cancers (see Chapter 31).

DNA REPLICATION ENZYMES HAVE MANY APPLICATIONS

Radioactive Labeling of DNA

Understanding the enzymology of DNA replication has been useful in developing important methodologies which have had a major impact on modern biology. Table 22.2 lists some of these enzymes and their applications. One of the first uses of DNA polymerases in recombinant DNA methodologies was for the radiolabeling of DNA using ^{32}P-dNTPs and *E. coli* DNA pol I in a reaction called **nick translation**. By providing a template of double-stranded DNA containing random single-strand nicks (produced by mild treatment with the endonuclease DNaseI), DNA pol I is able to remove deoxynucleotides with its 5′ to 3′ exonuclease activity,

Table 22.2. DNA Polymerases and Their Uses in Recombinant DNA Methodologies

Enzyme	3′–5′ Exonuclease	5′–3′ Exonuclease	Polymerase activity	Processivity	Application
E. coli DNA	Yes	Yes	Moderate	Low	Nick translation polymerase I
Klenow fragment	Yes	No	Moderate	Low	Blunting DNA ends
T4 DNA polymerase	Yes	No	Moderate	Low	Radiolabeling DNA
Reverse transcriptase	No	No	Low	Moderate	Synthesizing cDNA
Modified T7 DNA polymerase	No	No	High	High	Sanger dideoxy sequencing
Taq DNA polymerase	No	Yes	High	High	Polymerase chain reaction

and then replace them with radioactive ^{32}P-dNTPs using its polymerizing and proofreading activities. A modified form of DNA polymerase I, called the **Klenow fragment**, does not contain the 5′ to 3′ exonuclease activity and it can be used for adding nucleotides to recessed 3′ ends of double-stranded DNA. Table 22.2 also lists the characteristics of the bacteriophage **T4 DNA polymerase**, which has many of the same properties of the Klenow fragment, except that its 3′–5′ exonuclease activity is extremely efficient. This attribute makes T4 DNA polymerase useful for radiolabeling DNA in a reaction containing ^{32}P-dNTPs.

Synthesis of Complementary DNA (cDNA)

An unusual DNA polymerase that is described in more detail in Chapter 30 is the RNA-dependent DNA polymerase called **reverse transcriptase.** Reverse transcriptase was first identified in a class of eukaryotic viruses known as retroviruses. This enzyme is capable of synthesizing complementary DNA (cDNA) using an RNA or DNA primer annealed to a single-strand RNA template. Howard Temin and David Baltimore codiscovered reverse transcriptase and were awarded the Nobel Prize in Medicine in 1976. It is safe to say that modern biotechnology, which depends on the isolation of gene-coding sequences for many of its pharmaceutical applications, would not have been possible without the utilization of this remarkable DNA polymerase.

DNA Sequencing

DNA polymerases have also been exploited as tools for genome analysis and characterization. One of these enzymes is a modified form of bacteriophage T7 DNA polymerase (which lacks 3′ to 5′ exonuclease activity). The basic DNA sequencing reaction, which first used the Klenow fragment, as described by Sanger in 1977, was greatly improved by utilizing this modified T7 DNA polymerase. The Sanger dideoxy DNA sequencing method is described in Chapter 2.

Selective DNA Amplification

The second DNA polymerase, which has drastically changed the analysis of nucleic acids for molecular biological studies, is derived from the thermophilic bacteria *Thermus aquaticus.* This unusual DNA polymerase called **Taq polymerase,** is a single polypeptide chain with high processivity and a temperature optimum for polymerization of 80°C. Taq polymerase can be used in a procedure called the **polymerization chain reaction (PCR),** which can be used to amplify selected regions of DNA using oligonucleotide primers. Because Taq polymerase can be used at such high temperatures, the primers can be designed so that they anneal to target DNA with high specificity. A description of PCR and its applications is given in Box 22.1.

BOX 22.1. POLYMERASE CHAIN REACTION (PCR) USING TAQ DNA POLYMERASE

In the mid 1980s, a very simple, yet extremely powerful, variation of the in vitro DNA polymerization reaction was described by researchers at a biotechnology company in California. This technique is called the polymerase chain reaction (PCR) and it is based on the idea that if two primers are complementary to sequences which flanked a region of DNA, then successive rounds of primer annealing, DNA synthesis, and denaturation can produce a large quantity of DNA product with the two primers as the terminal sequences. This elegant strategy is now used worldwide in a large variety of research and diagnostic applications. It's major advantage over standard nucleic-acid techniques is that PCR can be used to amplify DNA sequences up to a million-fold, which makes it an extremely sensitive technique.

The key technological breakthrough in the application of PCR to molecular biology came when DNA polymerases from thermophilic bacteria were used in place of *E. coli* DNA polymerase. Since different temperature optimums are required for the three steps of annealing (~50°C), polymerization (~70°C), and denaturation (95°C), a DNA polymerase was needed that could function at temperatures well above 37°C. Automation of PCR by special temperature blocks that can quickly change the temperature of hundreds of reaction tubes at once has greatly facilitated the application of PCR to laboratory research. Figure 22.12 shows how successive rounds of PCR can quickly produce large amounts of a specific product. Some of the more spectacular uses of PCR include cloning of ancient DNA from extinct species such as the woolly mammoth ruling out suspects in criminal cases by forensic science, and providing genomic markers (STSs) in chromosome-mapping studies used in the Human Genome Project (see Chapter 21).

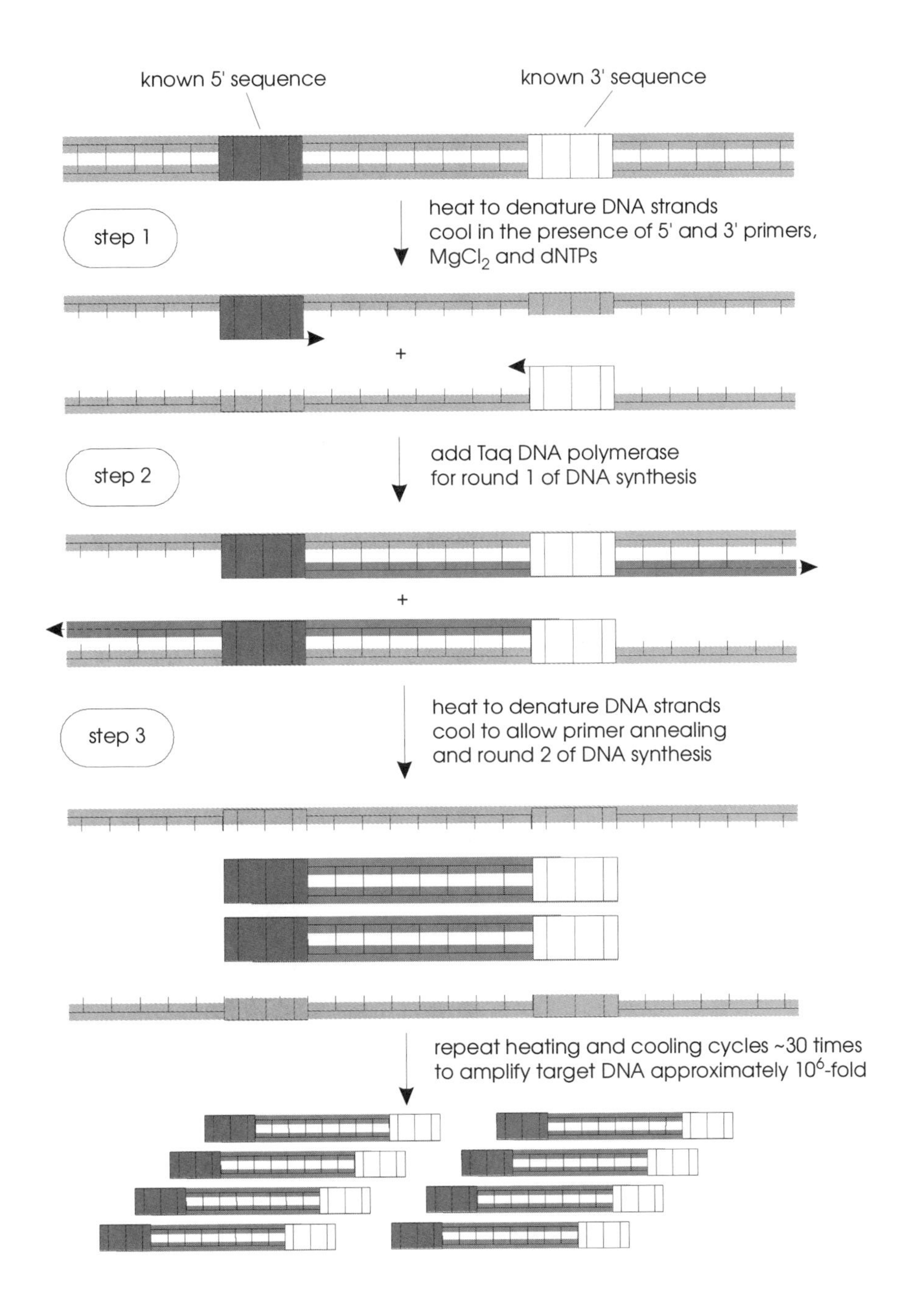

Figure 22.12.

Schematic drawing showing how PCR leads to amplification. Note that the original template DNA strands also participate in each round of amplification (not shown in step 3); however, the vast majority of amplified material is derived from amplified-target DNA templates, as depicted.

SUMMARY

1. DNA replication is semiconservative and proceeds bidirectionally from an origin of replication. DNA synthesis is always in the 5′ to 3′ direction relative to the template strand. A replication fork consists of a leading strand in which DNA synthesis is continuous, and a lagging strand characterized by the discontinuous formation of short Okazaki fragments.

2. There are at least three classes of DNA polymerases, based on their functional differences. In *E. coli,* DNA pol III is required for DNA synthesis at the replication fork, DNA pol I fills in gaps in the lagging strand, and DNA pol II is needed for DNA repair. Mammalian DNA polymerases have been characterized and shown to have similar functions.

3. All DNA polymerases require a free 3′ hydroxyl group to catalyze the elongation step. This 3′ hydroxyl can be provided by RNA or DNA primers.

4. The replisome is a large multiprotein complex required for replication fork movement. In ad-

dition to DNA polymerase, other proteins at the replication fork include helicases, topoisomerases, single-strand binding proteins, DNA primase, and DNA ligase.

5. Origins of DNA replication are sites of initiation of DNA synthesis. In *E. coli,* the DNA sequences designated OriC contain binding sites for multiple subunits of DnaA protein. The binding of DnaA proteins to OriC facilitates helix unwinding and leads to the establishment of bidirectional replication forks. Yeast have been shown to contain analogous DNA sequences at replication origins, and yeast origin-binding proteins have been isolated and characterized.

6. Many of the DNA polymerases have been exploited as tools in a variety of molecular biological methods. Most notably, Taq DNA polymerase is the key enzyme required for the polymerase chain reaction (PCR).

REFERENCES

Botchan, M. (1996): Coordinating DNA replication with cell division: Current status of the licensing concept. *Proc. Natl. Acad. Sci. USA* **93**:9997.

Burhans W. C., Huberman J. A. (1994): DNA replication origins in animal cells: A question of context? *Science* **263**:639.

Coverley D., Laskey R. A. (1994): Regulation of eukaryotic DNA replication. *Annu. Rev. Biochem.* **63**:745.

DePamphilis M. L. (1993): Eukaryotic DNA replication: Anatomy of an origin. *Annu. Rev. Biochem.* **62**:29.

Dutta A. (1993): DNA replication: Trans-plication factors? *Current Biol* **3**:709.

Erlich H. A., Gelfand D., Sninsky J. J. (1991): Recent advances in the polymerase chain reaction. *Science* **252**:1643.

Herendeen, D.R. and Kelly, T.J. (1996): DNA polymerase III: Running rings around the fork. *Cell* **84**:5.

Huberman J. A. (1995): Cell cycle: A license to replicate. *Nature* **375**:360.

Joyce C. M., Steitz T. A. (1994): Function and structure relationships in DNA polymerases. *Annu. Rev. Biochem.* **63**:777.

Kornberg A. (1991): Control of initiation of the *Escherichia coli* chromosome. Cold Spring Harbor *Symp. Quant. Biol.* **56**:275.

Krishna T. S. R., Kong X-P., Gary S., Burgers P. M., Kuriyan J. (1994): Crystal structure of the eukaryotic DNA polymerase processivity factor PCNA. *Cell* **79**:1233.

Liang P., Pardee A. B. (1992): Differential display of eukaryotic messenger RNA by means of the polymerase chain reaction. *Science* **257**:967.

Marahrens Y., Stillman B. (1992): A yeast chromosomal origin of DNA replication defined by multiple functional elements. *Science* **255**:817.

Newlon C. S. (1993): Two jobs for the origin replication complex. *Science* **262**:1830.

Reiss J., Cooper D. N. (1990): Application of the polymerase chain reaction to the diagnosis of human genetic disease. *Hum. Genet.* **85**:1.

Stillman B. (1994): Smart machines at the DNA replication fork. *Cell* **78**:725.

Stukenberg P. T., Turner J., O'Donnell M. (1994): An explanation for lagging strand replication: Polymerase hopping among DNA sliding clamps. *Cell* **78**:877.

Yuzhakov, A., Turner, J., O'Donnell, M. (1996): Replisome assembly reveals the basis for asymmetric function in leading and lagging strand replication. *Cell* **86**:877.

REVIEW QUESTIONS

1. How did Meselson and Stahl demonstrate that DNA replication is semiconservative?
   a) They biochemically labeled DNA and showed that DNA synthesis results in a 1:1 mixture of old and new DNA strands.
   b) They used a 1:1 mixture of ^{15}N and ^{14}N in bacterial media to obtain newly synthesized DNA that was uniformly labeled on both strands.
   c) They compared the replication of linear and circular DNA grown in vitro and in vivo.
   d) They used electron microscopy to physically separate ^{15}N-labeled DNA from ^{14}N-labeled DNA.
   e) The results of their experiments actually showed that DNA is replicated conservatively.

2. An Okazaki fragment is a(n)
   a) RNA primer
   b) Eco RI digestion product
   c) catalytic domain of DNA polymerase
   d) short segment of DNA synthesized on the lagging strand
   e) autonomously replicating segment of DNA from yeast

3. The low error rate of nucleotide misincorporation during DNA synthesis is due to
   a) low levels of processivity by DNA polymerases
   b) high levels of processivity by DNA polymerases
   c) 3′ to 5′ exonuclease activity in DNA polymerase and mismatch repair

d) mismatch repair only
e) the β subunit of DNA pol III

4. The DNA replication complex requires a variety of proteins to facilitate fork movement. Which answer best describes the minimum set of proteins required for *E. coli* DNA replication in an vitro reaction?
   a) DNA pol I, primase, SSB, and ligase
   b) SSB, helicase, and topoisomerase
   c) ligase, DNA pol I, and DNA pol III
   d) DNA pol III, helicase, SSB, and primase
   e) topoisomerase, helicase, and DNA pol II

5. The role of DnaA protein in controlling DNA synthesis in *E. coli* is
   a) functionally equivalent to the yeast origin recognition complex (ORC), which binds DNA and causes a local unwinding of the DNA helix
   b) to prevent torsional stress due to helicase activity at the fork
   c) to increase DNA pol III processivity
   d) to prime DNA synthesis of Okazaki fragments on the lagging strand
   e) to bind to a series of 13 bp A-T rich repeats in the origin and prevent DNA bending during fork movement

6. The polymerase chain reaction (PCR) is a(n)
   a) in vitro reaction mimicking oxidative phosphorylation
   b) form of DNA synthesis utilized by viruses
   c) series of reiterative DNA synthesis steps using oligonucleotide primers
   d) a way to enhance the specificity of DNA mismatch repair
   e) technique based on the DNA synthesis properties of reverse transcriptase

ANSWERS TO REVIEW QUESTIONS

1. ***a*** The experiment required a method to discriminate between newly synthesized DNA (light strands) and template DNA (heavy strands).

2. ***d*** An Okazaki fragment is a product of lagging-strand DNA synthesis.

3. ***c*** Both the proofreading activity of DNA polymerase and mismatch repair are required for high fidelity in DNA synthesis.

4. ***d*** DNA pol III, helicase, SSB, and primase would be sufficient for fork movement in vitro; topoisomerases are required in vivo.

5. ***a*** DnaA protein binding to the 9 bp repeats is required for DNA distortion at the A-T rich 13 bp repeats, similar to the function of ORC in ARS replication.

6. ***c*** DNA amplification by PCR requires that a cycle of DNA denaturation, primer annealing, and DNA synthesis by Taq DNA polymerase, be repeated ~30 times to achieve DNA target amplification of approximately a million-fold.

CHAPTER

23

DNA REPAIR AND RECOMBINATION

INTRODUCTION

Maintenance of DNA integrity by DNA repair mechanisms, and alterations in DNA organization by recombination, are two important processes which contribute to changes in the information stored in DNA. First, DNA damage that results in base changes can lead to permanent alterations in the DNA sequence if they are not repaired before the next round of replication. Second, DNA rearrangements through the process of recombination can be the consequences of random events that contribute to genetic diversity or the result of highly specialized mechanisms such as immunoglobulin gene rearrangements. In this chapter we describe these two general processes of DNA repair and recombination and emphasize the contribution of special DNA-dependent enzymes in maintaining the integrity of the genome.

MUTATIONS ARE THE RESULT OF DNA DAMAGE

Depurinations, Deaminations, and Thymine Dimers

Misincorporated nucleotides due to DNA replication are efficiently removed by the proofreading activity of DNA polymerases. But what about damaged DNA as a result of exposure to chemical mutagens or from ultraviolet radiation? A common form of DNA damage is the result of spontaneous **depurination** of guanine or adenine to deoxyribose through hydrolysis of the *N*-glycosyl linkage. It is estimated that there may be as many as 10^3 depurinations per 10^9 base pairs per day in human cells. A related type of spontaneous chemical alteration is the **deamination** of cytosine to form uracil. As shown in Figure 23.1, if this is not corrected, a C-G base pair will become an A-T base pair following DNA replication.

Another common type of DNA damage is due to the formation of **thymine dimers** induced by ultraviolet and ionizing radiation. These intrastrand covalent bonds form a cyclobutyl ring between the C-5 and C-6 of one thymine, and the same carbons on the adjacent thymine. The result is that the DNA backbone on the thymine-containing strand is structurally altered, which may cause disruption of the hydrogen bonds with the corresponding adenine residues on the complementary strand (Fig. 23.2.).

Natural Selection Requires a Background Level of DNA Damage

Chemical compounds present in the environment can also, under some circumstances, cause DNA damage. Two classes of such compounds are **deaminating**

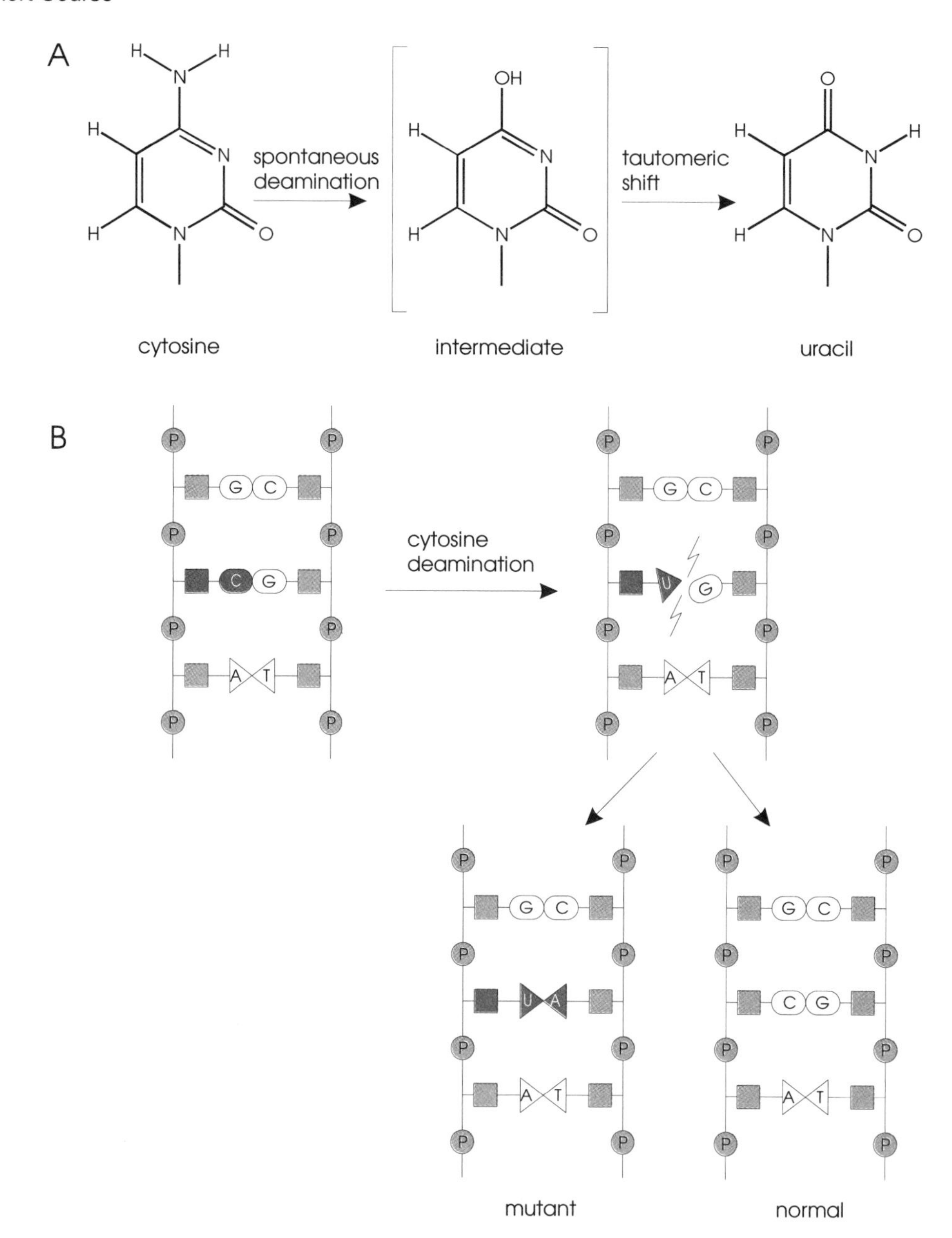

Figure 23.1.

Schematic drawing showing how cytosine deamination could cause a mutation. (A) Deamination reaction converting cytosine to uracil. (B) Cytosine deamination, if not corrected, will result in a mutation from G-C to A-T after subsequent rounds of DNA replication.

agents such as nitrous acid or nitrites, and **alkylating agents** like dimethylsulfate. A variety of laboratory tests have been designed to assess the potency of these compounds as mutagens. The **Ames test,** developed by Bruce Ames, uses bacteria as a genetic host to determine what effect various compounds have on the mutation frequency of a given gene. If a particular compound has chemical mutagenic properties, then colonies of mutated bacteria can readily be identified by the Ames test. The addition of protein extracts prepared from mammalian livers can alter the mutagenicity of some compounds in this assay due to the formation of metabolic intermediates by liver enzymes.

The vast majority of DNA mutations resulting from these natural and man-made sources are corrected by DNA repair systems in the cell. However, it is estimated that unrepaired mutations become incorporated into the genome at about 1 in every 10^9 base pair per generation, based on evolutionary studies of DNA sequences in noncoding regions of eukaryotic genomes. If this is extrapolated to a single gene, then the **mutation rate** would be about one mutation per gene in every 10^5–10^6 cell generations, depending on how big the gene is. The principle of **natural selection** is based on the assumption that some mutations are beneficial and provide a survival advantage to individuals within a

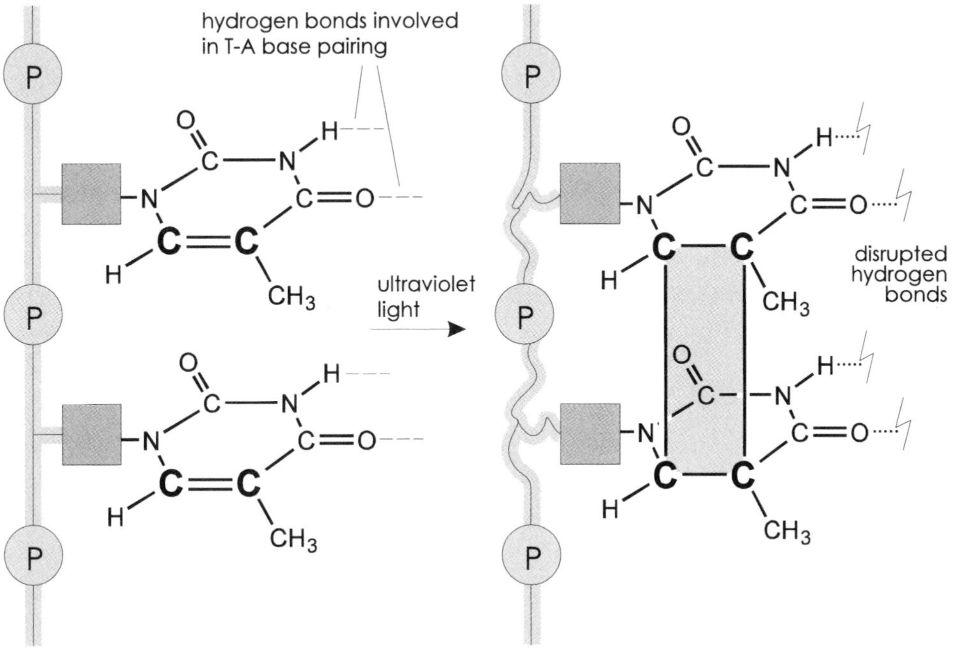

Figure 23.2.
Schematic drawing of thymine dimer formation.

species. Therefore, not all mutations are "bad," and in fact, the driving force of evolution depends on a delicate balance between DNA repair and genetic diversity.

DNA REPAIR MECHANISMS

How is a cell able to repair the $>10^4$ mutations per day per genome which are thought to occur? The answer is that the cell spends a lot of energy on enzymes that constantly monitor DNA for mutations. Fortunately, because of the duplex nature of the DNA helix, there are always two copies (one being the complement strand of the other) of genetic information. In many types of DNA damage, only one of the two strands needs to be repaired. Using mechanisms that distinguish between the damaged and undamaged strands, repair enzymes replace the mutated DNA, using information contained in the complementary strand.

As is the case in DNA replication, most of what we know about DNA repair mechanisms comes from the study of *E. coli*. However, recent studies of DNA repair in eukaryotes, primarily genetic analysis in yeast and genetic studies of human diseases, have revealed that similar processes are also found in these organisms. Table 23.1 lists the four basic types of DNA repair systems in *E. coli* and the proteins which have been characterized. The human genes for the corresponding proteins are also shown in cases where they have been identified.

The four basic steps in DNA repair are **recognize, remove, resynthesize,** and **religate**. The enzymes that recognize and remove damaged DNA are highly specific and can recognize a variety of structures resulting from different types of DNA alterations. The last two

Table 23.1. Table of the Four DNA Repair Systems in *E. coli* and the Human Homologues

Repair system	Type of damage	*E. coli* proteins	Human genes
Nucleotide excision	Pyrimidine dimers or other structural perturbations	ABC exonuclease, DNA pol I, ligase	XPA, XPB/ERCC3, XPC, XPD/ERCC2, XPE, XPF/ERCC1 and XPG/ERCC5, ERCC4, ERCC6
Mismatch	Nucleotide mismatches	MutH, MutL, MutS, DNA helicase, SSB, DNA pol III, ligase, exonuclease I	MLH1, PMS1, and PMS2 (MutL homologues) MSH2 (MutS)
Base excision	Removal of uracil, hypoxanthine, xanthine, and alkylated bases	DNA glycosylases, AP endonuclease, DNA pol I, ligase	Yeast DNA pol δ (mammalian DNA pol β), human DNA glycosylase
Direct	Chemical reversal of methylation or photoactivation	O^6-methylguanine DNA methyltransferases, DNA photolyases	O^6-methylguanine DNA methyltransferases

steps in DNA repair, resynthesize and religate, are carried out in *E. coli* by DNA polymerase I and DNA ligase, respectively.

Nucleotide Excision Repair

The removal of thymine dimers by **nucleotide excision repair** is shown in Figure 23.3. This is sometimes called bulky lesion repair because the structure recognized by the DNA repair enzymes is an abnormality in the DNA backbone, such as would result from the formation of the cyclobutyl ring in a thymine dimer (see Fig. 23.2).

The human genetic disease, **xeroderma pigmentosum** (XP), is a defect in DNA repair of thymine dimers. This condition results in extreme sensitivity to sunlight with a predisposition to skin cancer. Genetic

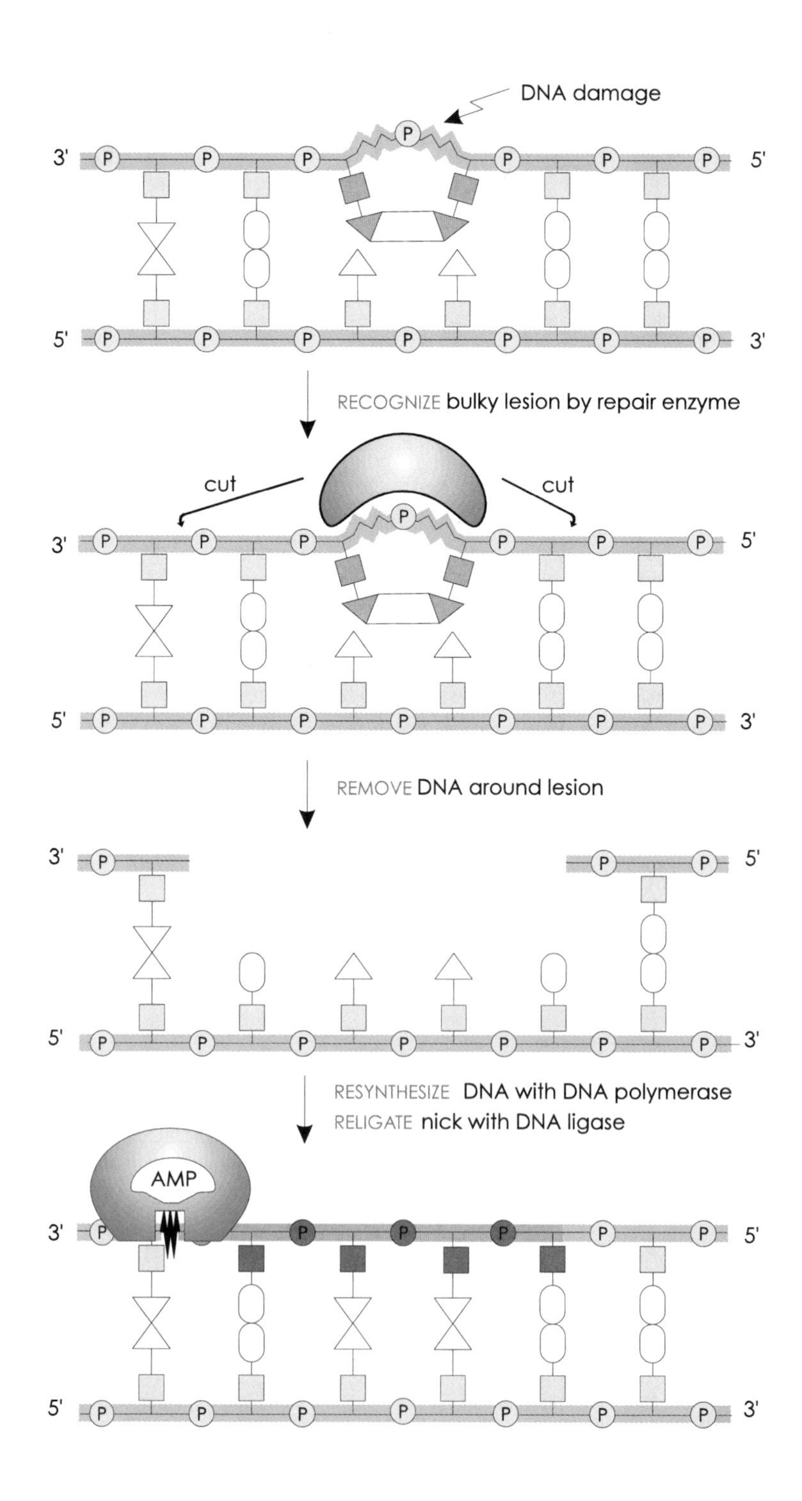

Figure 23.3.

Schematic drawing of the four steps in DNA repair: recognize, remove, resynthesize, and religate.

analyses of individuals with XP have identified at least seven different **complementation groups** (mutations in at least seven distinct genes, see Appendix). Two of the XP genes encode proteins that recognize bulky DNA (XPA and XPE) and at least two are helicases (XPB and XPD). Interestingly, the helicase encoded by the XPA gene (also called ERCC3), is bifunctional and found in RNA polymerase II complexes which are involved in gene transcription. It has been observed that transcribed DNA is preferentially repaired as compared to bulk DNA, and that RNA polymerases stall at sites of thymine dimers on the template strands of transcribed genes.

Mismatch Repair

A second important repair system is called **mismatch repair.** Occasionally, faulty DNA replication results in a mismatched base pair between the nascent and template strands. These errors are corrected by three proteins in *E. coli*; **MutS, MutH, and MutL.** This system is only able to correct newly synthesized DNA and it's fidelity is based on the fact that nascent DNA chains are initially unmethylated on adenine residues in the sequence GATC. This differential methylation pattern serves the purpose of distinguishing the new strand (unmethylated) from the template strand (methylated). This is important because the repair enzymes need to recognize which of the two nucleotides is the mismatch; otherwise, it could lead to a mutation if the wrong nucleotide is removed. Figure 23.4 illustrates how these three proteins correct a mismatch error in newly synthesized DNA.

Note that the unmethylated GATC sequence need not be immediately adjacent to the mismatch because exonucleases can remove the intervening DNA in the 3′ or 5′ direction, depending on the relative location of the incorrect nucleotide. Recently, the human homologues of MutS (hMSH2) and MutL (hMLH1) have been isolated based on the finding that a majority of patients with **hereditary nonpolyposis colon cancer** (HPCC) have mutations in genes which are functionally similar to the bacterial repair enzymes. Although there is still much to learn about the biochemistry of DNA repair in human cells by **hMSH2** and **hMLH1,** the identification of genetic defects in DNA repair genes in colon cancer cells may represent a key understanding about how cancer cells obtain a plethora of genomic alterations over a relatively short time (Chapter 31).

Base-Excision Repair

Base-excision repair involves the removal of nucleotides that have lost the base moiety as a result of depurination or by the action of DNA glycosylases (enzymes that remove abnormal bases such as deaminated cytosines). As in the other two repair systems, the first step is to remove the deoxyribose 5-phosphate through the action of an endonuclease and a phosphodiesterase which cleave the phosphate backbone and create a single nucleotide gap. DNA polymerase I then replaces the nucleotides and DNA ligase seals the nick. **DNA glycosylases** are enzymes that recognize inappropriate bases in DNA such as uracil, hypoxanthine, and xanthine, all of which result from the deamination of cytosine, adenine, and guanine, respectively. The aberrant base is removed by cleavage of the *N*-glycosyl bond by a DNA glycosylate which creates an "abasic" site. Importantly, uracil DNA glycosylases will not cleave uracil residues from RNA or thymine residues from DNA. Since uracil glycosylases act only on DNA, and uracil-containing nucleotides are not incorporated into DNA during DNA synthesis, then anytime a uracil is found in DNA it is removed.

Direct Repair

The fourth known repair system is referred to as **direct repair** because it does not involve the removal of any nucleotides and is independent of DNA polymerase I and DNA ligase. The most common example of this is the methylation of guanine at the O^6 position by alkylating agents. **O^6-methylguanine** can base pair with thymine and if uncorrected, can lead to a mutation from a G-C to an A-T base pair after two rounds of replication, similar to the case with cytosine deamination to uracil. The protein **O^6-methylguanine DNA methyltransferase** removes the CH_3 group by transferring it to cysteine residue in the enzyme and thus directly corrects the damage by regenerating guanine at that position. Direct repair is a highly specialized process, and indeed, very few examples of direct repair enzymes have been found.

TWO MECHANISMS OF DNA RECOMBINATION

Early genetic analyses of plants and bacteria indicated that simple patterns of heredity were not sufficient to

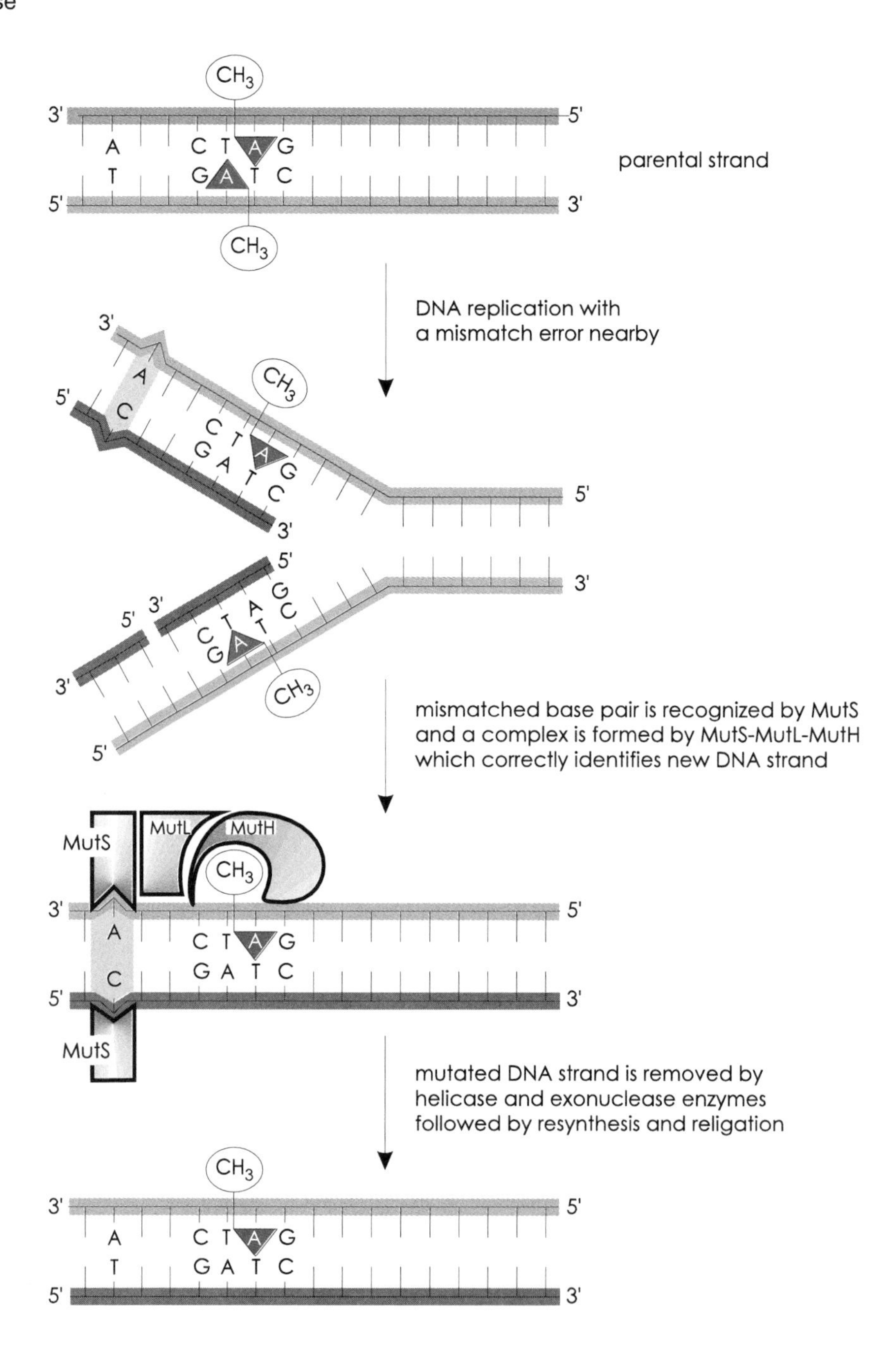

Figure 23.4.

Mismatch repair by MutH, MutS, and MutL proteins in *E. coli.*

explain the diversity of phenotypes observed in experimental genetic crosses. Following the elucidation of the structure of DNA, and the characterization of proteins involved in DNA replication and repair, it was proposed that DNA is capable of rearranging in either random or directed ways to create the genetic diversity first observed by geneticists. The basic mechanism of DNA recombination involves the breaking and rejoining of DNA strands in such a way that genetic information is exchanged or shuffled depending on the type of break (single or double strand) and the polarity of the DNA strands involved in the recombination. Two major types of recombination have been characterized at the molecular level, general or **homologous recombination** and **site-specific recombination.** One major difference between these two processes is that site-specific recombination requires proteins that recognize specific sequences at the recombination site, whereas homologous recombination can occur anywhere along the DNA strand, provided that the incoming DNA is complementary in sequence. Homologous recombination is driven by DNA base pairing and utilizes many of the same enzymes needed in other DNA-modifying reactions.

CROSSOVER FREQUENCIES AND GENETIC MAPS

The geneticists could provide evidence that homologous recombination must be taking place because genetic markers from paternal and maternal sources would sometimes be exchanged. The further two genes are physically apart from each other on the same chromosome, the more likely that homologous recombination events will be observed at the phenotypic level. If the genes are close to each other, it is less likely that homologous recombination will occur within the intervening DNA segment. The principal of using crossover frequencies to make genetic maps is shown in Figure 23.5.

MOLECULAR MECHANISM OF HOMOLOGOUS RECOMBINATION

If we look in more detail at how homologous recombination is accomplished, we find that it is a complicated process, and not too surprisingly, it presents a number of DNA structural and polarity problems that need to be solved. The two major steps in the reaction are (1) single-strand invasion to promote the formation of an in-

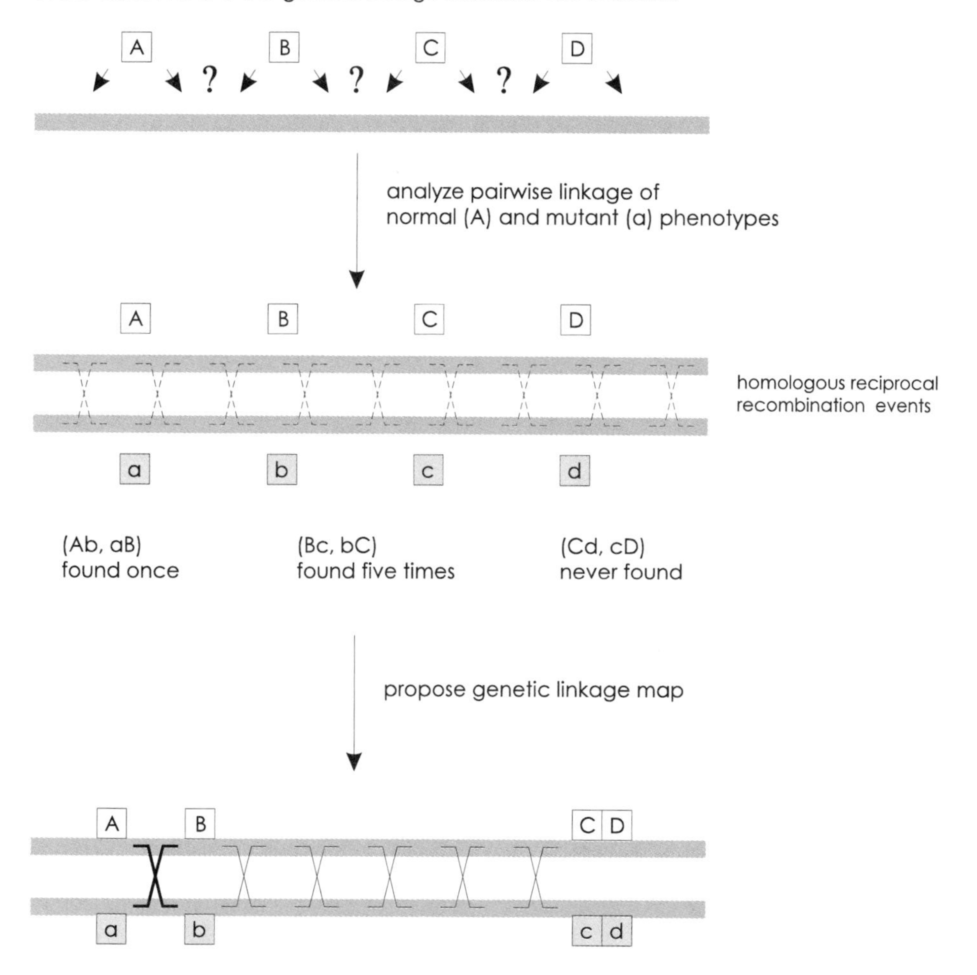

Figure 23.5.
Calculated crossover frequencies can be used to construct genetic maps.

termediate structure called a cross-strand exchange or Holliday junction, and (2) resolving this four-strand structure by isomerization and subsequent cleavage to release the recombined DNA molecules.

Step 1: Strand Invasion and Branch Migration

In the first step of homologous recombination, a single-strand nick in one of the two DNA molecules leads to the formation of a free single strand that displaces the corresponding strand in the homologous chromosome, as shown in Figure 23.6. The best example of this reaction is carried out by the **RecA protein** in *E. coli* which has a high affinity for single-stranded DNA and is also able to facilitate the "invasion" of one DNA strand into another. The second part of this reaction is the nicking and exchange of the displaced single strand so that it can pair with its complementary sequence present in the invading DNA molecule. This structure is called a **heteroduplex** because it consists of single strands from each of two donor molecules. Finally, through a process known as **branch migration,** the crossover point (where the invading DNA strand pairs with the target DNA) is able to travel up and down the heteroduplex DNA much like a zipper. Since each strand of the heteroduplex is base paired with its complementary strand on the other molecule, branch migration is really like two parallel zippers. Branch migration is a protein-directed event in *E. coli* where ATP hydrolysis by the RecA protein drives the reaction. The net result of branch migration is that it determines how much genetic information will be exchanged.

Step 2: Isomerization and Resolution of Holliday Junctions

The second critical step in homologous recombination is the **isomerization** and **resolution** of the two DNA molecules. The physical location where all four heteroduplex DNA strands meet, each base-paired with the complementary strand, is called a **Holliday junction** (named after Robin Holliday, who first proposed it). Figure 23.7 schematically illustrates how these DNA gymnastics are thought to occur, resulting in the formation of DNA characteristic of a homologous recombination event. Starting with the structure at the junction,

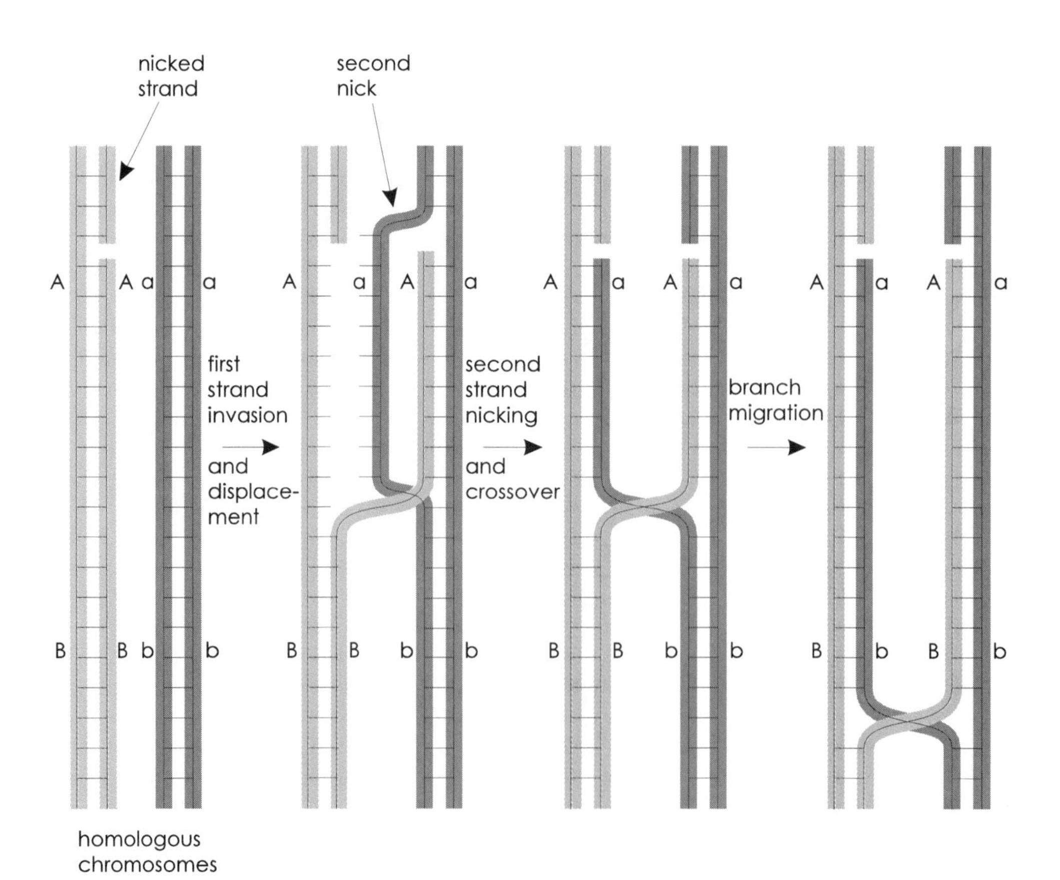

Figure 23.6.
Key initiating steps in homologous recombination are strand crossing over, strand nicking and invasion, and branch migration.

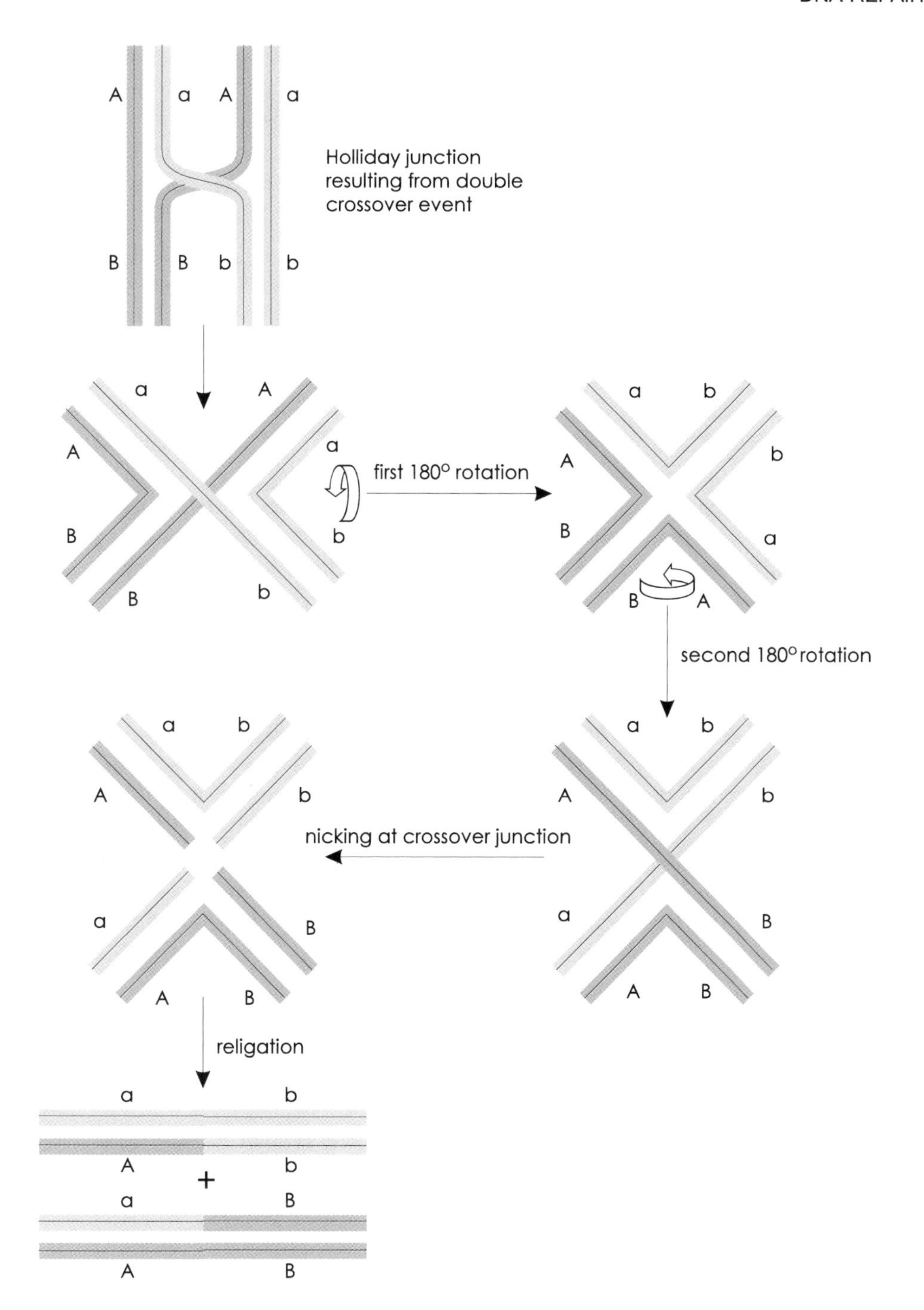

Figure 23.7.
Isomerization and resolution of the Holliday junction.

the DNA needs to be rotated through two 180° angles in first the vertical plane and then the horizontal plane to form the isomerized structure shown in Figure 23.7. Endonuclease cleavage of both strands, followed by ligation of the single-strand nicks resulting from the cleavage, will produce two DNA molecules containing a reciprocal crossover.

It is somewhat difficult to imagine how such a structural feat could be accomplished, but there is compelling evidence to support this recombination model. First, genetic analyses have demonstrated that genetic information is definitely exchanged between each strand. Second, DNA modifying enzymes that have the necessary functions for homologous recombination have been identified in several organisms. Third, high-resolution electron microscopy of recombination intermediates in *E. coli* were found to have the predicted structure of a Holliday junction.

SITE-SPECIFIC RECOMBINATION OF IMMUNOGLOBULIN GENES

Antibody diversity is crucial to a functional immune system. During development, 10^6–10^8 different antibodies are individually expressed by antibody-specific lymphocytes. Based on the number of genes estimated to be encoded by the human genome (~10^5), a single gene assignment for each antibody cannot explain how so many different antibody molecules are expressed. During the late 1970s, coding sequences for **immunoglobulin genes,** which encode antibody proteins, were isolated and analyzed. It was found that although some parts of the coding sequence were the same when immunoglobulin genes from clonally distinct cells were compared, other parts of the gene were very different. It was already known at the protein level that antibodies contain a **constant region** and a **variable region.** Amino-acid sequence differences in the variable region are what give antibodies their unique specificities. The key observation that explained how antibody proteins, and the corresponding genes, could have so many different sequence arrangements, came when it was found that the immunoglobulin genes in germ cells (sperm and egg cells) were different from those in mature lymphocytes. It was discovered that **site-specific recombination** was occurring in lymphocytes between variable-region gene segments (for which there are multiple copies in the genome) and constant-region gene segments.

Immunoglobulin Genes Consist of Variable and Constant Gene Segments

Figure 23.8 shows the relationship between variable regions (V), joining segments (D and J), and constant regions (C) in DNA specifying heavy-chain immunoglobulin molecules. Notice that in germ cells, these coding sequences are spread out across a large segment of DNA with the V regions 5′ of the D, J, and C regions. However, in lymphocytes, various parts of these gene segments are linked in combinations that reflect site-specific recombination events.

Antibody Diversity is Generated by Combinatorial V-D-J Recombination

The heavy-chain locus in mice has ~1000 V regions, 12 D regions, 4 J regions, and several constant regions. During a two-step recombination pathway, the D and J regions are rearranged, followed by a V region recombination with this D-J segment to create a heavy chain **V-D-J recombination** product. This would give $10^3 \times 12 \times 4 = 48{,}000$ different possible heavy-chain variable regions. Two genes, ***rag-1*** and ***rag-2,*** were found to be required for V(D)J recombination because lymphocytes with mutations in *rag-1* or *rag-2* are recombination defective. The Rag-1 and Rag-2 proteins have recently been shown to function together as a site-specific V(D)J recombinase using an in vitro recombination assay.

The mechanism of V-D-J recombination involves two repeating units of either 7 or 9 nucleotides, separated by 11–12 or 21–23 nucleotides, which act as "guides" in the recombination reactions. Homologous base pairing between these repeating sequences facilitates alignment and subsequent recombination events. The products of V-D-J recombination are the rearranged linear gene and a circular piece of DNA that contains the intervening deleted sequences. Importantly, the V(D)J recombination events are not very precise, which gives rise to additional variable-region diversity in the protein coding sequence.

The μ-constant region segment is a component of all newly rearranged immunoglobulin genes, but later in the life span of the cell, there is yet another recombination event called **heavy-chain switching** in which the μ chain is deleted and a γ chain is joined to the same V-D-J segment. The μ- and (δ-) chain antibodies are membrane-bound, whereas the γ-chain antibodies (as well as other heavy-chain antibodies) are secreted by the cell.

GENE TARGETING BY HOMOLOGOUS RECOMBINATION

In standard genetic analyses, mutations in genes that specify a selectable phenotype are obtained by screening large numbers of mutagenized cells, or whole organisms, searching for the desired phenotype. Genetic manipulations can eventually give rise to offspring with a single mutation in the gene of interest. These strategies work best with bacteria, lower eukaryotes (yeast and nematodes), and invertebrates (fruit fly) because the life spans of these organisms are short and the genetic linkage maps are extensive. However, to study mammalian genetic traits, for example, in mice, standard genetic approaches are laborious and difficult. With the advent of recombinant DNA and the availability of cloned-gene sequences, researchers began to de-

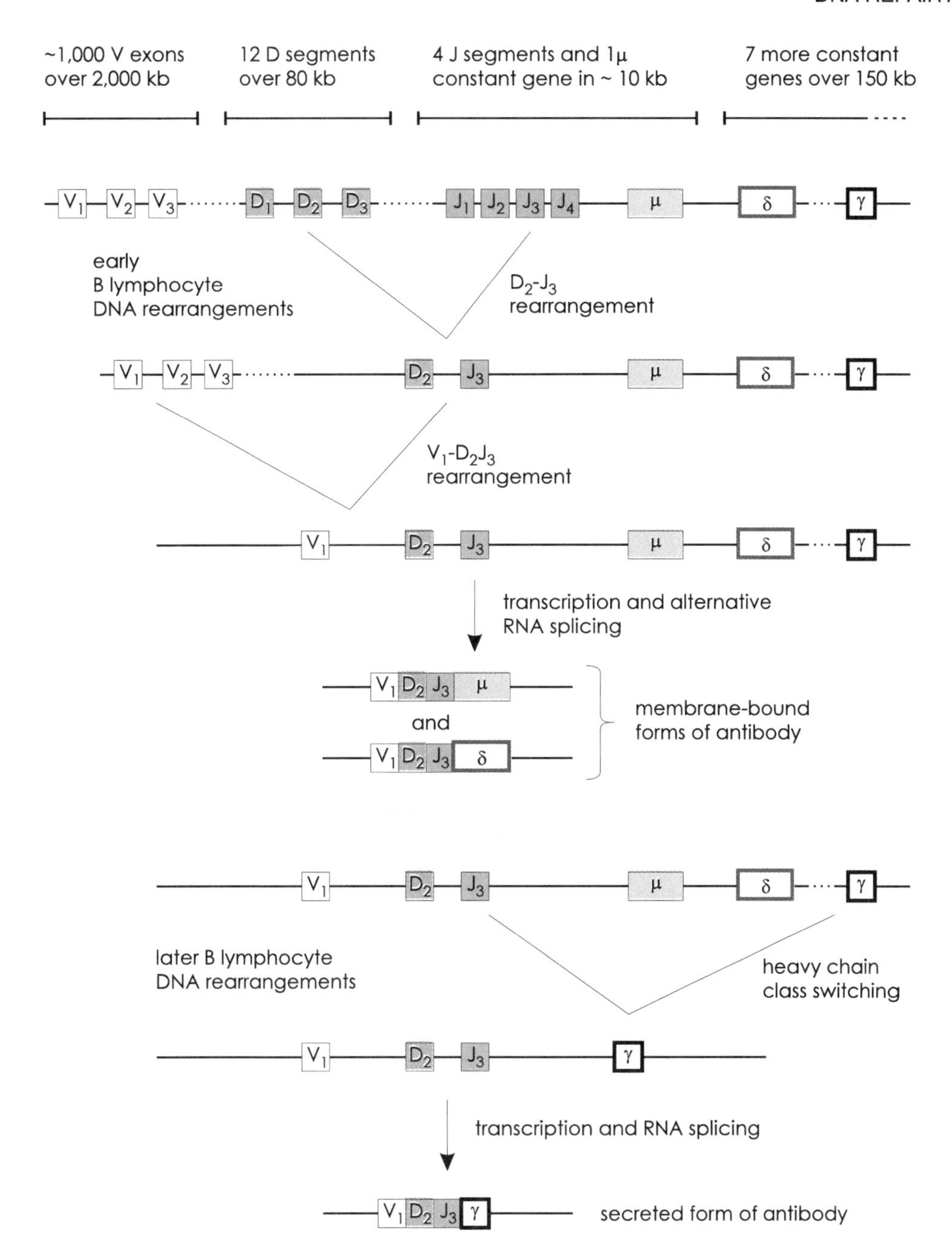

Figure 23.8.

Schematic of germ line and somatic cell heavy chain gene arrangements.

velop direct methods to manipulate the mouse genome as a means to develop animal models of human genetic diseases. This strategy, a form of **reverse genetics,** is based on the ability to cause specific mutations in selected genes by in vitro DNA manipulations using mouse **embryonic stem cell lines** (ES cells). By reintroducing these modified embryonic cells back into normal mouse embryos, it is possible to generate transgenic animals that contain germline mutations (see Box 23.1).

The molecular basis for reverse genetics in mice is the ability to harness homologous recombination to get the desired gene mutations. Some of the early pioneers of what has become known as **targeted gene transfer** were the two research groups led by Oliver Smithies and Mario Capecchi. Following what had been learned from homologous recombination in yeast, these groups, and others, eventually discovered ways to identify ES cells which had undergone homologous recombination between host genes and exogenously added DNA. Targeted gene transfer in mouse cells is dependent on selection schemes that identify rare cells in which the desired homologous recombination event has occured. With the methods used in the early 1980s,

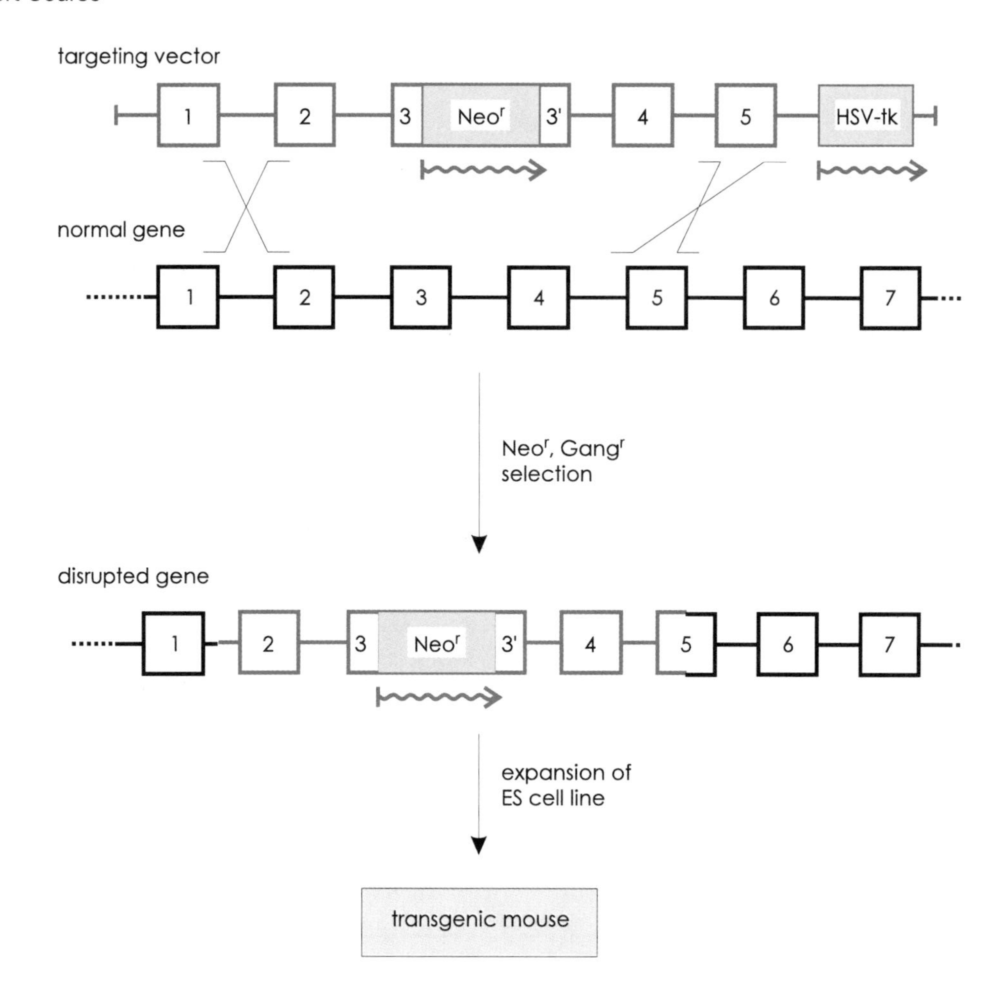

Figure 23.9.

One of the strategies used to identify homologous recombination events is based on selecting for cells that are resistant to both geneticin (neo^r) and gangciclovir. Selection for neo^r identifies cells that have stably integrated the targeting vector. Double crossover homologous recombination events should result in loss of the herpesvirus thymidine kinase (tk) gene from the vector, whereas random integrations would most likely include both the neo^r and tk genes.

BOX 23.1. UTILIZING HOMOLOGOUS RECOMBINATION AS A MEANS OF GENE TARGETING

The primary goal of developing targeted gene transfer techniques has been to develop animal models that will allow detailed studies of specific human diseases. As shown in Figure 23.9, homologous recombination can be utilized to disrupt genes by insertion of the neomycin resistance gene. To make use of this technology, developmental biologists optimized methods to introduce cells containing the targeted mutations back into mice so that the disrupted gene would be stably inherited. The technique utilizes mouse embryonic stem (ES) cells derived from the blastocyst of a pregnant female mouse. Once the mutated ES clonal line is isolated, these cells are injected into a second mouse embryo and this cell mass is then implanted into a pseudopregnant female for it to develop. When the mice are born, they contain mixtures of cells from the targeted ES cells and the cells derived from the host embryo. These chimeric mice are easy to identify because the ES cells came from a mouse with a different color coat than the female mouse providing the host embryo. As shown in Figure 23.10, if any of the ES cells were incorporated into the gametic cell lineage during development in utero, then when these first-generation mice are bred, it is possible to obtain mice that carry the mutation in the germ line. Inbreeding of these heterozygous mice can produce homozygous offspring.

A number of mouse strains have been developed that contain mutated or "knocked-out" genes as a result of homologous recombination in ES cells. One of the most exciting has been the development of a mouse strain with a disruption in the gene responsible for cystic fibrosis, the **cystic-fibrosis transmembrane-conductance regulator** (CFTR). These mice have several of the disease-associated problems that are found in cystic-fibrosis patients, the most severe of which is that nearly all of the homozygous CFTR knock-out mice die within the first 40 days because of intestinal obstruction, perforation, and fatal peritonitis. However, the debilitating respiratory problems that characterize the human form of the disease, are not manifested in these mice. Nevertheless, a variety of cell biological studies have been performed and research has begun to identify pharmacological agents that could be used to treat the altered-ion transport functions that are normally caused by loss of CFTR function. Another animal model of human disease made by targeted gene disruption is **Gaucher's disease,** due to disruption of the mouse glucocerebrosidase gene. Surprisingly, many of the knockout-mouse strains developed in the last several years have no recognizable phenotype. This has been interpreted to mean that there are likely to be more redundant biochemical, cellular, and physiological pathways than previously thought.

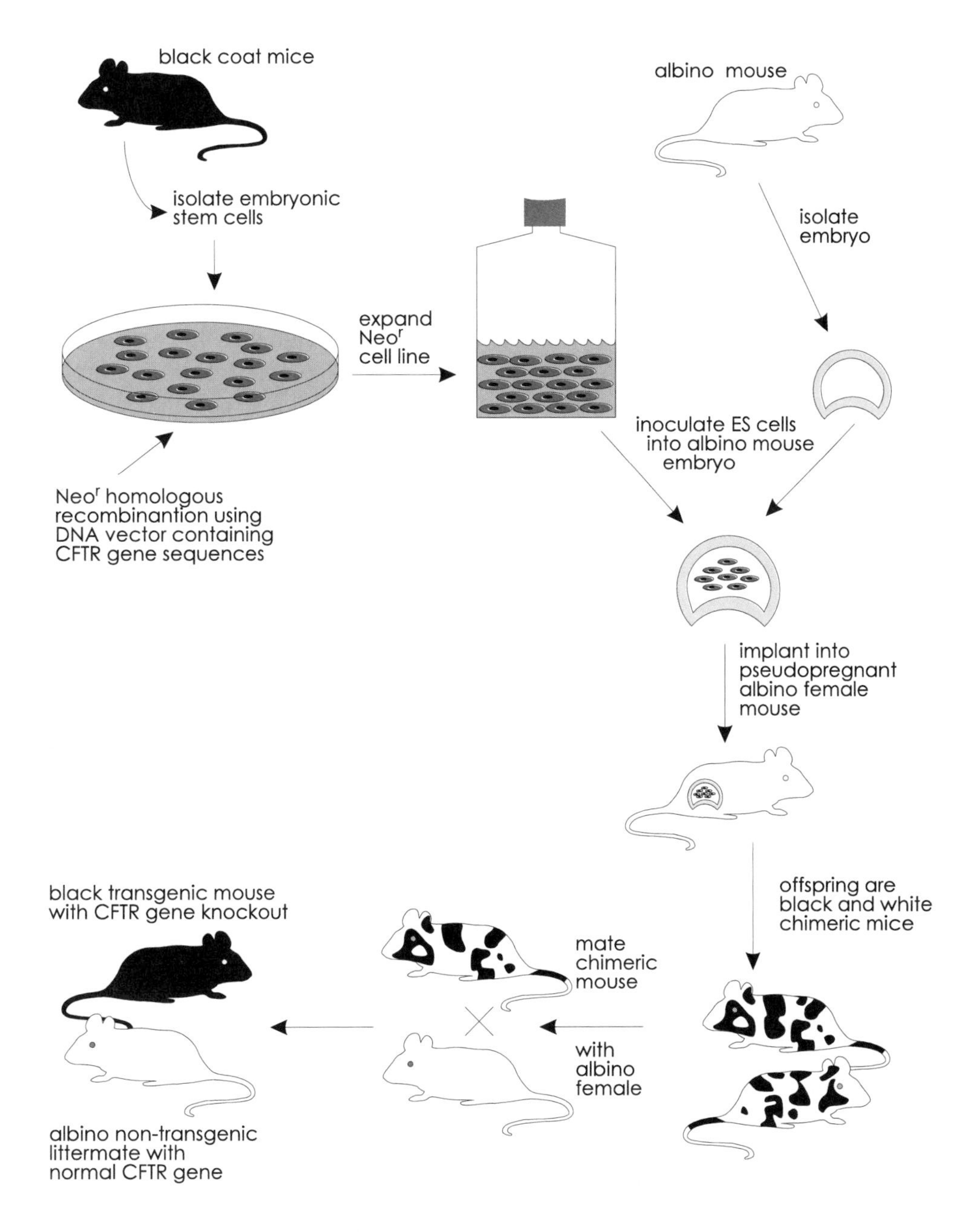

Figure 23.10.

Method of developing transgenic mouse strains that contain gene-targeted mutations. The strategy of selecting for homologous recombination events in ES cells is described in Figure 23.9. The critical step in producing transgenic mouse strains is to identify mice with the desired genetic alteration in their germ-line cells (gametes).

the chance of finding cells with the correct recombination event was only 1 in 1000, and elaborate screening techniques were required. However, in the 10 years since these experiments were performed, new selection schemes have been developed which bring the frequency of homologous recombination up to almost 10% of the cells surviving the selection.

Figure 23.9 illustrates one of the strategies often used to select for cells that have a targeted gene disruption as a result of homologous recombination. The double recombination event required for the replacement of target gene sequences with that of marker gene sequences is presumably mediated by cellular recombination enzymes that utilize the strand crossover mechanism shown in Figures 23.6 and 23.7. By selecting for cells that have stably integrated the neomycin-resistant gene (neo^r), but are resistant to the DNA synthesis inhibitor gangciclovir (herpesvirus thymidine kinase, but not mammalian thymidine kinase, is able to phosphorylate this nucleotide analogue), it is possible to enrich significantly for cells that contain the desired gene disruption (Figure 23.9). This would be called a **gene knockout,** and there are several examples where this type of reverse genetics has been used to develop animal models of human diseases (see Box 23.1).

SUMMARY

1. There are four basic steps in most DNA repair mechanisms; *recognize, remove, resynthesize, and religate.* The first two steps require special DNA repair proteins and the latter two steps are carried out by DNA polymerase and DNA ligase, respectively.

2. *Nucleotide excision repair* involves the removal of "bulky" lesions in DNA such as thymine dimers. The human genetic disease xeroderma pigmentosum is the result of defects in nucleotide excision repair proteins.

3. *Mismatch repair* is required to remove incorrect nucleotides from mismatched base pairs in newly synthesized DNA. It is critical that the repair proteins be able to distinguish between the template and nascent DNA strands to avoid "correcting" the wrong nucleotide. In *E. coli,* the nascent DNA strand is transiently unmethylated at adenines in the sequence GATC. Mutations in the human homologues (hMSH2 and hMLH1) of bacterial-mismatch repair enzymes have been associated with hereditary nonpolyposis colon cancer.

4. *Base-excision repair* requires the recognition and removal of inappropriate bases such as uracil, hypoxanthine, and xanthine from DNA. The DNA glycosylases are enzymes that remove the aberrant base by cleaving the *N*-glycosyl bond. *Direct repair* is different than the other repair mechanism in that the mutated base is repaired directly, without removal of the nucleotide. The protein O^6-methylguanine DNA methyltransferase, for example, carries our direct repair.

5. Homologous recombination in *E. coli* is a two-step process that requires the RecA protein. In step 1, a single-stranded DNA invades the target DNA molecule and base pairs with its homologous sequence. In step 2, the heteroduplex formed in step 1 is isomerized to form a Holliday junction which is resolved by subsequent DNA cleavage and ligation reactions.

6. Immunoglobulin gene rearrangement is an example of site-specific recombination in higher eukaryotes. This process involves the *nearly* precise recombination of V-D-J segments which then recombine with the constant-region segment of the heavy-chain gene. The Rag-1 and Rag-2 proteins present in lymphocytes function as a site-specific recombinase to mediate immunoglobulin gene rearrangements.

7. Gene targeting in mouse embryonic stem cells is an application of homologous-recombination principles. This approach utilizes "reverse genetics" and has been used in mice to develop animal models of several human genetic diseases.

REFERENCES

Aboussekhra A., Biggerstaff M., Shivji M. K. K. , et al. (1995): Mammalian DNA nucleotide excision repair reconstituted with purified protein components. *Cell* **80:**859.

Alani E., Chi N.-W., Kolodner R. (1995): The *Saccharomyces cerevisiae* Msh2 protein specifically binds to duplex oligonucleotides containing mismatched DNA base pairs and insertions. *Genes Dev.* **9:**234.

Ariyoshi M., Vassylyev D. G., Iwasaki H., Nakamura H., Shinagawa H., Morikawa K. (1994): Atomic structure of the RuvC resolvase: A Holliday junction-specific endonuclease from *E. coli. Cell* **78:**1063.

Bartl S., Baltimore D., Weissman I. L. (1994): Molecular evolution of the vertebrate immune system. *Proc. Natl. Acad. Sci. USA* **91:**10769.

Blunt T., Finnie N. J., Taccioli G. E., et al. (1995): Defective DNA-dependent protein kinase activity is linked to V(D)J recombination and DNA repair defects associated with the murine *acid* mutation. *Cell* **80:**813.

Capecchi M. R. (1994): Targeted gene replacement. *Sci. Am.* **270:**52.

Cleaver J. E., Layher S. K. (1995): "If the shoe fits:" Clues on structural recognition of DNA damage. *Cell* **80:**825.

Demple B., Harrison L. (1994): Repair of oxidative damage to DNA: Enzymology and biology. *Annu. Rev. Biochem.* **63:**915.

Dorin J. R., Dickinson P., Alton E. W. F. W., et al. (1992): Cystic fibrosis in the mouse by targeted insertional mutagenesis. *Nature* **359:**211.

Dubrova, Y.E., Nesterov, V.N., Krouchinsky, N.G., et al. (1996): Human minisatellite mutation rate after the Chernobyl accident. *Nature* **380:**683.

Kolodner, R. (1996): Biochemistry and genetics of eukaryotic mismatch repair. *Genes Dev.* **10:**1433.

Lindahl T. (1994): DNA repair: DNA surveillance defect in cancer cells. *Current Biol.* **4:**249.

Loeb L. A. (1994): Microsatellite instability: Marker of a mutator phenotype in cancer. *Cancer Res.* **54:**5059.

Modrich P. (1994): Mismatch repair, genetic stability, and cancer. *Science* **266:**1959.

Mol C. D., Arvai A. S., Slupphaug G., et al. (1995): Crystal structure and mutational analysis of human uracil–DNA glycosylase: Structural basis for specificity and catalysis. *Cell* **80:**869.

Reardon J. T., Thompson L. H., Sancar A. (1993): Excision repair in man and the molecular basis of xeroderma pigmentosum syndrome. Cold Spring Harbor, *Symp. Quant. Biol.* **58:**605.

Sancar A. (1994): Mechanisms of DNA excision repair. *Science* **266:**1954.

Shinagawa H., Iwasaki H. (1996): Processing the Holliday junction in homologous recombination. *Trends Biochem. Sci.* **21:**107.

Svejstrup J. Q., Wang Z., Feaver W. J., et al. (1995): Different forms of TFIIH for transcription and DNA repair: Holo-TFIIH and a nucleotide excision repairosome. *Cell* **80:**21.

Tybulewicz V. L. J., Tremblay M. L., LaMarca M. E., et al. (1992): Animal model of Gaucher's disease from targeted disruption of the mouse glucocerebrosidase gene. *Nature* **357:**407.

Van Gent, D.C., Ramsden, D.A. and Gellert, M. (1996): The RAG1 and RAG2 proteins establish the 12/23 rule in V(D)J recombination. *Cell* **85:**107.

van Vuuren A. J., Vermeulen W., Ma L., et al. (1994): Correction of xeroderma pigmentosum repair defect by basal transcription factor BTF2 (TFIIH). *EMBO J* **13:**1645.

Wu H., Liu X., Jaenisch R. (1994): Double replacement: Strategy for efficient introduction of subtle mutations into the murine *Col1a-1* gene by homologous recombination in embryonic stem cells. *Proc. Natl. Acad. Sci. USA* **91:**2819.

REVIEW QUESTIONS

1. The balance between DNA mutations and DNA repair is important in evolutionary terms because
   a) too much DNA damage will always lead to species extinction
   b) DNA repair requires energy from ATP to correct DNA damage
   c) DNA damage and DNA repair are unrelated cellular processes
   d) cytosine deamination is a rare event, but is often mutagenic
   e) DNA mutations in gamete cells, if unrepaired, allow for natural selection

2. The four sequential steps in most forms of DNA repair are
   a) recognize, remove, resynthesize, and religate
   b) religate, resynthesize, remove, and recognize
   c) remove, resynthesize, religate, and recognize
   d) restructure, remove, resynthesize, and religate
   e) restore, restructure, reutilize, and repair

3. An important difference between the mechanisms of mismatch repair and nucleotide-excision repair is that
   a) in mismatch repair, either strand could represent the mutation, whereas in nucleotide-excision repair, the damage creates a unique structure
   b) mismatch repair replaces only the damaged base, whereas nucleotide-excision repair replaces the entire nucleotide
   c) nucleotide-excision repair occurs only in bacteria
   d) nucleotide-excision repair is important in DNA replication and mismatch repair is required for the repair of environmental DNA damage
   e) these two DNA repair pathways are actually identical and both utilize the proteins MutS, MutH, and MutL

4. Uracil glycosylases
   a) remove thymine dimers
   b) remove uracil from RNA
   c) remove uracil from DNA
   d) remove thymine from DNA
   e) remove uridine from DNA

5. Homologous recombination is an important process to generate genetic diversity. Which of the following statements is NOT TRUE with regard to the mechanism of homologous recombination?

a) Branch migration determines the extent of crossover.
b) DNA strand breakage is not required for homologous recombination.
c) Strand invasion requires sequence homology.
d) Resolution of Holliday junctions requires two 180° rotations.
e) Religation is the last step in homologous recombination.

6. Immunoglobulin gene arrangements result in
   a) large regions of gene duplication
   b) insertional mutations
   c) a maximum of 48,000 different antibody genes
   d) the creation of unique variable-constant region gene fusions
   e) the deletion of mutated immunoglobulin genes

7. The principle of targeted gene transfer, as a means to create strains of mice with predetermined genetic mutations, depends on
   a) careful cross-breeding
   b) site-specific recombination
   c) the ability to distinguish coat colors
   d) higher rates of DNA damage compared to DNA repair
   e) selection for rare homologous recombination events

ANSWERS TO REVIEW QUESTIONS

1. ***e*** Natural selection eliminates or establishes germline mutations.

2. ***a*** The correct sequence is recognize, remove, resynthesize, and religate.

3. ***a*** Mismatch repair functions to monitor newly synthesized DNA, whereas nucleotide-excision repair removes damaged nucleotides such as thymine dimers.

4. ***c*** Uracil glycosylases remove only the base (uracil) from DNA, not the entire nucleotide (uridine).

5. ***b*** DNA strand breakage is required at three steps for homologous recombination to be completed.

6. ***d*** Because of additional mispairing during V-D-J recombination events, there is a much larger number of possible V-C combinations than 50,000.

7. ***e*** Strategies including selection for double-drug resistance, and the use of rapid DNA screening by PCR, have greatly improved the success rate of gene targeting.

CHAPTER

24

RNA SYNTHESIS

INTRODUCTION

Deoxyribonucleic acid encodes the information required for essentially every cellular process, yet these nucleotides by themselves, have no direct influence on cell viability. The genetic blueprints inscribed in the DNA sequence must be "read" to be useful. The first step in this process is called DNA transcription or RNA synthesis. Enzymatically, RNA and DNA synthesis (Chapter 22) are very similar. A large protein complex is required for initiation of nucleic acid synthesis on a DNA template and the direction of chain extension is in the 5′ to 3′ direction. There are, however, several important differences between DNA and RNA synthesis. First, RNA is synthesized from ribonucleotides rather than deoxyribonucleotides. Second, uracil replaces thymine as the base that pairs with adenine; this may have some evolutionary advantages for DNA repair (Chapter 23). Third, RNA synthesis does not require a preexisting primer. In fact, RNA synthesis is the priming event for DNA synthesis through the action of DNA primase. Fourth, RNA synthesis is very selective, with only small regions of the genome being transcribed at any one time.

There are four major classes of RNA in cells; **messenger RNA** (mRNA), **ribosomal RNA** (rRNA), **transfer RNA** (tRNA), and **small nuclear RNA** (snRNA). The relative abundance, and complexity (number of distinct types), of these RNA molecules is quite different. About 80% of RNA in a cell consists of rRNA, the RNA component of ribosomes. There are only four types of eukaryotic rRNA (28S, 18S, 5.8S, and 5S) and, therefore, the sequence complexity of this class is actually quite low. In contrast, mRNA constitutes just 5% of total cellular RNA, yet it is the most diverse, with an estimated 10^4–10^5 different species which correspond to the same number of protein coding genes. Both tRNA and snRNA (in eukaryotes) make up the remaining fraction of RNA in a cell (~15%), with ~50 different types of tRNA and ~10 different snRNAs. The total number of molecules per cell of each class of RNA is therefore based on the relationship between the total mass of each RNA class, the average length of RNA molecules in that class, and the sequence complexity.

GENE TRANSCRIPTION REQUIRES DNA-DEPENDENT RNA POLYMERASES

For the most part, RNA synthesis is restricted to DNA sequences contained within the informational unit called a gene. A gene is usually identified by one of two methods. It can be the physical location of DNA sequences that are responsible for a **genetic phenotype** (genetic mutations are used to identify genes), or as defined by "reverse genetics," a gene is a transcription unit. The Human Genome Project, for example, is utilizing strategies to map transcribed regions of the genome as a first step in identifying genes. As described in Chapter 21, a molecular definition of a gene is a segment of DNA that specifies, *and regulates,* a **transcription unit.** All genes contain transcription units, but not all sequences in a gene are transcribed. Figure 24.1 shows a schematic representation of a typical eukaryotic gene. Transcription is initiated at the 5′ end of the gene in a region called the promoter, proceeds through the coding region, and terminates at the 3′ end of the gene. The other functional components of a gene are **regulatory sequences,** often found 5′ of the initiation site, which control the frequency of gene transcription (Chapters 28 and 29).

E. coli RNA Polymerase Holoenzyme Is a Multisubunit Protein

The RNA polymerases are large multisubunit protein complexes that contain the catalytic activity necessary for RNA synthesis. The core enzymatic subunit of an RNA polymerase catalyzes the polymerization of RNA by "transcribing" one of the two strands of a DNA template. However, gene transcription requires that the RNA polymerase be directed to a specific site in the promoter at the 5′ end of the gene, as shown in Figure 24.1. This specificity requires additional RNA polymerase protein subunits. The *E. coli* **RNA polymerase core enzyme** consists of five protein subunits which are referred to as β, β', α (two subunits), and ω. The **β subunit** is thought to be the catalytic component and the M_r weight of the core complex is ~350,000. The *E. coli* **RNA polymerase holoenzyme** also contains a sixth polypeptide, called the **σ subunit** (70,000 M_r), which associates transiently with the core RNA polymerase to serve the function of promoter recognition. Once elongation has begun, the σ subunit dissociates.

In addition to the RNA polymerase holoenzyme and a double-strand DNA template containing a promoter, RNA synthesis also requires all four ribonucleotide triphosphates (ATP, GTP, CTP, and UTP) and Mg^{2+} as a cofactor. The RNA polymerase is a metalloenzyme and contains Zn^{2+}. The first nucleotide is often an ATP or GTP and the triphosphate is maintained unless the 5′ end of the transcript is posttranscriptionally modified, as is the case in eukaryotes (Chapter 25). The RNA polymerization reaction is chemically very similar to DNA synthesis in that the nascent RNA strand is synthesized in the 5′ to 3′ direction. In this reaction, the 3′ hydroxyl group of the existing RNA chain undergoes a **nucleophile attack** on the α-phosphate of the incoming nucleotide which then releases PPi.

$$(\text{NMP})_n + \text{NTP} \rightarrow (\text{NMP})_{n+1} + \text{PP}_i$$

Each base in the RNA product is complementary to the opposing base in the template DNA strand as shown below:

DNA 3′ A G C T C G A A G C C 5′
RNA 5′ U C G A G C U U C G G 3′

Unlike *E. coli* DNA polymerases, there is no 3′ to 5′ exonuclease proofreading activity in the *E. coli* RNA polymerase and, therefore, the error rate is relatively higher (~10^{-4}–10^{-5}). Since RNA represents only a transient copy of DNA, and is not inherited through the germ line, this error frequency is tolerable. As discussed in Chapter 30, viral RNA polymerases and another enzyme, reverse transcriptase, which also lacks proofreading activity, have error frequencies in the same range as the bacterial RNA polymerase. This mutation rate is most likely beneficial to some viruses because it results in frequent alterations in viral protein sequences and thus allows the virus to escape immune-system defenses in the host (Chapter 30).

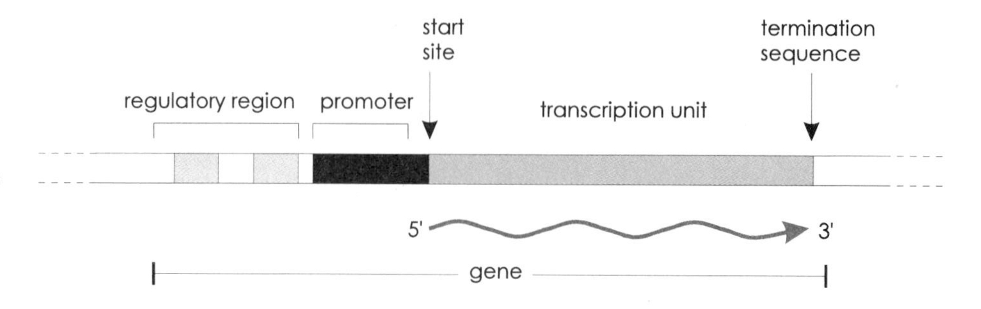

Figure 24.1.

Schematic drawing of a typical eukaryotic gene. The nascent RNA strand is shown as a wavy line and the rectangles in the gene regulatory region and gene promoter represent binding sites for proteins required for the initiation of RNA synthesis.

Eukaryotes Have Three Different RNA Polymerase Holoenzymes

Eukaryotic RNA polymerases are less well-characterized than the bacterial enzyme. Biochemical studies have so far identified three distinct nuclear RNA polymerase activities in eukaryotic cell extracts prepared from both yeast and human cells. These RNA polymerases, referred to as **RNA pol I, RNA pol II,** and **RNA pol III,** each contain a large number of subunits, some of which appeared to be shared between the different RNA polymerase subtypes. As shown in Table 24.1, RNA pol I transcribes ribosomal RNA genes, RNA pol II transcribes the majority of protein-coding genes into mRNA, and RNA pol III transcribes tRNA, small nuclear RNAs (snRNAs), and ribosomal 5S genes. Note also that the mitochondrial genome is transcribed by a nuclear-encoded mitochondrial RNA polymerase. The yeast mitochondrial RNA polymerase holoenzyme consists of a 140-kd catalytic subunit and a 43-kd promoter-recognition protein similar to the bacterial σ factor.

The fungal toxin α-amanitin, which is derived from a poisonous mushroom, is an inhibitor of RNA pol II and pol III (Table 24.1). As discussed later in this chapter, the gene specificity of these three RNA polymerases is due to their ability to recognize different promoter elements (DNA sequences) which are representative of each class of gene (rRNA, protein-coding, and tRNA/snRNA genes).

RNA SYNTHESIS CONSISTS OF THREE PHASES: INITIATION, ELONGATION, AND TERMINATION

Since RNA synthesis uses only one of the two strands of DNA for a template, it is important to identify which strand is being "read." By convention, the **template strand** has the complementary sequence of the RNA product (with the exception of thymine in place of uracil) and the **nontemplate strand** has the same sequence as the RNA. The nontemplate strand is also called the coding strand because the sequence of this strand can be used to decipher the primary amino-acid sequence of an encoded protein (see Chapter 26).

Initiation of Transcription in *E. coli*

The binding of *E. coli* RNA polymerase holoenzyme to bacterial promoters, and the subsequent initiation of RNA synthesis, has been studied in detail and a general scheme of these events is shown in Figure 24.2.

By performing detailed analyses of several different bacterial promoters, and then compiling the sequence information from a large number of promoters, it was found that two highly conserved DNA sequences are required for bacterial promoter function. One of these is called the **-35 region** because it is centered ~35 nucleotides "upstream" (5′) of the RNA synthesis-initiation site where the first ribonucleotide is paired with the corresponding base in the template strand. The other conserved sequence is called the **-10 region,** which lies ~10 nucleotides upstream of the start site. The RNA polymerase slides along the DNA until it encounters an accessible -35 region to form an initial protein–DNA complex called the **closed complex.** This complex (which includes the σ factor) then migrates a little further in the 3′ direction up to the -10 region, where it promotes the unwinding of the DNA double helix, exposing the start site of transcription on the template strand. This second protein–DNA conformation is called the **open complex,** which constitutes an active transcription unit. The A-T rich sequences in the -10 region may facilitate helix unwinding because of the lower hydrogen-bonding energy of A-T base pairs. In vitro transcription experiments using purified *E. coli* RNA polymerase have led to a kinetic model explaining the two-step mechanism of open and closed complexes:

Table 24.1. The Functional Properties of the Eukaryotic Nuclear RNA pol I, II, and III Enzymes Differ Based on the Class of Genes They Transcribe[a]

RNA polymerase	Genes transcribed	Relative activity (%)	α-Amanitin sensitivity
I	28S, 18S, 5.8S rRNA	~60	Insensitive
II	mRNA and snRNA	~30	Very sensitive
III	tRNA and 5S rRNA	~10	Moderately sensitive

[a]Note the relative contributions of each of these enzymes to the total RNA polymerase activity in the cell and their differential sensitivity to α-amanitin.

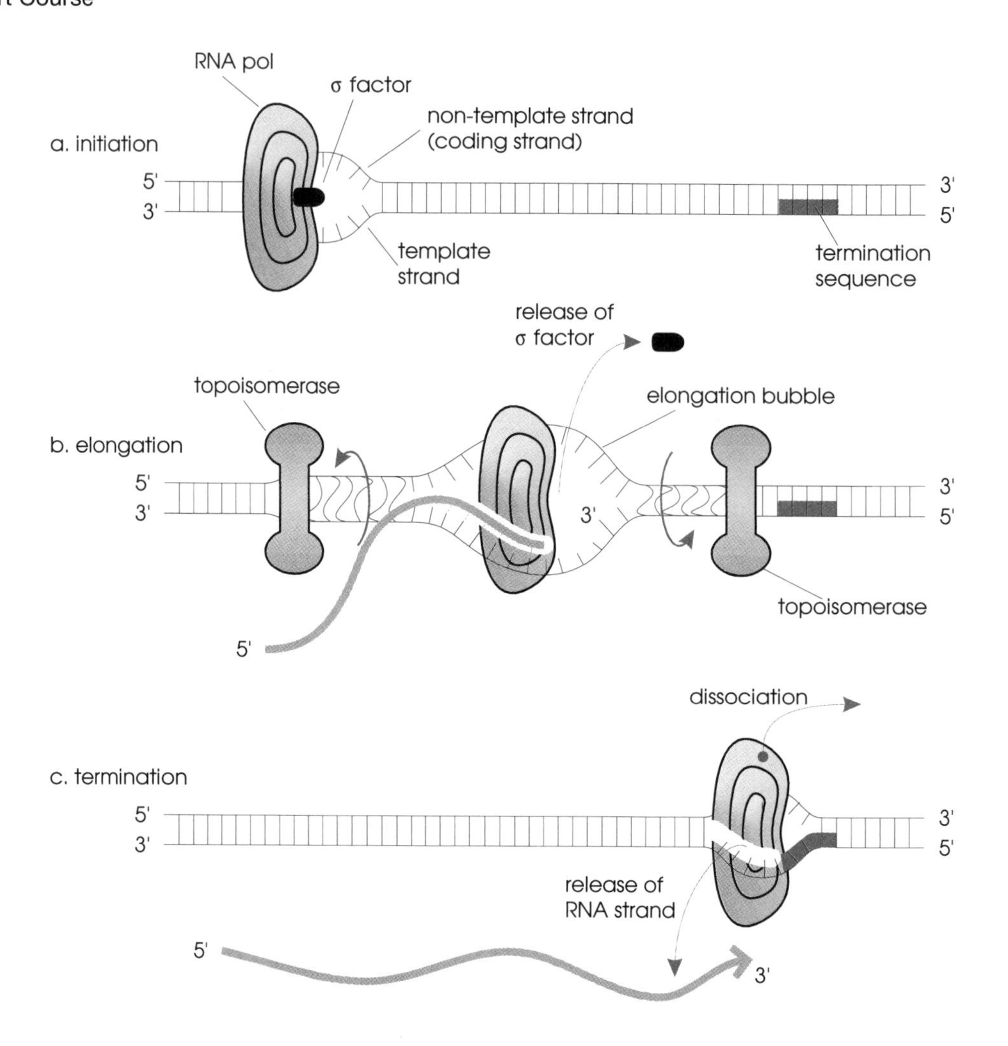

Figure 24.2.

The three phases of RNA synthesis are initiation, elongation, and termination of transcription.

$$R + P \overset{K_B}{\leftrightarrow} RP_c \overset{k_f}{\rightarrow} RP_O$$

In step one, the free RNA polymerase (R) interacts reversibly with the promoter (P) as defined by the equilibrium constant K_B. In the second step, DNA-bound RNA polymerase (RP_c) translocates to the -10 region, which initiates an isomerization reaction (defined by the rate constant k_f) that results in the irreversible formation of the open complex (RP_O). As will be discussed in Chapters 28 and 29, a prevailing model to explain the mechanism of the transcriptional *regulation* is that regulatory factors alter either K_B or k_f, and thereby influence RNA synthesis-initiation rates.

The Elongation Phase of Transcription

Ribonucleic acid polymerization, also called elongation, proceeds in the 5′ to 3′ direction partially through the energy released by cleavage of the phosphate bond in the incoming ribonucleotide, and by the subsequent hydrolysis of pyrophosphate to inorganic phosphate. The σ factor is known to dissociate immediately after elongation begins, which leaves the core RNA polymerase to complete gene transcription. Multiple RNA polymerase complexes actually load onto the promoter region in sequential fashion, thus allowing the gene to be transcribed continuously. The rate of elongation by *E. coli* RNA polymerase is about 30 nucleotides per second and a ~2 kb transcription unit is transcribed in a little over a minute. The RNA transcript is only base paired with the DNA template in the region of the **elongation bubble** (about 15 nucleotides), thus leaving the growing RNA chain trailing behind as the DNA helix reforms behind the elongation complex (Figure 24.3). Torsional stress ahead of the elongation bubble is produced by the unwinding activity of RNA polymerase, but just as in DNA synthesis, topoisomerases prevent a net increase in superhelicity. The completed RNA transcript is released once the RNA polymerase disengages from the DNA template at the site of transcriptional termination.

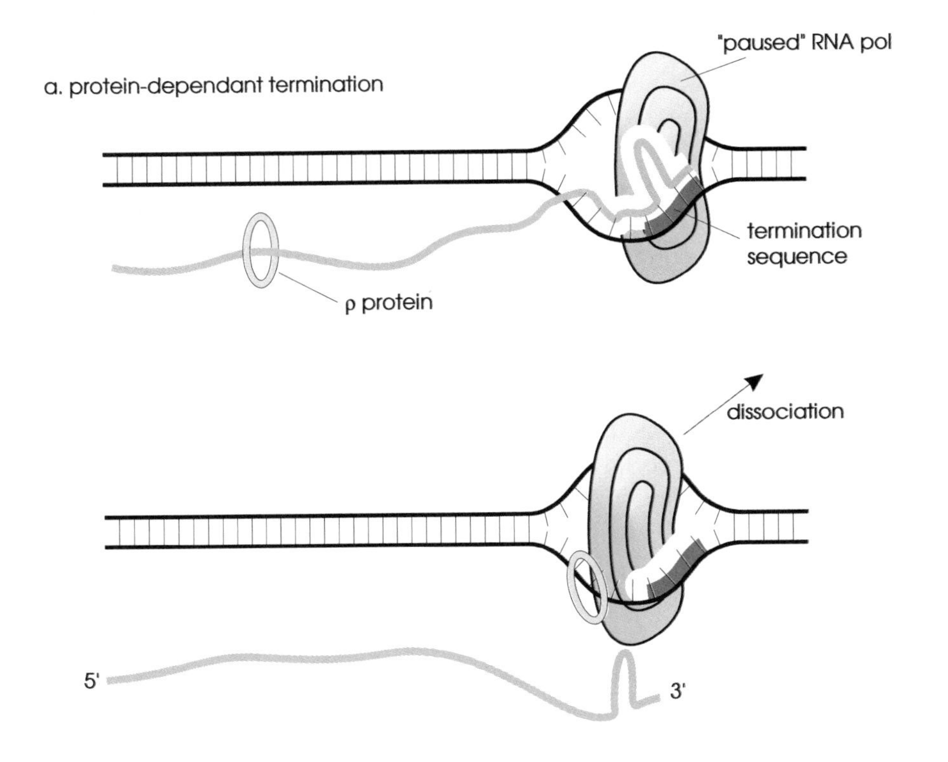

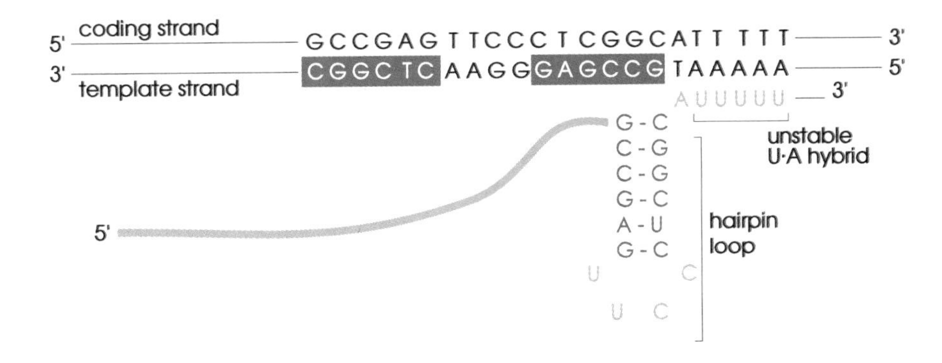

Figure 24.3.
Transcriptional termination by protein-dependent and protein-independent mechanisms.

Transcriptional Termination Mechanisms

In bacteria, the termination of transcription occurs by one of two mechanisms, as shown in Figure 24.3. **Protein-dependent termination** is due to the interaction of a terminator protein (ρ) with the RNA polymerase elongation complex just as it pauses at a hairpin loop which forms in the nascent transcript. The ρ protein is an ATP-dependent RNA–DNA helicase and it's mode of action is to disrupt the RNA–DNA hybrid, which leads to the disassembly of the elongation complex. **Protein-independent termination** is characterized by a similar hairpin-loop structure, but in this more general case of termination, there is a run of adenylates in the template strand (which specifies a string of uridylates in the transcript) just downstream of the hairpin. Chain termination probably occurs when the elongation complex pauses at the hairpin structure, causing the RNA–DNA hybrid of A-U base pairs to destabilize and the complex disengages. Eukaryotic transcriptional termination appears to be similar to protein-independent mechanisms and is coupled to a process involving the addition of a 3′ polyadenylate tail (Chapter 25).

SOME ANTIBIOTICS ARE INHIBITORS OF RNA SYNTHESIS

Actinomycin D is an antibiotic from *Streptomyces* that has an unusual structure which allows it to intercalate into double-stranded DNA. The phenoxazone ring intercalates into the DNA between adjacent G-C base pairs, and the cyclic peptides are thought to occupy space along the DNA helix. This structure effectively blocks transcriptional elongation in both prokaryotes and eukaryotes even at low concentrations. Actinomycin D is a useful biochemical tool to study transcriptional processes because it has little effect on DNA replication or protein translation. By adding low concentrations of actinomycin D to cells for a short period of time, it is possible to determine if the cellular process being studied requires gene transcription.

Rifamycin is another antibiotic isolated from *Streptomyces.* Rifamycin blocks RNA synthesis by binding directly to the beta subunit of bacterial RNA polymerases and specifically inhibiting formation of the first phosphodiester bond. This is in contrast to actinomycin D, which blocks elongation but not initiation. A synthetic derivative of rifamycin called rifampicin has been developed for clinical treatment of bacterial infections.

TRANSCRIPTIONAL INITIATION IN EUKARYOTES REQUIRES A LARGE MULTISUBUNIT COMPLEX

Gene transcription in eukaryotic systems relies on the same basic mechanisms first described in bacteria. However, eukaryotic transcription is more complex because of the increase in genetic information that must be selectively processed. In this section, we describe the involvement of a diverse array of auxiliary proteins that associate with eukaryotic RNA polymerases to form the transcriptional initiation complex. Since mechanisms of gene regulation involve modulating rates of transcriptional initiation (Chapters 28 and 29), it is important to have a basic understanding of the protein components that constitute a eukaryotic initiation complex. For clarity, we will ignore the influences of chromatin structure on eukaryotic gene promoter "architecture" in this section, and instead address this important concept in Chapter 29.

Protein Components in the RNA pol II Initiation Complex

Studies from mammalian, fruit fly, and yeast cells have shown that as many as eight protein fractions are required for accurate transcriptional initiation in vitro. One of these fractions contains the catalytic activity defined as **RNA pol II,** another fraction referred to as **TFIID,** contains a protein that binds an A-T rich sequence (this is described in more detail later). Six fractions, called **TFIIA, TFIIB, TFIIE, TFIIF, TFIIH, and TFIIJ,** need to be included for maximum rates of transcription. Many of these fractions contain multiple polypeptides, bringing the total number of protein subunits in the initiation complex to >50. The assembly of the RNA pol II **initiation complex** in vitro is often done by sequential addition of the various fractions to a reaction mixture containing the DNA template, rNTPs, and Mg^{2+}. It has recently been found that ~5–10% of yeast RNA pol II is present in a preassembled complex containing TFIIB, TFIIF, TFIIH, TFIIJ, and additional yeast proteins as defined by the SRB genes. This complex functions as a RNA pol II holoenzyme in in vitro transcription assays containing a DNA-bound TFIID/TFIIA complex and soluble TFIIE. Figure 24.4 illustrates a proposed assembly pathway for eukaryotic transcriptional initiation complexes.

RNA pol II Contains a Phosphorylated Carboxy-Terminal Repeat Domain

Genetic and biochemical studies of core RNA pol II have shown that it alone contains 14 protein subunits. The largest subunit of RNA pol II has been cloned from a variety of eukaryotic organisms and found to encode a ~200 kDa protein that retains homology to the β and β′ subunits of bacterial RNA polymerase. One unusual feature of this protein is the presence of a **heptamer repeat** at the carboxy terminus called the **C-terminal repeat domain** (CTD). This repeat has the consensus sequence Tyr-Ser-Pro-Thr-Ser-Pro-Ser, which varies in length, with yeast having 26 heptad repeats and mammals having 52. Two forms of this protein have been biochemically defined, one is RNA pol IIA which is unphosphorylated, and the other is RNA pol IIO, which is hyperphosphorylated primarily on serine and threonine residues in the CTD. Several different CTD kinases have been identified which are capable of phosphorylating CTD residues in vitro. Hyperphosphorylation of the CTD tail is correlated with the activated form of RNA pol II (IIO) both in vitro and in vivo. Genetic experiments in yeast and mice have demonstrated that the CTD is required for viability, indicating that it has an essential role in RNA pol II functions, possibly as a target for transcriptional control by regulatory proteins.

In addition to CTD kinases, the RNA pol II

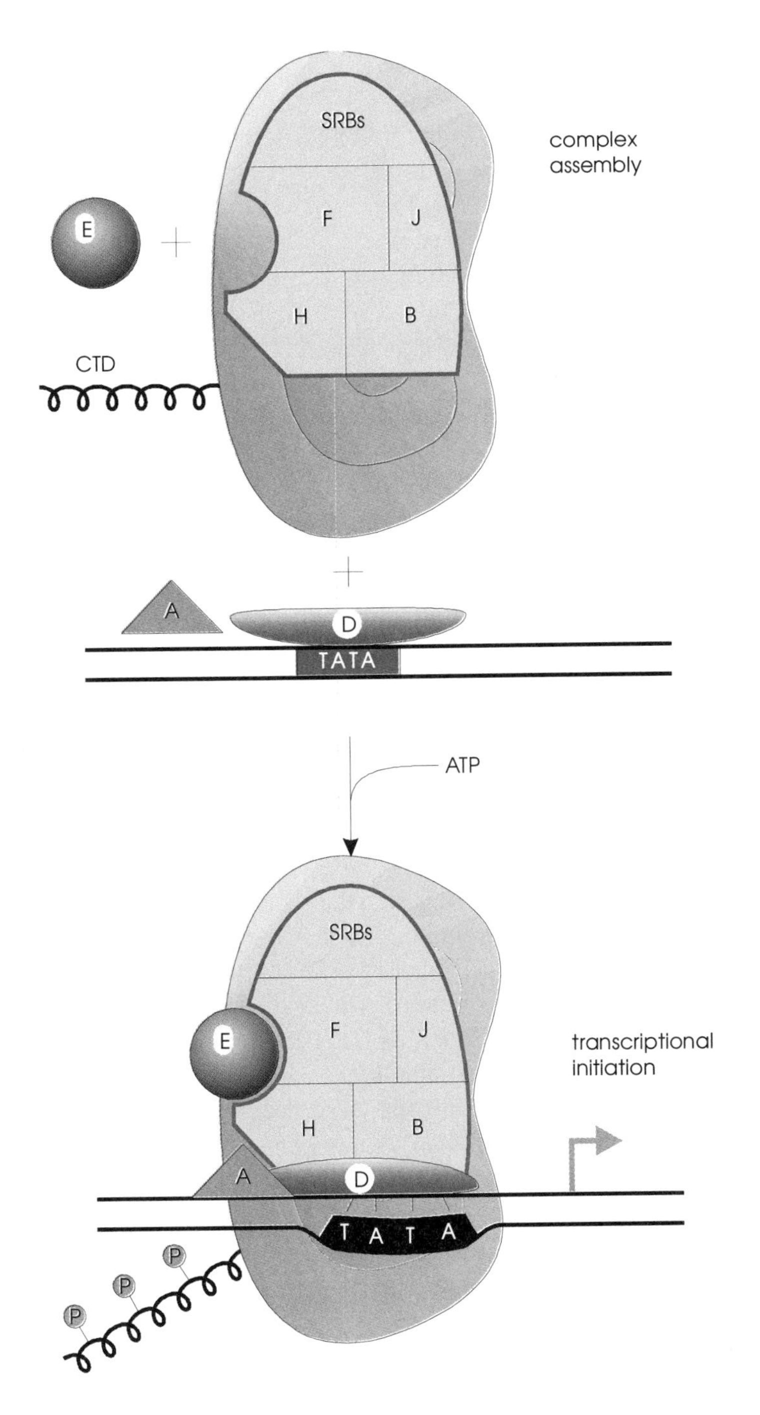

Figure 24.4.
Assembly of auxiliary transcription factors into a RNA pol II initiation complex in yeast by a pathway utilizing the preformed RNA pol II holoenzyme. Phosphorylation of the RNA pol II CTD by TFIIH kinase is thought to be a key initiating step in transcription. The bold arrow represents promoter activation and transcriptional initiation.

holoenzyme protein component TFIIH, contains a DNA-dependent kinase activity capable of phosphorylating serine and threonine residues in the CTD repeat using either ATP or GTP. Moreover, TFIIE, another required initiation factor, stimulates TFIIH CTD kinase activity ten-fold, suggesting a temporal role for TFIIE in complex formation. TFIIH is not only a kinase, it also contains an ATP-dependent helicase activity which is stimulated five-fold by TFIIE. As shown in Figure 24.4, CTD phosphorylation is thought to be involved in converting RNA pol II into an active conformation.

EUKARYOTIC GENE PROMOTERS CAN BE DIVIDED INTO CLASSES

Just as biochemical analyses had identified three separate nuclear RNA polymerase activities, molecular ge-

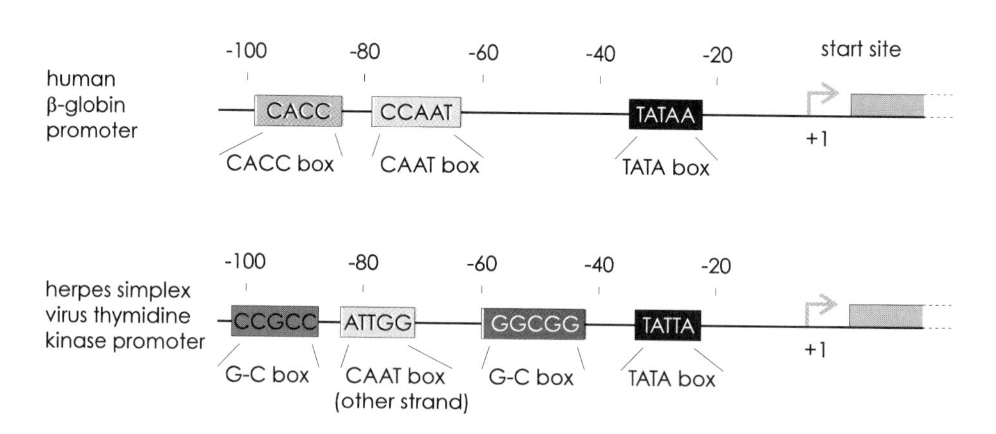

Figure 24.5.
Diagram showing functional elements within the β-globin and thymidine-kinase promoters. The CAAT box in the thymidine-kinase promoter is in the opposite orientation relative to the transcriptional start site as compared to the CAAT box in the β-globin promoter.

netic studies led to the compilation of a long list of promoter sequences corresponding to genes known to be transcribed by RNA pol I, II, or III. Two of the first eukaryotic promoters to be studied in detail were the mouse **β-globin gene promoter** and the **thymidine-kinase gene promoter** from herpes simplex virus. Both of these promoters are transcribed by RNA pol II and they share some common features that are reminiscent of bacterial promoters. Figure 24.5 summarizes the results from promoter mapping studies of the β-globin and thymidine-kinase genes. Each of these DNA sequence motifs represent high-affinity binding sites for eukaryotic transcription factors.

The DNA mutagenesis studies revealed that a sequence centered around -30 relative to the start site of transcription was required for accurate initiation of the β-globin and thymidine-kinase genes. Because of the A-T rich sequence in this region (usually TATAA), it has become known as the **TATA box.** Earlier studies on a variety of other RNA pol II transcribed genes had also implicated the TATA box as a critical element in promoter function and its fixed position at roughly 30 nucleotides upstream of the start site was reminiscent of the -35 and -10 regions of prokaryotic promoters. Box 24.1 describes the general approach of site-directed mutagenesis which has been used to identify essential DNA sequences required for transcriptional initiation of a variety of promoters.

As shown in Figure 24.5, additional sequences required for promoter function were also identified in the β-globin and thymidine-kinase genes. One of these elements, called the **CAAT box,** is commonly found in mRNA encoding genes, and the other, often called a **GC box,** serves as the binding site for the transcription factor SP-1. Box 24.2 describes how DNA binding assays can be used to purify proteins with sequence-specific DNA binding activities such as **C/EBP,** a protein with high affinity for the thymidine-kinase promoter CAAT box.

The basic approach of combining mutagenesis studies with functional assays of promoter activity in vitro and in vivo has now been performed on a large number of genes transcribed by all three RNA poly-

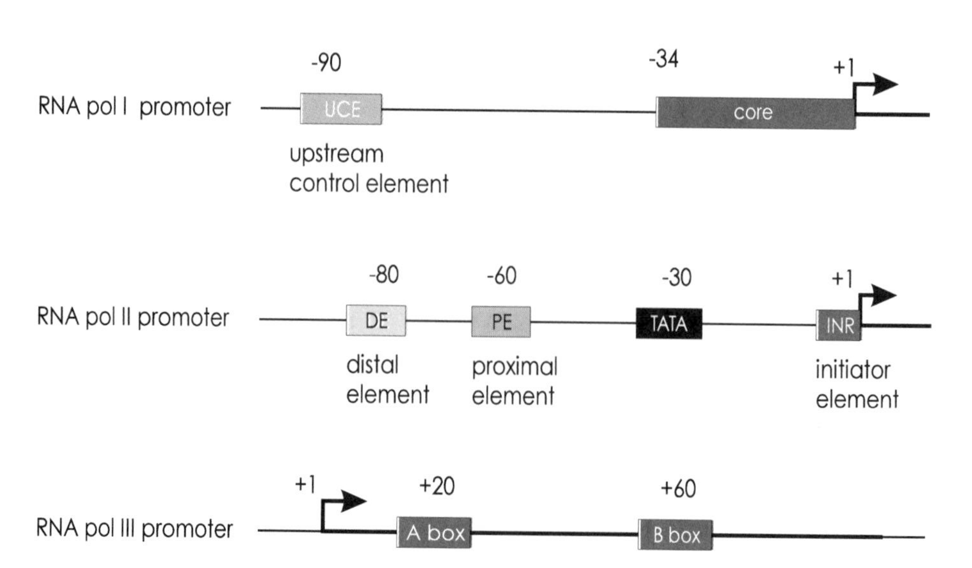

Figure 24.6.
Schematic representation of the three types of RNA pol I, II, and III eukaryotic promoters. The DNA sequence elements required for accurate initiation are shown as boxes. Note that the RNA pol III promoter elements (A box and B box) reside within the coding sequence of tRNA genes.

merases. It is clear from these studies that several recurring sequence motifs allow for the classification of most promoters into three general categories based on the species of RNA polymerase that transcribes them, as shown in Figure 24.6.

An important concept is that a relationship exists between specific DNA sequences and the proteins that bind to them. This is very similar to the way that the foundation of a building determines where the walls will be erected. In fact, the term "promoter architecture" has been coined to illustrate this point. As discussed in Chapter 29, selective eukaryotic gene regulation is the result of cell-specific promoter architectures.

TATA BINDING PROTEIN (TBP) IS A UNIVERSAL EUKARYOTIC TRANSCRIPTION FACTOR

To follow the analogy of erecting a building using the DNA sequence of a promoter as a foundation, it next becomes important to identify proteins that make up

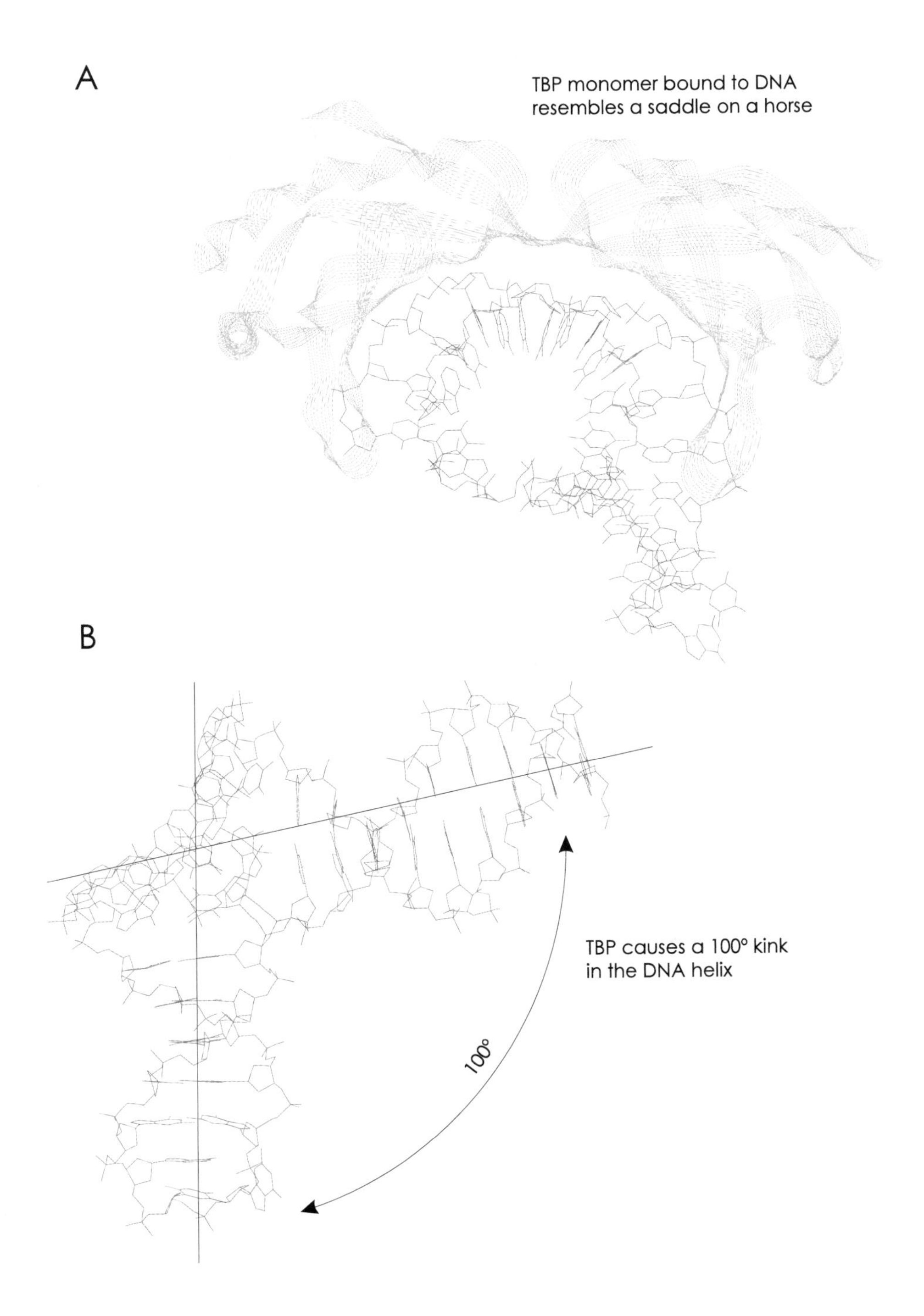

Figure 24.7.

The proposed structure of TBP bound to DNA based on X-ray crystallography data. (A) A monomer of TBP appears to form a protein "saddle" that sits atop the DNA helix. (B) TBP-binding to the TATA box causes a bend in the DNA double

the building materials. In the simplest case, this means purifying and characterizing the sequence-specific DNA binding proteins required for transcription. Although a large number of transcription factors have been identified using this scheme, we will focus our attention on the **TATA binding protein** (TBP), which is a required component of the transcriptional initiation complex for all three RNA polymerases.

Diversity and flexibility in transcriptional control is accomplished by assembling large, highly specific, protein complexes at promoter regions; these initiation complexes are sensitive to cellular modulating signals. The key to specificity is combinatorial determinants, that is, using the same building materials, in different combinations, to assemble unique transcription complexes. As mentioned earlier, there may be as many as 100,000 transcriptionally regulated human genes, but there are likely to be no more than a few hundred transcription factors. The TATA binding protein represents the best case so far of how a single, relatively small, transcription factor can be multifunctional as result of numerous discrete protein–protein interactions. The

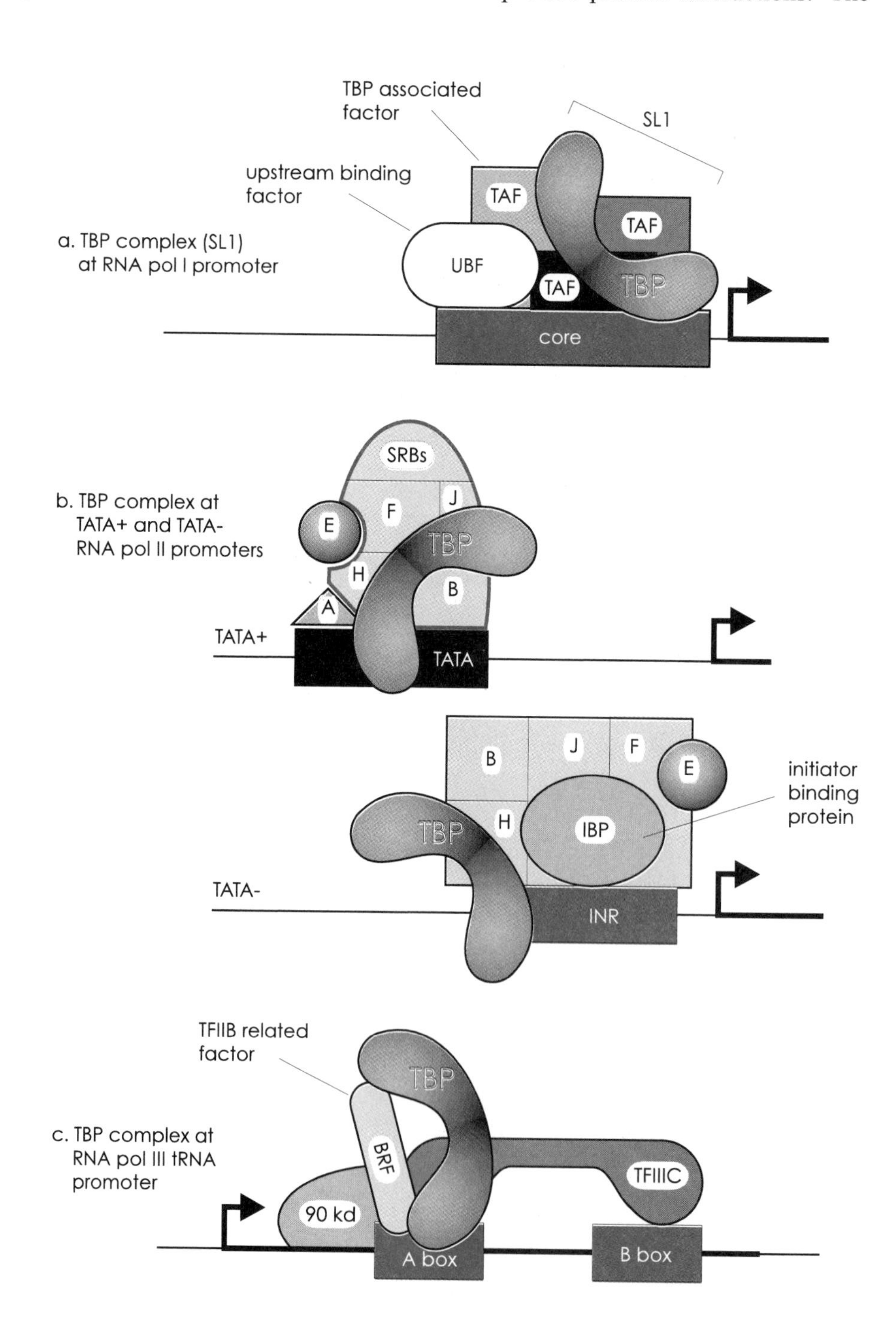

Figure 24.8.

TBP is a protein component of RNA pol I, II and III dependent initiation complexes. (a) TBP is part of the SL1 complex at RNA pol I dependent promoters, (b) TBP is bound to the TATA box in TATA+ RNA pol II-dependent promoters and to the initiator region (INR) in TATA-promoters and (c) TBP is associated with the TFIIB related factor (BRF) in initiation complexes at RNA pol III transcribed genes. The respective RNA polymerase core enzymes would also be present in these various initiation complexes.

TBP binds with high affinity to the TATA box present in many RNA pol II promoters (Figure 24.6), and amino-acid residues encoding the DNA binding domain of TBP have evolutionarily been conserved from archaebacteria to humans.

The biochemical structure of yeast TBP bound to DNA has been determined by X-ray crystallography and two very interesting observations have been made. The first is that TBP has an internal pseudo-dimer interaction between two adjacent protein domains, forming a "saddle" structure that binds DNA (Figure 24.7). The exposed surface of TBP in this configuration serves as a contact area for a variety of other transcription factors.

The second unusual feature of TBP DNA binding is that the protein causes a large kink in the DNA helix of ~100 degrees. This pronounced DNA bending may facilitate the interaction of auxiliary DNA binding proteins with TBP and other proteins in the transcriptional initiation complex. As would be expected by the polarity of RNA synthesis in the 5′ to 3′ direction, TBP binding to DNA is directional such that the same protein surface is always pointed toward the transcriptional start site.

Although TBP in yeast cells can be purified as a single polypeptide, TATA box–TBP-dependent in vitro transcription activities from mammalian cells purifies as a large multisubunit complex. For example, the human TFIID fraction contains the 38 kDa TBP polypeptide and eight other protein subunits, which together form a complex of >700 kDa. These additional **TBP-associated factors** (TAFs) are required for regulated transcription in vitro, although TAF functions in vivo are less clear. Similar to the TBP/TAF complex found in TFIID, other large protein complexes are required for the in vitro transcription of all RNA pol I, II, and III promoters (Figure 24.6). Surprisingly, TBP was found to be an integral component of the initiation complexes for all three types of promoters, even though they do not all contain TATA box binding sites (Figure 24.8).

The role of TBP in TATA-containing promoters, such as the β-globin and thymidine-kinase genes, may relate to its ability to bend DNA and cause local perturbations in DNA structure, as shown by X-ray crystallography data (Fig. 24.7). However, it is less clear why TBP is required for the transcription of TATA-less promoters. One explanation is that TBP is multifunctional; it binds DNA to alter DNA structure using the underside of the saddle, and it is a protein coordinator, making multiple contacts with a variety of TAFs through the top side of the saddle. In TATA-less promoters, this ability to hold proteins together in the proper configuration may be

BOX 24.1. SITE-DIRECTED MUTAGENESIS

There are a number of situations in which it is desirable to mutate DNA sequences that have already been cloned. To determine what role specific DNA sequences have in the function being studied (protein activity or promoter function), two general approaches are usually taken. In the first method, called **deletion mutagenesis,** both large and small fragments of the cloned DNA are removed by restriction-enzyme digestion or by exonuclease digestion of linear DNA. The remaining plasmid DNA is then religated and sequenced to determine the endpoint of the deletion.

In the second type of **site-directed mutagenesis,** point mutations or small deletions or insertions are created by annealing a mutated oligonucleotide to a complementary single-strand circular plasmid. As shown in Figure 24.9, the mutant oligonucleotide will be mispaired with the wild-type complementary sequence in the region of mutation. However, the 3′ end of the mutant oligonucleotide is correctly annealed to the template and therefore can serve as a primer for in vitro DNA synthesis. The completed circular plasmid DNA can then be introduced into bacterial cells and individual colonies containing single-plasmid species can be isolated.

A modification of this method has been developed by Thomas Kunkel to increase the yield of plasmids containing the desired mutations which can sometimes be <1% of the plasmid pool. In this scheme, the template DNA is "marked" with uridine (in place of thymidine) by isolating single-strand M13 bacteriophage DNA from a strain of bacteria that has elevated pools of dUTP owing to a deficiency in dUTPase (dut-) and a second deficiency in the enzyme uracil *N*-glycosylase (ung-) which normally removes uracil residues from DNA (Chapter 23). By transforming wild-type (dut+, ung+) bacteria with the double-stranded DNA prepared in vitro, using the standard mix dNTPs, plasmids derived from the unmutated DNA (dU+) will be eliminated preferentially. The majority of plasmids remaining after this selection process will be those synthesized in vitro using the mutant oligonucleotide as a primer and dNTPs (~60–80% of plasmid pool).

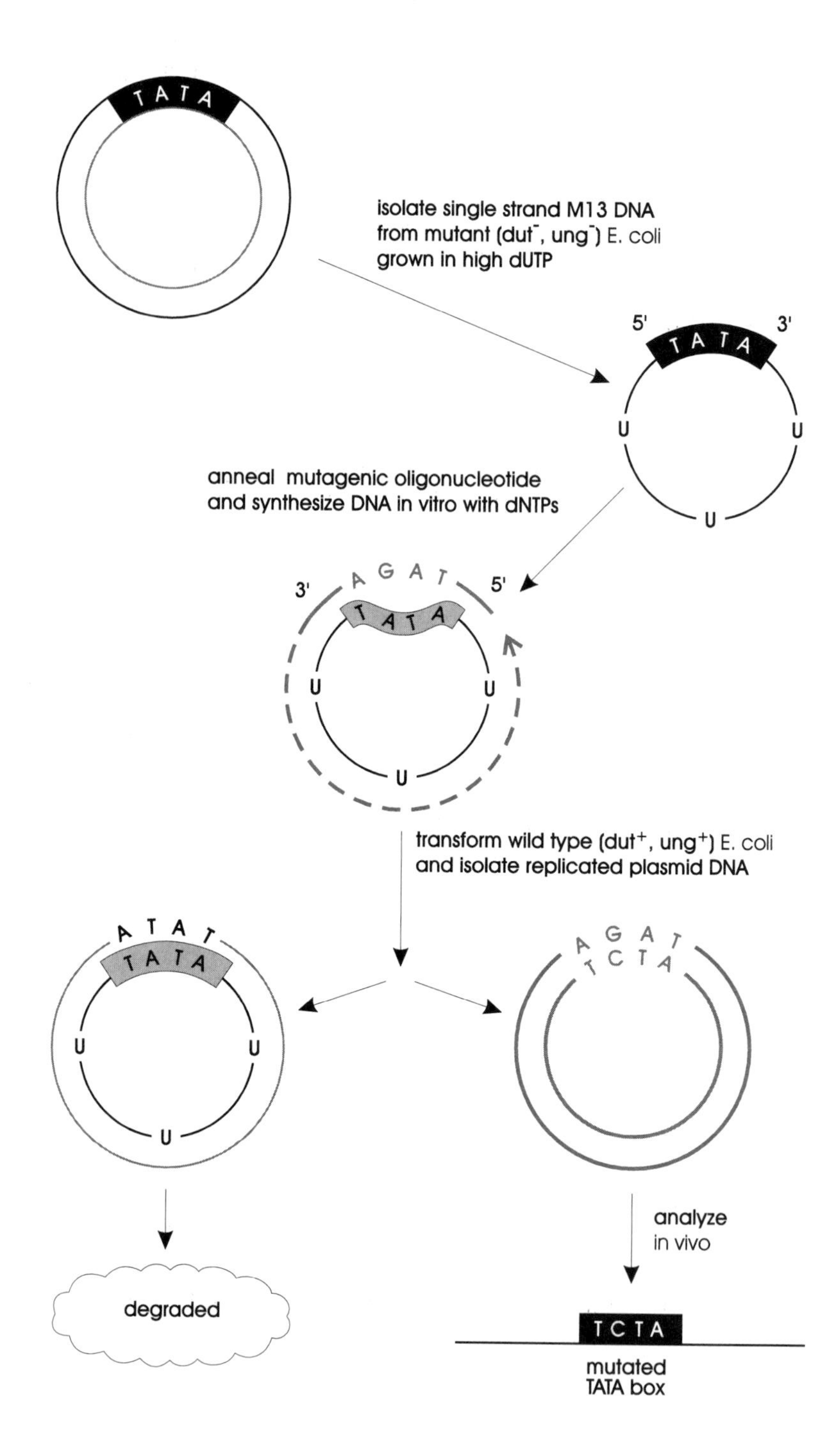

Figure 24.9.

Strategy for Kunkel oligonucleotide-directed mutagenesis using uridine incorporation as a means to eliminate unmutated plasmid DNA from the pool.

BOX 24.2. ANALYSIS OF SEQUENCE-SPECIFIC PROTEIN–DNA INTERACTIONS

Sequence-specific DNA binding activity can be assessed by several methods; two of the most sensitive and useful are **DNA footprinting** and **electrophoretic mobility shift assays** (EMSA). These procedures have been utilized to identify and purify various DNA binding proteins required for transcription, such as the **C/EBP protein,** which binds the 5′-CCAAT-3′ sequence (Fig. 24.5).

In DNA footprinting, a DNA fragment is prepared that contains the target binding site radioactively labeled at one end with ^{32}P. In one set of reactions, various amounts of protein extracts or purified protein are incubated with the end-labeled

DNA under conditions that promote DNA binding, and a control reaction is set up containing only the DNA fragment. Each of these reactions is treated with either chemicals that randomly cleave DNA or the endonuclease DNase I, under limiting conditions that produce on average only one cut per molecule. These treated samples are then electrophoresed through a polyacrylamide gel and the gel is autoradiographed to produce an image similar to the one shown in Figure 24.10a. The "footprint" represents the location on the DNA fragment where the majority of molecules were bound specifically by the protein, thus excluding the chemical or DNase I from attacking the DNA. The "ladder" of bands above and below the footprint are similar to what is produced in the dideoxy-termination method of DNA sequencing. Note that a uniform ladder is present in the lane containing DNA treated in the absence of added protein.

Gel-mobility shift assays are based on the principle that protein–DNA complexes migrate slower in an electrophoretic assay than DNA alone. In this procedure, a small double-stranded oligonucleotide (labeled with ^{32}P) is incubated with purified protein or nuclear extracts immediately before loading a low-ionic-strength agarose gel. The gel is then electrophoresed under low-temperature conditions and the migration of the unbound and protein-bound oligonucleotide are monitored by autoradiograhy. As shown in Figure 24.10b, protein binding retards migration of the oligonucleotide through the gel, producing a "shifted" band above the band corresponding to the unbound oligonucleotide. By including various amounts of unlabeled oligonucleotide in the binding reaction, or by using a labeled oligonucleotide containing a mutated sequence, it is possible to determine binding affinities and specificity, respectively.

a. DNA footprint

increasing [protein]

M

protected region

b. EMSA

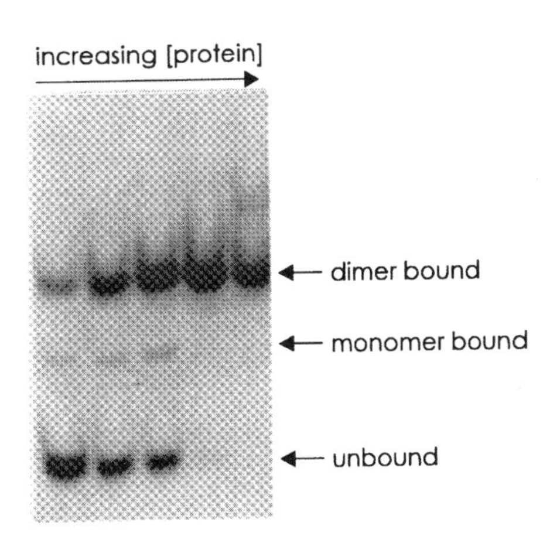

Figure 24.10.
Protein–DNA interactions can be characterized experimentally by the DNA footprinting technique and the electrophoretic mobility shift assay (EMSA). (a) In vitro DNAse I footprinting; (b) EMSA. Reactions contained an end-labeled radioactive fragment (a) or oligonucleotide (b) incubated with the zinc-finger DNA binding domain of the glucocorticoid receptor protein, a eukaryotic transcription factor.

critical for building a functional initiator complex. It is likely that some TAFs are shared between these different initiator complexes, but there are also additional "TAF-like" proteins that are found to be promoter selective.

SUMMARY

1. RNA synthesis, also called gene transcription, proceeds in the 5′ to 3′ direction and involves the nucleophilic attack of the incoming nucleotide on a free 3′ OH end. RNA synthesis does not require a preexisting primer. A transcription unit is that portion of a gene that is transcribed and is distinct from the 5′ promoter region.

2. RNA synthesis requires DNA-dependent RNA polymerase holoenzymes. The *E. coli* RNA polymerase consists of five subunits, one of which, the σ subunit, is only used during initiation for promoter recognition. Eukaryotes have three different RNA polymerase holoenzymes called RNA pol I, RNA pol II, and RNA pol III.

3. Transcription consists of three phases; initiation, elongation, and termination. Transcriptional initiation in *E. coli* involves the binding of the RNA polymerase holoenzyme to specific sequences in the promoter region called the -35 and -10 regions. It has been proposed that the RNA polymerase complex needs to undergo a conformation change from the inactive "closed" complex, to the active "open" complex, before RNA synthesis can begin.

4. Transcriptional termination in *E. coli* is either protein-dependent, which is mediated by a termination protein (ρ) that associates with the RNA polymerase elongation complex and cause template disengagement, or protein-independent, a mechanism that involves the formation of an inhibitory stem–loop structure.

5. Transcriptional initiation of RNA pol II-dependent genes in eukaryotes involves the assembly of a large protein complex at the promoter that requires as many as eight multisubunit components: RNA pol II, TFIID, TFIIB, TFIIE, TFIIF, TFIIH, TFIIJ, and TFIIA. Phosphorylation of the carboxy terminal repeat domain (CTD) of the RNA pol II large subunit is associated with initiation of RNA synthesis.

6. The TATA binding protein (TBP) is a universal eukaryotic transcriptional initiation protein that is required for the transcription of RNA pol I-dependent (rRNA), RNA pol II-dependent (mRNA and snRNA), and RNA pol III-dependent (tRNA and 55 rRNA) genes. Auxiliary proteins called TBP-associated factors (TAFs) provide promoter-selective properties to TBP and facilitate transcriptional regulation.

REFERENCES

Busby S., Ebright R. H. (1994): Promoter structure, promoter recognition, and transcription activation in prokaryotes. *Cell* **79**:743.

Berk A. J. (1995): Biochemistry meets genetics in the holoenzyme. *Proc. Natl. Acad. Sci. USA* **92**:11952.

Chao, D.M., Young, R.A. (1996): Activation without a vital ingredient. *Nature* **383**:119.

Cormack B. P., Struhl K. (1992): The TATA-binding protein is required for transcription by all three nuclear RNA polymerases in yeast cells. *Cell* **69**:685.

Gill G. (1994): Transcriptional initiation: Taking the initiative. *Curr. Biol.* **4**:374.

Greenblatt J. (1992): Transcription: Riding high on the TATA box. *Nature* **360**:16.

Hernandez N. (1993): TBP, a universal eukaryotic transcription factor? *Genes Dev.* **7**:1291.

Kainz M., Roberts J. (1992): Structure of transcription elongation complexes in vivo. *Science* **255**:838.

Kim Y., Geiger J. H., Hahn S., Sigler P. B. (1993): Crystal structure of a yeast TBP/TATA-box complex. *Nature* **365**:512.

Koleske A. J., Young R. A. (1994): An RNA polymerase II holoenzyme responsive to activators. *Nature* **368**:466.

Li, X-Y, Green, M.R. (1996): Tumorgenesis: Transcriptional elongation and cancer. *Curr Bio* **l6**:943.

Lu H., Zawel L., Fisher L., Egly J.-M., Reinberg D. (1992): Human general transcription factor IIH phosphorylates the C-terminal domain of RNA polymerase II. *Nature* **358**:641.

Martinez E., Chiang C.-M., Ge H., Roeder R. G. (1994): TATA-binding protein-associated factor(s) in TFIID function through the initiator to direct basal transcription from a TATA-less class II promoter. *EMBO J* **13**:3115.

McDowell J. C., Roberts J. W., Jin D. J., Gross C. (1994): Determination of intrinsic transcription termination efficiency by RNA polymerase elongation rate. *Science* **266**:822.

Oelgeschlager, T., Chiang, C-M, Roeder, R.G. (1996): Topology and reorganization of a human TFIID-promoter complex. *Nature* **382**:735.

Peterson M. G., Tjian R. (1992): Transcription: The tell-tail trigger. *Nature* **358**:620.

Reines, D., Conaway, J.A., Conaway, R.C. (1996): The RNA polymerase II general elongation factors. *Trends Biochem. Sci* **21**:351.

Roeder, R.G. (1996): The role of general initiation factors in transcription by RNA polymerase II. *Trends Biochem. Sci* **21**:327.

Rudloff U., Eberhard D., Tora L., Stunnenberg H., Grummt I. (1994): TBP-associated factors interact with DNA and govern species specificity of RNA polymerase I transcription. *EMBO J* **13**:2611.

Schaeffer L., Roy R., Humbert S., et al. (1993): DNA repair helicase: A component of BTF2 (TFIIH) basic transcription factor. *Science* **260**:58.

Severinov K., Mustaev A., Severinova E., et al. (1995): Assembly of functional *Escherichia coli* RNA polymerase containing β subunit fragments. *Proc. Natl. Acad. Sci. USA* **92**:4591.

Sheldon M, Reinberg D. (1995): Transcriptional activation: Tuning-up transcription. *Curr Biol* **5**:43.

Svejstrup, J.Q., Vichi, P., Egly, J-M. (1996): The multiple roles of transcription/repair factor TFIIH. *Trends Biochem. Sci* **21**:346.

REVIEW QUESTIONS

1. DNA and RNA synthesis are enzymatically very similar cellular processes, yet there are several key mechanistic differences. Which statement is NOT TRUE regarding RNA synthesis?
   a) RNA synthesis requires a free hydroxyl end for chain elongation.
   b) RNA polymerases utilize uridine rather than thymidine.
   c) DNA synthesis requires a preexisting primer and RNA polymerase does not.
   d) RNA synthesis proceeds in the 3′ to 5′ direction.
   e) Gene promoters are to RNA synthesis what origins are to DNA synthesis.

2. Transcriptional termination occurs as a result of
   a) RNA polymerase transcribing off the end of chromosomes
   b) formation of an RNA structure that promotes polymerase dissociation
   c) DNA binding proteins that intercalate into DNA and block transcription
   d) RNA degradation from the 3′ end
   e) dissociation of σ factor from RNA polymerase holoenzyme

3. Assembly of the RNA pol II initiation complex in eukaryotes requires a much larger number of proteins than what is needed to form a transcriptional initiation complex in prokaryotes. What answer best explains the reason for this difference?
   a) Cell-specific regulation in eukaryotes requires that transcriptional initiation be tightly controlled and multisubunit protein complexes facilitate this.
   b) Eukaryotes have a larger number of DNA binding proteins than prokaryotes.
   c) Eukaryotes have more genes than prokaryotes.
   d) The C-value paradox.
   e) Eukaryotic promoters contain a TATA box for TBP binding and prokaryotic promoters contain a -35 and a -10 region to which RNA polymerase binds.

4. What is the relationship between RNA pol I, II, and III and the three classes of genes that they transcribe?
   a) RNA pol I, II, and III transcribe tRNA, rRNA, and mRNA, respectively.
   b) The promoter architecture of each gene class reflects an optimal arrangement for the respective RNA polymerases.
   c) RNA pol I and RNA pol III both transcribe protein-coding genes.
   d) Gene class complexity is directly related to polymerase activity in the cell.
   e) Both b and d.

5. Why do you think genetic mutations in the yeast TBP gene are lethal?
   a) TBP is a required transcriptional termination protein.
   b) Yeast cells lacking functional TBP transcribe only rRNA genes.
   c) TBP-associated factors bind to and inhibit DNA polymerase α.
   d) TBP-deficient yeast cells become light-sensitive.
   e) TBP is required for the transcription of RNA pol I-, II-, and III-dependent genes.

ANSWERS TO REVIEW QUESTIONS

1. ***d*** RNA synthesis proceeds in the 5′ to 3′ direction just like DNA synthesis.

2. ***b*** Both protein-dependent and protein-independent transcriptional termination involve the for-

mation of a stem–loop structure in the nascent RNA.

3. ***a*** Controlling the initiation rates of transcription is a combinatorial process in eukaryotes that requires multisubunit protein complexes.

4. ***b*** RNA pol I, II, and III complexes recognize rRNA, mRNA, and tRNA/snRNA encoding genes, respectively; coevolution of promoter sequences and RNA polymerase subunits would explain this observation.

5. ***e*** Yeast cells unable to transcribe any gene due to lack of TBP will die.

CHAPTER

25

RNA PROCESSING

INTRODUCTION

Essentially all RNA molecules synthesized in eukaryotes require some form of posttranscriptional processing. In the simplest case, this means site-specific cleavage of the primary transcript into smaller, but functional, subfragments. Sometimes the bases are modified by methylation or other chemical reactions, as in tRNA molecules, or nucleotides are added to the transcript in the absence of DNA (addition of a 5′ methylguanosine cap to the 5′ end of the transcript or polyadenylation of the 3′ end of mRNA). However, by far the most prevalent RNA processing reaction in eukaryotes is the removal of introns through a mechanism called **RNA splicing.** As described in Chapter 21, most eukaryotic protein-encoding genes consist of exons that specify the amino-acid sequence of the protein product, interspersed with noncoding intron sequences. Therefore, before the mRNA in eukaryotes can be translated into protein, the primary RNA transcript (also called **hnRNA** for heterogeneous RNA), must be processed through a mechanism known as RNA splicing (Fig. 25.1). This is in contrast to the coding sequences of bacterial genes which are uninterrupted and transcribed as a single mRNA that can be directly translated into protein without RNA processing.

In this chapter we describe the biochemistry of RNA splicing and emphasize the catalytic role of RNA itself in promoting site-specific processing of RNA molecules. In addition, other RNA processing pathways are discussed, including the processing of rRNA and tRNA precursors, 3′ polyadenylation, regulated alternative mRNA splicing, and mRNA degradation. Together, these various modifications of primary RNA transcripts can control the quality and quantity of functional RNA in a cell.

MOST RNA SPLICING MECHANISMS REQUIRE AN RNA-CATALYZED TWO-STEP REACTION

When introns in eukaryotic genes were discovered over 20 years ago, it was clear that it would be necessary to physically isolate both the genomic DNA and RNA sequences (usually in the form of complementary DNA or "cDNA" as described in Box 25.1) to determine the sequence relationship between the genomic copy of the gene, and the fully processed mRNA. Figure 25.2 schematically shows how an **RNA–DNA hybrid,** formed between an mRNA and the template genomic DNA, results in the appearance of DNA segments being excluded from the hybrid. The single-strand loops of DNA represent the intron sequences that have been removed during the process of RNA splicing. Structures such as these were first observed using an electron microscope to examine the β-globin gene and β-globin mRNA.

Four general types of RNA splicing have so far

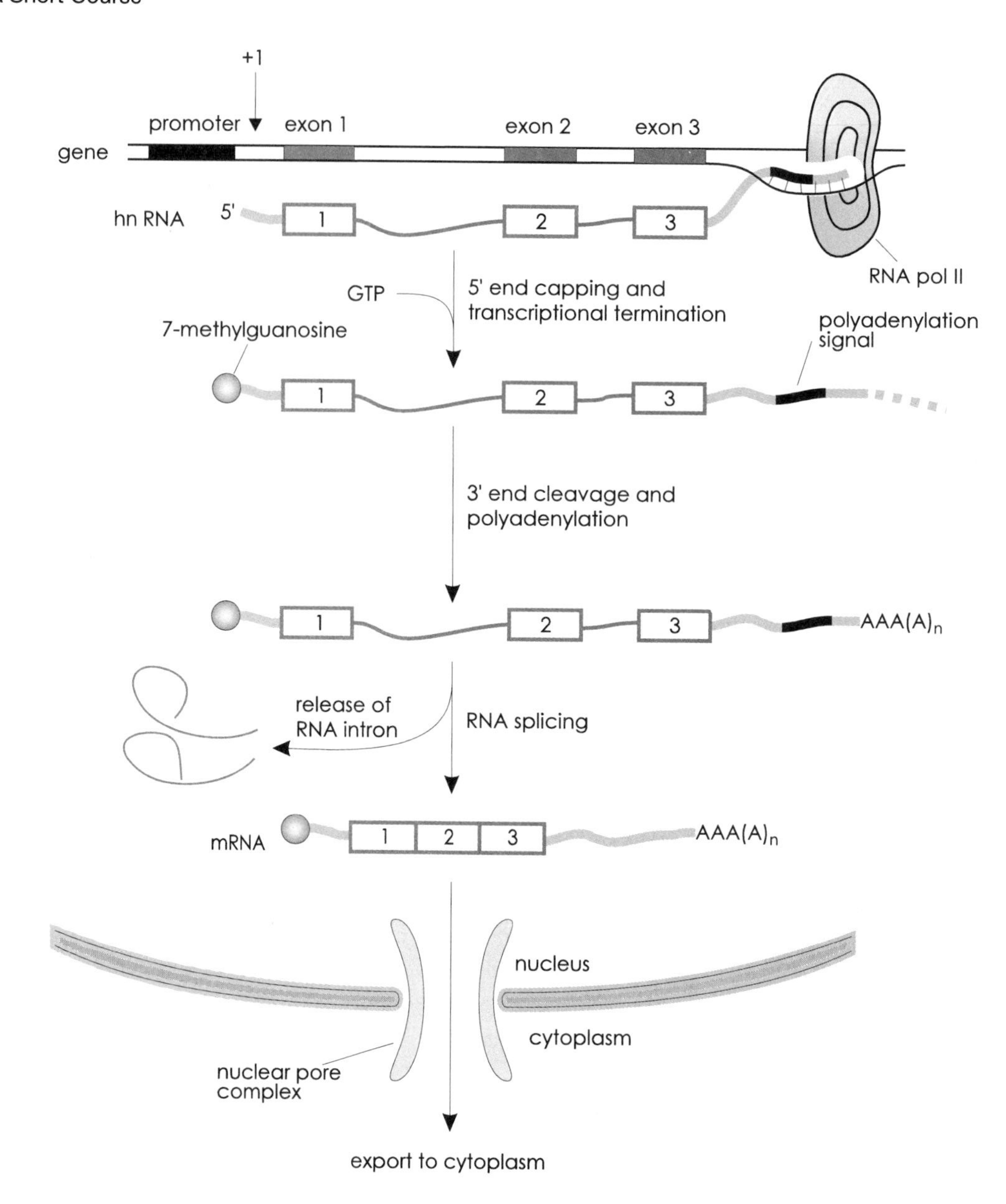

Figure 25.1.
Overview of RNA processing showing 7-methyl guanosine cap addition to the 5′ end, processing of hnRNA into mRNA by the process of RNA splicing, 3′ polyadenylation, and mRNA export to the cytoplasm. Note that 5′ capping takes place on the nascent hnRNA before transcription is terminated.

been discovered. The first three described in this section, group I and group II **self-splicing** and **hnRNA splicing,** have two distinct similarities: (1) the breaking and rejoining of phosphodiester bonds, which maintains energy balance and (2) the RNA itself acts as the catalyst either in cis (self-splicing of group I and II introns) or in trans (hnRNA splicing by ribonucleoprotein complexes called spliceosomes). Self-splicing is probably a more ancient mechanism in that it is found in the organelles of plants and fungi, whereas hnRNA splicing by **spliceosomes** may have evolved to provide increased flexibility and to control cell-specific alternative splicing. The splicing of some **yeast tRNAs** is a fourth type of RNA splicing reaction that requires a conventional protein-encoded endonuclease and an ATP-dependent ligase. For all four RNA processing reactions, the secondary and tertiary structure of RNA provides the required specificity and plays a significant role in lowering the energy barrier for catalysis by bringing reactive groups into close proximity.

Self-Splicing of Group I Introns Is Mediated by an External Guanine Nucleoside

The first step in self-splicing of **group I introns** is the nucleophilic attack on the 3′-5′ phosphodiester bond between the exon and intron at the 5′ end of the intron, as shown in Figure 25.3. The 3′ hydroxyl of a free guanosine acts as a nucleophile in the **first transesterification step** to form a new 3′-5′ phosphodiester bond with the 5′ nucleotide in the intron, thus releasing the upstream

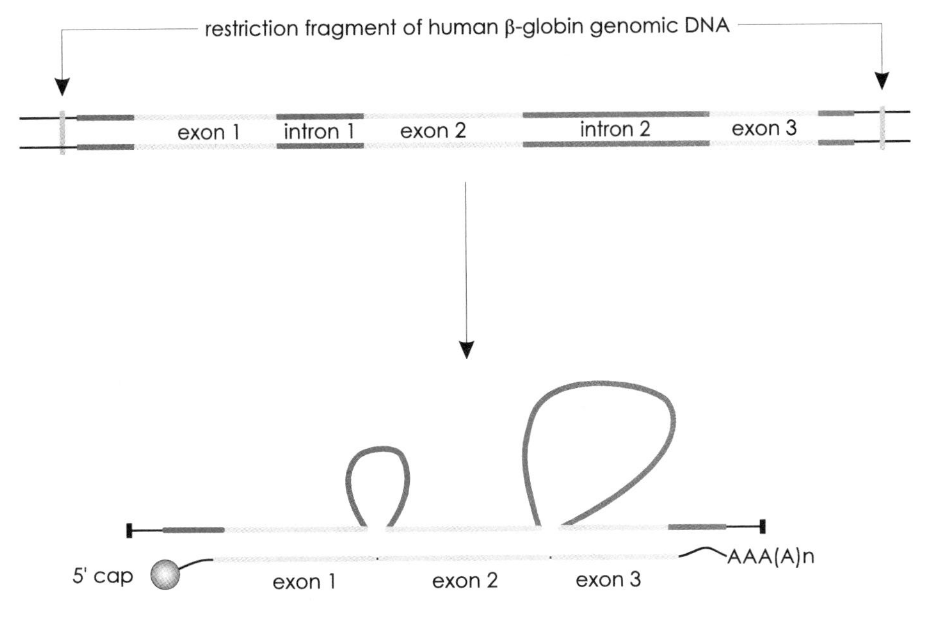

Figure 25.2.

Schematic drawing of an RNA–DNA hybrid showing the looping out of introns 1 and 2 in the DNA template strand of the human β-globin gene. The two purified components in the hybridization reaction are a heat-denatured fragment of β-globin genomic DNA and fully processed β-globin mRNA.

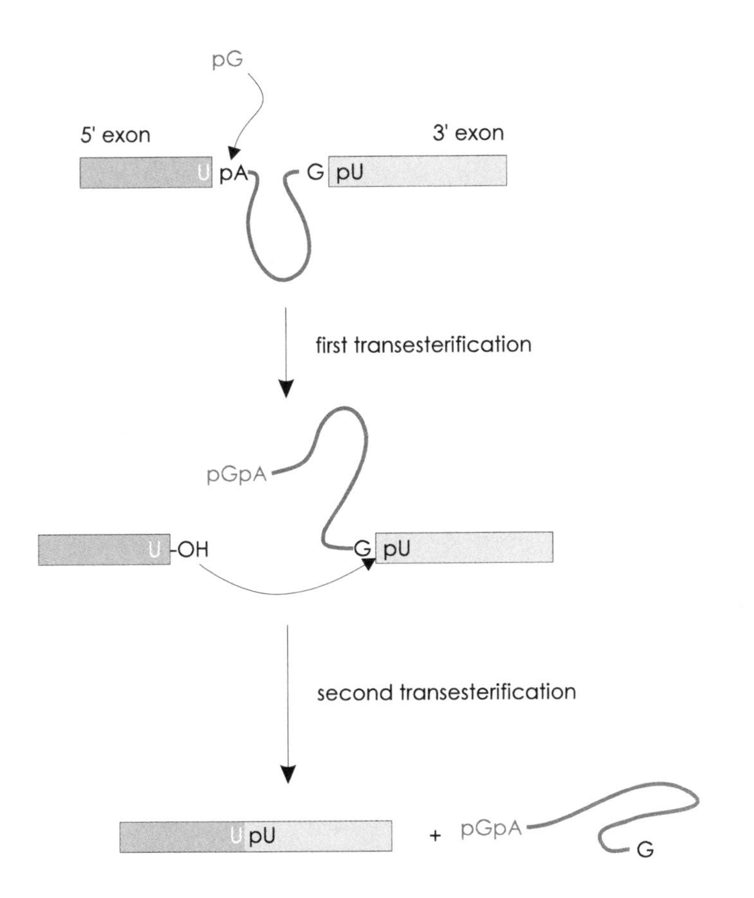

Figure 25.3.

Schematic drawing showing the two transesterification steps of group I self-splicing. In the first step there is a nucleophilic attack by the guanosine cofactor on the phosphodiester bond at the 5′ boundary of the intron. In the second transesterification, the free 3′ hydroxyl attacks the phosphodiester bond at the downstream intron–exon border, thus causing the release of the intron as a linear RNA fragment and the covalent linkage of the two exons.

exon as a free end (which is most likely held in place by the overall structure of the RNA molecule).

In the **second transesterification step,** the 3′ hydroxyl of the upstream exon acts as a nucleophile and attacks the 3′-5′ phosphodiester bond at the 3′ intron–exon boundary. This results in the precise removal of a linear intron sequence and the covalent linkage of the two exon sequences through a normal 3′-5′ phosphodiester bond. The guanosine acts as a cofactor in the reaction, but because the number of bonds is unchanged, this is considered an energy-neutral reaction.

Self-Splicing of Group II Introns Is Catalyzed by an Internal Adenylate Residue

Figure 25.4a shows the reaction steps for the removal of a **group II intron** by self-splicing. The two basic transesterification steps are the same as described for splicing of group I introns, except in this case, the nucleophile comes from the 2′ hydroxyl of an adenylate residue present in the intron itself. The first nucleophilic attack results in the formation of an unusual **2′-5′ phosphodiester bond** between the adenylate and the 5′ nucleotide in the intron (Fig. 25.4b). The 3′ hydroxyl of the upstream exon is then able to attack the 5′ phosphate at the downstream exon–intron border in the second transesterification step. This splicing reaction involves the formation of an intermediate structure called a **lariat** which is released after the second transesterification step. The unusual 2′-5′ phosphodiester bond in the lariat, at a position called the **branch site,** forms a structure with three phosphodiester bonds. As shown in Figure 25.4b, the adenylate residue remains attached to the rest of the intron through a standard 3′-5′ phosphodiester bond.

The importance of RNA structure in group I and

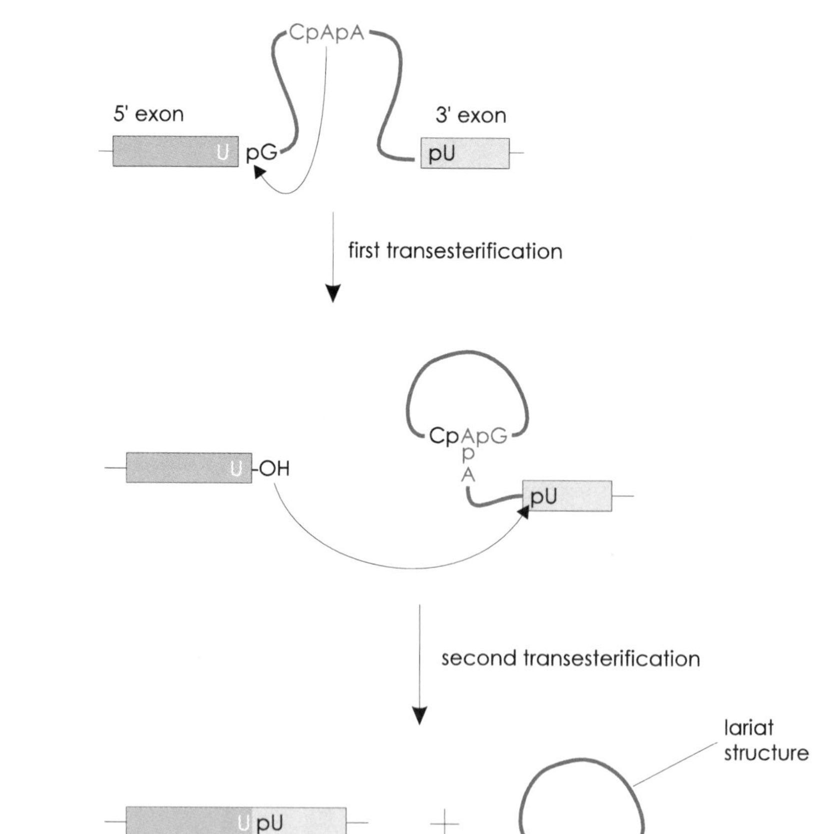

Figure 25.4.
Biochemistry of group II self-splicing. (a) RNA processing pathway of group II self-splicing introns.

(continued)

II self-splicing reactions can be shown by heating the RNA to denature intrastrand base pairing or by adding denaturing agents such as urea. Both of these treatments block in vitro self-splicing. The discovery in 1982 by Thomas Cech, that self-splicing reactions could proceed entirely in the absence of protein, supported the view that RNA is likely the ancestral macromolecule to DNA and proteins.

hnRNA Splicing Requires a snRNP Spliceosome Complex

The predominant form of RNA splicing in eukaryotes is spliceosome-mediated removal of introns from hnRNA. This reaction is chemically similar to group II splicing in that an adenylate residue in the intron acts as a nucleophile to attack the 5′ exon. This results in the formation of a lariat structure in the first step, which is followed by a second transesterification step that joins the 5′ and 3′ exons and releases the intron (Fig. 25.5).

The difference between self-splicing and spliceosome-mediated hnRNA splicing is that hnRNA splicing is dependent on **small nuclear RNA** (snRNA) molecules and associated proteins which form **small nucleoproteins** (snRNPs) often called **snurps.** One explanation for why snRNPs are required for hnRNA splicing is that the structural requirements for the reaction are imposed on the RNA substrate by the spliceosome. This would mean that the intron sequences themselves need not be conserved for splicing to be accurate (other than the conserved GU–AG residues at the intron boundaries, as shown in Fig. 25.5). In this context, hnRNA splicing would be considered to be a highly evolved form of group II self-splicing because it provides increased flexibility and a point of regulation.

Four snRNPs, **U1, U2, U5, and U4/U6,** associate with intron sequences to form the spliceosome complex

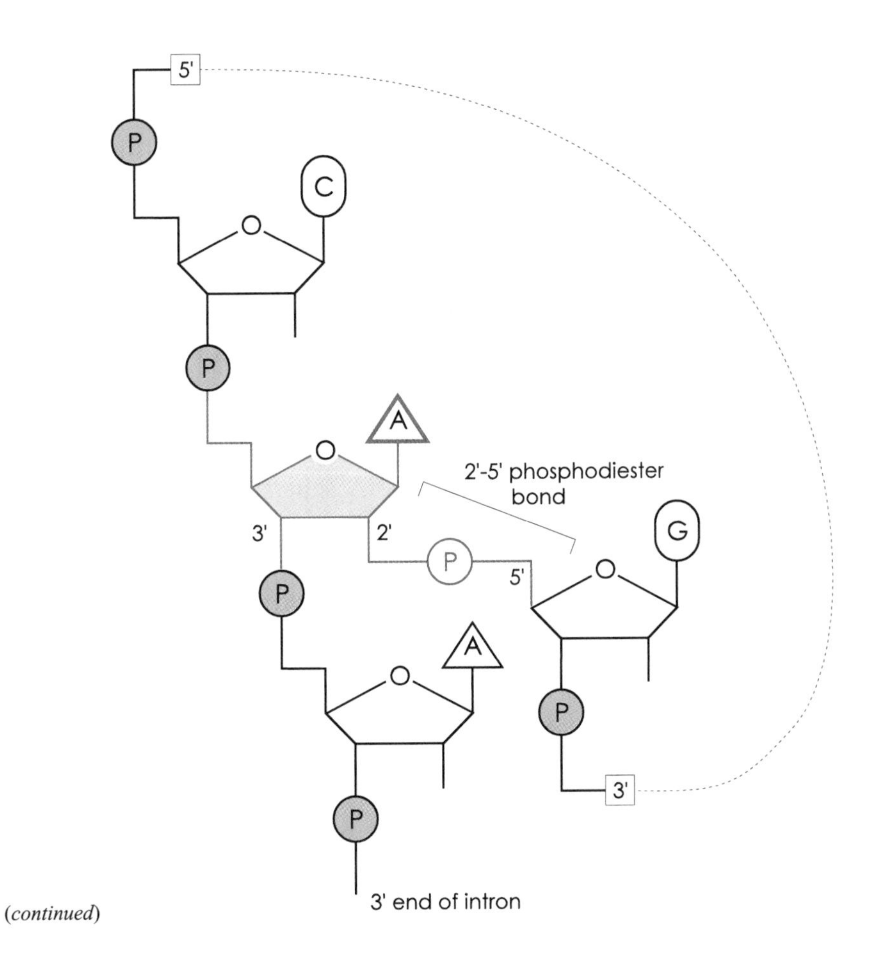

Figure 25.4. *(continued)*
(b) Structure of the 2′-5′ phosphodiester bond formed between the adenylate within the intron and the 5′ nucleotide at the intron–exon border.

(continued)

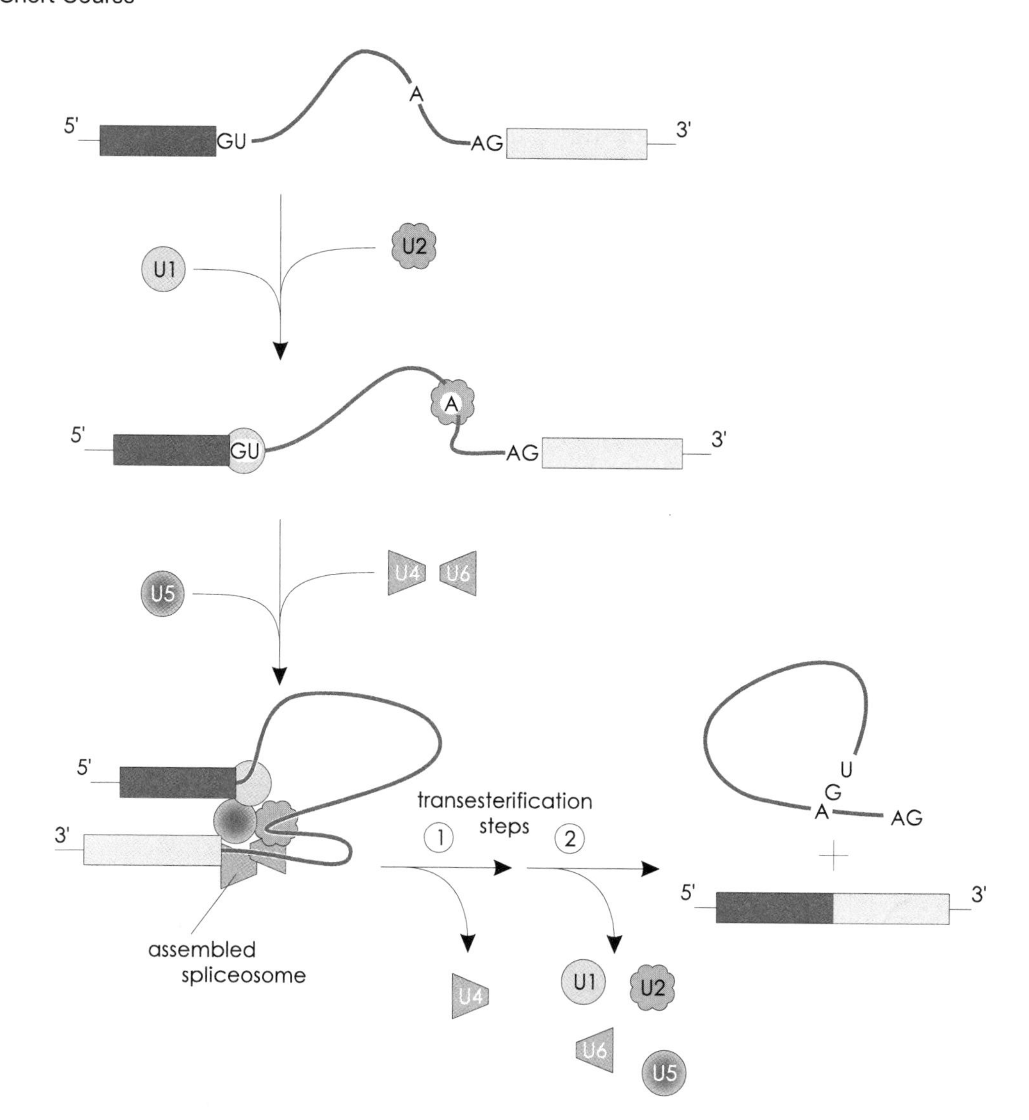

Figure 25.5.
Spliceosome assembly pathway showing the conserved GU–AG residues at the intron boundaries and the adenylate residue at the lariat branch site. Note that the dissociation of the U4–U6 complex after the first transesterification step leads to the formation of the U2–U6 complex which is critical to completion of the splicing reaction.

which contains five RNAs and >60 proteins. Figure 25.5 shows several steps in the hnRNA splicing reaction and the overall assembly of the spliceosome complex. Two critical snRNA interactions provide specificity in the spliceosome-directed pathway. The first is base pairing of **U1 RNA** with the 5′ exon–intron boundary, and the second is base pairing of **U2 RNA** with the branch site, causing an extrusion of the reactive adenylate residue. Once the **U4/U6 snRNP** rearranges in an ATP-dependent conformational change to dissociate U4, the U6 residues are free to interact with U2 and then U1. The precise alignment of the 5′ boundary and the branch point by U6 is complemented by the juxtaposition of the 3′ intron boundary with the reaction center of the **U5 snRNP.**

The role of U6 in the splicing reaction, as the central "organizer" of the active site, is supported by the finding that deletions of the **U6 gene** in yeast are lethal and that U6 sequences are highly conserved in evolution. Based on the similarity between group II self-splicing and the hnRNA splicing reaction, and on the conformation of the active site in the U6, U2, U5 complexes, it seems clear that hnRNA splicing is also an RNA-catalyzed event. However, there is still much to be learned about the molecular mechanism of spliceosome-mediated hnRNA splicing. For example, a class of non-snRNP proteins, called SR proteins, has also been shown to be required for 5′ splice-site recognition and spliceosome assembly.

RIBOZYMES ARE SITE-SPECIFIC ENDORIBONUCLEASES

The RNA-catalyzed cleavage of *Tetrahymena* rRNA introns represents a classic example of group I RNA self-splicing. While in the process of characterizing this reaction, Cech and co-workers found that the RNA itself behaves much like an enzyme in that it accelerates the rate of transesterification and is highly specific. Moreover, they discovered that the rRNA intron sequence, called **L-19 IVS** (intervening sequence), is capable of functioning as a true enzyme by promoting multiple nucleotidyl transfer reactions between oligonucleotides substrates such as pentacytidylic acid (C_5) and a guanosine residue at the 3′ end of L-19 IVS. In this reaction, nucleotides can be removed or added to oligonucleotide substrates with Michaelis–Menten kinetics having K_m=42 μM and a k_{cat}=2/min. Although this reaction is slower than most protein enzymes, it raises the catalysis rate by 10^{10}. Based on these properties, RNA molecules with catalytic function are called **ribozymes.** Another example of a ribozyme is the RNA component of **RNase P,** which cleaves a phosphodiester bond during the processing of *E. coli* tRNA transcripts. This reaction can be performed in vitro in the absence of any protein; however, in vivo, the 17.5 kDa RNase P protein is required, most likely to maintain the ribozyme in the active form.

Synthetic ribozymes have been developed and shown to act as site-specific endonucleases. These RNA enzymes contain a ~20 nucleotide sequence that functions as the catalytic site and flanking sequences that are complementary to sequences in the target RNA substrate. One type of trans-acting ribozyme is the **hammerhead** ribozyme originally found in plant virus-like particles called viroids. Empirical analysis of hammerhead ribozymes has revealed residues in both the ribozyme and the substrate that are required for function and several applications of ribozyme technology have now been described. Figure 25.6 illustrates an example of how a ribozyme has been used in vivo to inactivate an endogenous mRNA by endonucleolytic cleavage.

The application of ribozyme technology as a form of therapy in humans may be possible, and in vitro stud-

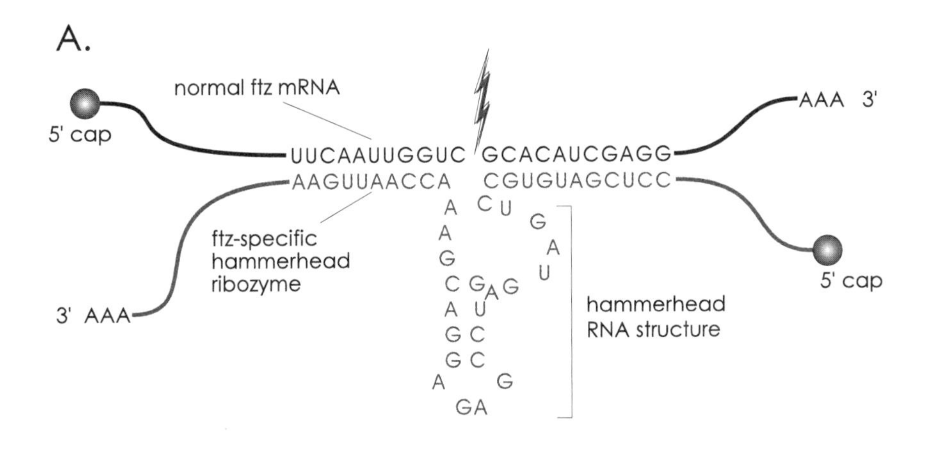

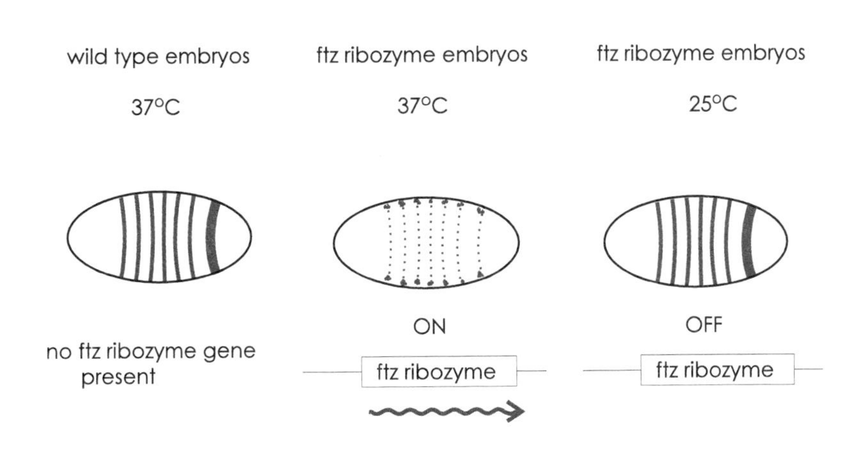

Figure 25.6.
Ribozymes have been used to specifically cleave RNA sequences in vivo. (A) Sequence of the relevant portion of the fushi taruzu (ftz) RNA that is a target for cleavage by the genetically engineered ftz hammerhead ribozyme. (B) Diagram showing the observed results in fly embryos containing the ftz hammerhead ribozyme gene. The normal pattern of ftz gene expression is seen as seven stripes in the wild-type embryo (left) or in ftz ribozyme embryos at low temperature when the ftz ribozyme gene is off (right). However, when the ftz gene is activated in the ftz ribozyme embryos by heat induction, ftz RNA is dramatically diminished (center), thus demonstrating the in vivo activity of the ftz ribozyme.

ies have shown that hammerhead ribozymes can be used to inactive human immunodeficiency virus (HIV) RNA in cultured human cells.

PROCESSING OF rRNA AND tRNA FROM PRECURSOR TRANSCRIPTS

rRNA Processing

In *E. coli*, there are seven copies of the rRNA genes contained within a single transcription unit. The primary transcript is ~6500 nucleotides long and encodes the **23S rRNA** and **16S rRNA.** Also encoded in this transcript are several tRNAs and the small 5S rRNA, which is a component of the bacterial ribosome. Initially, rRNAs were characterized by their sedimentation rates through sucrose density gradients as measured by **Svedberg units** (S). Molecular cloning and nucleic-acid sequencing has shown that the 23S *E. coli* rRNA is ~2900 nucleotides long and the 16S rRNA is ~1600 nucleotides in length. In bacteria, specific ribonucleases (**RNases**) are required to process the primary rRNA transcript into the 23S and 16S rRNA products.

Figure 25.7 shows the processing pathway of human **45S rRNA** into the two functional **28S rRNA** and **18S rRNA** products. As described in Chapter 24, eukaryotic rRNA genes are transcribed by RNA pol I. Less is known about the enzymology of eukaryotic rRNA processing than that of bacteria; however, it is presumed that site-specific RNases are also involved in the cleavage reactions.

The primary human 45S rRNA transcripts is ~14,500 nucleotides long and encodes the 28S, 18S, and **5.8S rRNA** species. Methylation of the rRNA precursor is preferential for sequences that encode the

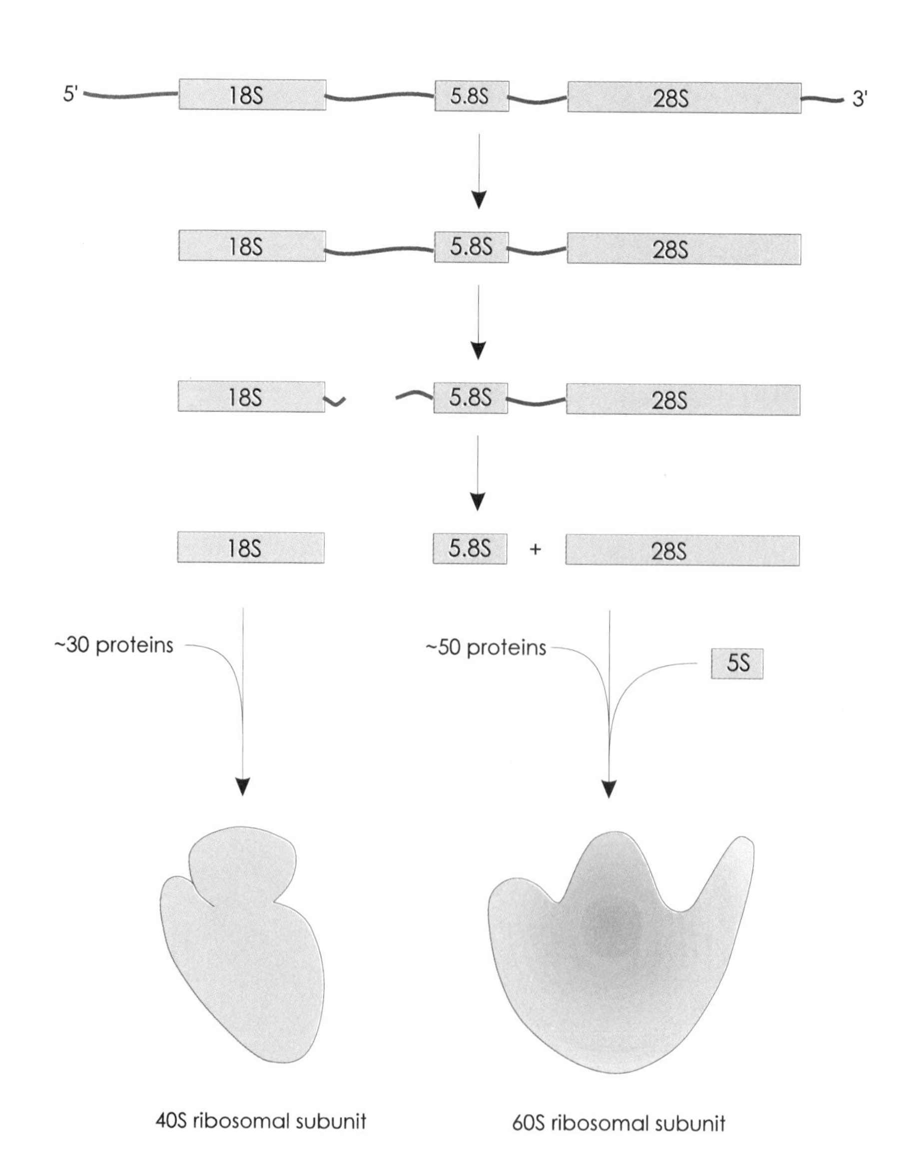

Figure 25.7.

Processing of human 45S rRNA into the functional 28S, 18S, and 5.8S products. Assembly of the 60S large ribosomal subunit requires the 28S, 5.8S, and 5S RNAs, whereas the 40S small ribosomal subunit contains the 18S rRNA.

rRNA products, strongly suggesting that methylation patterns guide the RNA processing reaction. The 5S rRNA component of eukaryotic ribosomes is encoded by 5S rRNA genes, which are transcribed by RNA pol III (Chapter 24).

tRNA Processing

Posttranscriptional processing of tRNA requires several distinct steps, as summarized in Figure 25.8. First, the 5′ and 3′ ends must be cleaved to release the tRNA sequence from the larger precursor transcript and introns must be removed if they are present. Second, the required **CCA charging sequence** at the 3′ end of tRNA must sometimes be added by a **nucleotidyl transferase.** Third, all tRNAs contain a large number of **modified bases** which result from reductions, methylations, and deaminations. These modifications can affect codon recognition by the tRNAs during protein synthesis (Chapter 26).

In yeast, about 40 of the ~400 nuclear tRNA genes contain an intron which is removed by a splicing reaction that depends on protein-dependent cleavage and ligation reactions (this is the fourth type of RNA splicing). The yeast tRNA introns are variable in length (~15–50 nucleotides long) and have no recognizable conserved sequence. Interestingly, in all of the yeast tRNA introns studied thus far, a trinucleotide sequence is found in the intron that is complementary to the anticodon in the same tRNA (see Fig. 25.8). For example, if the anticodon is 5′–CUU–3′, then the sequence 5′–AAG–3′ would be present in the intron. These sequences are paired in the precursor tRNA transcript to form a stem structure that is recognized by a specific RNA **endonuclease.**

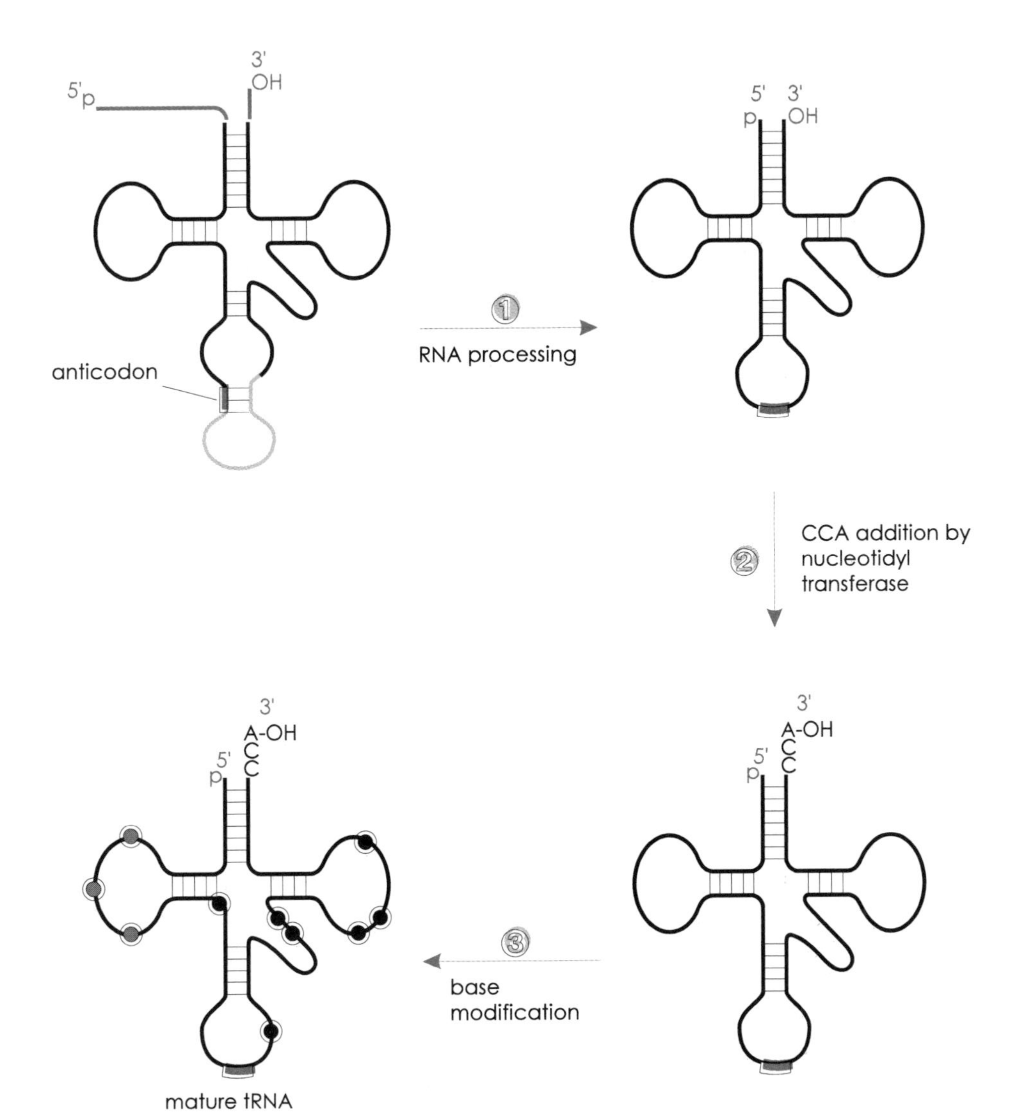

Figure 25.8.

tRNA processing steps. Step 1, RNase cleavage and removal of intron by tRNA splicing reaction. Step 2, for some tRNAs the terminal CCA is added by nucleotidyl transferase. Step 3, specific base modifications. Note that in yeast tRNAs with introns, the anticodon forms a stem structure with complementary nucleotides in the intron which may facilitate specificity in the splicing reaction.

PROCESSED mRNA CONTAINS A 5′-METHYL GUANOSINE CAP AND A 3′-POLYADENYLATED TAIL

Eukaryotic mRNAs contain an unusual **7-methyl guanosine** at the 5′ end of the transcript called a **5′ cap.** This guanosine is added to the transcript shortly after RNA synthesis is initiated by enzymes associated with the RNA pol II elongation complex. There is an unusual **5′,5′-triphosphate linkage** in the 5′ cap between the guanosine and the first nucleotide of the mRNA (usually adenine), as shown in Figure 25.9.

The 5′ cap on mRNA is important for initiation of protein synthesis and it also protects the mRNA transcript from exonucleolytic degradation. The RNA transcripts synthesized by RNA pol I and III do not have 5′ caps.

The 3′ end of eukaryotic mRNAs is posttranscriptionally modified in a two-step reaction that involves a large protein complex. In the first step, a **cleavage–polyadenylation specificity factor** (CPSF) recognizes and binds to the sequence **AAUAA** found in the 3′ end of most mRNAs (Fig. 25.10). Another protein complex, called the **cleavage stimulation factor** (CStF), interacts with CPSF to activate cleavage of the mRNA at a site approximately 20 nucleotides downstream of the AAUAA. In the second 3′ processing step, the enzyme **poly(A) polymerase** (PAP) binds to the free 3′ end of the transcript and initially adds a short **poly A tail** of ~20 nucleotides. This short poly A tail is subsequently extended to ~200 A residues following stimulation of PAP activity by an oligo(A) binding protein. The polyadenylation reaction is an important regulatory step because the length of poly A tails modulates both mRNA stability and translational efficiency. Interestingly, cell-cycle-regulated **histone mRNA** transcripts are not polyadenylated. Instead, the 3′ end of histone mRNA is cleaved at a sequence in the mRNA just downstream of a highly conserved stem–loop structure. This stem–loop structure most likely protects histone mRNA from 3′ degradation.

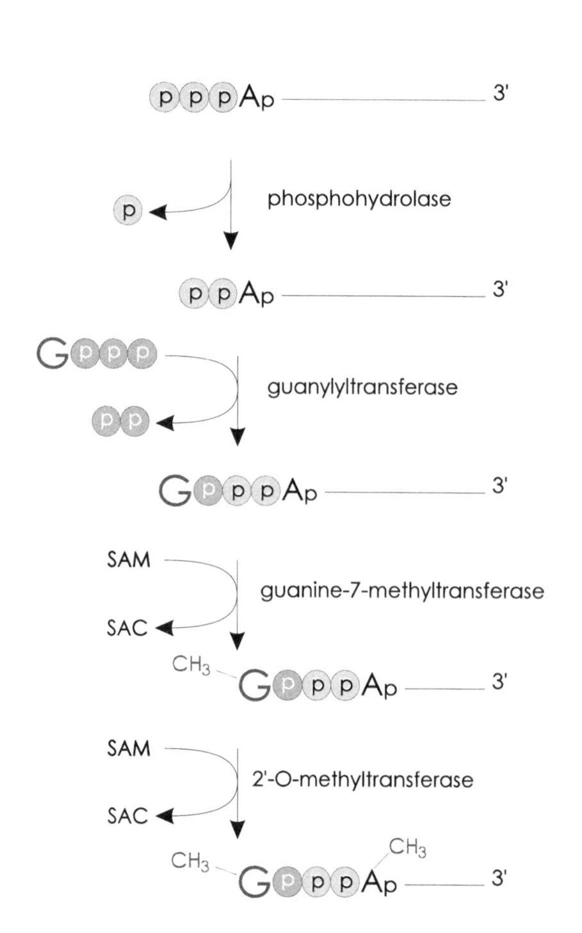

Figure 25.9.

5′ Capping of eukaryotic mRNA. (a) Enzymatic reactions required for 5′ capping; SAM is *S*-adenosyl methionine and SAC is *S*-adenosyl homocysteine. (b) Structure of 7-methylguanosine cap.

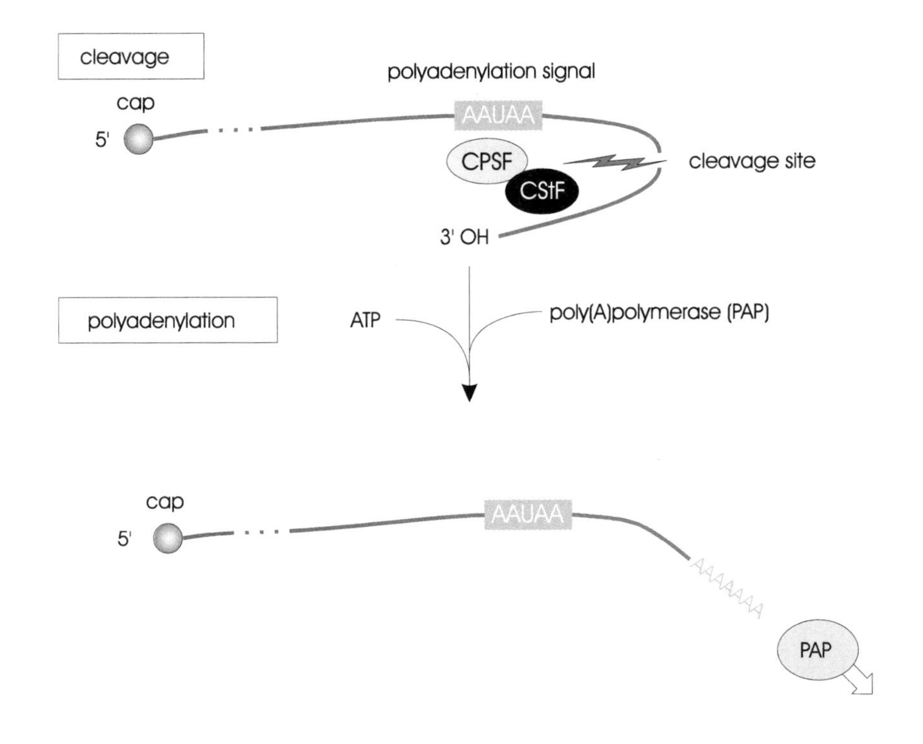

Figure 25.10.

Polyadenylation of the 3′ end of eukaryotic mRNA involves both a 3′ cleavage step at a site downstream of the sequence AAUAA and polydenylation of the free 3′ hydroxyl by poly(A) polymerase.

unspliced RNA

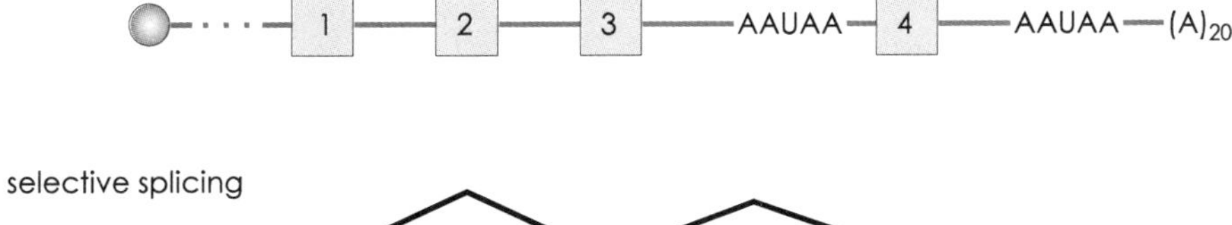

selective splicing

1 2 3 4 AAUAA—(A)200

alternate 5' site

alternate 3' site

alternate poly(A) site

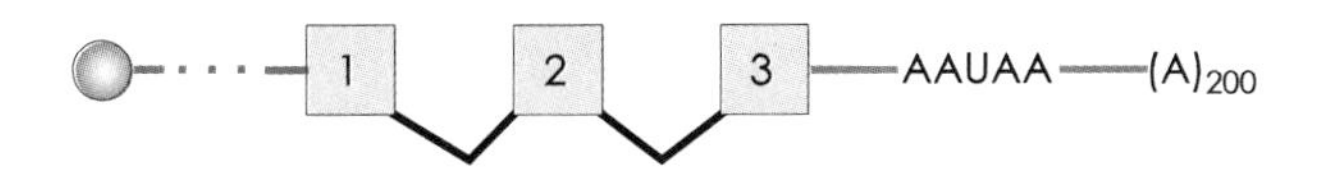

Figure 25.11.

The four basic modes of alternative RNA processing include selective splicing, use of an alternate 5′ donor site, use of an alternate 3′ acceptor site, and use of an alternate polyadenylation site.

mRNA PROCESSING CAN BE REGULATED BY ALTERNATIVE SPLICING AND POLYADENYLATION

One advantage of having genes split into multiple exons is that it provides an opportunity to mix and match small bits of genetic information. Most mammalian genes have 10 or more exons and there are a number of examples where the same gene specifies multiple protein products through the use of **alternative mRNA processing** pathways. Considering the role of secondary and tertiary RNA structure in determining the many properties of RNA, it is easy to imagine that the accessibility of 5′ and 3′ splice sites to the splicing machinery could be influenced by RNA binding proteins. For example, cell-specific RNA binding proteins could either mask preferred splice sites or change local RNA structure to promote splicing of alternate sites. Moreover, cell-specific regulation of cleavage and polyadenylation reactions have also been shown to produce a variety of altered transcripts. Figure 25.11 summarizes the various forms of alternative mRNA splicing that have been found.

One of the best examples of cell-specific mRNA processing is the alternative splicing and polyadenylation of **calcitonin/CGRP** transcripts in rat thyroid and brain cells. In thyroid cells, cell-specific cleavage and polyadenylation results in a truncated thyroid form of the transcript-encoding calcitonin. The brain form of this primary transcript (CGRP) is alternatively spliced to both remove calcitonin coding sequences and retain 3′ exons unique to CGRP. The pathway of cell-specific calcitonin/CGRP alternative polyadenylation and splicing was determined by isolating and characterizing cDNA sequences from thyroid and brain cells using cDNA cloning techniques such as those described in Box 25.1.

MECHANISMS OF RNA DEGRADATION INVOLVE 5′ AND 3′ EXONUCLEASES

All transcripts have a finite lifetime in the cell. The steady-state level of individual RNA species in a cell is determined by both the rate of transcription and the rate

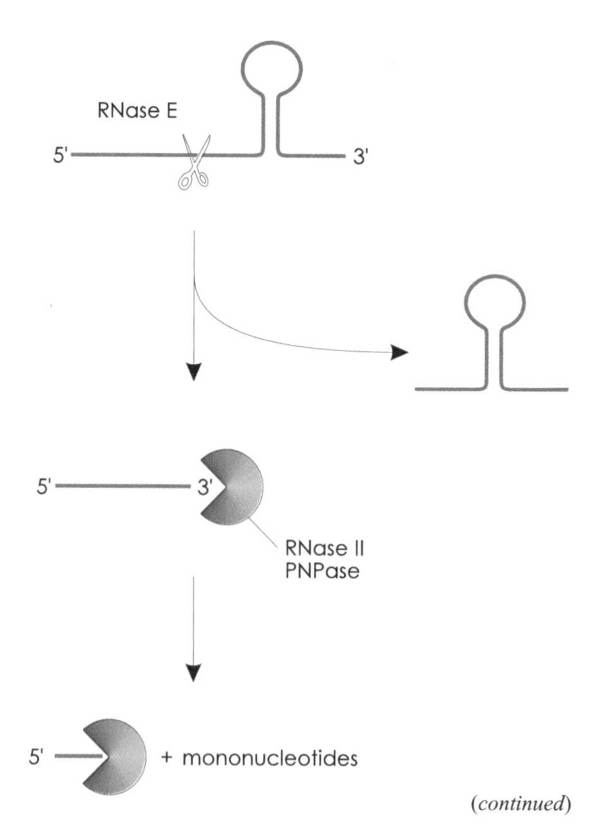

Figure 25.12.

Mechanisms of RNA degradation in (A) bacteria and (B) yeast. The scissors represent endonucleases and the circular objects depict exonucleases. The decapping degradation pathway in yeast is the predominate mechanism.

(continued)

of decay. Studies on bacterial and yeast systems have provided some clues about the role of RNA sequences in protecting transcripts from **exonucleolytic degradation,** and several RNAases have been identified in *E. coli* that are responsible for RNA turnover. The basic mechanism of **RNA decay** in bacteria involves the removal of RNA sequences from the 3′ untranslated region that normally protects the RNA from 3′ to 5′ exonucleolytic attack. Secondary structure predictions indicate that these protective RNA sequences form stable stem–loop structures. As shown in Figure 25.12A, RNaseE is an *E. coli* endonuclease which cleaves off the stem–loop structure, thus exposing the 3′ end of the RNA to exonucleolytic attack by polynucleotide phosphorylase (PNPase) and RNaseII.

Experiments in yeast have shown that progressive shortening of the poly(A) tail is a key step in mRNA degradation. In most cases, poly(A) tail shortening results in decapping at the 5′ end and subsequent degradation by 5′ to 3′ exonucleases. Alternatively, complete removal of the poly(A) tail can sometimes lead to degradation by 3′ to 5′ exonucleases (Fig. 25.12B). It is likely that still other RNA degradation pathways exist in eukaryotes because there is evidence that endonucleolytic cleavages by enzymes, similar to RNaseE in *E. coli,* also play a role.

Regulation of mRNA degradation would be one way to control the amount of protein being synthesized if the rate of transcription did not change. One of the best examples of this is the posttranscriptional regulation of **transferrin receptor** (TfR) mRNA levels. In higher eukaryotes, the control of iron homeostasis is dependent on the RNA binding activity of an iron-binding protein called **iron-responsive-element binding protein** (IRE-BP), which is now known to be the enzyme **cytoplasmic aconitase.** In the presence of high

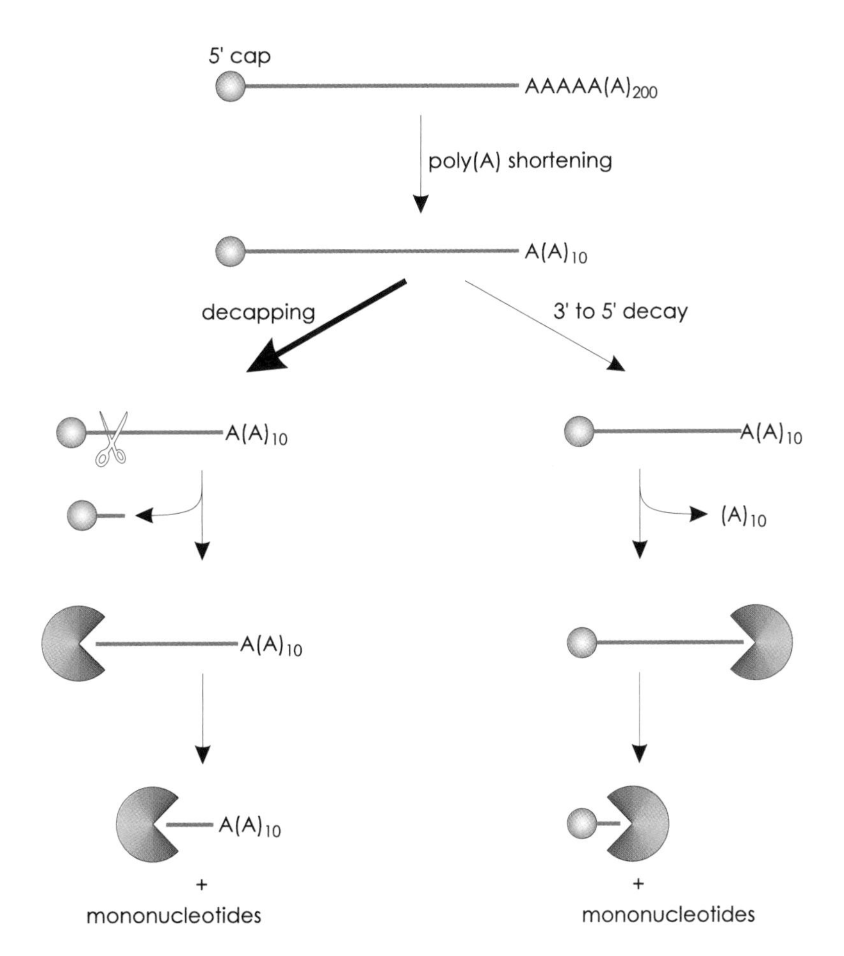

Figure 25.12. (*continued*)

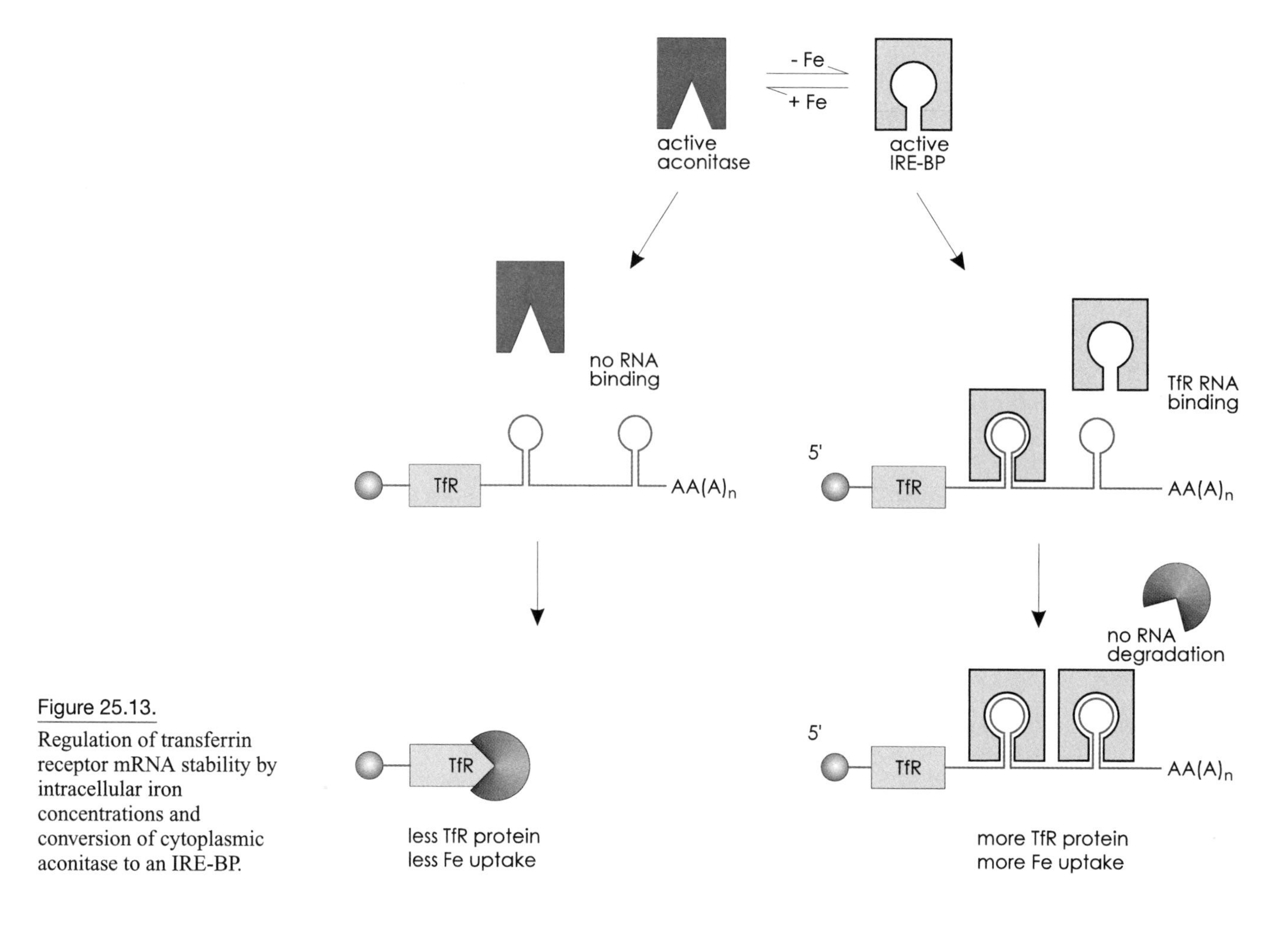

Figure 25.13.
Regulation of transferrin receptor mRNA stability by intracellular iron concentrations and conversion of cytoplasmic aconitase to an IRE-BP.

amounts of iron, cytoplasmic aconitase enzymatic function is active, but it is unable to bind TfR RNA. However, when iron stores are low, cytoplasmic aconitase loses its enzymatic activity and is converted to the active form of IRE-BP, which binds to TfR RNA stem–loop structures called **iron-responsive elements** (IREs). The TfR mRNA contains multiple IREs in the 3′ untranslated region. As shown in Figure 25.13, when cytoplasmic aconitase is bound to TfR IREs, a rate-limiting step in the degradation of TfR mRNA is inhibited.

Cytoplasmic aconitase binding results in an increase in the steady-state level of transferrin receptor mRNA, and subsequently TfR protein. Once the intracellular levels of iron return to normal, IRE binding activity of cytoplasmic aconitase is inactivated and TfR mRNA is rapidly degraded.

BOX 25.1. CHARACTERIZATION OF mRNA TRANSCRIPTS USING RECOMBINANT DNA TECHNIQUES

Isolation of complementary DNA (cDNA)

Ribonucleic acid is a difficult nucleic acid to work with because it is very sensitive to RNases that are prevalent in the lab (your hands are a good source of RNases!) and the techniques for sequencing RNA are indirect. To circumvent these problems, mRNA can be converted into cDNA using an enzyme called **reverse transcriptase** (Chapter 30). Reverse transcriptase is an RNA-dependent DNA polymerase that uses an RNA template to synthesize cDNA. The single-stranded cDNA product of the reverse transcriptase reaction (corresponding to the original template strand in the double-strand genomic DNA) is used in a second reaction with DNA polymerase to synthesize double-stranded DNA (Fig. 25.14). Two common ways to isolate the coding sequences of a gene are to screen a **cDNA library** using radioactive nucleic-acid probes, or to make antibodies against the encoded protein and screen a cDNA library that produces a fusion protein in the host bacteria (an expression library).

mRNA analysis by Northern blotting

Since gene expression is often cell-specific, and some transcripts are differentially spliced in various cell types, mRNA transcripts are often analyzed by a blotting technique similar to Southern blotting for DNA (see Chapter 2). In RNA blotting, called **Northern blotting,** the RNA samples are electrophoresed on agarose gels containing a denaturing reagent such as formaldehyde to disrupt intrastrand base pairing and RNA secondary structures. The RNA is transferred from the gel onto a solid support, such as a nitrocellulose membrane, so that the pattern f the RNA in the gel is preserved on the membrane support as shown in Figure 25.15. The filter membrane containing single-stranded RNA can then be hybridized with single-stranded cDNA probes that have been radioactively labeled in vitro using ^{32}P-dNTPs and DNA polymerase. The washed filter is exposed to film for autoradiography.

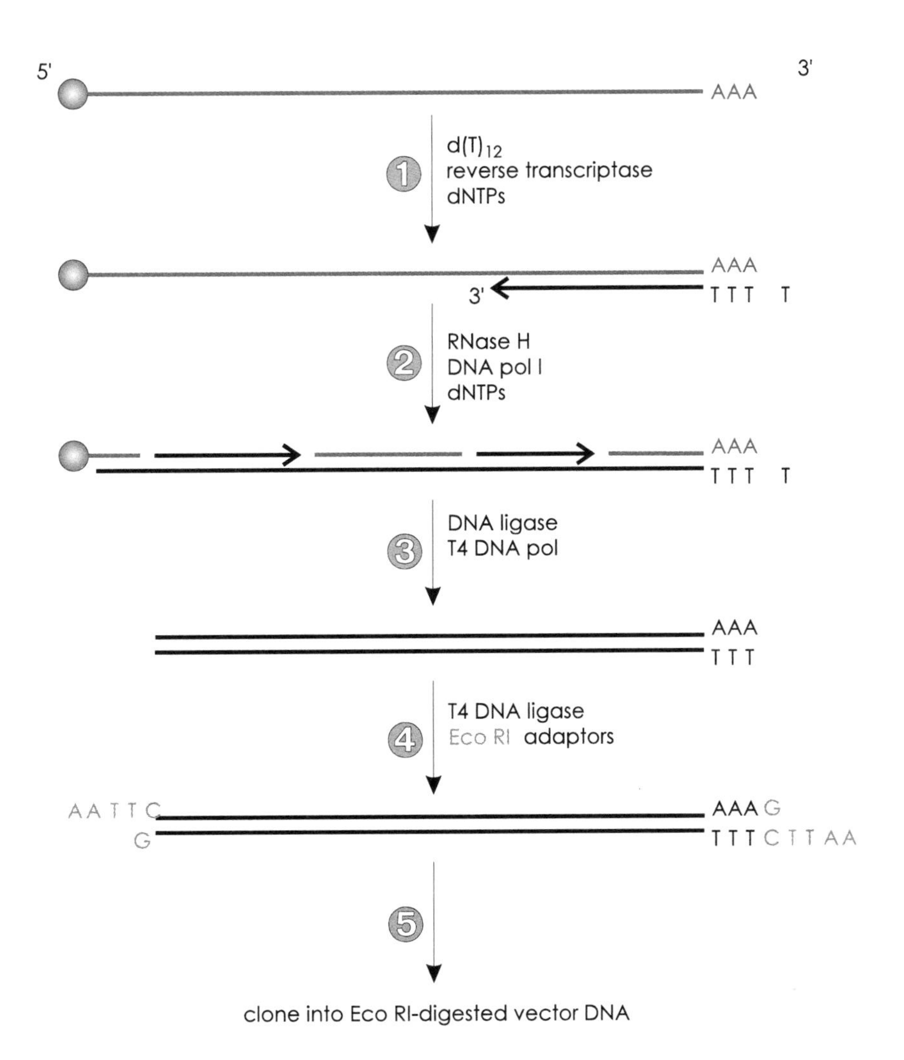

Figure 25.14.

Enzymatic reactions commonly used to synthesize double-stranded cDNA from mRNA and oligonucleotide primers such as oligo d(T). The enzyme reverse transcriptase is an RNA-directed DNA polymerase first discovered in eukaryotic retroviruses. The conversion of RNA to DNA in step 2 is accomplished by combining the endonucleolytic activity of *E. coli* RNase H to create nicks in the RNA–DNA hybrid with the DNA polymerizing activity of *E. coli* DNA pol I, which uses the residual RNA as a primer and the first-strand cDNA as a template. *E. coli* DNA ligase is used in step 3 to repair single strand nicks and T4 DNA polymerase creates blunt ends. Ligation of the Eco RI adaptors to the blunt-ended double-stranded cDNA facilitates cloning into plasmid or lambda DNA vectors that have been cleaved by the restriction enzyme Eco RI.

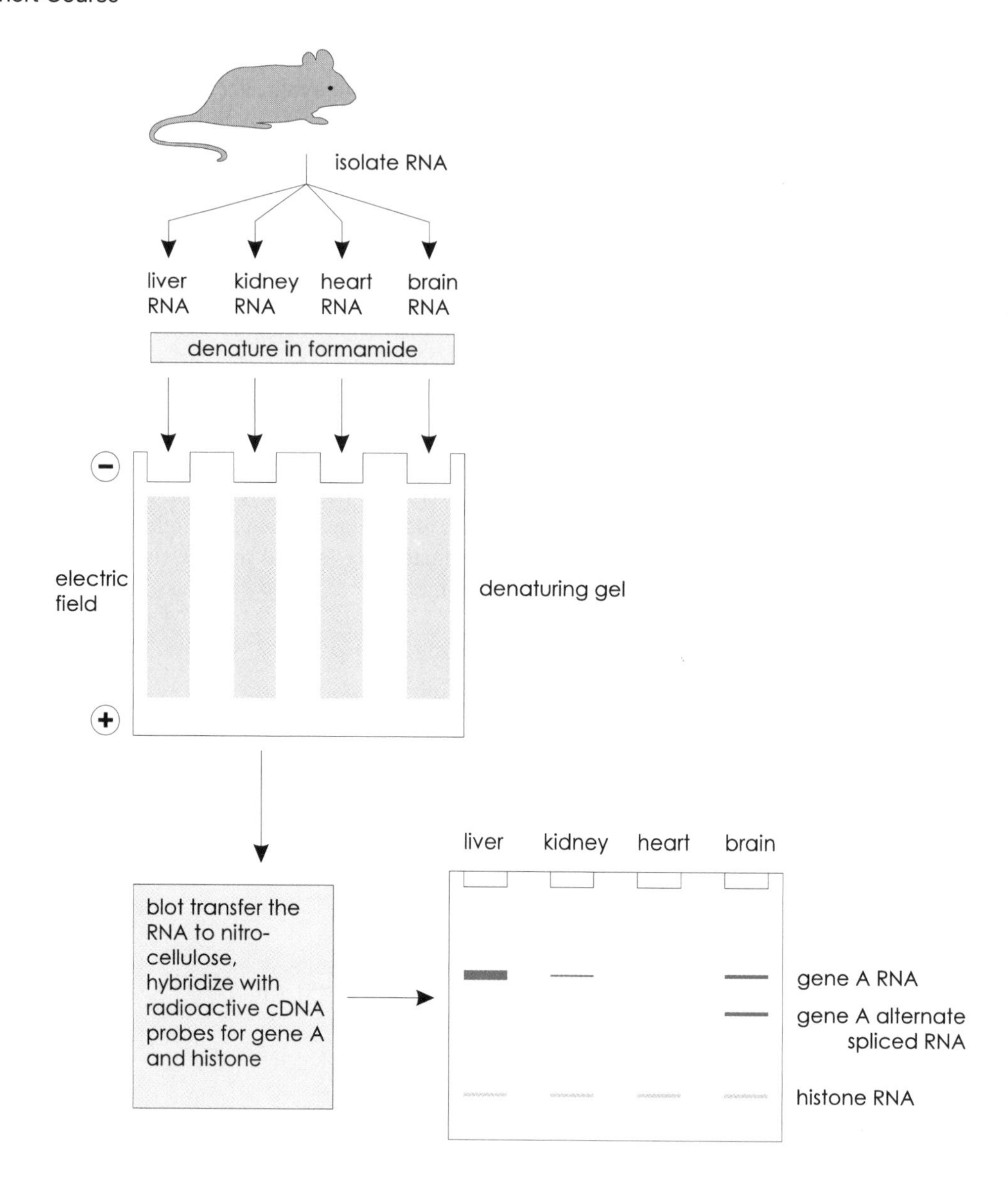

Figure 25.15.
Northern blotting to characterize RNA species in various cell types. The RNA is separated by gel electrophoresis in a denaturing formaldehyde agarose gel, transferred to a nitrocellulose filter, and hybridized with a gene-specific cDNA probe (gene A cDNA). Note that a second probe is also used to detect the presence of RNA molecules, such as histone RNA, that can be used to control for loading differences.

SUMMARY

1. Posttranscriptional processing of primary RNA transcripts includes RNA splicing, 5′ capping, 3′ polyadenylation, and tRNA base modifications. Many of these alterations increase RNA stability and enhance mRNA translation.

2. Group I intron self-splicing is mediated by an external guanine nucleoside which acts as a nucleophile in the first of two transesterification reactions. This protein-free reaction is dependent on intramolecular RNA structures.

3. Group II intron self-splicing utilizes an adenylate residue located in the intron as the nucleophile in the first transesterification step. In both group I and group II self-splicing reactions, the upstream exon–intron boundary is cleaved first and then the 3′ OH of the upstream exon carries out a nucleophilic attack on the downstream exon.

4. The process of hnRNA splicing requires a large snRNP spliceosome complex to facilitate essentially the same two-step mechanism as used for group II self-splicing. In hnRNA splicing, the U6 snRNA serves as the central organizer and is involved in catalyzing the second transesterification step, which releases the intron as a lariat.

5. Ribozymes are catalytic RNA molecules that are able to cleave at specific sites in target RNA mol-

ecules. The hammerhead ribozyme is a type of trans-acting ribozyme that may be useful for in vivo applications as a site-specific RNase.

6. Both rRNA and tRNA processing involves sequential cleavages of large precursor transcripts to yield the functional RNA products. The human 45S rRNA precursor transcript is processed into the mature 28S, 18S, and 5.8S species by poorly defined cleavage reactions. The tRNA processing includes numerous base modifications.

7. Eukaryotic mRNAs contain a 5′7-methylguanosine cap and a 3′ polyadenylate tail, both of which are added by specific enzymes localized to the nucleus. Alternative RNA splicing and polyadenylation contribute to transcript heterogeneity and specificity, and can result in the synthesis of cell-specific proteins.

8. Steady-state levels of RNA can be controlled by balancing rates of synthesis and degradation. Both endoribonucleases and exoribonucleases play a role in RNA degradation pathways. Control of transferrin-receptor mRNA levels by cytoplasmic aconitase/IRE-BP activities is a classic example of regulation of RNA turnover rates in response to physiological signals.

REFERENCES

Beelman C. A., Parker R. (1995): Degradation of mRNA in eukaryotes. *Cell* **81:**179.

Binder R, Horowitz J. A., Basilion J. P., Koeller D. M., Klausner R. D., Harford J. B. (1994): Evidence that the pathway of transferrin receptor mRNA degradation involves an endonucleolytic cleavage within the 3′ UTR and does not involve poly(A) tail shortening. *EMBO J* **13:**1969.

Cohen S. N. (1995): Surprises at the 3′ end of prokaryotic RNA. *Cell* **80:**829.

Flory C. M., Pavco P. A., Jarvis T. C., et al. (1996): Nuclease-resistant ribozymes decrease stromelysin mRNA levels in rabbit synovium following exogenous delivery to the knee joint. *Proc. Natl. Acad. Sci. USA* **93:**754.

Harris M. E., Nolan J. M. Malhotra A., Brown J. W., Harvey S. C., Pace N. R. (1994): Use of photoaffinity crosslinking and molecular modeling to analyze the global architecture of ribonuclease P RNA. *EMBO J* **13:**3953.

Kiss-László, Z., Henry, Y., Bachellerie, et al. (1996): Site-specific ribose methylation of preribosomal RNA: A novel function for small nucleolar RNAs. *Cell* **85:**1077.

Kohtz J. D., Jamison S. F., Will C. L., et al. (1994): Protein–protein interactions and 5′-splice-site recognition in mammalian mRNA precursors. *Nature* **368:**119.

Lou H., Cote G. J., Gagel R. F. (1994): The calcitonin exon and its flanking intron sequences are sufficient for the regulation of human calcitonin/calcitonin gene-related peptide alternative RNA splicing. *Mol. Endocrinol.* **8:**1618.

Manley J. L. (1995): Messenger RNA polyadenylation: A universal modification. *Proc. Natl. Acad. Sci. USA* **92:**1800.

Manley, J.L. and Tacke, R. (1996): SR proteins and splicing control. *Genes Dev.* **10:**1569.

Nilsen T. W. (1994): RNA–RNA interactions in the spliceosome: Unraveling the ties that bind. *Cell* **78:**1.

Pandey N. B., Williams A. S., Sun J.-H., Brown V. D., Bond U., Marzluff W. F. (1994): Point mutations in the stem-loop at the 3′ end of mouse histone mRNA reduce expression by reducing the efficiency of 3′ end formation. *Mol. Cell. Biol.* **14:**1709.

Proudfoot N. J. (1994): Post-transcriptional regulation: Chasing your own poly(A) tail. *Curr. Biol.* **4:**359.

Rossi J. J. (1994): Practical ribozymes: Making ribozymes work in cells. *Curr. Biol.* **4:**469.

Scott W. G., Finch J. T., Klug A. (1995): The crystal structure of an all-RNA hammerhead ribozyme: A proposed mechanism for RNA catalytic cleavage. *Cell* **81:**991.

Sharp P. A. (1994): Split genes and RNA splicing. *Cell* **77:**805.

Sun L.-Q., Warrilow D., Wang L., Witherington C., Macpherson J., Symonds G. (1994): Ribozyme-mediated suppression of Moloney murine leukemia virus and human immunodeficiency virus type I replication in permissive cell lines. Proc *Natl. Acad. Sci. USA* **91:**9715.

Wahle, E. and Keller, W. (1996): The biochemistry of polyadenylation. *Trends Biochem.Sci.* **21:**247.

Zhao J. J., Pick L. (1993): Generating loss-of-function phenotypes of the *fushi tarazu* gene with a targeted ribozyme in *Drosophila. Nature* **365:**448.

REVIEW QUESTIONS

1. The four sequential steps in eukaryotic mRNA processing are
   a) 5′ capping, nuclear export, 3′ polyadenylation, and RNA splicing
   b) RNA splicing, nuclear export, 5′ capping, and 3′ polyadenylation
   c) transcription, 3′ polyadenylation, RNA splicing, and nuclear export

d) 5′ capping, 3′ polyadenylation, RNA splicing, and nuclear export
e) RNA splicing, 5′ capping, 3′ polyadenylation, and nuclear export

2. What is mechanistically similar between group II intron self-splicing and spliceosome-mediated hnRNA splicing?
   a) Nucleophilic attack by an intrastrand adenylate in transesterification step 1
   b) Nucleophilic attack by a free guanine nucleotide in transesterification step 1
   c) Formation of a lariat structure and a 2′-5′ phosphodiester bond at the branch
   d) Nucleophilic attack by 3′ hydroxyl of the 5′ exon in transesterification step 2
   e) A, c, and d are all correct

3. One of the roles of snRNP-associated proteins in spliceosome function appears to be to
   a) perform the endonucleolytic cleavage in transesterification step 1
   b) protect the snRNA components from degradation
   c) form complementary base pairs with GU– AG sequences at splice boundaries
   d) promote proper folding of snRNAs to facilitate RNA-mediated catalysis
   e) facilitate nuclear export of processed mRNA molecules

4. Most all eukaryotic hnRNAs are posttranscriptionally modified by the
   a) addition of a CCA trinucleotide to the 3′ end
   b) exonucleolytic removal of the 5′ leader sequence
   c) addition of a 7-methyl guanosine to the 5′ end
   d) removal of exons and splicing together of introns
   e) addition of ~200 adenosines to the 5′ end

5. The steady-state level of RNA in the cell is controlled both by RNA synthesis rates and by RNA turnover. What answer best explains how an alternate polyadenylation site might be utilized to control RNA stability?
   a) RNA structures that protect 3′ ends may be included/excluded.
   b) 3′ Polyadenylation tails might be longer/ shorter.
   c) Transcriptional termination might be stimulated/inhibited.
   d) mRNAs may contain/lack a 7-methylguanosine cap.
   e) Nuclear export might be more random/selective.

6. The discovery and subsequent utilization of reverse transcriptase revolutionized modern molecular biology *most significantly* by
   a) facilitating the construction of YAC libraries for the Human Genome Project
   b) providing a therapeutic target for viral infections through drug discovery
   c) creating a means to synthesize RNA from DNA in vitro
   d) leading to the birth of the biotechnology industry through cDNA cloning
   e) providing an easy way to design longer ribozymes

ANSWERS TO REVIEW QUESTIONS

1. ***d*** Both 5′ capping and 3′ polyadenylation precede splicing and export.

2. ***e*** The biochemistry of group II intron self-splicing and spliceosome-mediated hnRNA splicing differs only in the requirement for cis or trans components, respectively.

3. ***d*** The snRNA molecules (e.g., U6, U2, and U5) form the catalytic center in the spliceosome and snRNP RNA structures are likely influenced by snRNP proteins.

4. ***c*** The 5′ capping is required for RNA stability and mRNA translation.

5. ***a*** The RNA stability is often controlled by the presence/absence of stem–loop structures located in the 3′ end of RNA transcripts.

6. ***d*** The ability to express proteins from cloned cDNA is a major focus of many biotechnology companies.

CHAPTER 26

PROTEIN SYNTHESIS

INTRODUCTION

Protein synthesis is the most-energy consuming process in the cell, requiring as much as 90% of the ATP a cell generates. To maintain the fidelity of protein synthesis, there are a number of important quality-control checkpoints that result in a high level of accuracy. In this chapter we describe information transfer from mRNA to the primary amino-acid sequence of proteins via "adaptor" tRNA molecules. We also review how the ribosome functions as an efficient ribonucleoprotein machine, moving from the 5′ end to the 3′ end of the mRNA. Although the basic mechanisms involved in protein synthesis have been known for almost 30 years, recent discoveries have shown that some of our earlier assumptions were wrong (e.g., the peptidyl transferase is not a protein but a ribozyme). In addition, molecular analyses of protein synthesis have shown that there are indeed exceptions to the general dogma of molecular biology, three examples of which are mRNA editing, translational frameshifting, and protein splicing.

OVERVIEW OF PROTEIN SYNTHESIS

There are two steps in protein synthesis where polarity of information is important. The first is the relationship between the 5′ to 3′ directionality of mRNA, and the NH_3^+ to COO^- terminal direction of protein synthesis. The utilization of tRNA as the adaptor is the second step where polarity of information is crucial. The tRNA has a bipolar function, it needs to correctly link each amino acid to the corresponding position encoded by the mRNA. Figure 26.1 shows an overview of how mRNA synthesis and protein translation share the same polarity. Moreover, similar to transcription, translation can also be broken down into three discrete components; initiation, elongation, and termination.

Soon after the structure of DNA was discovered, Francis Crick hypothesized that an adaptor molecule, such as the one shown in Figure 26.2, would be required for protein synthesis. It was discovered that **tRNA serves as the adaptor molecule** by linking the information stored in mRNA to the primary sequence of the polypeptide. The adaptor function of tRNA is mediated by base pairing between mRNA sequences called **codons,** and complementary sequences on tRNA called **anticodons.** For every codon on the mRNA, a single amino acid is delivered by the tRNA to the growing polypeptide chain. The term **genetic code** refers to the specific sequences in the mRNA codon that determine which tRNA molecule is going to have the complementary anticodon and, therefore, what amino acid is required at that position in the final polypeptide chain.

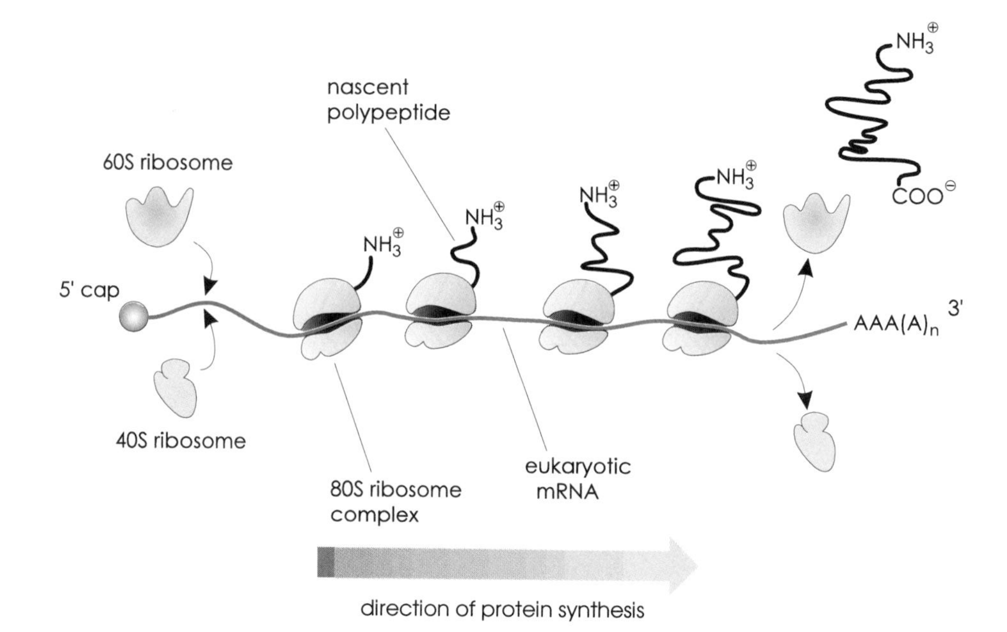

Figure 26.1.
Overview of the flow of information from mRNA to protein. The polarity of mRNA and protein sequences are analogous in that both are synthesized from left to right.

In addition to mRNA and tRNA, the third major class of RNA molecule required for protein synthesis is rRNA. Together with as many as 70 ribosomal proteins, rRNA folds into a two-subunit macromolecule complex called a **ribosome** (Chapter 5). In bacteria, the ribosomes attach to mRNA as it is being synthesized, thereby coupling transcription and translation. In eukaryotes, protein synthesis occurs in the cytoplasm, either by free ribosomes in the cytosol or by membrane-bound ribosomes associated with the endoplasmic reticulum. The differences between prokaryotic and eukaryotic protein synthesis are illustrated in Figure 26.3.

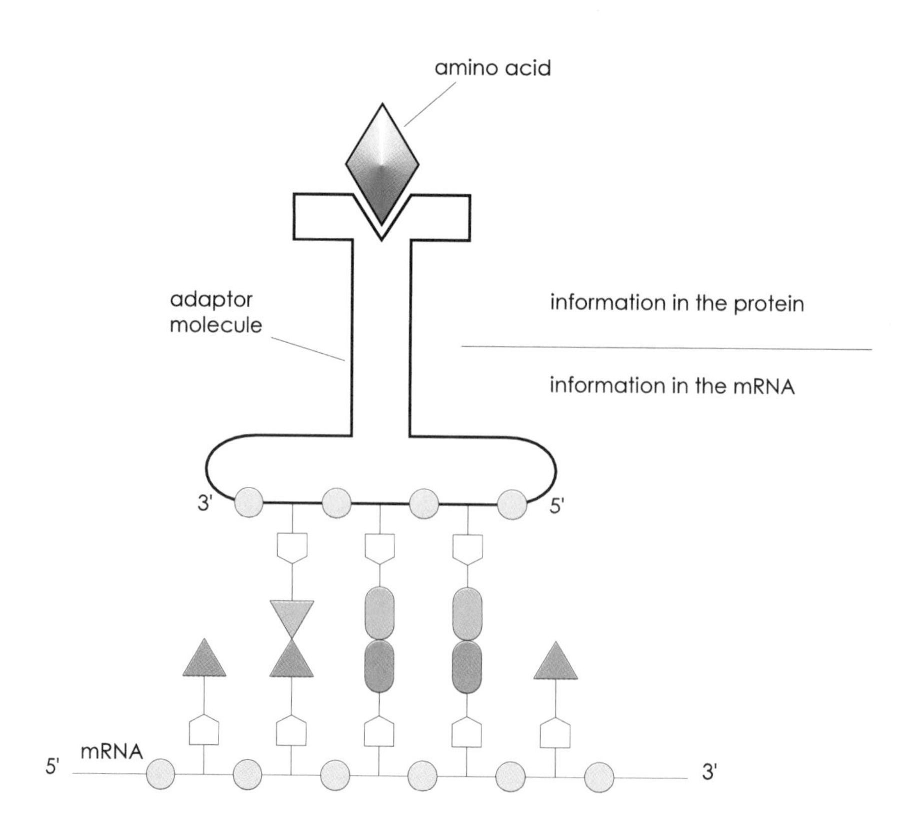

Figure 26.2.
Translation requires a bifunctional adaptor molecule which bridges the primary sequence information in mRNA and protein.

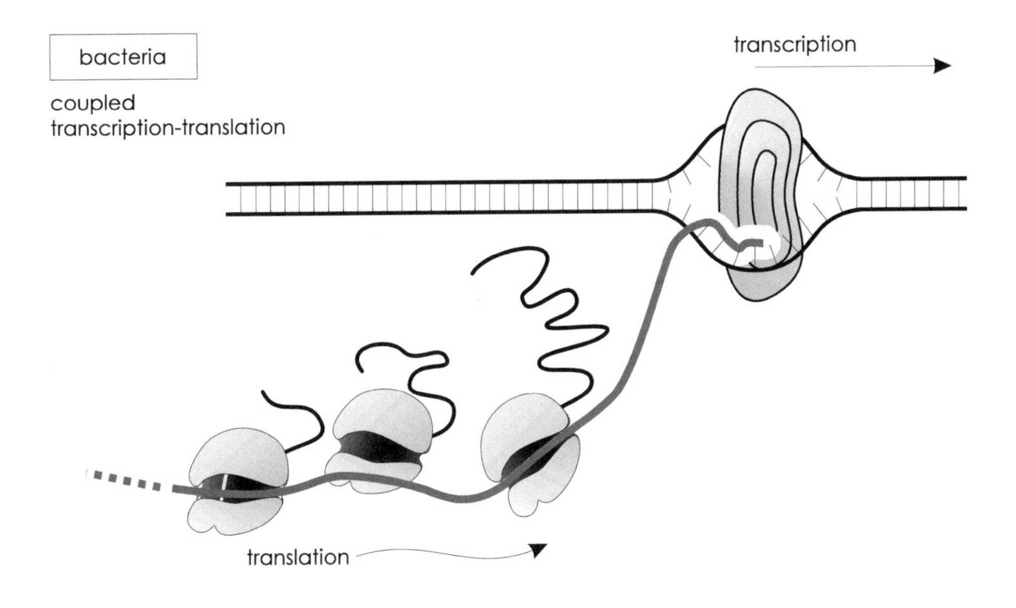

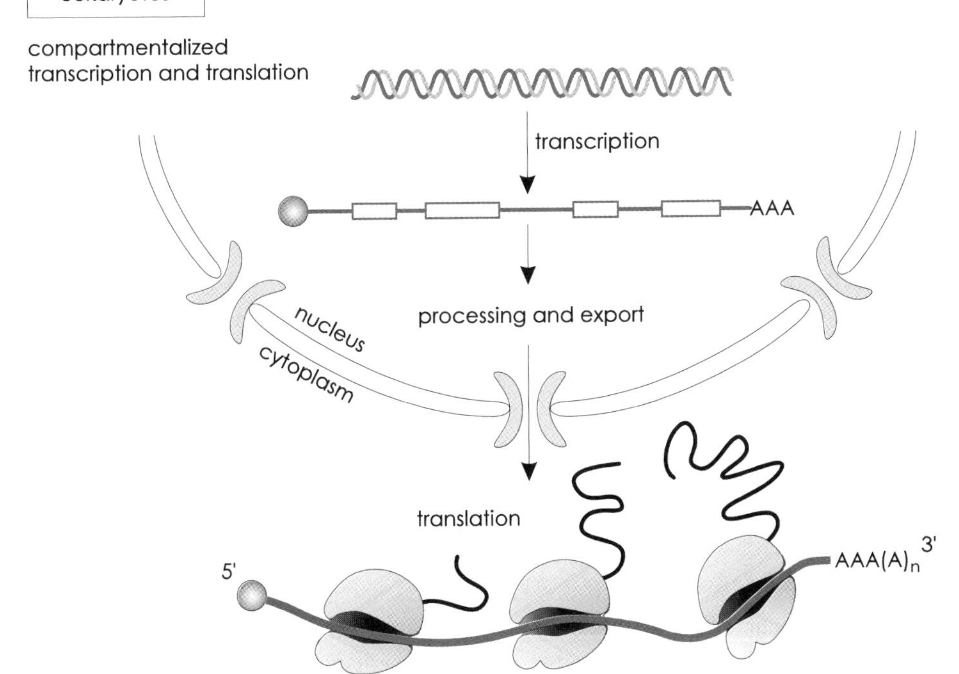

Figure 26.3.
Translation and transcription are coupled in bacteria, whereas in eukaryotes, these two processes are compartmentalized.

THE GENETIC CODE IS BASED ON TRIPLETS

Simple arithmetic indicates that the genetic code cannot be binary since there are only four bases in RNA (A, G, C, and U), yet 20 different amino acids (19 amino acids and 1 imino acid) are known to exist in cells. Therefore, the number of possible combinations using just two bases in the codon is not enough. For example, four bases in the first position and four bases in the second position would only be sufficient to encode 16 possible amino acids ($4 \times 4 = 16$). However, a ternary- (or triplet-) based code would be more than enough by specifying 64 possible amino acids ($4 \times 4 \times 4 = 64$).

Three different experimental approaches were taken in the early 1960s to prove that the genetic code is based on a continuous string of **triplet ribonucleotides** in the mRNA. The general idea for each experiment is illustrated in Figure 26.4.

In the first experiment, a synthetic **homopolymer**

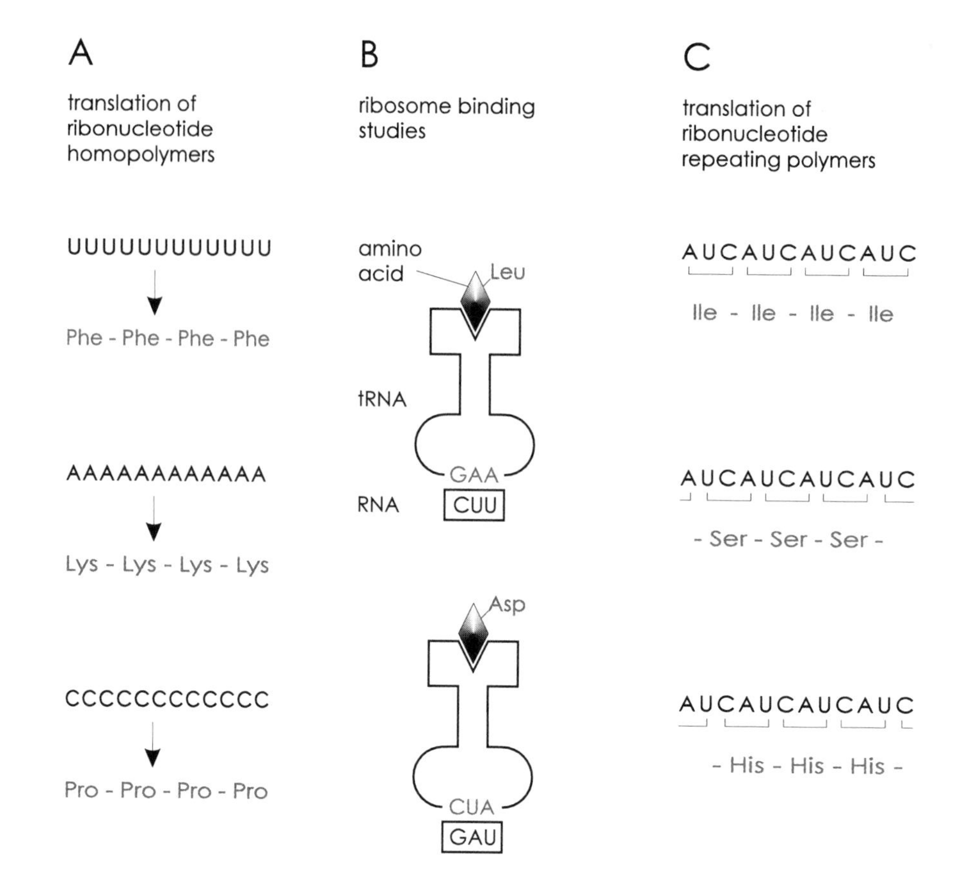

Figure 26.4.
Three different experiments were done to demonstrate that the genetic code is based on triplets. These include the use of (A) ribonucleotide homopolymers, (B) ribosome binding studies, and (C) translation of ribonucleotide repeating polymers.

of RNA was synthesized and used as a template for in vitro protein synthesis using *E. coli* protein extracts and GTP. It was found that poly(U) resulted in the synthesis of poly(Phe), poly(A) encoded poly(Lys), and poly(C) specified poly(Pro) (Fig. 26.4A). By mixing various ratios of the ribonucleotides together, it was possible to make randomized RNA heteropolymers which directed the synthesis of polypeptides that incorporated more than one amino acid.

In the second experiment, ribonucleotide trimers of a known sequence were synthesized and incubated with *E. coli* ribosomes and aminoacylated tRNAs. By knowing the trinucleotide sequence used in the experiment, and by identifying which amino acid was attached to the tRNA, it was possible to identify 50 trinucleotides that corresponded to all 20 possible amino acids (Fig. 26.4B). This was the first good evidence that multiple codons could be used to specify the same amino acids.

The third approach used **repeating ribonucleotide polymers** containing known repeating sequences (Fig. 26.4C). When these were used as templates for in vitro protein synthesis, it was found that each ribonucleotide polymer could specify as many as three different repeating polypeptide products. Of the 64 possible codons, 61 were found to specify amino acids and 3 were later defined as termination codons. Figure 26.5 shows the genetic code for protein synthesis in *E. coli,* which is for the most part applicable to mRNA translation in mammalian cells. Note that methionine and tryptophan are only specified by single codons, whereas almost all the other amino acids have from two to four codons (except leucine and serine which are encoded by six codons). Methionine is the first amino acid in essentially all proteins; however, methionine residues are also found within the polypeptide sequence. The amino-terminal methionine is called the initiator methionine.

THE WOBBLE POSITION IN CODON–ANTICODON BASE PAIRING

Close inspection of the genetic code suggested to Francis Crick that the third base in the triplet codon may be more flexible than the first two. For example, the four codons for proline all begin with two cytosines, but can end in U, C, A, or G (CCU, CCC, CCA, or CCG). In

first nucleotide	second nucleotide: U	second nucleotide: C	second nucleotide: A	second nucleotide: G
U	UU*U* Phe	UC*U* Ser	UA*U* Tyr	UG*U* Cys
	UU*C* Phe	UC*C* Ser	UA*C* Tyr	UG*C* Cys
	UU*A* Leu	UC*A* Ser	UA*A* stop	UG*A* stop
	UU*G* Leu	UC*G* Ser	UA*G* stop	UG*G* Trp
C	CU*U* Leu	CC*U* Pro	CA*U* His	CG*U* Arg
	CU*C* Leu	CC*C* Pro	CA*C* His	CG*C* Arg
	CU*A* Leu	CC*A* Pro	CA*A* Gln	CG*A* Arg
	CU*G* Leu	CC*G* Pro	CA*G* Gln	CG*G* Arg
A	AU*U* Ile	AC*U* Thr	AA*U* Asn	AG*U* Ser
	AU*C* Ile	AC*C* Thr	AA*C* Asn	AG*C* Ser
	AU*A* Ile	AC*A* Thr	AA*A* Lys	AG*A* Arg
	AU*G* Met	AC*G* Thr	AA*G* Lys	AG*G* Arg
G	GU*U* Val	GC*U* Ala	GA*U* Asp	GG*U* Gly
	GU*C* Val	GC*C* Ala	GA*C* Asp	GG*C* Gly
	GU*A* Val	GC*A* Ala	GA*A* Glu	GG*A* Gly
	GU*G* Val	GC*G* Ala	GA*G* Glu	GG*G* Gly

Figure 26.5.
Table of the genetic code showing the 61 amino acid codons and three termination codons (STOP). The codon AUG specifies methionine (Met) and is often referred to as the initiator codon.

addition, the tRNA anticodon often contains the nucleotide inosine (produced by deamination of adenine) which base pairs with the third nucleotide of the mRNA codon. Inosine is capable of forming hydrogen bonds with U, C, or A. Based on these observations, Crick proposed the **Wobble Hypothesis,** which states that the first two positions of the codon follow strict Watson–Crick base pairing (G-C and A-U), but the third position is able to form nonconventional base pairs if the tRNA has a I, G, or U in the corresponding anticodon position. The basis of the Wobble Hypothesis is illustrated in Figure 26.6.

The best way to conceptualize the principle of the Wobble Hypothesis is from the perspective of the tRNA anticodon, as this emphasizes the use of a single tRNA species to recognize more than one mRNA codon. The *first* base in the tRNA anticodon corresponds to the *last* base in the mRNA codon, which is referred to as the **wobble position.** For example, tRNAs with an inosine in the 5′ position of the anticodon have the greatest flexibility and can bind to as many as three different codons (Fig. 26.6). The Wobble Hypothesis was confirmed experimentally using **ribosome binding assays,** as shown in Figure 26.4b. Note that based on the Wobble Hypothesis, only 32 tRNAs would be necessary to translate all 61 codons. Most genetic systems are found to have more than the minimum of 32 different tRNAs, although usually less than the 61 possible tRNAs.

Figure 26.6.

The principle of the Wobble Hypothesis, showing the various codon/anticodon base pairings that are possible when the first position in the tRNA anticodon (5′ to 3′) is a U, G, or I.

PREDICTING OPEN-READING FRAMES BASED ON cDNA SEQUENCE

Just as sentences have capital letters to denote the first word, and periods are used to mark the end of a sentence, the genetic code specifies when a protein sequence should begin and when it should end. This is important because in the triplet code there are three possible **reading frames** which can be used, depending on the positioning of the tRNA. Figure 26.7A shows how only one of three triplet registers contains an initiator methionine codon in an **open-reading frame** which is not blocked by one of the three termination codons (UAA, UAG, UGA).

By using the genetic code, it is possible to infer an amino-acid sequence of a putative protein based on the analysis of a cloned cDNA sequence. As shown in Figure 26.7B, computer algorithms can rapidly scan a

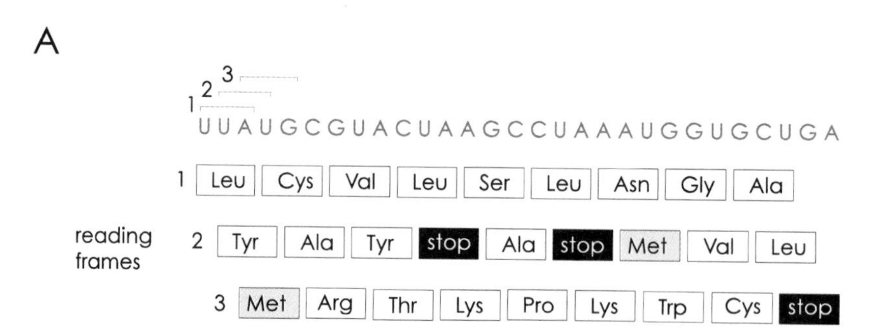

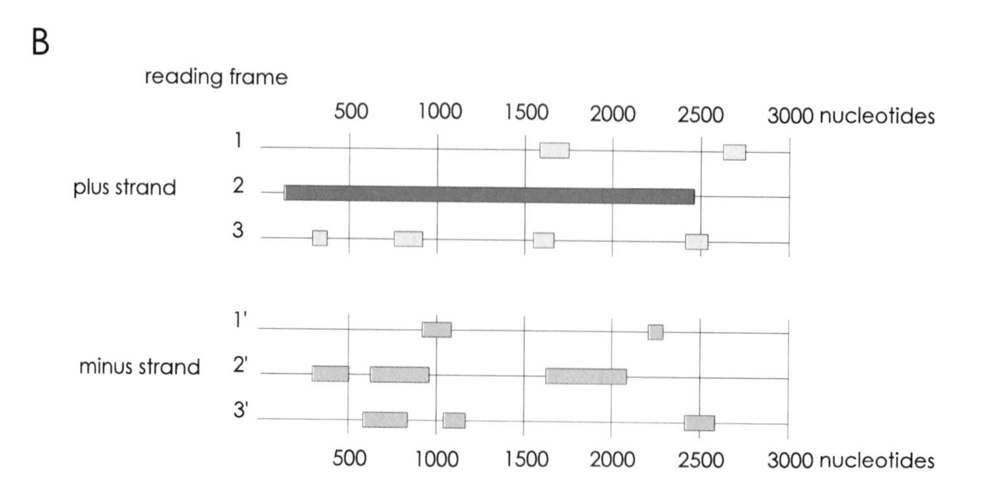

Figure 26.7.

The genetic code allows for three open-reading frames depending on the codon register. (A) Deciphering a single nucleotide sequence into the three possible open-reading frames demonstrates that only reading frame 3 encodes a protein beginning with a methionine and ending with a termination codon. (B) Computer analysis of cDNA sequences is an easy way to identify potential open-reading frames. In this example, only reading frame 2 on the plus cDNA strand of the human glucocorticoid receptor-coding sequence was found to contain the appropriate methionine and termination codons to specify a ~85 kDa protein (777 amino acids).

2000-nucleotide cDNA sequence to identify open reading frames that begin with an AUG codon and end with one of the three termination codons. Since most proteins are usually 15 kDa or larger, and terminators would occur about once every 20 codons by chance alone (3/64), open-reading frames in cloned cDNA sequences are usually defined as 100 codons or more. Note that searching for methionine residues is not a reliable way to find the bona fide initiator methionine codon because most proteins contain methionines within the coding sequence.

EXCEPTIONS TO THE CENTRAL DOGMA OF MOLECULAR BIOLOGY

Until recently, the only deviations from the genetic code in eukaryotes were found in mitochondria and chloroplast, where some codons are different and the wobble position is even more flexible (mitochondria use only 24 types of tRNA to translate 13 mitochondrial proteins). However, three phenomenon, translational frameshifting, RNA editing, and protein splicing, remind us that nature is full of surprises.

An example of **translational frameshifting** is found in the synthesis of the retroviral proteins gag and pol (see Chapter 30). Most of the time, the single retroviral transcript is translated to produce a gag protein which terminates at a UAG codon and is encoded in a reading frame offset by one position relative to the pol protein coding sequence. However, about 5% of the time, the ribosome "stutter steps" just prior to the gag UAG stop codon and then resumes translation in the -1 reading frame to create a large gag–pol fusion protein. This fusion protein is then cleaved by an endoprotease which releases the active reverse transcriptase. Since much more gag protein is needed for viral packaging than is the polymerase for viral replication, this occasional translational frameshifting is one way to regulate the relative levels of each protein with the most efficient use of coding information.

The term **RNA editing** refers to posttranscriptional alternations in the mRNA sequence prior to translation. In *trypanosome* mitochondrial mRNA, U residues are inserted or deleted from the cytochrome oxidase subunit II mRNA in a mechanism involving a guide RNA. This was found by comparing the sequence of the gene in the mitochondrial genome and the cloned cytochrome oxidase subunit II cDNA. Another type of RNA editing is in the cell-specific base modification of apolipoprotein B mRNA. In the intestine, a site-specific cytidine deaminase converts a CAA glutamine codon to a UAA termination codon. The full-length liver form of the protein (Apo B100) and the truncated intestinal form of apolipoprotein B (Apo B48) have distinct functional properties.

A third exception to the Central Dogma of molecular biology is a process known as **protein splicing** in which an internal polypeptide segment, called an intein, is removed by a peptide cleavage reaction to yield a branched intermediate. The polypeptide chains bordering the intein, are then covalently re-attached through a new peptide bond. The first reported example of protein splicing came from the analysis of a yeast ATPase gene which predicted an open reading frame that was 50 kDa larger than the 69 kDa ATPase. It was found that a 119 kDa precursor protein was first synthesized, but then underwent protein splicing to produce the active form of the ATPase.

tRNA IS THE MOLECULAR ADAPTOR IN PROTEIN SYNTHESIS

The attachment of amino acids to tRNAs by aminoacyl tRNA synthetases is called **tRNA charging.** A tRNA lacking an activated amino acid is rejected by the translational machinery and considered "uncharged." Several recurring themes have emerged from studying the structure of charged tRNAs. First, the anticodon and the amino acid are **physically separated** from each other and can be visualized as being at two ends of an inverted "L." Second, many of the bases in tRNAs are **modified posttranscriptionally**; some of these modifications are uniform among all tRNAs, whereas others are specific for each tRNA. Third, the structural data reveal that there are **four intrastrand RNA–RNA helices** which are formed by standard (G-C, A-U) and nonstandard (G-U) base pairing. Figure 26.8 shows the conventional two-dimensional "cloverleaf" configuration of a typical tRNA (see also the tertiary structure of tRNA in Chapter 2). Based on the characterization of a large number of tRNAs, it is clear that both the **anticodon sequence** (with base modifications), and the nucleotides present in the backbone of the tRNA molecule (also with modifications), are sufficiently different to distinguish the tRNAs from each other without grossly affecting the overall structure. Note that the 3′ end contains the **sequence CCA** and that the amino acid is linked to the adenine residue. As described in Chapter 25, this terminal CCA sequence can be added posttranscriptionally by the enzyme **tRNA**

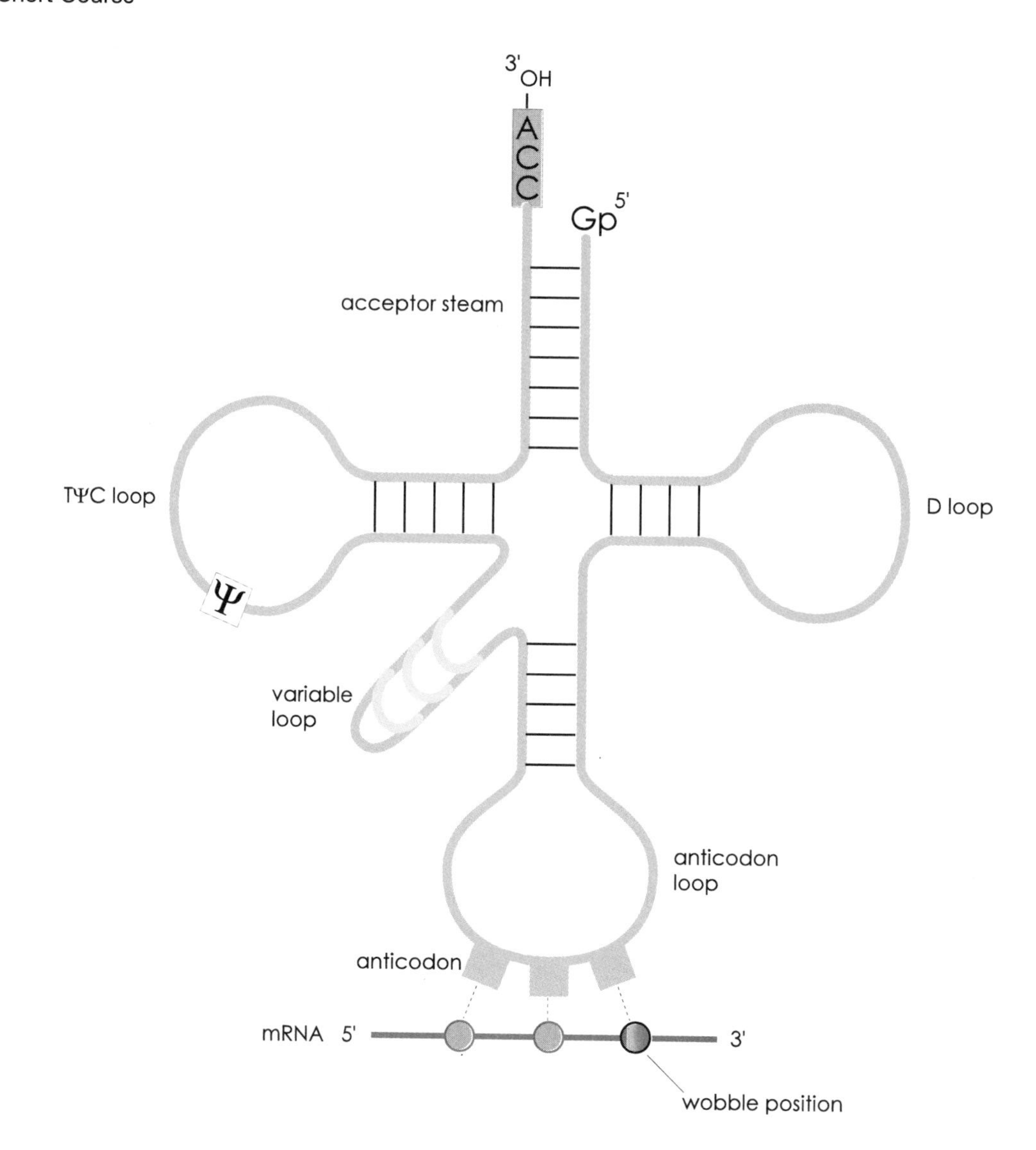

Figure 26.8.
A schematic drawing of prototypical tRNA in the two-dimensional clover leaf configuration. Most tRNAs are 74–95 nucleotides in length, and differ by the number of nucleotides in the variable loop. The TΨC loop is named for the presence of an invariant pseudo-uridine base modification (Ψ) and the D loop contains several dihydrouridine residues.

nucleotidyl transferase if it is not already present in the tRNA sequence.

AMINO ACIDS ARE ATTACHED TO tRNAs BY SPECIFIC AMINOACYL-tRNA SYNTHETASES

There are 20 different **aminoacyl-tRNA synthetases,** the enzymes that covalently link the amino acid to the correct tRNA. We will denote the tRNA species as **$tRNA^{AA}$** and the charged tRNA as **$AA\text{-}tRNA^{AA}$**. Charging must be a very accurate system because the measured error rate (when a tRNA is charged with the wrong amino acid) is less than 10^{-4}. Aminoacyl-tRNA synthetases are bifunctional in that they are capable of distinguishing between all 20 amino acids, and the full complement of cellular tRNAs. In *E. coli,* 20 aminoacyl-tRNA synthetases have been identified (one for each amino acid), indicating that related tRNAs, such as the six possible $tRNA^{Leu}$ species, must be recognized by the same enzyme. A two-step reaction is required for the **aminoacylation** of tRNAs:

(1) Amino acid + ATP $\leftrightarrow$ amino acid-AMP + PP_i ($PP_i \rightarrow 2P_i$)
(2) Amino acid-AMP + tRNA $\rightarrow$ amino acid-tRNA + AMP

The product of the first reaction, amino acid-AMP, remains bound to the enzyme until the second reaction is completed. This reaction is physiologically irreversible with a standard free energy of hydrolysis = –29 kJ/mol, owing to the hydrolysis of two high-energy phosphate bonds.

Figure 26.9 shows these two reactions in more detail. In the first step, the carboxyl group of the amino acid forms an anhydride bond by reacting with the α-phosphate of ATP and displacing pyrophosphate. This aminoacyl adenylate remains bound in the active site of the enzyme, where it is then used in the second reaction that transfers the aminoacyl group to either the 2′ or 3′ OH (depending on the amino acid) of the terminal adenylate residue on the incoming tRNA.

Since there is no "proofreading" step built in to the elongation reactions of protein synthesis, it is critical that the tRNA charging reaction be error free. There are basically two ways that the fidelity of tRNA charging is monitored by aminoacyl-tRNA synthetases. The first mechanism requires that the correct tRNA have a high enough affinity for the tRNA binding site that allows a conformational change in the enzyme to accelerate the aminoacylation step (reaction 2 in Fig. 26.9). Recent molecular genetic analyses of tRNAs have helped to define the tRNA bases that are critical for this binding specificity and affinity. This set of rules has be-

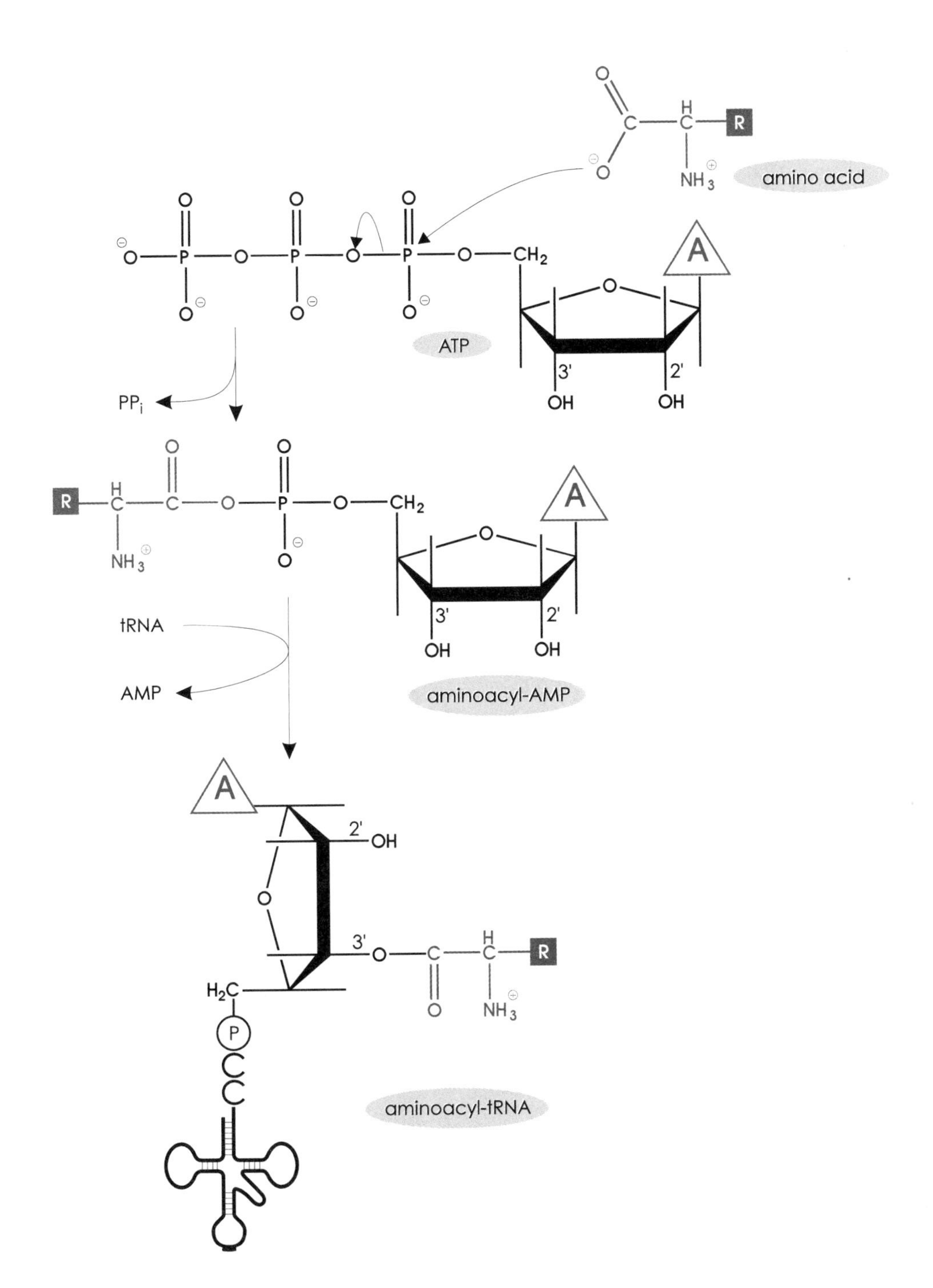

Figure 26.9.

The two-step reaction of tRNA aminoacylation by aminoacyl tRNA synthetases.

come known as the **second genetic code** because they are dependent on the sequence of specific bases in the tRNA backbone which are critical for enzyme tRNA interactions. Studies both in vitro and in vivo with all 20 of the *E. coli* aminoacyl-tRNA synthetases have shown that they require at least two separate regions of the tRNA for full recognition. For most of the aminoacyl-tRNA synthetases, this means that sequences in both the anticodon and in the acceptor stem are required for correct recognition. As with many examples of substrate or ligand recognition, multiple independent contact points are one way to insure the greatest level of specificity.

The second mechanism to eliminate errors is at the level of amino-acid discrimination. The incorrect amino acid is removed either by hydrolyzing the aminoacyl adenylate compound after step 1 but before step 2, or as is the case with some aminoacyl-tRNA synthetases, the incorrect aminoacyl-tRNA is formed, but then hydrolyzed immediately after it is formed.

THE THREE STAGES OF POLYPEPTIDE SYNTHESIS

Counting the two phosphate bonds that are hydrolyzed during the charging of each tRNA, and the two additional phosphate hydrolysis steps required for each cycle of elongation (hydrolysis of 2GTP $\rightarrow$ 2GDP + $2PP_i$), the minimum net energy expenditure of protein synthesis is –101 kJ/mol for every peptide bond formed during protein synthesis.

In this section, we describe the three basic stages of protein synthesis; initiation, elongation, and termination. These three processes are fairly similar between prokaryotes and eukaryotes, with the two exceptions being that more protein factors have been identified as necessary for eukaryotic protein synthesis, and that transcription and translation are physically linked in prokaryotes but not in eukaryotes. Note that the reactions will be schematized as a single ribosome transversing the mRNA, but as shown in Figure 26.3, translation actually occurs on polyribosomes.

Initiation

The first step in translation is the assembly of the initiation complex. In bacteria, this is done in a three-step process that involves the **initiation factors** IF-1, IF-2, and IF-3, the **30S and 50S ribosomal subunits,** and the **hydrolysis of GTP.** As shown in Figure 26.10, the first step is the formation of a preinitiation complex consisting of IF-1, IF-3, and the 30S ribosomal subunit. This preinitiation complex binds with high affinity to a translational start site on the mRNA transcript which is close to the 5′ end and includes the first codon of the open-reading frame. This codon is called the **initiation codon** and it is usually **AUG,** which specifies the amino-acid methionine in the genetic code (Fig. 26.6).

In the second step, a complex consisting of the appropriate charged tRNA and IF-2 bound with GTP together form the initiation complex by base pairing between the tRNA anticodon and the mRNA codon. The first tRNA utilized in bacterial protein synthesis is the initiator **fMet-tRNA$^{\mathbf{fMet}}$** (the aminoacylated form of tRNAf^{Met}). This special initiator tRNA is described in the next section. In the third step, GTP is hydrolyzed to GDP; IF-1, IF-2, and IF-3 dissociate; and the large 50S subunit binds to form the functional initiation complex. These three steps are summarized in Figure 26.10.

As shown in Figure 26.10, a purine-rich sequence, called the **Shine–Dalgarno sequence,** is centered ~10 nucleotides upstream of the AUG codon in most bacterial mRNAs. It was discovered that the 3′ end of bacterial 16S rRNA has a complementary sequence which can base pair with the Shine–Dalgarno sequence to position the 30S subunit relative to the initiator AUG codon. **Ribosome positioning** in eukaryotes is more difficult to predict, but there are two general rules for identifying the most likely eukaryotic AUG initiator codon: (1) it is often the first AUG encountered by the ribosome after it binds the 5′ mRNA cap and "scans" in the 3′ direction and (2) there are preferred contexts for the initiator AUG which have been defined by the consensus sequence GCCPuCCAUGG.

Is there anything special about the aminoacyl-tRNA that is used in the formation of the initiation complex? Studies in bacteria have shown that there is a difference between the initiator Met-tRNA^{Met} and a Met-tRNA^{Met} which inserts methionine residues during elongation. In *E. coli,* there are two tRNA^{Met} molecules, one for initiator methionines and one for internal methionines. The initiator tRNA, called tRNA^{fMet}, is first charged with methionine and then this aminoacyl-tRNA is recognized by the enzyme tRNA^{fMet} formyl transferase, which formylates the NH_2 group on methionine in a reaction using 10-formyl tetrahydrofolate.

The product of this reaction, fMet-tRNA^{fMet}, is

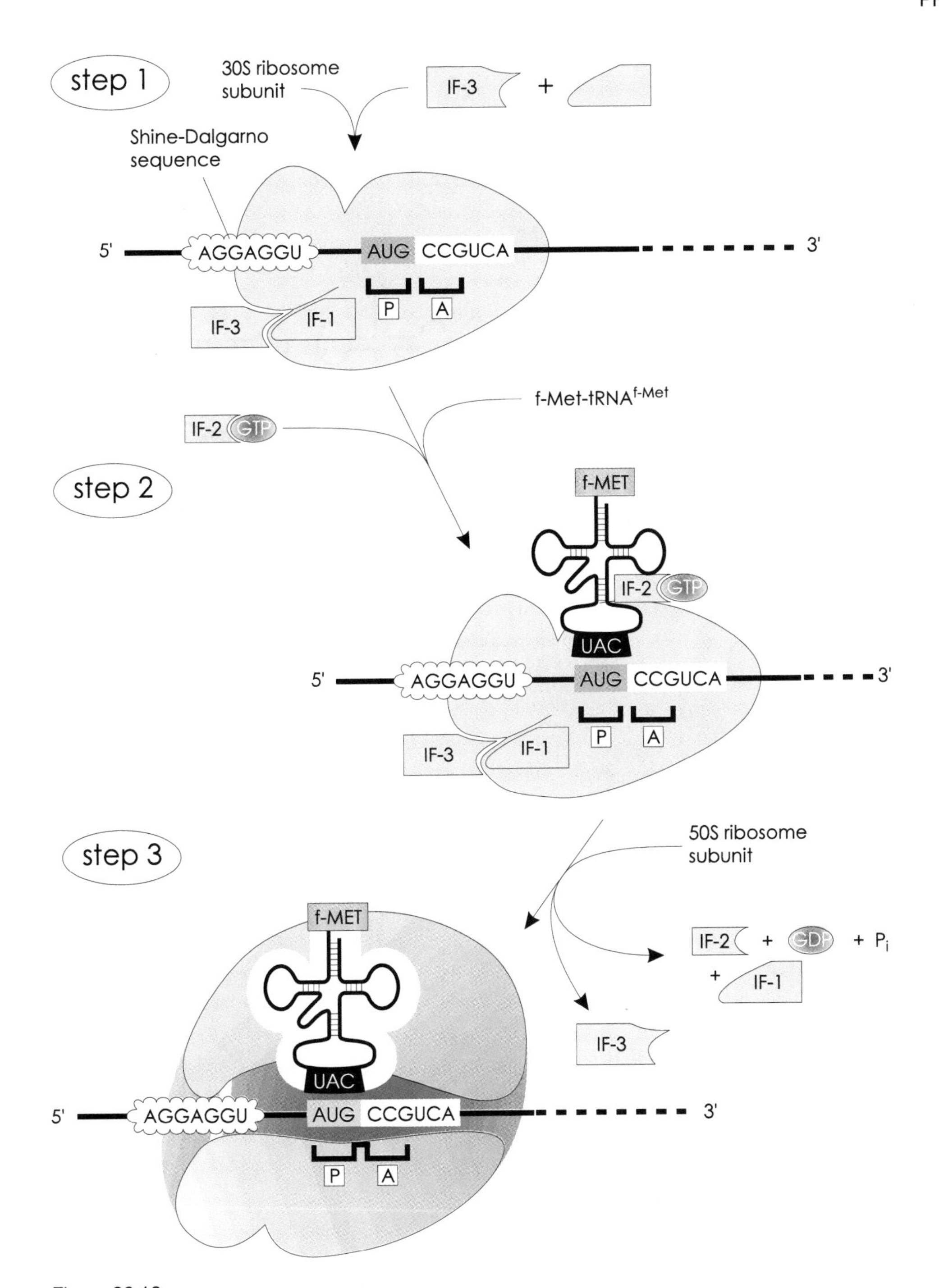

Figure 26.10.

Formation of the translational initiation complex in bacteria involves ribosome interactions with a sequence in the mRNA that is upstream (5′) of the AUG and called the Shine–Dalgarno sequence. The purine-rich Shine–Dalgarno sequence (AGGAGG) is complementary to a portion of the 16S rRNA. Step 1 involves the interaction of IF-1 and IF-3 with the 30S ribosome subunit to form a complex which is positioned precisely at the initiator codon (AUG). The addition of IF-2, GTP, and the f-Met-tRNA$^{f\text{-}Met}$ result in the formation of the preinitiation complex in step 2. In step 3, hydrolysis of GTP and the release of IF-2 and IF-1 allows for the binding of the 50S ribosome subunit which completes the assembly of the initiation complex. The P and A sites are also schematically shown (see text for details).

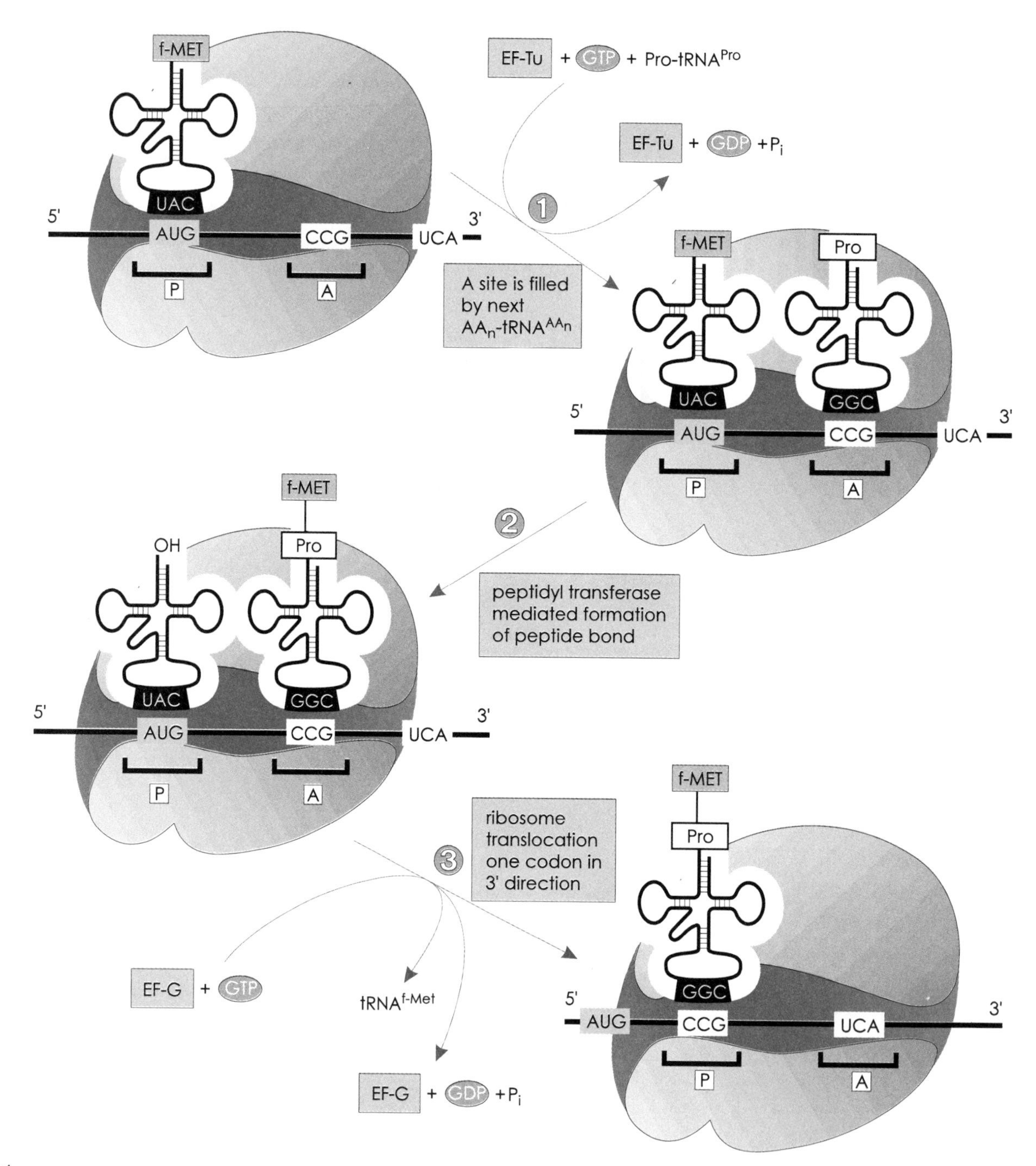

Figure 26.11.

The elongation reaction in bacterial protein synthesis requires three steps. Step 1 involves the entry of an aminoacyl-tRNA into the A site of the ribosome in a reaction that requires GTP hydrolysis and the elongation factor EF-Tu. In step 2, the amino group of the incoming amino acid attacks the carbonyl of the amino acid in the P site, a reaction catalyzed by the peptidyl transferase ribozyme encoded in the 23S rRNA of the 50S subunit. Step 3 is the GTP-catalyzed translocation of the 70S ribosome one codon position in the 3′ direction. This step requires the elongation factor EF-G and results in a vacancy in the A site, which is now ready to accept the next aminoacyl-tRNA. If the tRNA exit site (E site) were included in the figure, it would be positioned to the left side of the P site.

different than Met-$tRNA^{Met}$ and can only be used to form the initiation complex. In bacteria, the formyl group, or sometimes the entire fMet, is removed post-translationally from the amino terminus of the protein. In eukaryotes, the fMet moiety is not used; however, there are two different $tRNA^{Met}$ species, one for the initiator methionine and one for internal methionines. These two tRNAs are distinguishable by their nucleotide sequences and pattern of base modifications. Interestingly, protein synthesis in mitochondria and chloroplasts requires an organelle-encoded $tRNA^{fMet}$ as the initiator tRNA which supports the possible evolutionary relationship between eukaryotic organelles and bacteria.

Elongation

Three key steps in the elongation stage of protein synthesis are required for the addition of each amino acid. Because these steps are repeated for each peptide bond formed, this is sometimes called the **elongation cycle.** The central theme in elongation is that the fully assembled ribosomal complex functions as a **ribonucleoprotein machine** which rapidly moves 5′ to 3′ down the mRNA, much like a ratchet. At the center of this complex are two binding sites which line up over a pair of triplet codons, as shown in Figure 26.11. These two sites are called the **P site,** for peptidyl (or polypeptide), and the **A site,** for aminoacyl (or acceptor). A third site, called the **E site** for tRNA exit site, is also a functional component of the ribosome, but for reasons of clarity, it is not included in the figures. After the initiation complex is formed, the initiator fMet-$tRNA^{fMet}$ is positioned in the P site through base pairing with the AUG $codon_1$.

In the first step of elongation, the A site is filled by the appropriate aminoacyl-tRNA as specified by $codon_2$ in a reaction that requires **elongation factor EF-Tu** and GTP. Once the aminoacyl-tRNA is correctly positioned within the A site and base paired with $codon_2$, GTP is hydrolyzed and the EF-Tu-GDP complex is released.

In the second step of elongation, the α-amino group of the amino acid in the A site (AA_2) acts as a nucleophile and attacks the carbonyl group of AA_1 (in this case fMet). This reaction leads to the formation of a **dipeptidyl-tRNA** in the A site and a deacylated-$tRNA^{fMet}$ in the P site (Fig. 26.12). As shown by Harry Noller in 1992, this peptidyl transferase reaction is catalyzed by ribozyme activity present in the 23S rRNA of the 50S ribosome subunit (and the 28S rRNA in the 60S ribosome subunit in eukaryotes), rather than by ribosomal proteins, as originally thought.

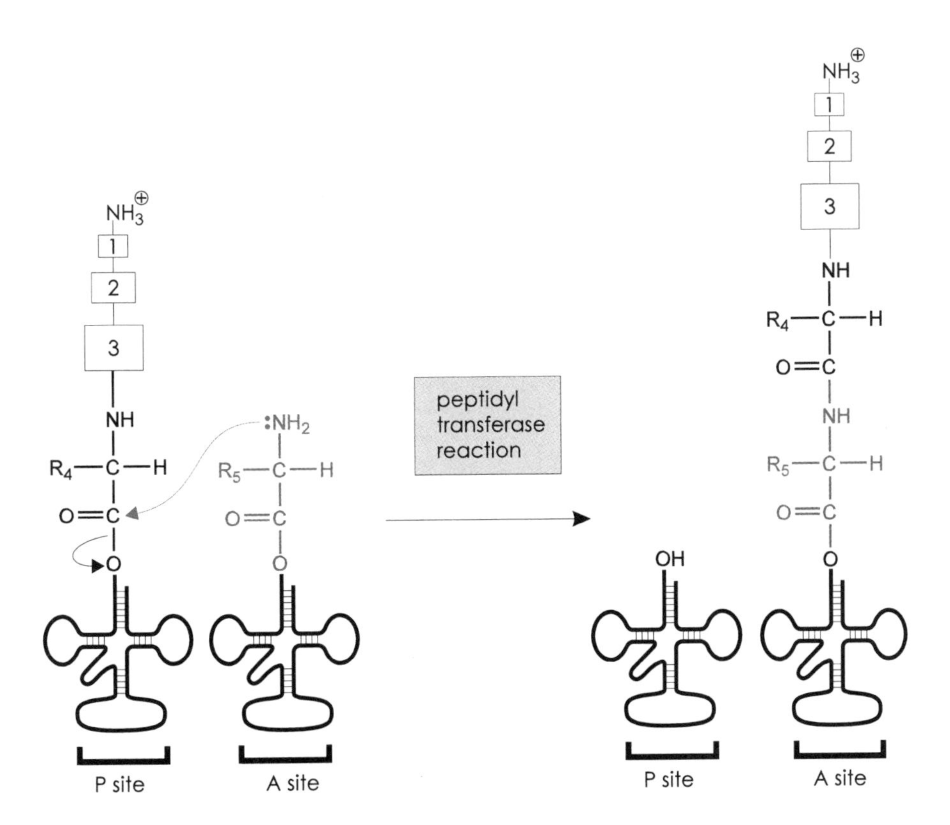

Figure 26.12.
Peptidyl transferase reaction in the elongation phase of protein synthesis.

The final step of elongation is the translocation of the ribosome a distance of one codon in the 3′ direction. This step requires the hydrolysis of another GTP and results in the displacement of the A site into the $codon_3$ position. The dipeptidyl-tRNA stays bound to the mRNA at $codon_2$, but since the ribosome is translocated, this tRNA is now within the P site. This translocation process also creates an empty A site for the next aminoacyl-tRNA and the E site (not shown in Fig. 26.11) now contains the deacylated-tRNA. The bacterial elongation factor EF-G is required for ribosome translocation and it is thought that hydrolysis of GTP results in a conformational change in the ribosome to facilitate the translocation step. This series of three steps is repeated for each amino acid added to the growing polypeptide chain (Fig. 26.13).

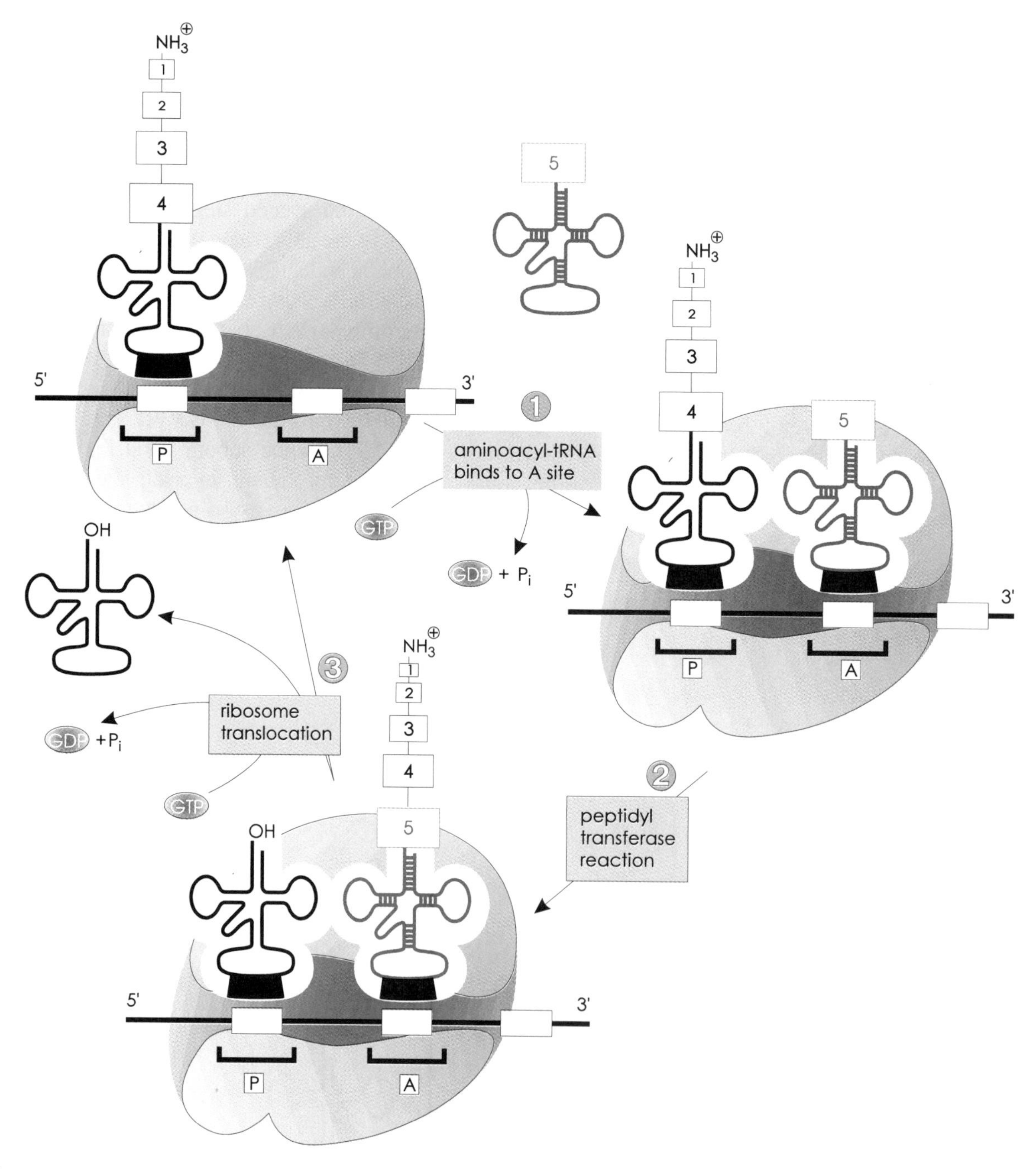

Figure 26.13.

The elongation cycle requires the hydrolysis of 2 GTPs for every additional amino acid added. This cycle continues as long as there are aminoacylated tRNAs in the A site that can function as acceptors in the peptidyl transferase reaction.

Termination

The last stage in protein synthesis is termination of polypeptide chain elongation. This occurs when one of the three **termination codons** (UAA, UAG, and UGA) is in the A site. Normally there are no tRNAs with the proper anticodon for a termination signal, so the A site remains empty. In this case, **termination factors** bind in the A site and promote chain termination by causing the peptidyl transferase reaction to add the growing peptide chain to a water molecule. As shown in Figure 26.14, bacteria contain three release factors, RF-1, RF-2, and RF-3. RF-1 recognizes the termination codons UAA and UAG, while RF-2 recognizes UAA and UGA. RF-3 is a GTP-binding protein that is needed to stimulate RF-1 and RF-2 binding to the A site when a termination codon is present. The only known release factor in eukaryotes is eRF, which is a GTP-hydrolyzing protein that appears to perform the same functions as RF-1, RF-2, and RF-3 combined.

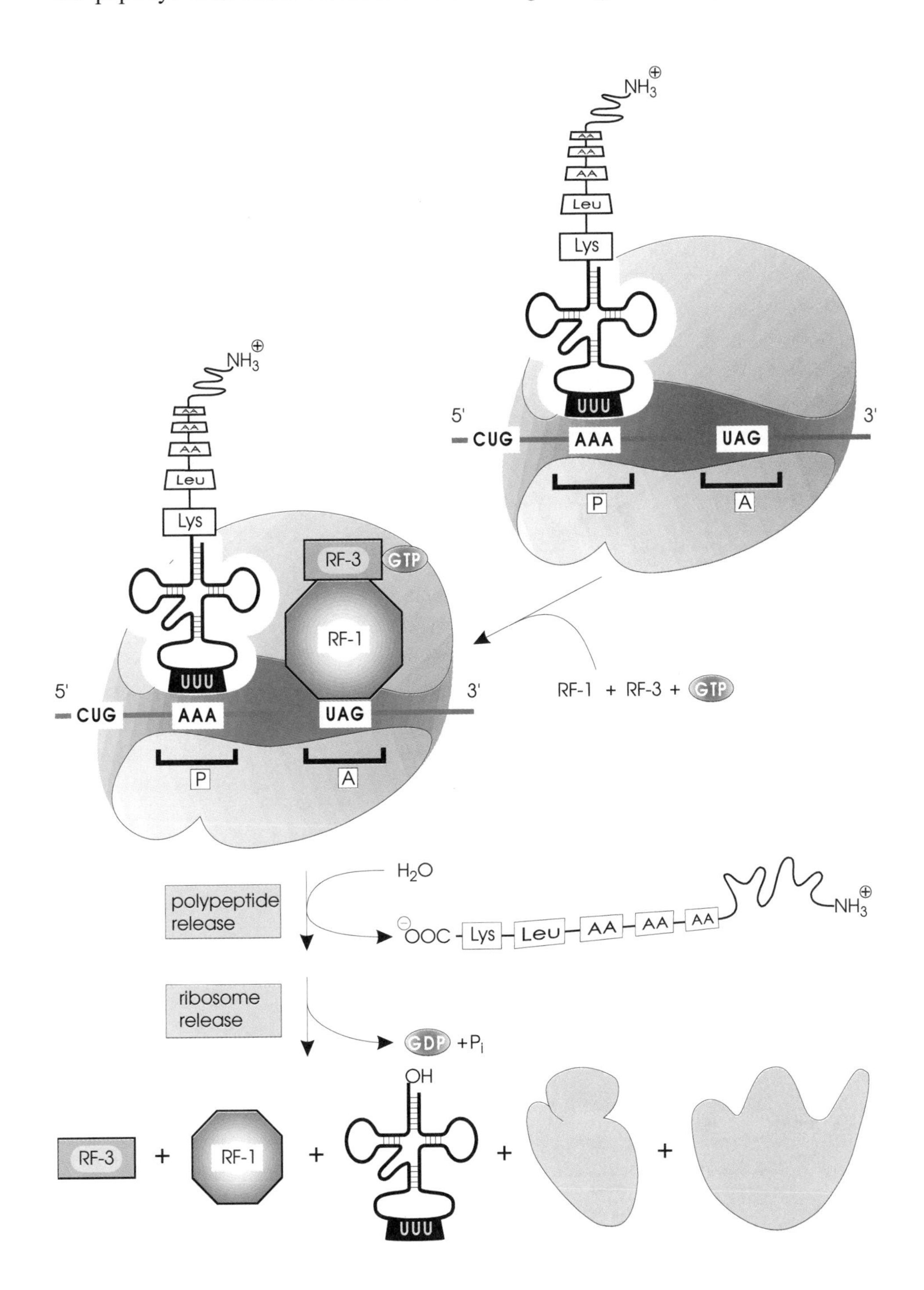

Figure 26.14.
Translational termination in bacteria. RF-1 or RF-2 bind in the A site following stimulation by RF-3-GTP. This protein complex causes the peptidyl transferase to add the polypeptide chain to H_2O, resulting in chain termination and release. Subsequent GTP cleavage by RF-3 causes the 70S ribosome to dissociate from the mRNA.

There are at least two situations where the reaction is not terminated even though a termination codon is in the A site. This process results in **translational read-through.** The first is translational frameshifting, which was described earlier as a stutter step by the ribosome. In this case, the termination codon in one reading frame is avoided by the ribosome shifting to one of the other two reading frames. The second case in which a termination codon does not function to terminate protein synthesis is if the cell contains an aminoacyl-tRNA with a mutation in the anticodon such that it now recognizes one of the termination codons as coding for its amino acid. These tRNAs were discovered by genetic analyses in bacteria and are called **suppressor tRNAs.**

TRANSLATIONAL CONTROL OF PROTEIN SYNTHESIS

Controlling the efficiency of productive translational initiation is an important mechanism of regulated gene

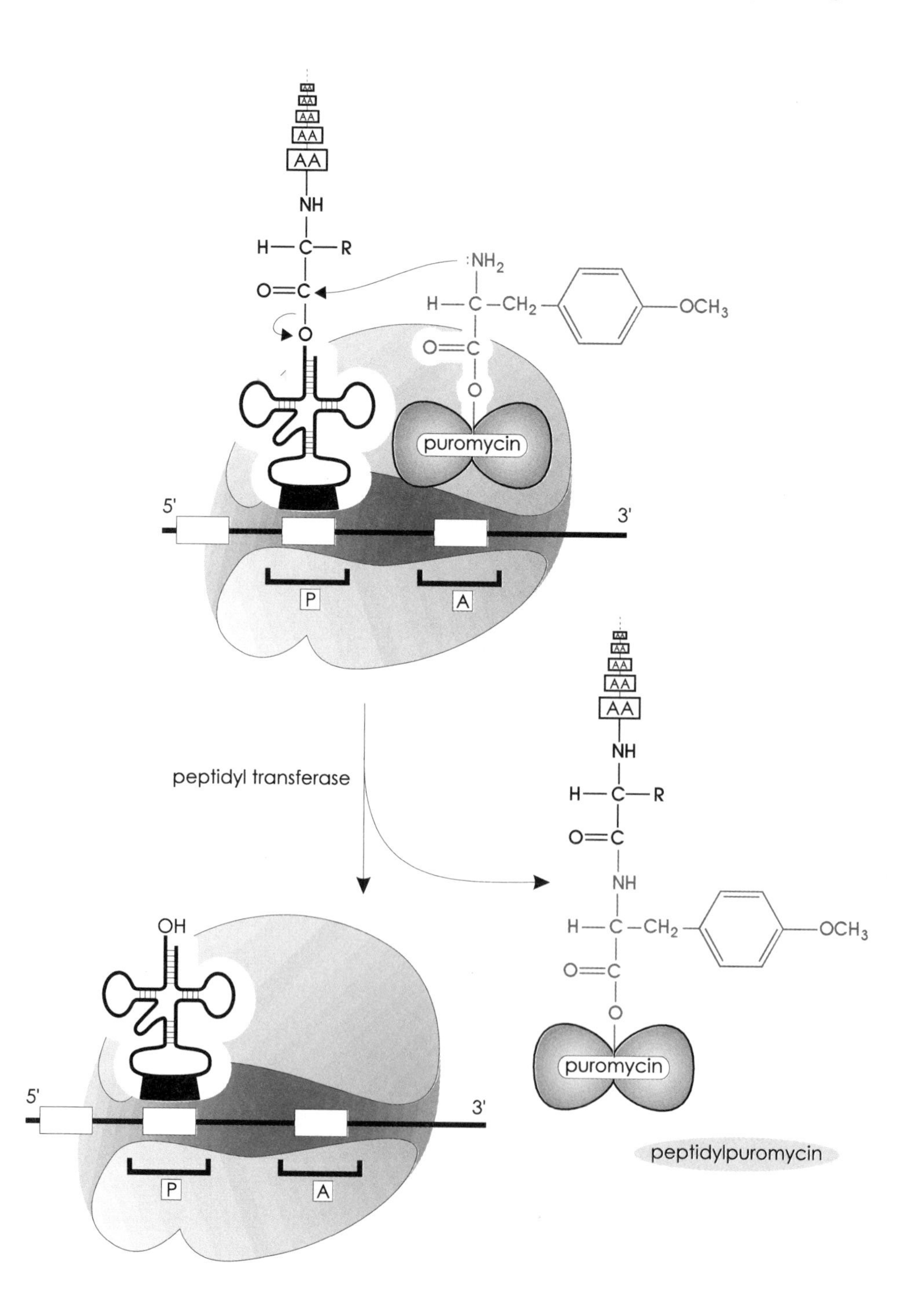

Figure 26.15.

Mechanism of action of puromycin. Puromycin is able to enter the ribosome A site and function as an aminoacyl tRNA analogue, resulting in polypeptide chain termination in both prokaryotes and eukaryotes.

Table 26.1. Bacterial and Eukaryotic Protein Synthesis Inhibitors

Inhibitor	Specificity	Mechanism of action
Puromycin	Prokaryotes, eukaryotes	Causes premature polypeptide chain termination by binding to the A site
Tetracycline	Prokaryotes	Binds to the A site and prevents entry of aminoacyl tRNA
Chloramphenicol	Prokaryotes	Inhibits the 50S ribosome peptidyl transferase activity
Streptomycin	Prokaryotes	Inhibits the fidelity of translational initiation
Diptheria toxin	Eukaryotes	Inactivates eukaryotic elongation factor eEF-2
Ricin	Eukaryotes	Inactivates 28S rRNA by *N*-glycolytic cleavage of an adenine
Cycloheximide	Eukaryotes	Inhibits the 60S ribosome peptidyl transferase activity

expression in prokaryotes and eukaryotes. One example in eukaryotic cells is the control of ferritin translation. **Ferritin** binds iron and functions as an iron-storage protein in the cell. A stem–loop structure called an **iron-responsive element** (IRE, see Chapter 25) resides in the 5′ untranslated region of ferritin mRNAs from a variety of organisms. Under conditions of high iron, the bifunctional IRE binding protein, cytoplasmic aconitase, is in a conformation that cannot bind to the ferritin mRNA IRE which allows for maximal ferritin protein synthesis. However, when iron concentrations are low, cytoplasmic aconitase readily binds to the ferritin IRE and inhibits the formation of the translation initiation complex, causing a decrease in ferritin protein synthesis. However, cytoplasmic aconitase binding to IREs in **transferrin receptor** (TfR) mRNA has the opposite effect on TfR protein synthesis (Chapter 25). When iron is limiting in the cell, TfR mRNA is protected from rapid RNA degradation by cytoplasmic aconitase binding and TfR synthesis is increased, resulting in the uptake of more extracellular iron.

Another example of translational control in eukaryotes is the inhibition of yeast GCN4 protein synthesis by stem–loop structures present in the 5′ end of the mRNA. GCN4 control, and an analogous situation in bacteria, links amino-acid biosynthesis to **ribosome pausing** in the 5′ end of the mRNA. This mechanism was first described for the tryptophan operon in *E. coli* and it is often referred to as **attenuation.** Transcriptional and translational control of the tryptophan biosynthetic enzymes are described in Chapter 28.

INHIBITORS OF PROTEIN SYNTHESIS

A good example of translational inhibition is the mechanism by which **puromycin** mimics the structure of the aminoacyl group of aminoacyl-tRNA, as shown in Figure 26.15. Puromycin is an **aminoacyl-tRNA analogue** that is able to bind to the A site of both prokaryotic and eukaryotic ribosomes, even without a corresponding tRNA, and thus act as an acceptor for the peptidyl transferase reaction.

The result is the immediate termination of translation and the release of a truncated protein. Two potent antibiotics that specifically inhibit bacterial translation, are **tetracycline,** which blocks the A site and prevents the entry of aminoacyl-tRNAs, and **chloramphenicol,** which inhibits the peptidyl transferase activity of the 23S rRNA. The mechanisms of action of these antibiotics, including **streptomycin,** which alters the fidelity of translation in bacteria, are listed in Table 26.1.

Diphtheria toxin is one of the most potent toxins known; a single molecule of diphtheria toxin is sufficient to inactivate enough eukaryotic elongation factor eEF-2 in a cell to cause death. **Ricin** is derived from castor beans and its mode of action is to inactivate eukaryotic 28S rRNA molecules by removing a single adenine residue by *N*-glycolytic cleavage. The other inhibitor of eukaryotic protein synthesis shown in Table 26.1 is **cycloheximide,** which inhibits peptidyl transferase activity of the 60S ribosome subunit.

SUMMARY

1. Protein synthesis requires mRNA, ribosomes, aminoacylated tRNAs, and the hydrolysis of four high-energy phosphate bonds per peptide linkage (counting the two phosphate bonds hydrolyzed during the aminoacylation step). Translation of the mRNA is in the 5′ to 3′ direction and the polypeptide is synthesized from the NH_2 to the COOH terminus one amino acid at a time.

2. The tRNAs serve as the molecular adaptors that read the genetic information encoded within the mRNA strand. The genetic code is based on trinucleotide sequences called codons; tRNAs contain the complementary anticodon sequences. Three types of experiments demonstrated that of the 64 possible triplets, 61 can specify aminoacylated tRNAs, each containing one of the 20 possible amino acids. Three codons have no corresponding tRNAs and serve as termination codons.

3. The Wobble Hypothesis states that only two of the three nucleotides in the codon need to be base paired with the tRNA anticodon for triplet recognition. Experimental tests of the Wobble Hypothesis confirmed this prediction and it was concluded that a minimum of only 32 tRNAs would be needed to specify all 61 codons; however, most cells contain more than this lower number of tRNAs.

4. Translational frameshifting, RNA editing, and protein splicing are three examples where the Central Dogma of information transfer from DNA → RNA → protein does not strictly apply. The amino-acid sequence of polypeptide products produced from these reactions cannot be predicted from the gene-encoded DNA or mRNA sequences.

5. There are 20 different aminoacyl tRNA synthetases, each of which recognizes the various tRNAs and "charges" them with the appropriate amino acid. Several recognition steps in this aminoacylation reaction help to reduce the error frequency due to incorrect tRNA charging. Aminoacylation requires the hydrolysis of ATP → AMP + $2P_i$.

6. Protein synthesis requires three discrete phases: initiation, elongation, and termination. Initiation involves the assembly of an initiation complex consisting of initiation factors, ribosomal subunits, mRNA, the initiator methionine tRNA, and a GTP hydrolysis step. Elongation is a continuous cycling of aminoacylated tRNAs into the A site, GTP hydrolysis, and a peptidyl transferase reaction catalyzed by the ribozyme activity encoded within the large rRNA. Termination occurs when one of the three termination codons is present in the A site and no tRNA is available to function as the acceptor in the peptidyl transferase reaction.

7. The protein synthesis inhibitors tetracycline, chloramphenicol, and streptomycin all block bacterial protein synthesis. Several eukaryotic translational inhibitors have also been found and they include diphtheria toxin, ricin, and cycloheximide. Puromycin causes premature chain termination in both prokaryotes and eukaryotes by functioning as an aminoacyl tRNA analog.

REFERENCES

Cammack R. (1993): A new use for an old enzyme. *Current Biol.* **3**:41.

Cattaneo R. (1991): Different types of messenger RNA editing. *Annu. Rev. Genet.* **25**:71.

Cooper A. A., Stevens T. H. (1995): Protein splicing: self-splicing of genetically mobile elements at the protein level. Trends Bioc. *Sciences* **20**:351.

Ibba, M., Hong, K. W., Sherman, et al. (1996): Interactions between tRNA identity nucleotides and their recognition sites in glutaminyl-tRNA synthetase determine the cognate amino acid affinity of the enzyme. *Proc.Natl.Acad.Sci.USA* **93**:6953.

Illangasekare M., Sanchez G., Nickles T., Yarus M. (1995): Aminoacyl-RNA synthesis catalyzed by an RNA. *Science* **267**:643.

Jacks T., Madhani H. D., Masiarz F. R., Varmus H. E. (1988): Signals for ribosomal frameshifting in the rous sarcoma virus gag-pol region. *Cell* **55**:447.

Kozak M. (1994): Features in the 5′ non-coding sequences of rabbit α- and β-globin mRNAs that affect translational efficiency. *J. Mol. Biol.* **235**:95.

Neupert W., Lill R. (1994): Protein synthesis: Cradle at the ribosome. *Nature* **370**:421.

Nierhaus K. H. (1996): An elongation factor turn-on. *Nature* **379**:491.

Nissen P., Kjeldgaard M., Thirup S., et al. (1995): Crystal structure of the ternary complex of Phe-$tRNA^{Phe}$, EF-Tu, and a GTP analog. *Science* **270**:1464.

Rodin, S., Rodin, A. and Ohno, S. (1996): The presence of codon-anticodon pairs in the acceptor stem of tRNAs. *Proc.Natl.Acad.Sci.USA* **93**:4537.

Saks M. E., Sampson J. R., Abelson J. N. (1994): The transfer RNA identity problem: A search for rules. *Science* **263**:191.

Samaha R. R., Green R., Noller H. F. (1995): A base pair between tRNA and 23S rRNA in the peptidyl transferase center of the ribosome. *Nature* **377**:309.

Schimmel P., Ribas de Pouplana L. (1995): Transfer RNA: From minihelix to genetic code. *Cell* **81**:983.

Schmidt E., Schimmel P. (1994): Mutational isolation of a sieve for editing in a transfer RNA synthetase. *Science* **264**:265.

Sollner-Webb, B. (1996): Trypanosome RNA editing: Resolved. *Science* **273**:1182.

Tuite M. F., Stansfield I. (1994): Translation: Knowing when to stop. *Nature* **372**:614.

Von Ahsen U., Noller H. F. (1995): Identification of bases in 16S rRNA essential for tRNA binding at the 30S ribosomal P site. *Science* **267**:234.

REVIEW QUESTIONS

1. tRNA is called the "adaptor" molecule in protein synthesis because it
   a) "links" the 30S and 50S ribosomal subunits together
   b) can "adapt" to different conformations during protein synthesis
   c) functions to "interpret" mRNA sequence into corresponding protein sequence
   d) "reads" the polypeptide sequence to direct RNA synthesis
   e) can be "recognized" by both aminoacyl tRNA synthetases and amino acids

2. If the genetic code were based on quadruplet codons rather than triplet codons, and the first two 5′ nucleotides in the tRNA anticodon were wobble positions, would there need to be more or less different tRNAs for protein synthesis?
   a) 256 different tRNAs would be needed.
   b) More than 150 but less than 250 tRNAs would be needed.
   c) Less than 20 different tRNAs would be needed.
   d) It depends on how many aminoacyl tRNA synthetases the cell contains.
   e) The same number (30–50 tRNAs) would be sufficient.

3. AUG and UAG specify start and stop codons in protein synthesis, respectively. Which open-reading frame(s) in the mRNA shown below would encode a short polypeptide?

   5′—UUAUGAAUGUACCGUGGUAGUU—3′

   a) reading frame 1
   b) reading frame 2
   c) reading frame 3
   d) reading frames 1 and 3
   e) reading frames 2 and 3

4. Which statement is NOT TRUE regarding the accuracy of protein synthesis?
   a) The ribosome contains an exopeptidase for proofreading.
   b) Aminoacyl tRNA synthetases hydrolyze incorrect AA-$tRNA^{AA}$.
   c) tRNA binding to aminoacyl tRNA synthetases is highly specific.
   d) The error rate in protein synthesis is about 1 in 10,000.
   e) Codon–anticodon interactions ensure that correct tRNAs enter the A site.

5. The two steps in the elongation cycle of protein synthesis where GTP hydrolysis is required are
   a) peptidyl transferase reaction and ribosome translocation
   b) aminoacyl tRNA binding to the A site and ribosome translocation
   c) fMet-$tRNA^{fMet}$ binding to the A site and ribosome translocation
   d) peptidyl transferase reaction and aminoacyl tRNA binding to the P site
   e) RF-1 + RF-3 binding and ribosome translocation

6. Puromycin
   a) is an inhibitor of DNA synthesis
   b) inactivates elongation factor eEF-2
   c) binds to the P site and prevents entry of aminoacyl tRNA
   d) causes termination of protein synthesis by blocking ribosome translocation
   e) binds to the A site and functions as an aminoacyl tRNA analogue

ANSWERS TO REVIEW QUESTIONS

1. *c* tRNA is bifunctional and binds both the mRNA and amino acids.

2. *e* Because the first two positions in the anticodon would be wobble positions, the fourth nucleotide in the code would not contribute specificity beyond what is already possible with a triplet codon.

3. ***c*** Only reading frame 3 makes a polypeptide.

4. ***a*** There is no exopeptidase activity in ribosomes; fidelity depends primarily on the accuracy of tRNA charging reactions by aminoacyl tRNA synthetases.

5. ***b*** The two energy-requiring steps in the elongation cycle are tRNA binding to the A site and ribosome translocation.

6. ***e*** Puromycin binds to the A site and causes premature termination by functioning as an aminoacyl tRNA analogue.

CHAPTER 27

PROTEIN TARGETING AND TURNOVER

INTRODUCTION

A variety of processing events follow protein synthesis. In eukaryotes, many of these modifications are associated with specific subcellular compartments. For example, glycosylation of proteins requires enzymes that are present both in the endoplasmic reticulum and the Golgi apparatus. Other proteins need to be localized to specific regions in the cell, for example, the localization of integral membrane proteins to the plasma membrane, or nuclear-encoded mitochondrial proteins that are synthesized in the cytosol but need to be imported into the mitochondria. In general, proteins cannot passively cross membranes; therefore, mechanisms have evolved to target proteins to various cell compartments, including membranes. In addition, many cells synthesize proteins for export from the cell, for example, the synthesis of digestive enzymes and zymogens by the pancreas, or the synthesis of hormones by exocrine glands. As described in Chapter 26, eukaryotic protein synthesis takes place in the cytoplasm. The net movement of newly synthesized proteins through the cytoplasm to different cell compartments is called protein trafficking or targeting.

Elimination of unwanted proteins is just as important as synthesis of desirable proteins. The stability of cellular proteins varies over a wide range. Some unstable proteins have a half life of minutes, while others survive almost as long as the cell. A short half life simplifies the rapid control of protein concentration by synthesis or degradation. For example, in nondividing cells, ornithine decarboxylase is normally synthesized and degraded rapidly. When such cells are stimulated to proliferate, ornithine decarboxylase degradation slows down, resulting in a net increase in the concentration of ornithine decarboxylase protein.

SIGNAL PEPTIDE SEQUENCES DIRECT PROTEIN TARGETING

Each protein destined for a specific subcellular compartment carries with it the signal that determines its eventual location, like the address on an envelope. The molecular signal for protein targeting is in the form of an amino-acid sequence. Often the signal resides in the amino-terminal sequence of the protein; however, a carboxy-terminal sequence or an internal amino-acid sequence can also be utilized. The term **signal peptide** may be used to refer specifically to the signal that directs proteins across the **endoplasmic reticulum** (ER) or it may be used in a more general sense to refer to any continuous amino-acid sequence whose function is to direct protein targeting. When the signal is provided by noncontiguous sequences, the signal is referred to as a **signal patch.**

Signal peptides with the same function may vary considerably in precise sequence, but they do appear to have structural features in common. Figure 27.1 lists the signal sequences of several eukaryotic proteins and their final destinations in the cell. The signal sequence for proteins targeted to the ER is the best characterized. The ER signal peptide is usually ~10 amino acids in length and consists of mostly **hydrophobic residues** (often including Leu, Val Ile, Tyr, or Trp). The ER signal peptide is typically flanked on the NH_2-terminal site by a positively charged amino acid such as Arg or Lys, and on the carboxy-terminal side by polar residues or Ala.

Signal peptides can be recognized by a functional test using recombinant DNA methods to move signal peptides from one protein to another. Signal peptides can be removed from a protein and the loss of correct localization observed. Even more striking, a signal peptide can be placed at the beginning of a test protein and its effect on the cellular localization of the protein determined. It has even been shown that the nuclear-localization signal peptides can carry colloidal gold particles into the nucleus.

If synthesis of the first ~10–20 amino acids of a protein *does not* produce a functional signal peptide directing the protein to the membrane compartment of the ER, then protein synthesis continues in the cytoplasm on **membrane-free ribosomes** and the protein is released into the cytosol. These newly synthesized cytosolic proteins have a variety of possible destinations, including the cytoplasm, the nucleus, mitochondria,

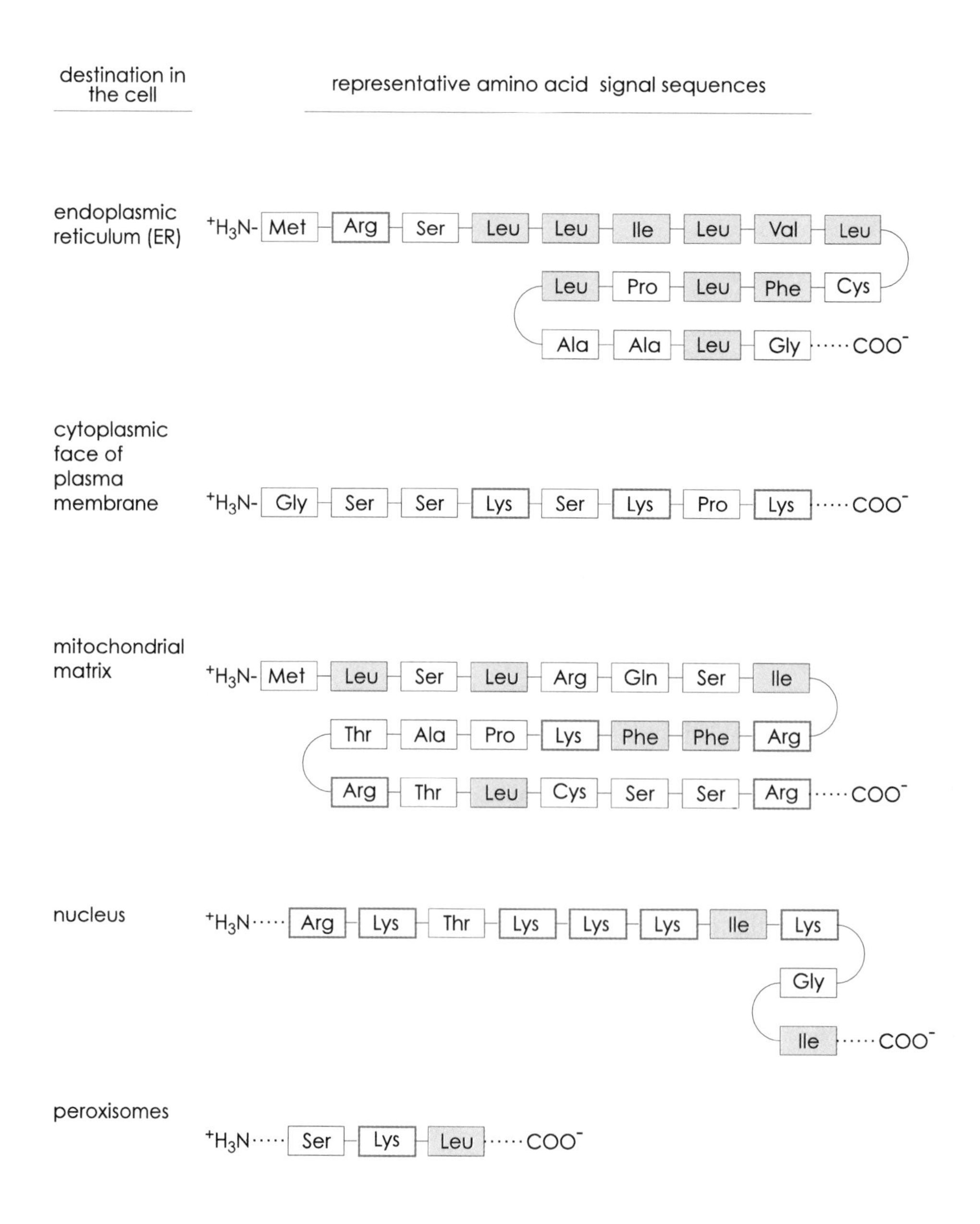

Figure 27.1.

List of signal peptides that function in eukaryotic proteins. Hydrophobic residues are shaded in light purple and positively charged residues are outlined in dark purple.

and the inner surface of cytoplasmic membranes. Each of these destinations is determined by the presence of specific classes of signal peptides. For each destination, there is a processing mechanism which recognizes the appropriate signal sequence. Proteins that *do* contain an ER signal peptide are targeted to the **ER membrane** where translation resumes as described in a later section. Figure 27.2 summarizes these two major protein-targeting pathways.

PROTEIN TARGETING THROUGH THE CYTOPLASM

Cytoplasmic Face of Membranes

To attach to membranes, proteins must be threaded through the membrane via the ER, as described below, or be modified by the attachment of a hydrophobic group that will act as an **anchor in the membrane.** In many cases, proteins that are released into the cytoplasm contain special signaling sequences which are recognized by enzymes that modify the protein, such as by attaching a fatty acid. The fatty acid then anchors the molecule in the membrane, leaving the protein on the cytoplasmic surface. Examples of this process include the attachment of **myristic acid** from myristyl CoA to the amino terminus of the Src protein tyrosine kinase, and the attachment of **palmitic acid** to a cysteine near the amino terminus of the Ras protein by the enzyme farnesyl protein transferase (Chapter 31).

Mitochondria

As described in earlier chapters, mitochondria are organelles in the cytoplasm that specialize in the synthesis of ATP by electron transport and oxidative phosphorylation. Mitochondria have their own DNA and ribosomes and synthesize some of their own proteins. Many of their proteins, however, are encoded by nuclear genes and are synthesized in the cytoplasm. A transport mechanism is required to get appropriate proteins from the cytoplasm into the mitochondria.

The **mitochondrial signal peptide** is located at the amino terminus and consists of up to 80 amino acids that can coil into an α-helix with polar residues on one side and nonpolar residues on the other. This protein structure is called an **amphipathic α-helix.** The process of protein import into mitochondria is not understood in detail. However, it appears to involve the following processes. The signal peptide binds to the outer mitochondrial membrane and passes through both outer and inner membrane, at a point where they meet. This requires the presence of the normal electrochemical gradient across the inner membrane. Once inside

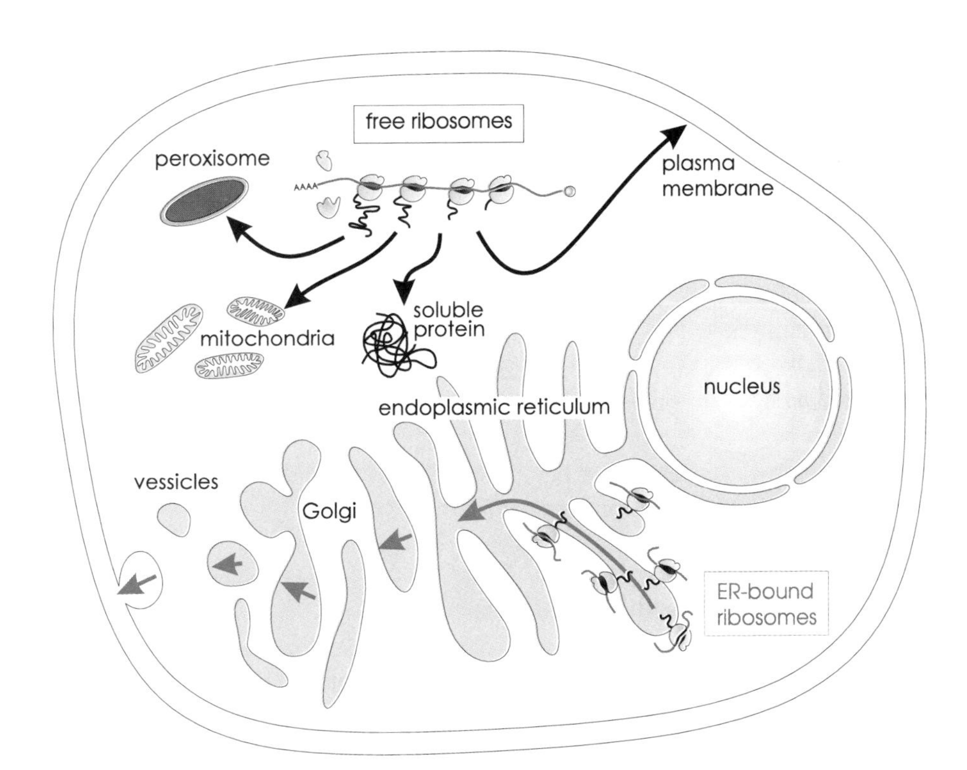

Figure 27.2.
Schematic drawing showing the two basic pathways of protein targeting—membrane-free protein synthesis and ER-bound protein synthesis.

the membrane, the signal peptide is cleaved off by a **matrix protease** and degraded. The remainder of the protein is unfolded by an ATP-dependent process and passes through the membrane into the **mitochondrial matrix** where it refolds.

The space between the inner and outer mitochondrial membrane can be reached by proteins that have two signal peptides. The first inserts the protein into the membrane and is cleaved in the matrix; the second remains and directs the protein to the intermembrane space.

Nucleus

The nuclear membrane contains pores that allow free diffusion of small molecules, but restrict the passage of large proteins. However, large proteins or protein complexes can pass through the pores by a selective transport mechanism. The signal that identifies a protein for nuclear import is called a **nuclear import signal.** It may occur anywhere in the protein, not just at the amino terminus. The sequence is usually short (4–8 amino acids), and is rich in Lys and Arg residues (Fig. 27.1). Transport across the nuclear membrane through **nuclear pores** is unusual in that the signal peptide is not cleaved after transport nor does the protein unfold during transport through the nuclear pore.

Peroxisomes

Peroxisomes are membrane-bounded organelles that are specialized for carrying out oxidative reactions. Their internal proteins are imported from the cytoplasm. The **peroxisome import signal** is the tripeptide Ser-Lys-Leu located near the carboxyl terminus of the protein. Peroxisome protein import is a critical cellular process, as evidenced by the severe tissue abnormalities found in patients with the inherited disease **Zelweyer syndrome.** These individuals are characterized as having protein-free peroxisomes.

PROTEIN TARGETING THROUGH MEMBRANE COMPARTMENTS

An ER signal peptide is highly hydrophobic, as shown in Figure 27.1. The hydrophobicity probably helps the signal peptide pass through the membrane. If a nascent polypeptide chain contains a signal peptide directing it to the ER, this portion of the protein will interact with a **signal recognition particle** (SRP) present in the cytoplasm. The SRP is a RNA–protein complex that binds to both the signal peptide of the nascent protein and the ribosome. It causes protein synthesis to temporarily pause, possibly by binding to the incoming tRNA binding site on the ribosome.

The SRP–peptide–ribosome complex in the cytoplasm binds to a SR receptor protein which is bound to the **cytoplasmic face of the ER.** This interaction causes the SRP to be released from the complex, leaving the signal peptide–ribosome complex associated with the ER membrane. Protein synthesis resumes once the SRP dissociates and the nascent protein feeds through the membrane to the interior of the ER (Fig. 27.3). If the protein manages to fold before passing through the membrane, it must be unfolded before the transport process can continue. Once inside the ER, the protein refolds in a process facilitated by a binding protein that recognizes incorrectly folded proteins, and by a protein disulfide isomerase and a prolyl isomerase.

The signal peptide may be cleaved on the inside of the ER membrane by an endopeptidase, or it may be left to guide the protein toward its destination. A third possibility is that the signal peptide remains in the ER membrane and anchors the protein there. Proteins may also be fixed on the inside of the ER membrane by covalent attachment to phosphatidylinositol in the membrane.

Inside the ER, many proteins undergo a variety of **modifications.** The most complex is the addition of sugars to form **glycoproteins.** Although some **glycosylation** can occur in the cytoplasm (e.g., addition of *N*-acetylglucosamine to Ser with an *O*-glycosidic link), most glycoproteins are modified in the ER by addition of a specific 14-residue **oligosaccharide.** This process occurs cotranslationally and is catalyzed by a membrane-bound **glycosyl transferase** utilizing Asn residues in the sequences -Asn-Xxx-Ser- and Asn-Any-Thr. Once on the luminal side of the ER membrane, proteins either remain in the ER or move to the Golgi apparatus for further distribution to the lysosomes, secretory vesicles, or plasma membrane, where they can remain or be exported from the cell. Figure 27.4 illustrates protein traffic out of the ER.

Golgi Apparatus

The Golgi apparatus takes proteins from the ER, continues their modification, particularly their glycosylation, and includes the formation of **proteoglycans.**

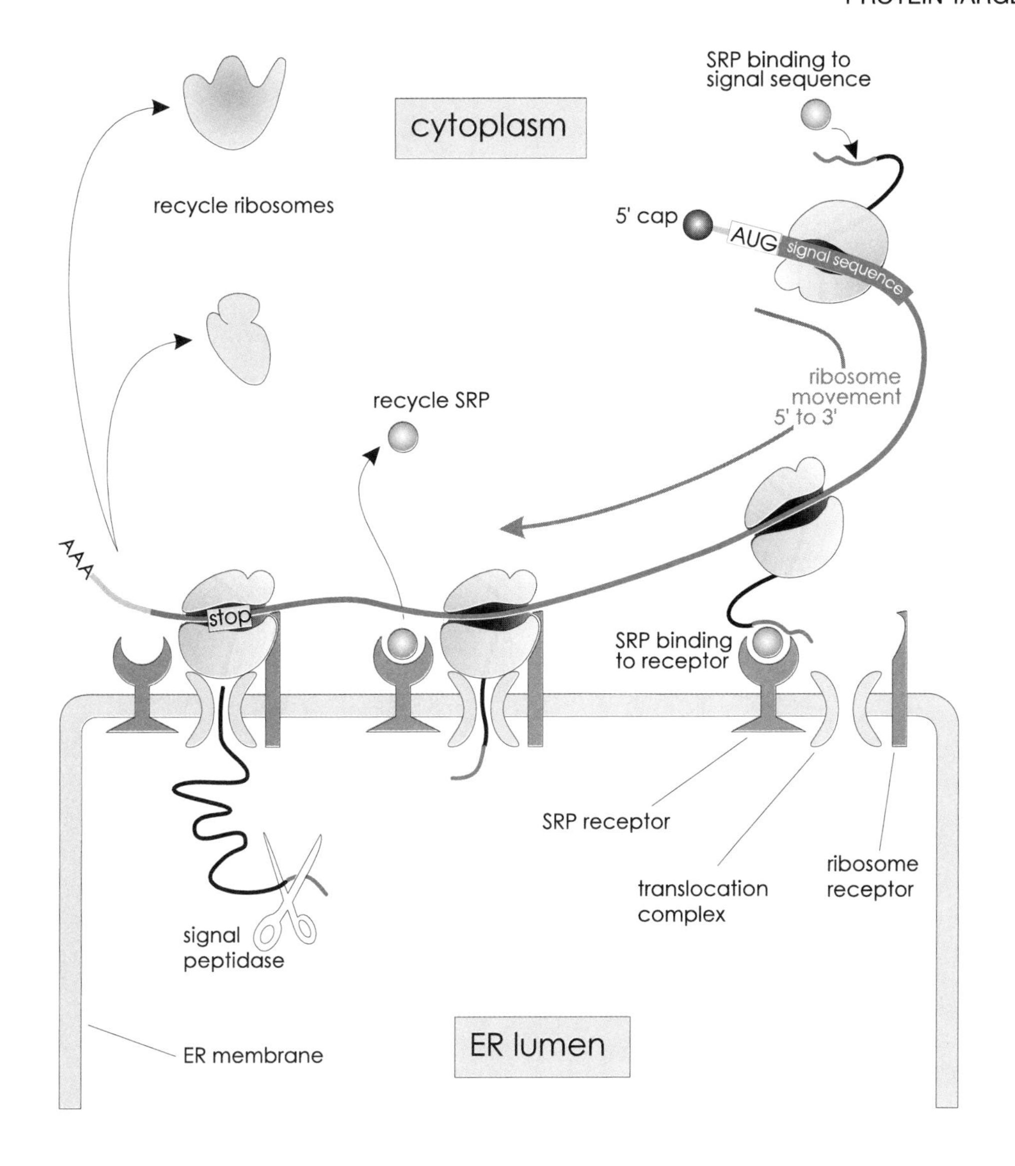

Figure 27.3.
Protein translocation through the ER by the SRP cycle. Nascent polypeptides that contain a signal sequence are directed to the ER by SRP binding interactions with the SRP receptor. The ribosome receptor and translocation complex facilitate protein translocation into the lumen of the ER, where a signal peptidase cleaves the signal peptide. Both the SRP and ribosomal components are recycled back into the cytoplasm.

Some proteolytic processing also takes place, especially of polyproteins. A polyprotein is a polypeptide that is cleaved at alternative sites after synthesis to form more than one mature protein. Some hormones are synthesized this way. The mature proteins are then distributed to the plasma membrane, the lysosomes, or the secretory vesicles.

Lysosomes

Lysosomes contain degradative enzymes that hydrolyze unwanted cellular molecules. Lysosomal enzymes reach the lysosomes via the Golgi. In the case of some lysosomal hydrolases, the Golgi adds a specific sugar residue, **mannose-6-phosphate,** to the enzymes which acts as a marker for lysosomal transport. Loss of the enzyme that adds mannose-6-phosphate gives rise to **I-cell disease,** a genetic abnormality that is usually fatal during childhood.

Secretory Vesicles

Proteins released by the Golgi are packaged into vesicles, which separate the proteins from the cytoplasm during transport. In the case of secretory vesicles, proteins destined for export from the cell are enclosed in secretory vesicles that bud off from the Golgi. These vesicles then diffuse to their destination, where they fuse with the membrane and release their contents.

Plasma Membrane

Proteins can be anchored to a membrane by attachment of a fatty acid (cytoplasmic face) or phosphatidylinosi-

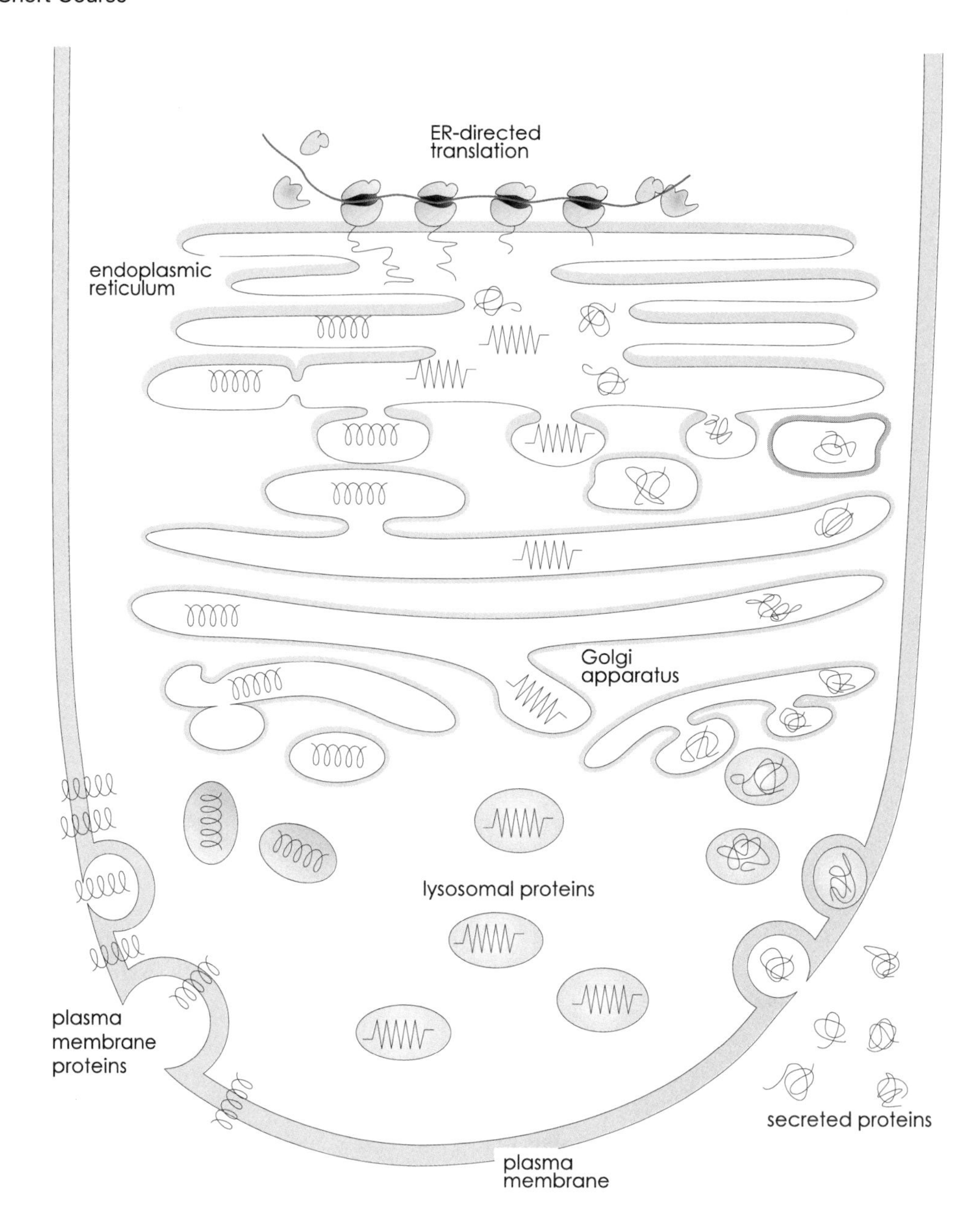

Figure 27.4.
Most proteins targeted to the ER by a signal peptide are directed to the Golgi apparatus for further sorting. Proteins leave the Golgi apparatus in membrane-coated vesicles and are destined for insertion in the plasma membrane, to become lysosomes or to fuse with the plasma membrane and release their contents as secreted proteins.

tol (luminal side of ER membrane). Proteins may also span a membrane, creating **transmembrane proteins.** The mechanism for threading a membrane-spanning protein through the membrane is probably analogous to the mechanism used to thread a protein through the ER membrane using a signal peptide.

PROTEIN TURNOVER

Several systems have been identified in cells to mediate protein turnover through degradation. **Lysosomes** contain a number of **proteases** and other **hydrolytic enzymes** (Fig. 27.5). All of these enzymes are **acid hydrolases** and have an optimum activity at ~pH 5. A primary role of lysosomes is to degrade proteins and other macromolecules that have been imported into the cell by endocytosis. **Endocytosis** is the process by which receptors on the cell surface bind to extracellular particles, invaginate the cell membrane, and form an intracellular vesicle that is delivered to the lysosome. Another source of protein for lysosomal-mediated protein degradation is through **autophagy,** for example, the turnover of mitochondria in liver cells. A third mecha-

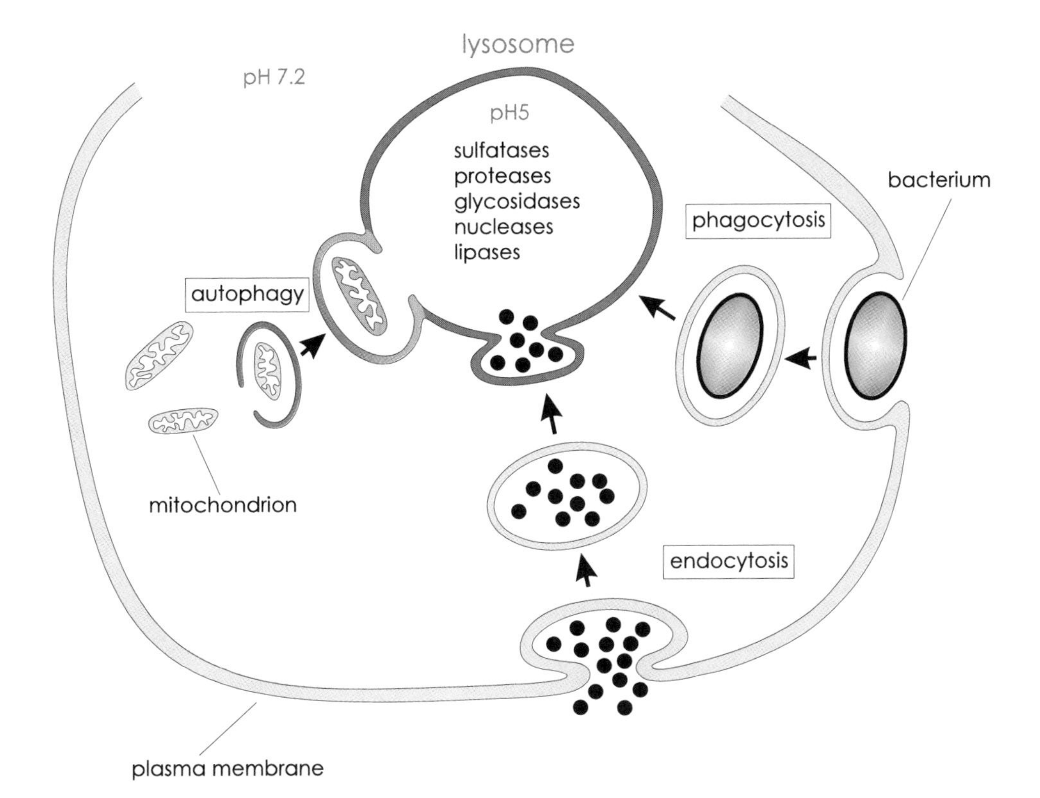

Figure 27.5.
The degradation of proteins by acid hydrolases in lysosomes involves vesicle fusion through three main pathways—autophagy, endocytosis, and phagocytosis. The low pH inside lysosomes is maintained by an ATP-driven proton pump.

nism is **phagocytosis,** such as the engulfment of bacteria by macrophages and neutrophils, two cell types present in the immune system.

Proteins, just like RNAs (Chapter 25), have a lifetime in the cell that is determined by their turnover rate. The half-life of a protein is the period of time it takes to degrade 50% of the protein and represents the stability of a protein. As shown in Table 27.1, protein half-lives are different and range from minutes to days. In general, regulatory proteins have short half-lives, such as those involved in metabolic control points, whereas proteins that provide "housekeeping" functions are long lived.

Protein Degradation in *E. coli*

In both bacterial and eukaryotic cells, cleavage of the peptide bond requires the hydrolysis of ATP. In *E. coli,* 70–80% of all ATP-dependent proteolysis is performed by two protease systems, the **Lon protease** and the **Clp A/P protease.** The Lon protease is composed of four identical subunits, whereas the Clp A/P protease is a regulated complex made up of the catalytic Clp C subunit and the ATP-dependent Clp A regulatory subunit. Signals that control the Lon and Clp A/C activities are not well understood, but one contribution may be the presence of certain **NH_2-terminal amino acids,** which

Table 27.1. Table of Protein Half-Lives

Protein	Half-life (hours)
α-2 repressor	0.1
c-Myc	0.5
RNA polymerase I	1.3
Tryptophan oxygenase	2.0
Phophoenolpyruvate carboxykinase	5.0
Glucocorticoid receptor	20
Chloramphenicol acetyltransferase	50
Aldolase	120

either confer protein stability or promote protein degradation depending on what amino acids are present. Alexander Varshavsky defined the **N-end rule** based on protein degradation studies in *E. coli* and yeast. By using the same target protein (β-galactosidase) and only changing the NH_2-terminal amino acids, they found that the amino acids Arg, Lys, Phe, Leu, and Trp caused β-galactosidase to be highly susceptible to degradation in both *E. coli* and yeast. It is likely that the Lon and Clp A/C protease systems in *E. coli* somehow recognize these amino acids as signals of proteolysis.

The Ubiquitin-Dependent Protein-Degradation Pathway

Varshavsky and colleagues have also characterized a nonlysosomal protein-degradation pathway in eukaryotic cells, primarily in yeast. In addition to confirming that the N-end rule applies to eukaryotic proteins as well, they identified a 76-amino-acid protein called **ubiquitin** as one of the major signals for protein degradation. Ubiquitin got its name because it is found in essentially every cell type ever examined. The sequence of ubiquitin is almost identical in yeast and humans,

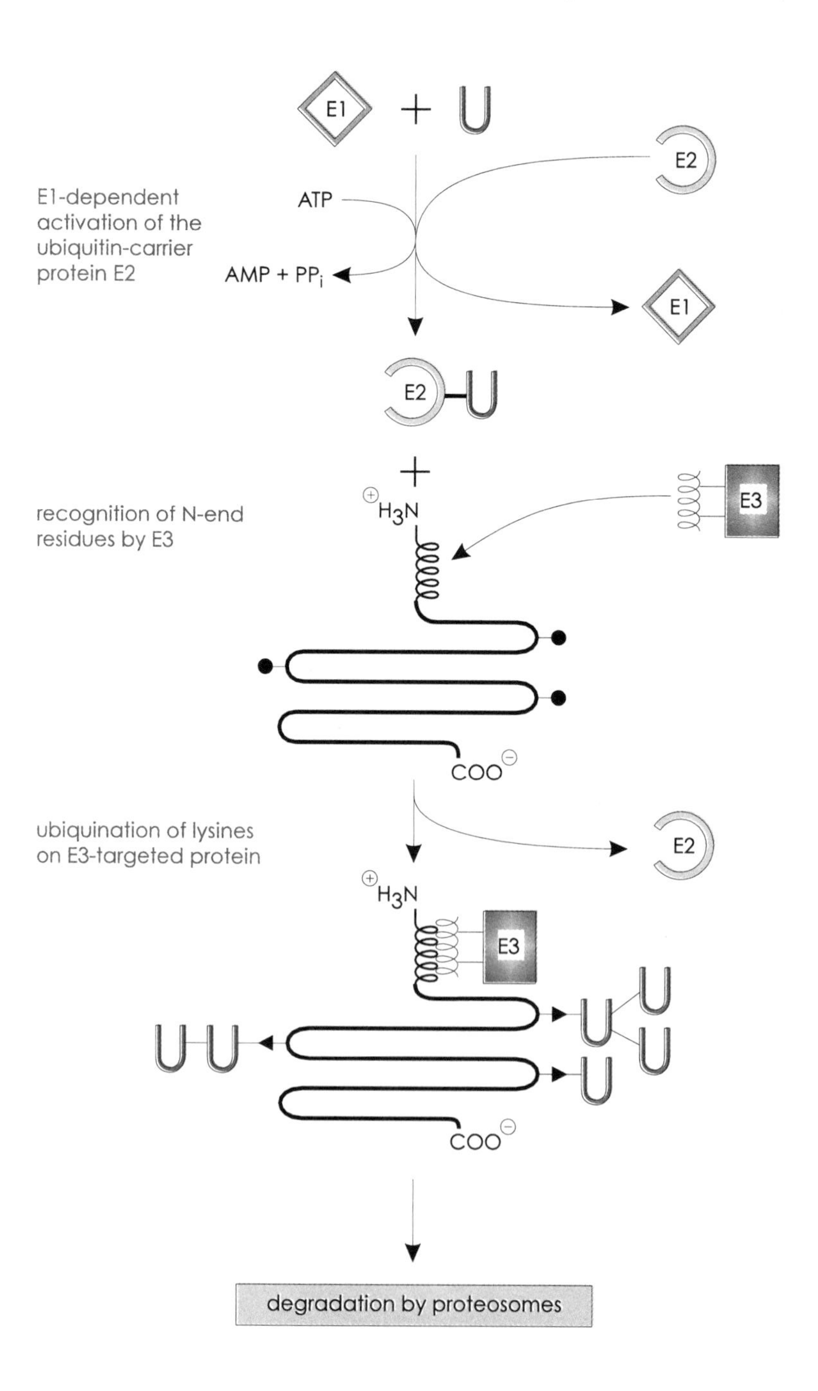

Figure 27.6.

Ubiquitin-dependent protein degradation pathway. ATP hydrolysis is required to attach ubiquitin to the carrier protein E1, which then passes the ubiquitin moiety to the activating protein E2. The protein E3 recognizes the NH_2-terminal residues and directs E2 to conjugate ubiquitin to lysine residues in the protein.

demonstrating intense evolutionary pressure to remain unchanged. Ubiquitin is a small, highly conserved protein which can be attached to proteins targeted for degradation through an **isopeptide linkage** of the *C*-terminal glycine of ubiquitin to specific lysine residues of the target protein.

A single ubiquitin molecule is not enough to signal degradation; for example, a small proportion of the stable histones H2A and H2B have a single attached ubiquitin. Rather, degradation is signaled by the presence of many ubiquitin molecules, usually as a **branched multiubiquitin chain.** As shown in Figure 27.6, ubiquitin is covalently attached to the epsilon-amino group of lysines through a series of steps involving at least three enzymes (E1, E2, and E3) and the hydrolysis of ATP.

According to Varshavsky's N-end rule, the presence of certain NH_2-terminal amino acids and polyubiquitin together act as a broadcast signal to the cells proteolytic machinery to degrade the protein. One of the physiological substrates of the N-end rule of protein degradation by the ubiquitin pathway is the α subunit of a yeast heterotrimeric guanine-nucleotide binding protein called Gα. A similar G-type Gα is also short lived in mouse cells, suggesting that similar mechanisms of G protein degradation occur in higher eukaryotes. Cyclins, proteins that control progression through the cell cycle, are regulated at the level of protein degradation by ubiquitination.

Proteosomes Are Cellular Garbage Disposals

Although the proteases involved in nonlysosomal ubiquitin-dependent protein degradation are not well characterized, the *site* of protein degradation in cells ap-

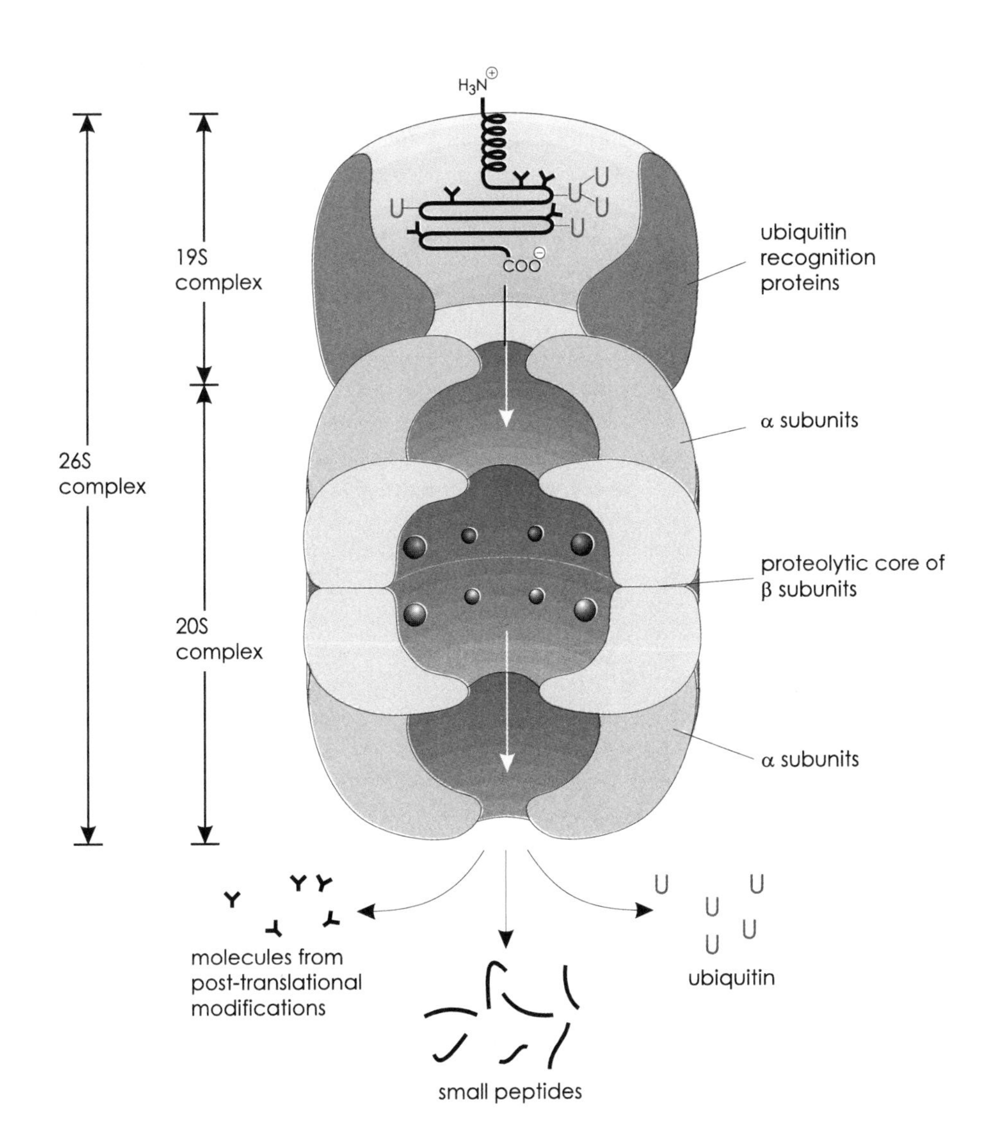

Figure 27.7.

Schematic drawing of an archaebacterium proteosome, illustrating its function as a cellular garbage disposal to degrade ubiquinated proteins.

pears to take place within large macromolecular structures called **proteosomes.** These large proteosome complexes require at least 10 different polypeptides to be functional. Structural data on proteosomes suggests that they are shaped like **hollow cylinders** with an opening on one end for the **ubiquinated protein,** a central region containing the catalytic sites of **multiple proteases,** and an exit port for the expulsion of **peptide fragments** (Fig. 27.7). The proteosome complex consists of two functional units, a 26S "holo" proteosome that contains a 19S ubiquitin-binding complex, and a 20S "core" proteosome that includes four stacks of seven-membered rings each consisting of seven α and seven β subunits. The 25.9 kDa β subunits form a central chamber that contains the peptidase activity required for protein degradation. Proteosomes have been called the trash cans of the cell, but perhaps garbage disposals would be a more accurate description!

PEST Sequences

Another signal for protein degradation, independent of ubiquitination, is the presence of **PEST sequences,** which are amino-acid sequences containing large amounts of Pro (P), Glu (E), Ser (S), and Thr (T). The presence of multiple PEST sequences is correlated with high turnover rates of proteins. Experiments have shown that, for some proteins, removal of the PEST sequences can result in protein stabilization. However, the mechanism by which PEST sequences determine protein turnover rates is still not known, although it is thought to be mediated by sequence-selective proteases.

SUMMARY

1. Signal peptides function as subcellular "addresses" to target proteins to specific compartments in the cell. These sequences are often at the NH_2 or COOH termini of the protein, but others are located within the sequence.

2. The ER signal peptide is ~10 amino acids long, is present on the NH_2 terminus of ER-targeted proteins, and consists of mostly hydrophobic residues. Nascent polypeptide chains on polyribosomes that contain the ER signal peptide are directed to the ER membrane to complete protein synthesis; polypeptides lacking this signal peptide are synthesized in the cytoplasm.

3. Proteins targeted to cytoplasmic compartments contain a variety of signaling sequences which direct them to the cytoplasmic face of membranes, mitochondria, the nucleus, or peroxisomes.

4. Protein targeting to membranes involves recognition of the ER signal peptide by the SRP, which facilitates attachment of the peptide–ribosome complex to the SRP receptor on the ER. Protein synthesis at the ER membrane results in translocation of the newly synthesized protein into the ER lumen. Once the protein is refolded inside the ER, it can be targeted to additional cellular locations such as the Golgi apparatus, lysosomes, secretory vesicles, or the plasma membrane, depending on what other signal sequences are present in the protein.

5. Protein turnover by degradative processes in the cell is an important regulatory mechanism because it controls protein concentrations. The Lon and Clp A/P proteases in *E. coli,* and the ubiquitin-dependent degradation pathway in eukaryotes, are two mechanisms of protein degradation that have been characterized. Proteosomes are large macromolecular structures that function as intracellular garbage disposals to hydrolyze proteins into small peptide fragments.

REFERENCES

Banfield D. K., Lewis M. J., Pelham HRB (1995): A SNARE-like protein required for traffic through the Golgi complex. *Nature* **375:**806.

Bennett M. K., Scheller R. H. (1994): A molecular description of synaptic vesicle membrane trafficking. *Annu. Rev. Biochem.* **63:**63.

Ciechanover A. (1994): The ubiquitin–proteasome proteolytic pathway. *Cell* **79:**13.

Craig E. A., Weissman J. S., Horwich A. L. (1994): Heat shock proteins and molecular chaperones: Mediators of protein conformation and turnover in the cell. *Cell* **78:**365.

Ellis, R. J. (1996): The "bio" in biochemistry: Protein folding inside and outside the cell. *Science* **272:**1448.

Fabre E., Hunt E. C. (1994): Nuclear transport. *Curr. Biol.* **6:**335.

Goldberg A. L. (1995): Functions of the proteasome: The lysis at the end of the tunnel. *Science* **268:**522.

Hilt W., Wolf D. (1996): Proteasomes: destruction as a program. *Trends Bioc. Science 21*:96.

Lord, J. M., (1996): Protein degradation: Go outside and see the proteasome. *Curr. Biol.* **6:**1067.

Löwe J., Stock D., Jap B., Zwicki P., Baumeister W., Huber R. (1995): Crystal structure of the 20S proteasome from the archaeon *T. acidophilum* at 3.4 Å resolution. *Science* **268:**533.

Mellman I., Simons K. (1992): The Golgi complex: In vitro veritas? *Cell* **68**:829.

Moore M. S. (1995): David and Goliath in nuclear transport. *Curr. Biology* **5**:1339.

Rechsteiner, M. and Rogers, S. W. (1996): PEST sequences and regulation by proteolysis. *Trends Biochem.Sci.* **21**:267.

Rothman J. E. (1994): Mechanisms of intracellular protein transport. *Nature* **372**:55.

Schatz G., Dobberstein B. (1996): Common principles of protein translocation across membranes. *Science* **271**:1519.

Scheffner M., Nuber U., Huibregtse J. M. (1995): Protein ubiquitination involving an E1–E2–E3 enzyme ubiquitin thioester cascade. *Nature* **373**:81.

Valle D., Gärtner J. (1993): Human genetics: Penetrating the peroxisome. *Nature* **361**:682.

Varshavsky A. (1992): The N-end rule. *Cell* **69**:725.

Wickner W. T. (1994): How ATP drives proteins across membranes. *Science* **266**:1197.

Wolin S. L. (1994): From the elephant to *E. coli*: SRP-dependent protein targeting. *Cell* **77**:787.

Zheng, N., Gierasch, L. M. (1996): Signal sequences: The same yet different. *Cell* **86**:849.

REVIEW QUESTIONS

1. The major function of signal peptide sequences in proteins is to
   a) direct protein folding into the correct tertiary structure
   b) mark proteins for subcellular targeting
   c) influence the half-life of a protein
   d) increase the DNA binding affinity of proteins
   e) target proteins to various tissues in the body

2. The presence of an ER signal peptide at the amino terminus of a protein often initiates
   a) protein translocation to the mitochondria
   b) ribosome binding to the nuclear membrane
   c) protein trafficking through the Golgi apparatus
   d) protein targeting to peroxisomes
   e) premature termination of protein synthesis

3. How does SRP binding to an ER signal peptide result in localization of that protein to the ER lumen?
   a) Bifunctional SRP interactions between the signal peptide and the SRP receptor result in ribosome binding to the ER membrane.
   b) The SRP receptor cleaves the signal sequence following SRP binding.
   c) SRP is an RNA binding protein that crosslinks polyribosomes to the SRP receptor.
   d) Following the completion of protein synthesis in the cytoplasm, SRP directs the protein to ER membrane pores.
   e) SRP blocks the signal peptide from interacting with the nuclear-pore complex.

4. The two primary mechanisms of protein degradation in eukaryotic cells are
   a) autophagy and phagocytosis
   b) cytosolic exopeptidases and endocytosis
   c) lysosomal degradation pathways and PEST-sequence-mediated degradation
   d) lysosomal degradation pathways and ubiquitin-targeting to proteosomes
   e) PEST-sequence-mediated degradation and the Clp A/P protease system

5. Proteosomes are
   a) cellular organelles involved in protein trafficking
   b) small single-cell aquatic organisms found in ponds
   c) ribonucleoprotein complexes required for protein synthesis
   d) macromolecules that autodigest mitochondria
   e) macromolecules that degrade ubiquinated proteins

ANSWERS TO REVIEW QUESTIONS

1. ***b*** Signal peptides function like an address on an envelope.
2. ***c*** Most ER-targeted proteins move through the Golgi apparatus.

3. ***a*** SRP binding to both the signal peptide and SRP receptor facilitates ribosome binding to the ribosome receptor and subsequent protein translocation.

4. ***d*** Lysosomes and proteosomes degrade the majority of cellular proteins.

5. ***e*** Proteosomes degrade proteins and release peptide fragments through a mechanism involving an internal core of proteases.

CHAPTER

28

MECHANISMS OF GENE REGULATION IN PROKARYOTES

INTRODUCTION

Controlling when a gene is transcribed is an efficient way to preserve limited cellular resources. If a cell does not require a particular gene product, it can save energy by not producing it. This logic would seem to explain why, in the many cases where mechanisms of gene expression have been examined, the critical control point is often at the level of **transcriptional initiation.** However, as described in Chapters 25–27, there are a variety of other mechanisms whereby posttranscriptional and posttranslational processes can be utilized to control the steady-state levels of cellular proteins.

In this chapter we begin by defining several conceptual terms which apply to transcriptional regulation using a well-characterized system in *E. coli,* the *lac* operon. Following this we describe the transcriptional regulation of the *ara* and *trp* operons in *E. coli* which represent two of the best examples of positive and negative transcriptional control in prokaryotes. These three paradigms of transcriptional control in *E. coli* were specifically chosen because they demonstrate important concepts in gene regulation which we build on in Chapter 29 when we describe the more complex transcriptional regulatory circuits found in eukaryotes.

Note that we use the nomenclature of "*gene*" to denote the segment of DNA corresponding to the gene, and "Gene" to signify the protein product of the gene. For example, the expression of *lacZ* results in the synthesis of LacZ.

TRANSCRIPTION FACTORS CAN FUNCTION AS ACTIVATORS OR REPRESSORS OF TRANSCRIPTION

As described in Chapter 24, a gene consists of all the DNA sequences required for the regulation and expression of a gene product. In a typical protein-coding gene, this means that the promoter and upstream **regulatory regions** are in the nontranscribed portion of a gene, whereas the mRNA transcript represents the transcription unit. What is the mechanism by which the regulatory region of a gene is able to control the rates of transcriptional initiation at the promoter? The answer is not yet completely clear; however, it is known that DNA binding proteins interact with specific DNA sequences in the regulatory region of genes. These proteins are commonly called **transcription factors** and are thought to interact directly or indirectly with proteins in the transcriptional initiation complex. The combined effect of transcription-factor DNA binding and protein–protein interactions leads to alterations in the transcriptional initiation complex which modulate RNA polymerase activity.

The descriptive terms **transcriptional induction** (increased rates of initiation), and **transcriptional repression** (decreased rates of initiation), are often used to describe the function of transcription factors. Transcription factors that induce transcription are called **activators,** whereas those that repress transcription are called **repressors** (Fig. 28.1). There are examples of transcription factors that can function either as activators or repressors, depending on the physiological state of the cell and DNA sequence differences between various promoters. Transcription factors bind to DNA sequences in the regulatory regions of genes that are commonly called **response elements.** Transcription-factor binding to response elements results in altered rates of transcriptional initiation through a combination of protein–protein and protein–DNA interactions that modulate RNA polymerase activity.

Genes with **low basal rates of transcription** are often responsive to activators (a common mechanism in eukaryotes), whereas genes with high basal rates of transcription usually respond to repressors (typical of prokaryotes). Moreover, many genes can respond to both activators and repressors at different times, depending on the cellular environment. Finally, some genes, called **housekeeping genes,** encode important structural and metabolic proteins and may not respond at all to typical transcriptional regulators, but instead are constitutively expressed.

A common mechanism used by cells to fine tune transcriptional regulation in response to extracellular signals is the modulation of transcription-factor DNA binding. For example, the interaction of a small **effector molecule** with a transcription factor could alter its DNA binding activity. Figure 28.2 illustrates how typical activators and repressors can be controlled in the absence and presence of an effector molecule.

In prokaryotes, the effector could be cAMP or some small **metabolic product,** whereas the activity of eukaryotic transcription factors is often modulated by phosphorylation (Chapter 29). It can be seen from the schematic models in Figure 28.2 that there are at least

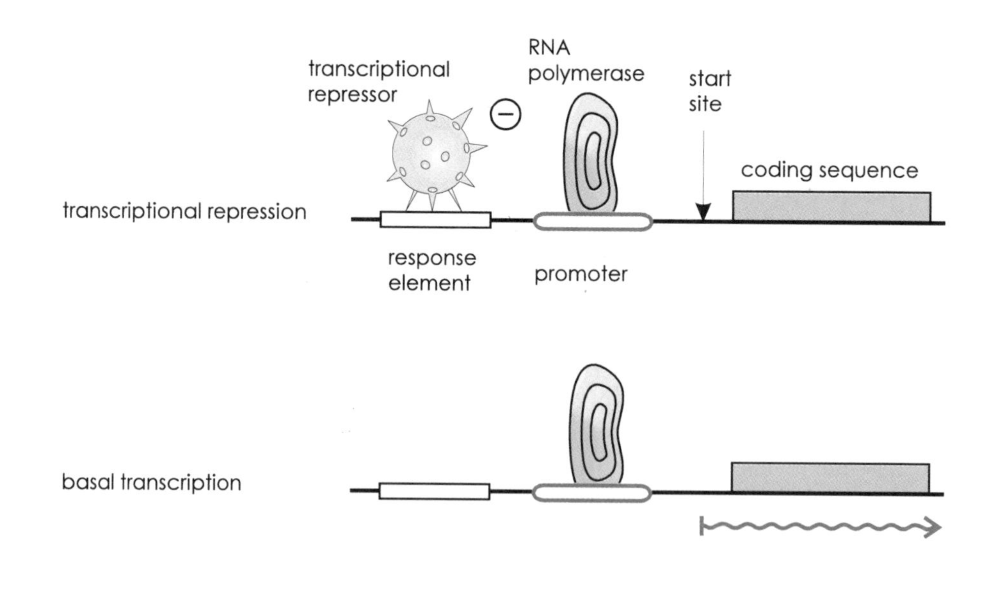

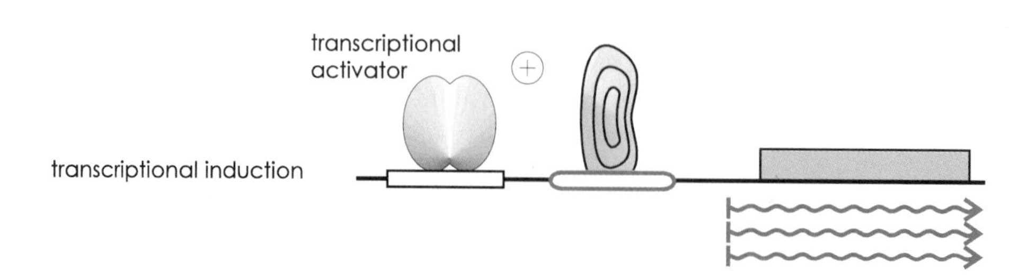

Figure 28.1.
Three rates of transcriptional initiation are described as repressed, basal, and induced. These relative promoter efficiencies result from transcription-factor modulation of RNA polymerase activity.

two mechanisms whereby transcription rates can be increased (**induction and derepression**) and decreased (**repression and deinduction**). In the examples shown, the key regulatory step is modulation of DNA binding activity. As described in Chapter 5, the molecular structure of a number of transcription factors, in the absence and presence of DNA, has been solved by X-ray crystallography.

TRANSCRIPTIONAL CONTROL OF THE *lac* OPERON IN *E. coli*

To understand how changes in the cellular environment can result in altered transcription rates of specific genes, we first examine the effect of **lactose** and **glucose** on the transcription of genes in the ***lac* operon** of *E. coli.* In bacteria, it is often found that genes encoding proteins that function together in a metabolic pathway are physically linked in the DNA segment, and in fact, these genes can share the same promoter. **Gene clusters** such as these in bacteria are called **operons.** The *lac* operon in *E. coli* is ~5000 base pairs in length and includes the *lac* promoter and regulatory region (called the ***lac* operator**) and the coding sequences for three genes (*lacZ, lacY, and lacA*) as shown in Figure 28.3.

When a single mRNA transcript contains the coding information for more than one protein, and those proteins are translated as independent polypeptides, it is called a **polycistronic mRNA.** Such polycistronic mR-

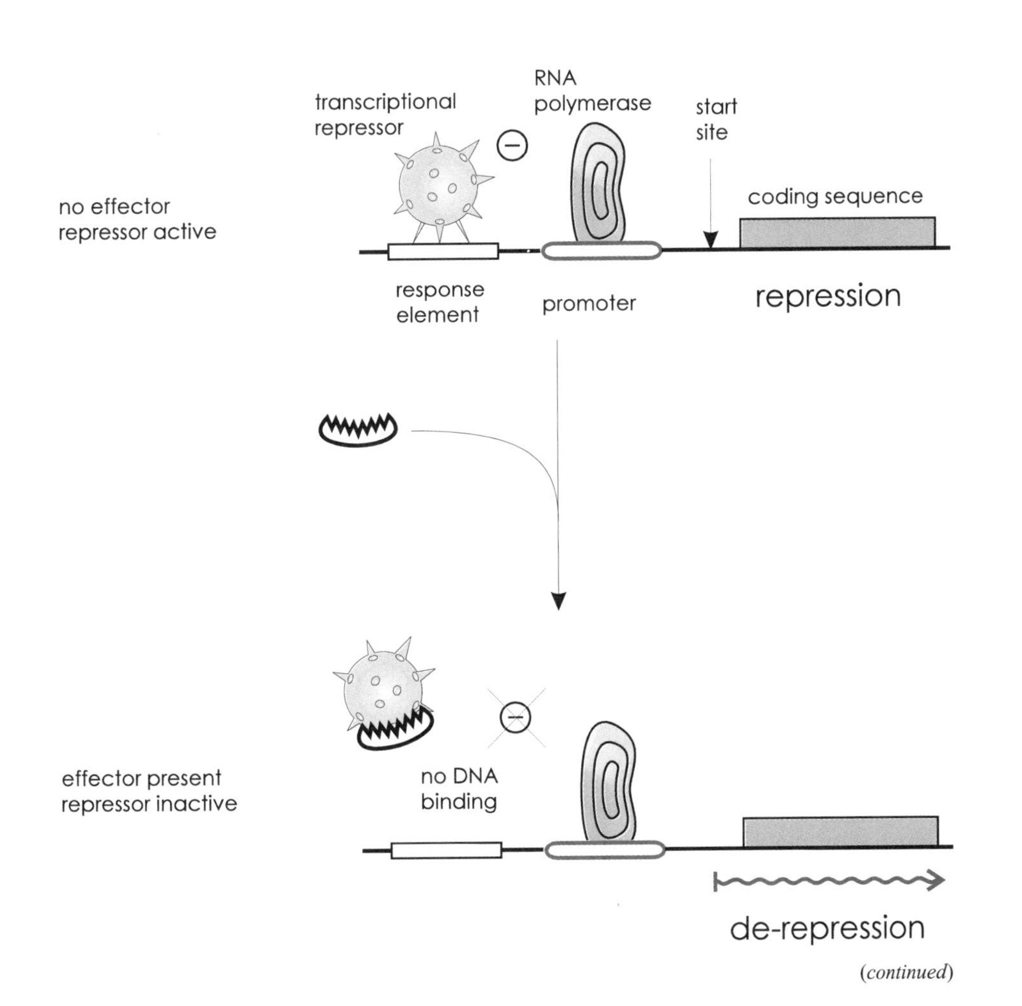

Figure 28.2.
Modulation of activator and repressor function by effector molecules. Repressors can either be inhibited (A) or stimulated (B) by effector-molecule interactions. Similarly, activators can either be stimulated (C) or inhibited (D) by effector-molecule interactions. Transcription-factor functions can also be modulated by posttranslational modification such as phosphorylation (see Fig. 29.10).

(*continued*)

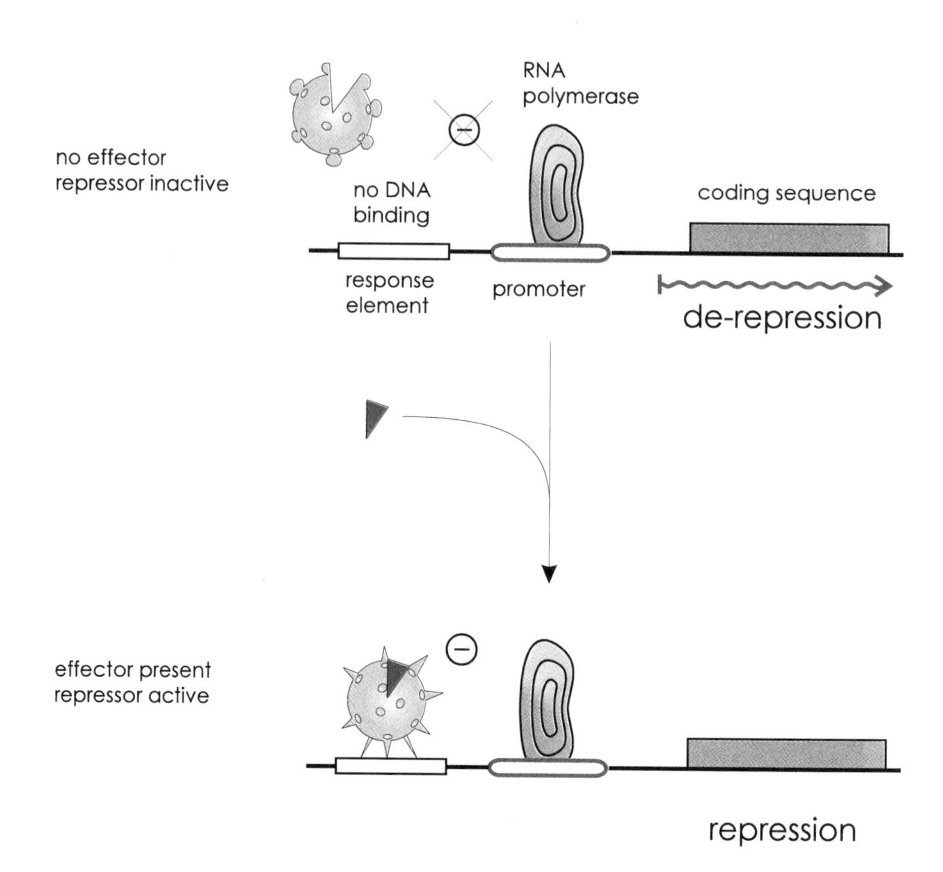

Figure 28.2. (*continued*)

NAs are common in bacteria. The polycistronic *lac* mRNA can be translated to produce β-galactosidase (LacZ), permease (LacY), and thiogalactoside transacetylase (LacA).

When lactose, or other **β-galactosides,** are present in the environment, transcription of the *lac* operon is induced. The result is an increase in all three enzymes, most importantly, β-galactosidase, which converts lactose to allolactose, and in a second reaction, rapidly cleaves allolactose to yield galactose and glucose (Fig. 28.4). Based on biochemical studies which revealed that allolactose is the active inducer of *lac* operon transcription, it was found that a synthetic nonhydrolyzable β-galactoside called isopropyl-β-D-thiogalactoside **(IPTG),** could function as a potent inducer of the *lac* operon. IPTG is used routinely to induce the expression of recombinant proteins in bacteria that contain plasmid vectors utilizing the *lac* promoter.

IPTG Causes Derepression of the *lac* Operon by Inactivating a Repressor

Genetics provided the first clue to the mechanism of *lac* operon regulation in the presence of IPTG. Several types of *E. coli* mutants were isolated that had defects in the regulatory pathway leading to elevated levels of *lac* mRNA. As shown in Figure 28.5A, the first class of mutants mapped to the *lac* operon itself and DNA sequence analysis revealed that the mutations were in the region of the *lac* promoter called the **lac operator** (functionally defined as a transcription factor-binding site or response element). The phenotype of these **operator mutants** indicated that the level of *lac* gene expression was constitutive, meaning that with or without IPTG, the gene was always on. These were called *lac* O^c mutants for operator **constitutive mutants.**

A second constitutive phenotype was found that

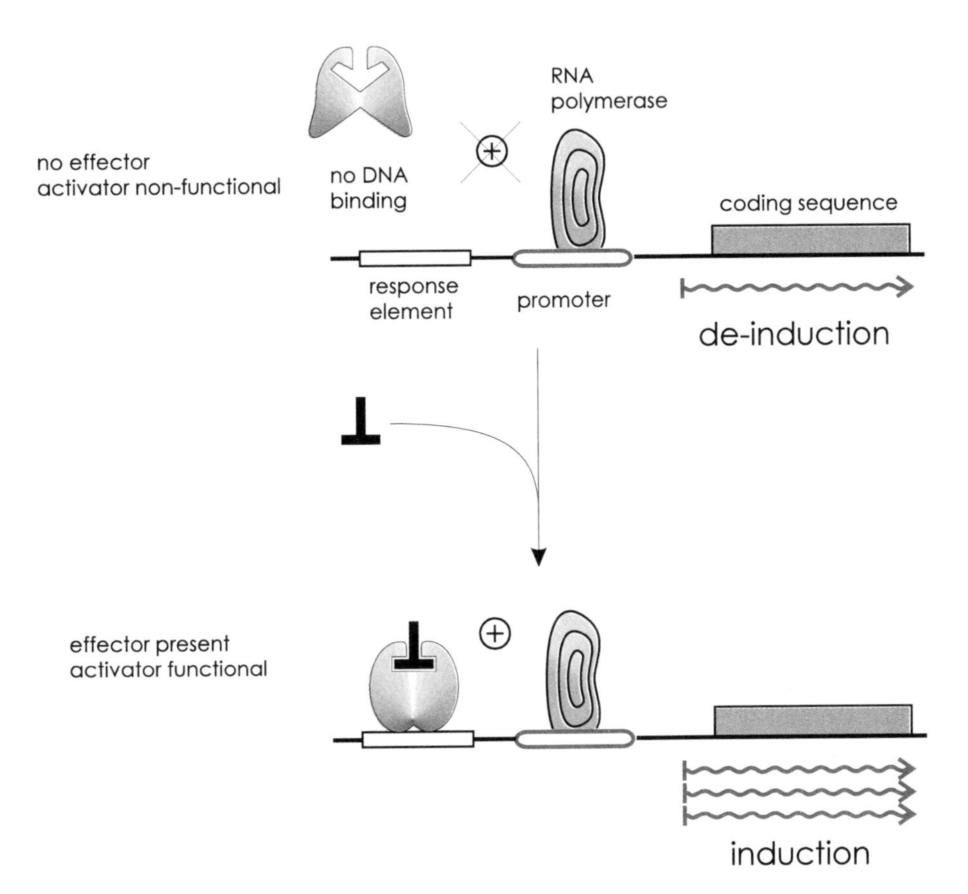

Figure 28.2. (*continued*)

mapped to the *lacI* gene, which lies just outside the *lac* operon. Molecular analysis of *lacI* mutations revealed that they were in the coding sequence of a **repressor protein** (Fig. 28.5B). In vitro DNA binding analysis demonstrated that some of the ***lacI* mutations** resulted in the synthesis of defective repressor protein that had **decreased DNA binding activity** owing to mutations in the functionally defined DNA binding domain. Genetic and biochemical studies demonstrated that the *lac* operon is controlled by the LacI repressor protein. In the absence of inducer (lactose or IPTG), the LacI repressor binds very tightly to the *lac* operator DNA sequences. The equilibrium constant for LacI repressor binding to the *lac* operator is $\sim 2 \times 10^{13}$/M. This specific binding affinity is $\sim 10^7$ times higher than the affinity of the LacI repressor for nonspecific DNA sequences.

The IPTG interacts with the LacI repressor and functions as a modulator of its DNA binding activity. In vitro DNA binding experiments reveal that the LacI repressor–IPTG complex has a 10^3-fold lower affinity for the *lac* operator sequence than the LacI repressor alone, even though IPTG has no effect on repressor binding to nonspecific sites. Taken together, the mechanistic effect of IPTG as an inducer of *lac* gene expression is **transcriptional derepression** of the *lac* operon. As shown in Figure 28.6, the *lac* repressor binding site (*lac* operator) overlaps the RNA polymerase binding site, which would support the role of LacI repressor as an inhibitor of transcriptional initiation. In vitro DNA footprinting studies demonstrate that RNA polymerase and LacI repressor binding to the operator sequence is **mutually exclusive.**

Catabolic Gene-Activator Protein (CAP) Positively Regulates the *lac* Operon

Control of the *lac* operon by derepression is only part of the story. Another regulatory system also plays a role

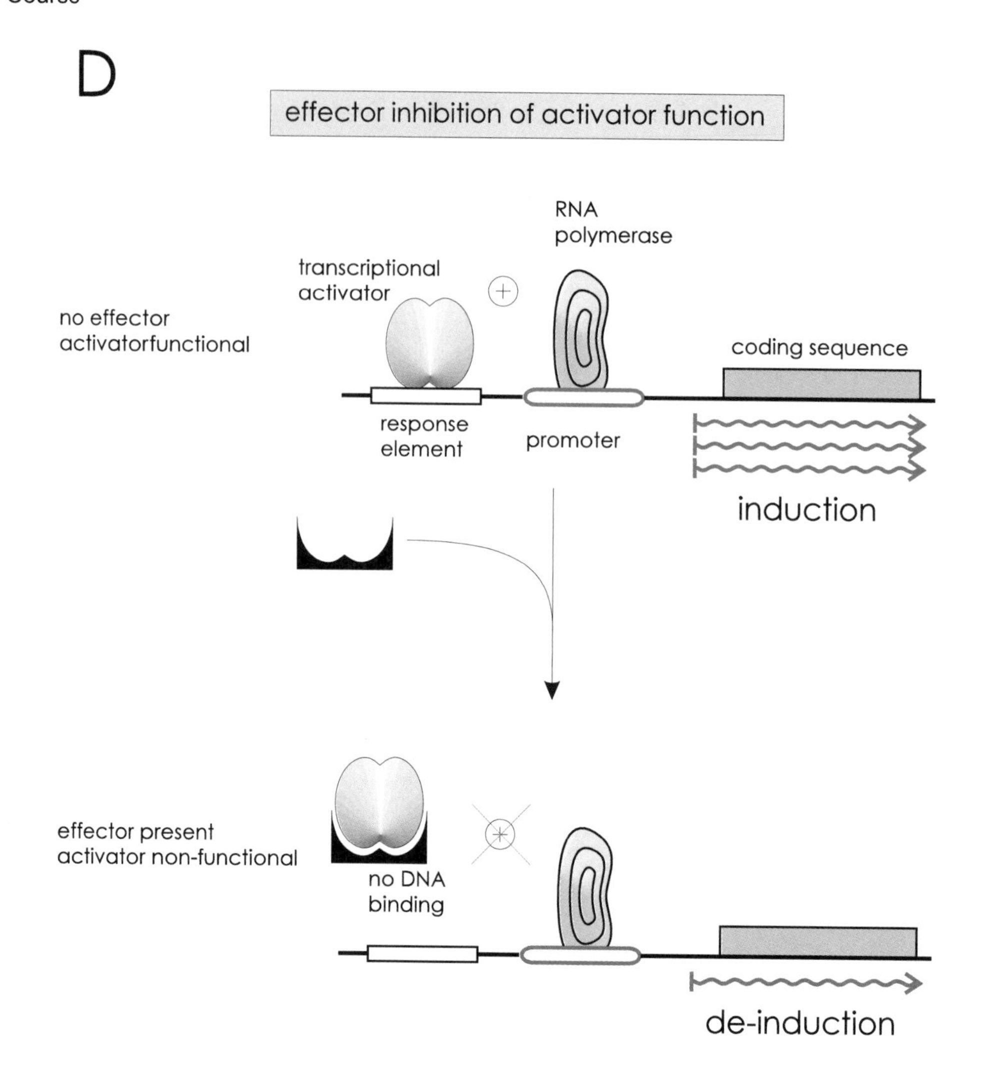

Figure 28.2. (*continued*)

in the transcriptional induction of the *lac* operon. In bacteria, glucose is the preferred carbon source, and when glucose and other sugars are present at the same time, glucose is metabolized first. This pathway is partially controlled by the **catabolic gene-activator protein** (CAP), which activates transcription of operons such as the *lac* operon, but only when lactose is present and glucose levels are low. The mechanism for **CAP** control of the *lac* operon is based on the finding that CAP is a transcription factor that binds with high affinity to DNA sequences found in the promoters of genes subject to catabolic control, for example, the lactose, galactose, and arabinose operons.

To understand transcriptional control by CAP, we

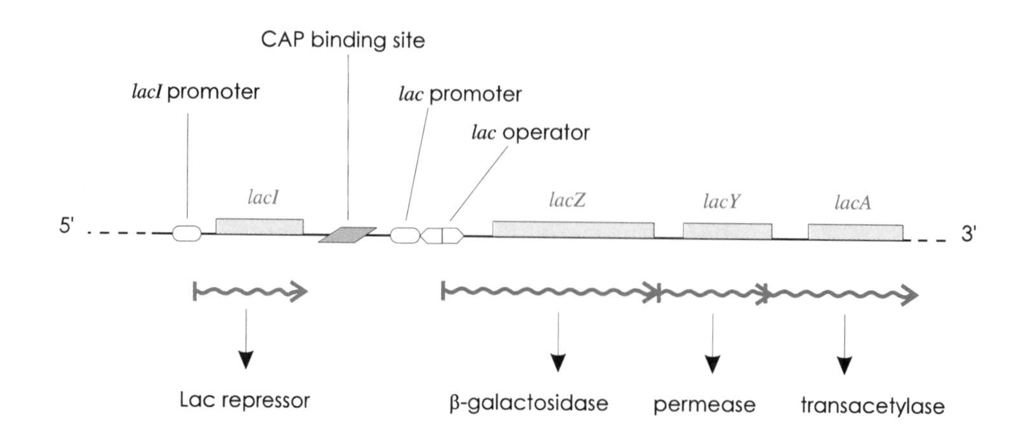

Figure 28.3.

Functional map of the *lac* operon in *E. coli.* Note that the *lacI* gene is physically linked to the *lac* operon, but it is not considered to be part of the *lac* operon because the *lacI* promoter is not regulated by factors that regulate the *lac* operon.

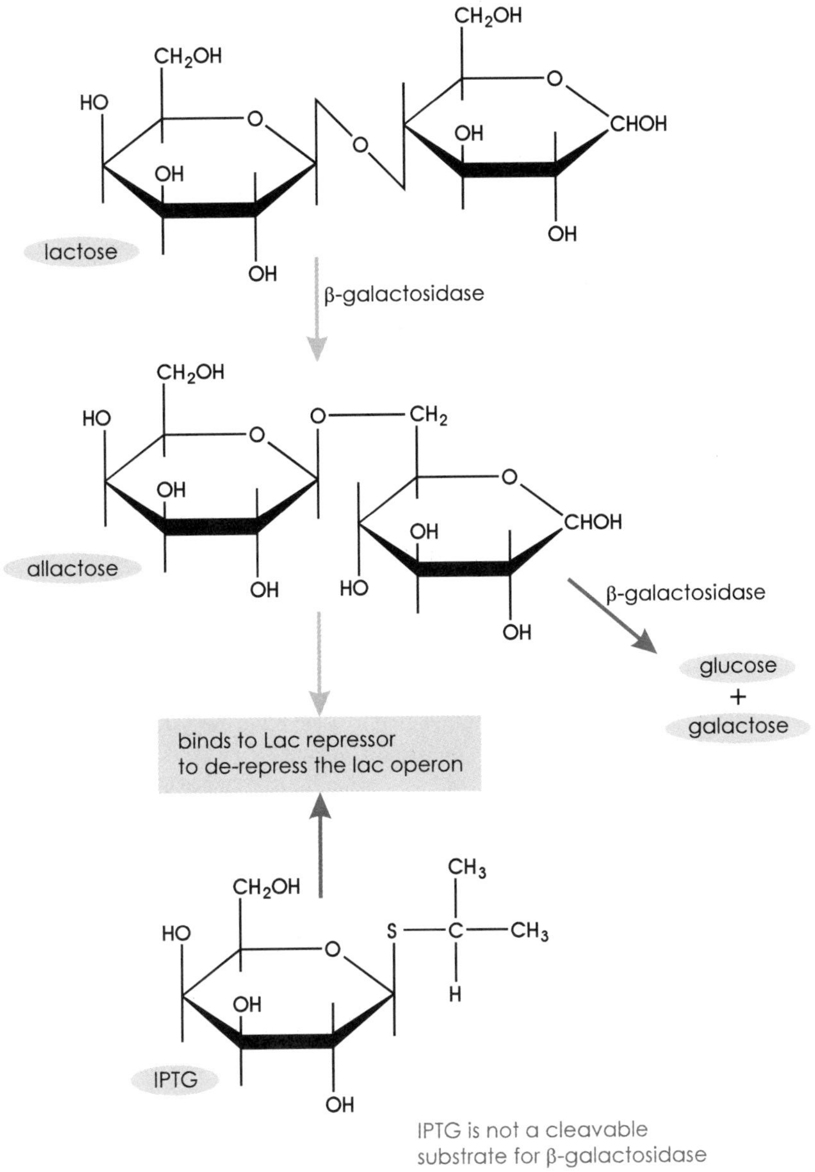

Figure 28.4.
Allactose and IPTG are inducers of the *lac* operon. β-Galactosidase rapidly produces galactose and glucose from lactose, but IPTG is a nonhydrolyzable β-galactoside and, therefore, a physiologically more potent inducer of the lac operon.

will examine CAP regulation of the *lac* operon. In the presence of both glucose and lactose, transcription from the *lac* promoter is fairly weak, even though the repressor is not bound. However, when the levels of glucose decrease, and lactose is still present, transcription from the *lac* operon increases dramatically. The reason for this is that **CAP** is an activator of the *lac* operon, and positively regulates transcriptional initiation. Biochemical crosslinking studies have shown that a region on CAP known to be required for transcriptional regulation is in **direct physical contact** with the carboxy terminus of the *E. coli* **RNA polymerase α subunit** when bound in a ternary complex at the *lac* promoter. These data, combined with genetic analyses, suggest that CAP facilitates RNA polymerase promoter binding and open-complex formation.

How does the presence or absence of glucose control CAP binding to the *lac* promoter? The answer is similar to what was shown for the *lac* repressor. In this case, cAMP functions as a modulator and binds to CAP to positively control its DNA binding activity. **Glucose transport** in *E. coli* leads to **deactivation of adenylyl cyclase,** which results in is low levels of intracellular cAMP. Therefore, when glucose is present, cAMP levels are low and CAP is inactive. However, when glucose levels are low, adenylyl cyclase is active, and intracellular cAMP levels increase, leading to an activation of CAP. Figure 28.7 summarizes the transcriptional ac-

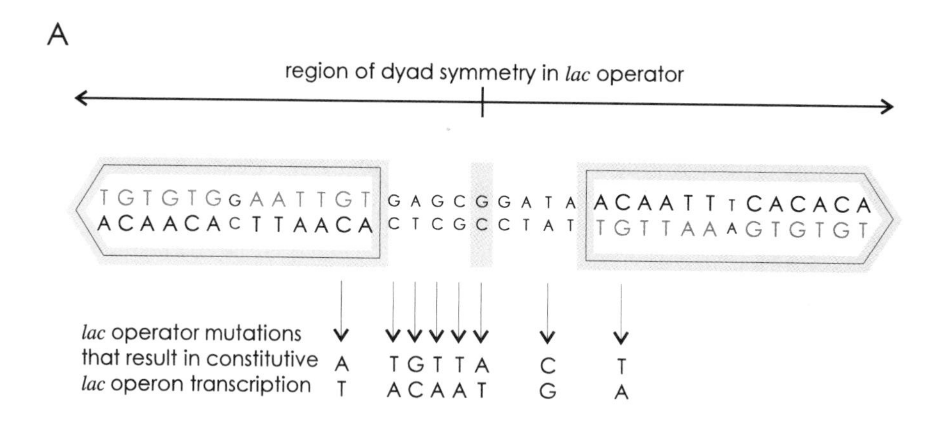

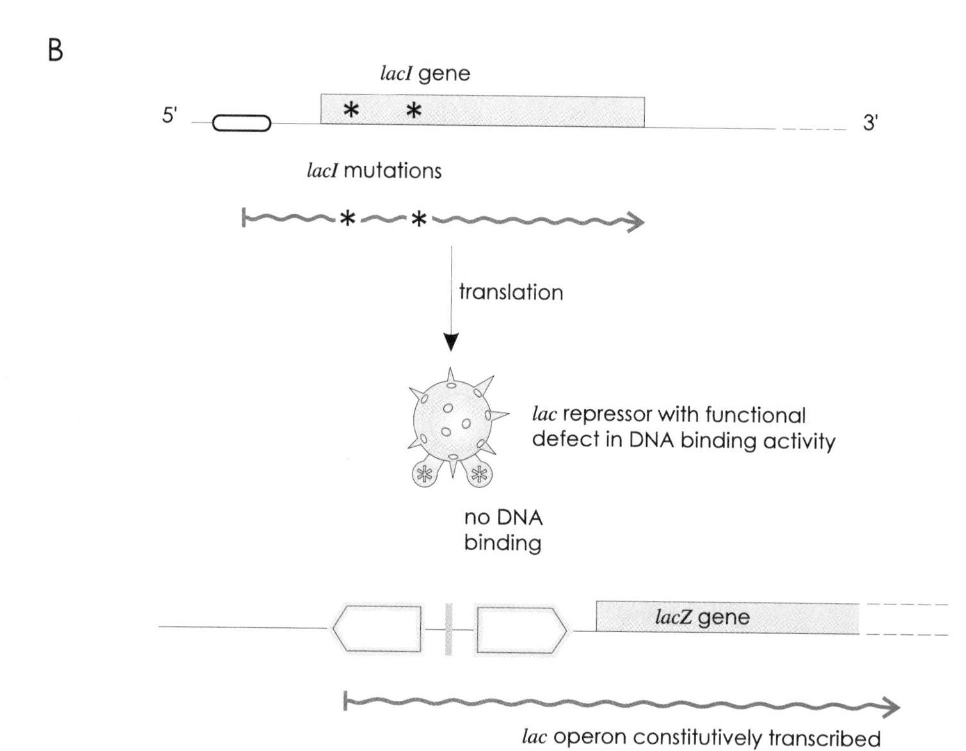

Figure 28.5.

Two classes of constitutive *lac* mutants that result in expression of the *lac* operon even in the absence of inducer. (A) The *lac* operator mutants that change the nucleotide sequence within the Lac repressor binding site, a region that contains an inverted repeat sequence. (B) The *lac* repressor (LacI) mutants that alter the amino-acid sequence within the DNA binding domain of the Lac repressor and thereby disrupt DNA binding function.

tivity of the *lac* operon under conditions of both low and high glucose and lactose levels. It is only when glucose levels are low, and lactose is present, that the *lac* operon is maximally transcribed, owing to the combined effects of LacI repressor-mediated derepression and CAP-dependent transcriptional induction through a direct interaction with the α subunit of RNA polymerase. What happens when both lactose and glucose levels are low? Experiments have shown that *lac* repressor binding to the *lac* operator is still able to block RNA polymerase from binding to the promoter, even though activated CAP is present (Fig. 28.7).

The following list summarizes several important concepts in regulated gene expression which are exemplified by the *lac* operon and apply to other systems:

1. Changes in the cellular environment lead to alterations in gene expression through modulation of transcription-factor DNA binding activity.
2. The binding of transcription regulatory factors to DNA is highly specific and often involves the formation of subunit multimers.
3. Transcription factors can either inhibit or enhance transcriptional initiation rates, most

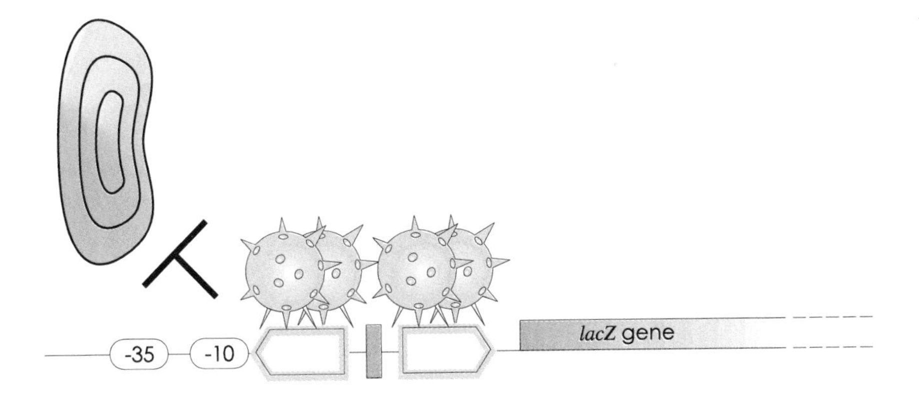

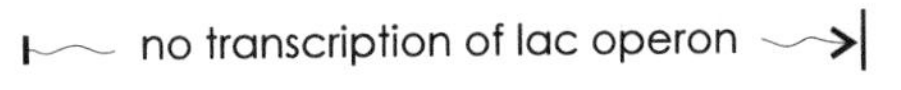

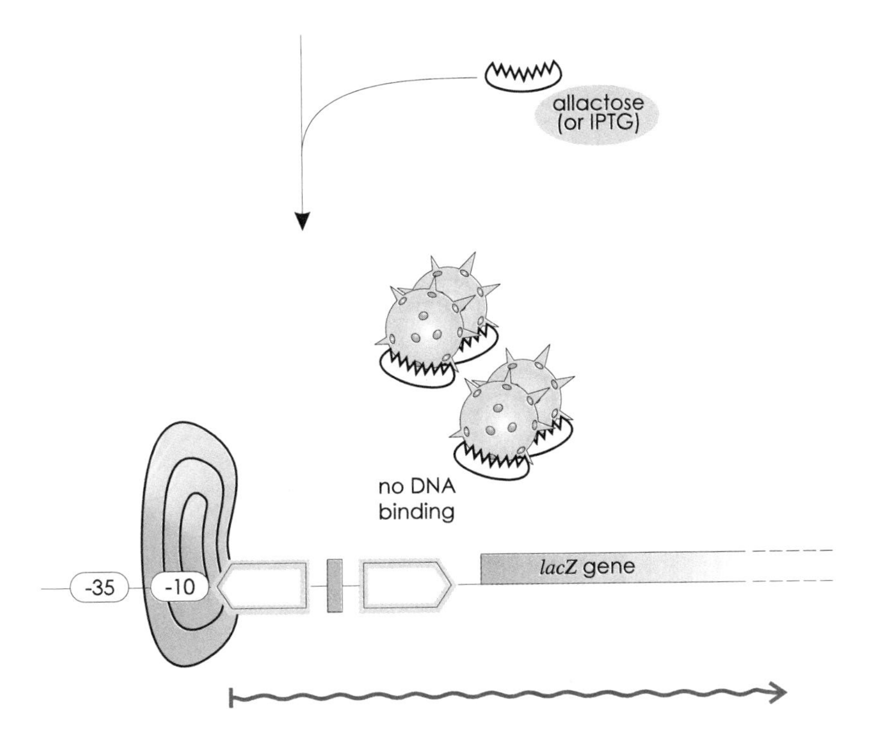

Figure 28.6.

Mechanism of transcriptional derepression of the *lac* operon by IPTG. Note that the Lac repressor binds to the *lac* operator as a tetramer, which is shown here schematically as two dimers binding the inverted repeat sequences.

likely through mechanisms that include direct or indirect protein–protein interactions with RNA polymerase.

AraC IS BOTH A REPRESSOR AND AN ACTIVATOR OF THE ARABINOSE OPERON

The arabinose (*ara*) operon in *E. coli* utilizes some of the same regulatory principles as the *lac* operon, but three additional mechanisms are also found. First, in the ***ara* operon,** the expression of a transcription factor gene is autoregulated by it's own gene product (**AraC** controls *araC* expression). Second, a transcription factor can function as a repressor or an activator in an effector-dependent manner (arabinose changes AraC from a repressor into an activator). Third, transcription factors often function together to effect a change in transcriptional initiation (AraC and CAP must both be bound to fully induce the *ara* operon).

The availability of arabinose as a carbon source for *E. coli* signals the induction of the *ara* operon, which encodes three proteins required for arabinose metabolism: ribulose kinase (*araB* gene), arabinose isomerase (*araA* gene), and ribulose-5-phosphate epimerase (*araD* gene).

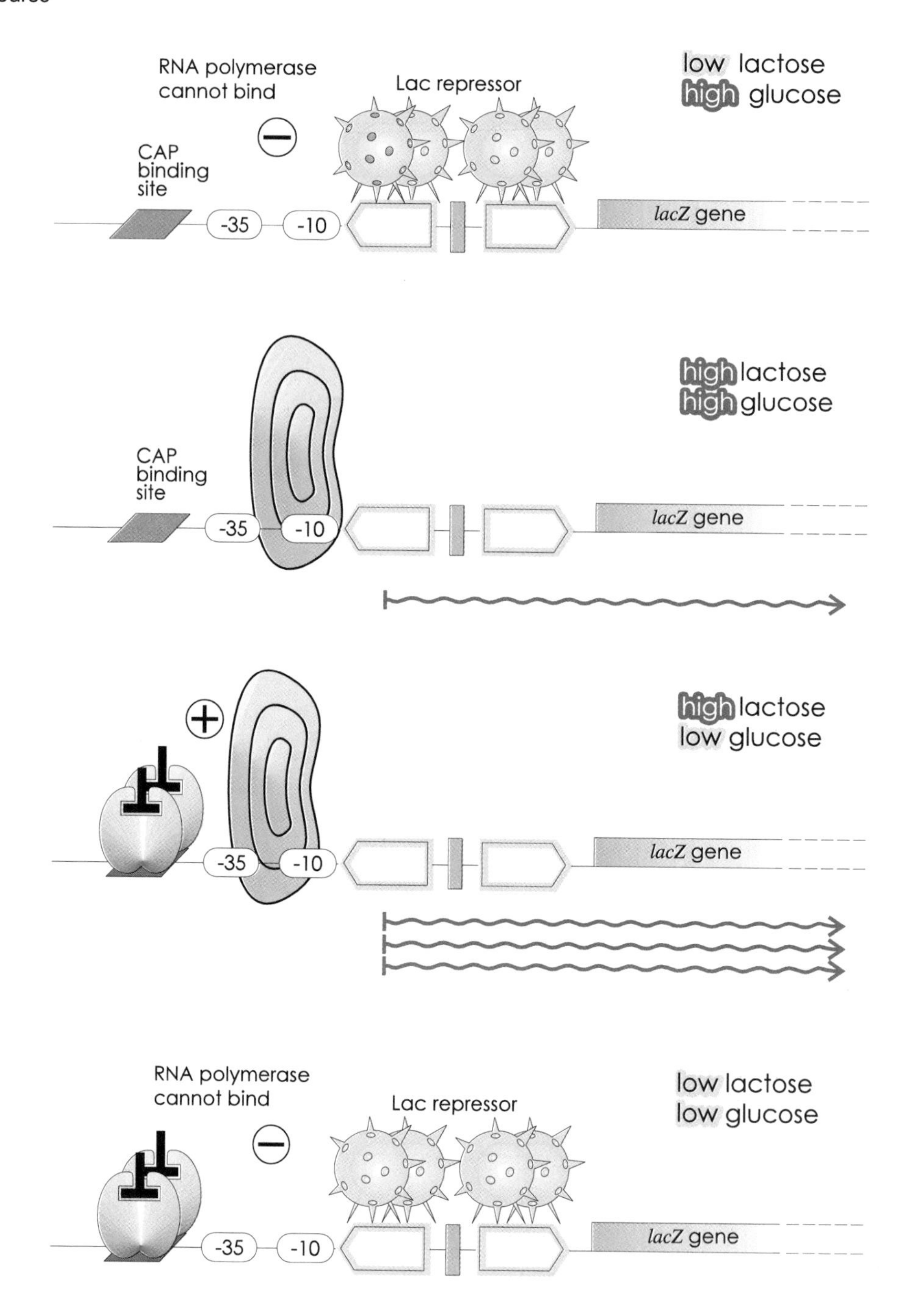

Figure 28.7.

Combined effects of derepression and activation of the *lac* operon under various conditions of lactose and glucose availability.

Expression of these three genes (***araBAD***) is regulated by AraC, which is encoded by the fourth gene in the *ara* operon, ***araC.*** Figure 28.8 illustrates three structural components of the *ara* operon which help explain how it can be so tightly regulated. First, the *araBAD* and *araC* transcription units are on opposite strands of the DNA and pointed away from each other, even though they are controlled by the same regulatory sequences (although the *araC* and *araBAD* genes are transcribed from independent promoters, P_C and P_{BAD}, respectively). Second, there are three distinct binding sites for AraC within the *ara* operon regulatory region which are called ***araO***$_1$, araO_2, and ***araI*** operator sites. Third, there is a CAP binding site situated between two *ara* operator sites. The promoter architecture of the *ara* operon is shown in Figure 28.9.

The expression of *araC* is controlled through an **autoregulatory** mechanism involving the AraC protein. In the absence of arabinose, AraC functions as a transcriptional repressor. Measurements of the steady-state levels of AraC protein in the cell indicate that it takes ~40 molecules of AraC to repress *araC* transcrip-

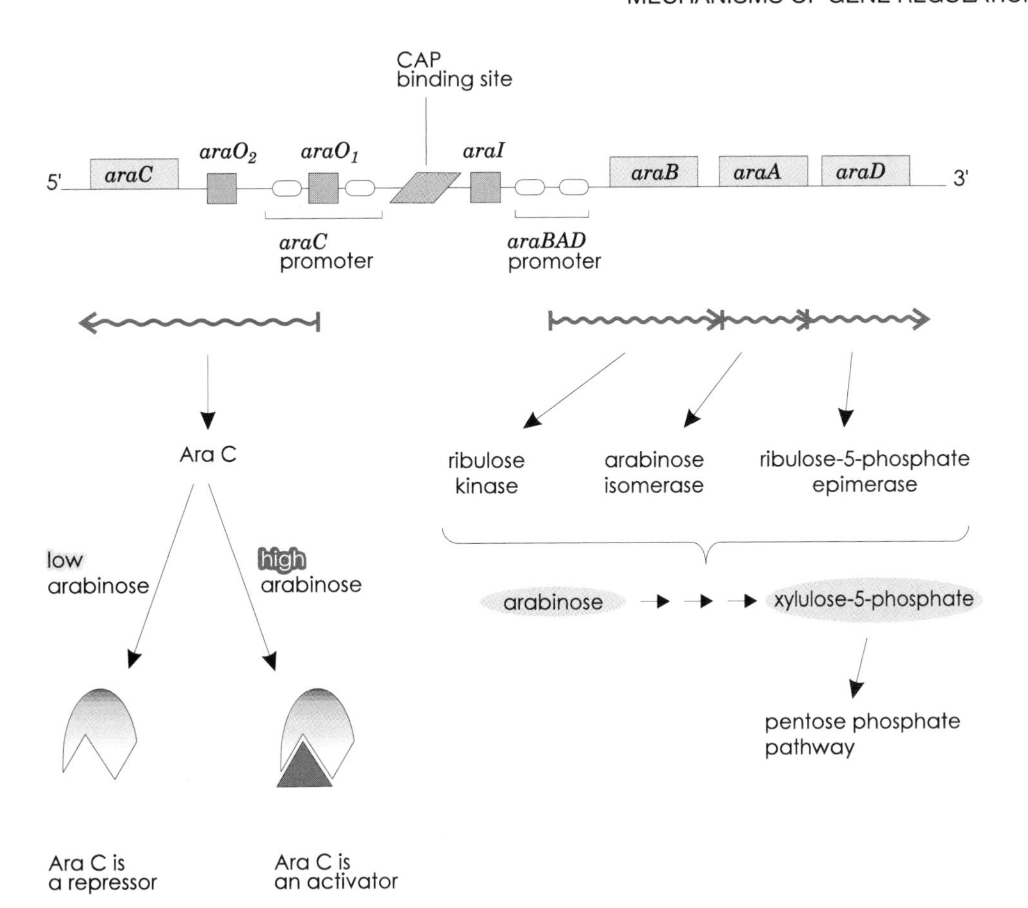

Figure 28.8.

The *ara* operon showing the AraC and CAP binding sites and the *araC, araB, araA,* and *araD* genes. Arabinose binding to AraC converts it from a transcriptional repressor into a transcriptional activator. The ***araC*** and ***araBAD*** promoters are specified by DNA sequences on the opposite strands.

tion. Therefore, when AraC protein levels are low (e.g., immediately after cell division), *araC* is expressed only until there is enough AraC to repress its transcription. There are many examples in which transcriptional autoregulatory mechanisms control the expression of transcription factor genes in both prokaryotes and eukaryotes.

When glucose levels are high in the cell, resulting in low cAMP concentrations, and arabinose levels are low, AraC binds as a dimer to the *araO*$_2$ and *araI* operator sites and represses transcription of **araBAD.** Physical, genetic, and biochemical studies have shown that under these conditions, the **DNA is looped out** between *araO*$_2$ and *araI,* as shown in Figure 28.9. However, when arabinose levels are high, and glucose levels are low (high cAMP), CAP–cAMP complexes are able to bind to the CAP binding site in the *ara* operon and disrupt the DNA loop. In addition, arabinose functions as a modulator of AraC and converts it from a repressor into an activator. Arabinose binding to AraC appears to alter AraC homodimerization properties and promote the formation of AraC–CAP complexes. Under these conditions, AraC-arabinose is functioning as a transcriptional activator because it is required for CAP–cAMP induction of *araBAD* transcription.

As arabinose levels drop owing to its metabolism by the *araBAD*-encoded proteins, AraC is converted back to a repressor and *araBAD* expression is decreased, even though CAP–cAMP complexes are still present in the cell. Interestingly, when glucose and arabinose are both abundant, the *ara* operon is repressed and, although the mechanistic basis for this is not totally understood, this result supports the model that *araBAD* induction is dependent on *both* AraC-arabinose and CAP–cAMP complexes.

Two of the mechanisms required for control of the *ara* operon, **DNA looping** and **multiple-protein interactions,** are hallmarks of regulated gene expression in eukaryotes. These mechanisms provide **increased flexibility** and opportunities for **combinatorial control** of transcription. For example, DNA looping allows response elements to be at various distances from the promoter, thus increasing the number of response elements that a gene can have (this also means that genomic evolution does not have to be precise). In addition, if more than one transcription factor is required for gene con-

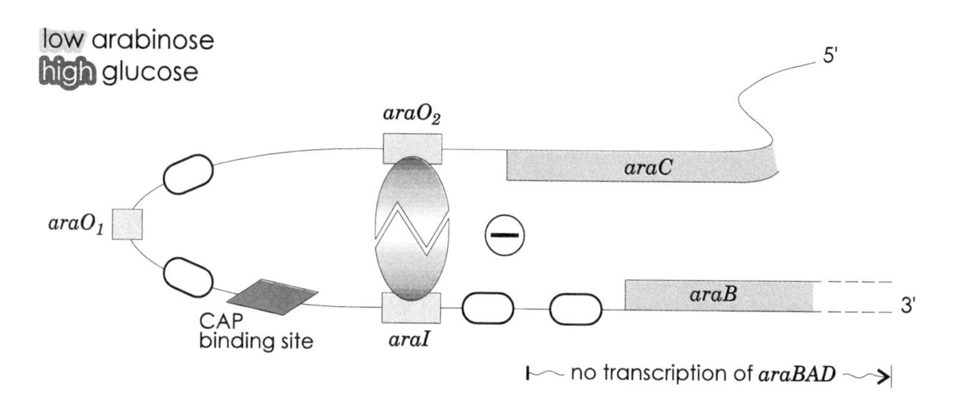

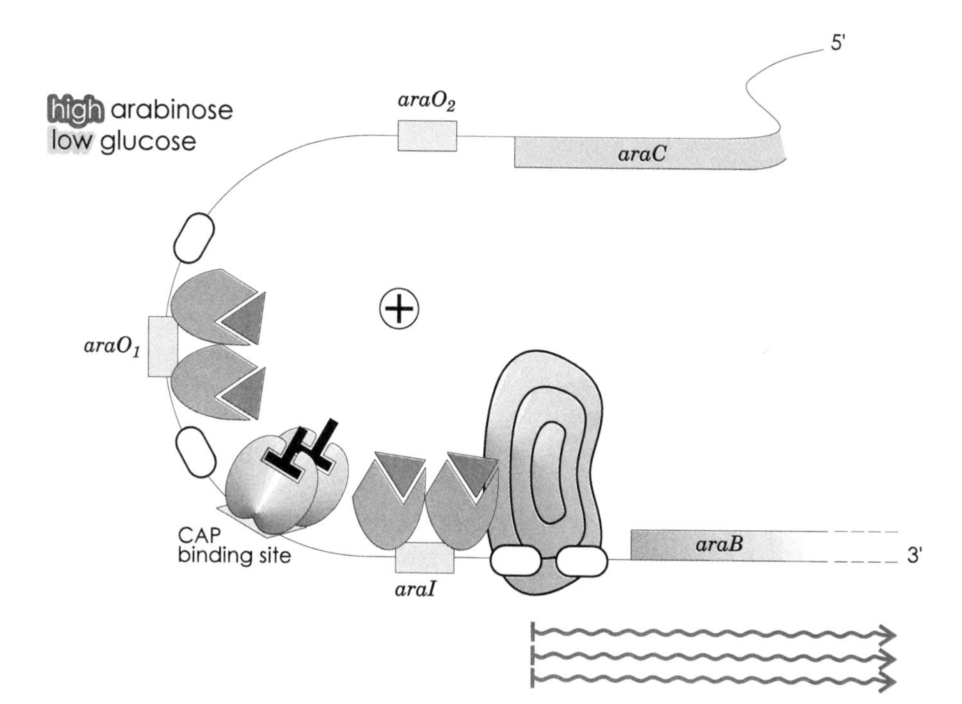

Figure 28.9.

AraC is both a repressor and an activator of the *ara* operon. When arabinose levels are low, AraC binds as a monomer to *araI* and *araO2* and causes a bend in the DNA which effectively blocks *ara* operon transcription. In the presence of arabinose, AraC binds as a dimer to both the *araO1* and *araI* sites, which can lead to an increase in transcription of the *ara* operon when activated CAP is bound to its site.

trol, multiple signals can be integrated and lead to transcriptional fine tuning. As exemplified by control of the *ara* operon, transcriptional regulation is rarely an all or none phenomenon, but rather a balance between optimal (induced) and suboptimal (repressed) promoter utilization.

DUAL CONTROL OF THE *trp* OPERON BY TRYPTOPHAN CONCENTRATIONS IN THE CELL

The physiological signal controlling the *lac* and *ara* operons is the utilization of carbon sources for metabolic energy. In contrast, the tryptophan (*trp*) operon is sensitive to the need for biosynthetic processes and is transcribed under conditions where intracellular concentrations of the amino acid tryptophan are below an optimal level for efficient protein synthesis. The ***trp* operon** consists of a promoter and operator region which controls the expression of a polycistronic mRNA encoding five proteins needed for **tryptophan biosynthesis.**

DNA Binding Activity of the Trp Repressor is Controlled by Tryptophan

Transcriptional control of the *trp* operon is mediated by the **Trp repressor** protein which binds to the ***trp* operator** sequence (Fig. 28.10). The DNA binding activity of the Trp repressor is controlled directly by tryptophan, which binds to the Trp repressor and functions as an effector molecule. In the presence of high concentrations

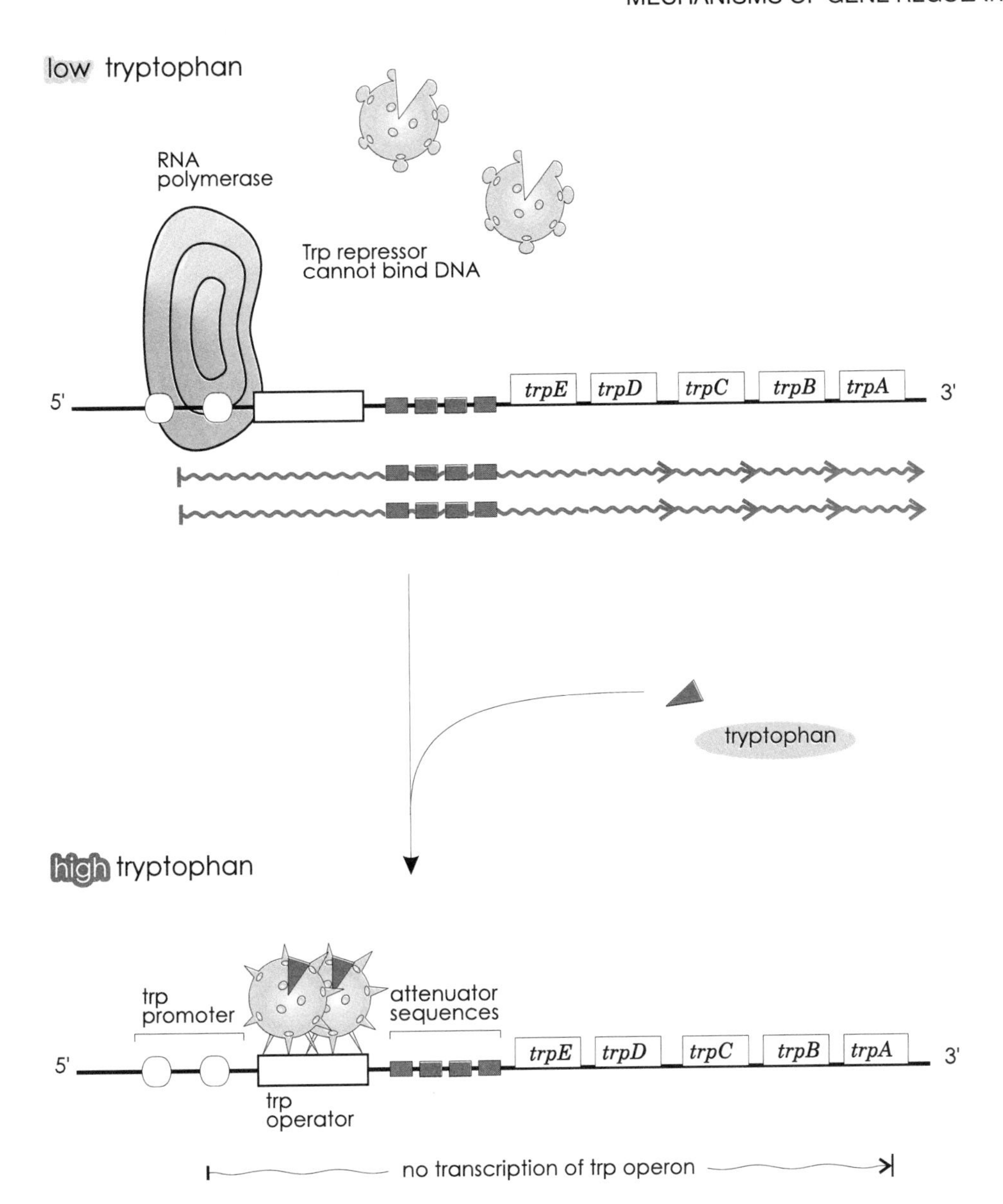

Figure 28.10.
Transcriptional control of the *trp* operon by activation of Trp repressor DNA binding activity.

of tryptophan, the Trp repressor–tryptophan complex forms a homodimer and binds tightly to the *trp* operator sequence, thus **blocking transcription.** However, when tryptophan levels are low, the tryptophan-free Trp repressor exists in an inactive form and cannot bind DNA. Under these conditions, the *trp* operon is transcribed by RNA polymerase and the tryptophan biosynthetic pathway is activated.

Attenuation Controls Transcriptional Elongation of *trp* Polycistronic mRNA

A second level of control of the tryptophan biosynthetic pathway was discovered by Charles Yanofsky when he characterized mutants in the *trp* operon that did not affect Trp repressor binding. Yanofsky and his colleagues characterized a novel form of transcriptional control they called **attenuation,** which depends on the unique linkage between transcription and translation in prokaryotes. As shown in Figure 28.11, the intracellular concentration of **TRP–tRNA$^{\mathbf{Trp}}$** determines if the ribosome will **pause** at a set of codons in the *trp* mRNA that specify consecutive Trp residues. When tryptophan levels are high, and TRP–tRNA^{Trp} is available, then the **transcriptional termination hairpin loop** forms and RNA polymerase disengages from the DNA template just downstream of a polyuridine track. This control mechanism linking events in translation with transcription is called attenuation. When TRP–tR-

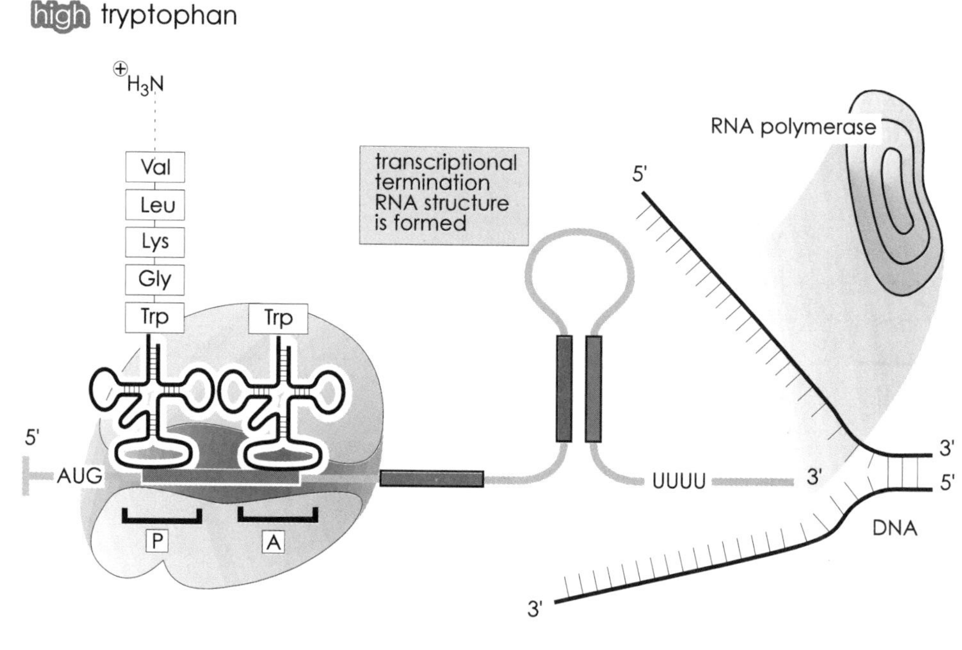

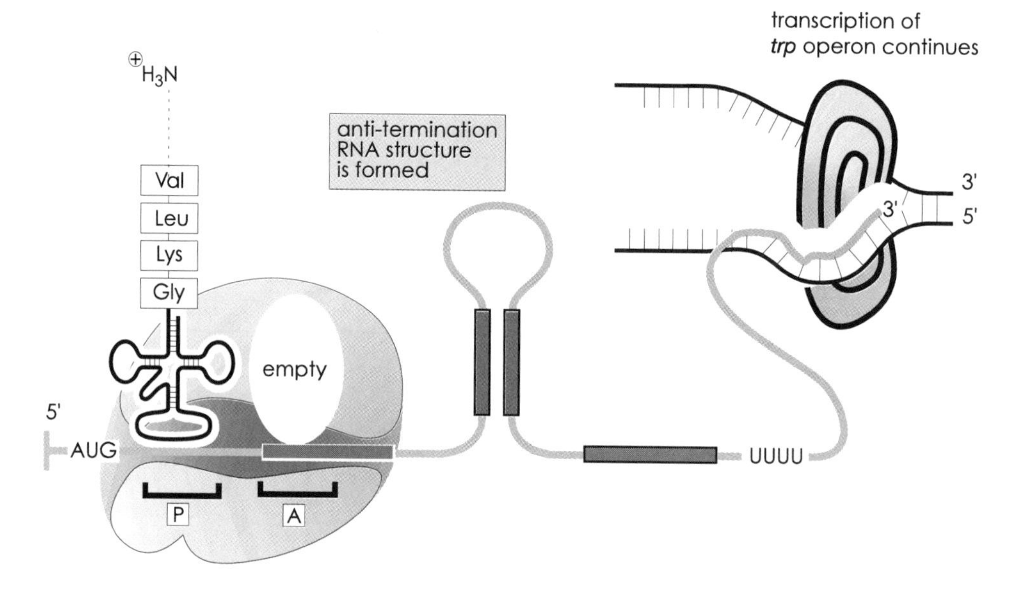

Figure 28.11.

Mechanism of transcriptional attenuation of the *trp* operon. Transcriptional attenuation, as exemplified by the *trp* operon, is possible in prokaryotes because of the functional link between transcription and translation.

NA^{trp} levels in the cell are low because tryptophan is limiting, then **ribosomes pause** at a pair of tryptophan codons in the RNA. This transient ribosomal pausing allows time for an alternate hairpin structure to form in the nascent RNA which disrupts the transcriptional termination signal and RNA polymerase is able to continue down the DNA template and complete transcription (Fig. 28.11).

Taken together, the combined effect of Trp repressor-mediated transcriptional repression, and TRP–tRNA^{Trp} sensitive transcriptional attenuation provides a highly sensitive mechanism to control utilization of the tryptophan biosynthetic pathway.

SUMMARY

1. Transcription factors are DNA binding proteins that interact with specific DNA sequences, called response elements, located in the regulatory or promoter regions of genes. In bacterial systems, these response elements are called operator sites.

Transcription factors can either activate or repress transcription depending on the promoter context and physiological state of the cell. The DNA binding activity of transcription factors can often be modulated by effector molecules.

2. The *lac* operon in *E. coli* consists of a set of clustered genes that encode enzymes needed for the metabolism of lactose. These genes are all transcribed as part of a single polycistronic mRNA. Transcription of the *lac* operon is controlled by both the LacI repressor and the catabolic gene-activator protein (CAP). Lactose and other β-galactosides, such as IPTG, bind to the LacI repressor and decrease its sequence-specific DNA binding activity by 1000-fold. In the presence of lactose, the *lac* operon is transcribed and the proteins required for lactose metabolism are synthesized.

3. CAP is a positive regulator of the *lac* operon and is able to greatly stimulate *lac* expression when lactose is present (LacI is inactive) and glucose levels are low (this leads to high cAMP levels, owing to the deactivation of adenylyl cyclase). Under these conditions, the CAP–cAMP complex binds to the *lac* operator and directly interacts with the α subunit of RNA polymerase. This interaction is thought to stimulate RNA polymerase binding to the promoter and formation of the open-initiation complex.

4. The *ara* operon is positively controlled by the activity of AraC, which can act both as a repressor and activator of *ara* operon transcription, depending on the intracellular concentrations of arabinose and glucose. In the absence of arabinose, AraC is a repressor and binds to two separate operator sites in the *ara* operon, leading to the formation of an inhibitory DNA loop. However, in the presence of arabinose, which binds to AraC, and in the presence of CAP–cAMP complexes, the inhibitory DNA loop is disrupted and CAP–cAMP complexes are able to stimulate transcription of the *ara* operon.

5. The *trp* operon controls the production of tryptophan biosynthetic enzymes at two levels, depending on the intracellular concentrations of tryptophan. When tryptophan levels are high, Trp repressor–tryptophan complexes are formed and Trp repressor binding to the *trp* operon inhibits transcription. A second level of control is called attenuation and it involves the premature termination of transcription when TRP–$tRNA^{Trp}$ is abundant and a termination structure is formed, resulting in the disengagement of RNA polymerase from the DNA.

REFERENCES

Ansari A. Z., Bradner J. E., O'Halloran T. V. (1995): DNA-bend modulation in a repressor-to-activator switching mechanism. *Nature* **374:**371.

Antson A. A., Otridge J., Brzozowski A. M., et al. (1995): The structure of *trp* RNA-binding attenuation protein. *Nature* **374:** 693.

Blatter E. E., Ross W., Tang H., Gourse R. I., Ebright R. H. (1994): Domain organization of RNA polymerase a subunit: C-terminal 85 amino acids constitute a domain capable of dimerization and DNA binding. *Cell* **78:**889.

Busby S., Ebright R. H. (1994): Promoter structure, promoter recognition, and transcription activation in prokaryotes. *Cell* **79:**743.

Bushman F. D. (1992): Activators, deactivators and deactivated activators. *Curr. Biol.* **2:**673.

Bustos S. A., Schleif R. F. (1993): Functional domains of the AraC protein. *Proc. Natl. Acad. Sci. USA* **90:**5638.

Chen Y., Ebright Y. W., Ebright R. H. (1994): Identification of the target of a transcription activator protein by protein–protein photocrosslinking. *Science* **265:**90.

Crothers D. M. (1994): Upsetting the balance of forces in DNA. *Science* **266:**1819.

Gralla J. D. (1991): Transcriptional control—Lessons from an *E. coli* promoter data base. *Cell* **66:**415.

Matthews, K. S. (1996): The whole lactose repressor. *Science* **271:**1245.

Kim B., Little J. W. (1992): Dimerization of a specific DNA-binding protein on the DNA. *Science* **255:**203.

Kustu S., North A. K., Weiss D. S. (1991): Prokaryotic transcriptional enhancers and enhancer-binding proteins. *Trends Biochem. Sci.* **16:**397.

Lewis M., Chang G., Horton N. C., et al. (1996): Crystal structure of the lactose operon repressor and its complexes with DNA and inducer. *Science* **271:**1247.

Ptashne M., Gann A. A. F. (1990): Activators and targets. *Nature* **346:**329.

Ross W., Gosink K. K., Salomon J., et al. (1993): A third recognition element in bacterial promoters: DNA binding by the a subunit of RNA polymerase. *Science* **262:**1407.

Shakked Z., Guzikevich-Guerstein G., Frolow F., Rabinovich D., Joachimiak A., Sigler P. B. (1994): Determinants of repressor/operator recognition from the structure of the *trp* operator binding site. *Nature* **368:**469.

Sogaard-Andersen L., Valentin-Hansen P. (1993): Protein–protein interactions in gene regulation: The cAMP–CRP complex sets the specificity of a second DNA-binding protein, the CytR repressor. *Cell* **75:**557.

Storz G., Tartaglia L. A., Ames B. N. (1990): Transcriptional regulator of oxidative stress-inducible genes: Direct activation by oxidation. *Science* **248:**189.

Zhou Y., Pendergrast P. S., Bell A., Williams R., Busby S., Ebright R. H. (1994): The functional subunit of a dimeric transcription activator protein depends on promoter architecture. *EMBO J.* **13:**4549.

REVIEW QUESTIONS

1. Transcription factors are
   a) small metabolic effector molecules that regulate DNA binding activity
   b) proteins that modulate transcriptional elongation rates
   c) proteins that modulate transcriptional initiation rates
   d) environmental stimuli that send signals to gene promoters
   e) DNA binding proteins that protect DNA from endonucleases

2. Which of the following statements is NOT TRUE regarding mechanisms of transcriptional regulation?
   a) Effector molecules can stimulate transcription factor DNA binding.
   b) Effector molecules can inhibit transcription factor DNA binding.
   c) Deinduction results in lower transcription rates.
   d) Derepression results in higher transcription rates.
   e) Transcription factors can only function as repressors.

3. IPTG-induction of β-galactosidase activity is the result of
   a) stimulation of Lac repressor function
   b) IPTG binding to the *lac* operon and inducing transcription
   c) IPTG binding to the *lacI* gene product and inhibiting its activity
   d) inhibition of β-galactosidase degradation
   e) cleavage of IPTG by β-galactosidase

4. Under what condition is RNA polymerase activity at the *lac* operon maximal?
   a) high lactose, low glucose
   b) high glucose, low lactose
   c) high glucose, high lactose
   d) low lactose, low glucose
   e) high IPTG, high glucose

5. The effect of arabinose on *araBAD* transcription is
   a) to decrease transcription by inhibiting binding of AraC to DNA
   b) to increase transcription by bending DNA
   c) to cause derepression by altering the DNA binding characteristics of AraC
   d) to activate CAP binding to DNA
   e) to increase transcription in the presence of high glucose

6. Attenuation at the *trp* operon results in
   a) premature termination of DNA replication
   b) the formation of an antitermination hairpin loop in the RNA
   c) the formation of a translational termination hairpin loop in the RNA
   d) the synthesis of enzymes that metabolize tryptophan
   e) the dissociation of RNA polymerase from *trp* operon DNA sequences.

ANSWERS TO REVIEW QUESTIONS

1. ***c*** Transcription factors are proteins that regulate transcription.

2. ***e*** Transcription factors can function as repressors and/or activators.

3. ***c*** IPTG binds to and inhibits the *lacI* gene product, that is, the Lac repressor.

4. ***a*** Under these conditions, Lac repressor is inhibited and CAP is active.

5. ***c*** AraC–arabinose complexes bind to different DNA sequences than AraC–AraC complexes.

6. ***e*** Under most growth conditions TRP-tRNA^{TRP} is abundant and transcriptional termination by attenuation is favored.

CHAPTER

29

MECHANISMS OF GENE REGULATION IN EUKARYOTES

INTRODUCTION

Regulated gene expression in eukaryotes is more complex than that of prokaryotes in at least two respects. First, the number of posttranscriptional steps subject to regulation is greater in eukaryotes because mRNA must be processed and exported to the cytoplasm for translation. Second, since multicellular organisms contain the same DNA in every cell, but only a portion of it is transcribed in any one cell type, a large number of transcription factors are needed to control the temporal and spatial expression of genes. In this chapter, we focus on the mechanisms by which transcriptional initiation rates of specific genes are altered in response to developmental or environmental cues in eukaryotes. This includes a description of various properties that make eukaryotic transcription factors so versatile. In addition, two specific examples of gene regulatory pathways are presented—steroid-regulated gene expression by ligand-activated transcription factors called steroid receptors, and signal transduction by phosphorylation events initiating at the cell membrane and resulting in posttranslational modification of transcription factors in the nucleus. The inhibitory effects of chromatin structure and DNA methylation on eukaryotic gene expression are also presented.

REGULATED GENE EXPRESSION IN EUKARYOTES REQUIRES A LARGE NUMBER OF PROTEINS

The central theme in eukaryotic gene expression is the requirement for integrating information through a **combinatorial mechanism.** Figure 29.1 shows a prototypical RNA polymerase II-dependent eukaryotic promoter with upstream regulatory sequences. As described in Chapter 24, the transcription-initiation complex includes **RNA polymerase II holoenzyme** (RNA pol II), the universal transcription factor **TBP** (TATA binding protein), and a large number of **auxiliary transcription factors,** some of which are **TBP-associated factors** (TAFs). This large initiation complex is bound to DNA in a region surrounding the start site of transcription. The DNA sequences that contain binding sites for transcription factors are located both near the promoter and at large distances (1–10 kb) upstream or downstream.

Transcription-factor binding sites on DNA are commonly called response elements. Some transcription factors act to induce transcription; others in the same regulatory region can repress transcription. by having DNA sequences corresponding to response elements for a large number of regulatory factors present

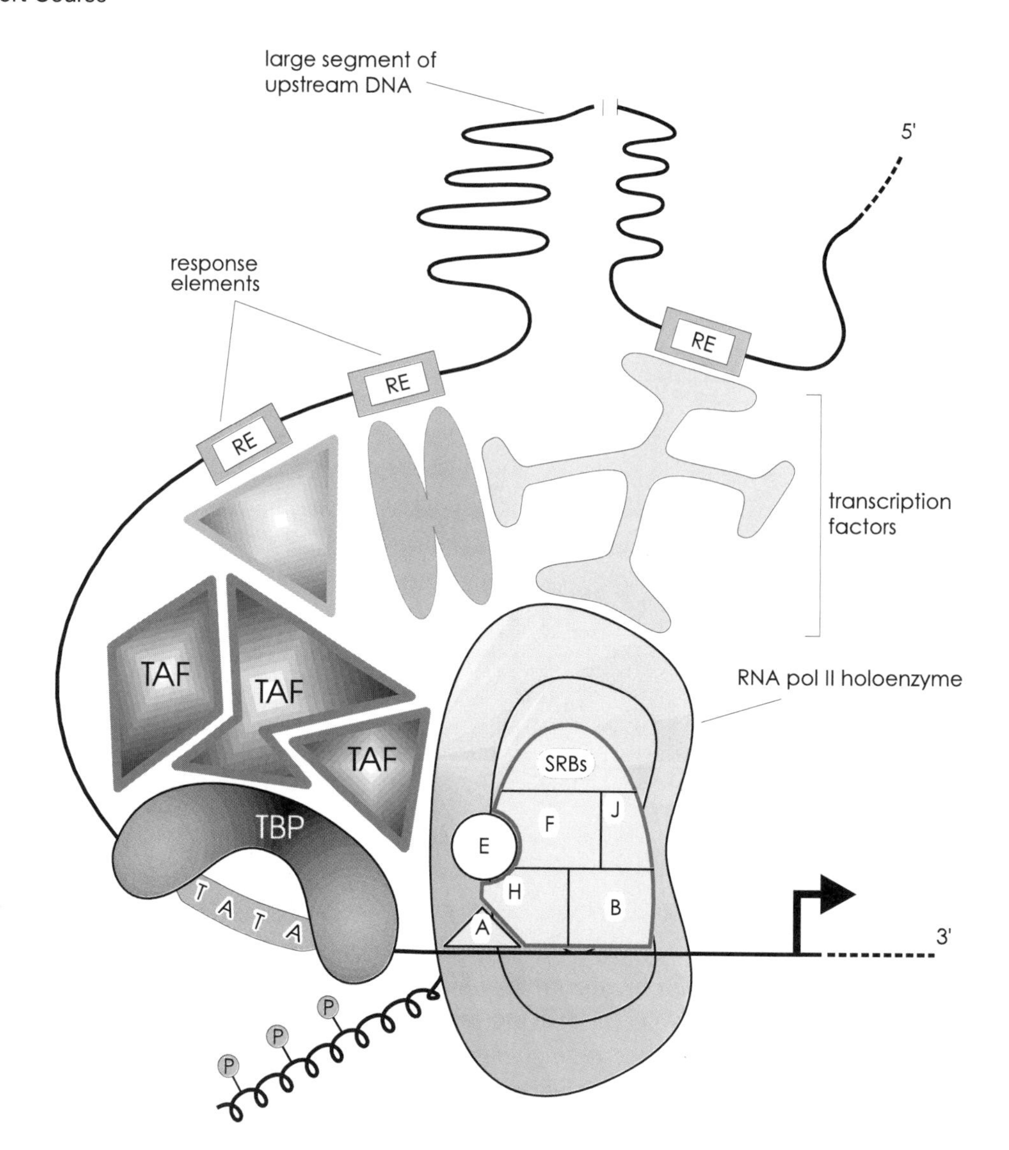

Figure 29.1.

A prototypical RNA pol II-dependent eukaryotic promoter. Three classes of transcription factors are schematically represented: the TFII auxiliary factors (TFIIF, TFIIJ, TFIIH, etc.), TBP with its associated factors (TAFs), and various transcriptional regulatory factors which bind to DNA sequences called response elements (RE). See Chapter 24 for more information about TBP and the RNA pol II holoenzyme.

in the same gene, combinatorial control of transcription is possible. For example, if transcriptional activators and repressors are bound in the regulatory region of the same gene, the level of expression can be fine tuned according to the relative "strength" of both positive and negative effects. In addition to these combinatorial interactions, some transcription factors are expressed in only selected cell types, which allows for cell-specific transcriptional control, as shown in Figure 29.2.

Enhancers are a class of DNA sequence response elements that function to modulate the transcription of heterologous promoters (they can be joined to unrelated promoter sequences). Enhancers are relatively insensitive to orientation and distance from the promoter. It is generally accepted that the binding of transcription factors to eukaryotic response elements, such as enhancers, often leads to **looping out** of the intervening DNA between the element and the transcription initiation complex. This arrangement is similar to that described for the transcriptional repression of the *ara* operon by AraC (Chapter 28). Since eukaryotic DNA is associated with chromosomal proteins (Chapter 3), and the initiation complex is known to contain a large number of transcription factors (Chapter 24), the actively transcribed eukaryotic promoter should be envisioned as a protein assembly center similar to the replication complex described in Chapter 22. To extend the analogy of a building foundation to describe the layout of a gene regulatory region (Chapter 24), the fully assembled eukaryotic promoter complex would be something like the Taj Mahal!

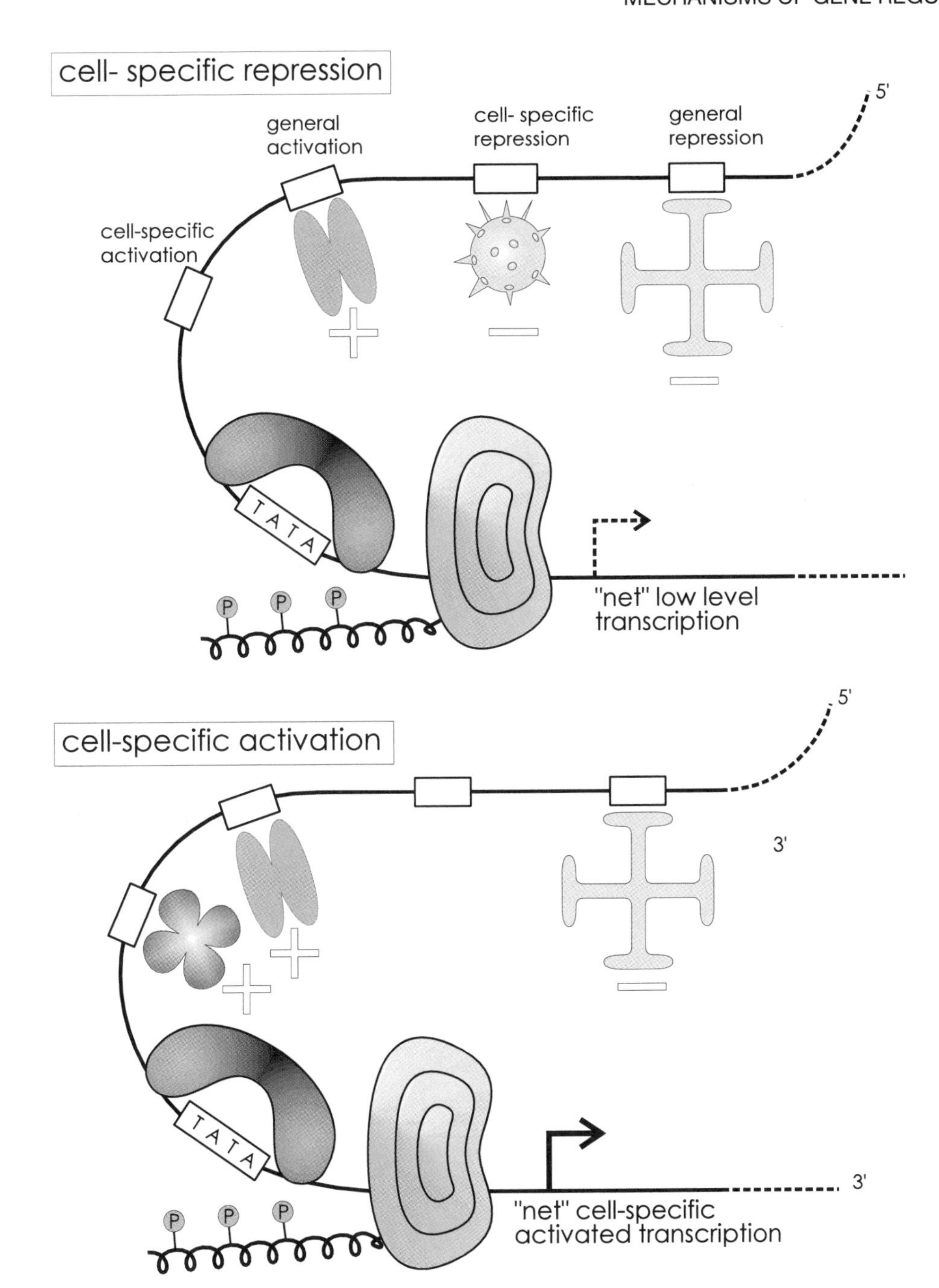

Figure 29.2.
Combinatorial control of eukaryotic transcription. The net effect of multiple transcription factors interacting with the transcriptional initiation complex can either be gene activation or gene repression. Since some transcription factors are cell-specific in their own expression, the presence or absence of these particular transcription factors can influence rates of transcription of individual genes.

MOST EUKARYOTIC TRANSCRIPTIONAL REGULATORS CONTAIN DISCRETE FUNCTIONAL DOMAINS

A large number of eukaryotic transcription factors have been biochemically characterized and the genes encoding them have been isolated. Many of these proteins are transcriptional activators that contain one of the DNA binding motifs described in Chapter 5 (zinc fingers, basic helix–loop–helix, or leucine zippers). In addition to specific DNA binding, transcriptional activation seems to require protein–protein contacts between transcription factors and some part of the transcription initiation complex. Using recombinant DNA techniques, **chimeric** transcription factors (a chimeric protein is one containing peptide sequences from two or more proteins) have been made by fusing together the coding sequences of different eukaryotic transcription factors.

As shown in Figure 29.3, these studies have demonstrated that it is possible to fuse the DNA bind-

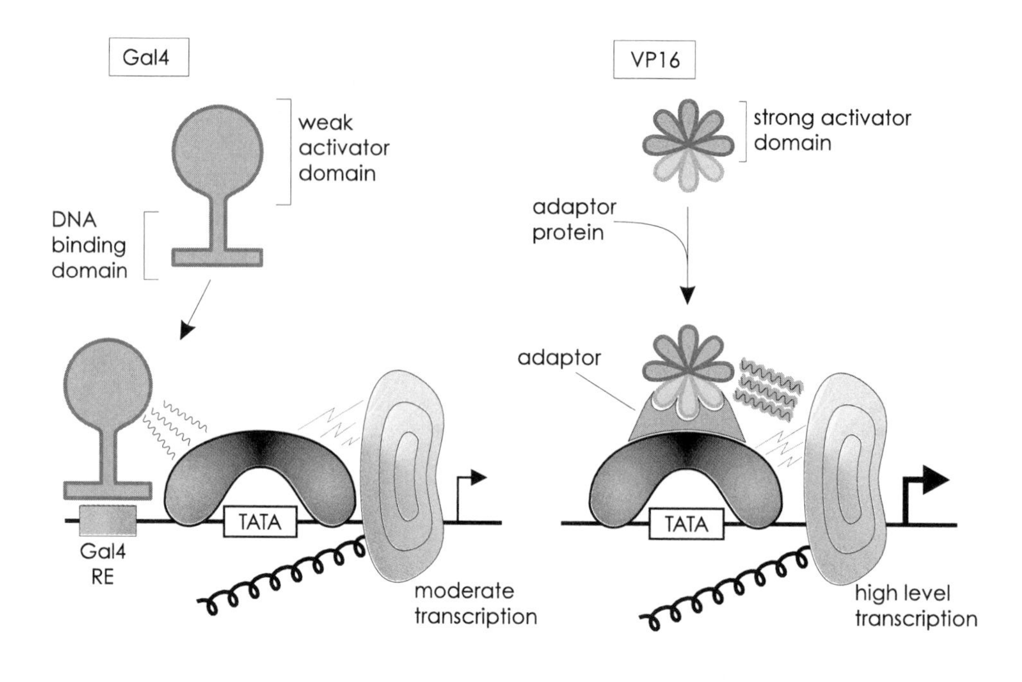

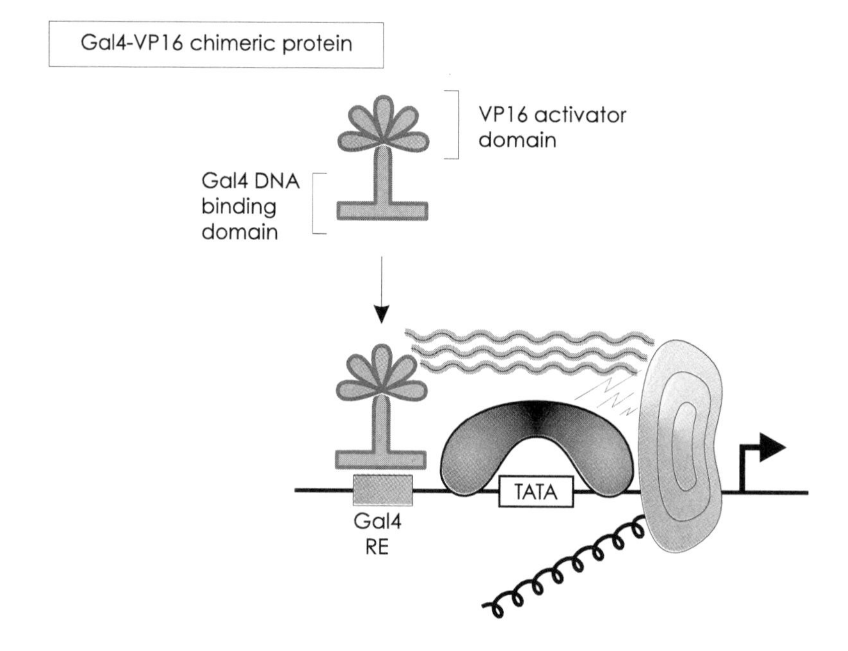

Figure 29.3.
Chimeric transcription factors can be made between heterologous proteins, demonstrating that DNA binding and transcriptional activation are often separable activities. Shown here are the results from experiments in which the DNA binding domain of a weak activator, Gal4, is fused to the potent activation domain of a nonDNA binding viral protein called VP16. The novel Gal4–VP16 chimeric protein is a strong activator of genes containing Gal4 response elements.

ing domain of the yeast **Gal4 protein** (a zinc-finger motif) with the transcriptional activation domain of the **Herpes Simplex virus VP16 protein,** to make a transcriptional activator that can function in yeast. Similarly, the Gal4 activation domain can be joined to the *E. coli* LexA repressor DNA binding domain to convert this transcriptional repressor into a transcription activator. Box 29.1 describes how reporter genes can be used to characterize DNA sequence response elements and transcription factors that control gene-specific transcriptional regulation.

In addition to independent DNA binding and activation sequences, many eukaryotic transcription factors also contain **dimerization domains,** which are sequences required for subunit interactions. Most prokaryotic DNA binding proteins bind to an inverted repeat DNA sequence as homodimers. However, many eukaryotic transcription factors are members of multigene families, and because of conserved structural properties in the dimer interface of the proteins, two related proteins can often form heterodimers. Since these related proteins may have subtle differences in their DNA binding domains, such **heterodimer pairings** can create discrete DNA binding specificities.

BOX 29.1. TRANSIENT COTRANSFECTION ASSAYS CAN BE USED TO STUDY TRANSCRIPTION IN VIVO

The mechanistic basis of rate-limiting steps in transcriptional initiation can best be understood by performing careful in vitro transcription experiments with purified components. However, in eukaryotic systems, this task is enormous because of all the factors involved and the effect of chromatin structure on transcription in vivo. An alternative and complementary approach is to introduce exogenous DNA into tissue-culture cell lines by the technique of **DNA transfection.** Methods for getting DNA into live cells by transfection are based on physical perturbations of the cell membrane, for example, by exposing cells to calcium-phosphate precipitates containing DNA or applying electric fields in the presence of DNA-containing buffers. By introducing an **expression gene** containing a cloned cDNA for a particular transcriptional factor, and combining it with a **reporter gene** having the appropriate response element, it is possible to rapidly identify functional domains in the transcription factor and to map DNA response elements in the reporter-gene promoter. Site-directed and deletion mutagenesis can be used to test various gene constructs and chimeric transcription-factor genes can be created using various recombinant-DNA strategies (see Box 24.1). Common reporter genes for the study of mammalian transcription are the bacterial **chloramphenicol acetyl transferase** (CAT) gene and the firefly **luciferase gene,** both of which are easy to assay for and are not found in mammalian cells. Figure 29.4 illustrates the principal of **transient cotransfection assays.**

One of the best examples of dimerization control of transcription-factor function is the interaction of **basic helix–loop–helix (bHLH) proteins.** Figure 29.5 shows that the interaction between various bHLH proteins can be reflected in the affinity of homo- and heterodimers for specific DNA sequences. Helix–loop–helix proteins lacking the basic region required for DNA binding, can dimerize with bHLH proteins and inhibit their DNA binding activity. These HLH proteins are said to have a **dominant negative** effect on bHLH proteins because they are dominant repressors of bHLH activity (Fig. 29.5).

TRANSCRIPTIONAL CONTROL BY LIGAND-DEPENDENT STEROID/NUCLEAR RECEPTORS

Steroid hormone receptors are members of a large family of ligand-regulated transcription factors which includes the **nuclear receptors.** These receptors are similar in their DNA binding domains, which consist of two **zinc fingers** and the inclusion of a carboxy-terminal segment that contains a **ligand-binding domain.** A functional map of the best-characterized nuclear receptors are shown in Figure 29.6. The **glucocorticoid receptor** is representative of the steroid receptors and it was the first gene in this family of transcription factors to be cloned.

When the amino-acid sequences of a number of the steroid/nuclear receptors are aligned through the DNA binding domain, there are several receptors with high similarity to the glucocorticoid receptor (progesterone and androgen), while others are much less similar, such as the retinoic acid, thyroid, and vitamin D receptors. The estrogen receptor is somewhat of a hybrid member of this family in that its DNA binding domain is more similar to the nuclear receptors, but the ligand binding domain (and its ligand) resembles that of the steroid receptors. The functional distinction between steroid and nuclear receptors is based on their mode of DNA binding. The *steroid receptors* bind to inverted-repeat **DNA response elements** as homodimers, whereas the *nuclear receptors* bind to **direct-repeat response elements** as heterodimers with the 9-cis retinoic acid receptor (RXR), as shown in Figure 29.7.

Steroid Receptors Function as Homodimers

The glucocorticoid subfamily and the estrogen receptor bind to related, but distinct, response elements (Fig. 29.7). Molecular genetic analysis of the glucocorticoid and estrogen receptors has shown that just three amino acids in the first zinc finger of these two receptors are responsible for DNA binding specificity (glucocorticoid receptors cannot bind estrogen response elements and vice versa). X-Ray crystallography data of the glucocorticoid-receptor DNA binding domain, complexed with a high-affinity glucocorticoid response element, confirmed that these three amino acids make important contacts with the DNA.

Both the **glucocorticoid** (GR) and **estrogen** (ER)

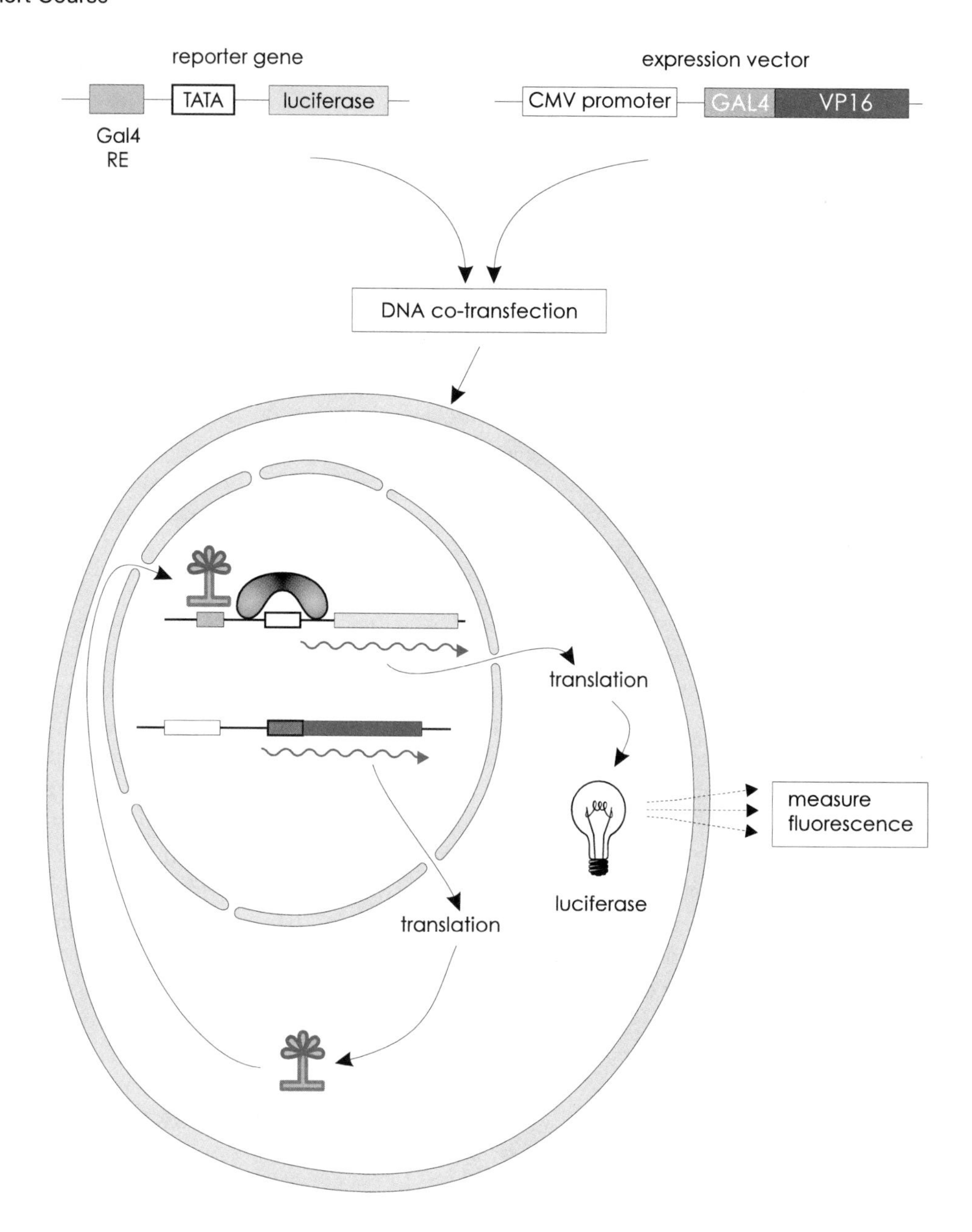

Figure 29.4.

The transient cotransfection assay, using a reporter gene and an appropriate expression vector. The reporter gene in this case contains a Gal4 response element linked to a minimal promoter located upstream of the firefly luciferase gene. The expression vector encodes the Gal4-VP16 chimeric gene described in Figure 29.3, which is downstream of a potent viral enhancer promoter from cytomegalovirus (CMV). These two DNA plasmids are introduced into a eukaryotic cell by electroporation which initiates a cascade of transcription and translation events resulting in the production of luciferase which can be measured by the amount of fluorescent product it generates. Luciferase levels are an accurate, albeit indirect, indicator of Gal4 reporter gene transcription.

receptors are steroid-activated transcription factors and each contains multiple **transcriptional activation** sequences. The most potent activation sequence in GR is in the amino terminus. As shown in Figure 29.8, GR is localized to the cytoplasm in the absence of hormone, as part of a protein complex which includes the **hsp 90 heat-shock protein.** This is an abundant protein found in all cells and functions as a molecular chaperone, facilitating protein folding. The name heat shock comes from the discovery that the level of hsp 90 in the cell increases dramatically after cell stresses such as heat. Studies have shown that hsp 90 is required to refold ligand-free GR into a conformation that is capable of binding glucocorticoids (Fig. 29.8).

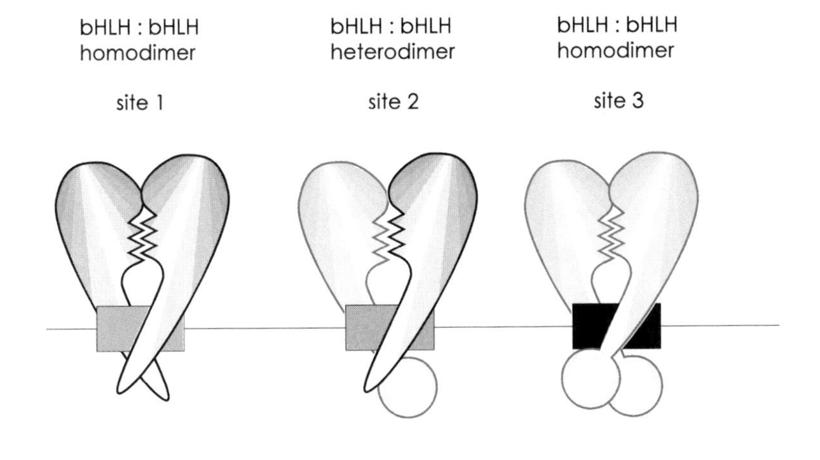

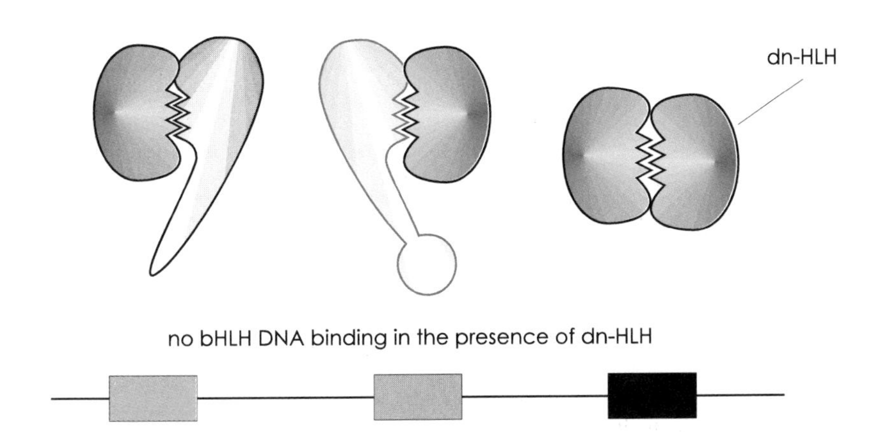

Figure 29.5.
The bHLH proteins can form heterodimeric complexes that recognize distinct binding sites. Two related bHLH proteins are shown which can bind to three different response elements as a result of forming homo- or heterodimers. An HLH protein lacking the basic region can function as a dominant negative inhibitor (dn-HLH) of bHLH proteins by forming heterodimers that are incapable of binding to DNA.

Nuclear Receptors Bind to DNA as Heterodimers

Three nuclear receptors, the **retinoic acid** (RAR), **thyroid** (TR), and **vitamin D** (VDR) receptors, were originally cloned by screening cDNA libraries for DNA sequences that are homologous to the zinc-finger DNA binding domain of the steroid receptors. Even though the ligands for nuclear receptors are chemically quite different than steroids, these proteins function as ligand-dependent transcription factors. The RAR, TR, and VDR bind to the estrogen–DNA response-element half-site (5′-AGGTCA-3′), but unlike ER, do not bind with high affinity to a palindromic sequence as homodimers. As shown in Figure 29.7, the 9-*cis*-retinoic acid nuclear receptor, commonly referred to as RXR, is required for DNA binding by forming a **heterodimer** with RAR, TR, and VDR.

How can the RAR, TR, and VDR nuclear receptors regulate distinct target genes if they all bind to the same DNA sequence? The answer is that each heterodimer has a high affinity for a response element that differs only in the spacing of the **direct repeat** sequence AGGTCA. The preference of each of these heterodimers for variable spacing between half-sites is a consequence of amino-acid differences in the DNA binding domain which includes one of two heterodimerization domains in the receptors (the other one is in the carboxy terminus). The dimerization domain in the carboxy terminus of nuclear receptors is symmetrical as it is in the steroid receptors; the dimerization interface located in the nuclear-receptor DNA binding domain is polarized to accommodate a "head-to-tail" interaction.

The **RXR-RAR, RXR-TR,** and **RXR-VDR** heterodimers all require the cognate ligand for the nuclear receptor partner (retinoic acid, thyroid hormone, or vitamin D, respectively) in order to function as transcrip-

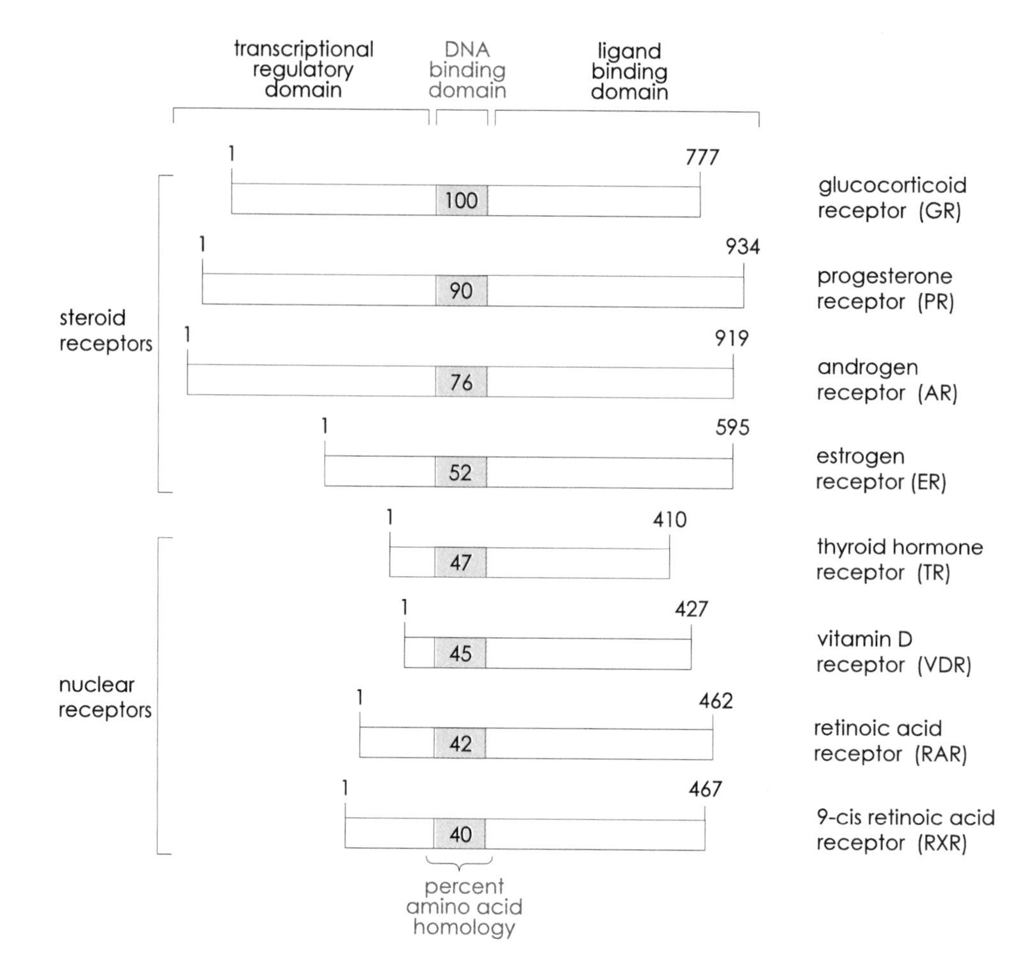

Figure 29.6.

Superfamily of steroid/nuclear receptors. Steroid/nuclear receptors are members of a large gene family of transcription factors that share significant amino-acid homology in the DNA binding domain. There is much less homology in the carboxy-terminal ligand-binding domain and a variable-length amino-terminal transcriptional regulatory domain. The total number of amino acids present in each receptor is shown.

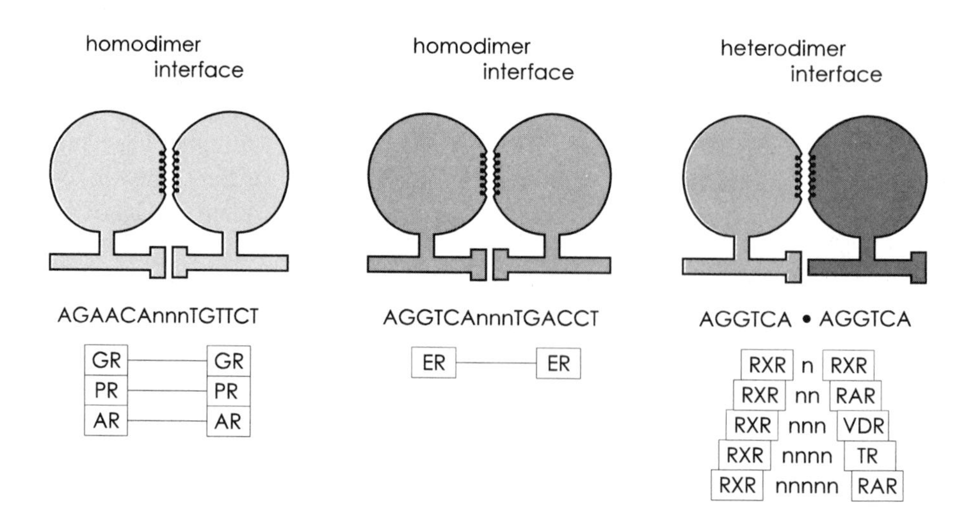

Figure 29.7.

Steroid and nuclear receptors are differentiated by their mode of DNA binding. GR, PR, AR, and ER all bind as homodimers to inverted-repeat sequences separated by three nucleotides (nnn), although the ER DNA binding domain recognizes a DNA sequence that is shared by the nuclear receptors. Nuclear receptors bind predominantly as heterodimers with RXR to direct-repeat sequences separated by one to five nucleotides. Structural studies have shown that RXR binds to the 5′ half-site of the response element, which may be important for ligand-dependent transcriptional regulatory activity of the heterodimeric partner.

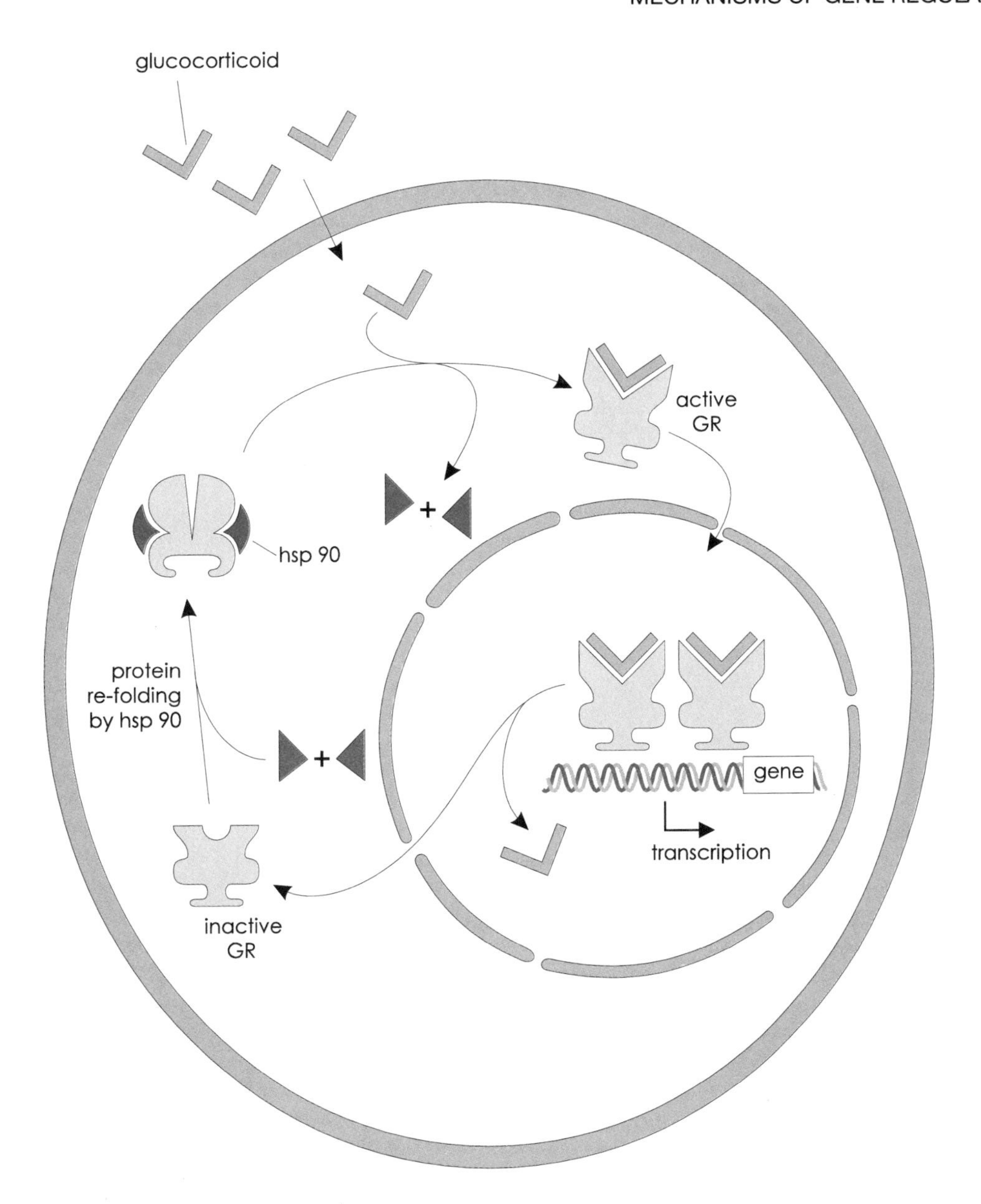

Figure 29.8.
Glucocorticoid activation of receptor function. Glucocorticoids enter the cell and bind to GR, forming the active complex, which releases two subunits of hsp 90. GR–hormone complexes are translocated to the nucleus where they interact with DNA and other transcription factors. When hormone levels drop, the ligand-free GR is recycled to the cytoplasm, where it is refolded by hsp 90 into a conformation capable of binding hormone.

tional activators. However, the RXR partner in these heterodimer pairs is ligand-free. This raises the question, is RXR a ligand-dependent transcriptional regulator, and if so, what is its ligand? Using a bioassay based on transient cotransfections (Box 29.1), it was found that **9-*cis*-retinoic acid** is a high-affinity ligand of the RXR nuclear receptor. Moreover, RXR-9-*cis*-retinoic acid homodimer complexes can activate transcription from a one-base-pair spacing of the direct repeat AGGTCA half-site (see Fig. 29.7).

This finding suggests a complex picture of ligand regulation of nuclear receptors, as shown in Figure 29.9. For example, in the presence of vitamin D, but the absence of 9-*cis*-retinoic acid, the RXR–VDR heterodimer could bind to vitamin D target genes (AGGTCA response elements containing a three-nucleotide spacer) and activate transcription. However, if 9-*cis*-retinoic acid levels increased, even if vitamin D is still present, RXR homodimers would form. This would have two consequences. First, vitamin D responsive genes would be deinduced because VDR requires RXR as a heterodimer partner to be fully active. Second, **RXR–RXR homodimers** could now regulate transcription from **RXR target genes** (response elements with a single nucleotide spacer). This dependence on RXR by RAR, TR, and VDR, coupled with the independent activity of RXR homodimers, allows for overlapping control of multiple gene networks in response to the relative ratios of the appropriate nuclear receptor ligands and 9-*cis*-retinoic acid.

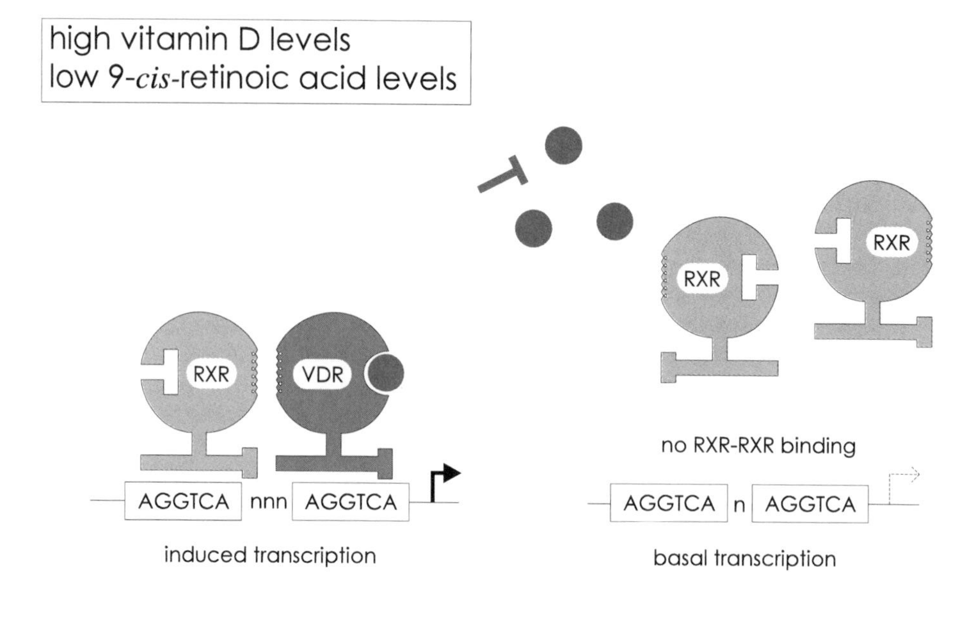

high vitamin D levels
high 9-*cis*-retinoic acid levels

VDR
VDR
RXR
RXR

no RXR-VDR or VDR-VDR binding

AGGTCA nnn AGGTCA

basal transcription

AGGTCA n AGGTCA

induced transcription

Figure 29.9.

Effect of 9-*cis*-retinoic acid on nuclear receptor activity. (Top) When vitamin D levels are high, but 9-*cis*-retinoic acid levels are low, vitamin D responsive genes would be activated preferentially by RXR–VDR heterodimers. (Bottom) When 9-*cis*-retinoic acid levels are high, ligand-bound RXR–RXR homodimers would form and preferentially activate 9-*cis*-retinoic acid responsive genes, while at the same time causing a decrease in the expression of vitamin D responsive genes.

PHOSPHORYLATION CASCADES TRANSMIT REGULATORY SIGNALS TO THE NUCLEUS

The cell membrane contains a large number of receptor proteins that bind **extracellular ligands.** In many cases, ligand-binding to the receptor stimulates an inherent kinase activity that phosphorylates other proteins directly. In addition, most **receptor kinases** undergo **autophosphorylation,** which promotes association with other cytosolic kinases through S_H2 protein domain interactions. As a result of multiple phosphorylation events, often including the phosphorylation and activation of other kinases, the extracellular signal is transduced through an intracellular **phosphorylation cascade.** The endpoint of this cascade can often be the modulation of transcription-factor activity by phosphorylation. The four basic mechanisms by which phosphorylation can alter transcription-factor activity are shown in Figure 29.10. These four mechanisms are regulation of (1) nu-

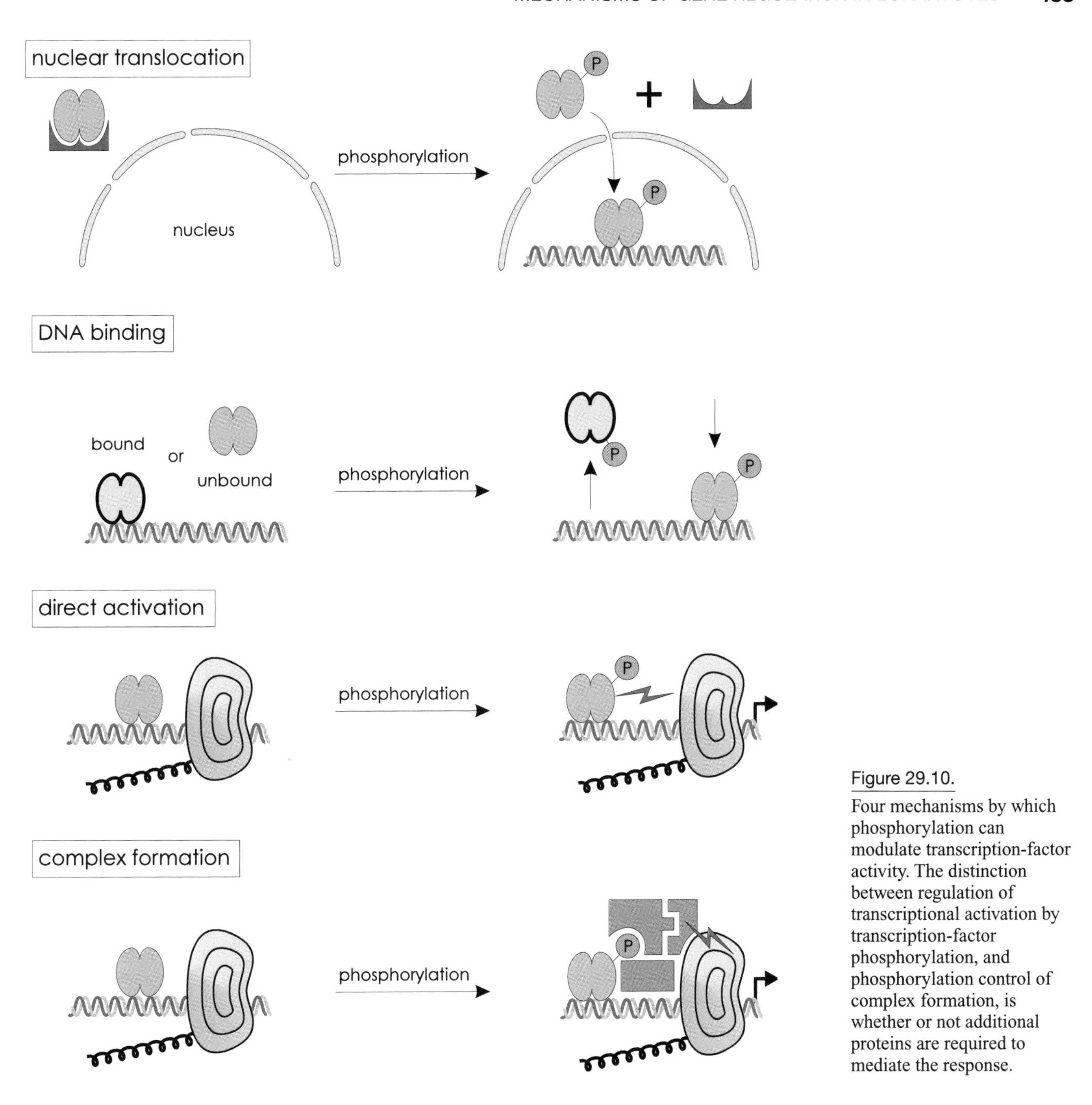

Figure 29.10.
Four mechanisms by which phosphorylation can modulate transcription-factor activity. The distinction between regulation of transcriptional activation by transcription-factor phosphorylation, and phosphorylation control of complex formation, is whether or not additional proteins are required to mediate the response.

clear translocation, (2) DNA binding, (3) transcriptional activation functions, and (4) multiprotein complex formation.

In this section we describe two signaling pathways in which the extracellular signal results in regulated gene expression through a phosphorylation cascade. The first example involves cAMP and its activation of a kinase, which translocates to the nucleus and phosphorylates a DNA-bound transcription factor. The second illustrates how phosphorylation of a cytosolic protein by a membrane-bound kinase modulates nuclear localization, and subunit interactions, of a transcription-factor complex.

Induction of CREB Activator Function by cAMP-Dependent Kinase A

Some ligand-activated membrane receptors transmit their signal by stimulating **adenylate cyclase** activity in the cell to produce **cAMP.** This activation pathway is mediated by a receptor-associated G protein called G_S (Chapter 16). In mammals, the most common mechanism by which cAMP serves as a **second messenger** involves cAMP binding to the regulatory subunit of **cAMP-dependent protein kinase A** (PKA). Dissociation of the regulatory subunit allows the catalytic subunit to be localized to the nucleus, where it phosphorylates the **cAMP-response-element binding protein** (CREB) on serine 133, a residue located in the transcriptional-activation domain. As shown in Figure 29.11, phosphorylated CREB is active and bound to cAMP response elements of **CREB target genes.**

The key control step in this signaling pathway is the activation of CREB transcriptional regulatory activity by PKA. It is not entirely clear how the phosphorylation on serine 133 by PKA causes transcriptional induction of CREB target genes. However, it has recently been found that another transcription factor, called CBP (**CREB binding protein**), is also phosphorylated by PKA, leading to its association with phosphorylated CREB. The CBP may be functioning as a **transcriptional adaptor protein,** or CBP and CREB together may contact the initiation complex and stimulate transcription.

Interferon Activation of STAT Nuclear Localization by Jak Tyrosine Kinases

Growth-factor signaling is mediated by membrane-bound receptors that transmit the signal to the cell as a result of **tyrosine-kinase activation.** The end result of this phosphorylation cascade can be altered expression of growth-regulatory genes through either induction or repression of transcription. Recent investigations studying **interferon** modulation of growth-regulatory genes

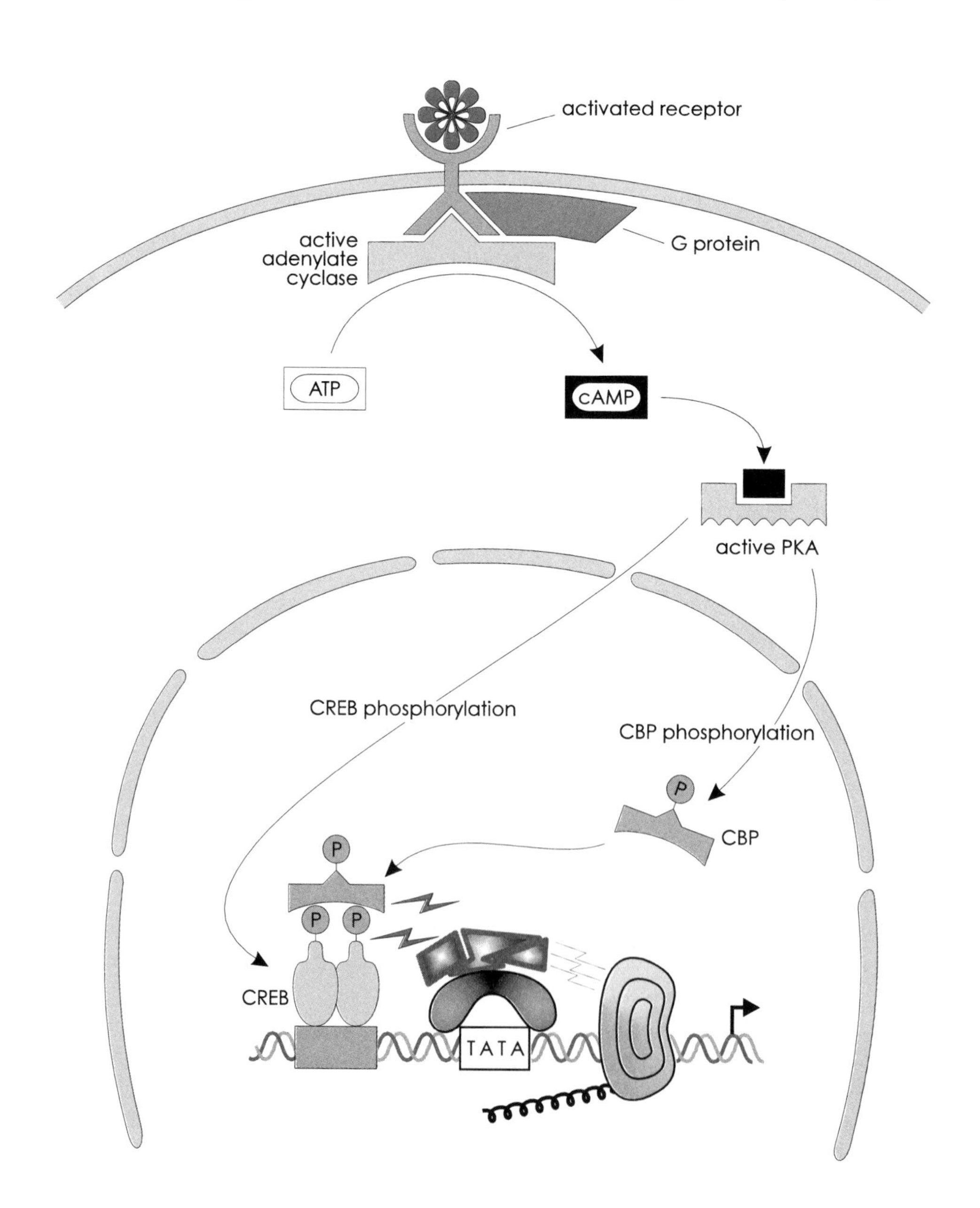

Figure 29.11.
Transcriptional regulation of cAMP responsive genes results from the phosphorylation of CREB and CBP by protein kinase A (PKA). When adenylate cyclase is activated by G-protein signaling through a membrane receptor, cAMP levels increase, PKA is activated, and CREB and CBP are phosphorylated, resulting in transcriptional induction of CREB-regulated genes.

revealed that this signaling pathway requires the activation of tyrosine kinases in the **Jak kinase family** through associations with the interferon receptor.

The **γ interferon receptor** activates Jak1 and Jak2 kinases through phosphorylation of their own tyrosine residues. Jak1 and Jak2 then phosphorylate two transcription factors called STAT1-α and STAT1-β (*s*ignal *t*ransducers and *a*ctivators of *t*ranscription), which normally reside in the cytoplasm of the cell in an inactive state. Once phosphorylated, STAT1-α and STAT1-β homodimers translocate to the nucleus and bind to **interferon response elements** located in growth-regulatory genes (Fig. 29.12). The STAT1-α and -β transcription factors are synthesized from alternatively spliced STAT1 transcripts encoded by the STAT1 gene.

The **α-interferon** pathway is more complex and demonstrates how kinase substrate specificity can also play a role. In this pathway, a third tyrosine kinase called Tyk2 is associated with the **α-interferon receptor,** and together with Jak1 (but not Jak2), phosphorylates the STAT1 proteins and another tyrosine-phosphorylated protein called STAT2. The STAT1 and STAT2 proteins form heterodimers, and in addition, associate with a 48-kDa DNA binding to form the active transcriptional regulatory complex.

The importance of STAT proteins as signal transducers and transcriptional regulators became clear when it was found that the **epidermal growth-factor receptor** also stimulates tyrosine kinases, which in turn, phosphorylate STAT1-α and another family member, STAT3. Since STAT proteins function as phosphorylation-dependent multisubunit transcription factors, the signaling pathways can be cell-type specific, depending on which Jak kinases are expressed and on the relative abundance of various STAT family members.

CHROMATIN STRUCTURE IS AN IMPORTANT MODULATOR OF GENE EXPRESSION

As described in Chapter 5, eukaryotic DNA is packaged in a nucleoprotein structure called **chromatin.** The lowest order of DNA packing is in the form of nucleosomes that contain ~140 bp of DNA wrapped around a complex of **histone proteins.** Higher orders of DNA packing were found to have an influence on gene transcription.

Two observations suggest that for some genes, **nucleosomes** may play a role in **inhibiting transcription.** First, some genes contain "phased" nucleosomes in the promoter region, meaning that nucleosomes are positioned in a specific, rather than random, arrangement. The **mouse mammary tumor virus** (MMTV) long-terminal repeat promoter is an example. The **glucocorticoid receptor** activates this promoter by facilitating the removal of a specific nucleosome that appears to be blocking the binding site of another transcription factor called NF1. After hormone withdrawal, the nucleosomes revert to their inhibitory state.

The second reason for suggesting that local chromatin structure limits transcriptional induction is the finding that some activator proteins often function better at specific gene promoters in the presence of **SNF/SWI nuclear proteins.** The SNF/SWI proteins are a related family of proteins that are evolutionarily conserved from yeast to humans. They do not bind DNA themselves, but can be coprecipitated with antibodies against activator proteins. It has been proposed that SNF/SWI proteins facilitate **derepression** of local chromatin structures by accompanying activator proteins to sites of action where they are able to **remodel chromatin structures.** Although the mechanism by which SNF/SWI proteins function is not yet clear, one of the proteins, SNF2, is a DNA-dependent ATPase, suggesting that ATP hydrolysis may be required for chromatin remodeling.

METHYLATION OF CpG DINUCLEOTIDES IS INHIBITORY TO TRANSCRIPTION

Another influence on gene transcription in eukaryotes is the presence of **5-methyl cytosine** in CpG dinucleotides. A large percentage of CpG residues in human cells contain 5-methyl cytosine, and several lines of evidence suggest that if these are clustered near a gene promoter, the expression of that gene is inhibited. One way to test for **CpG methylation** is to use restriction enzymes that cannot cut at sites containing 5-methyl cytosine to characterize specific promoter fragments by Southern blotting. A comparison of the DNA digestion pattern of two enzymes that recognize the same cleavage site, but differ in their ability to cut when 5-methyl cytosine is present (e.g., MspI cuts at all pCpCpGpG sites regardless of methylation at CpG, but HpaII can only cleave this site if the CpG is unmethylated) can be used to detect CpG methylation. These types of analyses have shown that for some genes, the level of CpG methylation near the promoter correlates with **lack of transcription.**

Another piece of evidence supporting the in-

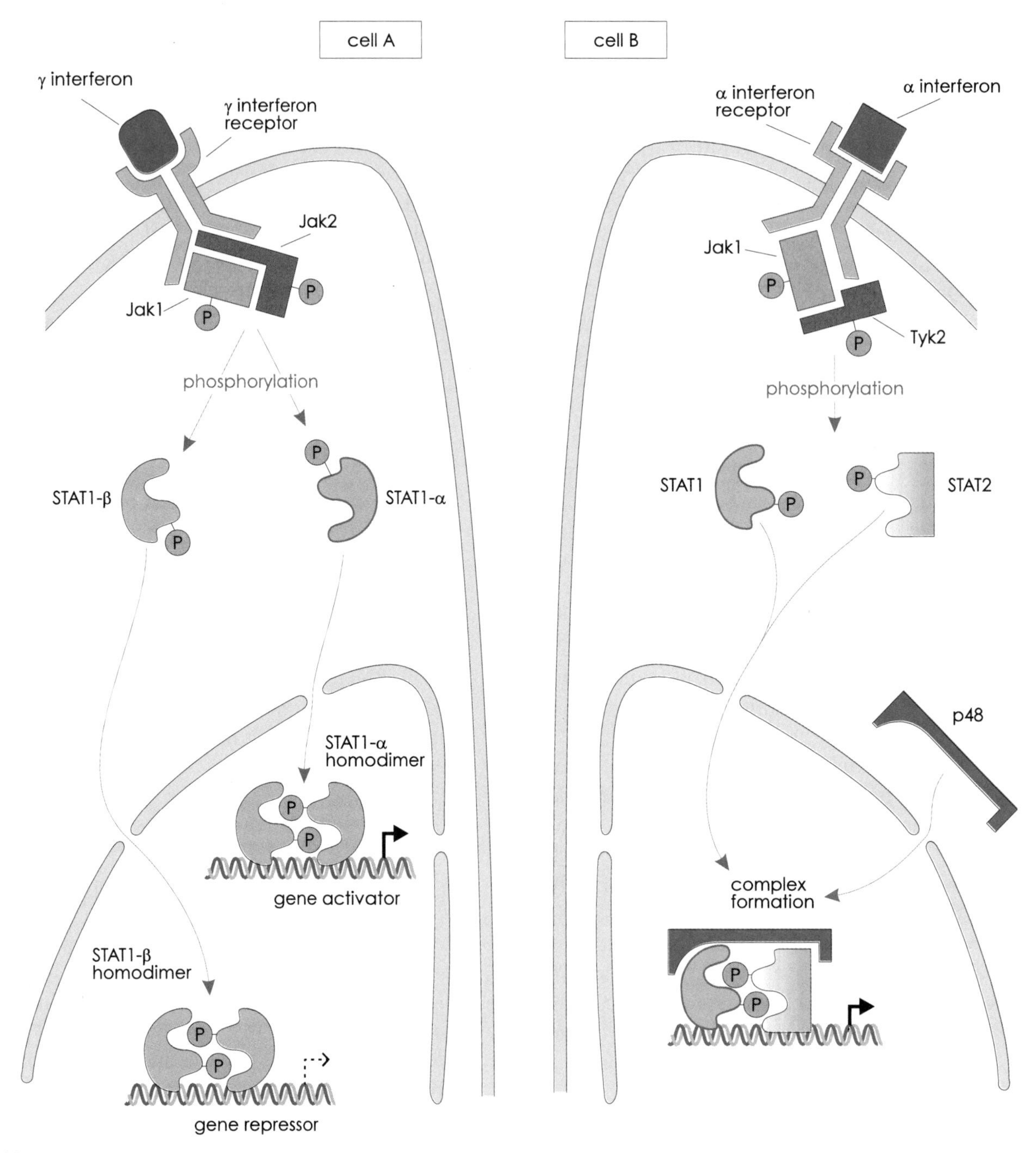

Figure 29.12.

Gamma and alpha interferon control of gene expression is mediated by Jak kinases and STAT transcription-factor proteins. Gamma interferon signaling in cell A (right) leads to the phosphorylation of STAT1-α and -β transcription factors which form homodimers and translocate to the nucleus to induce (STAT1-α homodimers) or repress (STAT1-β homodimers) transcription. Alpha interferon signaling in cell B (right) requires, in addition to Jak1 and STAT1, the Tyk2 kinase and the STAT2 transcription factor. A multiprotein complex consisting of phosphorylated STAT1 (α or β) and STAT2, along with a third protein, p48, is assembled at the α-interferon regulated promoter, resulting in gene activation.

hibitory effect of CpG methylation on transcription comes from studies in which human cells were grown in culture in media containing **5-azacytidine.** This compound replaces cytosine during replication and it cannot be methylated. Therefore, replicating DNA becomes hypomethylated under these conditions. The results from these experiments demonstrated an increase in the transcription of some genes, which were shown to have a reduced level of CpG methylation in the 5′ regulatory regions.

Although it is not known exactly how CpG methylation inhibits transcription, there is limited evidence that some transcription factors cannot bind to methylated DNA. Another possibility is that transcriptional inhibitory proteins exist in the cell which specifically bind to methylated CpG residues and thereby exclude transcription factors from gene-regulatory regions.

SUMMARY

1. Eukaryotic gene regulation requires a large number of transcription factors which function through a combinatorial mechanism to establish cell-specific and promoter-selective patterns of gene expression. The same basic principles of transcriptional control observed in prokaryotic cells is evident in eukaryotes, namely, transcription-factor binding to DNA response elements results in transcriptional induction or repression as a result of DNA looping and protein–protein interactions.

2. Experiments using transcription-factor chimeric proteins demonstrate that DNA binding and transcriptional regulatory activities are often organized within these proteins as discrete functional domains. These data suggest that sequence-specific DNA binding serves the role of localizing transcription factors to gene-regulatory regions to facilitate protein–protein interactions with the initiation complex.

3. Steroid/nuclear receptors are ligand-dependent transcription factors that transduce the hormonal signal directly to the nucleus by binding to target-gene response elements. These receptors can either induce or repress transcription, depending on ligand bioavailability and promoter sequence context. Steroid receptors bind to palindromic DNA sequences as homodimers, whereas nuclear receptors bind to direct-repeat DNA sequences as heterodimers with the RXR receptor.

4. A common transcriptional regulatory mechanism in eukaryotes is the modulation of transcription-factor activity by phosphorylation cascades. Receptor-associated G proteins transmit the extracellular signal by stimulating adenylyl cyclase activity and thus increasing intracellular levels of cAMP. Activation of cAMP-dependent protein kinase A leads to the phosphorylation and activation of two transcription factors, CREB and CBP, which together regulate transcription of CREB target genes.

5. Stimulation of the γ and α interferon receptors leads to the activation of Jak kinase family members which phosphorylate a variety of STAT transcription factors. Regulation of interferon-responsive genes is the result of phosphorylation-dependent nuclear translocation and dimerization of STAT transcription factors, which is required to activate their DNA binding and transcriptional regulatory functions.

6. Both chromatin structure and CpG methylation appear to negatively influence transcription of eukaryotic genes. The mechanisms by which these determinants modulate transcription are not clearly understood, but most data suggest that they may interfere directly or indirectly with transcription-factor binding to DNA response elements. It is likely that chromatin structure and DNA methylation play major roles in establishing gene-expression patterns during cell development.

REFERENCES

Bird A. P. (1993): Functions for DNA methylation in vertebrates. Cold Spring Harbor, *Symp. Quant. Biol.* **58:**281.

Briscoe J., Guschin D., Müller M. (1994): Signal transduction: Just another signalling pathway. *Curr. Biol.* **4:**1033.

Cowell I. G. (1994): Repression versus activation in the control of gene transcription. *TIBS* **1:**38.

Darnell J. E., Jr, Kerr I. M., Stark G. R. (1994): Jak–STAT pathways and transcriptional activation in response to IFNs and other extracellular signaling proteins. *Science* **264:**1415.

Farrell, S., Simkovich, N., Wu, Y., et al. (1996): Gene activation by recruitment of the RNA polymerase II holoenzyme. *Genes & Dev.* **10:**2359.

Felsenfeld, G., Boyes, J., Chumg, J., et al. (1996): Chromatin structure and gene expression. *Proc. Natl. Acad. Sci. USA* **93:**9384.

Glass C. K. (1994): Differential recognition of target genes by nuclear receptor monomers, dimers and heterodimers. *Endocrine Rev.* **15:**391.

Ihle, J. N. (1996): STATs: Signal transducers and activators of transcription. *Cell* **84:**331.

Janknecht, R., Hunter, T. (1996): A growing coactivator network. *Nature* **383:**22.

Kadesch T. (1993): Consequences of heteromeric interactions among helix–loop–helix proteins. *Cell. Growth Diff.* **4:**49.

Karin M. (1994): Signal transduction from the cell surface to the nucleus through the phosphorylation of transcription factors. *Curr. Opinion Cell Biology* **6:**415.

Ma P. C. M., Rould M. A., Weintraub H., Pabo C. O. (1994): Crystal structure of MyoD bHLH domain–DNA complex: Perspectives on DNA recognition and implications for transcriptional activation. *Cell* **77:**451.

Mangelsdorf D. J., Thummel C., Beato M., et al. (1995): The nuclear receptor superfamily: The second decade. *Cell* **83:**835.

McKnight S. L. (1996): Transcription revisited: A commentary of the 1995 Cold Spring Harbor Laboratory meeting "Mechanisms of Eukaryotic Transcription." *Genes & Dev.* **10:**367.

Miesfeld R. L. (1995): Biochemistry of glucocorticoid action. In DeGroot LJ, Ed: *Endocrinology* Philadelphia: W B Saunders.

Nordheim A. (1994): Transcription factors: CREB takes CBP to tango. *Nature* **370:**177.

Paranjape S. M., Kamakaka R. T., Kadonaga J. T. (1994): Role of chromatin structure in the regulation of transcription by RNA polymerase II. **Annu. Rev. Biochem. 63**:265.

Qureshi S. A., Salditt-Georgieff M., Darnell J. E., Jr (1995): Tyrosine-phosphorylated Stat1 and Stat2 plus a 48-kDa protein all contact DNA in forming interferon-stimulated gene factor 3. *Proc. Natl. Acad. Sci. USA 92:*3829.

Sadowski I., Ma J., Triezenberg S., Ptashne M. (1988): GAL4-VP16 is an unusually potent transcriptional activator. *Nature* **335:**563.

Sassone-Corsi P. (1994): Goals for signal transduction pathways: Linking up with transcriptional regulation. *EMBO J* **13:**4717.

Struhl K. (1996): Chromatin Structure and RNA polymerase II connection: Implications for transcription. *Cell* **84:**179.

Tjian R. (1995): Molecular machines that control genes. *Sci. Am.* **272:**54.

Tsai M.-J., O'Malley B. W. (1994): Molecular mechanisms of action of steroid/thyroid receptor superfamily members. *Annu. Rev. Biochem.* **63:**451.

REVIEW QUESTIONS

1. Why are the gene regulatory regions in eukaryotes often larger and more complex than that found in typical prokaryotic genes?
   a) Eukaryotic cells have nuclear membranes.
   b) Prokaryotic genes are always grouped in operons.
   c) Prokaryotic organisms generally have shorter life spans.
   d) Eukaryotes require mechanisms to control cell-specific gene expression.
   e) Eukaryotic genomes contain much more repetitive DNA.

2. What do the results from the experiments with the Gal4–VP16 chimeric transcription factor reveal about the mechanisms of transcriptional regulation?
   a) DNA binding and transcriptional activation can be distinct functions.
   b) Protein–DNA and protein–protein interactions are involved in gene activation.
   c) Yeast and viral transcription factors perform different functions in the cell.
   d) DNA response elements increase the local concentration of activator proteins.
   e) The statements in a, b, and d are all correct answers.

3. How do dimeric interactions between transcription factors contribute to combinatorial mechanisms of transcriptional regulation?
   a) More partners can bind to the same response element sequence.
   b) It allows, for example, three related proteins to bind to six different response elements.
   c) More proteins are required to bind fewer response elements.
   d) Less response elements are required to control more genes.
   e) Heterodimers and homodimers can bind to the same response elements.

4. The estrogen receptor and vitamin D receptor
   a) are both ligand-regulated transcription factors
   b) recognize different response element half-sites
   c) both bind as homodimers to direct-repeat sequences in DNA
   d) both form heterodimers with RXR
   e) share significant amino-acid homology in their hormone domains

5. Which statement is NOT TRUE regarding the effect of transcription-factor phosphorylation on protein activity?
   a) It can increase DNA binding activity.
   b) It can stimulate nuclear localization.
   c) It can activate dimerization properties.
   d) It can modulate transcriptional regulatory activities.
   e) The statements in a, b, c, and d are all not true.

6. Nucleosomal packing of DNA segments, and CpG methylation in gene regulatory regions, are both associated with
   a) increased rates of transcription
   b) transcriptional derepression
   c) decreased rates of transcription
   d) DNA damage by 5-methyl cytosine
   e) epidermal growth factor receptor activation

ANSWERS TO REVIEW QUESTIONS

1. ***d*** Cell-specific gene expression requires combinatorial mechanisms.

2. ***e*** Yeast and viral-transcription factors do perform similar functions.

3. ***b*** The proteins A, B, and C can bind to response elements recognized by AA, AB, AC, BB, BC, and CC.

4. ***a*** The mechanism of action of estrogen and vitamin D is receptor-dependent transcriptional regulation.

5. ***e*** This is a false statement because a, b, c, and d are all true.

6. ***c*** Chromatin condensation and DNA methylation inhibit transcription.

CHAPTER

30

MOLECULAR BIOLOGY OF ANIMAL VIRUSES

INTRODUCTION

The genomes of bacterial and animal cells consist of large DNA molecules that encode everything the cell needs to survive. In contrast, viruses contain much less nucleic acid, which can be in the form of single- or double-stranded DNA or RNA. Animal viruses must rely on the host cells they infect to assist them in completing their life cycle. In humans, viral infections can often be innocuous, as is true of rhinovirus (common cold) or most influenza virus infections. However, they can also be debilitating, or even deadly, as is the outcome of human immunodeficiency virus (HIV) infections. Fortunately, the combined efforts of virologists, immunologists, and biochemists have led to the production of viral vaccines which have been used to virtually eradicate at least a few virally transmitted human diseases (e.g., polio and smallpox). Nevertheless, there is still much we do not understand about these life forms, especially when it comes to designing safe and effective treatments for the most common (herpes virus), or most fatal (HIV), viral infections.

Before the age of modern biology and recombinant DNA techniques, the study of viruses was a convenient way to investigate mechanisms of gene expression. In fact, much of what we now know about information processing in eukaryotic cells was first discovered in animal viruses. More recently, the use of animal virus vectors in human gene therapy has provided potential new approaches to disease treatment. In this chapter we first describe the basic tenets of animal virus biology and biochemistry, and then describe various viruses and what we have learned from studying them. Finally, we discuss how some animal viruses are being used to develop biological tools for human gene therapy.

ANIMAL VIRUSES ARE STUDIED USING IMMORTALIZED CELL LINES

When cell biologists developed ways to propagate animal cells in culture, animal viruses could be studied at the molecular level. The most widely used in vitro cell models are those derived from immortalized differentiated cell types called cell lines. **Immortalized cell lines** are often established from tumors, usually from mouse or human tumors. Today, there are over 5000 animal cell lines available for animal cell studies. Table 30.1 lists several of the more commonly used animal cell lines.

These specialized cell lines are called immortal because they can divide indefinitely in tissue culture as

Table 30.1. Description of Several Common Mammalian Cell Lines

Cell line	Cell type	Origin	Year isolated
CHO	Ovary epithelial	Hamster	1957
3T3	Embryo fibroblast	Mouse	1963
MDCK	Kidney	Dog	1958
CV-1	Kidney	Monkey	1964
HeLa	Cervical carcinoma	Human tumor	1951
MCF7	Mammary epithelial	Human tumor	1973
LNCaP	Prostate epithelial	Human tumor	1977

long as they are provided optimal growth conditions and fresh media. However, we will make a distinction between immortal and **transformed** cell lines. A transformed cell line refers to an immortalized cell line that has the property of being **tumorgenic** when injected into animals. This means that it can form tumors in animals and has many of the attributes of a malignant cancer cell. We describe the transformed phenotype in more detail in Chapter 31 when we discuss how certain human oncogenes can transform immortalized cell lines.

VIRUSES EXPLOIT PROPERTIES OF THE CELL MEMBRANE TO ENTER AND EXIT THE HOST CELL

Unlike bacteriophage, which inject DNA into the bacterial cell after attaching to the cell surface, animal viruses usually enter cells by **endocytosis.** All viruses contain either RNA or DNA packaged within a protein coat called a capsid (Fig. 30.1). Virally encoded capsid proteins are assembled in the cell to form regular-shaped capsules that protect the viral genome from degradation. The viral-protein capsid and nucleic-acid genome form a structure called the **nucleocapsid.** Many animal viruses have nucleocapsids with a quasi-spherical shape called an **icosahedron.** This 20-sided shape solves two functional problems—packaging the nucleic acid so that it is protected from the environment, and providing a mechanism for viral self-assembly within the host cell.

Some viruses also have a lipid membrane, called a **viral envelope,** surrounding the nucleocapsid. The envelope is derived from the host-cell plasma membrane. Viruses have been classified based on both the type of viral-particle packaging (nucleocapsid, viral envelope, etc.), and on the nucleic-acid composition (RNA or DNA) of the viral genome.

Animal viruses can infect and lyse a cell, or they can reside in the cell for a long period of time and continuously release progeny to infect nearby cells. Regardless of the outcome, viruses first must enter the cell by binding to cellular-membrane proteins. Through the natural process of endocytosis, viruses are often brought into the cell by receptor-mediated events that involve **clatharin-coated pits,** as shown in Figure 30.1. As the cell membrane folds around the virus, coated vesicles are formed, which then enter the cytoplasm and fuse with other intracellular vesicles called **endosomes.**

During normal endocytosis, the contents of the endosome are passed onto the lysosome by membrane fusion where the lysosomal proteins can then degrade the captured proteins. Viruses like the **semilike forest virus** escape this fate by leaving the endosome before lysosomal membrane fusion. The low pH endosomes activates glycoproteins in the viral envelope which fuse with the endosomal membrane, resulting in release of the nucleocapsid into the cytoplasm. The capsid proteins and viral genome disassemble and viral production begins. Replication of viral genomes and viral-protein synthesis is described in more detail later.

The synthesis and release of enveloped viruses following a nonlytic infection is shown in Figure 30.2. The virus-assembly process takes advantage of normal protein trafficking in the cell (Chapter 27). Soluble capsid proteins are synthesized on free ribosomes and envelope proteins are targeted to the endoplasmic reticulum for synthesis and processing. The newly synthesized viral capsid proteins are transported through the Golgi apparatus and eventually become inserted into the plasma membrane of the cell. The capsid proteins and viral nucleic acid associate in the cytoplasm and then attach to the cytoplasmic side of the envelope proteins. Virus assembly is a thermodynamically favorable process which results in the exclusion of cellular membrane proteins and the budding off of viruses to the extracellular space.

Most viral vaccines are either made from viral-envelope or capsid proteins, or from whole neutralized viruses. These preparations are injected into the subject

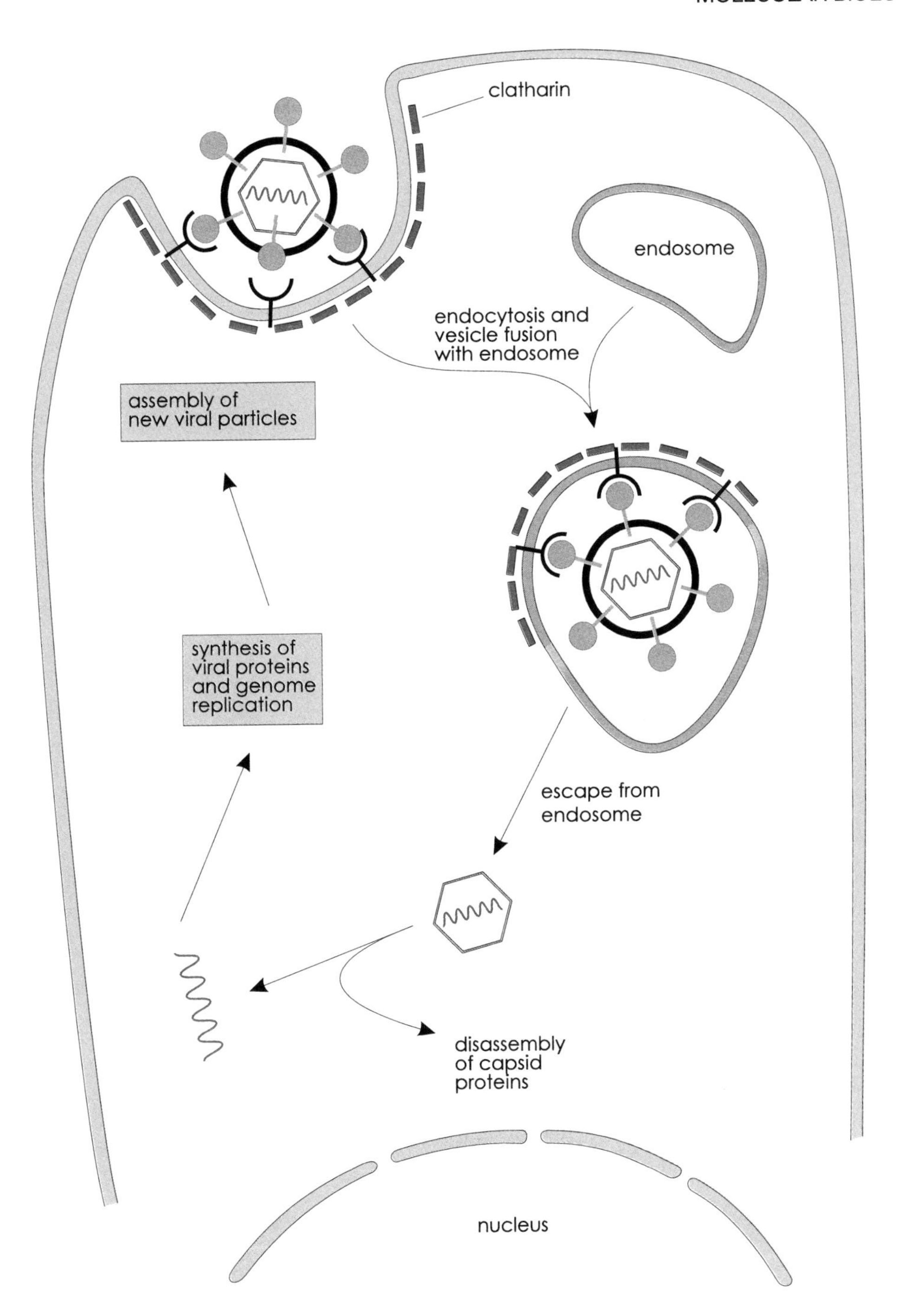

Figure 30.1.

Some eukaryotic viruses enter host cells by membrane fusion and endocytosis. Host-cell receptors interact with viral proteins, which initiates endocytosis. Once inside the cell, the virus escapes normal degradation pathways and releases its genetic material from the nucleocapsid.

over a period of time to elicit an immune response. If immune cells expressing antibodies specific for viral proteins are present in the individual's blood when they are later exposed to a virus, the virus will be recognized and destroyed. The most common reason why some vaccines are ineffective is that the genes encoding viral-coat proteins mutate frequently and thereby allow the virus to escape immune detection. For example, each seasonal strain of influenza virus requires a new vaccine.

ANIMAL VIRUSES CAN BE CLASSIFIED INTO FOUR MAJOR TYPES

Learning about animal virus life cycles is interesting because it not only teaches us something about important human pathogens, but also extends the basic principles of information transfer between DNA, RNA, and protein that we discussed in earlier chapters. Animal viruses have been discovered which seem to utilize

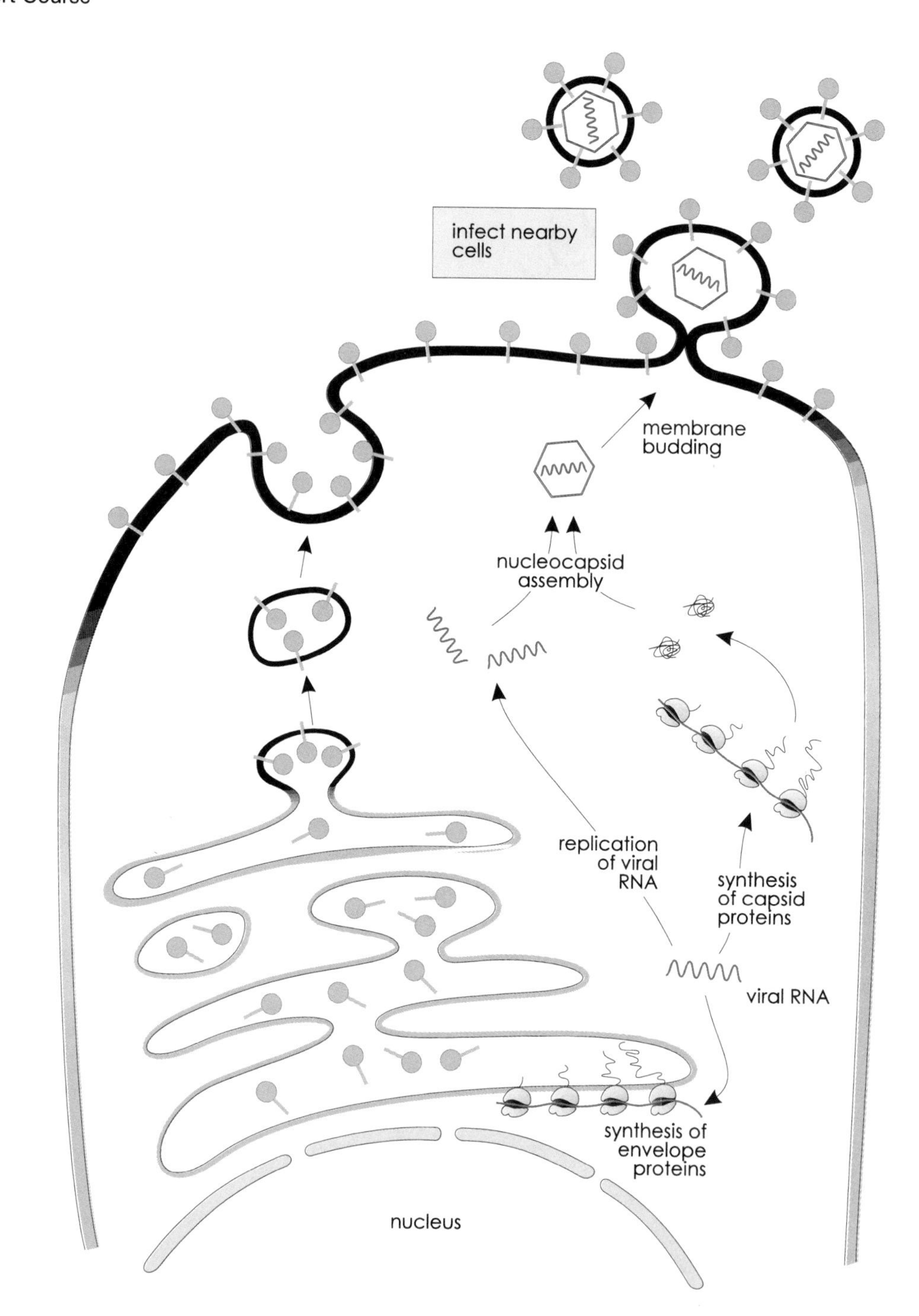

Figure 30.2.
Enveloped viruses are assembled at the host-cell membrane through normal protein-targeting pathways. Viral-envelope proteins become associated with the host-cell plasma membrane which facilitates viral packaging and cell-surface budding.

every mode of nucleic-acid replication at some point in their lifecycle; RNA → RNA, DNA → DNA, DNA → RNA, and RNA → DNA, as listed in Table 30.2.

Although more elaborate classification schemes have been developed, we will organize animal viruses into just four categories to reflect the nucleic-acid form of the virus and the first replicative product. For example, the classification "RNA–RNA" refers to an RNA virus that converts its RNA genome into a complementary RNA strand as a first step in replication. Similarly, a "RNA–DNA" virus is one that converts the RNA of the virion into DNA during replication.

RNA–RNA VIRUSES: POLIOVIRUS

There are two major classes of RNA–RNA viruses that differ in the strandedness of their RNA genomes, the (+) RNA and (–) RNA viruses. **Poliovirus** is referred to as a (+)-strand RNA virus because the RNA genome is the

Table 30.2. Four Classifications of Animal Viruses[a]

Classification	Type of viral genome	First replication product	Examples
RNA–RNA	(+) ss RNA	(–) ss RNA	Poliovirus, semilike virus, coronavirus, hepatitis A virus
	(–) ss RNA	(+) ss RNA	Rabies virus, influenza virus, measles virus
	ds RNA	(+) ss RNA	Reovirus
DNA–DNA	ds DNA (circle)	ds DNA	SV40 virus, polyoma virus
	ds DNA (linear)	ds DNA	Adenovirus, herpes virus
	ss DNA	ds DNA	Parvovirus
DNA–RNA	ds DNA	(+) ss RNA	Hepatitis B virus
RNA–DNA	ss RNA	ds DNA	HTLV-1 HIV-1

[a]ss denotes single-stranded and ds refers to double-stranded RNA or DNA; (+) is protein coding strand; (–) is noncoding.

mRNA for synthesis of poliovirus proteins soon after infection. **Influenza virus** is called a (–)-strand RNA virus because the infecting viral genome must first be copied into a (+)-strand mRNA by a viral-RNA transcriptase. Other well-known RNA–RNA viruses are **rhinovirus** and **coronavirus** which cause the common cold, **rhabdovirus,** which causes rabies, and **paramyxoviruses,** which are responsible for measles and mumps.

Poliovirus RNA Is Translated into a Single Polypeptide

Poliovirus is a member of the picornavirus family and contains a 7400-nucleotide (+)-strand RNA with a naked nucleocapsid. Figure 30.3 shows how the translation of a single large **poliovirus polypeptide** is processed by viral-encoded proteases to yield a variety of protein products. The two functional classes of viral proteins are the capsid proteins (VP1, VP2, VP3, and VP4) and the processing proteins (protease and replicase).

Poliovirus RNA in many respects looks like a typical eukaryotic mRNA with a polyadenylated 3′ tail and a protected 5′ terminus. However, the poly (A) tail is not added by the action of poly (A) polymerase, but instead is encoded in the viral genome. In addition, the 5′ end does not contain a 7-methylguanosine cap, but rather a covalently attached viral protein called **VPg** that has a **phosphodiester linkage** between a tyrosine residue in VPg and a uridine in the viral RNA.

The poliovirus RNA genome is translated into a single polypeptide called NCVP, which, when proteolytically processed by virally encoded proteases, results in the production of equimolar amounts of both capsid proteins and the replicase. This mode of protein synthesis is relatively inefficient because each repackaged virion requires 60 molecules of the capsid proteins, even though only a few replicase proteins are required to synthesize new viral genomes. Other viruses use more selective transcriptional and translation strategies. Perhaps one of the reasons poliovirus has not evolved further to streamline this process is that it already utilizes an effective method to direct the majority of cellular-protein synthesis machinery toward the translation of viral mRNA. For example, one of the poliovirus proteases specifically cleaves the cellular translation-initiation factor eIF-4B to inactivate it. The role of eIF-4B in the cell is to recognize 5′-methyl guanosine caps as a prerequisite to ribosome loading. Since poliovirus mRNA translation doesn't require this initiation factor, all the ribosomes in the cell are free for viral-protein synthesis.

Replication of the Poliovirus Genome

Another unusual feature of poliovirus is its mechanism of RNA replication. Once the amount of capsid proteins is sufficient to support virion packaging, the virus shifts into a mode of replicating its genome. **Replicase** is the viral RNA-dependent RNA polymerase needed for conversion of the (+) strand into the template (–) strand followed by synthesis of new (+) RNA strands for packaging (Fig. 30.4). Priming for (–)-strand synthesis requires the VPg protein and it involves the annealing of uridine residues attached to VPg to adenine residues in the 3′ terminus of the (+) strand. Replicase is able to begin synthesis by using the VPg protein–uridine as a primer and completing (–)-strand synthesis.

Once (–) RNA strands are synthesized, the production of (+)-strand poliovirus genomes is initiated.

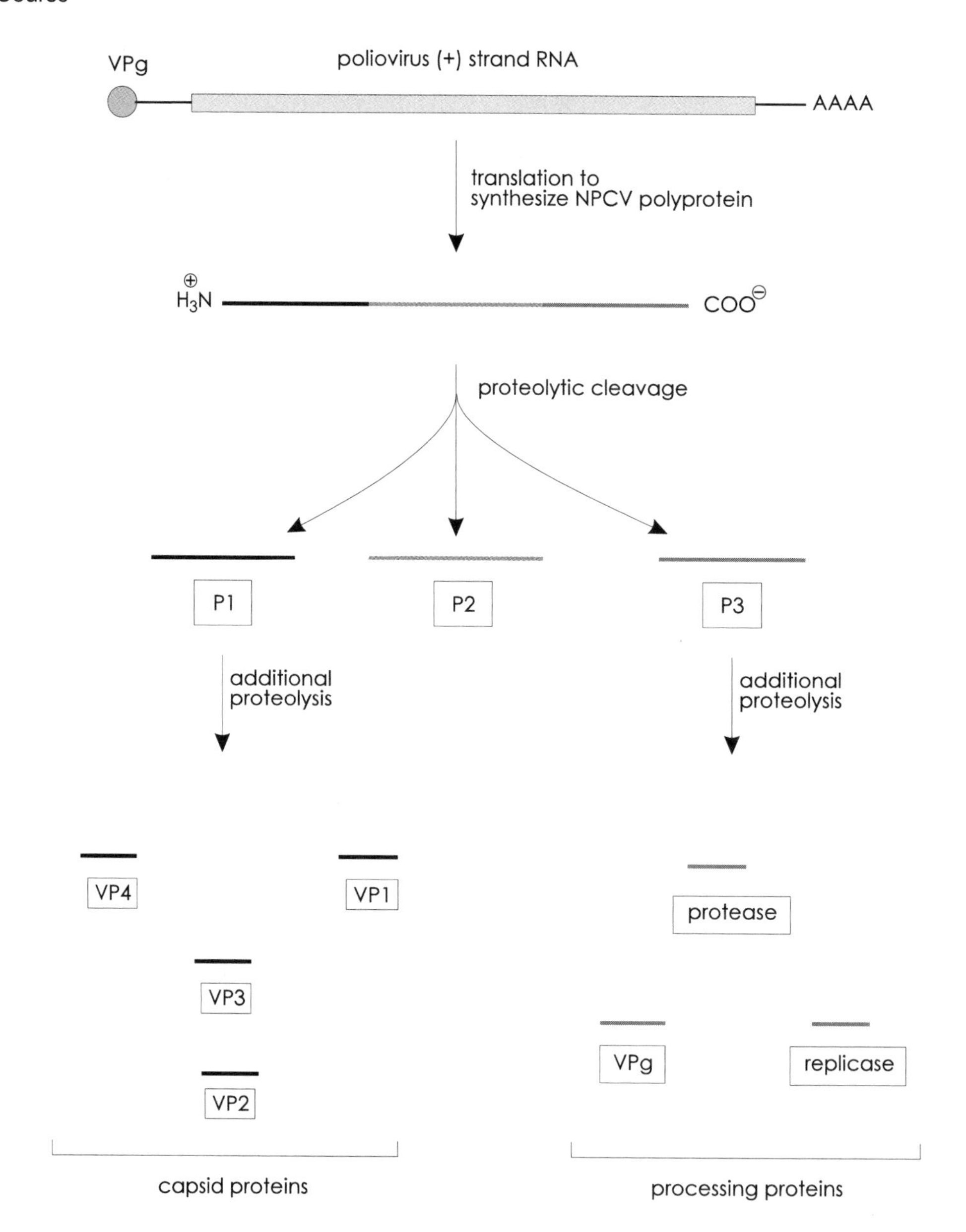

Figure 30.3.
Proteolytic processing of poliovirus proteins. A single polyprotein is synthesized from the infecting (+)-strand RNA. Subsequent proteolytic processing results in the accumulation of the various poliovirus proteins required for virus production.

Again, **VPg primer protein** is required to initiate (+)-strand synthesis. Some of the initial (+) strands made are used for protein synthesis, whereas later, after infection as capsid proteins begin to assemble, the majority of (+) strands are packaged into **virions** (infectious viral particles). The balance between using the (+) strands for translation of viral proteins, versus packaging the genome into virions, is determined by the availability of capsid proteins.

DNA–DNA VIRUSES: SV40 VIRUS AND ADENOVIRUS

The life cycle of DNA–DNA viruses is similar to that of a eukaryotic cell in that the genetic information is stored in the form of DNA. Moreover, RNA needs to be synthesized before viral proteins can be made. The DNA viruses have a wide range of genome sizes. Some DNA viruses have small genomes of less than 10 kb (**papilloma viruses** that cause warts), some have intermediate-sized genomes of ~40 kb (**adenoviruses** that infect the respiratory tract), and others are even larger, with genomes greater than 200 kb in size (**herpesvirus** and **smallpox virus**).

The smaller DNA viruses do not encode a DNA polymerase and can only replicate in cells that are actively cycling through the S phase, thus the host range of these viruses is limited. In contrast, the large DNA viruses, such as herpes simplex virus (HSV), encode a variety of proteins, which allows the virus to replicate in the nucleus of any cell it is able to infect. Herpes

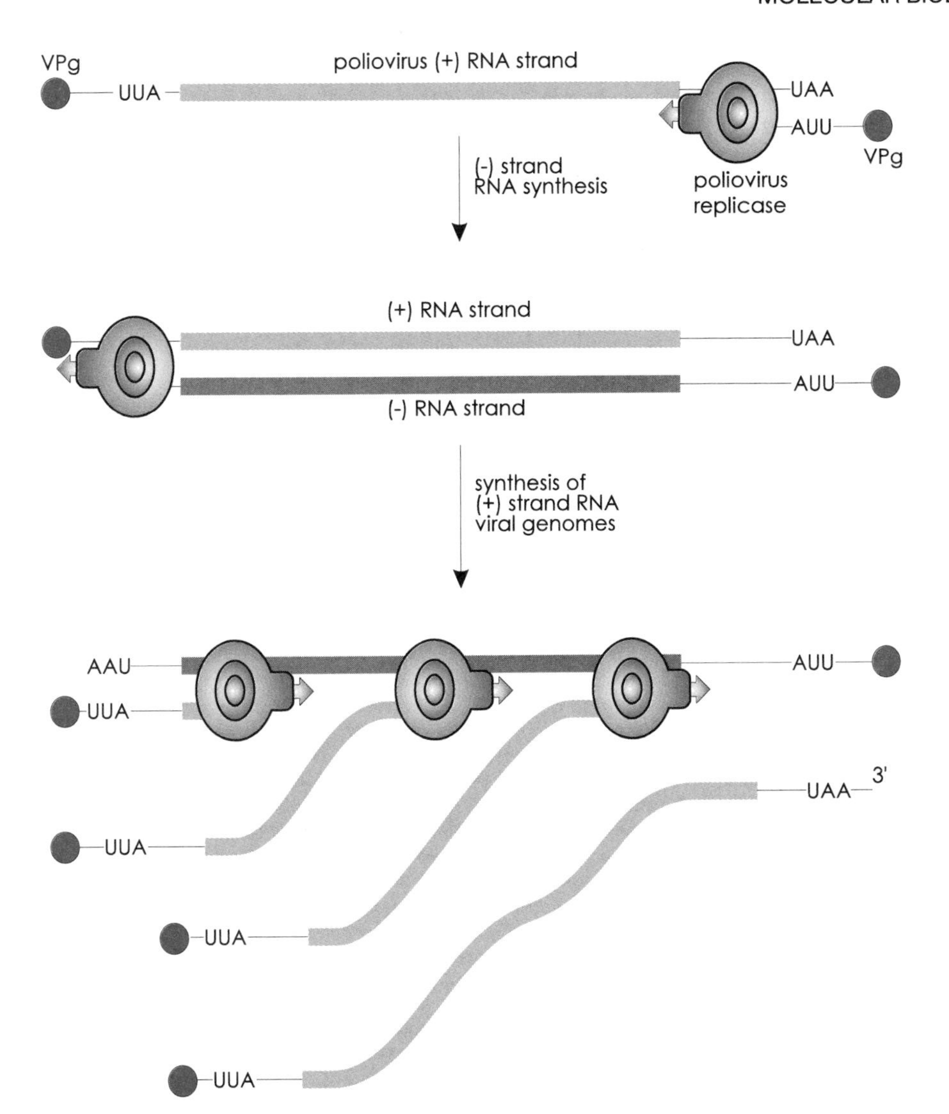

Figure 30.4.

Replication of the poliovirus genome. The poliovirus replicase utilizes the VPg priming protein to initiate synthesis of (–)-strand RNA, which becomes the template for (+)-strand RNA production.

viruses cause several human pathological conditions, such as oral cold sores (HSV type I) and genital blisters (HSV type II). Two common drugs used to treat HSV infections are **ganciclovir** and aciclovir, which are nucleotide analogues that specifically inhibit viral replication. Their mechanism of action is based on the fact that the HSV-encoded thymidine kinase uniquely phosphorylates these analogues and thereby activates them to function as chain terminators during replication. Since the human thymidine kinase cannot phosphorylate ganciclovir or aciclovir, DNA replication in uninfected cells is normal. In addition to its own thymidine kinase and DNA polymerase, the herpesvirus genome encodes a number of other proteins involved in DNA replication, including an origin-binding protein, a helicase/primase, single-strand binding protein and ribonucleotide reductase. Another well-characterized herpesvirus protein is the potent transcriptional-activator protein, VP16, which was described in Chapter 29.

In this section we briefly describe two of the most highly characterized DNA viruses—the monkey papovavirus SV40 and the adenoviruses. SV40 has taught us much about the functions of eukaryotic origins of DNA replication. It also encodes one of the best-suited **transforming** proteins known to cause tumors in animals, a multifunctional viral protein called the SV40 T antigen. Adenoviruses have also been extremely valuable biological reagents and have taught us much about transcriptional regulation and mRNA splicing in mammalian cells.

SV40 Virus Has Been Used to Study DNA Replication and Cell Transformation

SV40 is a circular, double-stranded DNA virus with a genome of 5243 base pairs. It was originally discovered as a contaminant in preparations of polio-virus vaccines prepared from monkey cells (simian virus). The **SV40 virus** normally kills monkey cells, but it can also infect a wide variety of mammalian tissue-culture cells. In rodent cells, and sometimes in human cells, SV40 infection does not cause cell lysis, but rather, at a low frequency is able to immortalize, and often transform, a variety of cell types. Many of the animal cell lines in use today were established by infecting primary cells with SV40 or introducing SV40 DNA directly into the cell by DNA transfection.

As shown in Figure 30.5, the SV40 genome contains a single **bidirectional origin of replication** that is coincident with the only transcriptional promoter in the viral genome. Transcription from the leftward side of the SV40 promoter (early promoter), in the counterclockwise direction, leads to the production of just two proteins, **small t antigen** and **large T antigen.** These two proteins share ~80 amino acids at their amino termini as a result of differential splicing of the transcript. Small t antigen protein is truncated as compared to large T antigen because of an in-frame stop codon in a portion of the intron which is removed from the T antigen transcript.

The "t/T" stands for **transforming protein,** which reflects how these proteins were originally characterized. It was found that the immortalization and transformation of infected rodent cells were due to T antigen. **SV40 T antigen** is a **multifunctional protein** which is required for **SV40 DNA replication** and binds to the viral replication origin. SV40 T antigen is also a **transcriptional activator** and is able to stimulate transcription of the SV40 promoter. Figure 30.6 shows the functional domains of SV40 T antigen and relevant DNA sequence elements in the SV40 origin of replication and transcriptional regulatory region.

The immortalizing and transforming properties of **SV40 T antigen** appear to be related to its ability to bind to and **inactivate** at least two different regulators of the host-cell cycle, the **Rb** (retinoblastoma) and **p53** proteins. In Chapter 31, we discuss the normal function of Rb and p53, which is to control the cell cycle by preventing cells from entering S phase and replicating their DNA at inappropriate times. However, if Rb or p53 become mutated, or when their functions are inhibited by viral proteins like SV40 T antigen, cell proliferation is allowed to continue and additional mutations can accumulate in the genome. Mutations in Rb and p53 are linked to a large number of human cancer, and, therefore, it seems likely that transformation by SV40 T antigen, through inactivation of Rb or p53, is a paradigm for early events in human oncogenesis.

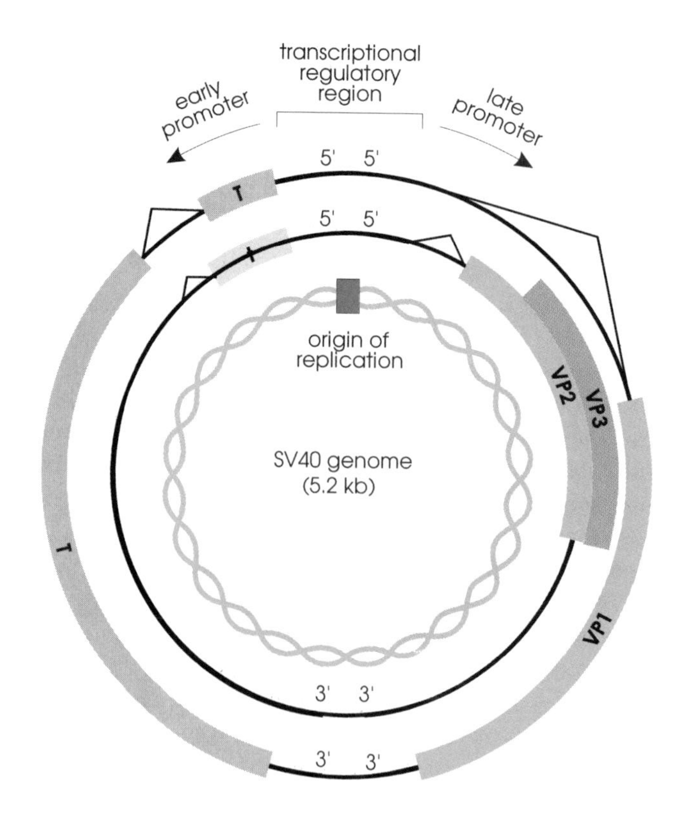

Figure 30.5.

SV40 genome organization showing the major viral protein-coding sequences and the replication and transcriptional regulatory regions.

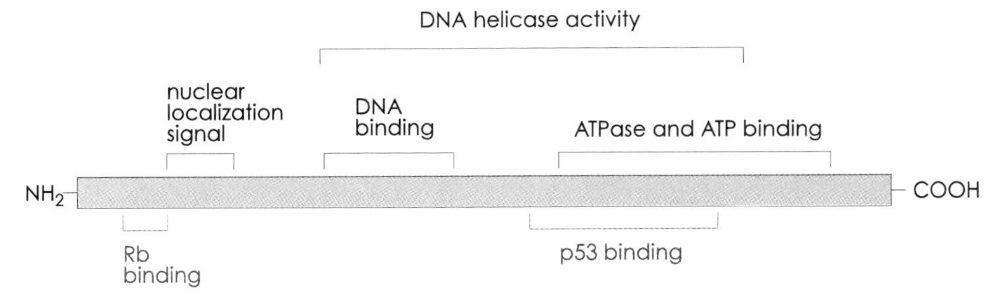

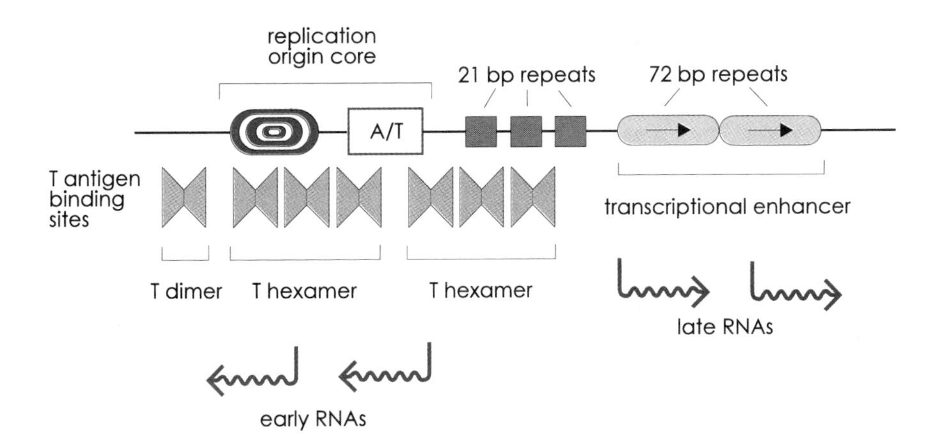

Figure 30.6.

Functional maps of the SV40 large T antigen and the region surrounding the origin of replication. (a) Genetic and biochemical analyses of T antigen coding sequences have revealed multiple overlapping functional regions. Some of these amino-acid sequences are required for the replication functions of T antigen and others are involved in protein interactions with host-cell regulatory proteins such as Rb and p53. (b) A ~300-bp region, which includes the SV40 origin of replication core, contains multiple binding sites for T antigen in addition to regulatory sequences (21- and 72-bp repeats) required for early and late gene transcription.

Adenovirus Utilizes Complex Transcription and RNA Splicing Strategies

Adenoviruses are linear, double-stranded DNA viruses that infect the respiratory tract of primates. The adenovirus genome contains over 36,000 base pairs and encodes at least 40 different viral proteins. Soon after infection, six different adenoviral transcription units, each with its own promoter, are transcribed using both host-cell and virally encoded transcription factors. The first **adenovirus protein** to be made is **E1A,** a viral **transcription factor** that is required for transcription of all other adenovirus promoters (including activation of the E1A promoter). E1A does not bind directly to DNA, but instead appears to function much like a transcriptional adaptor or coactivator protein (Chapter 29). Portions of the E1A transactivation domain interact with proteins in the RNA pol II initiation complex, such as TFIIB, and with other transcription factors. E1A expression in transfected cell lines can stimulate a number of cellular genes, suggesting that, in addition to its role in activating adenovirus promoters, it is also capable of altering host-cell gene expression.

Figure 30.7 shows the transcription pattern of early and late **adenoviral transcription units** which are transcribed from both strands. Adenoviral mRNAs are processed by the cellular machinery and contain **5′ methyl guanosine caps** and **poly(A) tails.** The adenovirus major late promoter is responsible for directing the transcription of a single precursor transcript, which gives rise to different mRNA classes that terminate at one of five polyadenylation signals. These mRNAs are then differentially spliced to remove internal portions of the coding sequence. Translation of these various major late mRNA species gives rise to a large number of adenoviral proteins required for viral replication and packaging.

DNA–RNA VIRUSES: HEPATITIS B VIRUS

The best-studied virus in this class in the **hepatitis B virus** (HBV), which is known to cause liver disease in humans. During its life cycle, the 3.2-kb DNA genome is converted into RNA by cellular RNA polymerase. This RNA is then used as a template by an HBV-encod-

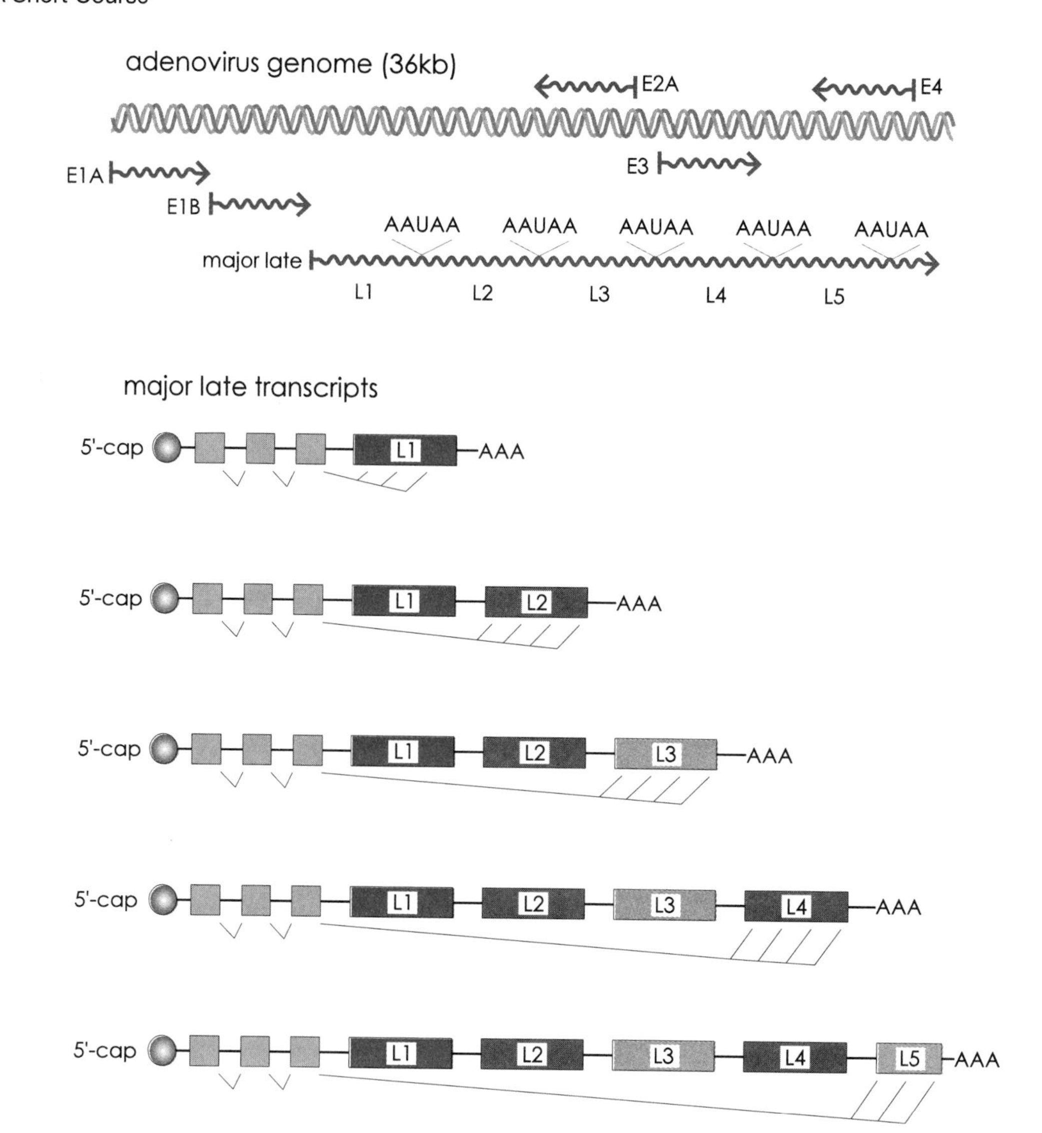

Figure 30.7.

Synthesis and processing of adenovirus transcripts. The adenovirus genome contains six promoters which give rise to multiple transcripts. The major late promoter is especially active and generates several classes of transcripts which terminate at any one of five polyadenylation signals. Through an elaborate alternative splicing pathway, each class of major late transcript is processed to include the same amino-terminal protein sequences linked to various carboxy terminal residues.

ed reverse transcriptase to synthesize new strands of viral DNA. The complete replication cycle is **DNA → RNA → DNA.** Hepatitis B virus has the opposite life cycle of a retrovirus, which replicates its genome in the order of RNA → DNA → RNA.

Two *different* viruses are associated with human liver disease. **Hepatitis A virus is a RNA picornavirus** which causes *infectious hepatitis*; **HBV is a DNA virus** responsible for *serum hepatitis.* Human liver disease, as a result of HBV infection, is a serious worldwide health problem, and it is estimated that over 200 million people are **chronically infected.** A substantial number of these people will die from liver cirrhosis or hepatocellular carcinoma. In Asia, which is endemic for HBV infection, hepatocellular carcinoma is one of the leading causes of death in males, and HBV-infected individuals are 100 times more likely to die in this way than uninfected people.

The mechanistic basis for HBV-related liver disease and **hepatocellular carcinoma** is not completely understood; however, two possibilities have been proposed. First, the virally encoded **HBV X protein** (HBx) has been shown to cause liver cancer in transgenic mice expressing only this HBV gene suggesting that HBx is directly responsible for transformation. Molecular studies of HBx have shown that it is capable of modulating host-gene expression by stimulating phosphorylation cascades that activate transcription factors through serine/threonine kinases. Second, chronic infection of HBV often results in **HBV integration** into the host genome and this could either inactivate genes by disruption, or abnormally activate genes as a result of HBV transcriptional regulatory sequences. The two mechanisms that could explain how HBV infection leads to liver-cell cancer are illustrated in Figure 30.8.

Fortunately, **HBV vaccines** are now available and most healthcare workers are encouraged to be immunized. One of the first widely used products of **biotechnology** was a second-generation HBV vaccine made from expressing the **HBV envelope-coat protein** gene in yeast and using the purified protein as an immunizing agent.

RNA–DNA VIRUSES: RETROVIRUSES

By far the most pathogenic of all human viruses are the RNA–DNA family of retroviruses. The life cycle of retroviruses is shown in Figure 30.9. **Retrovirus** virions are enveloped icosahedral nucleocapsids containing two copies of the single-stranded RNA genome, a specific host-cell tRNA for priming genome replication after infection, and two molecules of **reverse transcriptase,** which is the viral polymerase. After binding to a cell-surface receptor protein, the viral particle is

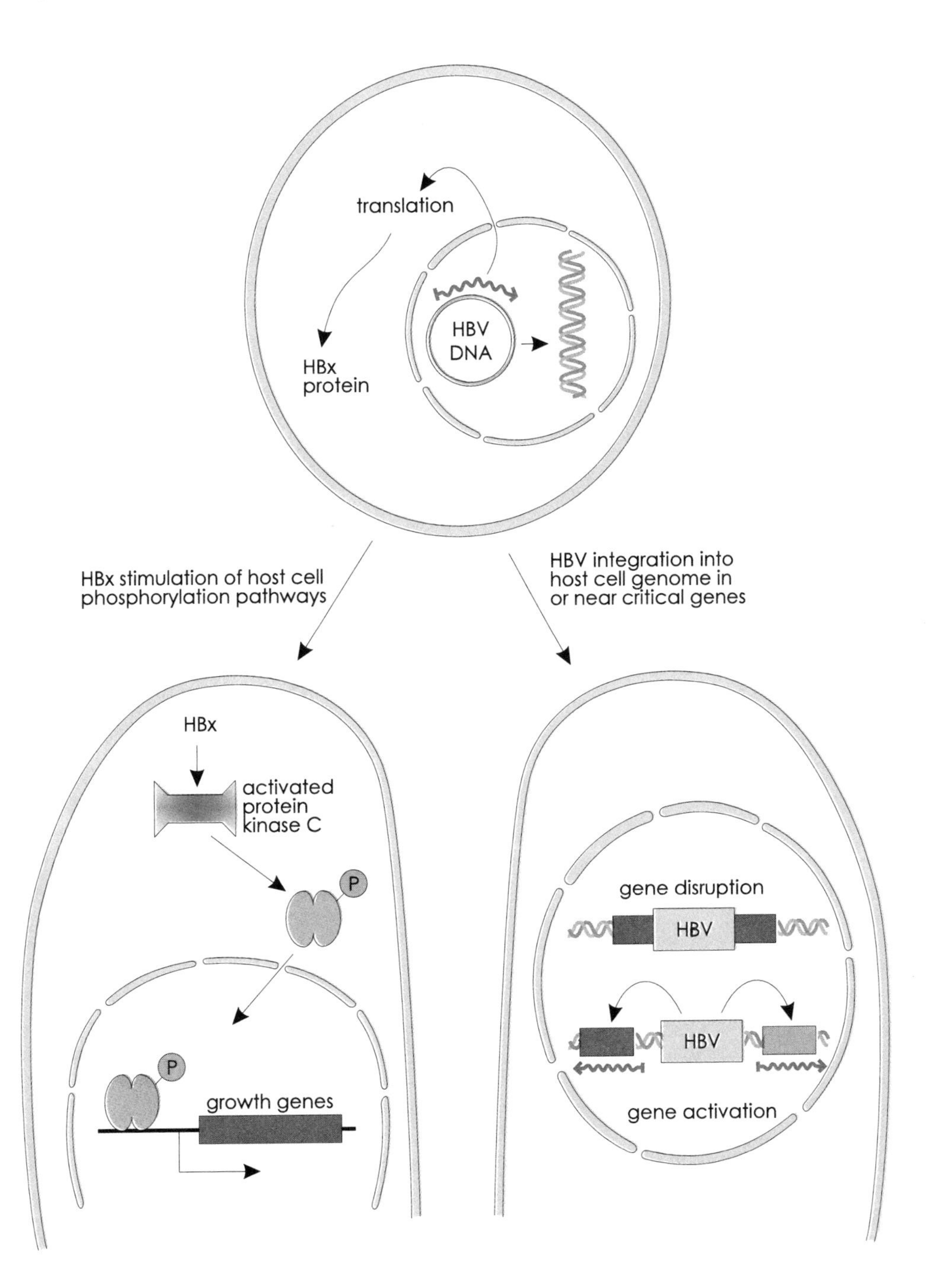

Figure 30.8.

Two modes by which chronic HBV infection could promote liver disease. (Left) The viral HBx protein functions as a stimulator of host-cell protein kinase C activity, which results in the phosphorylation of activator proteins such as AP-1. Induction of growth-regulatory gene transcription by AP-1 could lead to liver-cell proliferation. (Right) Integration of the HBV genome into host-cell DNA could result in the disruption of growth inhibitory genes or cause aberrant activation of growth-stimulatory genes.

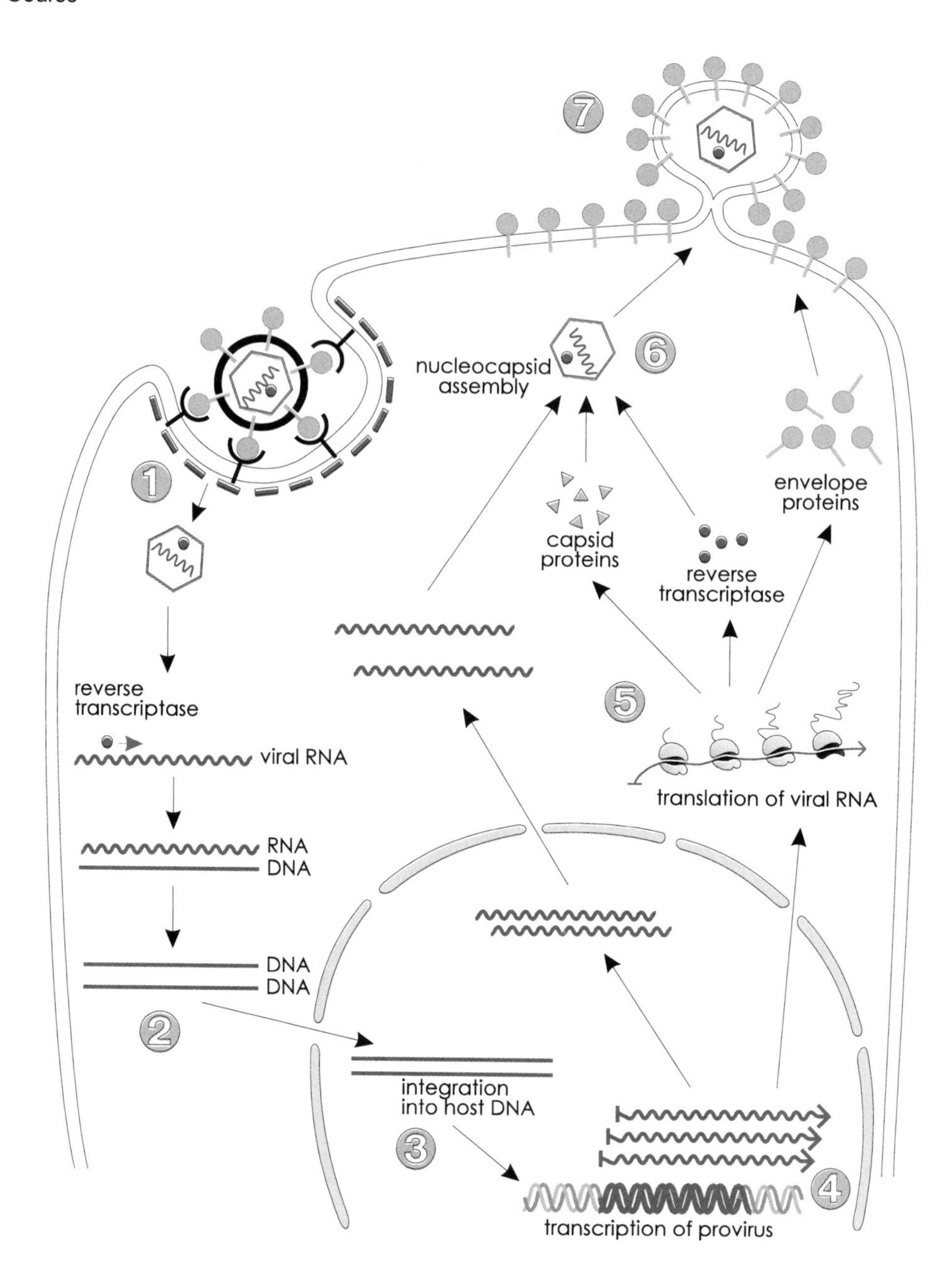

Figure 30.9.

Seven key steps in the retroviral life cycle. The infectious virus actually contains two copies of the RNA genome and two molecules of reverse transcriptase. The processes of endocytosis (step 1) and membrane budding (step 7) of retroviral virions is similar to that shown in Figures 30.1 and 30.2. The infected host cell is not immediately killed by the virus, but instead serves as a manufacturing plant for large-scale virion production.

endocytosed and the viral RNA genome is released into the cytoplasm, where it is immediately copied into double-stranded DNA. The double-stranded, linear retroviral DNA enters the nucleus and integrates in the host-cell genome, where it serves as transcription unit for the production of retroviral RNA. This integrated DNA copy of the retroviral genome is called a **provirus.** In the final step, the fully processed viral transcripts are exported to the cytoplasm, where they are translated to produce viral-packaging proteins and are also packaged into progeny virus. One of the important differences between retroviruses and other types of animal viruses is the dependence on **host-genome integration** for retroviral gene transcription and replication.

Four seminal discoveries in the last 20 years highlight the importance of retroviruses:

1. In the early 1970s, Howard Temin and David Baltimore independently discovered the retroviral enzyme **reverse transcriptase,** an RNA-directed DNA polymerase. This discovery not only proved once again that all rules in biology have exceptions (it was thought up to that point that genome information could only flow from DNA to RNA), but it also provided a

valuable tool for recombinant DNA technology. As described in Chapter 22, reverse transcriptase is a useful molecular biological reagent to clone mRNA transcripts by converting RNA into complementary DNA (cDNA).

2. Michael Bishop and Harold Varmus demonstrated in the mid-1970s that animal retroviruses can cause tumors by the altered expression of host genes known as **oncogenes.** This discovery led to our current understanding of the genetic basis of cancer (this principle is described in Chapter 31).
3. In 1980, researchers showed for the first time that a human virus can cause cancer. The virus was a retrovirus called the **human T-cell leukemia virus** (HTLV-1), and by the combined efforts of many researchers, it was found that a retrovirally encoded gene produce, the HTLV-1 Tax protein, caused adult T-cell leukemia through its action as a transcriptional transactivator of host-cell genes.
4. The most daunting claim to fame for a human retrovirus came in the early 1980s, when teams of researchers in France and the United States jointly identified **human immunodeficiency virus** (HIV-1) as the etiologic agent in the human acquired immunodeficiency syndrome called **AIDS.** HIV infection of human T cells eventually results in T-cell death and loss of these important cells from the immune system. AIDS patients with extremely low numbers of T cells are highly susceptible to opportunistic infectious diseases, which are often the underlying cause of death due to AIDS.

A worldwide effort to combat the fatal result of HIV-1 infection has led to some progress and we now have a better understanding of HIV-1 pathogenicity. Much of this research has focused on developing biochemical strategies to inhibit the activity of HIV-1 reverse transcriptase. Figure 30.10 shows the molecular structure of the HIV-1 reverse transcriptase obtained by X-ray crystallography. Data such as these have been used in an attempt to design better enzymatic inhibitors of viral DNA synthesis, for example, to improve the potency and viral specificity of the competitive inhibitor 3′-azido-2′,3′-dideoxythymidine (AZT). Unfortunately, high rates of mutation in the reverse-transcriptase coding sequence that accompanies retroviral genome replication (reverse transcriptases lacks a 3′-5′ exonuclease proofreading function), combined with the toxicity of these compounds due to their ability to cross-inhibition human DNA polymerases, severely limits their usefulness as therapeutic agents to treat AIDS.

THE USE OF VIRAL-EXPRESSION VECTORS IN HUMAN GENE THERAPY

One of the long-term goals of molecular genetic medicine is to use **gene replacement** as a therapeutic regimen to treat human diseases due to single gene defects. The two basic approaches under consideration at this point are eukaryotic viral vectors and nonviral DNA delivery. At this point, the development of viral vectors for **human gene therapy** is more advanced than nonviral DNA delivery strategies, although both approaches will likely be utilized as this technology improves.

The most promising viral vectors developed to

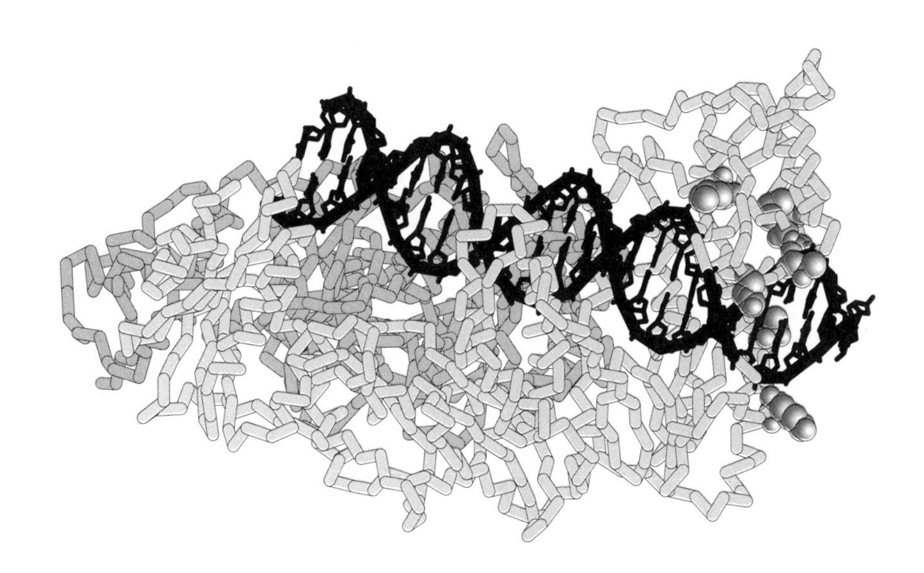

Figure 30.10.
Molecular structure of the HIV-1 reverse-transcriptase heterodimer (p66 and p51 subunits) complexed with an RNA–DNA nucleic-acid strand. The relative positions of amino-acid mutations in the reverse-transcriptase complex, which result in resistance to AZT and didexoyinosine (ddI), are shown as spheres.

date are derived from retroviruses and adenoviruses. As shown in Figure 30.11, by removing viral genes (gag-pol-env) and replacing them with the therapeutic gene plus a marker gene for drug resistance, it is possible to use retroviral-packaging cell lines to produce recombinant retroviruses that are capable of host-cell DNA integration as nonreplicating proviruses. In a pilot study, **retroviral vectors** were used to partially correct a deficiency in the **adenosine deaminase gene** in several patients with this inherited disorder. Lymphocytes from these patients were infected with the retroviral vector in vitro and then the reengineered cells were returned to the patients.

Recombinant **adenoviruses** have been used to deliver the **cystic fibrosis transmembrane conductance regulator** (CFTR) gene to the lungs of patients with cystic fibrosis. Adenoviruses are ideal for this approach because they naturally infect human respiratory cells. In these adenoviral vectors, the CFTR gene has been put in place of the adenovirus E1A gene. The viral vectors can be packaged in a specialized human cell line that expresses E1A protein to complement E1A deletion in the viral genome.

The advantage of using viral vectors for human gene therapy is their potential to deliver DNA to large numbers of target cells. Indeed, this attribute is what

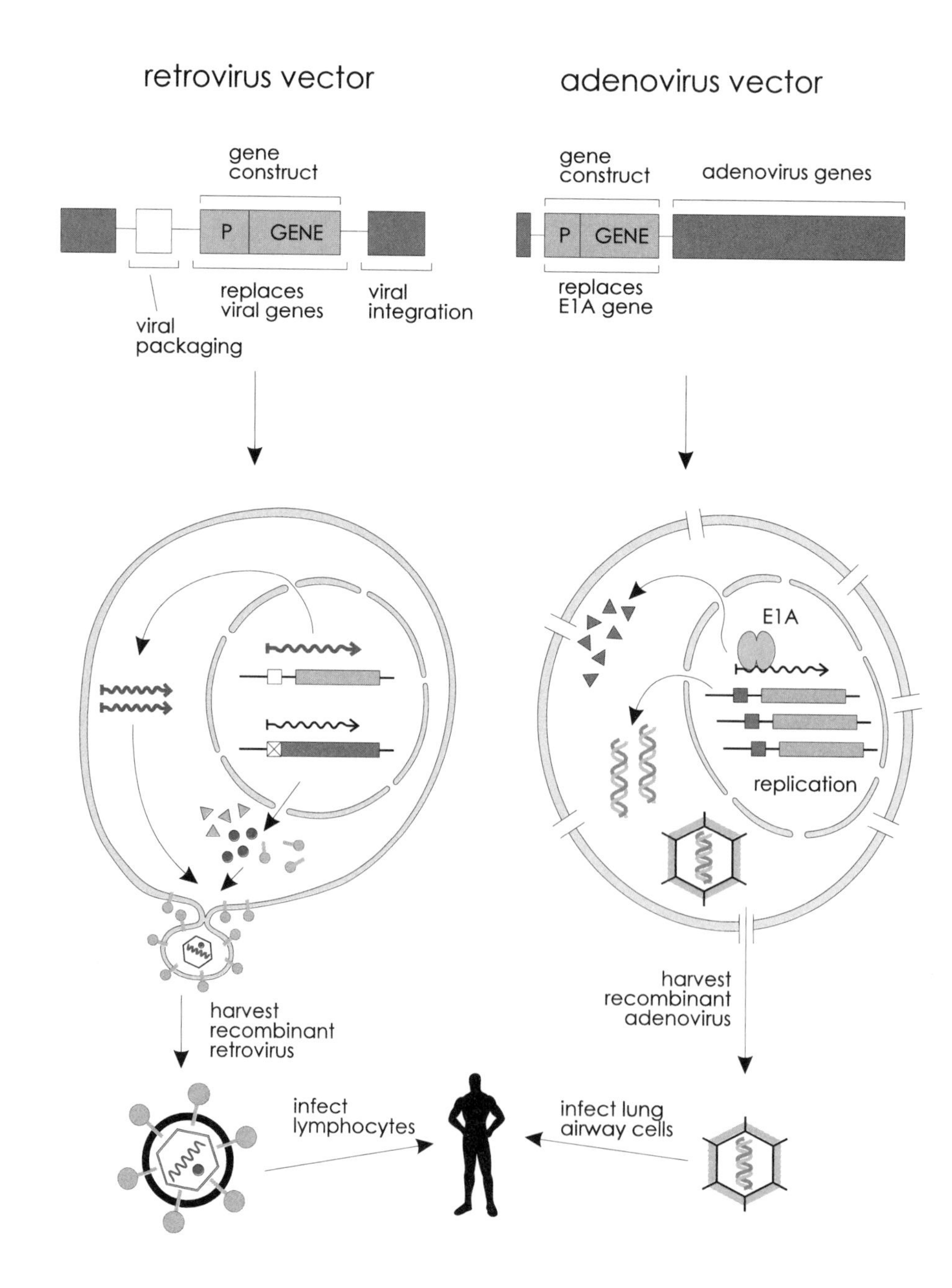

Figure 30.11.

Schematic drawing of retroviral and adenovirus vectors being developed for human gene therapy. The special viral "packaging cell lines" are needed to complete recombinant virus assembly by providing missing gene functions. Infection of host cells with these recombinant viruses should be nonproductive for viral replication because key viral genes have been deleted.

makes most animal viruses such a threat to their hosts. However, some concerns still need to be addressed, and several problems have to be solved, before viral vectors will become the first line of therapy for human disease treatment. For example, host-cell DNA integration by the viral vector could cause insertional mutations leading to loss of cell growth control (see Chapter 31). There is also a remote possibility that mutations in the viral vector could arise and thereby convert the otherwise harmless DNA reagent into an infectious virus with an altered host-cell range. For these reasons, other non-viral based human gene therapy approaches are also being developed.

SUMMARY

1. Animal viruses can be propagated and studied in the laboratory using immortalized cell lines. It has been shown that animal viruses exploit many of the host cell's normal processes, including DNA replication, RNA and protein synthesis, and protein targeting and transport. The cell membrane is utilized both as a port of entry through endocytosis and as a means of spreading infection to nearby cells by the process of membrane fusion and virus budding.

2. Based on the nucleic-acid content of the infectious virus, and the nucleic-acid product of the *first* replication phase, most common animal viruses can be classified into four groups. These four modes of replication are DNA → DNA, RNA → RNA, DNA → RNA, and RNA → DNA.

3. Poliovirus is an RNA–RNA virus that redirects the host-cell protein synthesis machinery toward translation of its own RNA genome. The poliovirus polypeptide is cleaved into active subunits by the proteolytic action of virally encoded proteases. Poliovirus RNA replication utilizes a novel protein–uridine primer to initiate RNA synthesis using its own RNA-dependent RNA polymerase.

4. SV40 is a DNA–DNA virus that normally infects monkey cells. The SV40 origin of replication has been extensively studied as a model for the initiation of eukaryotic DNA replication because it is dependent on host-cell proteins. The SV40 T antigen is a multifunctional protein that is required for SV40 replication and for the transcription of SV40 encoded genes. SV40 T antigen has been shown to cause *transformation* of rodent cell lines, which is at least partially due to its ability to inactivate cellular regulatory proteins such as Rb and p53.

5. Hepatitis B virus (HBV) is a DNA–RNA virus that infects human liver cells. Chronic HBV infection is a worldwide health problem because it causes liver cirrhosis and hepatocellular carcinoma. Two possible mechanisms for HBV-related liver disease are host-cell gene disruption by HBV integration and altered transcription of host-cell genes by virally encoded proteins such as HBx.

6. Retroviruses are RNA–DNA viruses that require host-cell genome integration to complete their replication cycle. The retroviral enzyme reverse transcriptase is an RNA-dependent DNA polymerase that has important uses in recombinant DNA methodologies such as the synthesis of complementary DNA (cDNA). The human immunodeficiency virus (HIV) is the etiologic agent that causes acquired immunodeficiency syndrome (AIDS) in humans. HIV infects and destroys human T cells, which are required for immune system functions.

7. One strategy being explored for human gene therapy is to use recombinant retrovirus and adenovirus vectors to deliver functional copies of specific genes to human cells. Both strategies involve the use of packaging cell lines which provide the missing gene functions necessary to produce infectious, but nonreplicating, viruses. Two examples of viral-mediated gene therapy are the use of a recombinant retrovirus to introduce the adenosine deaminase gene into lymphocytes and the inhalation of recombinant adenoviruses to deliver the cystic fibrosis transmembrane conductance regulator gene to lung epithelial cells.

REFERENCES

Amalfitano, A., Begy, C., Chamberlain, J. S. (1996): Improved adenovirus packaging cell lines to support the growth of replication-defective gene-delivery vectors. *Proc. Natl. Acad. Sci. USA* **93**:3352.

Benn J., Schneider R. J. (1994): Hepatitis B virus HBx protein activates Ras–GTP complex formation and establishes a Ras, Raf, MAP kinase signaling cascade. *Proc. Natl. Acad. Sci. USA* **91**:10350.

Bischoff, J. R., Kirn, D. H., Williams, A., et al. (1996): An adenovirus mutant that replicates selectively in p53-deficient human tumor cells. *Science* **274**:373.

Cohen J. (1994): Bumps on the vaccine road. *Science* **265**:1371.

Condra J. H., Schleif W. A., Blahy O. M., et al. (1995): *In vivo* emergence of HIV-1 variants resistant to multiple protease inhibitors. *Nature* **374:**569.

Fauci A. S. (1993): Multifactorial nature of human immunodeficiency virus disease: Implications for therapy. *Science* **262:**1011.

Hill, C. M., Littman, D. R. (1996): Natural resistance to HIV? *Nature* **382:**668.

Kasahara N., Dozy A. M., Kan Y. W. (1994): Tissue-specific targeting of retroviral vectors through ligand–receptor interactions. *Science* **266:**1373.

Katz R. A., Skalka A. M. (1994): The retroviral enzymes. *Annu. Rev. Biochem.* **63:**133.

Kitamura N. B., Semler B., Rothberg B. G., et al. (1981): Primary structure, gene organization and polypeptide expression of poliovirus RNA. *Nature* **291:**547.

Kohlstaedt L. A., Wang J., Friedman J. M., Rice P. A., Steitz T. A. (1992): Crystal structure at 3.5 Å resolution of HIV-1 reverse transcriptase complexed with an inhibitor. *Science* **256:**1783.

Liu F., Green M. R. (1994): Promoter targeting by adenovirus E1a through interaction with different cellular DNA-binding domains. *Nature* **368:**520.

Morgan R. A., Anderson W. F. (1993): Human gene therapy. *Annu. Rev. Biochem.* **62:**191.

Pear W. S., Nolan G. P., Scott M. L., Baltimore D. (1993): Production of high-titer helper-free retroviruses by transient transfection. *Proc. Natl. Acad. Sci. USA* **90:**8392.

Perales J. C., Ferkol T., Beegen H., Ratnoff O. D., Hanson R. W. (1994): Gene transfer *in vivo*: Sustained expression and regulation of genes introduced into the liver by receptor-targeted uptake. *Proc. Natl. Acad. Sci. USA* **91:**4086.

Price T. N. C., Moorwood K., James M. R., Burke J. F., Mayne L. V. (1994): Cell cycle progression, morphology and contact inhibition are regulated by the amount of SV40 T antigen in immortal human cells. *Oncogene* **9:**2897.

Rabinovich N. R., McInnes P., Klein D. L., Hall B. F. (1994): Vaccine technologies: View to the future. *Science* **265:**1401.

Spence R. A., Kati W. M., Anderson K. S., Johnson K. A. (1995): Mechanism of inhibition of HIV-1 reverse transcriptase by nonnucleoside inhibitors. *Science* **267:**988.

Tooze J. (1981): DNA tumor viruses. Part 2, Molecular biology of tumor viruses. Cold Spring Harbor Laboratory, Cold Spring Harbor, New York.

Varmus H. E. (1988): Retroviruses. *Science* **240:**1427.

REVIEW QUESTIONS

1. What technical advance in cell biology made it feasible to perform biochemical and molecular biological studies of animal viruses?
   a) The characterization of clatharin-coated pits
   b) The ability to propagate cell lines in culture
   c) The invention of the electron microscope
   d) The development of monoclonal antibodies
   e) The identification of human diseases caused by viruses

2. What macromolecular "organizing complex" in the cell directs the assembly of viral proteins and nucleic acids into an infectious virion?
   a) The Golgi apparatus
   b) The nuclear matrix
   c) Clatharin-coated pits
   d) A collection of special polyribosomes
   e) There is no "organizing complex"

3. Which of the pathways shown below is required for complete replication of the HIV-1 genome in infected human T cells?
   a) DNA → RNA
   b) DNA → RNA → DNA
   c) RNA → DNA
   d) RNA → DNA → RNA
   e) RNA → DNA → DNA

4. What property of viruses, such as the influenza virus and HIV-1, make it so difficult to develop *effective* vaccines?
   a) Viruses rapidly accumulate mutations in coat proteins during replication.
   b) Influenza and HIV-1 both have very large and complex genomes (>100 kb).
   c) RNA is unstable; vaccines against ssRNA viruses have never been successful.
   d) Viruses always lyse the cells they infect, making them difficult to study.
   e) So few people are infected with these viruses that there is very little research.

5. What has been learned about cellular processes by studying the obscure DNA–DNA monkey-kidney virus called SV40?
   a) Control of DNA synthesis at replication origins
   b) Relationship between DNA response elements and transcription
   c) The role of cell-cycle proteins in tumorgenesis

d) Amino-acid composition of nuclear localization-signal peptides
e) All of the above

6. Why have viral vectors been considered as potentially good biological reagents for human gene therapy?
a) They can randomly integrate into the middle of host-cell genes.
b) They are highly specific in selecting what cell types to infect.
c) They are very efficient vehicles for transferring genetic information.
d) They can replicate after infection and enter nearby cells.
e) There are no concerns that mutations could alter the viral vector.

ANSWERS TO REVIEW QUESTIONS

1. ***b*** Cell culture made it possible, and economical, to study animal viruses.

2. ***e*** Viral assembly is an energetically favorable set of molecular reactions, each of which utilizes some normal cellular process in the host.

3. ***d*** HIV-1 is a retrovirus.

4. ***a*** Amino-acid changes in coat and envelope proteins, as a result of mutations in the viral genome which occur during replication, allows viruses to escape immune detection.

5. ***e*** Biochemical and molecular analyses of the SV40 life cycle have been very informative, mainly because this simple virus utilizes a variety of host-cell proteins.

6. ***c*** Viral vectors can be very useful for getting DNA (or RNA) into cells; however, there are still a number of technical difficulties that need to be solved, such as host-cell specificity and viral genome stability.

CHAPTER

31

ONCOGENES AND HUMAN CANCER

INTRODUCTION

A dominant theme in biochemistry is understanding regulatory checkpoints, for example, how a cell senses changes in the environment at the molecular level and what events are set in motion to respond to those signals. We now know that most, if not all, cancer-causing mutations result from perturbations in regulatory mechanisms. These could be loss-of-function mutations in gene products that control when a cell divides or how it repairs its DNA, or they could be gain-of-function mutations in growth-stimulatory proteins. Cell-growth control is governed by positive signals, much like a gas pedal in a car, and negative signals, similar to a braking system. Cancer-cell growth could be due to loss-of-growth control by mutations that cause the gas pedal to stick in the "on" position, and/or, defects in the braking system, so that cell division cannot slow down. Recent progress in cell biology and cancer research has led to the identification of critical genes involved in controlling cancer-cell growth, and in some cases, it is even known how certain gene products function differently in a cancer cell as compared to its normal counterpart.

The observation that tumors are the result of rapidly dividing cells led to the term **oncogenesis** to describe events resulting in uncontrolled cell growth. "Onco" is Greek for bulk mass, and "genesis" means origination; therefore, oncogenesis refers to the initiation of tumor growth. In this chapter, we first give an overview of what cancer is and how it affects cellular phenotypes, and briefly describe factors that seem to contribute to its development. This is followed by a discussion on the molecular basis of cancer and the classification of **oncogenes.** Finally, we focus on two of the genes most often associated with cancer, *ras* and *p53,* and discuss how mutations in these genes contribute directly to oncogenesis.

EARLY EVIDENCE THAT DAMAGE TO DNA IS A PREREQUISITE TO CANCER

The first question that researchers tried to answer was, "What causes cancer?" Three lines of evidence strongly suggested that cancer-cell growth was the result of mutations in the DNA. First, incidence of cancer rises exponentially with age. As shown in Figure 31.1, before the age of 50, human cancer is a relatively rare disease, but after that, there is a dramatic increase in the incidence rate. **Epidemiological studies** indicated that **age** was a strong contributing factor to cancer, suggesting that accumulation of unrepaired **somatic mutations** in DNA could be the reason.

Second, there is a strong correlation between the potency of DNA-damaging agents to act as mutagens in

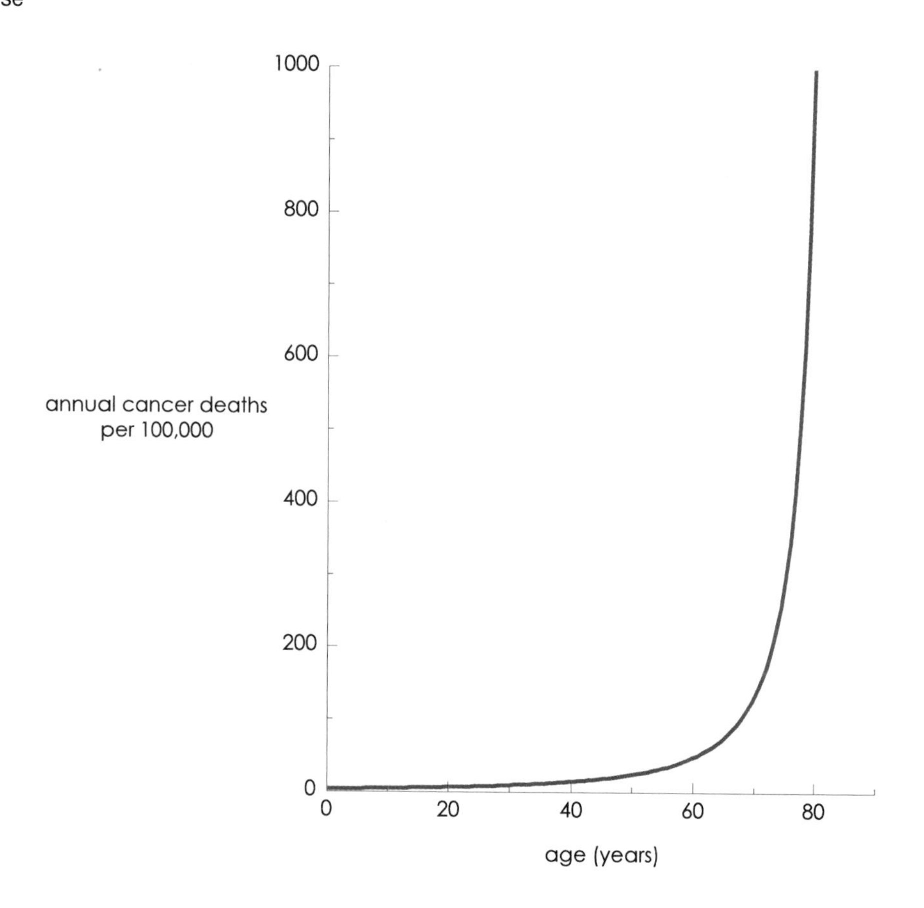

Figure 31.1.
Cancer rates in the population increase dramatically with age.

bacteria and their ability to cause cancer in animals when given at extremely high doses (carcinogens). Although tests such as these have contributed to our understanding of **carcinogens** in animals, it has been difficult to identify cancer-inducing agents in humans, because human exposure occurs at much lower levels and over a long period of time. For example, epidemiological studies have shown a link between smoking and lung cancer. However, not everyone who smokes gets lung cancer, even though more smokers get lung cancer than nonsmokers.

The third observation that supported a central role for DNA in cancer came from the study of chromosomes in human leukemia cells. **Karyotyping** is a cytological technique used to characterize the gross structure of chromosomes in mitotic cells. Researchers noticed that in cells from one type of leukemia, chronic myelogenous leukemia, there was almost always a **chromosome rearrangement** between chromosomes 9 and 22. This reciprocal translocation led to identification of a small 22:9 chromosome which came to be called the **Philadelphia chromosome** because it was discovered at the Wistar Institute in Philadelphia. Importantly, the karyotype of nonleukemic cells from patients with this form of leukemia is normal.

SEVERAL STAGES OF DEREGULATED CELL GROWTH PRECEDE THE MALIGNANT STATE

Most cells divide during development as precursor cells and then once they reach their predetermined fate, they stop dividing and **terminally differentiate.** A differentiated cell is the end point in development for any given cell type (e.g., hepatocytes, muscle cells, and neurons). Occasionally, however, changes occur in these cells and they **dedifferentiate** and begin dividing again to form what is known as a benign growth. A harmless type of benign growth is a wart, which is due to a localized papilloma virus infection. A different situation, however, can be seen by examining the various stage of **skin cancer.** As shown in Figure 31.2, **ultraviolet light-induced DNA mutations** in basal cells of the dermis could lead to an increase in the division rate of a single cell compared to the surrounding cells.

As a benign tumor enlarges, owing to cell proliferation, it may become obvious on the surface of the skin, and it could at this point be surgically removed. However, if not detected at this initial stage of deregulated cell division, some cells in the tumor could acquire another trait which allows them to grow more ag-

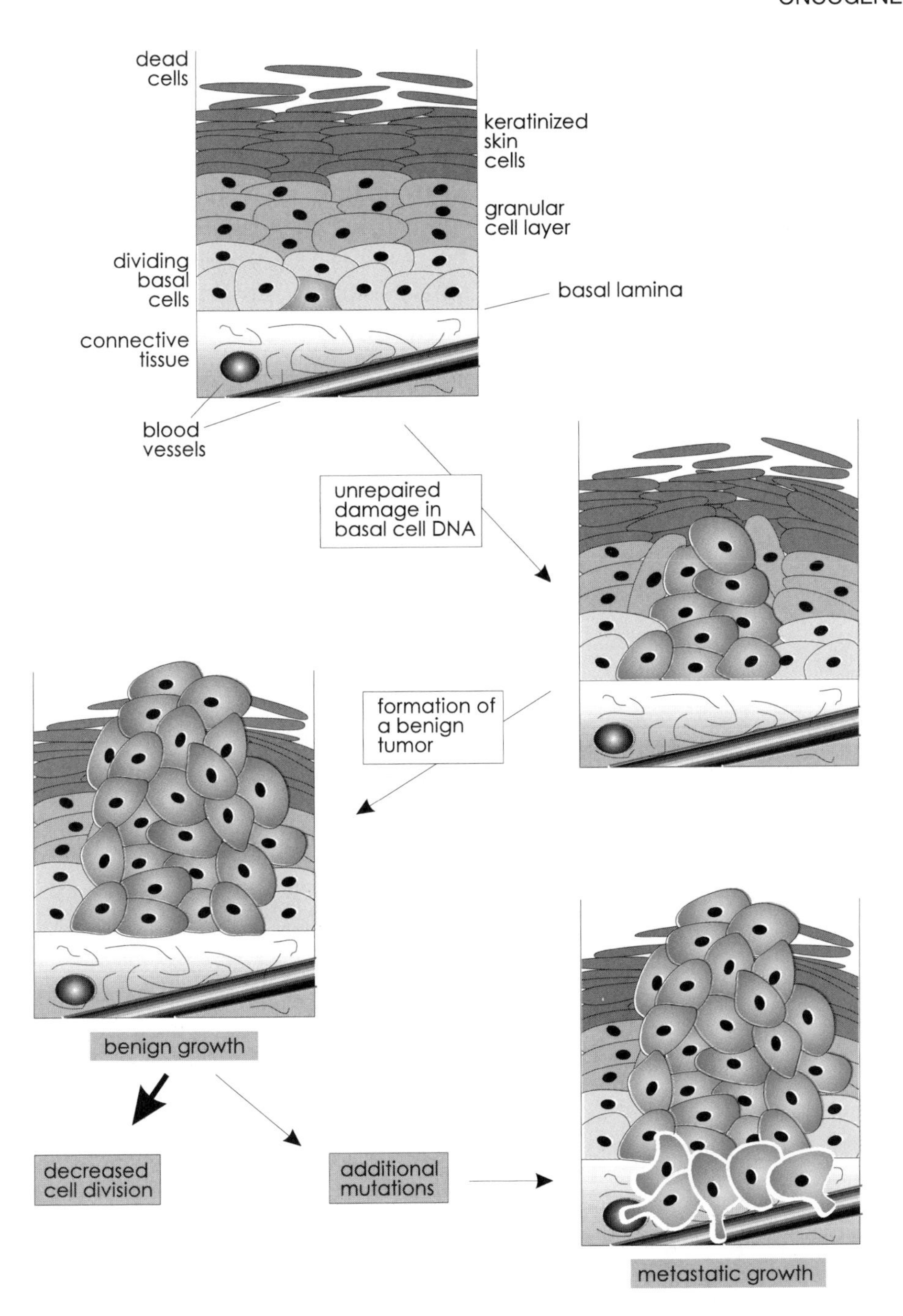

Figure 31.2.

Progressive stages of benign and malignant growth. Basal cells in the dermis normally divide at a fixed rate to replace cells that die and are lost at the skin surface. If unrepaired damage in a basal cell results in increased rates of cell division, a benign tumor will eventually appear on the surface. Additional rare genetic mutations are required to convert cells in the benign tumor into more aggressive malignant cells that escape through the basal lamina and enter the bloodstream.

gressively and to become **vascularized** (inundated with blood vessels). Benign tumor growth becomes **malignant** when the altered cells are able to degrade the basal lamina and escape into the circulatory system. Often times malignant cancer cells **migrate** to the lymph nodes or the liver, where second sites of cancer growth develop (Fig. 31.2).

A precancerous cell mass (benign growth) can be distinguished from advanced stages of malignancy by a pathologist who examines biopsy sections for evidence of **cell invasion.** Various stages of cancer are graded histologically, with grade 1 being the presence of dedifferentiated "precancerous" cells, whereas grade 4 is a tumor that is fully malignant with widespread invasion of other tissues. Breast and prostate cancer are examples of tumors that are evaluated by a graded score of benign to malignant.

What accounts for these different stages in cancer-cell growth? First, the benign cancer cell, with mutations in genes that allow the cell to divide locally, could accumulate additional mutations in genes that control cell mobility and attachment to surface receptors on neighboring cells. Some of the cancer cells could begin to express secreted proteases which accelerate basal lamina breakdown and invasiveness. Studies have shown that an initial increase in **cell division rates** often result in the accumulation of **additional mutations.** This makes sense, because with each round of DNA replication, there is a finite chance that new mutations will arise (see Chapter 22). Accumulation of multiple genetic mutations will eventually lead to the establishment of a subpopulation of malignant cells that cause the clinical signs of cancer (Fig. 31.3a).

The second way that a benign dividing cell could become more aggressive is by a process called **promotion** which aggravates an otherwise minor genetic defect by causing *inappropriate gene expression.* Studies have shown that **initiators** are mutagenic compounds that cause DNA damage and that promotion is only a contributing factor to cancer when cells have first been exposed to initiators, that is, sustained DNA damage. Figure 31.3b shows how a mutated cell can be stimulated to divide after being exposed to a promoting agent such as a hormone. For example, increased levels of the

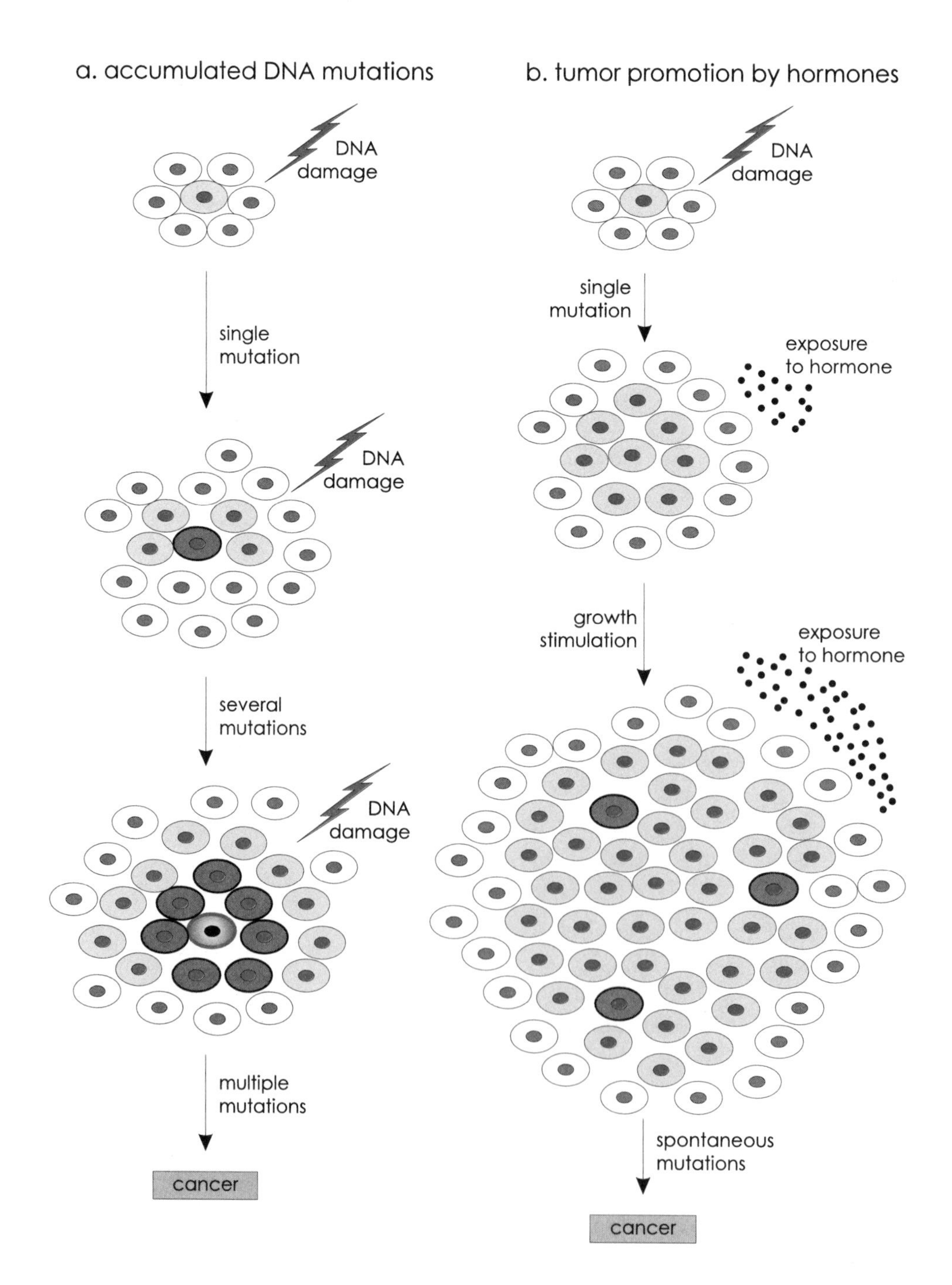

Figure 31.3.

Two models showing how cancer can be caused by unrepaired DNA damage. (a) Accumulation of multiple genetic mutations in a small number of cells can lead to the expression of the malignant phenotype. If the initiating mutations were in DNA repair enzymes, additional deleterious mutations would likely accumulate rapidly. (b) The combination of a genetic mutation in a key growth-regulatory gene, followed by prolonged exposure to promoting agents such as hormones, can also result in clinical cancer. Increased rates of cell division enhance the probability that a cell will acquire additional mutations.

steroid hormones estrogen and testosterone are associated with breast and prostate cancer, respectively. These steroids by themselves are not mutagens and, therefore, it is unlikely that they cause additional mutations in the cells. However, they do signal alterations in gene expression through activation of their respective hormone receptors. Other tumor-promoting agents, such as phorbol esters, activate cellular phosphorylation cascades, which, as we saw in Chapter 29, leads to the activation of a variety of transcription factors. Therefore, the combined affect of an initiating DNA mutation, and altered gene expression as a result of prolonged exposure to promoting agents, could accelerate the development of malignant growth.

ONCOGENESIS: GOOD GENES GONE BAD

We will refer to oncogenes as those human genes that have been shown to play a role in cancer when they are either mutated, deleted, or abnormally expressed. The term **protooncogene** is used to denote the normal unmutated form of the oncogene. To understand how these oncogenes were found, we must first review the difference between an immortalized and a transformed cell, two terms that were introduced in Chapter 30 when describing properties of SV40 T antigen. **Immortalization** refers to the ability of a cell line to grow indefinitely in tissue culture; however, in the simplest context, these cells are not cancerous because they do not form tumors in animals and often require special growth media to survive in culture. In contrast, **transformed cell lines** are immortalized cells that have acquired additional tumorgenic properties. Cellular phenotypes commonly associated with **transformation** include:

1. Decreased adherence to plastic; cells have round morphology
2. Growth in soft agar; decreased requirement for substrate attachment
3. Ability to attain high-density growth in tissue-culture systems
4. Decreased requirement for supplemental growth factors in the media
5. Ability to cause tumors when injected into immunodeficient mice

There are three basic ways that an immortalized cell line can become transformed: (1) if it is infected with a DNA tumor virus or oncogenic retrovirus, (2) by introduction of stably integrated DNA from a transformed cell or a primary tumor, or (3) culturing the cell in media containing mutagens. Figure 31.4 illustrates how transformed cells can be identified among a high background of nontransformed cells. This transformation assay was used to identify and isolate transformed cells containing "activated" oncogenes.

Composite Retroviruses Contain Oncogenes

The first clue that cells could be transformed by defects in normal genes came from studies of the **composite retroviruses.** Although this class of animal retroviruses have not been linked to human cancer, cell-culture studies demonstrated that the retroviral genome encoded all the information required for cell transformation.

With the advent of recombinant DNA technology, it was possible to isolate and sequence the genomes of a large number of **oncogenic retroviruses.** These studies revealed that the normal arrangement of the retroviral gag-pol-env genes was altered in oncogenic retroviruses and, more importantly, this class of retroviruses contained DNA that was unique to that particular viral strain. When these *atypical* viral DNA sequences were compared to the GENBANK nucleotide database using computer algorithms, it was found that many were homologous (similar but not identical) to **mouse and human genes** already present in the database. In addition, it was also shown by such computer analyses that many of the nonviral sequences shared amino-acid homologies with **cellular protein kinases.** Figure 31.5 shows the genome alignment of several oncogenic retroviruses that were shown to contain oncogenes.

The key experiment identifying these DNA sequences as oncogenes was to show that DNA transfection of just this portion of the viral genome, in the context of a eukaryotic expression vector, was sufficient to transform cell lines. In addition, when the protooncogene counterparts of these *captured* retroviral oncogenes were used in DNA transfection experiments, they were not oncogenic. By comparing the sequence of the two forms of the gene, it was found that the **retroviral oncogene** was either truncated or contained **small deletions** or **point mutations** in regions known to be required for protein function. Since the mutated form of the gene caused transformation even in cells expressing the corresponding protooncogene, they were called **dominant oncogenes.** These experiments suggested that the retroviral form of the protein had acquired an ac-

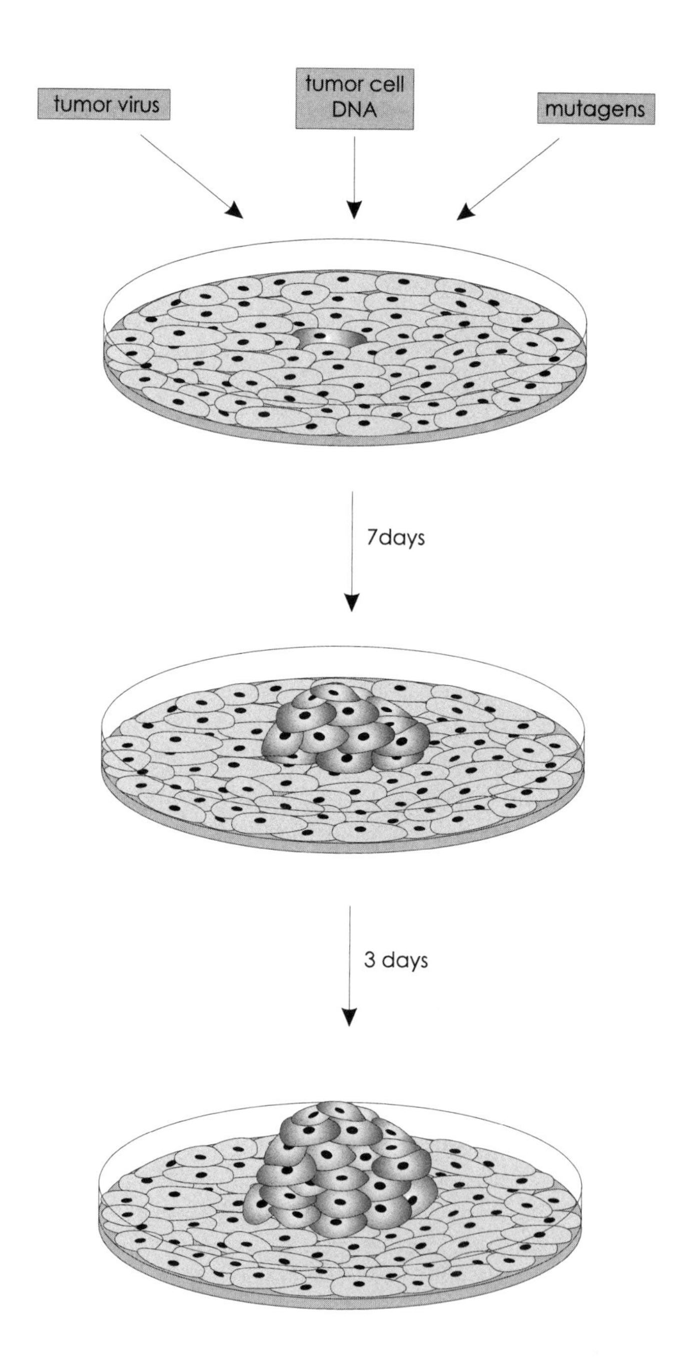

Figure 31.4.
The transformation assay is based on the finding that transformed cells are not contact-inhibited in tissue culture and can literally grow on top of each other. Colonies of transformed cells can be visualized after about 10 days in culture, physically isolated and expanded into a pure population of genetically similar cells.

tivity which allowed it to interfere with the function of the normal cellular protein. The molecular mechanism of retroviral oncogene capture most likely involves recombination with host-cell mRNA that is converted to cDNA by the virally encoded reverse transcriptase because retroviral oncogenes contain no introns.

Some Human Tumor Cells Contain Dominant Oncogenes

Since the animal oncogenic retroviruses could transform cells in culture, researchers set out to determine if DNA isolated from **primary humor tumors** could do the same thing. As shown in Figure 31.6, using a series of steps involving DNA transfection of randomly sheared genomic DNA, it was possible to identify human genes that could cause transformation of a mouse-immortalized cell line called **NIH 3T3.** When the oncogenic human DNA was sequenced, it was found to be homologous to some of the retroviral oncogenes previously identified.

The first human oncogene found by this approach was the ***ras* gene,** which encodes a monomeric GTP-binding protein involved in signal transduction. It was also possible to mutate NIH 3T3 cells with chemical

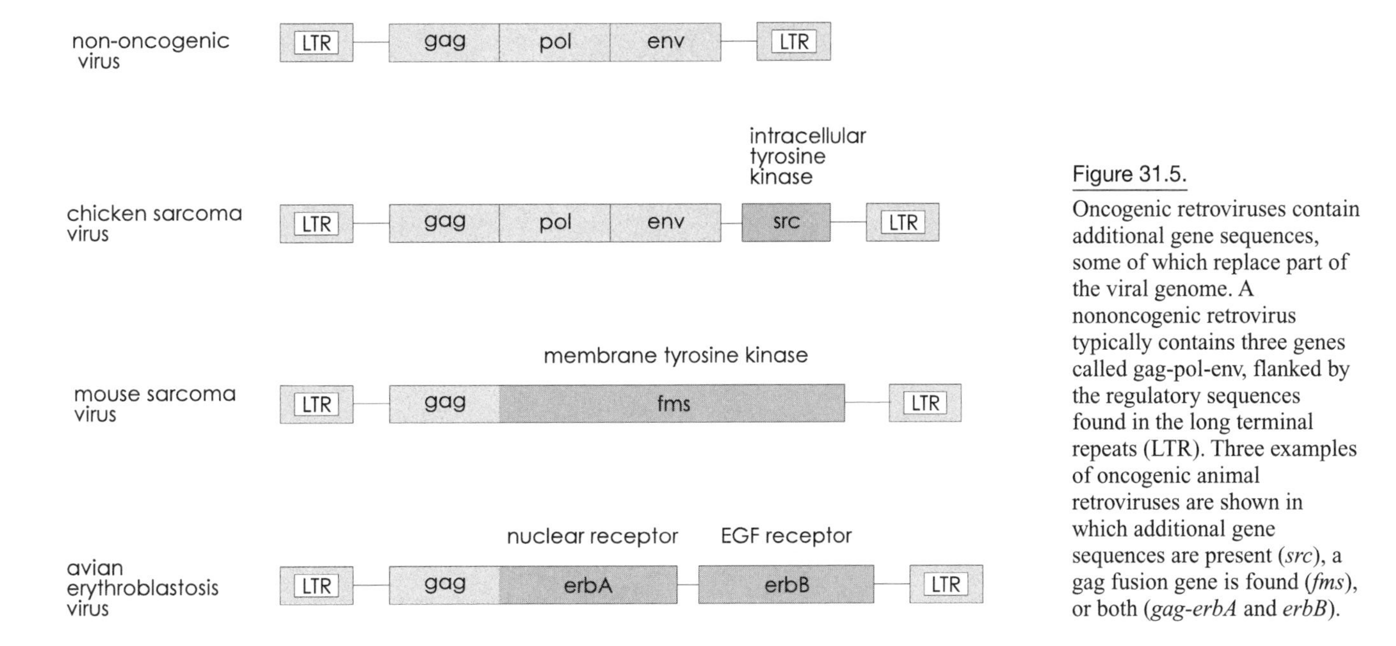

Figure 31.5.

Oncogenic retroviruses contain additional gene sequences, some of which replace part of the viral genome. A nononcogenic retrovirus typically contains three genes called gag-pol-env, flanked by the regulatory sequences found in the long terminal repeats (LTR). Three examples of oncogenic animal retroviruses are shown in which additional gene sequences are present (*src*), a gag fusion gene is found (*fms*), or both (*gag-erbA* and *erbB*).

carcinogens directly to cause transformation. These experiments led to the discovery that one of the most common mutations in the chemically treated NIH 3T3 cells was the *ras* gene. In fact, the ***ras* point mutations** identified by both these approaches, were almost always in codons 12, 13, or 61. Similar to the retroviral oncogenes, these *ras* mutations had a **dominant phenotype** in this assay. We discuss the functional significance of these "gain-of-function" *ras* mutations later in the chapter.

Oncogenes can be Found at Chromosomal Breakpoints

Cytogeneticists had known for years that many types of **leukemia** contained specific **chromosomal abnormalities.** As the search for oncogenes escalated, molecular biologists teamed with the leukemia researchers to clone the rearranged regions of specific chromosomes. Figure 31.7 shows two of the oncogenes found in this way: the ***myc* gene,** first cloned from a chromosome 8:14 translocation common to Burkitts lymphoma, and the ***abl* gene,** a tyrosine-kinase encoding gene which is disrupted on chromosome 9 by a 9:22 translocation, creating the Philadelphia chromosome in chronic myelogenous leukemia.

The normal Myc protein is expressed in Burkitts lymphoma cells, but because of the juxtaposition between the *myc* gene and the very active immunoglobulin promoter–enhancer region, the translocated *myc* gene is highly overexpressed. The chromosomal translocation in chronic myelogenous leukemia (the 22:9 Philadelphia chromosome), creates a different problem in that two gene-coding sequences are fused together. This results in a ***abl* gene** fusion protein containing the tyrosine-kinase domain of Abl fused to the amino terminus of the Bcr (breakpoint cluster region) protein. The **Bcr–Abl** protein is oncogenic, but the normal Bcr and Abl proteins are not.

Recessive Mutations in Tumor-Suppressor Genes Are Also Oncogenic

So far we have only described oncogenes that behave in a dominant manner when mutated. For example, the success of the transformation assay depended on the fact that the "activated" Ras protein is oncogenic in cells that have normal Ras. In contrast, there is another class of oncogenes that result from "loss-of-function" mutations and they have been called **tumor-suppressor genes.** Mutations or deletions of tumor-suppressor genes promote cancer because without their activity, the cell continues to divide even when it is not supposed to. The NIH 3T3 mouse-cell transformation assay cannot detect tumor-suppressor genes because this type of oncogene can only be identified by its absence. Most tumor-suppressor genes have been found using a combination of human genetic analyses and molecular cloning techniques.

Originally, tumor-suppressor genes were defined

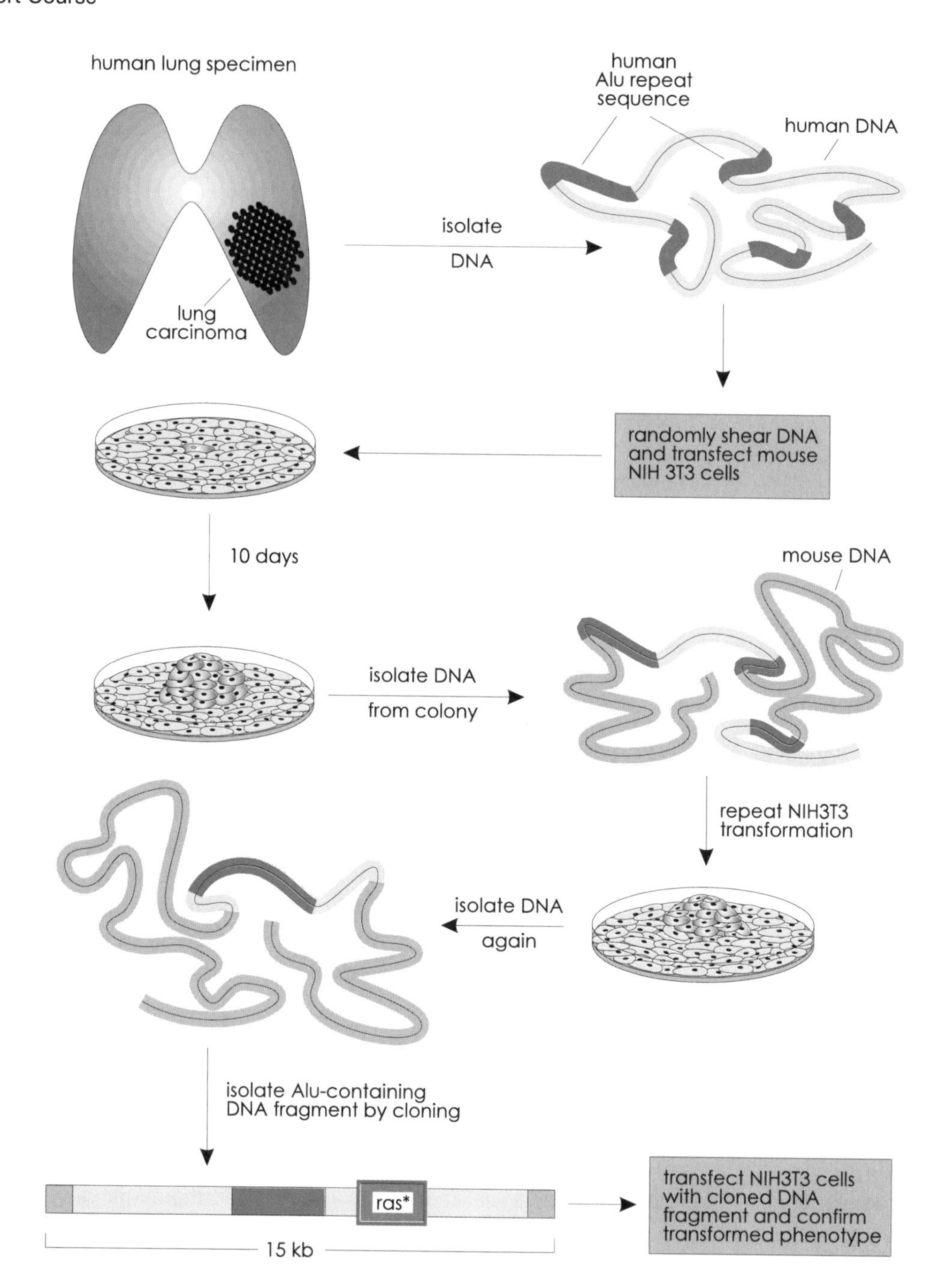

Figure 31.6.
Isolation of *ras* genes from human tumor DNA. Randomly sheared human DNA obtained from lung carcinoma cells was transfected into mouse NIH 3T3 cells and a colony (focus) of transformed cells was isolated. DNA from these cells contained a mixture of human and mouse DNA sequences. Following one or two more rounds of the transformation assay, a genomic library of the transformed cell DNA was constructed and lambda clones containing human-specific Alu-repetitive DNA were isolated. The *ras* gene was identified by mapping the functional segment of human DNA that could by itself transform mouse NIH 3T3 cells. *ras** refers to the oncogenic form of *ras* which contains one or more point mutations.

by their ability to suppress the transformed phenotype in cells following transfection of the unmutated gene. However, we will use the broader definition of tumor-suppressor/recessive oncogene here to refer to oncogenes that have been identified as loss-of-function mutations in cancer cells.

The first tumor-suppressor gene to be cloned was the **retinoblastoma (*Rb*) gene.** Alfred Knudson had been studying the predisposition of children from certain families to have a high rate of retinoblastoma eye tumors. He noticed that the rate of hereditary retinoblastoma in these children was as much as 1000 times higher than expected, compared to the normal population. In addition, hereditary retinoblastoma results in multiple independent tumors in both eyes of the patient, whereas the nonhereditary form is almost always only a single tumor in just one eye.

Based on his observations, Knudson proposed in 1971 that the rare nonhereditary form of retinoblastoma was due to two spontaneous mutations, one in each copy of a single *Rb* gene. This became known as Knudson's **two-hit model** of retinoblastoma. Knudson pre-

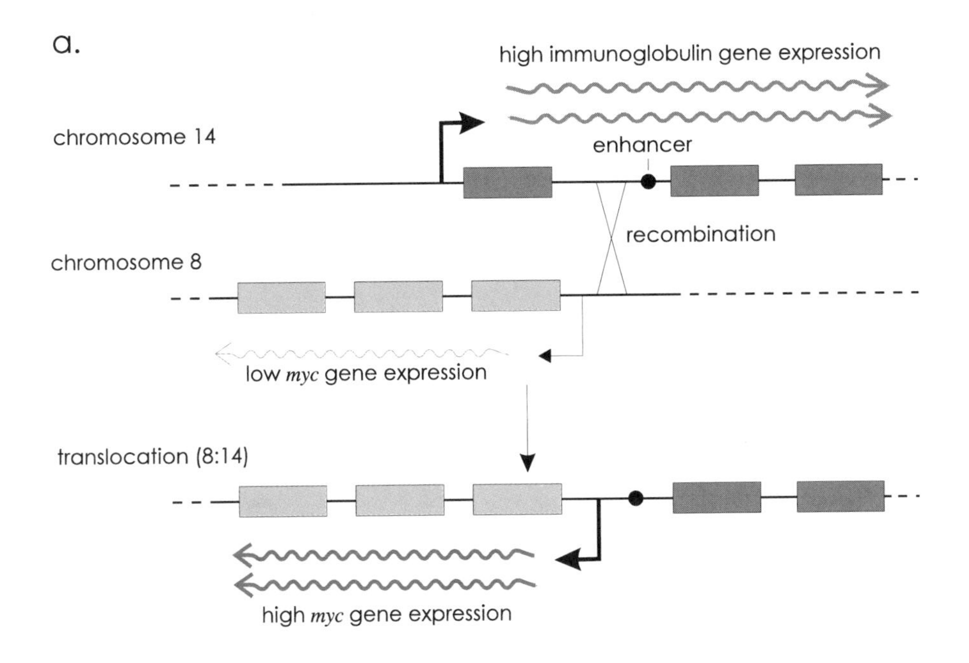

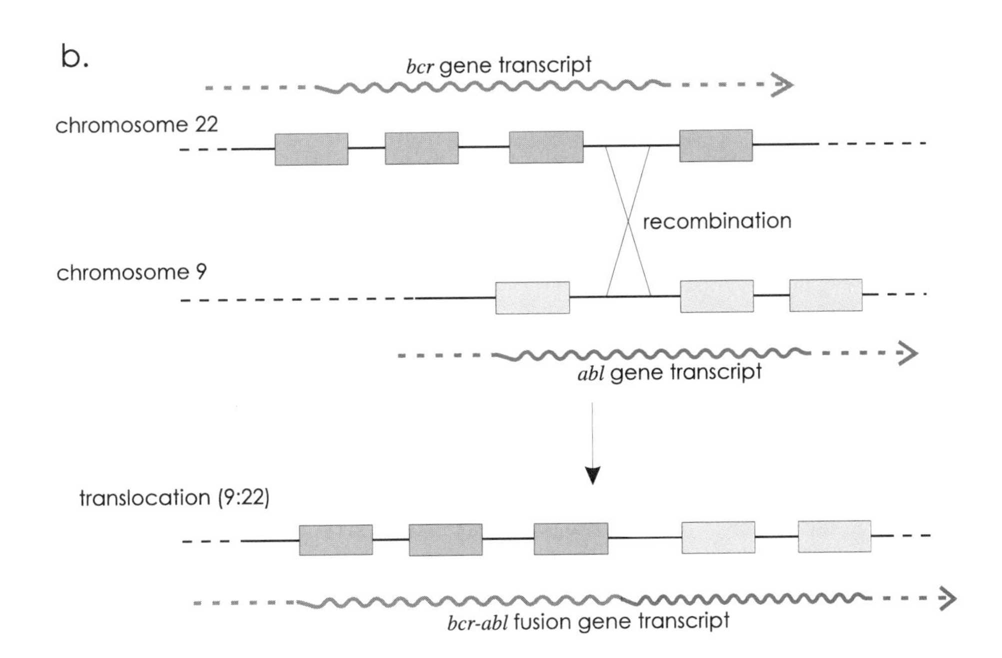

Figure 31.7.
Molecular analysis of genes found at chromosomal breakpoints in the DNA of cancer cells is another way to identify oncogenes. The transcription of *myc* is abnormally high in cancer cells from Burkitts lymphoma patients, owing to the translocation of the immunoglobulin gene-transcriptional enhancer into the vicinity of the *myc* gene promoter (a). Translocation between chromosomes 22 and 9 results in the creation of a fusion gene (and protein) containing parts of both the *bcr* and *abl* genes (b).

dicted that hereditary retinoblastoma was the result of a child inheriting one defective *Rb* gene from one parent and having a single spontaneous mutation in the *Rb* gene donated from the other parent. Figure 31.8 illustrates the principle of Knudson's two-hit model of retinoblastoma, which was later shown to be correct, using the cloned Rb gene to examine Rb DNA sequences in different cell types from retinoblastoma patients.

Tumor-cell karyotypes from a large number of retinoblastoma patients revealed that a small region on chromosome 13 was often deleted or near a chromosomal breakpoint. The *Rb* gene was eventually cloned and sequenced and found to encode a **cell-cycle control protein** which normally prevents cells from replicating their DNA in the S phase. The Rb activity is regulated by phosphorylation and one of its functions is to inhibit the transcription genes required for cell proliferation. Cells with mutations or deletions of *Rb* would therefore bypass the normal braking mechanism that prevents cells from entering the cell cycle.

Table 31.1 lists other tumor-suppressor/recessive

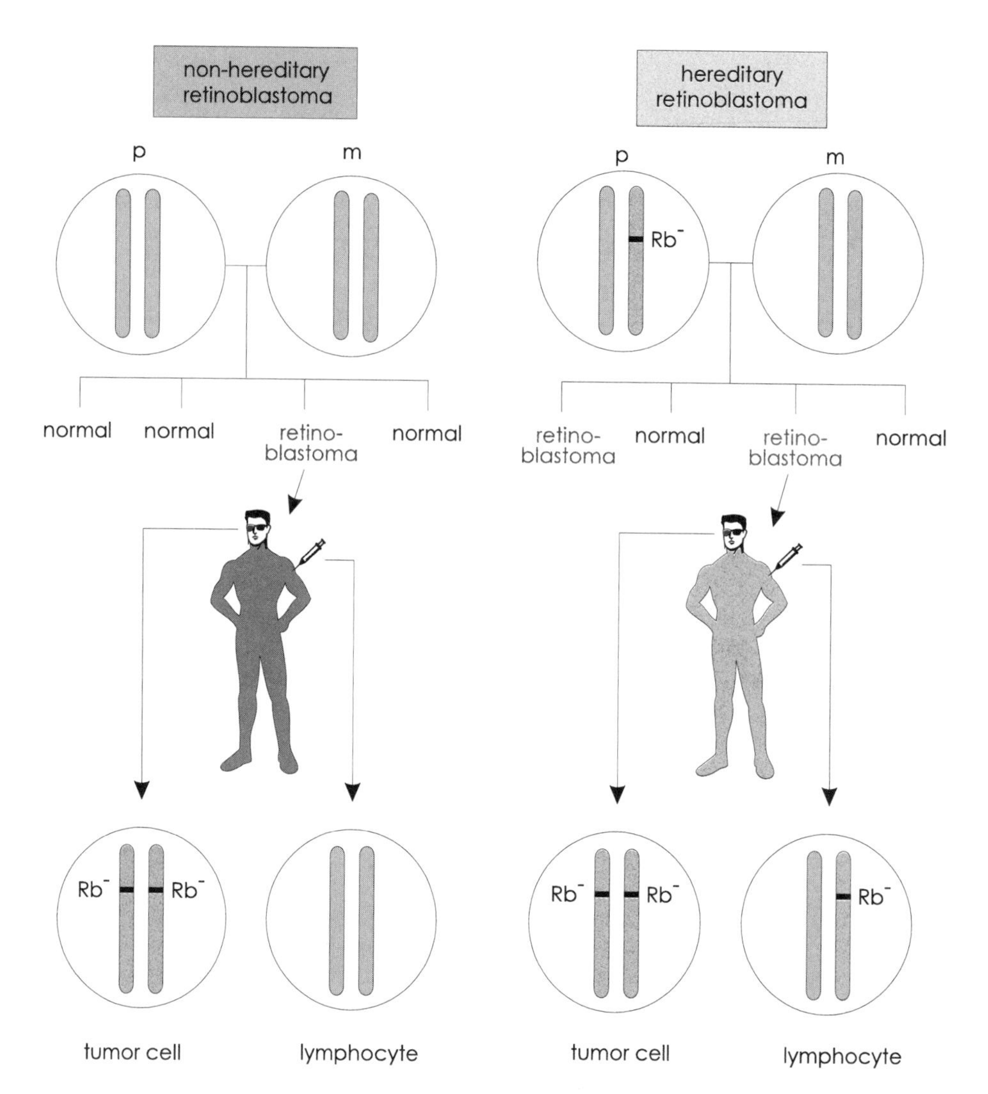

Figure 31.8.

Knudson's two-hit model of carcinogenesis is best illustrated by the pattern of *Rb* gene mutations in various cell types from retinoblastoma patients. Molecular analysis of the *Rb* gene in retinoblastoma cells reveals that both copies of the gene are mutated or deleted, confirming the two-hit model. In patients with nonhereditary retinoblastoma, which is a very rare cancer, nontumor cells such as lymphocytes contain two functional copies of *Rb*. However, normal lymphocytes from patients with hereditary retinoblastoma have been shown to contain one mutated copy of *Rb*, which would explain why this type of retinoblastoma afflicts more individuals in the family and at a younger age.

oncogenes that have been identified, including two DNA mismatch repair genes, *MLH1* and *MSH2*, which are found to be defective in a large percentage of patients with **hereditary nonpolyposis colon cancer.** Obviously, mutations in either of the genes encoding these two DNA mismatch repair enzymes would lead to a rapid buildup of additional mutations in the genome, and subsequent progression to the transformed phenotype. There are now several other examples of candidate tumor-suppressor/recessive oncogenes which have been identified by screening for genetic markers associated with certain hereditary forms of cancer. For example, mutations in the *BRCA1* and *BRCA2* genes have been found in cases of **hereditary breast and ovarian**

Table 31.1. Human Tumor-Suppressor/Recessive Oncogenes[a]

Gene	Biochemical function	Chromosomal location	Disease
Rb	Modulates transcription	13q14	Retinoblastoma
p53	Transcription factor	17p13	Many tumor types
WT-1	Transcription factor	11p13	Wilm's tumor
APC	Cell adhesion	5q31	Adenomatous polyposis
DCC	Cell adhesion	18q21	Colon carcinoma
hMLH1	DNA repair	3p21	Nonpolyposis colon cancer
hMSH2	DNA repair	2p16	Nonpolyposis colon cancer

[a]Loss-of-function mutations or deletions of these genes have been associated with cellular transformation.

cancers, and defects in the *ATM* gene, which causes **ataxia telangiectasia,** may predispose individuals to a variety of cancers. Molecular genetic studies indicate that BRCAI is a transcription factor and that ATM functions to modulate DNA repair processes.

ONCOGENE GENE PRODUCTS ALTER MULTIPLE-GROWTH CONTROLLING STEPS

To date, well over 60 dominant and recessive oncogenes have been identified. Moreover, the normal functions of the protooncogenes have in a number of cases, been broadly defined. In fact, as with many molecular genetic approaches, the characterization of a defective gene product, in this case the oncogene, has elucidated the role of that protein in normal cell processes. For example, the *Rb* gene was first identified in retinoblastoma, but its characterization has led to a greater understanding of cell-cycle control. The collaborative research between basic science and cancer research have been one of the most immediate benefits resulting from the discovery of oncogenes.

Figure 31.9 summarizes the wide variety of cellular processes in which oncogenes have been shown to act. As can be seen, virtually every level of cellular regulation controlling cell proliferation is subject to disruption by dominant or recessive oncogenes. By analyzing mutations in both dominant and recessive oncogenes in

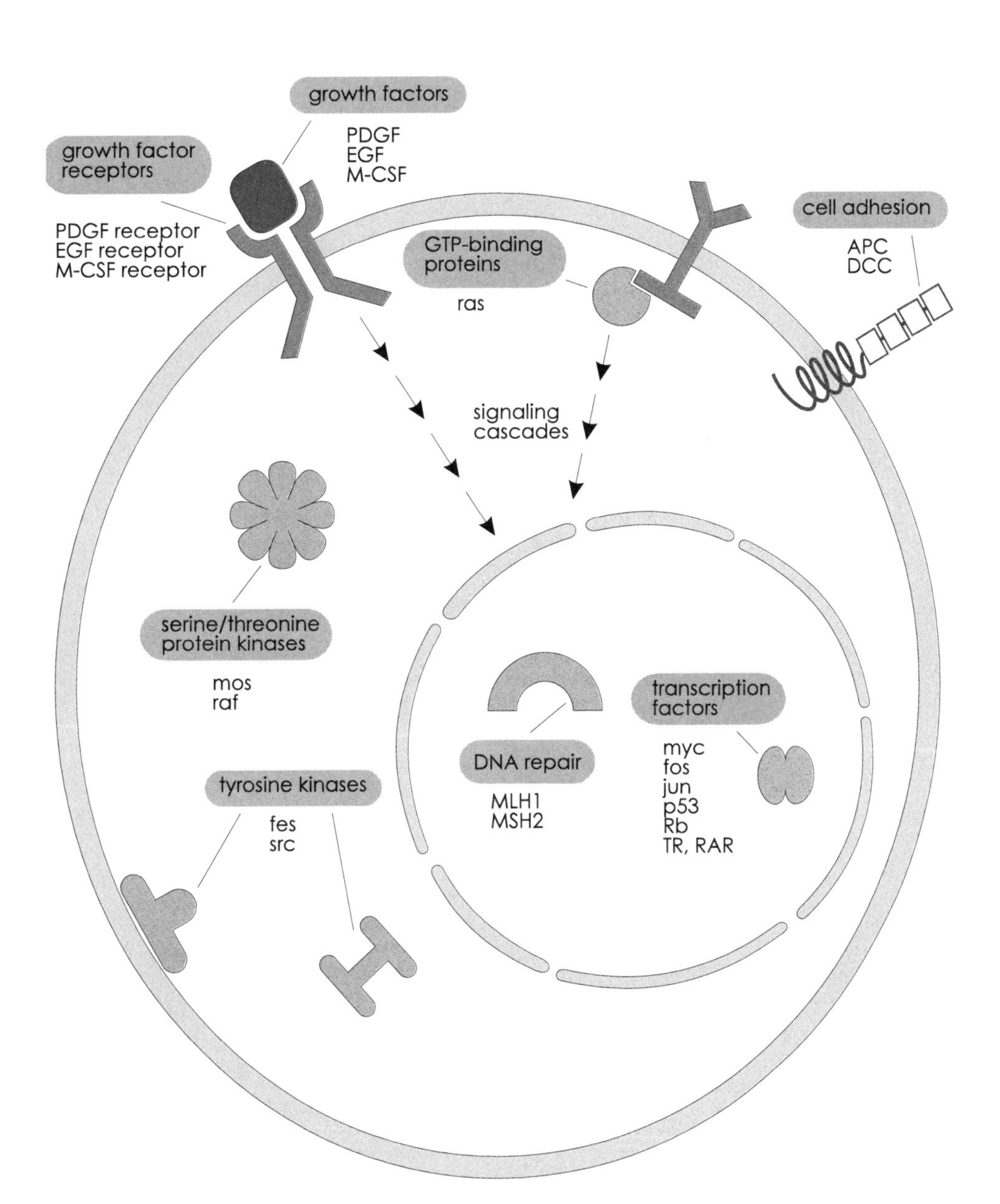

Figure 31.9.
Oncogenes have been found to control a variety of cell processes. Gain-of-function (dominant oncogenes) or loss-of-function (recessive oncogenes) mutations in these critical genes results in the disruption of normal cell function.

a large number of tumor types, it has been suggested that multiple independent oncogenes need to be mutated during the progression of precancerous cells to the malignant phenotype. Such a multistep mechanism of carcinogenesis is consistent with the observed age-related bias of cancer cell deaths (Fig. 31.1).

MUTATIONS IN THE ONCOGENES *RAS* AND *p53* ARE EXTREMELY COMMON

Mutations in the dominant oncogene *ras,* and the tumor-suppressor/recessive oncogene *p53,* represent the most commonly found gene mutations in human cancers, and in fact many transformed cells contain mutations in both of these oncogenes. What accounts for the high frequency of *ras* and *p53* mutations? There are at least two possible explanations. First, it could be that these oncogenes directly control a **critical growth-regulatory step** which is the central modulator of cell proliferation and common to all cancer cells, that is, a cell is not transformed *unless* it has a mutation in one of these two genes. Second, Ras or p53 could control genome stability, or initiate a signaling cascade from the membrane to the nucleus. Therefore, they **act early in the multistep pathway** such that most cancer cells start out with a mutation in one of these genes, even though these gene products themselves do not directly cause transformation. It seems likely that both of these possibilities contribute to the high incidence of *ras* and *p53* mutations. In this section, we first describe what is known about Ras and its central role in signal transduction and then discuss the function of p53 in controlling the cell cycle.

Ras Connects Growth-Factor Signaling to the Intracellular Phosphorylation Cascade

Ras is a membrane-bound GTPase that is required for a variety of signal-transduction pathways. Although it was first thought to be a G protein because of these features, Ras is distinct from the heterotrimeric G proteins α, β, and γ (Chapter 16). Ras acts positively to stimulate the cascade of kinase-driven phosphorylation events that culminate in the activation of nuclear transcription factors. Figure 31.10 shows that the active form of Ras has a molecule of GTP bound in the nucleotide binding site, whereas the inactive form of Ras contains a GDP in that site. The intrinsic GTPase activity of Ras therefore converts the **active Ras–GTP form** into the **inactive Ras–GDP moiety.** Two types of proteins, GAPs (GTPase activating proteins) and GNRFs (guanine nucleotide releasing factors) control the activity of Ras such that GAPs are negative regulators and GNRFs are positive regulators of Ras.

The most prevalent dominant mutations in *ras* (codons 12, 13, and 61) cause a defect in the Ras GTPase activity such that it is always in the active Ras–GTP form. Under these circumstances, the cell receives a "false" signal to activate growth-promoting genes, even in the absence of growth factors. Therefore, one reason why *ras* mutations may be so common is that Ras serves a central role in membrane-to-nucleus signal transduction, and a dominant mutation in *ras* will still be deleterious, even if a normal *ras* gene is present.

Growth-factor receptors such as the **epidermal growth factor receptor** (EGFR) are membrane-bound tyrosine kinases which are positioned upstream of Ras in this signal-transduction pathway. When epidermal growth factor binds to EGFR, its **tyrosine-kinase activity** is stimulated and the cytoplasmic tail of EGFR is autophosphorylated on tyrosine residues, resulting in the formation an $\mathbf{S_H2}$ protein binding site. The adaptor protein Grb2 has one S_H2 domain which binds to the phosphorylated EGFR tail and two S_H3 domains which bind to S_H3 binding sites on SOS (a GNRF). Thus, when EGFR is activated by growth-factor binding, Grb2 brings Sos to the membrane to activate Ras. As described earlier, ErbB is a retroviral oncogene-encoded protein that is homologous to EGFR and is constitutively activated, even in the absence of growth factor. Cells expressing ErbB (or constitutive mutants of EGFR) would therefore constitutively activate Ras through the **Grb2–SOS adaptor complex.**

Once Ras is in its active form, it binds to another oncogene product called **Raf,** which localizes Raf to the membrane and activates its serine/threonine specific kinase activity. Among the many phosphorylation targets of Raf is another kinase called **MAP kinase kinase** (also called MEK), which is activated by Raf and in turn phosphorylates cytosolic MAP kinases on threonine and tyrosine residues. Activated **MAP kinases** (also called ERK1 and ERK2) are serine/threonine kinases that, once activated, can localize to the nucleus where they **phosphorylate transcription factors** such as **Fos** and **Jun.** MAP kinases also phosphorylate and activate other kinases (such as Rsk) which then pass the signal on to additional transcription factors (e.g., SRF, serum response factor). Therefore, downstream of Ras, a series of phosphorylation reactions culminate in the activation or inactivation of a variety of transcription

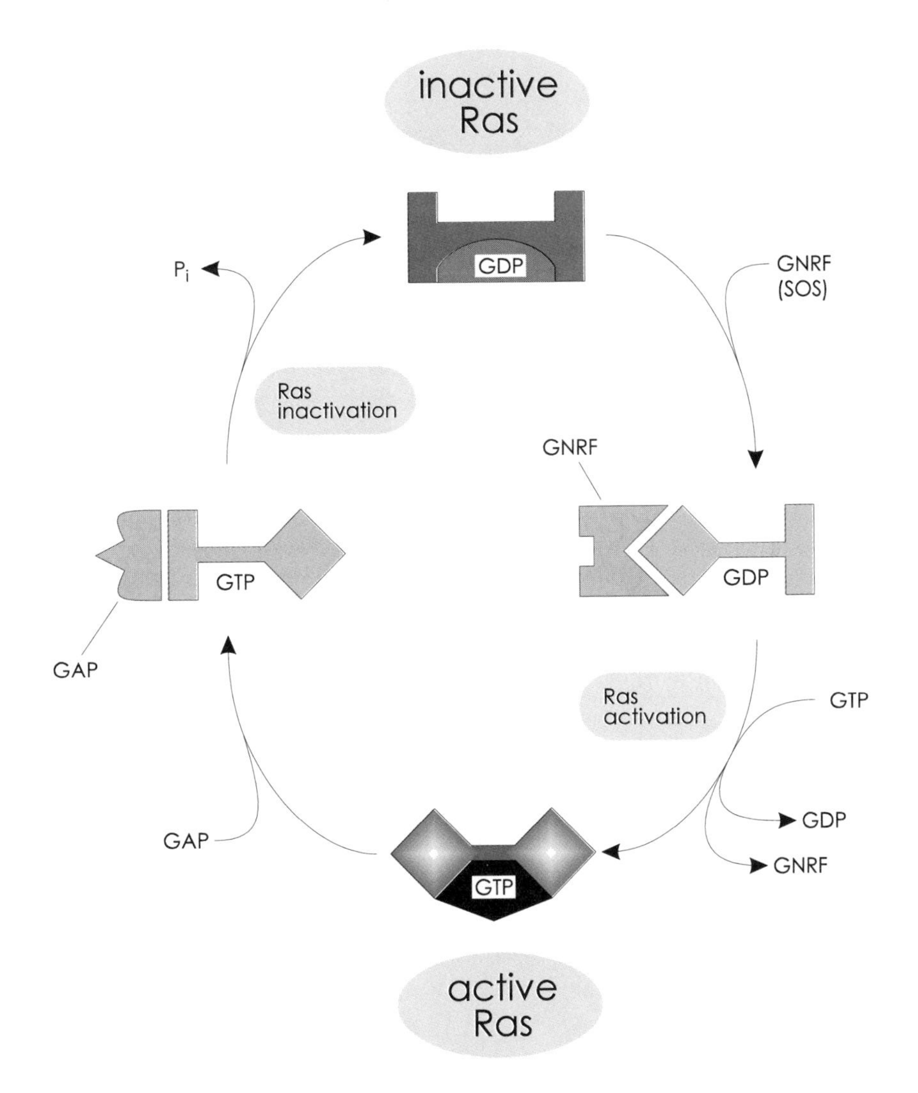

Figure 31.10.
Ras activation by proteins that modulate Ras GTPase activity. Ras functions in cell signaling are governed by the presence of GTP or GDP in the nucleotide binding site of Ras. GTP-bound Ras is active and GDP-bound Ras is inactive. GAP proteins stimulate Ras GTPase activity, which inactivates Ras signaling functions; GNRF proteins such as SOS stimulate GDP–GTP nucleotide exchange and therefore Ras activation.

factors. *Raf, fos, jun,* and *myc* are all oncogenes, and dominant mutations in any of these results in growth stimulation, even if there is no upstream signal from Ras or EGFR.

Although elucidation of the **Ras-signaling pathway** shown in Figure 31.11 was only recently accomplished, researchers are already devising ways to short circuit the dominant activity of mutated *ras.* One of these strategies is to inhibit the **farnesylation** of Ras on a cysteine residue near the carboxy terminus. Farnesylation is required to anchor Ras to the **cell membrane** which is apparently necessary for Ras-mediated activation of Raf kinase (Fig. 31.11). Several Ras inhibitors have been developed which target the enzyme farnesyltransferase. One idea is that by inhibiting enzymes required for Ras farnesylation, it may be possible to disrupt the dominant activity of mutated Ras in cancer cells. Initial studies using cell lines have been promising, but it is too early to tell yet how efficacious this approach will be in humans, because cellular proteins other than Ras also need to be farnesylated.

The Normal Function of *p53* Is To Inhibit the Cell Cycle when DNA is Damaged

The retinoblastoma gene is the classic example of a tumor suppressor/recessive oncogene. Also, *p53* is considered a tumor-suppressor/recessive oncogene because the wild-type p53 protein can be shown to suppress the growth of transformed cells containing a *p53* deletion. However, the oncogenic properties of *p53* is more complicated because many *p53* mutations actually result in the expression of an aberrant form of p53 protein that is able to inhibit the function of normal p53 protein present in the cell. This type of mutation is called a **dominant negative mutation** because it is a loss-of-function mutation that behaves as if it is an *inhibitory* gain-of-

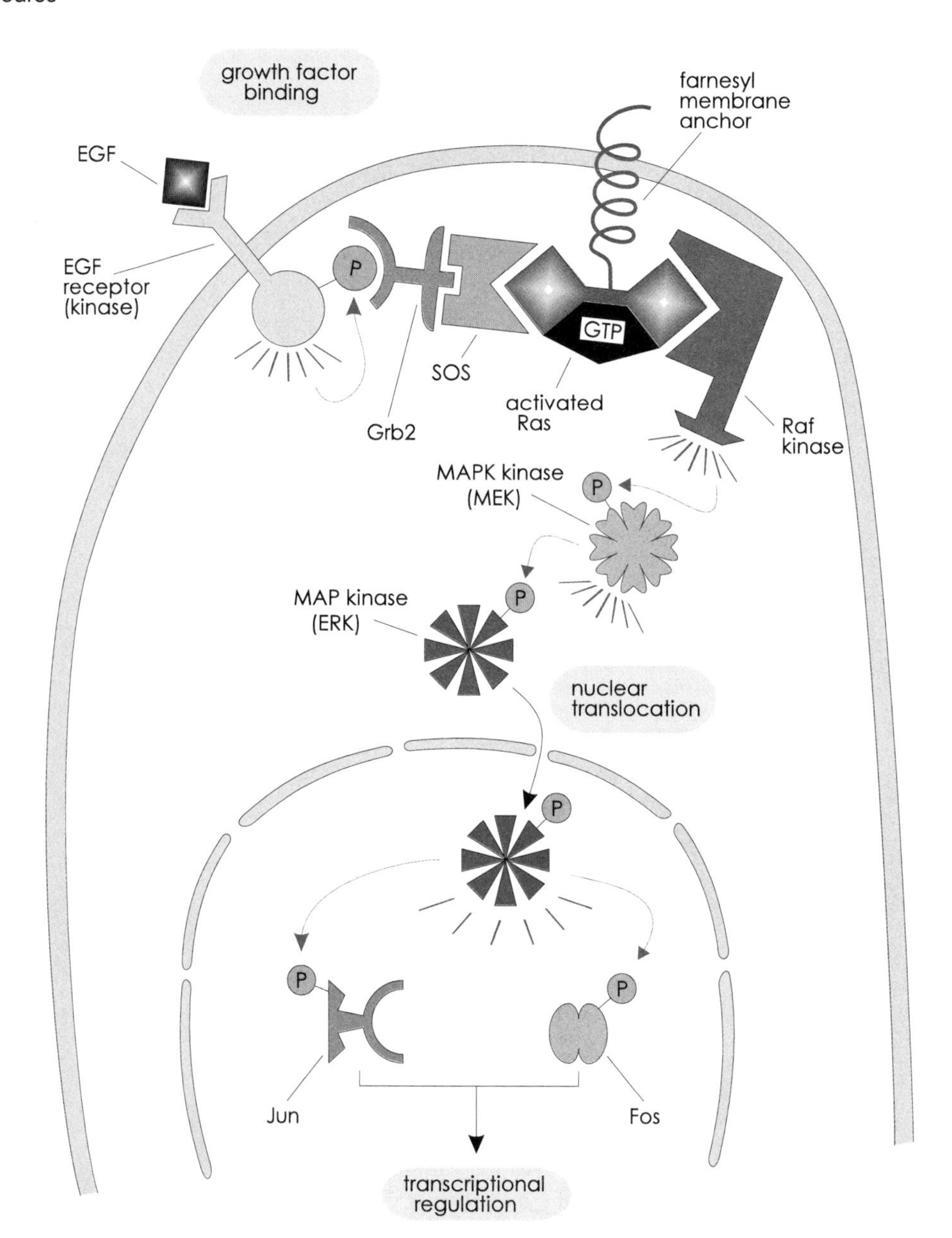

Figure 31.11.
Proposed phosphorylation-signaling cascade from EGF receptor to Fos and Jun through a sequential series of kinase-activation steps. The role of Ras in stimulating Raf kinase activity is thought to involve Ras-mediated membrane localization of Raf, which would explain the requirement for Ras farnesylation.

function mutation. The mechanistic basis for *p53* dominant negative mutations is thought to be that p53 function is dependent on the formation of multisubunit protein complexes. Inclusion of one or more dominant negative p53 molecules into this complex results in loss of function to the entire complex. This would be a biochemical example of "one bad apple spoils the whole bunch"!

The p53 has been dubbed the "Guardian of the Genome" because it surveys for **DNA damage,** such as would result from irradiation, and then relays the signal to the cell-cycle machinery to stop in G1 and wait for the damage to be repaired. Moreover, under some circumstances, p53 is able to induce damaged cells to die by a form of cell suicide called **apoptosis.** This type of p53-mediated programmed cell death would eliminate the possibility that such a cell could go on to become tumorgenic. However, a mutation in *p53* that abrogates these important *guardian* functions would allow cells to enter S phase and attempt to **replicate damaged DNA,** rather than repairing it first or undergoing apoptosis (Fig. 31.12). Cells that do proceed through S phase under these conditions are found to accumulate a variety of genomic lesions including **gross chromosomal rearrangements.**

Why are *p53* mutations so common? First, p53 binds to DNA as a tetramer, and **dominant negative mutants** are able to disrupt the formation or activity of

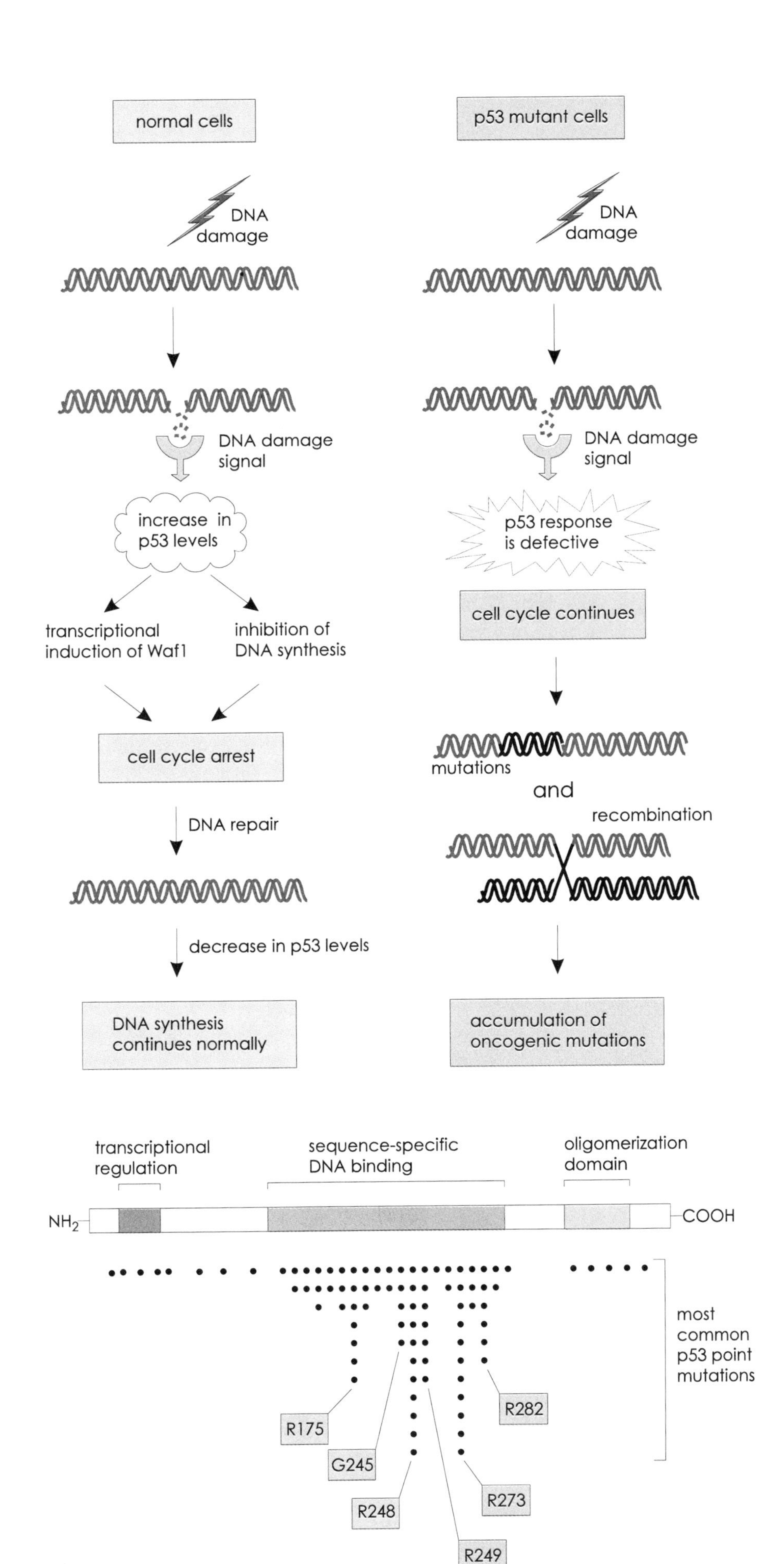

Figure 31.12.

Model of p53-mediated control of cell-cycle progression. The proposed functions of p53 in normal cells are shown, and in cells lacking p53 or expressing a dominant form of p53, both of which result in a defect in p53 functions.

Figure 31.13.

The vast majority of p53 mutations are localized to the DNA binding domain and effect specific protein–DNA interactions. The dots show the distribution of *p53* mutations across the entire protein-coding sequence and the boxes identify the most frequently mutated amino acids.

these multisubunit complexes. Therefore, a cell only needs one mutated copy of *p53* to cause a defect in cell-cycle control. Second, a significant portion of the *p53* coding sequence is **susceptible to mutations.** Sequencing of *p53* genes in a large number of cancer cells has shown that mutations which alter over 200 different amino-acid residues are deleterious to p53 function. Most of these mutations are in the p53 DNA binding domain (Fig. 31.13). Third, p53 has the important function of protecting cells from replicating damaged DNA. A cell that loses this ability will quickly accumulate additional mutations and **chromosome rearrangements** which ultimately could result in cell transformation. Therefore, similar to Ras, p53 function is central to cellular homeostasis and a key player in cell-cycle control.

SUMMARY

1. Three observations led early cancer researchers to propose that mutations in DNA were central to the mechanism of oncogenesis. First, cancer is most prevalent in older people, suggesting an accumulation of somatic mutations. Second, mutagens can also function as carcinogens in animals. Third, human leukemias are often associated with abnormal karyotypes showing chromosomal rearrangements. The discovery of human oncogenes 20 years later confirmed this proposal.

2. Tumor progression from the benign to malignant state suggests that multiple mutations are required for the development of metastatic cancer. Initiators are DNA-damaging agents and tumor promoters are growth stimulators. It is thought that cell proliferation enhances the probability of accumulating additional DNA mutations, which may explain the effect of tumor promoters.

3. Composite retroviruses contain mutated copies of genes acquired from the genomes of host cells. This discovery led to the direct isolation of mutated human genes shown to have "oncogenic" activity. Dominant oncogenes such as *ras* are able to function even in cells that express unmutated forms of the same protein. Many of these dominant oncogenes interfere with kinase-signaling pathways by deregulating phosphorylation pathways.

4. Tumor-suppressor genes are recessive oncogenes that normally function to inhibit cell proliferation. Loss-of-function mutations or deletions of tumor-suppressor genes leads to uncontrolled cell division. Rb and p53 are two such tumor suppressor/recessive oncogenes which have cell-cycle regulatory functions and are mutated in a wide variety of cancers.

5. Ras is a cell-signaling molecule that plays a critical role in linking extracellular signals to intracellular phosphorylation cascades which control the activity of nuclear transcription factors. Mutations in Ras that lead to constitutive activity have been found in a large percentage of human cancers. It might be possible to neutralize Ras by inhibiting farnesylation of the Ras-carboxy terminus.

6. Loss of function or dominant negative mutations in p53 have been found in a majority of human cancers. It is thought that the normal function of p53 is to cause cell-cycle arrest when DNA damage exists in cells about to enter S phase and undergo DNA replication. A defect in p53 function would allow damaged DNA to be replicated, which could result in profound genetic alterations.

REFERENCES

Ames B. N., Gold L. S., Willett W. C. (1995): The causes and prevention of cancer. *Proc. Natl. Acad. Sci. USA* **92**:5258.

Armstrong J. F., Kaufman M. H., Harrison D. J., Clarke A. R. (1995): High frequency developmental abnormalities in p53-deficient mice. *Curr. Biol.* **5**:931.

Baserga R. (1994): Oncogenes and the strategy of growth factors. *Cell* **79**:927.

Bishop J. M. (1995): Cancer: The rise of the genetic paradigm. *Genes Dev.* **9**:1309.

Black D. (1994): Familial breast cancer: *BRCA1* down, *BRCA2* to go. *Curr. Biol.* **4**:1023.

Chapman, M. S., Verma, I. M. (1996): Transcriptional activation by BRCAI. *Nature* **382**:678.

Cho Y., Gorina S., Jeffrey P. D., Pavletich N. P. (1994): Crystal structure of a p53 tumor suppressor–DNA complex: Understanding tumorigenic mutations. *Science* **265**:346.

Denissenko, M. F., Pao, A., Tang, M-S, Pfeifer, G. P. (1996): Preferential formation of benzo[a]pyrene adducts at lung cancer mutational hotspots in p53. *Science* **274**:430.

Fischer S. G., Cayanis E., Bonaldo M. F., et al. (1996): A high resolution annotated physical map of the human chromosome 13q12-13 region containing the breast cancer susceptibility locus BRCA2. *Proc. Natl. Acad. Sci. USA* **93**:690.

Gibbs J. B., Oliff A., Kohl N. E. (1994): Farnesyltransferase inhibitors: Ras research yields a potential cancer therapeutic. *Cell* **77**:175.

Hartwell, L. H., Kastan M. B. (1994): Cell cycle control and cancer. *Science* **266**:1821.

Hawley, R. S., Friend, S. H. (1996): Strange bedfellows in even stranger places: The role of ATM in meiotic cells, lymphocytes, tumors and its functional links to p53. *Genes & Dev.* **10:**2383.

Hunter T. (1995): Protein kinases and phosphatases: The yin and yang of protein phosphorylation and signaling. *Cell* **80:**225.

Karlovich C. A., Bonfini L., McCollam L., et al. (1995): In vivo functional analysis of the Ras exchange factor Son of Sevenless. *Science* **268:**576.

Karp J. E., Broder S. (1994): New directions in molecular medicine. *Cancer Res.* **54:**653.

Ko, L. J., Prives, C. (1996): p53: Puzzle and paradigm. *Genes & Dev.* **10:**1054.

Levine A. J. (1993): The tumor suppressor genes. *Annu. Rev. Biochem.* **62:**623.

Maignan S., Guilloteau J.-P., Fromage N., Arnoux B., Becquart J., Ducruix A. (1995): Crystal structure of the mammalian Grb2 adaptor. *Science* **268:**291.

Marshall, C. J. (1996): Raf gets it together. *Nature* **383:**127.

Pawson T. (1995): Protein modules and signaling networks. *Nature* **373:**573.

Rabbitts T. H. (1994): Chromosomal translocations in human cancer. *Nature* **372:**143.

Varmus, H., Wienberg, R. A. (1993): Genes and the biology of cancer. Scientific American Library, W.H. Freeman Co., New York.

Verma I. M., Vogt P. K. (1995): Oncogenes: 20 years later. *Genes Dev.* **9:**1289.

Weinberg R. A. (1995): The retinoblastoma protein and cell cycle control. *Cell* **81:**323.

White M. A., Nicolette C, Minden A, et al. (1995): Multiple Ras functions can contribute to mammalian cell transformation. *Cell* **80:**533.

Wynder E. L., Hoffmann D. (1994): Smoking and lung cancer: Scientific challenges and opportunities. *Cancer Res.* **54:**5284.

REVIEW QUESTIONS

1. The most direct evidence linking DNA damage to cancer is
   a) epidemiological studies that associate smoking with lung cancer
   b) the identification of abnormal karyotypes in leukemic cells
   c) the finding that high levels of mutagenic compounds in rats is carcinogenic
   d) the finding that mutated genes isolated from tumors can transform cells
   e) pedigree studies showing that predisposition to cancer can be inherited

2. Why might increased rates of cell proliferation in dedifferentiated cells be a contributing factor to tumor progression?
   a) The more cells there are, the larger the tumor will be.
   b) Proliferating cells require increased blood supplies and growth factors.
   c) Mutation rates will be higher owing to replication of unrepaired DNA.
   d) Hormones could be released and cause proliferating cells to undergo apoptosis.
   e) The dedifferentiated cells might start to differentiate again.

3. How did the characterization of retroviral genomes lead to the oncogenic theory of cancer initiation?
   a) It was found that some retroviruses could cause human lung cancer.
   b) Retroviral *ras* genes were identical in DNA sequence to the human *ras* gene.
   c) Retroviral genomes contained segments of the Philadelphia chromosome.
   d) Oncogenic retroviral genes were shown to be homologous to human genes.
   e) All of the above.

4. *ras* is classified as a dominant oncogene because
   a) mutant Ras* proteins activate signal transduction in cells with normal Ras
   b) mutant Ras* is a much larger than normal Ras
   c) *ras* mutations are passed from one generation to the next
   d) mutant Ras* is a potent GTPase and it depletes energy stores in the cell
   e) *ras* is actually a tumor-suppressor gene, not a dominant oncogene

5. How can some *p53* mutations lead to its classification as a tumor-suppressor/recessive oncogene, and other *p53* mutations result in it being labeled a dominant oncogene?
   a) The *p53* gene is very large and mutations are rare.
   b) p53 is a tyrosine kinase involved in signal transduction.
   c) *p53* is a member of a multigene family and several mutations are possible.
   d) Decreased levels of normal p53, and too much mutant p53, are both deleterious.

e) p53 is a bifunctional protein and can have a double mutation in the same gene.

6. If cancer is due to mutations in key regulatory genes, how do humans remain relatively free of cancer for up to 50 years?
   a) DNA repair mechanisms are extremely efficient.
   b) Cells with damaged DNA are often removed by apoptosis.
   c) It takes time to accumulate multiple mutations in the same cell.
   d) Most precancerous conditions are benign and go undetected for years.
   e) All of the above.

ANSWERS TO REVIEW QUESTIONS

1. ***d*** The most direct evidence came from the characterization of oncogenes.

2. ***c*** Dedifferentiated cells likely contain some DNA mutations and replication will increase the rate at which additional mutations will accumulate.

3. ***d*** Oncogenic retroviral genes such as *ras** are similar, but not identical, to *ras*.

4. ***a*** Oncogenic Ras* is constitutively activated and unaffected by normal Ras.

5. ***d*** p53 functions as a tetrameric protein and, therefore, either insufficient levels of p53 in the cell, or expression of aberrant forms, will result in a p53-deficient phenotype.

6. ***e*** Eukaryotic cells have a number of highly evolved mechanisms to avoid catastrophic mutations. Moreover, cancer is a clinical definition of a life-threatening disease and most precancerous lesions are benign.

APPENDIX

MENDELIAN GENETICS

Genetics provides a powerful approach to the study of living organisms that is complementary to the molecular approach; it is essential for understanding and applying biochemistry and molecular biology. *Genetics* is based on the observation that offspring inherit characteristics from their parents. *Inheritance* is observed both at the *organismal* level and at the level of individual cells and can be described in terms of discrete inheritable units called ***genes.*** In simple cases, a particular ***trait,*** such as eye color or a *genetic disease,* is due to a single inheritable unit or gene. In other cases, combinations of genes determine more complex traits.

Almost all of the genes of a eukaryotic cell are found in the chromosomes in the nucleus. In most eukaryotic cells—the *somatic cells*—chromosomes occur in pairs and the cells are said to be *diploid.* During reproduction, the chromosomes are separated into sets, with each set—called the *haploid genome*—containing one copy of each chromosome. In the female, chromosomes are packaged into *eggs,* which also contain cytoplasm and mitochondria; in the male, chromosomes are packaged into *sperm,* which are little more than DNA in a molecular swimsuit with flippers. Each egg contains only one set of genes, as does each sperm. Sperm and egg fuse in a random collision process to produce a fertilized egg that contains the normal two copies of each chromosome—one from each parent. Thus, the inheritance of genes is governed by a random process that must be described statistically.

The quantitative aspects of inheritance are very important for mapping the positions of genes. Consider parents with traits, A and B, that differ. Let's say that one parent has forms a and b and the other has forms a′ and b′. If A and B are on different chromosomes, the offspring will be just as likely to inherit different combinations of A and B as they are to inherit either of the parents' combinations. However, if A and B are on the same chromosome, the offspring will be much more likely to resemble either parent than to have a different combination of A and B forms. However, changes in the combination of A and B are possible even when they are on the same chromosome through the process of ***recombination,*** in which pieces of paired *chromosomes exchange* with one another. In this case, the closer A and B are to each other, the more likely it is that they will stay together during *reproduction.* By measuring the *frequency* with which the combination of A and B changes through multiple generations, geneticists determine the relative positions of genes on chromosomes.

The process of converting a *diploid cell* into *haploid cells* is called *meiosis.* Diploid cells may also increase in number—proliferate—by growth, including replicating each chromosome to give four copies of each, and then splitting, or dividing, to produce two diploid cells by mitosis.

Different forms of a particular gene are called *alleles* of that gene. When the two copies of a gene present in one cell are significantly different, the cell can take on the characteristic determined by one allele, the other allele, or a mixture of the two. Some alleles are inherently *recessive,* as when a mutation results in loss of function. Other alleles are inherently *dominant,* as when a mutation dominates the normal gene product through loss of regulation so that the mutant gene product is active at all times—constitutively active—instead of being regulated by the cell. A cell containing identical alleles of a particular gene is said to be ***homozygous*** for that gene, whereas a cell containing different alleles of a gene is said to be ***heterozygous*** for that gene. Thus, a dominant allele will result in expression of the characteristic it describes, whether the allele is present in one copy or two. The characteristic described by a recessive allele will only appear when that allele is present in both copies of the gene. In other words, a dominant characteristic will occur in a cell that is either heterozygous or homozygous for the cognate (i.e., matching) gene, while a recessive characteristic only appears in the homozygous situation. The behavior and appearance of a cell or organism—the net result of all the *expressed* characteristics—is the ***phenotype***; the total complement of genes present in a cell is its ***genotype*** or ***genome.***

SUMMARY

1. Several genetic concepts are important for understanding biochemistry and molecular biology. The inherited traits of an individual are due to the genes carried in the cells of that individual. Different versions, alleles, of a gene give rise to different traits.

2. Higher organisms inherit one copy of each gene from each parent. The total complement of genes in an individual is the genotype or genome. Some genes are expressed in a dominant manner, others in a recessive manner. The expressed traits of an individual are the phenotype.

GLOSSARY

A-, B-, and Z-DNA: DNA is found as a base-paired duplex. A-, B-, and Z-DNA are different duplex structures with the same base-pairing scheme, but different three-dimensional structures. A- and B-DNA are right-handed helices; Z-DNA is left-handed. B-DNA is the predominant structure of DNA in cells; A-DNA is found in duplex structures containing RNA.

Activator: A transcription factor that stimulates transcriptional initiation rates.

Active site: The portion of an enzyme that binds the substrate and participates directly in catalysis is the active site of the enzyme.

Active transport: Molecules and ions can be transported across membranes by a variety of mechanisms. Transport mechanisms that require the expenditure of energy—usually the hydrolysis of ATP—are known as active-transport mechanisms.

α-helix: Protein chains fold into complex shapes. The α-helix is a recurring structural motif in proteins. It has 3.6 residues per turn of the helix, which is stabilized by hydrogen bonds parallel to the helix axis between backbone carboxyl and amide groups.

Allosteric: Enzyme regulation may occur by the binding of small molecule regulators. When the regulator binds to a site separate from the enzyme-active site, the enzyme is said to be allosteric.

Antibiotic resistance gene: A gene that encodes an enzyme capable of inactivating a specific antibiotic; the β-lactamase gene confers resistance to ampicillin.

Anticodon: The three nucleotides in tRNA that are complementary to a corresponding codon in mRNA.

ATP: Adenosine 5′-triphosphate (ATP) is the major molecule that carries energy from storage forms to cellular reactions that require energy. It is like an energy currency in the cell.

Autosomally inherited: A genetic trait that is encoded on a chromosome other than the sex chromosomes and, therefore, is inherited equally in males and females.

Bacteriophage: Bacterial virus; bacteriophage λ is a biological reagent used in recombinant DNA cloning.

Beta oxidation: A process of fatty-acid breakdown producing acetylCoA.

B-form DNA: See A-, B-, and Z-DNA.

bp: Base pairs in a nucleic acid duplex.

β-structure: Proteins fold into complex shapes. The β-structures are a family of structures that recur frequently in proteins. In β-structures, the polypeptide backbone is relatively extended with a rotation per residue of 180°. The structure contains more than one polypeptide strand stabilized in a parallel or antiparallel conformation by hydrogen bonds between backbone carboxyl and amide groups, perpendicular to the direction of the polypeptide chains.

Calmodulin: A protein that binds calcium and modulates the activities of other proteins in a calcium-dependent manner.

cAMP: Cyclic AMP, adenosine 3′,5′-cyclic monophosphate; cAMP is a second messenger-signaling molecule in both prokaryotes and eukaryotes.

Carcinogen: Any biological, chemical, or physical agent that causes cancer.

Cascade: In several situations in the body, a series of similar enzyme reactions occurs in which a signal is passed and often amplified between the beginning and end of the series. Such a series is called a cascade. The most common types are proteolytic and protein phosphorylation cascades.

cDNA: Complementary DNA; made by synthesizing the complementary strand of mRNA using the enzyme-reverse transcriptase.

cDNA library: A collection of double-stranded cDNA sequences representing a given pool of mRNA; cDNA libraries can be maintained in bacteriophage or plasmid vectors.

Chemotherapy: A form of therapy in which chemicals—or drugs—are used to treat cancer.

Chromatin: Is primarily a DNA–protein complex containing histones and other proteins.

Chromatography: A method for separating mixtures of different molecules into pure or partly pure fractions. The molecules in the mixture bind to a stationary phase—paper, thin-layer, or particles packed into a column—and are differentially eluted by a mobile phase.

Chromosome: A particle that contains part of the genetic material of a cell nucleus—DNA—complexed with histone and nonhistone proteins. During cell division, chromosomes condense and become visible in the light microscope.

Citric-acid cycle: A metabolic process by which "acetate" is converted to CO_2 and reduced coenzymes are formed, as well as GTP. Requires sparking by intermediates, thus the cyclic nature.

Clathrin-coated pits: Regions of plasma membrane containing the structural protein clathrin; molecules that bind to receptors in clathrin-coated pits are internalized as a result of vesicle formation.

Codon: Three consecutive nucleotides in mRNA that specify a given amino acid during protein synthesis; the anticodon of tRNA is complementary to the mRNA codon.

Coenzyme, cofactor: Many enzyme reactions involve molecules that cooperate with the enzyme in carrying out the reaction. Such molecules—coenzymes or cofactors—may provide energy, as through the hydrolysis of ATP, or provide a chemical group required by the reaction, as in the oxidation of NADH. The coenzyme or cofactor is regenerated by a separate reaction and used again. Many coenzymes or cofactors are derived from vitamins.

Collagen: Collagens form a family of proteins with a tissue-specific distribution, including types I, II, and III—found in connective tissue such as filaments—and types IV and V—found in basal laminae—forming sheets of tissue.

Competitive inhibition: An inhibition where the K_M for a reaction is changed but the V_{max} is not. This can occur by competition with the substrate for the binding site on the enzyme, in which case the inhibitor resembles the substrate chemically. Physiologically, this occurs more frequently by action at a different site, in which case the inhibitor need not structurally resemble the substrate.

Complementary mutation: In the context of double-stranded RNA structure, a complementary mutation is a second mutation that restores base pairing lost in the primary mutation.

Conservative replacement: In the context of protein structure, a conservative replacement occurs when an amino acid in a protein is replaced—during evolution or by genetic manipulation—by an amino acid with similar chemical and structural properties of its side chains.

Cooperativity: In the context of ligand binding, cooperativity means that the binding of one molecule of the ligand affects the binding of a second molecule. For example, the binding of one molecule of oxygen to hemoglobin increases the affinity for the binding of subsequent molecules of oxygen.

Covalent modification: The additions of groups to proteins or deletions from proteins, usually causing a change in the protein properties.

C-value paradox: The observation that organismal complexity is not always directly related to genome size because of differences in the amount of repetitive DNA.

Cytoplasm: A compartment of eukaryotic cells. The cytoplasm is the solution between the nucleus and the plasma membrane. It contains other organelles such as mitochondria and the cytoskeleton.

Cytoskeleton: A protein structure that controls the shape of the cell and directs molecular transport.

DNA: Deoxyribonucleic acid, a polymer of deoxyribonucleotides. DNA contains the genetic information of the cell in the sequence of nucleotides along its length.

DNA polymorphisms: Alterations in DNA sequence that occur in more than 1% of the population; these DNA rearrangements or nucleotide changes are useful in RFLP mapping.

DNAase-1, DNAse-1: An enzyme that cuts DNA, DNAase-1 cuts nonspecifically and is used to determine regions of DNA that are bound to proteins and distinguish between different chromatin structures.

Dominant mutation: This class of mutations results in a phenotype that is observed in a heterozygous genetic background; dominant oncogenes like *ras** are active in a cell that contains wild-type *ras* because Ras* is a gain of function mutant.

Effector: A small molecule that regulates the function of a protein; cAMP is a positive effector of CAP transcription-factor function and protein kinase A activity.

Electrostatic interaction: The force between charged atoms or groups: attractive for opposite charges; repulsive for charges of the same sign.

Embryonic stem (ES) cell: Karyotypically normal cells derived from a mouse blastocyst; ES cells are used to make transgenic mice by serving as pluripotent progenitors.

Endoplasmic reticulum (ER): A membrane found within the cell that separates the cytoplasm from the lumen of the endoplasmic reticulum. Proteins destined to cross the plasma membrane, or to enter lysozomes, initially pass through the ER on their way to the Golgi apparatus.

Enhancer: A DNA sequence corresponding to a high-affinity binding site for a specific transcription factor; enhancers are a class of response elements that can function with heterologous promoters and from large distances 5′ or 3′ of a promoter.

Epithelial cells, epithelium: Epithelial tissues are sheets of cells forming containers or barriers in the body. The skin, for example, is an epithelial tissue.

Eukaryotic: Eukaryotic cells have a separate nucleus that contains almost all of the cellular DNA. Yeasts and higher organisms, including humans, have eukaryotic cells and are termed eukaryotic organisms. Lower organisms are prokaryotes.

Extracellular matrix: The proteoglycan-rich structures found between cells, in tissues.

Extrachromosomal DNA: DNA molecules, usually circular, that replicate autonomously, and often independently, of chromosomes; plasmids and mitochondrial DNA are examples of extrachromosomal DNA.

Fibroblasts: Cells of the skin, found below the outer, epithelial layer, or connective tissue.

5′ to 3′ Direction: In a nucleic acid, the 5′ end normally has a free 5′-phosphate and the 3′ end has a free 3′-OH. Thus, the 5′ to 3′ direction goes from the 5′ to the 3′ position through the sugar and then the phosphate links the 3′ on one sugar to the 5′ position on the next sugar.

G protein: A protein that binds GTP and hydrolyzes it to GDP. G proteins are molecular switches that change their structural state depending on whether GTP or GDP is bound. There are two classes of G proteins, the small G proteins such as Ras and the heterotrimeric G proteins such as the G protein that couples adenyl cyclase to the epinephrine receptor, providing the link between the primary messenger, epinephrine, outside the cell and the second messenger, cyclic AMP, formed by adenyl cyclase inside the cell.

Gametes: Haploid cells such as sperm and egg cells that are capable of fusing to form a diploid zygotic cell; mutations in gamete cells are inherited.

Gel electrophoresis (including SDS, 2D): A method for separating molecules—for example, proteins or nucleic acids—that depends on their mobility in an electric field. A gel is used to prevent mixing of the solution by convection during electrophoresis and to restrict diffusion. An SDS gel contains the detergent SDS, which acts as denaturing agent and shields the protein charge so that proteins separate according to their molecular weight on SDS gel electrophoresis. In two-dimensional gel electrophoresis, proteins are separated in two dimensions on a rectangular gel, usually using isoelectric focusing in one direction and SDS electrophoresis in the perpendicular direction.

Gene knockout: An insertional mutation or deletion in a specific gene which eliminates the expression of a functional gene product.

Genes: A gene determines an inherited trait. Genes are encoded in the sequence of the cellular DNA.

Genetic code: The set of codons in mRNA, specifying each of 61 possible tRNA anticodons and 3 nonsense (STOP) codons; it is based on 4^3 triplet combinations.

Genome: The entire genetic information of a single cell.

Genotype: The genetic traits present in an organism. The traits may or may not be expressed. The expressed traits are the phenotype.

Gluconeogenesis: The formation of glucose from nonglucose precursors.

Glycogenolysis: The breakdown of glycogen to glucose phosphates and glucose.

Glycolysis: The process of converting glucose or glucose phosphate to pyruvate or lactate and producing ATP. This can occur in the absence of oxygen.

Glycoproteins: Proteins containing polysaccharide residues.

Heteroduplex: A double-strand hybrid of complementary nucleic acid; DNA: DNA heteroduplexes are formed during recombination.

Heterogeneous RNA (hnRNA): The primary RNA transcript of a gene, this usually refers to the unspliced precursor of mRNA.

Heterozygous: In genetics, a cell is heterozygous for a particular gene when the two copies of the gene found in one diploid cell are different.

Histones: Small basic proteins that organize the DNA of chromosomes into nucleosomes.

Holliday junction: A DNA intermediate formed during homologous recombination; Robin Holliday first proposed the existence of this structure.

Homozygous: In genetics, a cell is homozygous for a particular gene when the two copies of the gene found in one diploid cell are identical.

Human gene therapy: A form of treatment designed to introduce DNA into cell types harboring genetic defects. Human gene therapy is only in the experimental stages but would offer an alternative to standard pharmaceutical treatments.

Hydrogen bonds: Weak noncovalent bonds in which a hydrogen is shared between the donor and acceptor atoms.

Hydropathy plot: A hydropathy plot shows the average hydrophobicity of successive short sequences of amino acids as a function of position along the protein sequence.

Hydrophobic, hydrophobic interactions: A hydrophobic residue is one that will not interact with water. The term "hydrophobic interactions" refers to the tendency of hydrophobic groups to aggregate in aqueous solution.

Hyperbolic curve: A graph of a class of functions of the form $y = ax/(b + x)$. Near the origin, the graph increases almost linearly and then the slope decreases until, at high x values, it approaches a maximum value.

Hypersensitive site: A site on chromatin that has a sensitivity to DNAase-1 similar to that of free DNA.

Immortalized cell lines: Tissue-culture cell lines that can be propagated indefinitely in the laboratory; some immortalized cell lines have a transformed phenotype.

Intermediate filaments: Intermediate filaments are assemblies of intermediate filament proteins that provide mechanical strength to animal cells.

Inverted repeat: In the context of nucleic-acid sequences, an inverted repeat of a sequence is a copy of the sequence with the direction reversed.

Ionic bonds: Electrostatic interactions.

Isoelectric point, pI: The pH at which the net charge on a molecule is zero.

Isozyme, isoenzyme: Isozymes, or isoenzymes, are different enzymes that carry out the same chemical reaction. They are different proteins usually with different kinetic properties.

Karyotype: A visual display of metaphase chromosomes from somatic cells which collectively represents the chromosomal makeup of an individual; a normal human karyotype is 44 autosomes (2 each of 22) and 2 sex chromosomes.

kb: Kilobase pairs, 10^3 nucleotides.

kDa: Kilodalton, a measurement of molecular weight.

Ketone bodies: Small lipid-like molecules, usually produced from partial fatty-acid metabolism by the liver. They can be used for energy by most tissues, including the brain.

K_M: K_M is the Michaelis–Menten constant defined by the Michaelis–Menten equation: $V = V_{max}[S]/(K_M + [S])$. In many cases, K_M is related to the affinity of an enzyme for its substrate.

Krebs cycle: See *Citric-acid cycle.*

Lagging strand: The template strand of DNA which is copied discontinuously during DNA synthesis to produce Okazaki DNA fragments.

Lariat: The closed, circular RNA structure formed as an intermediate in RNA splicing of group II self-spliced introns and spliceosome-mediated hnRNA splicing.

Leading strand: The template strand of DNA which is copied continuously in the 5′ to 3′ direction during DNA synthesis.

Ligand: A molecule that binds to another molecule. For example, when oxygen binds to hemoglobin, oxygen is a ligand.

Lipid bilayer: The structure of biological membranes in which the hydrophobic tails of lipids aggregate in two layers.

Lipids: Very hydrophobic molecules that can be released from cells with hydrophobic solvents.

Lipogenesis: The formation of lipid compounds. Also refers to fatty-acid synthesis from nonlipid precursors.

Lipoproteins: A complex of triacylglycerols, cholesterol, cholesterol esters, phospholipids, and protein. These include very-low-density lipoproteins (VLDL), low-density lipoproteins (LDL), intermediate-density lipoproteins (IDL), high-density lipoproteins (HDL), and chylomicrons.

Major groove: In B-form DNA, the phosphate backbone forms a ridge running around the stacked bases. The spaces between the two backbones in the double strand are the major and minor grooves, the major groove being wider and deeper.

Malignant: Tumor cells that migrate from the tumor origin to other parts of the body through the process of metastasis.

Megabase; mb: A million base pairs of DNA (1000 kb).

Meiosis: The process by which haploid germ cells are produced from diploid cells.

Messenger RNA (mRNA): The RNA transcript that serves as the template for protein synthesis; in eukaryotes, mRNA is fully processed hnRNA.

Metabolic pathway: A series of enzymatic reactions in which the product of one reaction is the substrate for the next reaction in the pathway.

Michaelis–Menten: Michaelis and Menten—the latter is one of the few women to be recognized in this way—are responsible for the Michaelis–Menten equation, which describes the dependence of the reaction velocity of an enzyme-catalyzed reaction on the substrate concentration.

Micrococcal nuclease: An enzyme that cleaves double-stranded DNA. It is used to map nucleosome positions due to its strong preference for cutting linker DNA over DNA in nucleosome core particles.

Microtubules: Long thin aggregates of tubulin found in the cytoplasm as part of the cytoskeleton.

Minor groove: In B-form DNA, the phosphate backbone forms a ridge running around the stacked bases. The spaces between the two backbones in the double strand are the major and minor grooves, the minor groove being shallower and narrower.

Mitochondria: Organelles found in the cytoplasm that organize the enzymes that use molecular oxygen to produce ATP for use as an energy source in the cell.

Mitosis: The process of cell division in which the chromosomes are segregated and daughter cells are formed.

Mitotic: A mitotic cell is undergoing mitosis. A mitotic chromosome is a chromosome in the condensed state found during mitosis.

Mixed inhibition: An inhibition where both V_{max} and K_M are changed.

Multigene family: A related set of genes that arose from gene duplication during evolution; the DNA sequences of these genes are homologous but not identical.

Mutation frequency: The frequency within a population that a specific genetic mutation occurs per generation.

Mutation rate: The number of mutation events in a unit of time, often expressed as mutations per gene per generation.

Natural selection: The differential selection, either positive or negative, of a genetic trait over evolutionary time as reflected in reproductive capacity; this mechanism is the central tenet of Darwinian evolution.

Noncompetitive inhibition: An inhibition where the V_{max} is changed but not the K_M.

Northern blotting: An experimental technique in which RNA molecules, after gel electrophoresis, are blotted to a membrane and then incubated with a radioactive probe that hybridizes to complementary sequences, demonstrating their presence on the blot. Northern blotting is used to detect specific RNA sequences in a mixture of RNA molecules.

Nucleocapsid: The complex of protein and nucleic acid which forms a virus particle; enveloped viruses, such as HIV-1, have a membrane-coated nucleocapsid.

Nucleolus: An organelle, within the cell nucleus, where the genes for the large rRNA precursor are found and transcribed.

Nucleoplasm: The solution within the nuclear membrane, not including the nucleolus or the nuclear matrix.

Nucleoside: A component of nucleic acids and other biological molecules. A nucleoside is a sugar, such as ribose, connected to an organic base, such as adenine.

Nucleosome: The major repeating subunit structure found in chromosomes.

Nucleotide: A component of nucleic acids and other biological molecules. A nucleotide is a sugar, such as ribose, connected to an organic base, such as adenine, and to one or more phosphate groups.

Nucleus: The cellular compartment where nucleic acid synthesis takes place and almost all the cellular DNA is stored.

Okazaki fragments: Short DNA segments generated during discontinuous replication of the lagging strand; these were described by Reiji Okazaki, based on experiments using radioactive deoxynucleotides as precursors in DNA synthesis.

Oncogenes: Any gene that can promote tumorigenesis; dominant oncogenes are gain-of-function mutants, recessive oncogenes are loss-of-function mutations, and protooncogenes are the unmutated form of an oncogene.

Oncogenesis: The initiation of cancer as a result of mutations in protooncogenes.

Open-reading frame (ORF): A segment of DNA or RNA that predicts a continuous set of codons devoid of termination codons; the relationship between predicted ORFs and actual proteins, however, requires direct experimental evidence.

Origin of DNA replication: The physical location in a chromosome or genome where DNA replication is initiated by the formation of a replication bubble.

Oxidative phosphorylation: The formation of ATP from ADP plus inorganic phosphate during the transfer of hydrogen or electrons down the electron-transport system.

Palindromes: A palindrome is a sequence, of base pairs in DNA for example, that is identical when read in the 5′ to 3′ direction on opposite DNA strands.

Phenotype: The genetic traits expressed in an organism.

Phosphorylation cascade: The sequence of events in which the activity of kinases/phosphatases is controlled by phosphorylation/dephosphorylation events; phosphorylation cascades function in signal-transduction pathways.

Phylogenetic: Related to the study of evolutionary relationships.

Physiological conditions: Conditions of temperature, ionic strength, pH, and so on that exist in living cells.

Plasmids: A circular DNA molecule, usually small in comparison with a chromosomal DNA molecule. A plasmid usually functions inside a cell alongside the chromosomal DNA, providing additional genes.

Polymerase chain reaction (PCR): A method for selecting and amplifying a specific DNA sequence starting from a mixture of DNA sequences.

Probe: In the context of hybridization, a probe is used to determine whether a specific nucleotide sequence is present in a mixture of sequences. Normally, the probe has a sequence complementary to the required sequence.

Prokaryotic: Prokaryotic cells lack a separate nucleus. Lower organisms such as bacteria are prokaryotes.

Protein kinase: An enzyme that transfers the γ-phosphate from ATP to an amino-acid side chain on a protein substrate.

Protein phosphatase: An enzyme that removes a phosphate group from a protein substrate, releasing inorganic phosphate.

Provirus: The intermediate life form of a retrovirus which is integrated into the host genome as a double-stranded DNA molecule.

Pseudogene: A gene segment that is similar to a known gene, but is nonfunctional because of genetic drift (mutations over time); pseudogenes are thought to arise by gene duplication and by mechanisms involving reverse transcription of mRNA.

Receptor: In the context of the interaction of hormones and other factors with cells, a receptor is the cellular protein that interacts directly with the hormone or other factor.

Receptor protein kinases: A receptor that has protein kinase activity.

Recombinant DNA technology: A term referring to the biochemical manipulation of DNA fragments using DNA modifying enzymes for the purposes of gene cloning.

Recombination: The exchange of genetic information between chromosomes that occurs during meiosis.

Regulatory region: A segment of DNA that controls gene expression; the regulatory region of a gene and the transcription unit together constitute a functional gene.

Replication: A process that begins with one molecule and ends with two molecules, each identical with the original molecule.

Restriction enzyme: An endonuclease that cleaves DNA at a specific site; restriction enzymes are the molecular "scissors" of recombinant DNA technology.

Reverse genetics: The series of manipulations in which genetic phenotypes are studied by first isolating genes and then characterizing their function in vivo using recombinant DNA technology; classic genetics begins with a known phenotype and subsequently determines the genotype.

Reverse transcriptase: An RNA-dependent DNA polymerase first characterized in retroviruses; this enzyme is used to synthesize cDNA in vitro.

Ribosome: A large ribonucleoprotein complex that is required for all protein synthesis in the cell.

Ribozyme: A catalytic RNA molecule capable of modifying another molecule in trans; ribozymes have been identified which can either cleave or synthesize RNA.

RNA: Ribonucleic acid, a polymer of ribose nucleotides. RNA plays many roles in gene expression.

RNA splicing: The primary transcript produced from DNA contains both introns and exons. The process by which introns are removed and the exons joined together is called RNA splicing.

rRNA: RNA molecules found in ribosomes.

Salt links: Molecular electrostatic interactions.

Sanger sequencing: A method, named after Fred Sanger, for determining the nucleotide sequence of DNA.

Sarcoplasmic reticulum: The equivalent in muscle cells of the endoplasmic reticulum.

Second messenger: Activation of membrane-bound hormone receptors, in many cases, leads to the release, inside the cell, of small molecules—including cAMP, inositol trisphosphate, or diacylglycerol—that carry the hormonal signal to other parts of the cell where target enzymes are regulated. These small intracellular molecules are known as second messengers.

Sequence divergence: The dissimilarity between two DNA sequences as a result of genetic changes at the nucleotide level. Sequence divergence occurs over evolutionary time and reflects the outcome of natural selection.

Sigmoidal curve: A curve, sometimes called S-shaped, that starts from the origin as a straight line of zero slope that gradually increases and then decreases again to approach zero slope.

Signal peptide: A linear sequence of amino acids that specifies a particular subcellular targeting pathway; the presence of an ER signal peptide targets a protein to the ER through a process involving the signal recognition particle (SRP).

Signal-recognition particle: A ribonucleoprotein particle that helps proteins cross the endoplasmic reticulum.

Signal transduction: The process by which a signal moves from one place to another and from one molecule to another.

snRNP: Small nuclear ribonucleoprotein particle. Many snRNPs are involved in RNA splicing in the cell nucleus.

Somatic cell: All cells other than gamete cells; a somatic-cell mutation is not inherited.

Southern blotting: A method commonly used to detect the presence of a specific DNA sequence in a mixture of DNA sequences.

Spliceosome: A large ribonucleoprotein complex that mediates hnRNA splicing.

Substrate: A compound that is converted into product by an enzyme-catalyzed reaction.

Substrate cycle: Series of reactions (also known as futile cycle) where the original substrate is reformed, but with the loss of one or more high-energy phosphates.

Sugars: Carbohydrate molecules made from CH_2OH units.

Supercoiling: In a coiled-coil structure, the second or subsequent hierarchies of coiling are called supercoils.

Tandem repeat: In a sequence, a tandem repeat is a repeat of a particular sequence, in the same direction.

Taq polymerase: A DNA-dependent DNA polymerase isolated from the bacterium *Thermus aquaticum*; Taq polymerase functions at high temperatures and is the key enzyme in the polymerase chain reaction (PCR).

Template strand: The strand of DNA that is being copied during nucleic-acid synthesis. The template strand in RNA synthesis represents the noncoding strand and is complementary in sequence to the RNA.

Tight junction: Direct contacts between the plasma membranes of adjacent cells, particularly in epithelial tissues, are called *tight junctions.*

Topological domains: Topological domains are separated from one another in such a way that a topological change, such as supercoiling of a DNA molecule, cannot propagate from one domain to another.

Trait: In genetic studies, a trait is a property of an organism, such as eye color, that can be used to follow the segregation of genes.

Transamination: The movement of nitrogen from one carbon skeleton to another carbon skeleton, usually by the action of amino transferases.

Transcript: This refers to the product of RNA synthesis; mRNA is often called a gene transcript.

Transcription: The process of synthesizing RNA on a DNA template.

Transcription factor: A protein that functions to initiate or modulate transcription. Most, but not all, transcription factors are sequence-specific DNA binding proteins.

Transcription unit: The portion of a gene that is transcribed; this includes 5′ and 3′ non protein-coding sequences.

Transduction: In the context of cell signalings conversion of messages (usually from hormones) across membranes to form effector molecules (usually second messengers).

Transformed: The phenotype of immortalized cells that have tumorigenic properties; this term is also used in recombinant DNA technology to describe the introduction of exogenous DNA into bacteria.

Transition state: In an enzyme reaction, the enzyme and substrate combine and then pass through a transitory state of high-energy—the transition state—on the way to forming the enzyme–product complex.

Translation: In the context of gene expression, translation refers to the process of synthesizing protein from an mRNA template.

Tricarboxylic-acid cycle: See *Citric-acid cycle.*

tRNA: Transfer RNA; used on the ribosome as an adaptor molecule to translate the nucleotide sequence of mRNA into the amino-acid sequence of a protein.

tRNA charging: Aminoacylation of tRNA by aminoacyl tRNA transferases; this is a prerequisite step in protein synthesis.

Tropomyosin, troponin: Proteins found together with actin in muscle fibers.

Tumorgenic cell: A cell that can form tumors in animals; transformed cells are tumorgenic and can grow to high densities in tissue culture.

Urea cycle: The formation of urea from amino group and ammonia. Also elucidated by H. A. Krebs, the discoverer of the citric-acid cycle.

V_{max}: The reaction velocity for an enzyme-catalyzed reaction at very-high-substrate concentration.

Virion: An infectious virus particle.

Vitamin: A coenzyme or cofactor or a precursor of a coenzyme or cofactor.

Wobble hypothesis: As proposed by Francis Crick, the Wobble hypothesis explains why one tRNA anticodon can be complementary to more than one mRNA codon.

X-Ray crystallography: A method for determining the three-dimensional structures of molecules, including proteins tRNAs and oligodeoxyribonucleotides.

Yeast artificial chromosome (YAC): A genomic cloning vector which is propagated in yeast as an autonomously replicating chromosome.

Zymogen: A protein that becomes an active enzyme when a portion of the protein is removed by proteolysis.

INDEX